# FUNDAMENTAL EQUATIONS OF STATICS

## Moments:

### Moment of a force about a point:

$$\mathbf{M}_O = \mathbf{r} \times \mathbf{F} = \begin{vmatrix} \mathbf{i} & \mathbf{j} & \mathbf{k} \\ r_x & r_y & r_z \\ F_x & F_y & F_z \end{vmatrix}$$

$$= (r_y F_z - r_z F_y)\mathbf{i} + (r_z F_x - r_x F_z)\mathbf{j} + (r_x F_y - r_y F_x)\mathbf{k}$$

### Moment of a force about a line:

$$M_{OB} = \mathbf{M}_O \cdot \mathbf{e}_n = (\mathbf{r} \times \mathbf{F}) \cdot \mathbf{e}_n = \begin{vmatrix} \mathbf{i} & \mathbf{j} & \mathbf{k} \\ r_x & r_y & r_z \\ F_x & F_y & F_z \end{vmatrix} \cdot \mathbf{e}_n$$

$$= \begin{vmatrix} e_{nx} & e_{ny} & e_{nz} \\ r_x & r_y & r_z \\ F_x & F_y & F_z \end{vmatrix}$$

## Center of Gravity, Center of Mass, Centroid:

### Center of mass:

$$x_G = \frac{1}{m}\int x\, dm \qquad y_G = \frac{1}{m}\int y\, dm \qquad z_G = \frac{1}{m}\int z\, dm$$

### Centroid of a plane area:

$$x_C = \frac{1}{A}\int x\, dA \qquad y_C = \frac{1}{A}\int y\, dA$$

## Beams:

### Load-shear relations:

$$\frac{dV}{dx} = w \qquad \text{or} \qquad dV = w\, dx$$

$$\int_{x_1}^{x_2} w\, dx = \int_{V_1}^{V_2} dV = V_2 - V_1$$

### Shear-moment relations:

$$\frac{dM}{dx} = V \qquad \text{or} \qquad dM = V\, dx$$

$$\int_{x_1}^{x_2} V\, dx = \int_{M_1}^{M_2} dM = M_2 - M_1$$

## Second Moments of Area and Moments of Inertia:

### Second moments of plane areas and radius of gyration:

$$I_x = \int_A y^2\, dA = A k_x^2 \qquad k_x = \sqrt{\frac{I_x}{A}}$$

$$J_z = \int_A r^2\, dA = A k_z^2 \qquad k_z = \sqrt{\frac{J_z}{A}}$$

### Parallel-Axis Theorem for Second Moments:

$$I_{x'} = I_{xC} + d_x^2 A \qquad J_{z'} = J_{zC} + d_x^2 A$$

### Principal Second Moments:

$$I_{x'} = I_x \cos^2\theta + I_y \sin^2\theta - 2I_{xy}\sin\theta\cos\theta$$

$$I_{u,v} = \frac{I_x + I_y}{2} \pm \left[\left(\frac{I_x - I_y}{2}\right)^2 + I_{xy}^2\right]^{1/2}$$

$$\tan 2\theta_\rho = -\frac{2I_{xy}}{(I_x - I_y)}$$

### Moments of Inertia:

$$I_x = \int_m r_x^2\, dm = \int_m (y^2 + z^2)\, dm$$

### Parallel-Axis Theorem for Moments of Inertia:

$$I_{x'} = I_{xG} + (y_G^2 + z_G^2)m = I_{xG} + d_x^2 m$$

### Product of Inertia:

$$I_{xy} = \int_m xy\, dm \qquad I_{x'y'} = I_{xyG} + x_G y_G\, m$$

## Miscellaneous:

### The law of sines:

$$\frac{a}{\sin\alpha} = \frac{b}{\sin\beta} = \frac{c}{\sin\gamma}$$

### The law of cosines:

$$c^2 = a^2 + b^2 - 2ab\cos\gamma$$

### Newton's Law of Gravitation:

$$F = G\frac{m_1 m_2}{r^2}$$

$G = 3.439(10^{-8})\ \text{ft}^3/(\text{slug}\cdot\text{s}^2) = 6.673(10^{-11})\ \text{m}^3/(\text{kg}\cdot\text{s}^2)$

$m_e = 4.095(10^{23})\ \text{slug} = 5.976(10^{24})\ \text{kg}$

$r_e = 2.090(10^7)\ \text{ft} = 6.371(10^6)\ \text{m}$

# ENGINEERING MECHANICS
# STATICS

## SECOND EDITION

## WILLIAM F. RILEY

*Professor Emeritus*
*Iowa State University*

## LEROY D. STURGES

*Iowa State University*

### JOHN WILEY & SONS, INC.
**New York · Chichester · Brisbane · Toronto · Singapore**

ACQUISITIONS EDITOR    Charity Robey
DEVELOPMENTAL EDITOR    Madalyn Stone
MARKETING MANAGER    Debra Riegert
OUTSIDE PRODUCTION    York Production Services
TEXT DESIGN    Ann Marie Renzi and Karin Kincheloe
COVER DESIGN    Karin Kincheloe
MANUFACTURING MANAGER    Mark Cirillo
PHOTO EDITOR    Hilary Newman
ILLUSTRATION EDITOR    Sigmund Malinowski
ILLUSTRATION DEVELOPMENT    Boris Starosta
ELECTRONIC ILLUSTRATIONS    Precision Graphics
COVER PHOTOS    Tim Brown/ProFILES West, Inc.; (*inset*) Ken Rogers/WestLight

This book was set in Palatino by York Graphic Services, and printed and bound by Von Hoffmann Press. The cover was printed by Phoenix Color.

*Library of Congress Cataloging in Publication Data:*
Riley, William F. (William Franklin), 1925–
    Engineering mechanics : statics / William F. Riley, Leroy D.
Sturges. — 2nd ed.
        p.    cm.
    Includes bibliographical references.
    ISBN 0-471-05333-3 (cloth : alk. paper)
    1. Statics.   I. Sturges, Leroy D.   II. Title.
TA351.R55   1996
620.1′03—dc20                                            95-40833
                                                         CIP

Printed in the United States of America

10 9 8 7 6 5 4 3 2 1

# PREFACE

Our purpose in writing this statics book, together with the companion dynamics book, was to present a fresh look at the subject and to provide a more logical order of presentation of the subject material. We believe our order of presentation will give students a greater understanding of the material and will better prepare students for future courses and later professional life.

## INTRODUCTION

This text has been designed for use in undergraduate engineering programs. Students are given a clear, practical, comprehensible, and thorough coverage of the theory normally presented in introductory mechanics courses. Application of the principles of statics to the solution of practical engineering problems is demonstrated. This text can also be used as a reference book by practicing aerospace, automotive, civil, mechanical, mining, and petroleum engineers.

Extensive use is made in this text of prerequisite course materials in mathematics and physics. Students entering a statics course that uses this book should have a working knowledge of algebra, geometry, and trigonometry, and should have taken an introductory course in calculus and vector algebra.

Vector methods do not always simplify solutions of two-dimensional problems in statics; therefore, they are used only in instances where they provide an efficient solution to a problem. For three-dimensional problems, however, vector algebra provides a systematic procedure that often eliminates errors that might occur with a less systematic approach. Students are encouraged to develop the ability to select the mathematical tools most appropriate for the particular problem that they are attempting to solve.

## FEATURES

### Engineering Emphasis

Throughout this book, strong emphasis has been placed on the engineering significance of the subject area in addition to the mathematical methods of analysis. Many illustrative example problems have been integrated into the main body of the text at points where the presentation of a method can be best reinforced by the immediate illustration of the method. Students are usually more enthusiastic about a subject if they can see and appreciate its value as they proceed into the subject.

We believe that students can progress in a mechanics course only by understanding the physical and mathematical principles jointly, not by mere memorization of formulas and substitution of data to obtain answers to simple problems. Furthermore, we think that it is better to teach a few fundamental principles for solving problems than to teach a large number of special cases and trick procedures. Therefore the text aims to develop in the student the ability to analyze a given problem in a simple and logical manner and to apply a few fundamental, well-understood principles to its solution.

A conscientious effort has been made to present the material in a simple and direct manner, with the student's point of view constantly in mind.

## Free-body Diagrams

Most engineers consider the free-body diagram to be the single most important tool for the solution of mechanics problems. Mastering the concept of the free-body diagram is fundamental to success in this course. Students frequently have difficulty with the concept, and coverage in this book has been carefully designed to ensure student understanding. A step-by-step procedure walks the student through the process of developing a complete and correct free-body diagram. Whenever an equation of equilibrium is written, we recommend that it be accompanied by a complete, proper free-body diagram.

## Problem-solving Procedures

Success in engineering mechanics courses depends, to a surprisingly large degree, on a well-disciplined method of problem solving and on the solution of a large number of problems. The student is urged to develop the ability to reduce problems to a series of simpler component problems that can be easily analyzed and combined to give the solution of the initial problem. Along with an effective methodology for problem decomposition and solution, the ability to present results in a clear, logical, and neat manner is emphasized throughout the text. A first course in mechanics is an excellent place to begin development of this disciplined approach that is so necessary in most engineering work.

## Homework Problems

This book contains a large selection of problems that illustrate the wide application of the principles of statics to the various fields of engineering. The problems in each set represent a considerable range of difficulty. And, in this edition, we have grouped the homework problems according to this range of difficulty. We believe that a student gains mastery of a subject through application of basic theory to the solution of problems that appear somewhat difficult. Mastery, in general, is not achieved by solving a large number of simple but similar problems. The problems in this text require an understanding of the principles of statics without demanding excessive time for computational work. We have replaced approximately 400 homework problems with new ones.

## Significant Figures

Results should always be reported as accurately as possible. However, results should not be reported to 10 significant figures merely because the calculator displays that many digits. One of the tasks in all engineering work is to determine the accuracy of the given data and the expected accuracy of the final answer. Results should reflect the accuracy of the given data.

In a textbook, however, it is not possible for students to examine or question the accuracy of the given data. It is also impractical for the authors to place error bounds on every number. An accuracy greater than 0.2 percent is seldom possible in engineering work, since physical data is seldom known with any greater degree of accuracy. A practical rule for "rounding off" numbers, that provides approximately this degree of accuracy, is to retain four significant figures for numbers beginning with the figure 1 and three significant figures for numbers beginning with any figure from 2 through 9. In this book, all given data, regardless of the number of figures shown, are assumed to be sufficiently accurate to permit application of this practical rule. Therefore, answers are given to three significant figures, unless the number lies between 1 and 2 or any decimal multiple thereof, in which case four significant figures are reported.

## SI vs. US Units

Most large engineering companies deal in an international marketplace. In addition, the use of the International System of Units (SI) is gaining acceptance in the United States. As a result, most engineers must be proficient in both the SI system and the U.S. Customary System (USCS) of units. In response to this need, both U.S. Customary units and SI units are used in approximately equal proportions in the text for both illustrative examples and homework problems. As an aid to the instructor in problem selection, all odd-numbered problems are given in USCS units and even-numbered problems in SI units.

## Chapter Summaries and Key Equations

As an aid to students we have written a summary that appears at the end of each chapter. These sections provide a synopsis of the major concepts that are explained in the chapter and can be used by students as a review or study aid. In this edition, we have highlighted key equations in color to help students distinguish them.

## Answers Provided

Answers to about half of the problems are included in the back of the book. We believe that the first assignment on a given topic should include some problems for which the answers are given. Since the simpler problems are usually reserved for this first assignment, answers are provided for the first few problems of each article and thereafter are given for approximately half of the remaining problems. The problems whose answers are provided are indicated by an asterisk after the problem number.

## DESIGN

### Use of Color

In this edition we have continued to use a variety of colors. We believe that color will help students learn mechanics more effectively for two reasons: First, today's visually oriented students are more motivated by texts that depict the real world more accurately. Second, the careful color coding makes it easier for students to understand the figures and text.

Following are samples of figures found in the book. As you can see, force and moment vectors are depicted as red arrows; velocity and acceleration vectors are depicted as green arrows. Position vectors appear in blue; unit vectors in bold black; and dimensions as a thin black line. This pedagogical use of color is standard throughout this book and its companion dynamics book.

We have also used color to help students identify the most important study elements. For instance, example problems are always outlined in red and important equations appear in a green box.

## Illustrations

One of the most difficult things for students to do is to visualize engineering problems. Over the years, students have struggled with the lack of realism in mechanics books. We think that mechanics illustrations should be as colorful and three-dimensional as life is. To hold students' attention, we developed the text illustrations with this point in mind.

We started with a basic sketch. Then a specialist in technical illustration added detail. Then the art studio created the figures using *Adobe Illustrator©*. All of these steps enabled us to provide you with the most realistic and accurate illustrations on the market.

## Accuracy

After many years of teaching, we appreciate the importance of an accurate text. We have made an extraordinary effort to provide you with an error-free book. Every problem in the text has been worked out at least twice independently; many of the problems have been worked out a third time independently.

## DEVELOPMENT PROCESS

This book is the most extensively developed text ever published for the engineering market. The development process involved several steps.

1. **Input from Students.** We have made a special effort to involve our end-users, the students, in the development of this text. An extensive market research survey was sent to 525 current or prior users of the text. We also asked for suggestions from students on student comment cards, and we held two student focus groups, one at the University of Maryland and one at the State University of New York at Stony Brook.

2. **Input from Faculty**. Professors from the United States and Canada carefully reviewed each draft of this manuscript. In addition to these reviews, we asked 7 professors to keep diary reviews of the first edition as they were actually using it in class. To supplement the traditional reviewing process, we held a teleconference focus group with 8 professors across the country. The Wiley sales staff reported hundreds of comments and suggestions from professors across the country. All of these suggestions were carefully considered by the author and editorial team and incorporated whenever feasible.

3. **Manuscript and Illustration Development.** A development editor worked with the authors to hone both the manuscript and the art sketches to their highest potential. A special art developer worked with the authors and the art studio to enhance the illustrations.

## SUPPLEMENTS FOR THE BOOK

After years of teaching, we realize the importance of an accurate solution manual that matches the quality of the text. For that reason, we have prepared the manual ourselves. The manual includes a complete solution for every problem in the book, and especially challenging problems are marked with an asterisk. Each solution appears with the original problem statement and, where appropriate, the problem figure. We do this for the convenience of the instructor, who no longer will have to refer to both book and solution manual in preparing for class. The manual also contains transparency masters for use in preparing overhead transparencies. In this edition we are making the solutions manual and overheads available in digital form (CD-ROM IBM compatible) as well as in print form.

## ACKNOWLEDGMENTS

Many people participated directly and indirectly in the preparation of this book. In addition to the authors, many present and former colleagues and students contributed ideas concerning methods of presentation, example problems, and homework problems. Final judgments concerning organization of material and emphasis of topics, however, were made by the authors. We will be pleased to receive comments from readers and will attempt to acknowledge all such communications.

We'd like to thank the following people for their suggestions and encouragement throughout the reviewing process.

| | |
|---|---|
| Jeff Arrington | Abilene Christian University |
| Nancy Austin | Scottsdale Community College |
| Leonard B. Baldwin | University of Wyoming |
| Donald Bunk | Rensselaer Polytechnic Institute |
| Tim Burns | Leeward Community College |
| Thomas W. Butler | U.S. Naval Academy |
| Julie Chen | Boston University |
| Ravinder Chona | Texas A&M University |
| Turgay Erturk | University of Massachusetts-Lowell |
| Habib Eslami | Embry Riddle Aeronautical University |
| Jack Forrester | University of Tennessee-Knoxville |
| Bill Hauser | California Polytechnic State University-Pomona |
| John Kosmatka | University of California-San Diego |
| Normal Laws | University of Pittsburgh |
| Donald Lemke | University of Illinois-Chicago |
| J. Michael McCarthy | University of California-Irvine |
| Gaby Neunzert | Colorado School of Mines |
| Dr. C. M. North | Rose-Hulman Institute of Technology |
| Dr. R. A. Raouf | U.S. Naval Academy |
| Lisa Spainhour | Florida State University |
| George Staab | Ohio State University |
| Major Curtis Thalken | U.S. Military Academy at West Point |
| Francis M. Thomas | University of Kansas |
| Wallace Venable | West Virginia University |
| Carlos Ventura | University of British Columbia (Canada) |
| Charles White | Northeastern University |
| Shahram Zanganeh | Howard University |

We also want to thank the following students who took the time to complete a survey of the first edition. We used their input to help shape the second edition.

| | |
|---|---|
| Sam Fortner, Bill Michel | California Maritime Academy |
| Russ Ahlberg, Jerry Allen, Patricia Johns, Shih-Hao Kuo | California State-Long Beach |
| Gregory J. Arserio, Paul Critchley, Mark Thomas | Clarkson |
| Philip J. Hruska | Cleveland State |
| Scott Francis | Colorado State |
| Annie Wing | DMACC |
| A. Marasco | Drexel |
| Matthew Ruminski | EEC North |
| Troy Brodstrom, Aaron Sanders, Dean Skinner | GMI-Engineering and Tech. Institute |
| Russ Ladley | GRCC |
| Matthew Greenlaw | Greenville Tech |
| Jenny Johannsen, Mitch Wagner, Yoke Theng Yoon | Iowa State |
| Charles Njendu, Jason Winebester | Mississippi State |
| Duane Bredel, Fred Jones, Tommy McBane, Ryan Pardee | Oklahoma State |
| David Boyd, Robert Harder, Roy Hedges, Thomas Smith | Oregon State |
| Jeff Weatheril | Pacific Lutheran |
| Albert Abramson, Tiffany M. Bonner, Budhi Sayopno, Amber L. Master, Parag Patel, Charlie Roth, Matt L. Walkeb, A. Zimmerman | Rensselaer Polytechnic Institute |
| Jason Gold | Roger Williams |
| Rob Allman | Ryerson |
| D. Jefferson | Spellman College |
| Valerie Mercer, Christine Rudakewycz | Stevens Institute of Technology |
| Wan Ci Lei | SUNY-Fredonia |
| Dawson Belobrajdic, Roger Canto, Christine Landry, Keli Musick, Eric Monkoe, Heath Penland, Adam Samuels, Julio Toro | Texas A&M |
| Renee Bailey, Althea F. Woodson | Tuskegee University |
| Julia Fernandez | Universidad de Turnabo |
| Kevin Bates, Charles Burgess, Tom Richardson | University of Alabama-Huntsville |
| Adrienne Cooper, Da-Yu Chang, Dioa A. Manly, Hoang Vu | University of California-Irvine |
| Salem Alshibani, R. Chapman, Jacqueline Long | University of Hartford |
| Eric Lee | University of Hawaii-Manoa |
| Manuel Bonilla, Kevin Boomsona, Danny H. Cruz, Anthony Duran, Brian Houston | University of Illinois-Chicago |
| Marcus Brewer | University of Kansas |
| Stacy Brown, James Knight | University of Maine |
| Stephen Ulhorn | University of Miami |
| Cuyler Larson, Ray Meyers, Jennifer L. Shryer | University of Missouri-Rolla |

| | |
|---|---|
| Ban Lin Lau | University of Oklahoma |
| Neit J. Nieves, Luis Ortiz, Javier Rodriguez, Vanessa Rosas, Hector Soto | University of Puerto Rico |
| Rainone Chin | University of Toledo |
| John Treybal | University of Vermont |
| Young Mee Cho | University of Virginia |
| Chad Helgerson, Bishwa Shrestha | University of Wyoming |
| Mike Capelli, Charles Guerrero, Kevin Wallace | U.S. Coast Guard Academy |
| Ken Nelson, Robert Ozanich, Brad Reisinger | U.S. Military Academy-West Point |
| Mark Elliott | Western Kentucky University |
| Peter Arangoudakis | Western New England College |

William F. Riley
Leroy D. Sturges

# A NOTE FROM THE PUBLISHER TO STUDENTS

We wish to acknowledge the extraordinary contributions of a number of engineering students to the development of the second edition of *Engineering Mechanics*. All of our textbooks depend on the experience with and understanding of students' needs that our authors bring to the work. This book benefited from the direct suggestions of students themselves.

Among the many important recommendations these students made were the following:

- They liked the idea of having homework problems grouped according to difficulty. This is a feature we added to this edition.
- They liked the idea of highlighting important expressions. They found this to be a helpful study aid, especially when preparing for a test or reviewing the chapter. This is also a new feature to this edition.
- They liked the idea of helpful hints in the margins of worked-out example problems. They didn't want them at the end of the problem, but next to the specific step that they referred to. We added this feature to our text, and we made a concerted effort to place them at the appropriate place that would be the most helpful to students.
- The students wanted to see more photos. Photos make the concepts seem more realistic to them. When asked if they preferred an illustration to a photo on the chapter opener, they overwhelmingly preferred photos. Therefore, we are using photos for chapter openers.
- When the idea of a real-world conceptual example was explained to them, most of the students thought it was a good idea. We have about 2 of these qualitative examples per chapter. This is a new feature we have added to this edition.

The following students participated in a focus group that helped guide the development of this book. We are grateful for their intelligent and thoughtful contributions.

Gavin Appel

Noel Aquino

Ney de Vasconcellos

Frank Fischer

Rosa Genao

Patrick Haran

Robert Ieraci

Paresh Rana

Anita Ristorucci

Anand Sarkar

Doug Stubbe

Charity Robey
Executive Editor

Madalyn Stone
Senior Development Editor

# A GUIDE TO THE LEARNING FEATURES OF THIS BOOK

**CONCEPTUAL EXAMPLES**
relate engineering principles to real-world situa-
tions using a minimum of
math. They help students
develop qualitative rea-
soning and problem-
solving skills. We have
created 23 Conceptual
Examples for this edition,
which are highlighted in
the Table of Contents.

## CONCEPTUAL EXAMPLE 2-1: CHARACTERISTICS OF A FORCE

*Why must all three characteristics of a force be known before its effect on a body can be established?*

### SOLUTION

The characteristics needed to describe a force are its magnitude, its direction, and its point of application. The effects of a force on a body are determined by these characteristics.

The force acting at point *A* of the garbage can shown in Fig. CE2-1*a* has both a magnitude (100 lb) and a direction (horizontal and to the right). If the 100-lb force is sufficient to overcome the frictional resistance at the contact sur- face, the garbage can will begin to slip and move to the right. If the same force (magnitude and direction) acts at point *B* as shown in CE2-1*b*, the garbage can may tip before it slips. Thus, all three characteristics of a force (the magnitude, the direction, and the point of application) must be known before the effects of the force (slipping or tipping) on the garbage can can be determined.

*(a)*

*(b)*
**Fig. CE2-1**

## CONCEPTUAL EXAMPLE 4-1: MOMENT OF A FORCE

*Experience indicates that it is easier to turn a wheel if the force is applied near the rim of the wheel rather than near the axle. Explain why the moment of a force depends on the magnitude of the force, the direction of the force, and where the force is applied relative to the axis of rotation.*

*(a)*      *(b)*      *(c)*
**Fig. CE4-1**

### SOLUTION

The concept of a moment of a force can be illustrated by considering the door shown in Fig. CE4-1. The force **F** of Fig. CE4-1*a* is perpendicular to the door and tends to rotate the door about a line through the axes of the hinges. Greater force is required to rotate the door when the force is applied near the hinge line rather than at the knob. In Fig. CE4-1*a*, the moment of force **F** about the hinge line is defined as $M = Fd$, where $M$ is the moment of the force, $F$ is the magnitude of the force, and $d$ is the perpendicular distance from the hinge line to the line of action of the force. The distance $d$ is called the moment arm of the force.

If the force **F** is not perpendicular to the door (see Fig. CE4-1*b*), a larger force **F** is required to rotate the door. The tendency of an inclined force **F** to ro- tate the door can be seen by resolving force **F** into components as shown in Fig. 4-1*c*. Component $\mathbf{F}_y$ does not cause rotation because its line of action passes through the hinge line ($d = 0$). Component $\mathbf{F}_z$ does not cause rotation because its line of action is parallel to the hinge line. Only component $\mathbf{F}_x$, which is per- pendicular to the plane of the door and at a distance $d$ from the hinge line, tends to rotate the door.

## EXAMPLE PROBLEM 6-14

A beam is loaded and supported as shown in Fig. 6-44a. Determine the components of the reactions at supports A and B.

(a)                          (b)

**Fig. 6-44**

### SOLUTION

A free-body diagram of the beam is shown in Fig. 6-44b. The action of the pin support at A is represented by force components $A_x$ and $A_y$. The action of the roller support at B is represented by the force $B_y$, which acts perpendicular to the horizontal surface at B. The distributed loads can be represented temporarily on the free-body diagram by fictitious resultants $R_1$, $R_2$, and $R_3$ with lines of action at distances $x_{C1}$, $x_{C2}$, $x_{C3}$, respectively, from the left support. Thus,

> The resultant **R** of a distributed load is given by the area under the load diagram and its line of action passes through the centroid of the area of the diagram.

$$R_1 = A_1 = \frac{1}{2}(6)(3) = 9.00 \text{ kN} \qquad x_{C1} = \frac{2}{3}(3) = 2.0 \text{ m}$$

$$R_2 = A_2 = \frac{1}{2}(3)(3) = 4.50 \text{ kN} \qquad x_{C2} = 3 + \frac{1}{3}(3) = 4.0 \text{ m}$$

$$R_3 = A_3 = 3(3) = 9.00 \text{ kN} \qquad x_{C3} = 3 + \frac{1}{2}(3) = 4.5 \text{ m}$$

> The distributed load on the right half of the beam is divided into a rectangle and a triangle for which areas and centroid locations are known.

The beam is subjected to a coplanar system of parallel forces in the $y$-direction; therefore, $A_x = 0$. The two remaining equilibrium equations are available to solve for $A_y$ and $B_y$.

Determination of $B_y$:

$$+\circlearrowleft\Sigma M_A = 0: \quad B_y(6) - 9.00(2.0) - 4.50(4.0) - 9.00(4.5) = 0$$
$$B_y = +12.75 \text{ kN} \quad \mathbf{B}_y = 12.75 \text{ kN}\uparrow \qquad \text{Ans.}$$

Determination of $A_y$:

$$+\circlearrowleft\Sigma M_B = 0: \quad -A_y(6) + 9.00(4.0) + 4.5(2.0) + 9.00(1.5) = 0$$
$$A_y = +9.75 \text{ kN} \quad \mathbf{A}_y = 9.75 \text{ kN}\uparrow \qquad \text{Ans.}$$

> Summing moments about point B yields a value for force $A_y$ that is independent of the previous determination for force $B_y$.

Alternatively: (or as a check)

$$+\uparrow\Sigma F_y = 0: \quad A_y - 9.00 - 4.50 - 9.00 + 12.75 = 0$$
$$A_y = +9.75 \text{ kN} \quad \mathbf{A}_y = 9.75 \text{ kN}\uparrow$$

## EXAMPLE PROBLEM 9-16

A car is prevented from moving by pulling on a rope that is wrapped $n + \frac{1}{4}$ times around a tree (Fig. 9-41). The coefficient of friction between the rope and the tree is 0.35, and the force exerted by the car is 3600 N. If it is desired that the force exerted on the rope be no more than 125 N, determine $n$, the number of times the rope must be wrapped around the tree.

**Fig. 9-41**

### SOLUTION

The angle of twist necessary to hold the car is found from

$$3600 = 125 \, e^{0.35\beta}$$

or

$$\beta = \frac{\ln(3600/125)}{0.35} = 9.60 \text{ radians}$$

> If the rope is wrapped $n$ times around the tree, the angle $\beta$ equals $2\pi n$ radians.

which is 1.528 times around the tree. Any angle less than this will require a resisting force greater than 125 N, whereas any angle greater than this will require less force. Thus

$$n = 2$$        Ans.

will be sufficient.

In this edition we have grouped the homework problems according to the degree of difficulty.

At the end of every chapter, **REVIEW PROBLEMS** test students on all the concepts covered in the chapter. They often integrate topics and thus can employ realistic applications.

**COMPUTER PROBLEMS** allow students to tackle more realistic problems that would be difficult to solve without the use of a computer.

# CONTENTS

*A satellite is weightless in space; nevertheless, an astronaut must exert considerable force in order to move the satellite because of its mass.*

# Chapter 1

# GENERAL PRINCIPLES

## 1-1  INTRODUCTION TO MECHANICS

 **Mechanics** is the branch of the physical sciences that deals with the response of bodies to the action of forces. The subject matter of this field constitutes a large part of our knowledge of the laws governing the behavior of gases and liquids as well as the laws governing the behavior of solid bodies. The laws of mechanics find application in astronomy and physics as well as in the study of the machines and structures involved in engineering practice. For convenience, the study of mechanics is divided into three parts: namely, the mechanics of rigid bodies, the mechanics of deformable bodies, and the mechanics of fluids.

A study of the mechanics of rigid bodies can be further subdivided into three main divisions: statics, kinematics, and kinetics. **Statics** is concerned with bodies that are acted on by balanced forces and hence are at rest or have uniform motion. Such bodies are said to be in equilibrium. Statics is an important part of the study of mechanics because it provides methods for the determination of support reactions and relationships between internal force distributions and external loads for stationary structures. Many practical engineering problems involving the loads carried by structural components can be solved using the relationships developed in statics. The relationships between internal force distributions and external loads that are developed in statics play an important role in the subsequent development of deformable body mechanics.

**Kinematics** is concerned with the motion of bodies without considering the manner in which the motion is produced. Kinematics is sometimes referred to as the geometry of motion. Kinematics forms an important part of the study of mechanics, not only because of its application to problems in which forces are involved, but also because of its application to problems that involve only motions of parts of a machine. For many motion problems, the principles of kinematics, alone, are sufficient for the solution of the problem. Such problems are discussed in "Kinematics of Machinery" books, where the motion of machine elements such as cam shafts, gears, connecting rods, and quick-return mechanisms are considered.

**Kinetics** is concerned with bodies that are acted on by unbalanced forces; hence, they have nonuniform or accelerated motions. A study of kinetics is an important part of the study of mechanics because it provides relationships between the motion of a body and the forces and moments acting on the body. Kinetic relationships may be obtained by direct application of Newton's laws of motion or by using the integrated forms of the equations of motion that result in the principles of work–energy or impulse–momentum. Frequently the term **dynamics** is used in the technical literature to denote the subdivisions of mechanics with which the idea of motion is most closely associated, namely, kinematics and kinetics.

The branch of mechanics that deals with internal force distributions and the deformations developed in actual engineering structures and machine components when they are subjected to systems of forces is known as **mechanics of deformable bodies.** Books covering this part of mechanics commonly have titles like "Mechanics of Materials" or "Mechanics of Deformable Bodies."

The branch of mechanics that deals with liquids and gases at rest or in motion is known as **fluid mechanics.** Fluids can be classified as compressible or incompressible. A fluid is said to be compressible if the density of the fluid varies with temperature and pressure. If the volume of a fluid remains constant during a change in pressure, the fluid is said to be incompressible. Liquids are considered incompressible for most engineering applications. A subdivision of fluid mechanics that deals with incompressible liquids is commonly known as hydraulics.

In this book on statics and in the companion volume on dynamics only rigid-body mechanics will be considered. The two books will provide the foundation required for follow-on courses in many fields of engineering.

## 1-2   HISTORICAL BACKGROUND

The portion of mechanics known as statics developed early in recorded history because many of the principles are needed in building construction. Ancient Egyptian and Assyrian monuments contain pictorial representations of many kinds of mechanical implements. The builders of the pyramids of Egypt probably understood and used such devices as the lever, the sled, and the inclined plane. An early history of mechanics was published by Dr. Ernst Mach of the University of Vienna in 1893.[1]

---

[1]Dr. Ernst Mach, "Die Mechanik in ihrer Entwickelung historisch-kritisch dargestellt," Professor an der Universitat zu Wien. Mit 257 Abbildungen. Leipzig, 1893. First translated from the German by Thomas J. McCormack in 1902. *The Science of Mechanics*, 9th ed. The Open Court Publishing Company, LaSalle, Ill., 1942.

The milestone contributions to mechanics presented in this brief review were obtained from this source.

Archytas of Tarentum (circa 400 B.C.) founded the theory of pulleys. The writings of Archimedes (287–212 B.C.) show that he understood the conditions required for equilibrium of a lever and the principle of buoyancy. Leonardo da Vinci (1452–1519) added to Archimedes' work on levers and formulated the concept of moments as they apply to equilibrium of rigid bodies. Copernicus (1473–1543) proposed that the Earth and the other planets of the solar system revolve about the sun. From the time of Ptolemy in the second century A.D., it had been assumed that the Earth was the center of the universe. Stevinus (1548–1620) first described the behavior of a body on a smooth inclined plane and employed the parallelogram law of addition for forces. Varignon (1654–1722) was the first to establish the equality between the moment of a force and the moment of its components. Both Stevinus and Galileo (1564–1642) appear to have understood the principle of virtual displacements (virtual work), but the universal applicability of the principle to all cases of equilibrium was first perceived by John Bernoulli (1667–1748), who communicated his discovery to Varignon in a letter written in 1717.

The portion of mechanics known as dynamics developed much later, since velocity and acceleration determinations require accurate time measurements. Galileo experimented with blocks on inclined planes, pendulums, and falling bodies; however, he was handicapped by his inability to accurately measure the small time intervals involved in the experiments. Huygens (1629–1695) continued Galileo's work with pendulums and invented the pendulum clock. He also investigated the motion of a conical pendulum and made an accurate determination of the acceleration of gravity. Sir Isaac Newton (1642–1727) is generally credited with laying the true foundation for mechanics with his discovery of the law of universal gravitation and his statement of the laws of motion. Newton's work on particles, based on geometry, was extended to rigid-body systems by Euler (1707–1793). Euler was also the first to use the term *moment of inertia* and developed the parallel-axis theorem for moment of inertia. D'Alembert (1717–1783) introduced the concept of an inertia force. The sum of the forces on a body in motion is zero when an inertia force is included. Previous work in mechanics, which had been based largely on astronomical observations and geometrical concepts, was formalized by Lagrange (1736–1813), who derived the generalized equations of motion, analytically, by using energy concepts. The derivation of these equations, which are known as Lagrange's equations, represented a significant advancement in the development of classical mechanics. Another major advancement was made by Coriolis (1792–1843), who showed how the introduction of additional terms could validate Newton's laws when the reference frame is moving. More recent contributions to mechanics include Max Planck's (1858–1947) formulation of quantum mechanics and Albert Einstein's (1879–1955) formulation of the theory of relativity (1905). These new theories do not repudiate Newtonian mechanics; they are simply more general. Newtonian mechanics is applicable to the prediction of the motion of bodies where the speeds are small compared to the speed of light.

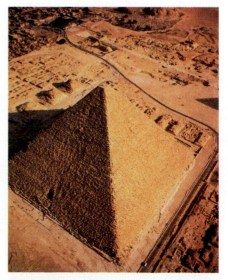

The builders of the pyramids of Egypt probably understood and used most of the basic principles of statics.

## 1-3  FUNDAMENTAL QUANTITIES OF MECHANICS

The fundamental quantities of mechanics are space, time, mass, and force. Three of the quantities—space, time, and mass—are absolute quantities. This means that they are independent of each other and cannot be expressed in terms of the other quantities or in simpler terms. The quantity known as a force is not independent of the other three quantities but is related to the mass of the body and to the manner in which the velocity of the body varies with time. A brief description of these and other important concepts follows.

**Space** is the geometric region in which the physical events of interest in mechanics occur. The region extends without limit in all directions. The measure used to describe the size of a physical system is known as length. The position of a point in space can be determined relative to some reference point by using linear and angular measurements with respect to a coordinate system whose origin is at the reference point. The basic reference system used as an aid in solving mechanics problems is an inertial reference system.[2] Measurements relative to this system are called absolute.

**Time** can be defined as the interval between two events. Measurements of this interval are made by making comparisons with some reproducible event such as the time required for the earth to orbit the sun or the time required for the earth to rotate on its axis. Solar time is earth rotation time measured with respect to the sun and is used for navigation on earth and for daily living purposes.

Any device that is used to indicate passage of time is referred to as a clock. Reproducible events commonly used as sensing mechanisms for clocks include the swing of a pendulum, oscillation of a spiral spring and balance wheel, and oscillation of a piezoelectric crystal. The time required for one of these devices to complete one cycle of motion is known as the period. The frequency of the motion is the number of cycles occurring in a given unit of time.

**Matter** is any substance that occupies space. A body is matter bounded by a closed surface. The property of a body that causes it to resist any change in motion is known as **inertia. Mass** is a quantitative measure of inertia. The resistance a body offers to a change in translational motion is independent of the size and shape of the body. It depends only on the mass of the body. The resistance a body offers to a change in rotational motion depends on the distribution of the mass of the body. Mass is also a factor in the gravitational attraction between two bodies.

A **force** can be defined as the action of one body on another body. Our concept of force comes mainly from personal experiences in which we are one of the bodies and tension or compression of our muscles results when we try to "pull" or "push" the second body. This is an example of force resulting from direct contact between bodies. A force can also be exerted between bodies that are physically separated. Gravitational forces exerted by the earth on the moon and on artificial satellites to keep them in earth orbit are examples. Since a body cannot exert a force on a second body unless the second body offers a resis-

The study of dynamics requires accurate time measurement. In the 16th and 17th centuries, experimenters such as Galileo used their pulse to measure small time intervals. The pendulum clock, invented by Huygens in the 17th century, made accurate and repeatable experimental measurements possible.

---

[2]An inertial reference system is defined as a coordinate system within which a body subject to no net force will either stay at rest or move with a constant velocity. For most engineering problems, axes fixed on the earth provide an adequate approximation of an inertial reference system. Velocity changes resulting from the earth's rotation and orbital motion about the sun are small enough to neglect in most problems.

tance, a force never exists alone. Forces always occur in pairs, and the two forces have equal magnitude and opposite sense. Although a single force never exists, it is convenient in the study of motions of a body to think only of the actions of other bodies on the body in question without taking into account the reactions of the body in question. The external effect of a force on a body is either acceleration of the body or development of resisting forces (reactions) on the body.

A **particle** is an object whose size and shape can be ignored when studying its motion. Only the position of the mass center of the particle need be considered. The orientation of the object or rotation of the object plays no role in describing the motion of a particle. Particles can be very small or very large. Smallness does not always guarantee that an object can be modeled as a particle; largeness does not always prevent modeling an object as a particle. For example, for studies of the orbital motion of the earth about the sun, the earth can be considered to be a particle because the rotation and internal structure of the earth have only insignificant effects on its orbital motion. When a body (large or small) in a mechanics problem can be treated as a particle, the analysis is greatly simplified since the mass can be assumed to be concentrated at a point and the concept of rotation is not involved in the solution of the problem.

A **rigid body** can be represented as a collection of particles. The size and shape of the body as well as the relative position of the particles comprising the body remain constant at all times and under all conditions of loading. The rigid-body concept represents an idealization of the true situation since all real bodies will change shape to a certain extent when subjected to a system of forces. Such changes are small for most structural elements and machine parts encountered in engineering practice; therefore, they have only a negligible effect on the acceleration produced by the force system or on the reactions required to maintain equilibrium of the body. The bodies dealt with in this book, with the exception of deformable springs, will be considered to be rigid bodies.

## 1-3.1 Newton's Laws

The foundations for studies in engineering mechanics are the laws formulated and published by Sir Isaac Newton in 1687. In a treatise called *The Principia*, Newton stated the basic laws governing the motion of a particle as follows:[3]

---

**Newton's Laws of Motion**

*Law 1.* Every body perseveres in its state of rest or of uniform motion in a straight line, except in so far as it is compelled to change that state by impressed forces.

*Law 2.* Change of motion is proportional to the moving force impressed and takes place in the direction of the straight line in which such force is impressed.

*Law 3.* Reaction is always equal and opposite to action; that is to say, the actions of two bodies upon each other are always equal and directly opposite.

---

[3]As stated in Dr. Ernst Mach, *The Science of Mechanics*, 9th ed. The Open Court Publishing Company, LaSalle, Ill., 1942.

These laws, which have come to be known as "Newton's Laws of Motion," are commonly expressed today as follows:

*Law 1.*   A particle originally at rest will remain at rest; a particle originally moving with a constant velocity will continue to move with a constant velocity along a straight line, unless the particle is acted on by an unbalanced force.

*Law 2.*   When a particle is acted on by an unbalanced force, the particle will be accelerated in the direction of the force; the magnitude of the acceleration will be directly proportional to the force and inversely proportional to the mass of the particle.

*Law 3.*   For every action there is an equal and opposite reaction. The forces of action and reaction between contacting bodies are equal in magnitude, opposite in direction, and collinear.

Newton's three laws were developed from a study of planetary motion (the motion of particles); therefore, they apply only to the motion of particles. During the eighteenth century, Leonhard Euler (1707–1783) extended Newton's work on particles to rigid-body systems.

The first law of motion is a special case of the second law and covers the case where the particle is in equilibrium. Thus, the first law provides the foundation for the study of statics. The second law of motion provides the foundation for the study of dynamics. The mathematical statement of the second law that is widely used in dynamics is

$$\mathbf{F} = m\mathbf{a} \tag{1-1}$$

where

$\mathbf{F}$ is the external force acting on the particle,

$m$ is the mass of the particle, and

$\mathbf{a}$ is the acceleration of the particle in the direction of the force.

The third law of motion provides the foundation for an understanding of the concept of a force since in practical engineering applications the word *action* is taken to mean force. Thus, if one body exerts a force on a second body, the second body exerts an equal and opposite force on the first.

## CONCEPTUAL EXAMPLE 1-1: ACTION AND REACTION

*When a force is being applied to a body, does the ability of the body to resist the force have any effect on the amount of force that can be applied?*

(a)

### SOLUTION

Newton's third law states that for every action (force) there is an equal and opposite reaction. Thus, when we try to "pull" or "push" a second body, we can exert the force on the second body only if the second body offers a resistance. Forces always exist in pairs, and the two forces (action and reaction) have equal magnitude and opposite sense.

For example, the action of a person's hand on the rubber band of Fig. CE1-1a is to stretch the band. In this case, the resistance offered by the band as it stretches is relatively small and the force that ultimately can be applied to the band before it breaks is small. In the example shown in Fig. CE1-1b, the action of the hands is to tend to slide or tip the crate. In this case the resistance offered by the crate can be very large, and it may be impossible for a single person to apply sufficient force to slide or tip the crate. In either case, two bodies are involved, and there is an action of one body (the person) on another body (the rubber band or the crate). The resistance that is offered by one body to the action of another body arises out of the ability of a body (1) to resist a change of shape (the rubber band) or (2) to resist a change of motion (the crate).

(b)

**Fig. CE1-1**

The law that governs the mutual attraction between two isolated bodies was also formulated by Newton and is known as the "Law of Gravitation." This law can be expressed mathematically as

$$F = G\frac{m_1 m_2}{r^2} \qquad \text{(1-2)}$$

where

F is the magnitude of the mutual force of attraction between the two bodies.
G is the universal gravitational constant,
$m_1$ is the mass of one of the bodies,
$m_2$ is the mass of the second body, and
r is the distance between the centers of mass of the two bodies.

Approximate values for the universal gravitational constant that are suitable for most engineering computations are

$G = 3.439(10^{-8})$ ft³/(slug · s²)      in the U.S. Customary system of units
$G = 6.673(10^{-11})$ m³/(kg · s²)      in the SI system of units

The mutual forces of attraction between the two bodies represent the action of one body on the other; therefore, they obey Newton's third law, which requires that they be equal in magnitude, opposite in direction, and collinear (lie along the line joining the centers of mass of the two bodies). The law of gravitation is very important in all studies involving the motion of planets or artificial satellites.

Some of the quantities and constants that may be of interest in applying the law of universal gravitation are listed in Table 1-1.

**TABLE 1-1    SOLAR SYSTEM MASSES AND DISTANCES**

| Mass: | | |
|---|---|---|
| of the earth | $= 4.095(10^{23})$ slug | $= 5.976(10^{24})$ kg |
| of the moon | $= 5.037(10^{21})$ slug | $= 7.350(10^{22})$ kg |
| of the sun | $= 1.364(10^{29})$ slug | $= 1.990(10^{30})$ kg |
| Mean or average radius: | | |
| of the earth[a] | $= 2.090(10^{7})$ ft | $= 6.371(10^{6})$ m |
| of the moon | $= 5.702(10^{6})$ ft | $= 1.738(10^{6})$ m |
| of the sun | $= 2.284(10^{9})$ ft | $= 6.960(10^{8})$ m |
| Mean or average distance from the earth: | | |
| to the moon | $= 1.261(10^{9})$ ft | $= 3.844(10^{8})$ m |
| to the sun | $= 4.908(10^{11})$ ft | $= 1.496(10^{11})$ m |

[a]Radius of a sphere of equal volume.

| | | |
|---|---|---|
| Polar radius | $= 2.0856(10^{7})$ ft | $= 6.357(10^{6})$ m |
| Equatorial radius | $= 2.0925(10^{7})$ ft | $= 6.378(10^{6})$ m |

### 1-3.2    Mass and Weight

The mass $m$ of a body is an absolute quantity that is independent of the position of the body and independent of the surroundings in which the body is placed. The weight $W$ of a body is the gravitational attraction exerted on the body by the earth or by any other massive body such as the moon. Therefore, the weight of the body depends on the position of the body relative to some other body. Thus for Eq. 1-2, at the surface of the earth:

$$W = G\frac{m_e m}{r_e^2} = mg \qquad (1\text{-}3)$$

where

$m_e$ is the mass of the earth,
$r_e$ is the mean radius of the earth, and
$g = G\dfrac{m_e}{r_e^2}$ is the gravitational acceleration.

## CONCEPTUAL EXAMPLE 1-2: WEIGHT AND MASS

*A claw hammer is used to drive a nail into a board as shown in Fig. CE1-2. Could the same claw hammer be used to drive a nail into the board in a spacecraft, far from any star or planet, where the hammer is weightless?*

### SOLUTION

Mass and weight are different quantities. Mass is an intrinsic property of a body that does not change as the body is moved from one location to another. Weight is the gravitational force exerted on a body by the earth or any other second body.

On the surface of the earth, the block, hammer, and nail have weights that result from the gravitational attraction of the earth. In a spacecraft far from any star or planet, the block, hammer, and nail are essentially weightless.

On earth, a force $\mathbf{F} = m\mathbf{a}$ (which depends on the mass $m$ of the hammer) is required to change the velocity of the hammer as it strikes the nail. The reaction to this force $\mathbf{F}$ drives the nail into the block. In space, the same force $\mathbf{F}$ is required to change the velocity of the hammer because the hammer has the same mass in space as on earth. Therefore, the claw hammer could be used to drive a nail into a board both on earth and in a spacecraft. Note also that even though the hammer has no weight in space, the same pain would be felt in space as on earth if the hammer misses the nail and strikes the astronaut's thumb.

Fig. CE1-2

The internationally adopted standard value for the gravitational acceleration at sea level and at a latitude of 45° is

$$g = 9.80665 \ \text{m/s}^2$$

or

$$g = 32.1740 \ \text{ft/s}^2$$

Approximate values for the gravitational acceleration that are suitable for most engineering computations are

$$g = 32.17 \ \text{ft/s}^2 = 9.807 \ \text{m/s}^2$$

A source of some confusion arises because the pound is sometime used as a unit of mass and the kilogram is sometimes used as a unit of force. In grocery stores in Europe, weights of packages are marked in kilograms. In the United States, weights of packages are often marked in both pounds and kilograms. Similarly, a unit of mass called the pound or the pound mass, which is the mass whose weight is one pound under standard gravitational conditions, is sometimes used.

Throughout this book on statics and the companion book on dynamics, without exception, the pound (lb) will be used as the unit of force and the slug will be used as the unit of mass for problems and examples when the U.S. Customary System of units is used. Similarly, the newton (N) will be used as the unit of force and the kilogram (kg) will be used as the unit of mass for problems and examples when the SI System of units is used.

A body weighs 250 lb at the earth's surface. Determine

a. The mass of the body.
b. The weight of the body 500 mi above the earth's surface.
c. The weight of the body on the moon's surface.

### SOLUTION

a. The weight of a body at the earth's surface is given by Eq. 1-3 as

$$W = mg$$

Thus,

$$m = \frac{W}{g} = \frac{250}{32.17} = 7.77\frac{\text{lb} \cdot \text{s}^2}{\text{ft}} = 7.77 \text{ slug} \qquad \text{Ans.}$$

b. The force of attraction between two bodies is given by Eq. 1-2 as

$$W = F = G\frac{m_1 m_2}{r^2}$$

or

$$Wr^2 = Gm_1 m_2 = \text{constant}$$

The mean radius of the earth (see Table 1-1) is $r_e = 2.090(10^7)$ ft = 3958 mi. Thus, for the two positions of the body

$$Wr_e^2 = W_{500}(r_e + 500)^2 = Gm_1 m_2 = \text{constant}$$

$$W_{500} = \frac{Wr_e^2}{(r_e + 500)^2}$$

$$= \frac{250(3958)^2}{(3958 + 500)^2} + 197.1 \text{ lb} \qquad \text{Ans.}$$

> When $h = 500$ mi, the body is $(500 + r_e)$ mi from the center of the earth.

c. On the moon's surface, the weight of the body is given by Eq. 1-2 as

$$W = G\frac{m m_m}{r_m^2}$$

The mean radius and mass of the moon (see Table 1-1) are $r_m = 5.702(10^6)$ ft and $m_m = 5.037(10^{21})$ slug. Also, $G = 3.439(10^{-8})$ ft$^3$/(slug $\cdot$ s$^2$). Thus,

$$W = G\frac{m m_m}{r_m^2}$$

$$= 3.439(10^{-8})\frac{7.77(5.037)(10^{21})}{[5.702(10^6)]^2} = 41.4 \text{ lb} \qquad \text{Ans.}$$

At what distance $d$, in kilometers, from the surface of the earth on a line from center to center would the gravitational force of the earth on a rocket be three times the gravitational force of the moon on the rocket?

### SOLUTION

Newton's "Law of Gravitation" (Eq. 1-2) states that the mutual force of attraction between two isolated bodies is

$$F = \frac{Gm_1m_2}{r^2}$$

when the force $F_e$ exerted on the rocket by the earth is three times the force $F_m$ exerted on the rocket by the moon, as shown in Fig. 1-1.

**Fig. 1-1**

$$\frac{Gm_em_r}{a^2} = \frac{3Gm_mm_r}{b^2}$$

$$a = \left[\frac{m_eb^2}{3m_m}\right]^{1/2} = \left[\frac{5.976(10^{24})b^2}{3(7.350)(10^{22})}\right]^{1/2} = 5.206b$$

From Table 1-1,

$$a + b = 3.844(10^8)\text{ m}$$

Therefore,

$$a = 3.225(10^8)\text{ m}$$
$$b = 0.619(10^8)\text{ m}$$
$$d = a - r_e = 3.225(10^8) - 6.371(10^6)$$
$$= 3.16(10^8)\text{ m} = 3.16(10^5)\text{ km} \qquad \text{Ans.}$$

A distance $d$ from the surface of the earth is a distance $a = d + r_e$ from the center of the earth as shown in Fig. 1-1.

# PROBLEMS

## Introductory Problems

**1-1\*** Calculate the mass $m$ of a body that weighs 600 lb at the surface of the earth.

**1-2\*** Calculate the weight $W$ of a body at the surface of the earth if it has a mass $m$ of 675 kg.

**1-3** Calculate the gravitational constant $g$, in U.S. Customary units, for a location on the surface of the moon.

**1-4** Calculate the gravitational constant $g$, in SI units, for a location on the surface of the sun.

**1-5\*** The planet Venus has a diameter of 7700 mi and a mass of $3.34(10^{23})$ slug. Determine the gravitational acceleration at the surface of the planet.

**1-6\*** The gravitational acceleration at the surface of Mars is $3.73 \text{ m/s}^2$, and the mass of Mars is $6.39(10^{23})$ kg. Determine the radius of Mars.

**1-7** Determine the gravitational force, in kips, exerted by the sun on the earth.

**1-8** Determine the gravitational force, in kilonewtons, exerted by the earth on the moon.

**1-9\*** The equatorial radius of the earth is $2.0925(10^7)$ ft, and the polar radius is $2.0856(10^7)$ ft. Determine the gravitational acceleration $g$ at the two locations.

**1-10\*** Two spherical bodies have masses of 60 kg and 80 kg, respectively. Determine the gravitational force of attraction between the spheres if the distance from center to center is 600 mm.

**1-11** Two solid spherical bodies have 12 in. and 10 in. diameters and are made of a material that weighs $0.284 \text{ lb/in.}^3$. Determine the gravitational force of attraction between the two spheres when they are touching each other.

**1-12** A satellite is placed in orbit $1.6(10^6)$ m above the surface of the moon. If the mass of the satellite is $3.0(10^4)$ kg, determine the gravitational force exerted on the satellite by the moon.

## Intermediate Problems

**1-13\*** Determine the weight $W$ of a satellite when it is in orbit 8500 mi above the surface of the earth if the satellite weighs 7600 lb at the surface.

**1-14\*** Determine the weight $W$ of a satellite when it is in orbit $20.2(10^6)$ m above the surface of the earth if the satellite weighs 8450 N at the surface.

**1-15** If a woman weighs 135 lb when standing on the surface of the earth, how much would she weigh when standing on the surface of the moon?

**1-16** Determine the weight $W$ of a body that has a mass of 1000 kg

a. At the surface of the earth.
b. At the top of Mt. McKinley (6193 m above sea level).
c. In a satellite at an altitude of 250 km.

**1-17\*** If a man weighs 210 lb at sea level, determine the weight $W$ of the man

a. At the top of Mt. Everest (29,028 ft above sea level).
b. In a satellite at an altitude of 200 mi.

**1-18** A space traveler weighs 800 N on earth. A planet having a mass of $5(10^{25})$ kg and a diameter of $30(10^6)$ m orbits a distant star. Determine the weight $W$ of the traveler on the surface of this planet.

## Challenging Problems

**1-19\*** The planet Jupiter has a mass of $1.302(10^{26})$ slug and a visible diameter (top of cloud layer) of 88,700 mi. Determine the gravitational acceleration $g$

a. At a point 100,000 miles above the top of the clouds.
b. At the top of the cloud layers.

**1-20\*** The planet Saturn has a mass of $5.67(10^{26})$ kg and a visible diameter (top of cloud layer) of $12.00(10^7)$ m. The weight $W$ of a planetary probe on earth is 4.50 kN. Determine

a. The weight of the probe when it is $6(10^8)$ m above the top of the clouds.
b. The weight of the probe as it begins its penetration of the cloud layers.

**1-21** The first U.S. satellite, *Explorer I*, had a mass of approximately 1 slug. Determine the force exerted on the satellite by the earth at the low and high points of its orbit, which were 175 mi and 2200 mi, respectively, above the surface of the earth.

**1-22** A neutron star has a mass of $2(10^{30})$ kg and a diameter of $10(10^3)$ m. Determine the gravitational force of attraction on a 10-kg space probe

a. When it is $10^6$ m from the center of the star.
b. At the instant of impact with the surface of the star.

**1-23\*** At what distance from the surface of the earth, in miles, is the weight of a body equal to one-half of its weight on the earth's surface?

**1-24** At what distance, in kilometers, from the surface of the earth on a line from center to center would the gravitational force of the earth on a body be exactly balanced by the gravitational force of the moon on the body?

## 1-4 UNITS OF MEASUREMENT

The building blocks of mechanics are the physical quantities used to express the laws of mechanics. Some of these quantities are mass, length, force, time, velocity, and acceleration. Physical quantities are often divided into fundamental quantities and derived quantities. **Fundamental quantities** cannot be defined in terms of other physical quantities. The number of quantities regarded as fundamental is the minimum number needed to give a consistent and complete description of all the physical quantities ordinarily encountered in the subject area. Examples of quantities viewed as fundamental in the field of mechanics are length and time. **Derived quantities** are those whose defining operations are based on measurements of other physical quantities. Examples of derived quantities in mechanics are area, volume, velocity, and acceleration. Some quantities may be viewed as either fundamental or derived. Mass and force are examples of such quantities. In the SI system of units, mass is regarded as a fundamental quantity and force as a derived quantity. In the U.S. Customary System of units, force is regarded as a fundamental quantity and mass as a derived quantity.

The magnitude of each of the fundamental quantities is defined by an arbitrarily chosen unit or "standard." The familiar yard, foot, and inch, for example, come from the practice of using the human arm, foot, and thumb as length standards. However, for any sort of precise calculations, such units of length are unsatisfactory. The first truly international standard of length was a bar of platinum-iridium alloy, called the standard meter[4] which was kept at the International Bureau of Weights and Measures in Sèvres, France. The distance between two fine lines engraved on gold plugs near the ends of the bar is defined to be one meter. Historically, the meter was intended to be one ten-millionth of the distance from the pole to the equator along the meridian line through Paris. Accurate measurements made after the standard meter bar was constructed show that it differs from its intended value by approximately 0.023 percent.

In 1961 an atomic standard of length was adopted by international agreement. The wavelength in vacuum of the orange-red line from the spectrum of isotope krypton 86 was chosen. One meter (m) is now defined to be 1,650,763.73 wavelengths of this light. The choice of an atomic standard offers advantages other than increased precision in length measurements. Krypton 86 is available everywhere, the material can be obtained relatively easily and cheaply, and all atoms of the material are identical and emit light of the same wavelength. The particular wavelength chosen is uniquely characteristic of krypton 86 and is very sharply defined. The definition of the yard, by international agreement, is 1 yard = 0.9144 m, exactly.[5] Thus, 1 in. = 25.4 mm, exactly; and 1 ft = 0.3048 m, exactly.

Similarly, time can be measured in a number of ways. Since the earliest times, the length of a day has been an accepted standard of time measurement. The internationally accepted standard unit of time, the second (s), has been defined in the past as 1/86,400 of a mean solar day or 1/31,557,700 of a mean solar year. Time defined in terms of

The international standard of length is the meter.

---

[4]The United States has accepted the meter as a standard of length since 1893.
[5]*Guide for the Use of the International System of Units,* National Institute of Standards and Technology (NIST) Special Publication 811, September 1991.

The cesium clock.

The standard kilogram.

the rotation of the earth must be determined by astronomical observations. Since these observations require at least several weeks, a good secondary terrestrial measure, calibrated by astronomical observations, is used. Quartz crystal clocks, based on the electrically sustained natural periodic vibrations of a quartz wafer, have been used as secondary time standards. The best of these quartz clocks have kept time for a year with a maximum error of 0.02 seconds.

To meet the need for a better time standard, an atomic clock has been developed that uses the periodic atomic vibrations of isotope cesium 133. The second based on this cesium clock was adopted as the time standard by the Thirteenth General Conference on Weights and Measures in 1967. The second is defined as the duration of 9,192,631,770 cycles of vibration of isotope cesium 133. The cesium clock provides an improvement over the accuracy associated with astronomical methods by a factor of approximately 200. Two cesium clocks will differ by no more than one second after running 3000 years.

The standard unit of mass, the kilogram (kg), is defined by a bar of platinum-iridium alloy that is kept at the International Bureau of Weights and Measures in Sèvres, France.

### 1-4.1 The U.S. Customary System of Units

Until very recently, almost all engineers in the United States used the U.S. Customary System of Units (sometimes called the British gravitational system) in which the base units are foot (ft) for length, the pound (lb) for force, and the second (s) for time. In this system, the foot is defined as 0.3048 m, exactly. The pound is defined as the weight at sea level and at a latitude of 45 degrees of a platinum standard, which is kept at the Bureau of Standards in Washington, D.C. This platinum standard has a mass of 0.453,592,43 kg. The second is defined in the same manner as in the SI system.

In the U.S. Customary System, the unit of mass is derived and is called a slug. One slug is the mass that is accelerated one foot per second squared by a force of one pound, or 1 slug equals $1 \text{ lb} \cdot \text{s}^2/\text{ft}$. Since the weight of the platinum standard depends on the gravitational attraction of the earth, the U.S. Customary System is a gravitational system of units rather than an absolute system of units.

### 1-4.2 The International System of Units (SI)

The original metric system provided a set of units for the measurement of length, area, volume, capacity, and mass based on two fundamental units: the meter and the kilogram. With the addition of a unit of time, practical measurements began to be based on the meter–kilogram–second (MKS) system of units. In 1960 the Eleventh General Conference on Weights and Measures formally adopted the Système international d'unités (International System of Units), for which the abbreviation is SI in all languages, as the international standard. Thirty-six countries, including the United States, participated in this conference.

The International System of Units adopted by the conference includes three classes of units: (1) base units, (2) supplementary units, and (3) derived units. The system is founded on the seven base units listed in Table 1-2.

**TABLE 1-2 BASE UNITS AND THEIR SYMBOLS**

| Quantity | Name of Unit | Symbol |
|---|---|---|
| Length | meter | m |
| Mass | kilogram | kg |
| Time | second | s |
| Electric current | ampere | A |
| Thermodynamic temperature | kelvin | K |
| Amount of substance | mole | mol |
| Luminous intensity | candela | cd |

Certain units of the international system have not been classified under either base units or derived units. These units, listed in Table 1-3, are called supplementary units and may be regarded as either base units or derived units.

Derived units are expressed algebraically in terms of base units and/or supplementary units. Their symbols are obtained by means of the mathematical signs of multiplication and division. For example, the SI unit for velocity is meters per second (m/s) and the SI unit for angular velocity is radians per second (rad/s). In the SI system, the

**TABLE 1-3 SUPPLEMENTARY UNITS AND THEIR SYMBOLS**

| Quantity | Name of Unit | Symbol |
|---|---|---|
| Plane angle | radian | rad |
| Solid angle | steradian | sr |

**TABLE 1-4 DERIVED UNITS AND THEIR SYMBOLS AND SPECIAL NAMES**

| Quantity | Derived SI Unit | Symbol | Special Name |
|---|---|---|---|
| Area | square meter | $m^2$ | — |
| Volume | cubic meter | $m^3$ | — |
| Linear velocity | meter per second | m/s | — |
| Angular velocity | radian per second | rad/s | — |
| Linear acceleration | meter per second squared | $m/s^2$ | — |
| Frequency | (cycle) per second | Hz | hertz |
| Density | kilogram per cubic meter | $kg/m^3$ | — |
| Force | kilogram · meter per second squared | N | newton |
| Moment of force | newton · meter | N · m | — |
| Pressure | newton per meter squared | Pa | pascal |
| Stress | newton per meter squared | Pa | pascal |
| Work | newton · meter | J | joule |
| Energy | newton · meter | J | joule |
| Power | joule per second | W | watt |

unit of force is derived and is called a newton. One newton is the force required to give one kilogram of mass an acceleration of one meter per second squared. Thus, $1 \text{ N} = 1 \text{ kg} \cdot \text{m/s}^2$. For some of the derived units, special names and symbols exist; those of interest in mechanics are listed in Table 1-4.

Prefixes are used to form names and symbols of multiples (decimal multiples and submultiples) of SI names. The choice of the appropriate multiple is governed by convenience and should usually be chosen so that the numerical values will be between 0.1 and 1000. Only one prefix should be used in forming a multiple of a compound unit, and prefixes in the denominator should be avoided. Approved prefixes with their names and symbols are listed in Table 1-5.

As the use of the SI system becomes more commonplace in the United States, engineers will be required to be familiar with both the SI system and the U.S. Customary System in common use today. As an aid to interpreting the physical significance of answers in SI units for those more accustomed to the U.S. Customary System, some conversion factors for the quantities normally encountered in mechanics are provided in Table 1-6.

For the foreseeable future, engineers in the United States will be required to work with both the U.S. Customary and SI systems of units; therefore, we have used both sets of units in examples and problems in this book on statics and the companion volume on dynamics.

**TABLE 1-5 MULTIPLES OF SI UNITS**

| Factor by Which Unit Is Multiplied | Prefix Name | Prefix Symbol |
|---|---|---|
| $10^{18}$ | exa | E |
| $10^{15}$ | peta | P |
| $10^{12}$ | tera | T |
| $10^{9}$ | giga | G |
| $10^{6}$ | mega | M |
| $10^{3}$ | kilo | k |
| $10^{2}$ | hecto[a] | h |
| 10 | deca[a] | da |
| $10^{-1}$ | deci[a] | d |
| $10^{-2}$ | centi[a] | c |
| $10^{-3}$ | milli | m |
| $10^{-6}$ | micro | $\mu$ |
| $10^{-9}$ | nano | n |
| $10^{-12}$ | pico | p |
| $10^{-15}$ | femto | f |
| $10^{-18}$ | atto | a |

[a]To be avoided when possible.

**TABLE 1-6    CONVERSION FACTORS BETWEEN THE SI AND U.S. CUSTOMARY SYSTEMS**

| Quantity | U.S. Customary to SI | SI to U.S. Customary |
|---|---|---|
| Length | 1 in. = 25.40 mm | 1 m = 39.37 in. |
| | 1 ft = 0.3048 m | 1 m = 3.281 ft |
| | 1 mi = 1.609 km | 1 km = 0.6214 mi |
| Area | 1 in.$^2$ = 645.2 mm$^2$ | 1 m$^2$ = 1550 in.$^2$ |
| | 1 ft$^2$ = 0.0929 m$^2$ | 1 m$^2$ = 10.76 ft$^2$ |
| Volume | 1 in.$^3$ = 16.39($10^3$) mm$^3$ | 1 mm$^3$ = 61.02($10^{-6}$) in.$^3$ |
| | 1 ft$^3$ = 0.02832 m$^3$ | 1 m$^3$ = 35.31 ft$^3$ |
| | 1 gal = 3.785 L[a] | 1 L = 0.2642 gal |
| Velocity | 1 in./s = 0.0254 m/s | 1 m/s = 39.37 in./s |
| | 1 ft/s = 0.3048 m/s | 1 m/s = 3.281 ft/s |
| | 1 mi/h = 1.609 km/h | 1 km/h = 0.6214 mi/h |
| Acceleration | 1 in./s$^2$ = 0.0254 m/s$^2$ | 1 m/s$^2$ = 39.37 in./s$^2$ |
| | 1 ft/s$^2$ = 0.3048 m/s$^2$ | 1 m/s$^2$ = 3.281 ft/s$^2$ |
| Mass | 1 slug = 14.59 kg | 1 kg = 0.06854 slug |
| Second moment of area | 1 in.$^4$ = 0.4162($10^6$) mm$^4$ | 1 mm$^4$ = 2.402($10^{-6}$) in.$^4$ |
| Force | 1 lb = 4.448 N | 1 N = 0.2248 lb |
| Distributed load | 1 lb/ft = 14.59 N/m | 1 kN/m = 68.54 lb/ft |
| Pressure or stress | 1 psi = 6.895 kPa | 1 kPa = 0.1450 psi |
| | 1 ksi = 6.895 MPa | 1 MPa = 145.0 psi |
| Bending moment or torque | 1 ft · lb = 1.356 N · m | 1 N · m = 0.7376 ft · lb |
| Work or energy | 1 ft · lb = 1.356 J | 1 J = 0.7376 ft · lb |
| Power | 1 ft · lb/s = 1.356 W | 1 W = 0.7376 ft · lb/s |
| | 1 hp = 745.7 W | 1 kW = 1.341 hp |

[a]Both L and 1 are accepted symbols for liter. Because "1" can be easily confused with the numeral "1", the symbol "L" is recommended for U.S. use by the National Institute of Standards and Technology (see NIST special publication 811, September 1991).

## EXAMPLE PROBLEM 1-3

A foreign manufacturer lists the fuel consumption for a new automobile as 15 kilometers per liter (km/L). Determine the fuel consumption in miles per gallon.

### SOLUTION

One accepted procedure for converting units is to write the associated units in abbreviated form with each of the numerical values used in the conversion. Like unit symbols can then be canceled in the same manner as algebraic symbols. The conversion factors (see Table 1-6) needed for this exercise are

$$1 \text{ km} = 0.6214 \text{ mi}$$
$$1 \text{ gal} = 3.785 \text{ L}$$

Thus,

$$15\frac{\text{km}}{\text{L}} \times 0.6214\frac{\text{mi}}{\text{km}} \times 3.785\frac{\text{L}}{\text{gal}} = 35.3 \text{ mi/gal} \qquad \text{Ans.}$$

## EXAMPLE PROBLEM 1-4

The value of G (universal gravitational constant) used for engineering computations in the U.S. system of units is G = $3.439(10^{-8})$ ft$^3$/(slug · s$^2$). Use the conversion factors listed in Table 1-6 to determine a value of G with units of m$^3$/(kg · s$^2$) suitable for computations in the SI system of units.

### SOLUTION

The conversion factors (see Table 1-6) needed for this example are

$$1 \text{ ft}^3 = 0.02832 \text{ m}^3$$
$$1 \text{ kg} = 0.06854 \text{ slug}$$

Thus,

$$G = 3.439(10^{-8})\frac{\text{ft}^3}{\text{slug} \cdot \text{s}^2} \times 0.02832\frac{\text{m}^3}{\text{ft}^3} \times 0.06854\frac{\text{slug}}{\text{kg}}$$
$$= 6.675(10^{-11})\frac{\text{m}^3}{\text{kg} \cdot \text{s}^2} \qquad \text{Ans.}$$

# PROBLEMS

## Introductory Problems

**1-25*** Determine the weight $W$, in U.S. Customary units, of an 85-kg steel bar under standard conditions (sea level at a latitude of 45 degrees).

**1-26*** Determine the mass $m$, in SI units, for a 600-lb steel beam under standard conditions (sea level at a latitude of 45 degrees).

**1-27** The velocity of light in space is approximately 186,000 mi/s. The velocity of light in units of kilometers per hour is?

**1-28** Using the fact that 1 m = 39.37 in., convert 5 m$^3$ of concrete to units of cubic yards of concrete.

**1-29*** Using the fact that 1 in. = 25.40 mm, convert a speed of 75 mi/h to units of meters per second.

**1-30*** The fuel consumption of an automobile is 14 km/L. The fuel consumption, in miles per gallon, is?

**1-31** An automobile has a 350-in.$^3$ engine displacement. The engine displacement, in liters, is?

**1-32** How many barrels of oil are contained in 100 kL of oil? One barrel (petroleum) equals 42.0 gal.

**1-33*** Express a speed of 20 nm/h (1 nautical mile = 6076 ft) in units of kilometers per minute.

**1-34*** One acre equals 43,560 ft$^2$. One hectare equals $10^4$ m$^2$. Determine the number of acres in 500 hectares.

**1-35** Verify the conversion factors listed in Table 1-6 for converting the following quantities from U.S. Customary units to SI units by using the values listed for length as defined values:

**a.** Velocity **b.** Acceleration

**1-36** Verify the conversion factors listed in Table 1-6 for converting the following quantities from SI units to U.S. Customary units by using the values listed for length as defined values:

**a.** Area **b.** Volume

Use 1 gal = 231 in.$^3$ and 1 L = 0.001 m$^3$.

## Intermediate Problems

**1-37** Verify the conversion factors listed in Table 1-6 for converting the following quantities from U.S. Customary units to SI units by using the values listed for length and force as defined values:

**a.** Mass **b.** Distributed load

**1-38** Verify the conversion factors listed in Table 1-6 for converting the following quantities from SI units to U.S. Customary units by using the values listed for length and mass as defined values:

**a.** Pressure or stress **b.** Bending moment or torque

**1-39** One acre equals 43,560 ft$^2$. One gallon equals 231 in.$^3$. Determine the number of liters of water in 2500 acre · ft of water.

**1-40** The viscosity of crude oil under conditions of standard temperature and pressure is $7.13(10^{-3})$ N · s/m$^2$. The viscosity of crude oil in U.S. Customary units (lb · s/ft$^2$) is?

**1-41*** The air pressure in an automobile tire is 35 psi. Express the pressure in appropriate SI units (kPa) by using the values listed in Table 1-6 for length and force as defined values.

**1-42** The stress in a steel bar is 150 MPa. Express the stress in appropriate U.S. Customary units (ksi) by using the values listed in Table 1-6 for length and force as defined values.

## Challenging Problems

**1-43*** Express the density, in SI units (kg/m$^3$), of a specimen of material that has a specific weight of 0.284 lb/in.$^3$.

**1-44*** Express the specific weight, in U.S. Customary units (lb/in.$^3$), of a specimen of material that has a density of 4500 kg/m$^3$.

**1-45** By definition, 1 hp = 33,000 ft · lb/min and 1 W = 1 N · m/s. Verify the conversion factors listed in Table 1-6 for converting power from U.S. Customary units to SI units by using the values listed for length and force as defined values.

**1-46** The specific heat of air under standard atmospheric pressure, in SI units, is 1003 N · m/kg · K. The specific heat of air under standard atmospheric pressure, in U.S. Customary units (ft · lb/slug · °R), is?

## 1-5  DIMENSIONAL CONSIDERATIONS

All the physical quantities encountered in mechanics can be expressed dimensionally in terms of the three fundamental quantities: mass, length, and time, denoted respectively by $M$, $L$, and $T$. The dimensions of quantities other than the fundamental quantities follow from definitions or from physical laws. For example, the dimension of velocity, $L/T$, follows from the definition of velocity, rate of change of position with time. Similarly, acceleration is defined as the rate of change of velocity with time and has the dimension $L/T^2$. From Newton's second law, force is defined as the product of mass and acceleration; therefore, force has the dimension $ML/T^2$. The dimensions of a number of other physical quantities commonly encountered in mechanics are given in Table 1-7.

### 1-5.1  Dimensional Homogeneity

An equation is said to be dimensionally homogeneous if the form of the equation does not depend on the units of measurement. For example, the equation describing the distance $h$ a body released from rest has fallen is $h = gt^2/2$, where $h$ is the distance traveled, $t$ is the time since release, and $g$ is the gravitational acceleration. This equation is valid whether length is measured in feet, meters, or inches and whether time is measured in hours, years, or seconds, provided $g$ is measured in the

**TABLE 1-7  DIMENSIONS OF THE PHYSICAL QUANTITIES OF MECHANICS**

| Physical Quantity | Dimension | Common Units | |
|---|---|---|---|
| | | SI System | U.S. Customary System |
| Length | $L$ | m, mm | in., ft |
| Area | $L^2$ | m², mm² | in.², ft² |
| Volume | $L^3$ | m³, mm³ | in.³, ft³ |
| Angle | $1\ (L/L)$ | rad, degree | rad, degree |
| Time | $T$ | s | s |
| Linear velocity | $L/T$ | m/s | ft/s |
| Linear acceleration | $L/T^2$ | m/s² | ft/s² |
| Angular velocity | $1/T$ | rad/s | rad/s |
| Angular acceleration | $1/T^2$ | rad/s² | rad/s² |
| Mass | $M$ | kg | slug |
| Force | $ML/T^2$ | N | lb |
| Moment of a force | $ML^2/T^2$ | N · m | ft · lb |
| Pressure | $M/LT^2$ | Pa, kPa | psi, ksi |
| Stress | $M/LT^2$ | Pa, MPa | psi, ksi |
| Energy | $ML^2/T^2$ | J | ft · lb |
| Work | $ML^2/T^2$ | J | ft · lb |
| Power | $ML^2/T^3$ | W | hp |
| Linear impulse | $ML/T$ | N · s | lb · s |
| Momentum | $ML/T$ | N · s | lb · s |
| Specific weight | $M/L^2T^2$ | N/m³ | lb/ft³ |
| Density | $M/L^3$ | kg/m³ | slug/ft³ |
| Second moment of area | $L^4$ | m⁴, mm⁴ | in.⁴, ft⁴ |
| Moment of inertia | $ML^2$ | kg · m² | slug · ft² |

same units of length and time as $h$ and $t$. Therefore, by definition, the equation is dimensionally homogeneous.

If the value $g = 32.2$ ft/s² is substituted in the previous equation, the equation obtained is $h = 16.1t^2$ ft/s². This equation is not dimensionally homogeneous since the equation applies only if length is measured in feet and time is measured in seconds. Dimensionally homogeneous equations are usually preferred because of the potential confusion connected with the unknown units of constants appearing in dimensionally inhomogeneous equations. Dimensionally homogeneous equations also eliminate the explicit use of unit conversion factors.

All like dimensions in a given equation should be measured with the same unit. For example, if the length dimension of a beam is measured in feet and the cross-sectional dimensions are measured in inches, all measurements should be converted to either feet or inches before they are used in a given equation. If this is done, the terms of the equation can be combined after the numerical values for the variables are substituted.

## EXAMPLE PROBLEM 1-5

The angle of twist for a circular shaft subjected to a twisting moment is given by the equation $\theta = TL/GJ$. What are the dimensions of $J$ if $\theta$ is an angle in radians, $T$ is the moment of a force, $L$ is a length, and $G$ is a force per unit area?

## SOLUTION

The dimensions of the given quantities are

$$\theta = 1 \text{ (dimensionless)}$$

$$T = FL = \frac{ML}{T^2}(L) = \frac{ML^2}{T^2}$$

$$G = \frac{F}{A} = \frac{ML/T^2}{L^2} = \frac{M}{LT^2}$$

All physical quantities can be expressed in terms of three fundamental quantities mass: ($M$), length ($L$), and time ($T$).

Therefore,

$$J = \frac{TL}{G\theta} = \frac{(ML^2/T^2)(L)}{(M/LT^2)(1)} = L^4 \qquad \text{Ans.}$$

Determine the dimensions of $I$, $R$, $w$, $M$, and $C$ in the dimensionally homogeneous equation

$$EIy = Rx^3 - P(x - a)^3 - wx^4 + Mx^2 + C$$

in which $x$ and $y$ are lengths, $P$ is a force, and $E$ is a force per unit area.

## SOLUTION

The equation can be written dimensionally as

$$\frac{F}{L^2}(I)(L) = R(L^3) - F(L - a)^3 - w(L^4) + M(L^2) + C$$

For this equation to be dimensionally homogeneous $a$ must be a length; hence, all terms must have the dimensions $FL^3$. Thus,

$$(I)\frac{F}{L} = (R)L^3 = (w)L^4 = (M)L^2 = C = FL^3$$

> Every term in a dimensionally homogeneous equation must have the same dimension.

The dimensions for each of the unknown quantities are obtained as follows:

$$I = \frac{L}{F}(FL^3) = L^4 \qquad \text{Ans.}$$

$$R = \frac{1}{L^3}(FL^3) = F \qquad \text{Ans.}$$

$$w = \frac{1}{L^4}(FL^3) = \frac{F}{L} \qquad \text{Ans.}$$

$$M = \frac{1}{L^2}(FL^3) = FL \qquad \text{Ans.}$$

$$C = FL^3 \qquad \text{Ans.}$$

# PROBLEMS

### Introductory Problems

**1-47*** Newton's law of gravitation can be expressed in equation form as

$$F = G\frac{m_1 m_2}{r^2}$$

If $F$ is a force, $m_1$ and $m_2$ are masses, and $r$ is a distance, determine the dimensions of $G$.

**1-48*** The elongation of a bar of uniform cross section subjected to an axial force is given by the equation

$$\delta = \frac{PL}{AE}$$

What are the dimensions of $E$ if $\delta$ and $L$ are lengths, $P$ is a force, and $A$ is an area?

**1-49** An important parameter in certain types of fluid flow problems when a free surface is present is the Froude number (Fr), which can be expressed in equation form as

$$\text{Fr} = \left[\frac{\rho v^2}{Lw}\right]^{1/2}$$

where $\rho$ is the density of the fluid, $v$ is a velocity, $L$ is a length, and $w$ is the specific weight of the fluid. Show that the Froude number is dimensionless.

**1-50** An important parameter in fluid flow problems involving thin films is the Weber number (We) which can be expressed in equation form as

$$\text{We} = \frac{\rho v^2 L}{\sigma}$$

**23**

where $\rho$ is the density of the fluid, $v$ is a velocity, $L$ is a length, and $\sigma$ is the surface tension of the fluid. If the Weber number is dimensionless, what are the dimensions of the surface tension $\sigma$?

**1-51*** The period of oscillation of a simple pendulum is given by the equation

$$T = k(L/g)^{1/2}$$

where $T$ is in seconds, $L$ is in feet, $g$ is the acceleration due to gravity, and $k$ is a constant. What are the dimensions of $k$ for dimensional homogeneity?

**1-52*** In the equation

$$y = y_0 + vt + \frac{1}{2}at^2$$

$y$ and $y_0$ are distances, $v$ is a velocity, $a$ is an acceleration, and $t$ is time. Is the equation dimensionally homogeneous?

**1-53** The modulus $k$ of a coil spring (force required to stretch the spring a unit distance) can be expressed in equation form as

$$k = \frac{Gr^4}{4R^3n}$$

in which $r$ and $R$ are lengths and $n$ is a dimensionless number. Determine the dimensions of $G$ (a property of the spring material).

**1-54** In the dimensionally homogeneous equation

$$U = Fd - \frac{Wv^2}{2g}$$

$F$ is a force, $W$ is a force, $d$ is a length, and $v$ is a linear velocity. Determine the dimensions of $U$ and $g$.

## Intermediate Problems

**1-55*** In the dimensionally homogeneous equation

$$\sigma = \frac{P}{A} + \frac{Mc}{I}$$

$\sigma$ is a stress, $A$ is an area, $M$ is a moment of a force, and $c$ is a length. Determine the dimensions of $P$ and $I$.

**1-56*** In the dimensionally homogeneous equation

$$Pd = \frac{1}{2}mv^2 + \frac{1}{2}I\omega^2$$

$d$ is a length, $m$ is a mass, $v$ is a linear velocity, and $\omega$ is an angular velocity. Determine the dimensions of $P$ and $I$.

**1-57** In the dimensionally homogeneous equation

$$\tau = \frac{Tr}{J} + \frac{VQ}{Ib}$$

$\tau$ is a stress, $T$ is a torque, $V$ is a force, $r$ and $b$ are lengths, and $I$ is a second moment of an area. Determine the dimensions of $J$ and $Q$.

**1-58** In the dimensionally homogeneous equation

$$\tau = \frac{P}{A} + \frac{Tr}{J}$$

$\tau$ is a stress, $A$ is an area, $T$ is a torque, and $r$ is a length. Determine the dimensions of $P$ and $J$.

## Challenging Problems

**1-59*** The equation $x = Ae^{-t/b} \sin(at + \alpha)$ is dimensionally homogeneous. If $A$ is a length and $t$ is time, determine the dimensions of $x$, $a$, $b$, and $\alpha$.

**1-60*** In the dimensionally homogeneous equation $w = x^3 + ax^2 + bx + a^2b/x$, if $x$ is a length, what are the dimensions of $a$, $b$, and $w$?

**1-61** Determine the dimensions of $a$, $b$, $c$, and $y$ in the dimensionally homogeneous equation

$$y = Ae^{-bt} \cos\left[\sqrt{1 - a^2}\, bt + c\right]$$

in which $A$ is a length and $t$ is time.

**1-62** Determine the dimensions of $c$, $\omega$, $k$, and $P$ in the differential equation

$$m\frac{d^2x}{dt^2} + c\frac{dx}{dt} + kx = P\cos\omega t$$

in which $m$ is mass, $x$ is length, and $t$ is time.

## 1-6   METHOD OF PROBLEM SOLVING

The principles of mechanics are few and relatively simple; however, the applications are infinite in their number, variety, and complexity. Success in engineering mechanics depends to a large degree on a well-disciplined method of problem solving. Experience has shown that the development of good problem-solving methods and skills results from solving a large variety of problems. Professional problem solving consists of three phases: problem definition and identification, model development and simplification, and mathematical solution and result interpretation. The problem-solving method outlined in this section will prove useful for the engineering mechanics courses that follow and for most situations encountered later in engineering practice.

Problems in engineering mechanics (statics, dynamics, and mechanics of deformable bodies) are concerned with the external effects of a system of forces on a physical body. Since the relationships between the external forces applied to a body and the motions or deformations that they produce are stated in mathematical form, the true physical situation must be represented by a mathematical model in order to obtain the required solution. Often it is necessary to make assumptions or approximations in setting up this model in order to simplify the solution. The most common approximation is to treat most of the bodies in statics and dynamics problems as rigid bodies. No real body is absolutely rigid; however, the changes in shape of a real body usually have a negligible effect on the acceleration produced by a force system or on the reactions required to maintain equilibrium of the body. Considerations of changes in shape under these circumstances would be an unnecessary complication of the problem. Similarly, the weights of many members can be neglected since they are small with respect to the applied loads, and a distributed force that acts over a small area can be considered to be concentrated at a point.

Usually, an actual physical problem cannot be solved exactly or completely. However, even in complicated problems, a simplified model can provide good qualitative results. Appropriate interpretation of such results can lead to approximate predictions of physical behavior or be used to verify the "reasonableness" of more sophisticated analytical, numerical, or experimental results. The engineer must always be aware of the actual physical problem under consideration and of any limitations associated with the mathematical model used. Assumptions must be continually evaluated to ensure that the mathematical problem solved provides an adequate representation of the physical process or device of interest.

As stated previously, the most effective way to learn the material contained in engineering mechanics courses is to solve a variety of problems. In order to become an effective engineer, the student must develop the ability to reduce complicated problems to simple parts that can be easily analyzed and to present the results of the work in a clear, logical, and neat manner. This can be accomplished by following the sequence of steps listed below.

1. Read the problem carefully.
2. Identify the result requested.
3. Identify the principles to be used to obtain the result.
4. Prepare a scaled sketch and tabulate the information provided.
5. Apply the appropriate principles and equations.
6. Report the answer with the appropriate number of significant figures and the appropriate units.
7. Study the answer and determine if it is reasonable.

The development of an ability to apply an orderly approach to problem solving constitutes a significant part of an engineering education. Also, the problem identification, model simplification, and result interpretation phases of engineering problem solving are often more important than the mathematical solution phase. No matter how elegantly or cleverly or exactly the mathematical solution is, it is of no value if it does not accurately reflect the physical problem which was to be solved.

## 1-7   SIGNIFICANCE OF NUMERICAL RESULTS

The speed and accuracy of pocket electronic calculators facilitate the numerical computations involved in the solution of engineering problems. However, the number of significant figures that can easily be obtained should not be taken as an indication of the accuracy of the solution. Rather, final results should be reported with a number of significant figures that is indicative of the accuracy of the solution. The accuracy of solutions to real engineering problems depends on three factors:

1.   The accuracy of the known physical data
2.   The accuracy of the physical model
3.   The accuracy of the computations performed

**Accuracy of the known physical data:** Engineering data is seldom known to an accuracy greater than 0.2 percent; therefore, calculated results should always be rounded off to the number of significant figures that will yield the same degree of accuracy as the data on which they are based. A practical rule for rounding off final numbers obtained in the computations involved in engineering analysis, which provides answers to approximately this degree of accuracy, is to retain four significant figures for numbers beginning with the figure "1" and to retain three significant figures for numbers beginning with any figure from "2" through "9".

**Accuracy of the physical model:** In order to obtain a solution to a problem, the true physical situation must be represented by a mathematical model. Of course, the final result is only as accurate as the physical model used. Care must exercised to ensure that the model and the associated mathematical problem being solved provide an adequate representation of the physical process or device they represent.

Often it is necessary to make simplifying assumptions or approximations in setting up a mathematical model of a real physical situation. Common assumptions are that bodies are rigid and do not deform, that air resistance has a negligible effect on the motion of a projectile, and that gravity is constant and independent of latitude or altitude. Any assumption that is not obvious should be clearly stated at the beginning of the problem. Once the solution is completed, the appropriateness of these assumptions should be checked. If the results are not consistent with the assumptions used to produce the results, then the assumptions must be modified and the problem solved again.

Pocket digital calculators.

**Accuracy of the computations performed:** Even if the physical model and the physical data are exact, computations almost always involve some round-off error. For example, calculators and computers round off fractions like "1/3" and values of functions like "sin 60°". The number of significant figures in the result depends on the exact machine used but is generally between 7 and 10 significant figures.

Arithmetic operations on these approximate numbers reduces the number of significant figures even further. A common practice is to treat the result as having one less significant figure than either of the numbers used to produce it. However, when two approximately equal numbers are subtracted, the result may have only one significant digit.

Extreme accuracy is not the goal of every problem solution. A conceptual study may require only a crude simple model with approximate data and a final accuracy of 50 to 100 percent. A feasibility study would require a more accurate model with more representative data and a final accuracy of 5 to 20 percent. A final, detailed design would seek the most accurate model and the best data available. It is the engineer's task to estimate how accurate the model and the data are and to report the results accordingly. Results reported with too many significant digits are not only misleading as to the real significance of the answer, they may also lead to an unexpected failure and tragedy.

In a book, of course, students have no way of knowing how accurate the given data are. When a length is given as 200 mm, there is no way of knowing whether the real measurement was 245 mm and rounded down to 200 mm (2 significant digits), or 200.4 mm and rounded down to 200 mm (3 significant digits), and so on. Therefore, the example problems in this book are solved (and the homework problems should be solved) with the assumption that the data provided are sufficiently accurate to justify at least three or four significant figures in the final answer. Intermediate calculations are (should be) worked out to one or two significant figures more than are reported in the final answer.

An error can be defined as the difference between two quantities. The difference, for example, might be between an experimentally measured value and a computed theoretical value. An error may also result from the rounding off of numbers during a calculation. One method for describing an error is to state a percent difference ($\%D$). Thus, for two numbers $A$ and $B$, if it is desired to compare number $A$ with number $B$, the percent difference between the two numbers is defined as

$$\%D = \frac{A - B}{B}(100)$$

In this equation, $B$ is the reference value with which $A$ is to be compared. The percent difference resulting from round-off error is illustrated in the following example.

Round off the number 12,345 to two, three, and four significant figures. Find the percent difference between the rounded-off numbers and the original number by using the original number as the reference.

### SOLUTION

Rounding off the number 12,345 to two, three, and four significant figures yields 12,000, 12,300, and 12,350. The percent difference for each of these numbers is

$$\%D = \frac{A - B}{B}(100)$$

For 12,000,

$$\%D = \frac{12{,}000 - 12{,}345}{12{,}345}(100) = -2.79\% \qquad \text{Ans.}$$

For 12,300,

$$\%D = \frac{12{,}300 - 12{,}345}{12{,}345}(100) = -0.36\% \qquad \text{Ans.}$$

For 12,350,

$$\%D = \frac{12{,}350 - 12{,}345}{12{,}345}(100) = +0.041\% \qquad \text{Ans.}$$

The minus signs associated with the above percent differences indicate that the rounded-off numbers are smaller than the reference number. Similarly, a positive percent difference indicates that the rounded-off number is larger than the reference number.

## PROBLEMS

Round off the numbers in the following problems to two significant figures. Find the percent difference between each rounded-off number and the original number by using the original number as the reference.

| | | | | | | | |
|---|---|---|---|---|---|---|---|
| 1-63* | a. | 0.0153 | b. | 0.0347 | c. | 0.0566 |
| 1-64* | a. | 0.8374 | b. | 0.4729 | c. | 0.6644 |
| 1-65 | a. | 1.8394 | b. | 3.4629 | c. | 6.7523 |
| 1-66 | a. | 3.6544 | b. | 7.5638 | c. | 8.9223 |

Round off the numbers in the following problems to three significant figures. Find the percent difference between each rounded-off number and the original number by using the original number as the reference.

| | | | | | | | |
|---|---|---|---|---|---|---|---|
| 1-67* | a. | 26.394 | b. | 74.829 | c. | 55.336 |
| 1-68* | a. | 374.93 | b. | 826.48 | c. | 349.33 |
| 1-69 | a. | 6471.9 | b. | 3628.7 | c. | 7738.2 |
| 1-70 | a. | 8521.4 | b. | 6748.3 | c. | 9378.7 |

Round off the numbers in the following problems to four significant figures. Find the percent difference between each rounded-off number and the original number by using the original number as the reference.

| | | | | | | | |
|---|---|---|---|---|---|---|---|
| 1-71* | a. | 63,746.2 | b. | 27,382.6 | c. | 55,129.9 |
| 1-72* | a. | 937,284 | b. | 274,918 | c. | 339,872 |
| 1-73 | a. | 918,273 | b. | 284,739 | c. | 342,691 |
| 1-74 | a. | 624,373 | b. | 785,239 | c. | 936,491 |

## Computer Problems

**C1-75** A common practice in rounding answers is to report numbers whose leading digit is 1 to an accuracy of 4 significant figures and all other numbers to an accuracy of 3 significant figures. Although this practice probably started with the accuracy with which slide rules could be read, it also reflects the fact that an accuracy of greater than 0.2 percent is seldom possible. This project will examine the error introduced by this and some other rounding schemes. For each of the rounding schemes below,

1. Generate 20,000 random numbers between 1 and 10.
2. Round each number to the specified number of significant figures. (Note that 3 significant figures is equivalent to 2 decimal places, 4 significant figures is equivalent to 3 decimal places, etc., since all numbers are between 1 and 10.)
3. Calculate the percent relative error for each number.

$$\text{PercentRelError} = \left| \frac{\text{Number} - \text{RoundNumber}}{\text{Number}} \right| *100$$

4. Plot `PercentRelError` versus `Number`.
5. Comment on the maximum round-off error and the distribution of round-off error.

a. Round all numbers to an accuracy of 3 significant figures.
b. Round numbers less than 2 to an accuracy of 4 significant figures and numbers greater than 2 to an accuracy of 3 significant figures.
c. Round numbers less than 3 to an accuracy of 4 significant figures and numbers greater than 3 to an accuracy of 3 significant figures.
d. Round numbers less than 5 to an accuracy of 4 significant figures and numbers greater than 5 to an accuracy of 3 significant figures.

**C1-76** When engineers deal with angles, they are usually more interested in the sine or cosine of the angle than they are with the angle itself. Since

$$\sin 5° = \cos 85° = \sin 175° = \sin 1085° = \cdots = 0.08176$$

the rounding of angles requires a different scheme than that described in Problem C1-75. That is, angles should be rounded to a specified number of decimal places rather than a specified number of significant figures. This project will examine the error introduced by rounding angles to various numbers of decimal places. For each of the cases below,

1. Generate 20,000 random angles between 1° and 89°. (Use a random number generator that produces decimal numbers and not just integers.) Calculate the sine and cosine of each angle.

2. Round each angle to the specified number of decimal places and calculate the sine and cosine of the rounded angle.
3. Calculate the percent relative error for each angle.

$$\text{PercentRelError} = \left| \frac{\sin(\text{Angle}) - \sin(\text{RoundAngle})}{\sin(\text{Angle})} \right| *100$$

or

$$\text{PercentRelError} = \left| \frac{\cos(\text{Angle}) - \cos(\text{RoundAngle})}{\cos(\text{Angle})} \right| *100$$

4. Plot `PercentRelError` versus `Angle`.
5. Comment on the maximum round-off error and the distribution of round-off error.

a. Round all angles to an accuracy of 1 decimal place.
b. Round all angles to an accuracy of 2 decimal places.
c. Round angles less than 10° to an accuracy of 3 decimal places and angles greater than 10° to an accuracy of 2 decimal places.

**C1-77** When two numbers are added or multiplied together, the result is always less accurate than the original numbers. This project will examine the error introduced by rounding two numbers before they are multiplied together.

a. Generate 80 random numbers between 4.51 and 5.49 (that is, $5 \pm 0.1*\text{RND}$). If any pair of these numbers are rounded to the nearest integer (5) and then multiplied together, the result will be 25. How does this result compare with the correct product obtained by multiplying the original two numbers together? Is the result accurate to the nearest integer? Is the result accurate to less than 10 percent?
b. Repeat part *a* for numbers between 49.51 and 50.49 ($50 \pm 0.01*\text{RND}$). Is the result accurate to the nearest integer? Is the result accurate to less than 1 percent?
c. Generate 20,000 random integers between 1 and 49. For each integer N, generate two random numbers that will round to that integer

$$N1 = N \pm 0.5*\text{RND} \qquad N2 = N \pm 0.5*\text{RND}$$

Plot the percent relative difference in the products

$$\text{ProdDiff} = \left| \frac{N1*N2 - N*N}{N*N} \right| *100$$

versus N. Compare this with the percent relative difference in the original numbers

$$\text{NumDiff} = \left| \frac{N1 - N}{N} \right| *100$$

# SUMMARY

The foundations for studies in mechanics are the laws formulated by Sir Isaac Newton in 1687. The first law deals with conditions for equilibrium of a particle; therefore, it provides the foundation for the study of statics. The second law, which establishes a relationship between the force acting on a particle and the motion of the particle, provides the foundation for the study of dynamics. The third law provides the foundation for understanding the concept of a force. In addition to the basic laws of motion, Newton also formulated the law of gravitation that governs the mutual attraction between two isolated bodies.

Physical quantities used to express the laws of mechanics can be divided into fundamental quantities and derived quantities. The magnitude of each fundamental quantity is defined by an arbitrarily chosen unit or "standard." The units used in the SI system are the meter (m) for length, the kilogram (kg) for mass, and the second (s) for time. The unit of force is a derived unit called a newton (N). In the U.S. Customary System of units, the units used are the foot (ft) for length, the pound (lb) for force, and the second (s) for time. The unit of mass is a derived unit called a slug. The U.S. Customary System is a gravitational rather than an absolute system of units.

The terms of an equation used to describe a physical process should not depend on the units of measurement (should be dimensionally homogeneous). If an equation is dimensionally homogeneous, the equation is valid for use with any system of units provided all quantities in the equation are measured in the same system. Use of dimensionally homogeneous equations eliminates the need for unit conversion factors. Success in engineering depends on a well-disciplined method of problem solving. Professional problem solving consists of three phases:

1. Problem definition and identification
2. Model development and simplification
3. Mathematical solution and result interpretation

In obtaining a solution to most problems, the true physical situation must be represented by a mathematical model. A common approximation made in setting up this model is to treat the body as a rigid body. Even though no real body is absolutely rigid, changes in shape usually have a negligible effect on the accelerations produced by a force system or on the reactions required to maintain equilibrium of the body; therefore, considerations of changes in shape usually result in an unnecessary complication of the problem. Anytime a mathematical model is used in solving a problem, care must be exercised to ensure that the model and the associated mathematical problem being solved provides an adequate representation of the physical process or device that it represents.

No matter how elegantly or cleverly or exactly the mathematical solution is, it is of no value if it does not accurately reflect the physical problem which was to be solved.

Accepted engineering practice requires that all work be neat and orderly. Sloppy solutions that cannot be understood by others have little or no value. Final results should always be reported with a number of significant figures that is indicative of the accuracy of the solution.

# REVIEW PROBLEMS

**1-78*** The planet Neptune has a mass of 1.03 $(10^{26})$ kg and a visible diameter (top of cloud layers) of $4.86 \times (10^7)$ m. Determine the gravitational acceleration $g$ at the top of the cloud layers.

**1-79*** The weight of the first Russian satellite, *Sputnik I*, on the surface of the earth was 184 lb. Determine the force exerted on the satellite by the earth at the low and high points of its orbit which were 149 mi and 597 mi, respectively, above the surface of the earth.

**1-80** The planet Jupiter has a mass of $1.90(10^{27})$ kg and a radius of $7.14(10^7)$ m. Determine the force of attraction between the earth and Jupiter when the minimum distance between the two planets is $6(10^{11})$ m.

**1-81** On the surface of the earth the weight of a body is 200 lb. At what distance from the center of the earth would the weight of the body be (a) 100 lb? (b) 50 lb?

**1-82*** At what distance from the center of the earth would the force of attraction between two spheres 1 m in diameter in contact equal the force of attraction of the earth on one of the spheres? The mass of each sphere is 250 kg.

**1-83*** The weight of a satellite on the surface of the earth prior to launch is 250 lb. When the satellite is in orbit 5000 miles from the surface of the earth, determine the force of attraction between the earth and the satellite.

**1-84** A fluid has a dynamic viscosity of $1.2(10^{-3})$ N · s/m². Express its dynamic viscosity in U.S. Customary units (lb · s/ft²).

**1-85** The stress equation for eccentric loading of a short column is

$$\sigma = -\frac{P}{A} - \frac{Pey}{I}$$

If $P$ is a force, $A$ is an area, and $e$ and $y$ are lengths, what are the dimensions of stress $\sigma$ and second moment of area $I$?

**1-86*** Determine the dimension of $c$ in the dimensionally homogeneous equation

$$v = \frac{mg}{c}\left[1 - e^{-ct/m}\right]$$

in which $v$ is a velocity, $m$ is a mass, $t$ is time, and $g$ is the gravitational acceleration.

**1-87*** In the dimensionally homogeneous equation

$$R = cv + ag$$

$R$ is a force, $v$ is a velocity, and $g$ is an acceleration. Determine the dimensions of $a$ and $c$.

**1-88** When a body moves through a fluid it experiences a resistance to its motion which can be represented by the equation $F = \frac{1}{2}C_{D}\rho V^2 A$ where $F$ is a force, $\rho$ is the density of the fluid, $V$ is the velocity of the body relative to the fluid, and $A$ is the cross-sectional area of the body. Show that the drag coefficient $C_D$ is dimensionless.

**1-89** Develop an expression for the change in gravitational acceleration $\Delta g$ between the surface of the earth and a height $h$ when $h << R_e$.

*T*he ropes at the left and right sides of the parachute form a concurrent force system.

# CONCURRENT FORCE SYSTEMS

## 2-1 INTRODUCTION

 A force was defined in Section 1-3 as the action of one body on another. The action may be the result of direct physical contact between the bodies or it may be the result of gravitational, electrical, or magnetic effects for bodies that are separated.

A force exerted on a body has two effects on the body: (1) the external effect, which is the tendency to change the motion of the body or to develop resisting forces (reactions) on the body, and (2) the internal effect, which is the tendency to deform the body. In many problems, the external effect is significant but the internal effect is not of interest. This is the case in many statics and dynamics problems when the body is assumed to be rigid. In other problems, when the body cannot be assumed to be rigid, the internal effects are important. Problems of this type are discussed in textbooks on "Mechanics of Materials" or "Mechanics of Deformable Bodies."

When a number of forces are treated as a group, they are referred to as a force system. If a force system acting on a body produces no external effect, the forces are said to be in balance and the body that experiences no change in motion is said to be in equilibrium. If a body is acted on by a force system that is not in balance, a change in motion of the body must occur. Such a force system is said to be unbalanced or to have a resultant.

*Why must all three characteristics of a force be known before its effect on a body can be established?*

**F = 100 lb**

*(a)*

**F = 100 lb**

*(b)*

**Fig. CE2-1**

### SOLUTION

The characteristics needed to describe a force are its magnitude, its direction, and its point of application. The effects of a force on a body are determined by these characteristics.

The force acting at point *A* of the garbage can shown in Fig. CE2-1*a* has both a magnitude (100 lb) and a direction (horizontal and to the right). If the 100-lb force is sufficient to overcome the frictional resistance at the contact surface, the garbage can will begin to slip and move to the right. If the same force (magnitude and direction) acts at point *B* as shown in CE2-1*b*, the garbage can may tip before it slips. Thus, all three characteristics of a force (the magnitude, the direction, and the point of application) must be known before the effects of the force (slipping or tipping) on the garbage can can be determined.

Two force systems are said to be equivalent if they produce the same external effect when applied in turn to a given body. Two systems that are equivalent are called replacements for each other. The resultant of a force system is the simplest equivalent system to which the original system will reduce. The process of reducing a force system to a simpler equivalent system is called reduction. The process of expanding a force or a force system into a less simple equivalent system is called resolution. A component of a force is one of the two or more forces into which the given force may be resolved.

## 2-2 FORCES AND THEIR CHARACTERISTICS

The properties needed to describe a force are called the characteristics of the force. The characteristics of a force are as follows:

1. Its magnitude
2. Its direction (orientation and sense)
3. Its point of application

The magnitude (positive numerical value) of a force is the amount or size of the force. In this book, the magnitude of a force will be expressed in newtons (N) or kilonewtons (kN) when the SI System of units is used and in pounds (lb) or kilopounds (kip) when the U.S. Customary System of units is used.

The direction of a force is the orientation and sense of the line segment used to represent the force. In a two-dimensional problem, the orientation can be specified by providing an angle as shown in Fig. 2-1*a* or by providing two dimensions as shown in Fig. 2-1*b*. In a three-dimensional problem, the orientation can be specified by providing three angles as shown in Fig. 2-2*a* or by providing three dimensions as shown in Fig. 2-2*b*. The sense of the force can be specified by placing an arrowhead on the appropriate end of the line segment used to represent the force. Alternatively, a plus or minus sign can be used with the magnitude of a force to indicate the sense of the force.

The point of application of a force is the point of contact between the two bodies. A straight line extending through the point of application in the direction of the force is called its line of action.

The three characteristics of a force are illustrated on the sketch of a block shown in Fig. 2-3. In this case, the force applied to the block can be described as a 100-lb (magnitude) force acting 30° upward and to the right (direction: orientation and sense) through point $A$ (point of application). A discussion of the manner in which these characteristics influence the reactions developed in holding a body at rest forms an important part of the study of statics. In a similar manner, a discussion of the manner in which these characteristics influence the change in motion of a body forms an important part of the study of kinetics.

### 2-2.1   Scalar and Vector Quantities

Scalar quantities can be completely described with a magnitude (number). Examples of scalar quantities in mechanics are mass, density, length, area, volume, speed, energy, time, and temperature. In mathematical operations, scalars follow the rules of elementary algebra.

A vector quantity has both a magnitude and a direction (line of action and sense) and obeys the parallelogram law of addition. Examples of vector quantities in mechanics are force, moment, displacement, velocity, acceleration, impulse, and momentum. Vectors can be classified into three types: free, sliding, or fixed.

1.  A **free vector** has a specific magnitude and direction (orientation and sense), but its line of action does not pass through a unique point in space. For example, consider a sailboat on a small lake. The velocity of the wind is a vector quantity, and both its magnitude and direction are important to the operator of the sailboat. The wind velocity does not have a line of action that passes through a unique point on the boat. Thus, the wind velocity vector is a free vector. Other quantities that can be represented by free vectors are displacement and acceleration. In Section 4-4 it will be shown that the vector used to represent the moment of a couple is a free vector.
2.  A **sliding vector** has a specific magnitude and direction, and its line of action passes through a unique point in space. The point of application of a sliding vector can be anywhere along its line of action. (See Section 2-2.2 for a discussion of the effect of a force on a rigid body.)
3.  A **bound vector** has a specific magnitude and direction, and its line of action passes through a unique point in space. The point of application of a bound vector is confined to a fixed point on its line of action. (See Section 2-2.2 for a discussion of the effect of a force on a deformable body.)

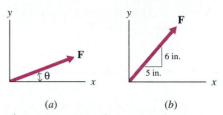

(a)                      (b)

**Figure 2-1**   Two-dimensional representations of the direction of a force.

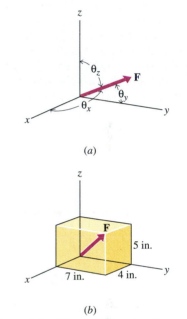

(a)

(b)

**Figure 2-2**   Three-dimensional representations of the direction of a force.

**Figure 2-3**   Characteristics of a force.

**Figure 2-4**   Principle of transmissibility.

Vector quantities are indicated in typeset print by the use of boldface type (**A**). Since it is not feasible to produce boldface characters with pencil or chalk, vector quantities are frequently indicated in these instances by using an arrow over the vector quantity ($\vec{A}$). When text material is prepared with a typewriter, vector quantities are frequently indicated by underling the vector quantity (<u>A</u>).

The use of scalars and vectors to represent physical quantities is a simple example of the modeling of physical quantities by mathematical methods. An engineer must be able to construct good mathematical models and to correctly interpret their physical meaning.

### 2-2.2   Principle of Transmissibility

In most statics and dynamics problems, the assumption is made that the body is rigid. As a result, only the external effects of any force applied to the body are of interest. When this is the case, the force can be applied at any point along its line of action without changing the external effects of the force. For example, a stalled automobile (see Fig. 2-4) can be moved by pushing on the back bumper or pulling on the front bumper. If the magnitude, direction, and the line of action of the two forces are identical, the external effect will be the same. Clearly, in this case, the point of application of the force has no effect on the external effect (motion of the automobile).

This fact is formally expressed by the principle of transmissibility as, "The external effect of a force on a rigid body is the same for all points of application of the force along its line of action." It should be noted that only the external effect remains unchanged. The internal effect of a force (stress and deformation) may be greatly influenced by a change in the point of application of the force along its line of action. In the case of the automobile, pushing on the rear bumper will tend to deform the rear bumper and pulling on the front bumper will tend to deform the front bumper. Clearly, in this case, the point of application of the force produces totally different internal effects (deformation of the two different bumpers of the automobile). For those cases where the principle of transmissibility applies (rigid-body mechanics), force can be treated as a sliding vector.

### 2-2.3   Classification of Forces

A force has been defined as the action of one physical body on another. Since the interaction can occur when there is contact between the bodies or when the bodies are physically separated, forces can be classified under two general headings: (1) contacting or surface forces, such as a push or a pull produced by mechanical means, and (2) noncontacting or body forces, such as the gravitational pull of the earth on all physical bodies.

Concrete bridge floor

Uniformly distributed load

**Figure 2-5**   Example and representation of a distributed load.

*Why is the principle of transmissibility valid only for the external effects of a force on a rigid body?*

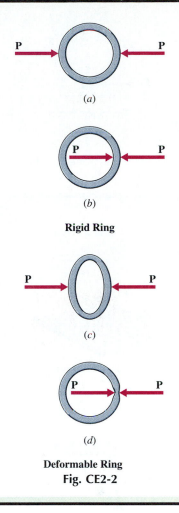

(a)

(b)

**Rigid Ring**

(c)

(d)

**Deformable Ring**
**Fig. CE2-2**

## SOLUTION

The principle of transmissibility states that "The *external effect* of a force on a *rigid* body is the same for all points of application of the force along its line of action."

In mechanics problems, which deal with rigid bodies, the external effects of a force are a tendency to change the motion of the body (dynamics problems) or a tendency to develop resisting forces, known as reactions, on the body (statics problems). The principle can be illustrated by considering the rigid ring of Fig. CE2-2. In Fig. CE2-2a, the ring is at rest under the action of two equal, opposite, collinear forces. If one of the forces is moved along its line of action, as shown in Fig. CE2-2b, the ring remains at rest. In both cases, the effect of the forces on the rigid ring is the same; namely, the ring remains at rest since no unbalanced force is present to produce an acceleration.

In mechanics of materials problems, which deal with deformable bodies, the internal effects of a force (a tendency to deform the body) are also important. In Fig. CE2-2c, the deformable ring is at rest under the action of two equal, opposite, collinear forces; however, the ring suffers an overall deformation under the action of the forces. If one of the forces is moved along its line of action, as shown in Fig. CE2-2d, the ring remains at rest, but the overall deformation of the ring changes to a very localized deformation in the region of the ring near the forces. Thus, moving one of the forces along its line of action does not produce motion of the ring but does produce a significant change in the deformation of the ring.

Forces may also be classified with respect to the area over which they act. A force applied along a length or over an area is known as a distributed force. The distribution can be uniform or nonuniform. The weight of a concrete bridge floor of uniform thickness (see Fig. 2-5) is an example of a uniformly distributed load. Any force applied to a relatively small area compared with the size of the loaded member can be assumed to be a concentrated force. For example, the force applied by a car wheel to the longitudinal members of a bridge (see Fig. 2-6) can be considered to be a concentrated load.

Any number of forces treated as a group constitute a force system. Force systems may be one-, two-, or three-dimensional. A force system is said to be concurrent if the action lines of all forces intersect at a

**Figure 2-6** Example and representation of a concentrated load.

Concurrent Forces

(a)

Coplanar Forces

(b)

Parallel Forces

(c)

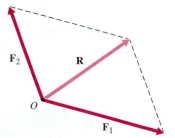

Collinear Forces

(d)

**Figure 2-7**   Different force systems.

common point (see Fig. 2-7*a*) and coplanar when all the forces lie in the same plane (see Fig. 2-7*b*). A parallel force system is one in which the action lines of the forces are parallel (see Fig. 2-7*c*). In a parallel force system, the senses of the forces do not have to be the same. If the forces of a system have a common line of action, the system is said to be collinear (see Fig. 2-7*d*).

## 2-3   RESULTANT OF TWO CONCURRENT FORCES

Any two concurrent forces $\mathbf{F}_1$ and $\mathbf{F}_2$ acting on a body can be replaced by a single force, called the resultant $\mathbf{R}$, which will produce the same effect on the body as the original two forces. The resultant of the two forces can be determined by adding the two forces vectorially using the parallelogram law. Mathematically, the sum of the two forces is given by the vector equation

$$\mathbf{F}_1 + \mathbf{F}_2 = \mathbf{R}$$

The process by which two forces are added graphically using the parallelogram law is illustrated in Fig. 2-8.

The resultant $\mathbf{R}$ of two forces can also be determined graphically by using one-half of a parallelogram. Since one-half of a parallelogram is a triangle, the method is called the triangle law for addition of vectors. When the triangle law is used to determine the resultant $\mathbf{R}$ of two forces $\mathbf{F}_1$ and $\mathbf{F}_2$, force $\mathbf{F}_1$ is drawn to scale first; then force $\mathbf{F}_2$ is drawn to scale from the tip of force $\mathbf{F}_1$ in a head-to-tail fashion. The closing side of the triangle, drawn from the origin $O$ to the tip of force $\mathbf{F}_2$, is the resultant $\mathbf{R}$ of the two forces. The process by which two forces are added using the triangle law is illustrated in Fig. 2-9. The triangle constructed using this method is called a force triangle.

Alternatively, force $\mathbf{F}_2$ can be drawn first; then force $\mathbf{F}_1$ can be drawn from the tip of force $\mathbf{F}_2$ in a head-to-tail fashion. Again, the resultant $\mathbf{R}$ of the two forces is the closing side of the triangle. As indicated in Fig. 2-9,

$$\mathbf{F}_1 + \mathbf{F}_2 = \mathbf{F}_2 + \mathbf{F}_1 = \mathbf{R}$$

The results shown in Fig. 2-9 indicate that the resultant $\mathbf{R}$ does not depend on the order in which forces $\mathbf{F}_1$ and $\mathbf{F}_2$ are selected. Figure 2-9 is a graphical illustration of the commutative law of vector addition.

Graphical methods for determining the resultant of two forces require an accurate scaled drawing if accurate results are to be obtained.

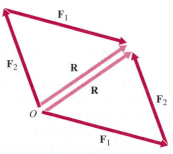

**Figure 2-8**   Parallelogram law for addition of forces.

**Figure 2-9**   Triangle law for addition of forces.

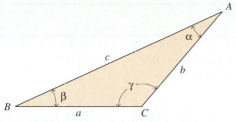

**Figure 2-10**  Lengths and angles for an oblique triangle.

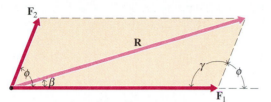

**Figure 2-11**  Resultant **R** of two forces **F**$_1$ and **F**$_2$.

In practice, numerical results are obtained by using trigonometric methods based on the law of sines and the law of cosines in conjunction with sketches of the force system. For example, consider the triangle shown in Fig. 2-10, which is similar to the force triangles illustrated in Figs. 2-8 and 2-9. For this general triangle, the law of sines is

$$\frac{a}{\sin \alpha} = \frac{b}{\sin \beta} = \frac{c}{\sin \gamma}$$

and the law of cosines is

$$c^2 = a^2 + b^2 - 2ab \cos \gamma$$

Application of the law of cosines to the determination of the magnitude of the resultant of the two forces **F**$_1$ and **F**$_2$ shown in Fig. 2-11 yields

$$R^2 = F_1^2 + F_2^2 - 2F_1F_2 \cos \gamma$$

Since $\cos \gamma = \cos (180° - \phi) = -\cos \phi$, this equation can be written in alternate form as

$$R^2 = F_1^2 + F_2^2 + 2F_1F_2 \cos \phi \qquad \text{(2-1)}$$

where $\phi$ is the angle between the two forces. Note also from Fig. 2-11 that the orientation of the line of action of the resultant force **R** with respect to the force **F**$_1$ is given by the expression

$$\sin \beta = \frac{F_2 \sin \phi}{R}$$

or

$$\beta = \sin^{-1}\left(\frac{F_2 \sin \phi}{R}\right) \qquad \text{(2-2)}$$

The procedure for determining the resultant **R** of a force system by using the law of sines and the law of cosines is demonstrated in the following example.

Two forces are applied at a point in a body as shown in Fig. 2-12a. Determine the magnitude of the resultant **R** of the two forces and the angle $\theta$ between the x-axis and the line of action of the resultant.

(a)

(b)

**Fig. 2-12**

## SOLUTION

The resultant **R** of the two forces is determined by adding the forces vectorially using the parallelogram law as illustrated in Fig. 2-12b. The magnitude $R$ of the resultant is obtained by using the alternate form of the law of cosines as expressed by Eq. 2-1. Thus,

$$R^2 = F_1^2 + F_2^2 + 2F_1F_2 \cos \phi$$
$$= 500^2 + 800^2 + 2(500)(800) \cos (180° - 56°)$$

From which

$$R = 665.32 \text{ lb} \cong 665 \text{ lb} \qquad \text{Ans.}$$

Since the line of action of force **F₁** coincides with the x-axis, the line of action of resultant **R** with respect to the x-axis is given by Eq. 2-2 as

$$\theta = \beta = \sin^{-1} \frac{F_2 \sin \phi}{R}$$

$$= \sin^{-1} \frac{800 \sin (180° - 56°)}{665.32} = 85.46° \cong 85.5° \qquad \text{Ans.}$$

> A parallelogram of forces is constructed by drawing lines parallel to the two known forces as shown in Fig. 2-12b.

Two forces are applied to an eye bracket as shown in Fig. 2-13a. Determine the magnitude of the resultant **R** of the two forces and the angle $\theta$ between the x-axis and the line of action of the resultant.

(a)

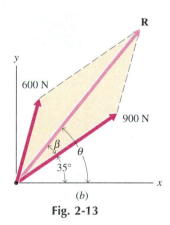

(b)

**Fig. 2-13**

## SOLUTION

The resultant **R** of the two forces is determined by adding the forces vectorially using the parallelogram law as illustrated in Fig. 2-13b. The magnitude $R$ of the resultant is obtained by using the alternate form of the law of cosines as expressed by Eq. 2-1. Thus,

$$R^2 = F_1^2 + F_2^2 + 2F_1F_2 \cos \phi$$
$$= 900^2 + 600^2 + 2(900)(600) \cos 40°$$

A parallelogram of forces is constructed by drawing lines parallel to the two known forces as shown in Fig. 2-13b.

From which

$$R = 1413.3 \, N \cong 1413 \, N \qquad \text{Ans.}$$

The line of action of resultant **R** with respect to the 900-N force is given by Eq. 2-2 as

$$\beta = \sin^{-1}\frac{F_2 \sin \phi}{R} = \sin^{-1}\frac{600 \sin 40°}{1413.3} = 15.84°$$

Thus, as shown in Fig. 2-13b,

$$\theta = \beta + 35° = 15.84° + 35° = 50.84° \cong 50.8° \qquad \text{Ans.}$$

# PROBLEMS

Use the law of sines and the law of cosines, in conjunction with sketches of the force triangles, to solve the following problems. Determine the magnitude of the resultant **R** and the angle $\theta$ between the x-axis and the line of action of the resultant for the following:

## Introductory Problems

2-1*   The two forces shown in Fig. P2-1.

**Fig. P2-1**

2-2*   The two forces shown in Fig. P2-2.

**Fig. P2-2**

2-3   The two forces shown in Fig. P2-3.

**Fig. P2-3**

2-4   The two forces shown in Fig. P2-4.

**Fig. P2-4**

2-5*   The two forces shown in Fig. P2-5.

**Fig. P2-5**

2-6*   The two forces shown in Fig. P2-6.

**Fig. P2-6**

2-7   The two forces shown in Fig. P2-7.

**Fig. P2-7**

2-8   The two forces shown in Fig. P2-8.

**Fig. P2-8**

**2-9*** The two forces shown in Fig. P2-9.

**Fig. P2-9**

**2-10** The two forces shown in Fig. P2-10.

**Fig. P2-10**

## Intermediate Problems

**2-11** The two forces shown in Fig. P2-11.

**Fig. P2-11**

**2-12*** The two forces shown in Fig. P2-12.

$210\left(\dfrac{5}{13}\right)$

$210\left(\dfrac{12}{13}\right)$

**Fig. P2-12**

r3240

**2-13** The two forces shown in Fig. P2-13.

**Fig. P2-13**

**2-14** The two forces shown in Fig. P2-14.

**Fig. P2-14**

## Challenging Problems

**2-15*** The two forces shown in Fig. P2-15.

**Fig. P2-15**

**2-16** The two forces shown in Fig. P2-16.

**Fig. P2-16**

**Figure 2-14**   Three coplanar, concurrent forces.

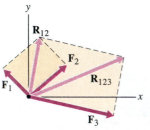

**Figure 2-15**   Use of the parallelogram law to sum three concurrent forces.

## 2-4   RESULTANT OF THREE OR MORE CONCURRENT FORCES

In the previous section, use of the parallelogram and triangle laws to determine the resultant **R** of two concurrent forces **F**$_1$ and **F**$_2$ was discussed. The method can easily be extended to cover three or more forces. As an example, consider the case of three coplanar, concurrent forces acting on an eye bolt as shown in Fig. 2-14. Application of the parallelogram law to forces **F**$_1$ and **F**$_2$, as shown graphically in Fig. 2-15, yields resultant **R**$_{12}$. Then, combining resultant **R**$_{12}$ with force **F**$_3$ through a second graphical application of the parallelogram law yields resultant **R**$_{123}$, which is the vector sum of the three forces.

In practice, numerical results for specific problems involving three or more forces are obtained algebraically by using the law of sines and the law of cosines in conjunction with sketches of the force system similar to those shown in Fig. 2-16. The sketches shown in Fig. 2-16 are known as force polygons. The order in which the forces are added can be arbitrary, as shown in Figs. 2-16a and 2-16b, where the forces are added in the order **F**$_1$, **F**$_2$, **F**$_3$ in Fig. 2-16a and in the order **F**$_3$, **F**$_1$, **F**$_2$ in Fig. 2-16b. Although the shape of the polygon changes, the resultant force remains the same. The fact that the sum of the three vectors is the same regardless of the order in which they are added illustrates the associative law of vector addition.

If there are more than three forces (as an example, Fig. 2-17 shows the eye bolt of Fig. 2-14 with four forces), the process of adding additional forces can be continued, as shown in Fig. 2-18, until all of the forces are joined in head-to-tail fashion. The closing side of the polygon, drawn from the tail of the first vector to the head of the last vector is the resultant of the force system. If the force polygon closes, the resultant of the force system is zero and the body to which the forces are applied is said to be in equilibrium.

Application of the parallelogram law to more than three forces requires extensive geometric and trigonometric calculation. Therefore, problems of this type are usually solved by using the rectangular-component method, which is developed in Section 2-7.

(a)

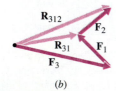

(b)

**Figure 2-16**   Illustration of associative law of vector addition.

**Figure 2-17**   Four coplanar, concurrent forces.

**Figure 2-18**   Force polygon for four coplanar, concurrent forces.

Determine the magnitude of the resultant **R** and the angle $\theta$ between the x-axis and the line of action of the resultant for three forces shown in Fig. 2-19a.

(a)

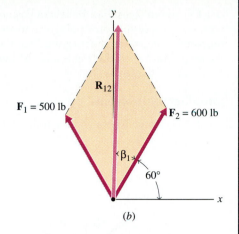

(b)

## SOLUTION

First, the resultant **R**$_{12}$ of forces **F**$_1$ and **F**$_2$ is determined by adding the two forces vectorially using the parallelogram law as illustrated in Fig. 2-19b. The magnitude $R_{12}$ (obtained using Eq. 2-1) is

$$R_{12}^2 = F_1^2 + F_2^2 + 2F_1F_2 \cos \phi_1$$
$$= 500^2 + 600^2 + 2(500)(600) \cos 60°$$

From which

$$R_{12} = 953.9 \text{ lb} \cong 954 \text{ lb}$$

The line of action of resultant **R**$_{12}$ with respect to force **F**$_2$ (see Fig. 2-19b) is given by Eq. 2-2 as

$$\beta_1 = \sin^{-1} \frac{F_1 \sin \phi_1}{R_{12}}$$
$$= \sin^{-1} \frac{500 \sin 60°}{953.9} = 27.00°$$

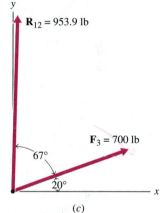

(c)

Next, the resultant **R** of forces **R**$_{12}$ and **F**$_3$ (see Fig. 2-19c) is determined by adding the two forces vectorially using the parallelogram law as illustrated in Fig. 2-19d. The magnitude $R$ (obtained using Eq. 2-1) is

$$R^2 = R_{12}^2 + F_3^2 + 2R_{12}F_3 \cos \phi_2$$
$$= 953.9^2 + 700^2 + 2(953.9)(700) \cos 67.0°$$

From which

$$R = 1386.3 \text{ lb} \cong 1386 \text{ lb} \qquad \text{Ans.}$$

The line of action of resultant **R** with respect to force **F**$_3$ (see Fig. 2-19d) is given by Eq. 2-2 as

$$\beta_2 = \sin^{-1} \frac{R_{12} \sin \phi_2}{R}$$
$$= \sin^{-1} \frac{953.9 \sin 67.0°}{1386.3} = 39.30°$$

Thus, as shown in Fig. 2-19d,

$$\theta = \beta_2 + 20° = 39.30° + 20° = 59.30° = 59.3° \qquad \text{Ans.}$$

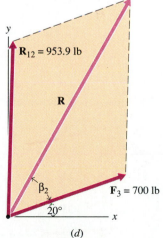

(d)

**Fig. 2-19**

Adding more than two forces requires successive applications of the parallelogram law.

# PROBLEMS

Use the law of sines and the law of cosines, in conjunction with sketches of the force polygons, to solve the following problems. Determine the magnitude of the resultant **R** and the angle $\theta$ between the $x$-axis and the line of action of the resultant for the following:

## Introductory Problems

**2-17*** The three forces shown in Fig. P2-17.

**Fig. P2-17**

**2-18*** The three forces shown in Fig. P2-18.

**Fig. P2-18**

**2-19** The three forces shown in Fig. P2-19.

**Fig. P2-19**

**2-20** The three forces shown in Fig. P2-20.

**Fig. P2-20**

**2-21*** The three forces shown in Fig. P2-21.

**Fig. P2-21**

**2-22*** The three forces shown in Fig. P2-22.

**Fig. P2-22**

**2-23**  The three forces shown in Fig. P2-23.

**Fig. P2-23**

**2-24**  The three forces shown in Fig. P2-24.

**Fig. P2-24**

## Intermediate Problems

**2-25\***  The three forces shown in Fig. P2-25.

**Fig. P2-25**

**2-26\***  The three forces shown in Fig. P2-26.

**Fig. P2-26**

**2-27**  The three forces shown in Fig. P2-27.

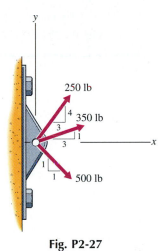

**Fig. P2-27**

**2-28**  The three forces shown in Fig. P2-28.

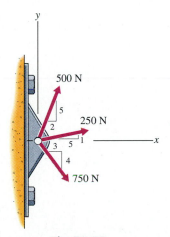

**Fig. P2-28**

**2-29*** The four forces shown in Fig. P2-29.

**2-30** The four forces shown in Fig. P2-30.

**Fig. P2-29**

**Fig. P2-30**

## 2-5 RESOLUTION OF A FORCE INTO COMPONENTS

In the previous two sections of this chapter, use of the parallelogram and triangle laws to determine the resultant **R** of two concurrent forces $\mathbf{F}_1$ and $\mathbf{F}_2$ or three or more concurrent forces $\mathbf{F}_1, \mathbf{F}_2, \ldots, \mathbf{F}_n$ was discussed. In a similar manner, a single force **F** can be replaced by a system of two or more forces $\mathbf{F}_a, \mathbf{F}_b, \ldots, \mathbf{F}_n$. Forces $\mathbf{F}_a, \mathbf{F}_b, \ldots, \mathbf{F}_n$ are called components of the original force. In the most general case, the components of a force can be any system of forces that can be combined by the parallelogram law to produce the original force. Such components are not required to be concurrent or coplanar. Normally, however, the term *component* is used to specify either one of two coplanar concurrent forces or one of three noncoplanar concurrent forces that can be combined vectorially to produce the original force. The point of concurrency must be on the line of action of the original force. The process of replacing a force by two or more forces is called resolution.

The process of resolution does not produce a unique set of vector components. For example, consider the four coplanar sketches shown in Fig. 2-20. It is obvious from these sketches that

$$\mathbf{A} + \mathbf{B} = \mathbf{R} \qquad \mathbf{E} + \mathbf{F} = \mathbf{R}$$
$$\mathbf{C} + \mathbf{D} = \mathbf{R} \qquad \mathbf{G} + \mathbf{H} + \mathbf{I} = \mathbf{R}$$

where **R** is the same vector in each expression. Thus, an infinite number of sets of components exist for any vector.

Use of the parallelogram and triangle laws to resolve a force into components along any two oblique lines of action is illustrated in the following examples.

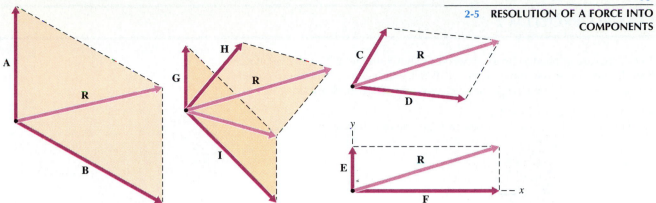

**Figure 2-20** Several different sets of components of a force **R**.

## EXAMPLE PROBLEM 2-4

Determine the magnitudes of the $u$- and $v$-components of the 900-N force shown in Fig. 2-21$a$.

### SOLUTION

The magnitude and direction of the 900-N force are shown in Fig. 2-21$b$. The components $\mathbf{F}_u$ and $\mathbf{F}_v$ along the $u$- and $v$-axes can be determined by drawing lines parallel to the $u$- and $v$-axes through the head and tail of the vector used to represent the 900-N force. From the parallelogram thus produced, the law of sines can be applied to determine the forces $\mathbf{F}_u$ and $\mathbf{F}_v$ since all the angles for the two triangles that form the parallelogram are known. Thus

$$\frac{F_u}{\sin 45°} = \frac{F_v}{\sin 25°} = \frac{900}{\sin 110°}$$

from which

$$F_u = |\mathbf{F}_u| = \frac{900 \sin 45°}{\sin 110°} = 677 \text{ N} \qquad \text{Ans.}$$

$$F_v = |\mathbf{F}_v| = \frac{900 \sin 25°}{\sin 110°} = 405 \text{ N} \qquad \text{Ans.}$$

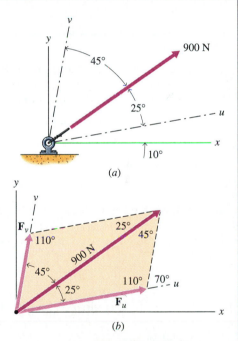

**Fig. 2-21**

Since the $u$- and $v$-axes are not perpendicular, normal sine and cosine relationships do not apply. Therefore, a parallelogram of forces together with the law of sines is used to solve for the components.

Two forces are applied to an eye bracket as shown in Fig. 2-22a. The resultant **R** of the two forces has a magnitude of 1000 lb and its line of action is directed along the x-axis. If the force **F**$_1$ has a magnitude of 250 lb, determine

a. The magnitude of force **F**$_2$.
b. The angle $\alpha$ between the x-axis and the line of action of the force **F**$_2$.

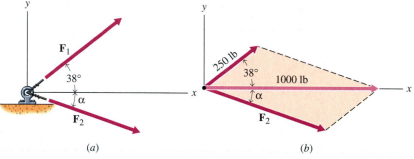

Fig. 2-22

---

## SOLUTION

The two forces **F**$_1$ and **F**$_2$, the resultant **R**, and the angle $\alpha$ are shown in Fig. 2-22b. The force triangle was drawn by using **F**$_1$, **R**, and the 38° angle. Completing the parallelogram identifies force **F**$_2$ and the angle $\alpha$.

a. Applying the law of cosines to the top triangle of Fig. 2-22b yields

$$F_2^2 = 250^2 + 1000^2 - 2(250)(1000)\cos 38°$$

from which

$$F_2 = |F_2| = 817.6 \cong 818 \text{ lb} \qquad \text{Ans.}$$

b. Applying the law of sines to the top triangle yields

$$\frac{250}{\sin \alpha} = \frac{817.6}{\sin 38°}$$

Thus,

$$\sin \alpha = \frac{250}{817.6}\sin 38° = 0.18825$$

from which

$$\alpha = 10.85° \qquad \text{Ans.}$$

> A force triangle can be drawn when two forces and the angle between them are known.

A 100-kN force is resisted by an eye bar and a strut as shown in Fig. 2-23a. Determine the component $\mathbf{F}_u$ of the force along the axis of eye bar $AB$ and the component $\mathbf{F}_v$ of the force along the axis of strut $AC$.

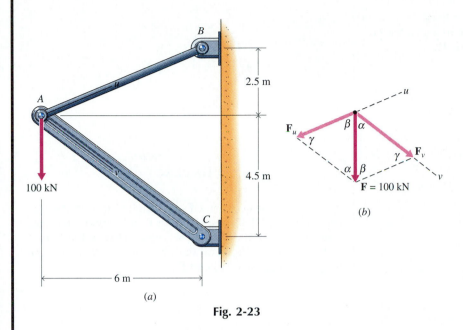

(a)

(b)

**Fig. 2-23**

---

## SOLUTION

The components $\mathbf{F}_u$ and $\mathbf{F}_v$ along the $u$- and $v$-axes can be determined by drawing lines parallel to the $u$- and $v$-axes through the head and tail of the vector used to represent the 100-kN force. From the parallelogram thus produced (see Fig. 2-23b), the law of sines can be applied to determine the forces $\mathbf{F}_u$ and $\mathbf{F}_v$ since all of the angles for the two triangles that form the parallelogram can be determined. Thus,

> A force triangle can be drawn when a force and the directions of its two components are known.

$$\alpha = \tan^{-1}\frac{6.0}{4.5} = 53.13° \qquad \beta = \tan^{-1}\frac{6.0}{2.5} = 67.38°$$
$$\gamma = 180 - 53.13 - 67.38 = 59.49°$$

From the law of sines:
$$\frac{F_u}{\sin \alpha} = \frac{F_v}{\sin \beta} = \frac{100}{\sin \gamma}$$

$$F_u = \frac{100}{\sin \gamma} \sin \alpha = \frac{100}{\sin 59.49°} \sin 53.13° = 92.86 \text{ kN} \cong 92.9 \text{ kN}$$
$$\mathbf{F}_u = 92.9 \text{ kN} \nearrow 22.6° \qquad\qquad \text{Ans.}$$

$$F_v = \frac{100}{\sin \gamma} \sin \beta = \frac{100}{\sin 59.49°} \sin 67.38° = 107.14 \text{ kN} \cong 107.1 \text{ kN}$$
$$\mathbf{F}_v = 107.1 \text{ kN} \searrow 36.9° \qquad\qquad \text{Ans.}$$

Note that the force in eye bar $AB$ is directed away from the bar (tension) and that the force in strut $AC$ is directed toward the strut (compression).

# PROBLEMS

Use the law of sines and the law of cosines in conjunction with sketches of the force triangles to solve the following problems.

## Introductory Problems

**2-31\*** Determine the magnitudes of the $u$- and $v$-components of the 750-lb force shown in Fig. P2-31.

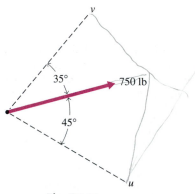

**Fig. P2-31**

**2-32\*** Determine the magnitudes of the $u$- and $v$-components of the 1000-N force shown in Fig. P2-32.

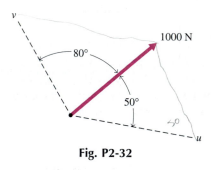

**Fig. P2-32**

**2-33** Determine the magnitudes of the $u$- and $v$-components of the 850-lb force shown in Fig. P2-33.

**Fig. P2-33**

**2-34** Determine the magnitudes of the $u$- and $v$-components of the 1500-N force shown in Fig. P2-34.

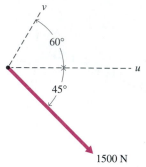

**Fig. P2-34**

**2-35\*** Two forces $\mathbf{F}_u$ and $\mathbf{F}_v$ are applied to a bracket as shown in Fig. P2-35. The resultant $\mathbf{R}$ of the two forces has a magnitude of 600 lb, and the angle between the line of action of the resultant and the $y$-axis is 28°. Determine the magnitudes of forces $\mathbf{F}_u$ and $\mathbf{F}_v$.

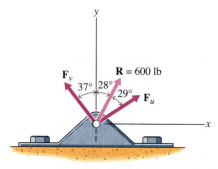

**Fig. P2-35**

**2-36\*** Two forces $\mathbf{F}_u$ and $\mathbf{F}_v$ are applied to a circular plate as shown in Fig. P2-36. The resultant $\mathbf{R}$ of the two forces has a magnitude of 900 N, and the angle between the line of action of the resultant and the $x$-axis is 30°. Determine the magnitudes of forces $\mathbf{F}_u$ and $\mathbf{F}_v$.

**Fig. P2-36**

**2-37** Two ropes are used to tow a boat upstream as shown in Fig. P2-37. The resultant **R** of the rope forces $F_u$ and $F_v$ has a magnitude of 400 lb and its line of action is directed along the axis of the boat. Determine the magnitudes of forces $F_u$ and $F_v$.

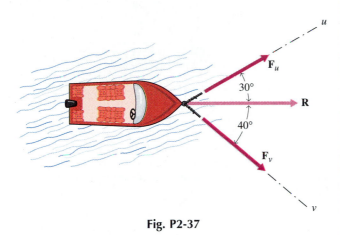

**Fig. P2-37**

**2-38** Two cables are used to support a stoplight as shown in Fig. P2-38. The resultant **R** of the cable forces $F_u$ and $F_v$ has a magnitude of 1350 N and its line of action is vertical. Determine the magnitudes of forces $F_u$ and $F_v$.

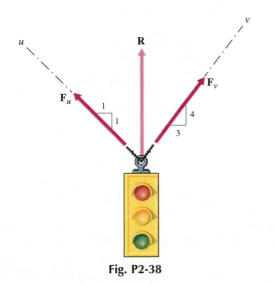

**Fig. P2-38**

**Intermediate Problems**

**2-39\*** Two forces $F_u$ and $F_v$ are applied to a bracket as shown in Fig. P2-39. If the resultant **R** of the two forces has a magnitude of 725 lb and a direction as shown on the figure, determine the magnitudes of forces $F_u$ and $F_v$.

**Fig. P2-39**

**2-40\*** Two forces $F_u$ and $F_v$ are applied to a bracket as shown in Fig. P2-40. If the resultant R of the two forces has a magnitude of 375 N and a direction as shown on the figure, determine the magnitudes of forces $F_u$ and $F_v$.

**Fig. P2-40**

**2-41** A 2000-lb force is resisted by two pipe struts as shown in Fig. P2-41. Determine the component $F_u$ of the force along the axis of strut $AB$ and the component $F_v$ of the force along the axis of strut $BC$.

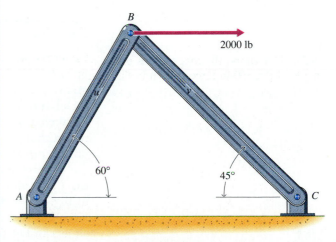

**Fig. P2-41**

53

**2-42** A 750-N force is resisted by two pipe struts as shown in Fig. P2-42. Determine the component $\mathbf{F}_u$ of the force along the axis of strut $AC$ and the component $\mathbf{F}_v$ of the force along the axis of strut $BC$.

**Fig. P2-42**

### Challenging Problems

**2-43\*** Three forces are applied to a bracket as shown in Fig. P2-43. The magnitude of the resultant $\mathbf{R}$ of the three forces is 5000 lb. If force $\mathbf{F}_1$ has a magnitude of 3000 lb, determine the magnitudes of forces $\mathbf{F}_2$ and $\mathbf{F}_3$.

**Fig. P2-43**

**2-44** A gusset plate is used to transfer forces from three bars to a beam as shown in Fig. P2-44. The magnitude of the resultant $\mathbf{R}$ of the three forces is 5000 N. If force $\mathbf{F}_1$ has a magnitude of 1000 N, determine the magnitudes of forces $\mathbf{F}_2$ and $\mathbf{F}_3$.

**Fig. P2-44**

## 2-6 RECTANGULAR COMPONENTS OF A FORCE

General oblique components of a force are not widely used for solving most practical engineering problems. Mutually perpendicular (rectangular) components, on the other hand, find wide usage. The process for obtaining rectangular components is greatly simplified since the parallelogram used to represent the force and its components reduces to a rectangle and the law of cosines used to obtain numerical values of the components reduces to the Pythagorean theorem.

A force $\mathbf{F}$ can be resolved into a rectangular component $\mathbf{F}_x$ along the $x$-axis and a rectangular component $\mathbf{F}_y$ along the $y$-axis as shown in Fig. 2-24a. The forces $\mathbf{F}_x$ and $\mathbf{F}_y$ are the vector components of the force $\mathbf{F}$. The $x$- and $y$-axes are usually chosen horizontal and vertical as shown in Fig. 2-24a; however, they may be chosen in any two perpendicular directions. The choice is usually indicated by the geometry of the problem.[1]

The force **F** and its two-dimensional vector components $\mathbf{F}_x$ and $\mathbf{F}_y$ can be written in Cartesian vector form by using unit vectors **i** and **j** directed along the positive $x$- and $y$-coordinate axes as shown in Fig. 2-24b. Thus,

$$\mathbf{F} = \mathbf{F}_x + \mathbf{F}_y = F_x\mathbf{i} + F_y\mathbf{j} \tag{2-3}$$

where the scalars $F_x$ and $F_y$ are the $x$ and $y$ scalar components of the force **F**. The scalar components $F_x$ and $F_y$ are related to the magnitude by $F = |\mathbf{F}|$ and the angle of inclination $\theta$ (direction) of the force **F** by the following expressions

$$
\begin{array}{ll}
F_x = F\cos\theta & F = \sqrt{F_x^2 + F_y^2} \\
F_y = F\sin\theta & \theta = \tan^{-1}\dfrac{F_y}{F_x}
\end{array}
\tag{2-4}
$$

The scalar components $F_x$ and $F_y$ of the force **F** can be positive or negative depending on the sense of the vector components $\mathbf{F}_x$ and $\mathbf{F}_y$. A scalar component is positive when the vector component has the same sense as the unit vector with which it is associated and negative when the vector component has the opposite sense.

Similarly, for problems requiring analysis in three dimensions, a force **F** in space can be resolved into three mutually perpendicular rectangular components $\mathbf{F}_x$, $\mathbf{F}_y$, and $\mathbf{F}_z$ along the $x$-, $y$-, and $z$-coordinate axes as shown in Fig. 2-25. The force **F** and its three-dimensional vector components $\mathbf{F}_x$, $\mathbf{F}_y$, and $\mathbf{F}_z$ can also be written in Cartesian vector form by using unit vectors **i**, **j**, and **k** directed along the positive $x$-, $y$-, and $z$-coordinate axes as shown in Fig. 2-26a. Thus,

$$
\begin{aligned}
\mathbf{F} &= \mathbf{F}_x + \mathbf{F}_y + \mathbf{F}_z = F_x\mathbf{i} + F_y\mathbf{j} + F_z\mathbf{k} \\
&= F\cos\theta_x\mathbf{i} + F\cos\theta_y\mathbf{j} + F\cos\theta_z\mathbf{k} = F\mathbf{e}_F
\end{aligned}
\tag{2-5}
$$

where $\mathbf{e}_F = \cos\theta_x\mathbf{i} + \cos\theta_y\mathbf{j} + \cos\theta_z\mathbf{k}$ is a unit vector along the line of action of the force. The scalar components $F_x$, $F_y$, and $F_z$ are related to

[1]Coordinate systems and axes are tools that may be used to advantage by the analyst. Machine components and structural elements do not come inscribed with $x$- and $y$-axes; therefore, the analyst is free to select directions that are convenient for his or her work.

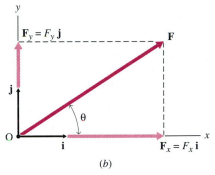

**Figure 2-24** Rectangular components $\mathbf{F}_x$ and $\mathbf{F}_y$ of a force **F**.

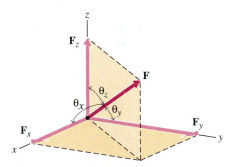

**Figure 2-25** Rectangular components $\mathbf{F}_x$, $\mathbf{F}_y$, and $\mathbf{F}_z$ of a force **F**.

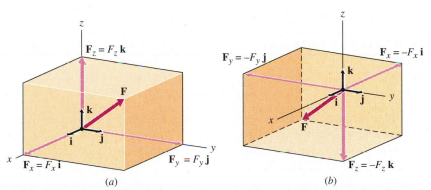

**Figure 2-26** Representing the rectangular components of a force **F** with unit vectors **i**, **j**, and **k**.

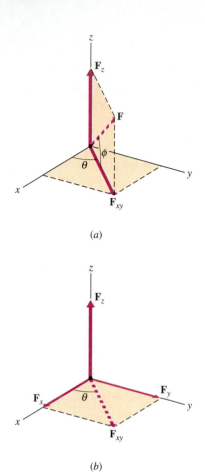

(a)

(b)

**Figure 2-27** Specifying the direction
of a force **F** with an azimuth angle $\theta$
and an elevation angle $\phi$.

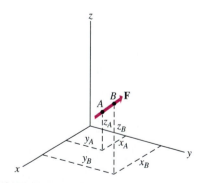

**Figure 2-28** Specifying the direction
of a force **F** with the coordinates of
two points along its line of action.

the magnitude and direction of the force **F** by the following expressions

$$F_x = F \cos \theta_x \qquad F_y = F \cos \theta_y \qquad F_z = F \cos \theta_z$$

$$\theta_x = \cos^{-1}\frac{F_x}{F} \qquad \theta_y = \cos^{-1}\frac{F_y}{F} \qquad \theta_z = \cos^{-1}\frac{F_z}{F} \quad \text{(2-6)}$$

$$F = \sqrt{F_x^2 + F_y^2 + F_z^2}$$

The angles $\theta_x$, $\theta_y$, and $\theta_z$ are the angles ($0 \leq \theta \leq 180°$) between the force **F** and the positive coordinate axes. The cosines of these angles, called direction cosines, must satisfy the equation

$$\cos^2 \theta_x + \cos^2 \theta_y + \cos^2 \theta_z = 1$$

If an angle is greater than 90° (see Fig. 2-26b), the cosine is negative, indicating that the sense of the component is opposite to the positive direction of the coordinate axis. Thus, Eqs. 2-6 provide the sign as well as the magnitude of the scalar components of the force and hold for any value of the angle.

In many engineering problems, the direction of the force is specified in terms of an azimuth angle $\theta$ and an elevation angle $\phi$, as shown in Fig. 2-27a. When the direction of the force is specified in this manner, the magnitudes of the three rectangular components $F_x$, $F_y$, and $F_z$ of the force can be determined by first resolving the force into a component $F_{xy}$ in the $xy$-plane and a second component $F_z$ perpendicular to the $xy$-plane. Thus,

$$F_{xy} = F \cos \phi$$
$$F_z = F \sin \phi$$

The component in the $xy$-plane can then be further resolved into components along the $x$- and $y$-axes as shown in Fig. 2-27b. Thus,

$$F_x = F_{xy} \cos \theta = F \cos \phi \cos \theta$$
$$F_y = F_{xy} \sin \theta = F \cos \phi \sin \theta$$

Finally, the direction of the force can be indicated by specifying the coordinates of two points $A(x_A, y_A, z_A)$ and $B(x_B, y_B, z_B)$ along the line of action of the force as shown in Fig. 2-28. For this method of specifying the direction,

$$\cos \theta_x = \frac{x_B - x_A}{\sqrt{(x_B - x_A)^2 + (y_B - y_A)^2 + (z_B - z_A)^2}}$$

$$\cos \theta_y = \frac{y_B - y_A}{\sqrt{(x_B - x_A)^2 + (y_B - y_A)^2 + (z_B - z_A)^2}}$$

$$\cos \theta_z = \frac{z_B - z_A}{\sqrt{(x_B - x_A)^2 + (y_B - y_A)^2 + (z_B - z_A)^2}}$$

The rectangular component of a force **F** along an arbitrary direction $n$ can be obtained by using the vector operation known as the dot or scalar product (see Appendix A). For example, the scalar component $F_x$ of a force **F** is obtained as

$$F_x = \mathbf{F} \cdot \mathbf{i} = (F_x\mathbf{i} + F_y\mathbf{j} + F_z\mathbf{k}) \cdot \mathbf{i}$$
$$= F_x(\mathbf{i} \cdot \mathbf{i}) + F_y(\mathbf{j} \cdot \mathbf{i}) + F_z(\mathbf{k} \cdot \mathbf{i}) = F_x$$

since

$$i \cdot i = j \cdot j = k \cdot k = 1$$

and

$$i \cdot j = j \cdot i = i \cdot k = k \cdot i = j \cdot k = k \cdot j = 0$$

In more general terms, if $e_n$ is a unit vector in a specified direction $n$ along a line $AB$ (see Fig. 2-29) and $e_F$ is a unit vector in the direction of the force $F$, then the rectangular component $F_n$ of the force $F$ is

$$F_n = \mathbf{F} \cdot \mathbf{e}_n = (F\mathbf{e}_F) \cdot \mathbf{e}_n = F(\mathbf{e}_F \cdot \mathbf{e}_n)$$

The rectangular component $F_n$ of the force $F$ can be expressed in vector form by multiplying the scalar component $F_n$ by the unit vector $e_n$. Thus,

$$\mathbf{F}_n = (\mathbf{F} \cdot \mathbf{e}_n)\mathbf{e}_n = F(\mathbf{e}_F \cdot \mathbf{e}_n) = F_n\mathbf{e}_n \qquad (2\text{-}7)$$

The angle $\alpha$ between the line of action of the force $F$ and the direction $n$ can be determined by using the dot-product relationship and the definition of a rectangular component of a force. Thus

$$F_n = F \cos \alpha = \mathbf{F} \cdot \mathbf{e}_n$$

therefore,

$$\alpha = \cos^{-1} \frac{\mathbf{F} \cdot \mathbf{e}_n}{F} = \cos^{-1} \frac{F_n}{F} \qquad (2\text{-}8)$$

Equation (2-8) can be used to determine the angle $\alpha$ between any two vectors $A$ and $B$ or between any two lines by using the unit vectors $e_1$ and $e_2$ associated with the lines. Thus,

$$\alpha = \cos^{-1} \frac{\mathbf{A} \cdot \mathbf{B}}{AB} \qquad (2\text{-}9)$$

or

$$\alpha = \cos^{-1} (\mathbf{e}_1 \cdot \mathbf{e}_2) \qquad (2\text{-}10)$$

The dot-product relationships represented by Eqs. 2-8, 2-9, and 2-10 apply to nonintersecting vectors as well as to intersecting vectors.

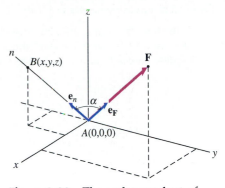

**Figure 2-29**  The scalar product of two vectors.

Two forces are applied at a point in a body as shown in Fig. 2-30a.

**a.** Determine the $x$ and $y$ scalar components for force $\mathbf{F}_1$.
**b.** Determine the $x$ and $y$ scalar components for force $\mathbf{F}_2$.
**c.** Express forces $\mathbf{F}_1$ and $\mathbf{F}_2$ in Cartesian vector form.

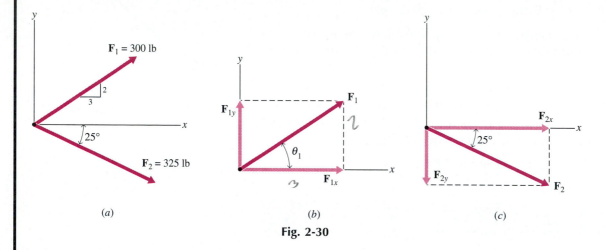

(a)        (b)        (c)

**Fig. 2-30**

## SOLUTION

**a.** The $x$ and $y$ components for force $F_1$ are shown in Fig. 2-30b. Since the direction of the force is given as a slope, the angle between the line of action of the force and the $x$-axis is

$$\theta_1 = \tan^{-1}\frac{2}{3} = 33.69°$$

Once the angle is known, the $x$ and $y$ components of the force are given by the expressions

$$F_{1x} = F_1 \cos \theta_1 = 300 \cos 33.69° = 249.6 \text{ lb} \cong 250 \text{ lb} \qquad \text{Ans.}$$
$$F_{1y} = F_1 \sin \theta_1 = 300 \sin 33.69° = 166.41 \text{ lb} \cong 166.4 \text{ lb} \qquad \text{Ans.}$$

> Rectangular components of a force are obtained by using normal sine and cosine relationships since the parallelogram of forces is a rectangle.

**b.** Similarly, for force $\mathbf{F}_2$, after noting that the $y$ component of the force is negative since its sense is opposite to the positive $y$ direction as shown in Fig. 2-30c,

$$F_{2x} = F_2 \cos \theta_2 = 325 \cos 25° = 294.6 \text{ lb} \cong 295 \text{ lb} \qquad \text{Ans.}$$
$$F_{2y} = -F_2 \sin \theta_2 = -325 \sin 25° = -137.35 \text{ lb} \cong -137.4 \text{ lb} \qquad \text{Ans.}$$

**c.** The two forces can be expressed in Cartesian vector form as

$$\mathbf{F}_1 = F_{1x}\mathbf{i} + F_{1y}\mathbf{j} = 250\mathbf{i} + 166.4\mathbf{j} \text{ lb} \qquad \text{Ans.}$$
$$\mathbf{F}_2 = F_{2x}\mathbf{i} + F_{2y}\mathbf{j} = 295\mathbf{i} - 137.4\mathbf{j} \text{ lb} \qquad \text{Ans.}$$

> Note that scalar components of a force may be positive or negative depending on the sense of the component.

A force **F** is applied at a point in a body as shown in Fig. 2-31.

**a.** Determine the $x$ and $y$ scalar components of the force.
**b.** Determine the $x'$ and $y'$ scalar components of the force.
**c.** Express the force **F** in Cartesian vector form for the $xy$- and $x'y'$-axes.

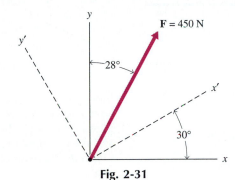

**Fig. 2-31**

---

**SOLUTION**

**a.** The magnitude $F$ of the force is 450 N. The angle $\theta_x$ between the $x$-axis and the line of action of the force is

$$\theta_x = 90° - 28° = 62°$$

Thus

$$F_x = F \cos \theta_x = 450 \cos 62° = +211 \text{ N} \qquad \text{Ans.}$$
$$F_y = F \sin \theta_x = 450 \sin 62° = +397 \text{ N} \qquad \text{Ans.}$$

**b.** The magnitude $F$ of the force is 450 N. The angle $\theta_{x'}$ between the $x'$-axis and the line of action of the force is

$$\theta_{x'} = \theta_x - 30° = 62° - 30° = 32°$$

Thus

$$F_{x'} = F \cos \theta_{x'} = 450 \cos 32° = +382 \text{ N} \qquad \text{Ans.}$$
$$F_{y'} = F \sin \theta_{x'} = 450 \sin 32° = +238 \text{ N} \qquad \text{Ans.}$$

As a check note that

$$F = \sqrt{F_x^2 + F_y^2} = \sqrt{F_{x'}^2 + F_{y'}^2} = \sqrt{211^2 + 397^2} = \sqrt{382^2 + 238^2} = 450 \text{ N}$$

**c.** The force **F** expressed in Cartesian vector form for the $xy$- and $x'y'$-axes is

$$\mathbf{F} = 211\mathbf{i} + 397\mathbf{j} \text{ N}$$
$$\mathbf{F} = 382\mathbf{e}_{x'} + 238\mathbf{e}_{y'} \text{ N} \qquad \text{Ans.}$$

The triangle formed by a force **F** and one of its rectangular components is a right triangle.

## EXAMPLE PROBLEM 2-9

A force **F** is applied at a point in a body as shown in Fig. 2-32.

**a.** Determine the $x$, $y$, and $z$ scalar components of the force.
**b.** Express the force in Cartesian vector form.

**Fig. 2-32**

### SOLUTION

**a.** The magnitude F of the force is 1500 lb. Thus,

$$F_x = F \cos \theta_x = 1500 \cos 72.0° = +464 \text{ lb} \quad \text{Ans.}$$
$$F_y = F \cos \theta_y = 1500 \cos 31.6° = +1278 \text{ lb} \quad \text{Ans.}$$
$$F_z = F \cos \theta_z = 1500 \cos 65.0° = +634 \text{ lb} \quad \text{Ans.}$$

As a check note that

$$F = \sqrt{F_x^2 + F_y^2 + F_z^2} = \sqrt{464^2 + 1278^2 + 634^2} = 1500 \text{ lb}$$

**b.** The force **F** expressed in Cartesian vector form is

$$\mathbf{F} = F_x\mathbf{i} + F_y\mathbf{j} + F_z\mathbf{k} = 464\mathbf{i} + 1278\mathbf{j} + 634\mathbf{k} \text{ lb} \quad \text{Ans.}$$

> In either two or three dimensions, a rectangular component of a force is equal to the product of the force and the cosine of the angle between the force and the component.

---

## EXAMPLE PROBLEM 2-10

A force of 50 kN is applied to an eye bolt as shown in Fig. 2-33.

**a.** Determine the angles $\theta_x$, $\theta_y$, and $\theta_z$.
**b.** Determine the $x$, $y$, and $z$ scalar components of the force.
**c.** Express the force in Cartesian vector form.

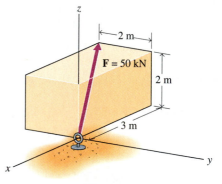

**Fig. 2-33**

### SOLUTION

**a.** The angles $\theta_x$, $\theta_y$, and $\theta_z$ can be determined from the geometry of the box shown in Fig. 2-33. The length of the diagonal of the box is

$$d = \sqrt{x^2 + y^2 + z^2} = \sqrt{(-3)^2 + (-2)^2 + (2)^2} = 4.123 \text{ m}$$

Thus,

$$\theta_x = \cos^{-1}\frac{x}{d} = \cos^{-1}\frac{-3}{4.123} = 136.69° \cong 136.7° \quad \text{Ans.}$$

$$\theta_y = \cos^{-1}\frac{y}{d} = \cos^{-1}\frac{-2}{4.123} = 119.02° \cong 119.0° \quad \text{Ans.}$$

$$\theta_z = \cos^{-1}\frac{z}{d} = \cos^{-1}\frac{2}{4.123} = 60.98° \cong 61.0° \quad \text{Ans.}$$

As a check:

$$\cos^2 \theta_x + \cos^2 \theta_y + \cos^2 \theta_z = \left(\frac{-3}{4.123}\right)^2 + \left(\frac{-2}{4.123}\right)^2 + \left(\frac{2}{4.123}\right)^2 = 1.000$$

> A force **F** and its three rectangular components can be represented with a rectangular parallelepiped (a box). Since the line of action of the force is along a diagonal of the box, the length of the diagonal and the lengths of the three sides of the box can be used to determine the three direction angles for the force.

**b.** Once the angles $\theta_x$, $\theta_y$, and $\theta_z$ have been determined, the three scalar components are obtained from the expressions

$$F_x = F \cos \theta_x = 50 \cos 136.69° = -86.38 \text{ kN} \cong -36.4 \text{ kN} \qquad \text{Ans.}$$
$$F_y = F \cos \theta_y = 50 \cos 119.02° = -24.26 \text{ kN} \cong -24.3 \text{ kN} \qquad \text{Ans.}$$
$$F_z = F \cos \theta_x = 50 \cos 60.98° = 24.26 \text{ kN} \cong 24.3 \text{ kN} \qquad \text{Ans.}$$

**c.** The force **F** expressed in Cartesian vector form is

$$\mathbf{F} = F_x\mathbf{i} + F_y\mathbf{j} + F_z\mathbf{k} = -36.4\mathbf{i} - 24.3\mathbf{j} + 24.3\mathbf{k} \text{ kN} \qquad \text{Ans.}$$

## EXAMPLE PROBLEM 2-11

A force **F** is applied at a point in a body as shown in Fig. 2-34a.

**a.** Determine the $x$, $y$, and $z$ scalar components of the force.
**b.** Express the force in Cartesian vector form.

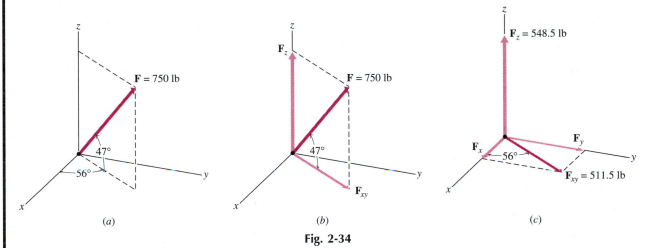

**Fig. 2-34**

## SOLUTION

**a.** When the direction of a force **F** in space is specified with an azimuth angle $\theta$ and an elevation angle $\phi$, as shown in Fig. 2-34a, the magnitudes of the three rectangular components $F_x$, $F_y$, and $F_z$ of the force are usually determined by first resolving the force into a component $F_{xy}$ in the $xy$-plane and a second component $F_z$ perpendicular to the $xy$-plane as shown in Fig. 2-34b. Thus,

> When the direction of a force **F** is specified with azimuth and elevation angles, successive applications of the parallelogram law can be used to determine the three rectangular components of the force.

$$F_{xy} = F \cos \phi = 750 \cos 47° = 511.5 \text{ lb} \cong 512 \text{ lb}$$
$$F_z = F \sin \phi = 750 \sin 47° = 548.5 \text{ lb} \cong 549 \text{ lb} \qquad \text{Ans.}$$

The component in the $xy$-plane can then be further resolved into components along the $x$- and $y$-axes as shown in Fig. 2-34c. Thus,

$$F_x = F_{xy} \cos \theta = 511.5 \cos 56° = 286.0 \text{ lb} = 286 \text{ lb} \qquad \text{Ans.}$$
$$F_y = F_{xy} \sin \theta = 511.5 \sin 56° = 424.1 \text{ lb} \cong 424 \text{ lb} \qquad \text{Ans.}$$

**b.** The force **F** expressed in Cartesian vector form is

$$\mathbf{F} = F_x\mathbf{i} + F_y\mathbf{j} + F_z\mathbf{k} = 286\mathbf{i} + 424\mathbf{j} + 549\mathbf{k} \text{ lb} \qquad \text{Ans.}$$

## EXAMPLE PROBLEM 2-12

A force **F** is applied at a point in a body as shown in Fig. 2-35.

**a.** Determine the $x$, $y$, and $z$ scalar components of the force.
**b.** Express the force in Cartesian vector form.

**Fig. 2-35**

### SOLUTION

**a.** When the direction of a force **F** in space is indicated by specifying the coordinates of two points $A(x_A, y_A, z_A)$ and $B(x_B, y_B, z_B)$ along the line of action of the force, as shown in Fig. 2-35, a position vector $\mathbf{r}_{B/A}$ can be written as

$$\mathbf{r}_{B/A} = (x_B - x_A)\mathbf{i} + (y_B - y_A)\mathbf{j} + (z_B - z_A)\mathbf{k}$$

The direction cosines associated with the position vector $\mathbf{r}_{A/B}$ and thus with the force **F** are

$$\cos \theta_x = \frac{x_B - x_A}{\sqrt{(x_B - x_A)^2 + (y_B - y_A)^2 + (z_B - z_A)^2}}$$

$$= \frac{475 - 250}{\sqrt{(475 - 250)^2 + (750 - 350)^2 + (500 - 400)^2}} = 0.3907$$

> When two points $A$ and $B$ on the line of action of a force **F** are used to specify the direction of the force, the position vector $\mathbf{r}_{B/A}$ can be used to determine the three direction cosines (or angles) for the force.

$$\cos \theta_y = \frac{y_B - y_A}{\sqrt{(x_B - x_A)^2 + (y_B - y_A)^2 + (z_B - z_A)^2}}$$

$$= \frac{750 - 350}{\sqrt{(475 - 250)^2 + (750 - 350)^2 + (500 - 400)^2}} = 0.8930$$

$$\cos \theta_z = \frac{z_B - z_A}{\sqrt{(x_B - x_A)^2 + (y_B - y_A)^2 + (z_B - z_A)^2}}$$

$$= \frac{500 - 400}{\sqrt{(475 - 250)^2 + (750 - 350)^2 + (500 - 400)^2}} = 0.2233$$

The scalar components $F_x$, $F_y$, and $F_z$ are obtained from the magnitude of the force and the direction cosines. Thus,

$$F_x = F \cos \theta_x = 1000(0.3907) = 390.7 \text{ N} \cong 391 \text{ N} \qquad \text{Ans.}$$
$$F_y = F \cos \theta_y = 1000(0.8930) = 893.0 \text{ N} = 893 \text{ N} \qquad \text{Ans.}$$
$$F_z = F \cos \theta_z = 1000(0.2233) = 223.3 \text{ N} \cong 223 \text{ N} \qquad \text{Ans.}$$

**b.** The force **F** expressed in Cartesian vector form is

$$\mathbf{F} = F_x\mathbf{i} + F_y\mathbf{j} + F_z\mathbf{k} = 391\mathbf{i} + 893\mathbf{j} + 223\mathbf{k} \text{ N} \qquad \text{Ans.}$$

## EXAMPLE PROBLEM 2-13

A force $\mathbf{F} = 218\mathbf{i} + 436\mathbf{j} + 873\mathbf{k}$ lb is applied at a point in a body as shown in Fig. 2-36. Determine

**a.** The rectangular scalar component $F_n$ of force $\mathbf{F}$ along line $OA$.
**b.** The angle $\alpha$ between force $\mathbf{F}$ and line $OA$.

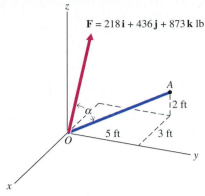

**Fig. 2-36**

### SOLUTION

**a.** The rectangular scalar component $F_n$ of force $\mathbf{F}$ is obtained by using the vector dot product $F_n = \mathbf{F} \cdot \mathbf{e}_n$, where $\mathbf{e}_n$ is a unit vector in a direction $n$ from point $O$ to point $A$. The distance $d$ from point $O$ to point $A$ is

$$d = \sqrt{x_A^2 + y_A^2 + z_A^2} = \sqrt{(-3)^2 + (5)^2 + (2)^2} = 6.164 \text{ ft}$$

$$\mathbf{e}_n = \frac{-3}{6.164}\mathbf{i} + \frac{5}{6.164}\mathbf{j} + \frac{2}{6.164}\mathbf{k} = -0.4867\mathbf{i} + 0.8112\mathbf{j} + 0.3245\mathbf{k}$$

$$F_n = \mathbf{F} \cdot \mathbf{e}_n = (218\mathbf{i} + 436\mathbf{j} + 873\mathbf{k}) \cdot (-0.4867\mathbf{i} + 0.8112\mathbf{j} + 0.3245\mathbf{k})$$

$$= 218(-0.4867) + 436(0.8112) + 873(0.3245)$$

$$= 530.9 \text{ lb} \cong 531 \text{ lb} \qquad \qquad \text{Ans.}$$

**b.** The magnitude $F$ of force $\mathbf{F}$ is

$$F = \sqrt{F_x^2 + F_y^2 + F_z^2} = \sqrt{(218)^2 + (436)^2 + (873)^2} = 999.9 \text{ lb}$$

$$\alpha = \cos^{-1}\frac{F_n}{F} = \cos^{-1}\frac{530.9}{999.9} = 57.93° \cong 57.9° \qquad \text{Ans.}$$

> The dot or scalar product of two intersecting vectors is defined as the product of the magnitudes of the two vectors and the cosine of the angle between them. Thus, the dot product can be used to obtain the rectangular scalar component of any vector in a specified direction.

## EXAMPLE PROBLEM 2-14

A force $\mathbf{F}$ is applied at a point in a body as shown in Fig. 2-37.

**a.** Express the force $\mathbf{F}$ in Cartesian vector form.
**b.** Determine the rectangular scalar component of force $\mathbf{F}$ along line $OA$.

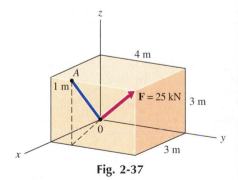

**Fig. 2-37**

### SOLUTION

**a.** The direction cosines for force $\mathbf{F}$ can be determined from the geometry of the box shown in Fig. 2-37. The length of diagonal $d$ of the box is

$$d = \sqrt{x^2 + y^2 + z^2} = \sqrt{(3)^2 + (4)^2 + (3)^2} = 5.831 \text{ m}$$

Thus,

$$F_x = F \cos \theta_x = 25\left(\frac{3}{5.831}\right) = 12.862 \text{ kN} \cong 12.86 \text{ kN}$$

$$F_y = F \cos \theta_y = 25\left(\frac{4}{5.831}\right) = 17.150 \text{ kN} \cong 17.15 \text{ kN}$$

$$F_z = F \cos \theta_z = 25\left(\frac{3}{5.831}\right) = 12.862 \text{ kN} \cong 12.86 \text{ kN}$$

**63**

The force **F** expressed in Cartesian vector form is

$$\mathbf{F} = F_x\mathbf{i} + F_y\mathbf{j} + F_z\mathbf{k} = 12.86\mathbf{i} + 17.15\mathbf{j} + 12.86\mathbf{k} \text{ kN} \qquad \text{Ans.}$$

**b.** The rectangular scalar component $F_n$ of force **F** is obtained by using the vector dot product $F_n = \mathbf{F} \cdot \mathbf{e}_n$, where $\mathbf{e}_n$ is a unit vector in a direction $n$ from point $O$ to point $A$. The distance $d_A$ from point $O$ to point $A$ is

$$d_A = \sqrt{x_A^2 + y_A^2 + z_A^2} = \sqrt{(3)^2 + (1)^2 + (3)^2} = 4.359 \text{ m}$$

$$\mathbf{e}_n = \frac{3}{4.359}\mathbf{i} + \frac{1}{4.359}\mathbf{j} + \frac{3}{4.359}\mathbf{k} = 0.6882\mathbf{i} + 0.2294\mathbf{j} + 0.6882\mathbf{k}$$

$$F_n = \mathbf{F} \cdot \mathbf{e}_n = (12.86\mathbf{i} + 17.15\mathbf{j} + 12.86\mathbf{k}) \cdot (0.6882\mathbf{i} + 0.2294\mathbf{j} + 0.6882\mathbf{k})$$

$$= 12.86(0.6882) + 17.15(0.2294) + 12.86(0.6882)$$

$$= 21.63 \text{ kN} \cong 21.6 \text{ kN} \qquad \text{Ans.}$$

> The dot product can be used to obtain the rectangular scalar component of any vector in a specified direction.

# PROBLEMS

## Introductory Problems

Determine the $x$ and $y$ scalar components of

**2-45\*** The force shown in Fig. P2-45.

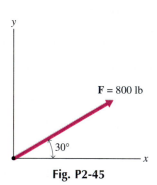

**Fig. P2-45**

**2-46\*** The force shown in Fig. P2-46.

**Fig. P2-46**

**2-47** The force shown in Fig. P2-47.

**Fig. P2-47**

**2-48** The force shown in Fig. P2-48.

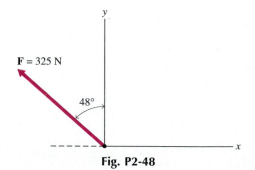

**Fig. P2-48**

**2-49\*** The force shown in Fig. P2-49.

**Fig. P2-49**

**2-50\*** The force shown in Fig. P2-50.

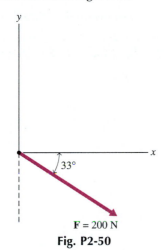

**Fig. P2-50**

**2-51** The force shown in Fig. P2-51.

**Fig. P2-51**

**2-52** The force shown in Fig. P2-52.

**Fig. P2-52**

**2-53\*** The force shown in Fig. P2-53.

**Fig. P2-53**

**2-54\*** The force shown in Fig. P2-54.

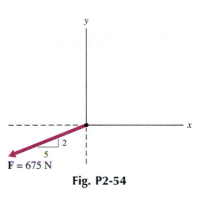

**Fig. P2-54**

## Intermediate Problems

**2-55\*** For the force shown in Fig. P2-55,

**a.** Determine the $x$, $y$, and $z$ scalar components of the force.
**b.** Express the force in Cartesian vector form.

**Fig. P2-55**

**2-56\*** For the force shown in Fig. P2-56,

**a.** Determine the $x, y,$ and $z$ scalar components of the force.
**b.** Express the force in Cartesian vector form.

**Fig. P2-56**

**2-57** For the force shown in Fig. P2-57,

**a.** Determine the $x, y,$ and $z$ scalar components of the force.
**b.** Express the force in Cartesian vector form.

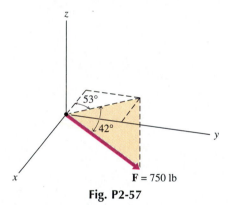

**Fig. P2-57**

**2-58** For the force shown in Fig. P2-58,

**a.** Determine the $x, y,$ and $z$ scalar components of the force.
**b.** Express the force in Cartesian vector form.

**Fig. P2-58**

**2-59\*** For the force shown in Fig. P2-59,

**a.** Determine the $x, y,$ and $z$ scalar components of the force.
**b.** Express the force in Cartesian vector form.

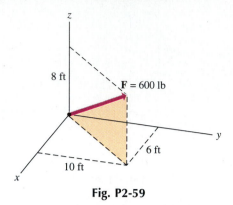

**Fig. P2-59**

**2-60\*** For the force shown in Fig. P2-60,

**a.** Determine the $x, y,$ and $z$ scalar components of the force.
**b.** Express the force in Cartesian vector form.

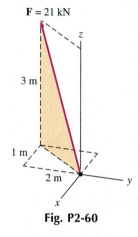

**Fig. P2-60**

**2-61** For the force shown in Fig. P2-61,

**a.** Determine the $x, y,$ and $z$ scalar components of the force.
**b.** Express the force in Cartesian vector form.

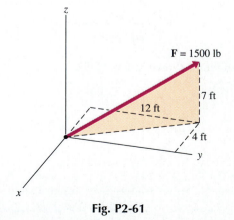

**Fig. P2-61**

**2-62** For the force shown in Fig. P2-62,

a. Determine the $x$, $y$, and $z$ scalar components of the force.
b. Express the force in Cartesian vector form.

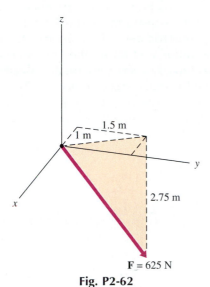

**Fig. P2-62**

## Challenging Problems

**2-63\*** Two forces are applied to an eye bolt as shown in Fig. P2-63.

a. Determine the $x$, $y$, and $z$ scalar components of force $\mathbf{F}_1$.
b. Express force $\mathbf{F}_1$ in Cartesian vector form.
c. Determine the angle $\alpha$ between forces $\mathbf{F}_1$ and $\mathbf{F}_2$.

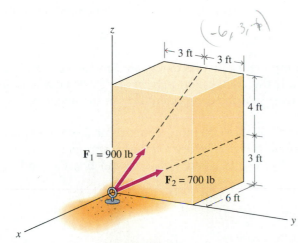

**Fig. P2-63**

**2-64\*** Two forces are applied to an eye bolt as shown in Fig. P2-64.

a. Determine the $x$, $y$, and $z$ scalar components of force $\mathbf{F}_1$.
b. Express force $\mathbf{F}_1$ in Cartesian vector form.
c. Determine the angle $\alpha$ between forces $\mathbf{F}_1$ and $\mathbf{F}_2$.

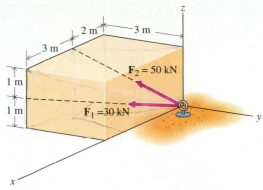

**Fig. P2-64**

**2-65** Two forces are applied to an eye bolt as shown in Fig. P2-65.

a. Determine the $x$, $y$, and $z$ scalar components of force $\mathbf{F}_2$.
b. Express force $\mathbf{F}_2$ in Cartesian vector form.
c. Determine the magnitude of the rectangular component of force $\mathbf{F}_2$ along the line of action of force $\mathbf{F}_1$.

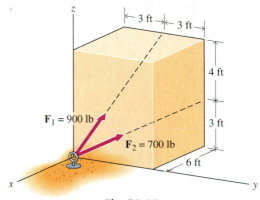

**Fig. P2-65**

**2-66** Two forces are applied to an eye bolt as shown in Fig. P2-66.

a. Determine the $x$, $y$, and $z$ scalar components of force $\mathbf{F}_2$.
b. Express force $\mathbf{F}_2$ in Cartesian vector form.
c. Determine the magnitude of the rectangular component of force $\mathbf{F}_2$ along the line of action of force $\mathbf{F}_1$.

**Fig. P2-66**

**67**

(a)

(b)

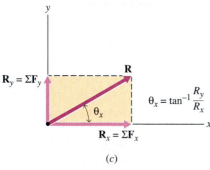

(c)

**Figure 2-38**   Rectangular components
and resultants **R** for three coplanar,
concurrent forces.

## 2-7   RESULTANTS BY RECTANGULAR COMPONENTS

Previous sections of this chapter discussed the use of the parallelogram
and triangle laws to determine the resultant **R** of two or more concur-
rent coplanar forces $\mathbf{F}_1, \mathbf{F}_2, \mathbf{F}_3, \ldots, \mathbf{F}_n$. Using the parallelogram law to
add more than two forces is time-consuming and tedious since the pro-
cedure requires extensive geometric and trigonometric calculation in
order to determine the magnitude and locate the line of action of the
resultant **R**. Problems of this type, however, are easily solved using the
rectangular components of a force discussed in Section 2-6.

For any system of coplanar concurrent forces, such as the three
shown in Fig. 2-38a, rectangular components $\mathbf{F}_{1x}$ and $\mathbf{F}_{1y}$, $\mathbf{F}_{2x}$ and $\mathbf{F}_{2y}$,
$\mathbf{F}_{3x}$ and $\mathbf{F}_{3y}$, ..., and $\mathbf{F}_{nx}$ and $\mathbf{F}_{ny}$ can be determined as shown in Fig.
2-38b. Adding the respective $x$ and $y$ components yields

$$\mathbf{R}_x = \Sigma\mathbf{F}_x = \mathbf{F}_{1x} + \mathbf{F}_{2x} + \mathbf{F}_{3x} + \cdots + \mathbf{F}_{nx}$$
$$= (F_{1x} + F_{2x} + F_{3x} + \cdots + F_{nx})\mathbf{i} = R_x\mathbf{i}$$
$$\mathbf{R}_y = \Sigma\mathbf{F}_y = \mathbf{F}_{1y} + \mathbf{F}_{2y} + \mathbf{F}_{3y} + \cdots + \mathbf{F}_{ny}$$
$$= (F_{1y} + F_{2y} + F_{3y} + \cdots + F_{ny})\mathbf{i} = R_y\mathbf{j}$$

By the parallelogram law

$$\mathbf{R} = \mathbf{R}_x + \mathbf{R}_y = R_x\mathbf{i} + R_y\mathbf{j}$$

The magnitude $R$ of the resultant can be determined from the
Pythagorean theorem:

$$R = \sqrt{R_x^2 + R_y^2} = \sqrt{(\Sigma F_x)^2 + (\Sigma F_y)^2}$$

The angle $\theta_x$ between the $x$-axis and the line of action of the resultant
**R**, as shown in Fig. 2-38c, is

$$\theta_x = \tan^{-1}\frac{R_y}{R_x} = \tan^{-1}\frac{\Sigma F_y}{\Sigma F_x}$$

The angle $\theta_x$ can also be determined, if it is more convenient, from the
equations

$$\theta_x = \cos^{-1}\frac{\Sigma F_x}{R} \qquad \text{or} \qquad \theta_x = \sin^{-1}\frac{\Sigma F_y}{R}$$

The sense of each component must be designated in the summations
by using a plus sign if the component acts in the positive $x$- and $y$-
direction, and a minus sign if the component acts in the negative $x$- or
$y$-direction.

In the general case of three or more concurrent forces in space,
such as the three shown in Fig. 2-39, rectangular components $\mathbf{F}_{1x}$, $\mathbf{F}_{1y}$,
and $\mathbf{F}_{1z}$; $\mathbf{F}_{2x}$, $\mathbf{F}_{2y}$, and $\mathbf{F}_{2z}$; $\mathbf{F}_{3x}$, $\mathbf{F}_{3y}$, and $\mathbf{F}_{3z}$; ...; and $\mathbf{F}_{nx}$, $\mathbf{F}_{ny}$, and $\mathbf{F}_{nz}$

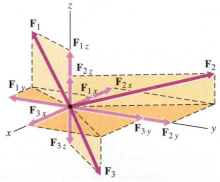

**Figure 2-39**   Rectangular components for three concurrent forces in space.

can be determined. Adding the respective $x$, $y$, and $z$ components yields

$$R_x = \Sigma F_x = F_{1x} + F_{2x} + F_{3x} + \cdots + F_{nx}$$
$$= (F_{1x} + F_{2x} + F_{3x} + \cdots + F_{nx})\mathbf{i} = R_x\mathbf{i}$$
$$R_y = \Sigma F_y = F_{1y} + F_{2y} + F_{3y} + \cdots + F_{ny}$$
$$= (F_{1y} + F_{2y} + F_{3y} + \cdots + F_{ny})\mathbf{j} = R_y\mathbf{j}$$
$$R_z = \Sigma F_z = F_{1z} + F_{2z} + F_{3z} + \cdots + F_{nz}$$
$$= (F_{1z} + F_{2z} + F_{3z} + \cdots + F_{nz})\mathbf{k} = R_z\mathbf{k}$$

The resultant **R** is then obtained from the expression

$$\mathbf{R} = \mathbf{R}_x + \mathbf{R}_y + \mathbf{R}_z = R_x\mathbf{i} + R_y\mathbf{j} + R_z\mathbf{k}$$

Once the scalar components $R_x$, $R_y$, and $R_z$ are known, the magnitude $R$ of the resultant and the angles $\theta_x$, $\theta_y$, and $\theta_z$ between the line of action of the resultant and the positive coordinate axes can be obtained from the expressions

$$R = \sqrt{R_x^2 + R_y^2 + R_z^2}$$

and

$$\theta_x = \cos^{-1}\frac{R_x}{R} \qquad \theta_y = \cos^{-1}\frac{R_y}{R} \qquad \theta_z = \cos^{-1}\frac{R_z}{R}$$

A unit vector $\mathbf{e}_R$ along the line of action of the resultant force **R** is given by the expression

$$\mathbf{e}_R = \cos\theta_x\,\mathbf{i} + \cos\theta_y\,\mathbf{j} + \cos\theta_z\,\mathbf{k}$$

Thus, the resultant force **R** can be expressed in Cartesian vector form as

$$\mathbf{R} = R\,\mathbf{e}_R = R(\cos\theta_x\,\mathbf{i} + \cos\theta_y\,\mathbf{j} + \cos\theta_z\,\mathbf{k})$$

## EXAMPLE PROBLEM 2-15

Determine the magnitude $R$ of the resultant and the angle $\theta_x$ between the line of action of the resultant and the $x$-axis for the three forces shown in Fig. 2-40.

**Fig. 2-40**

### SOLUTION

The magnitude $R$ of the resultant and the angle $\theta_x$ between the $x$-axis and the line of action of the resultant will be determined by using rectangular components. The sense of each component in the summations will be designated by a plus sign if the component acts in a positive $x$- or $y$-direction and by a minus sign if its acts in a negative $x$- or $y$-direction. Thus,

$$R_x = F_1\cos\theta_1 + F_2\cos\theta_2 + F_3\cos\theta_3$$
$$= -80\cos 35° + 40\cos 62° + 60\cos 28° = 6.244\text{ lb}$$
$$R_y = F_1\sin\theta_1 + F_2\sin\theta_2 + F_3\sin\theta_3$$
$$= 80\sin 35° + 40\sin 62° + 60\sin 28° = 109.372\text{ lb}$$

The x- and y-components of the resultant of any number of coplanar forces are the algebraic sums of the x- and y-components of the individual forces.

Once the rectangular scalar components have been determined, the magnitude $R$ of the resultant and the angle $\theta_x$ are determined by using the expressions

$$R = \sqrt{R_x^2 + R_y^2} = \sqrt{(6.224)^2 + (109.372)^2}$$
$$= 109.549 \cong 109.5\text{ lb} \qquad\qquad \text{Ans.}$$

$$\theta_x = \tan^{-1}\frac{R_y}{R_x} = \tan^{-1}\frac{109.372}{6.224} = 86.74° \cong 86.7° \qquad\qquad \text{Ans.}$$

Three forces are applied at a point on a body as shown in Fig. 2-41. Determine the magnitude $R$ of the resultant of the three forces and the angles $\theta_x$, $\theta_y$, and $\theta_z$ between the line of action of the resultant and the positive $x$-, $y$-, and $z$-coordinate axes.

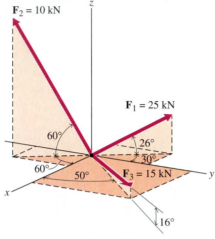

**Fig. 2-41**

## SOLUTION

The resultant $R$ of the three forces will be determined by using a scalar analysis. A scalar analysis is more efficient in this case since the directions of the forces are specified with azimuth and elevation angles. The magnitudes of the $x$-, $y$-, and $z$-components of each of the forces are as follows:
For force $\mathbf{F}_1$:

$$F_{1xy} = F_1 \cos \phi_1 = 25 \cos 26° = 22.470 \text{ kN}$$
$$F_{1z} = F_1 \sin \phi_1 = 25 \sin 26° = 10.959 \text{ kN}$$

The component in the $xy$-plane can be further resolved into components along the $x$- and $y$-axes. Thus,

$$F_{1x} = F_{1xy} \cos \theta_1 = 22.470 \cos 60° = 11.235 \text{ kN}$$
$$F_{1y} = F_{1xy} \sin \theta_1 = 22.470 \sin 60° = 19.460 \text{ kN}$$

For force $\mathbf{F}_2$:

$$F_{2xy} = F_2 \cos \phi_2 = 10 \cos 60° = 5.000 \text{ kN}$$
$$F_{2z} = F_2 \sin \phi_2 = 10 \sin 60° = 8.660 \text{ kN}$$
$$F_{2x} = F_{2xy} \cos \theta_2 = 5.000 \cos 60° = 2.500 \text{ kN}$$
$$F_{2y} = F_{2xy} \sin \theta_2 = 5.000 \sin 60° = 4.330 \text{ kN}$$

For force $\mathbf{F}_3$:

$$F_{3xy} = F_3 \cos \phi_3 = 15 \cos 16° = 14.419 \text{ kN}$$
$$F_{3z} = F_3 \sin \phi_3 = 15 \sin 16° = 4.135 \text{ kN}$$
$$F_{3x} = F_{3xy} \cos \theta_3 = 14.419 \cos 50° = 9.268 \text{ kN}$$
$$F_{3y} = F_{3xy} \sin \theta_3 = 14.419 \sin 50° = 11.046 \text{ kN}$$

Once the magnitudes of the rectangular components of the three forces are known, the scalar components $R_x$, $R_y$, and $R_z$ of the resultant are obtained

When the direction of a force $\mathbf{F}$ is specified with azimuth and elevation angles, successive applications of the parallelogram law can be used to determine the three rectangular components of the force.

by summing the appropriate components after a plus or minus sign has been assigned to designate the sense of the component.

$$R_x = \Sigma F_x = F_{1x} + F_{2x} + F_{3x}$$
$$= -11.235 + 2.500 + 9.268 = +0.533 \text{ kN}$$
$$R_y = \Sigma F_y = F_{1y} + F_{2y} + F_{3y}$$
$$= +19.460 - 4.330 + 11.046 = +26.176 \text{ kN}$$
$$R_z = \Sigma F_z = F_{1z} + F_{2z} + F_{3z}$$
$$= +10.959 + 8.660 + 4.135 = +23.754 \text{ kN}$$

The magnitude R of the resultant is

$$R = \sqrt{R_x^2 + R_y^2 + R_z^2}$$
$$= \sqrt{(0.533)^2 + (26.176)^2 + (23.754)^2} = 35.351 \text{ kN} \cong 35.4 \text{ kN} \qquad \text{Ans.}$$

The angles $\theta_x$, $\theta_y$, and $\theta_z$ are obtained from the expressions

$$\theta_x = \cos^{-1} \frac{R_x}{R} = \cos^{-1} \frac{+0.533}{35.351} = 89.14° \cong 89.1° \qquad \text{Ans.}$$

$$\theta_y = \cos^{-1} \frac{R_y}{R} = \cos^{-1} \frac{+26.175}{35.351} = 42.23° \cong 42.2° \qquad \text{Ans.}$$

$$\theta_z = \cos^{-1} \frac{R_z}{R} = \cos^{-1} \frac{+23.754}{35.351} = 47.78° \cong 47.8° \qquad \text{Ans.}$$

> The $x$-, $y$-, and $z$-components of the resultant of any number of forces in a three-dimensional system are the algebraic sums of the $x$-, $y$-, and $z$-components of the individual forces.

## EXAMPLE PROBLEM 2-17

Determine the magnitude $R$ of the resultant of the three forces shown in Fig. 2-42 and the angles $\theta_x$, $\theta_y$, and $\theta_z$ between the line of action of the resultant and the positive $x$-, $y$-, and $z$-coordinate axes.

### SOLUTION

In addition to the origin of coordinates, the lines of action of forces $\mathbf{F}_1$, $\mathbf{F}_2$, and $\mathbf{F}_3$ pass through points $(3, -2.5, 3.5)$, $(-3, -4.5, 3.5)$, and $(3, 5, 4)$, respectively. Since the coordinates of these points in space are known, it is easy to determine the unit vectors associated with each of the forces:

$$\mathbf{e}_1 = \frac{3\mathbf{i} - 2.5\mathbf{j} + 3.5\mathbf{k}}{\sqrt{(3)^2 + (-2.5)^2 + (3.5)^2}} = +0.5721\mathbf{i} - 0.4767\mathbf{j} + 0.6674\mathbf{k}$$

$$\mathbf{e}_2 = \frac{-3\mathbf{i} - 4.5\mathbf{j} + 3.5\mathbf{k}}{\sqrt{(-3)^2 + (-4.5)^2 + (3.5)^2}} = -0.4657\mathbf{i} - 0.6985\mathbf{j} + 0.5433\mathbf{k}$$

$$\mathbf{e}_3 = \frac{3\mathbf{i} + 5\mathbf{j} + 4\mathbf{k}}{\sqrt{(3)^2 + (5)^2 + (4)^2}} = +0.4243\mathbf{i} + 0.7071\mathbf{j} + 0.5657\mathbf{k}$$

Once the unit vectors $\mathbf{e}_1$, $\mathbf{e}_2$, and $\mathbf{e}_3$ are known, the three forces can be expressed in Cartesian vector form as

$$\mathbf{F}_1 = F_1\mathbf{e}_1 = 20(+0.5721\mathbf{i} - 0.4767\mathbf{j} + 0.6674\mathbf{k})$$
$$= +11.442\mathbf{i} - 9.534\mathbf{j} + 13.348\mathbf{k} \text{ kip}$$
$$\mathbf{F}_2 = F_2\mathbf{e}_2 = 30(-0.4657\mathbf{i} - 0.6985\mathbf{j} + 0.5433\mathbf{k})$$
$$= -13.971\mathbf{i} - 20.955\mathbf{j} + 16.299\mathbf{k} \text{ kip}$$
$$\mathbf{F}_3 = F_3\mathbf{e}_3 = 40(+0.4243\mathbf{i} + 0.7071\mathbf{j} + 0.5657\mathbf{k})$$
$$= +16.972\mathbf{i} + 28.284\mathbf{j} + 22.628\mathbf{k} \text{ kip}$$

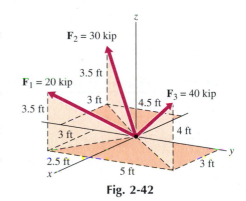

**Fig. 2-42**

> A force $\mathbf{F}$ can be written in Cartesian vector form by multiplying the magnitude of the force (a scalar) by a unit vector along the line of action of the force.

The resultant **R** of the three forces is

$$\mathbf{R} = \mathbf{F}_1 + \mathbf{F}_2 + \mathbf{F}_3 = R_x\mathbf{i} + R_y\mathbf{j} + R_z\mathbf{k} \text{ kip}$$

where

$$
\begin{aligned}
R_x &= \Sigma F_x = F_{1x} + F_{2x} + F_{3x} \\
&= 11.442 - 13.971 + 16.972 = 14.443 \text{ kip} \\
R_y &= \Sigma F_y = F_{1y} + F_{2y} + F_{3y} \\
&= -9.534 - 20.955 + 28.284 = -2.205 \text{ kip} \\
R_z &= \Sigma F_z = F_{1z} + F_{2z} + F_{3z} \\
&= 13.348 + 16.299 + 22.628 = 52.275 \text{ kip}
\end{aligned}
$$

> The $x$-, $y$-, and $z$-components of the resultant of any number of forces in a three-dimensional system are the algebraic sums of the $x$-, $y$-, and $z$-components of the individual forces.

Thus,

$$\mathbf{R} = 14.443\mathbf{i} - 2.205\mathbf{j} + 52.275\mathbf{k} \text{ kip}$$

The magnitude $R$ of the resultant is

$$R = \sqrt{R_x^2 + R_y^2 + R_z^2}$$

$$= \sqrt{(14.443)^2 + (-2.205)^2 + (52.275)^2} = 54.278 \cong 54.3 \text{ kip} \qquad \text{Ans.}$$

The angles $\theta_x$, $\theta_y$, and $\theta_z$ are obtained from the expressions

$$\theta_x = \cos^{-1}\frac{R_x}{R} = \cos^{-1}\frac{14.443}{54.278} = 74.57° \cong 74.6° \qquad \text{Ans.}$$

$$\theta_y = \cos^{-1}\frac{R_y}{R} = \cos^{-1}\frac{-2.205}{54.278} = 92.33° \cong 92.3° \qquad \text{Ans.}$$

$$\theta_z = \cos^{-1}\frac{R_z}{R} = \cos^{-1}\frac{52.275}{54.278} = 15.614° \cong 15.61° \qquad \text{Ans.}$$

# PROBLEMS

### Introductory Problems

Use the rectangular component method to solve the following problems. Determine the magnitude $R$ of the resultant of the forces and the angle $\theta_x$ between the line of action of the resultant and the $x$-axis for

**2-67*** The two forces shown in Fig. P2-67.

**Fig. P2-67**

**2-68*** The two forces shown in Fig. P2-68.

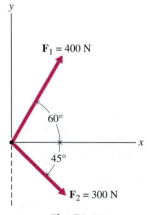

**Fig. P2-68**

**2-69**  The two forces shown in Fig. P2-69.

**Fig. P2-69**

**2-70**  The two forces shown in Fig. P2-70.

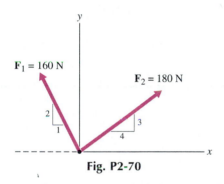

**Fig. P2-70**

**2-71***  The three forces shown in Fig. P2-71.

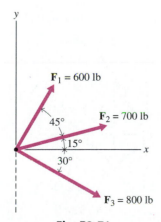

**Fig. P2-71**

**2-72***  The three forces shown in Fig. P2-72.

**Fig. P2-72**

**2-73**  The three forces shown in Fig. P2-73.

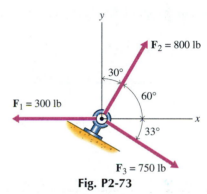

**Fig. P2-73**

**2-74***  The three forces shown in Fig. P2-74.

**Fig. P2-74**

**2-75** The three forces shown in Fig. P2-75.

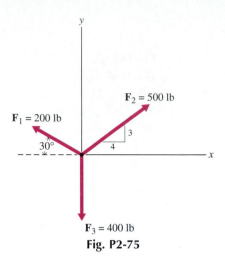

Fig. P2-75

**2-76*** The four forces shown in Fig. P2-76.

Fig. P2-76

**2-77** The four forces shown in Fig. P2-77.

Fig. P2-77

## Intermediate Problems

Use the rectangular component method to solve the following problems. Determine the magnitude $R$ of the resultant of the forces and the angles $\theta_x$, $\theta_y$, and $\theta_z$ between the line of action of the resultant and the positive $x$-, $y$-, and $z$-coordinate axes for

**2-78*** The two forces shown in Fig. P2-78.

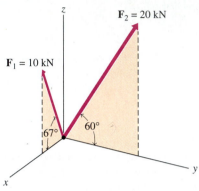

Fig. P2-78

**2-79*** The two forces shown in Fig. P2-79.

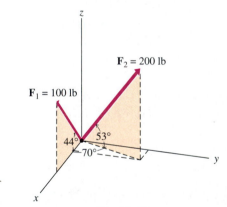

Fig. P2-79

**2-80** The two forces shown in Fig. P2-80.

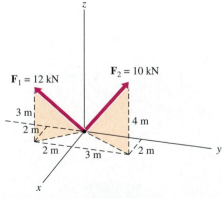

Fig. P2-80

**2-81** The two forces shown in Fig. P2-81.

Fig. P2-81

## Challenging Problems

Use the rectangular component method to solve the following problems. Determine the magnitude $R$ of the resultant of the forces and the angles $\theta_x$, $\theta_y$, and $\theta_z$ between the line of action of the resultant and the positive $x$-, $y$-, and $z$-coordinate axes for

**2-82*** The three forces shown in Fig. P2-82.

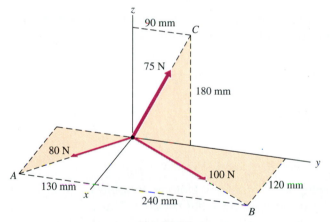

Fig. P2-82

**2-83*** The three forces shown in Fig. P2-83.

Fig. P2-83

**2-84** The three forces shown in Fig. P2-84.

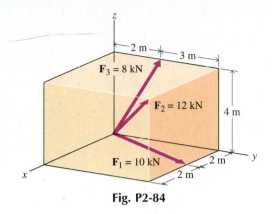

Fig. P2-84

**2-85*** The three forces shown in Fig. P2-85.

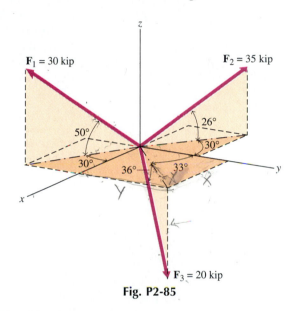

Fig. P2-85

**2-86** The three forces shown in Fig. P2-86.

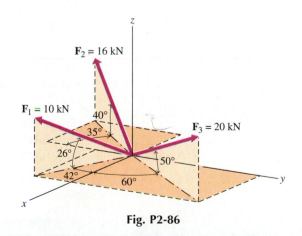

Fig. P2-86

# SUMMARY

A force is defined as the action of one physical body on another. Since the interaction can occur when there is contact between the bodies or when the bodies are physically separated, forces are classified as either surface forces (a push or a pull produced by mechanical means) or body forces (the gravitational pull of the earth). The characteristics of a force are its magnitude, its direction (orientation and sense), and its point of application. Since both a magnitude and a direction are needed to characterize a force and forces add according to the parallelogram law of addition, forces are vector quantities. A number of forces treated as a group constitute a force system.

The components of a force are any system of forces that can be combined by the parallelogram law to produce the original force. For most engineering problems, rectangular (mutually perpendicular) components of a force are more useful than general oblique components. The three rectangular components of a force $\mathbf{F}$ in space are $\mathbf{F}_x$, $\mathbf{F}_y$, and $\mathbf{F}_z$ along the $x$-, $y$-, and $z$-coordinate axes, respectively. The force $\mathbf{F}$ and its scalar components $F_x$, $F_y$, and $F_z$ can be written in Cartesian vector form by using unit vectors $\mathbf{i}$, $\mathbf{j}$, and $\mathbf{k}$ directed along the positive $x$-, $y$-, and $z$-coordinate axes as

$$\mathbf{F} = \mathbf{F}_x + \mathbf{F}_y + \mathbf{F}_z = F_x\mathbf{i} + F_y\mathbf{j} + F_z\mathbf{k}$$
$$= F\cos\theta_x\mathbf{i} + F\cos\theta_y\mathbf{j} + F\cos\theta_z\mathbf{k} \tag{2-5}$$

where the scalar components $F_x$, $F_y$, and $F_z$ are related to the magnitude $F$ and the direction of the force $\mathbf{F}$ by the expressions

$$F_x = F\cos\theta_x \qquad F_y = F\cos\theta_y \qquad F_z = F\cos\theta_z$$
$$F = \sqrt{F_x^2 + F_y^2 + F_z^2} \tag{2-6}$$

$$\theta_x = \cos^{-1}\frac{F_x}{F} \qquad \theta_y = \cos^{-1}\frac{F_y}{F} \qquad \theta_z = \cos^{-1}\frac{F_z}{F}$$

and

$$\cos^2\theta_x + \cos^2\theta_y + \cos^2\theta_z = 1$$

The rectangular component of a force $\mathbf{F}$ along an arbitrary direction $n$ can be obtained by using the vector dot product. Thus, if $\mathbf{e}_n$ is a unit vector in the specified direction $n$, the magnitude of the rectangular component $\mathbf{F}_n$ of the force $\mathbf{F}$ is

$$F_n = \mathbf{F}\cdot\mathbf{e}_n$$

The rectangular component $\mathbf{F}_n$ of the force $\mathbf{F}$ is expressed in Cartesian vector form as

$$\mathbf{F}_n = (\mathbf{F}\cdot\mathbf{e}_n)\mathbf{e}_n = F_n\mathbf{e}_n \tag{2-7}$$

The angle $\alpha$ between the line of action of the force $\mathbf{F}$ and the direction $n$ is determined by using the definition of a rectangular component of a force ($F_n = F\cos\alpha = \mathbf{F}\cdot\mathbf{e}_n$). Thus

$$\alpha = \cos^{-1}\frac{\mathbf{F}\cdot\mathbf{e}_n}{F} = \cos^{-1}\frac{F_n}{F} \tag{2-8}$$

A single force, called the resultant $\mathbf{R}$, will produce the same effect on a body as a system of concurrent forces. The resultant can be determined by adding the forces using the parallelogram law; however, this procedure is time-consuming and tedious when more than two forces comprise the system. Resultants are easily obtained, however, by using rectangular components of the forces. For the general case of two or more concurrent forces in space,

$$\mathbf{R}_x = \Sigma\mathbf{F}_x = R_x\mathbf{i} \qquad \mathbf{R}_y = \Sigma\mathbf{F}_y = R_y\mathbf{j} \qquad \mathbf{R}_z = \Sigma\mathbf{F}_z = R_z\mathbf{k}$$

The magnitude $R$ of the resultant and the angles $\theta_x$, $\theta_y$, and $\theta_z$ between the line of action of the resultant and the positive coordinate axes are

$$R = \sqrt{R_x^2 + R_y^2 + R_z^2}$$

$$\theta_x = \cos^{-1}\frac{R_x}{R} \qquad \theta_y = \cos^{-1}\frac{R_y}{R} \qquad \theta_z = \cos^{-1}\frac{R_z}{R}$$

# REVIEW PROBLEMS

**2-87\*** Two forces are applied to a bracket as shown in Fig. P2-87. Determine the magnitude of the resultant **R** of the two forces and the angle $\theta$ between the x-axis and the line of action of the resultant.

**Fig. P2-87**

**2-88\*** Two forces are applied to a bracket as shown in Fig. P2-88. Determine the magnitude of the resultant **R** of the two forces and the angle $\theta$ between the x-axis and the line of action of the resultant.

**Fig. P2-88**

**2-89** An 800-lb force is applied to the post shown in Fig. P2-89. Determine

**a.** The magnitudes of the x- and y-components of the force.

**b.** The magnitudes of the u- and v-components of the force.

**Fig. P2-89**

**2-90** Two forces $\mathbf{F}_1$ and $\mathbf{F}_2$ are applied to an eye bolt as shown in Fig. P2-90. Determine

**a.** The magnitude and direction (angle $\theta_x$) of the resultant **R** of the two forces.

**b.** The magnitudes of two other forces $\mathbf{F}_u$ and $\mathbf{F}_v$ that would have the same resultant.

**Fig. P2-90**

**2-91*** Three forces are applied to a bracket mounted on a post as shown in Fig. P2-91. Determine

a. The magnitude and direction (angle $\theta_x$) of the resultant **R** of the three forces.
b. The magnitudes of two other forces $\mathbf{F}_x$ and $\mathbf{F}_y$ that would have the same resultant.

**Fig. P2-91**

**2-92** Four forces act on a small airplane in flight, as shown in Fig. P2-92; its weight, the thrust provided by the engine, the lift provided by the wings, and the drag resulting from its motion through the air. Determine the resultant of the four forces and its line of action with respect to the axis of the plane.

**Fig. P2-92**

**2-93** Three forces are applied to a stalled automobile as shown in Fig. P2-93. Determine the magnitude of force $\mathbf{F}_3$ and the magnitude of the resultant **R** if the line of action of the resultant is along the $x$-axis.

**Fig. P2-93**

**2-94*** Four forces are applied at a point in a body as shown in Fig. P2-94. Determine the magnitude of the resultant **R** of the four forces and the angle $\theta$ between the $x$-axis and the line of action of the resultant.

**Fig. P2-94**

**2-95** Four forces are applied at a point in a body as shown in Fig. P2-95. Determine the magnitude and direction (angles $\theta_x$, $\theta_y$, and $\theta_z$) of the resultant **R** of the four forces.

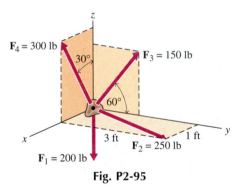

**Fig. P2-95**

**2-96*** Four forces are applied at a point in a body as shown in Fig. P2-96. Determine the magnitude and direction (angles $\theta_x$, $\theta_y$, and $\theta_z$) of the resultant **R** of the four forces.

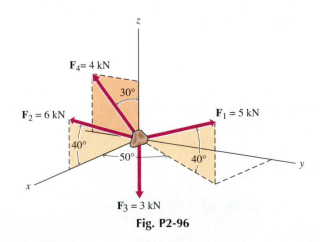

**Fig. P2-96**

**2-97** Two forces are applied at a point in a body as shown in Fig. P2-97. Determine

a. The magnitude and direction (angles $\theta_x$, $\theta_y$, and $\theta_z$) of the resultant **R** of the two forces.
b. The magnitude of the rectangular component of force **F**₁ along the line of action of force **F**₂.
c. The angle $\alpha$ between forces **F**₁ and **F**₂.

**Fig. P2-97**

**2-98** Three forces are applied with cables to the anchor block shown in Fig. P2-98. Determine

a. The magnitude and direction (angles $\theta_x$, $\theta_y$, and $\theta_z$) of the resultant **R** of the three forces.
b. The magnitude of the rectangular component of force **F**₁ along the line of action of force **F**₂.
c. The angle $\alpha$ between forces **F**₁ and **F**₃.

$F_1 = 272$ N
$F_2 = 500$ N
$F_3 = 650$ N

**Fig. P2-98**

## Computer Problems

**C2-99** Two forces **A** and **B** are applied to an eye bolt as shown in Fig. P2-99. If the magnitudes of the two forces are A = 50 lb and B = 100 lb, calculate and plot the magnitude of the resultant R as a function of the angle $\theta_A$ ($0° < \theta_A < 180°$). Also calculate and plot the angle $\theta_R$ that the resultant makes with the force **B** as a function of the angle $\theta_A$. When is the resultant a maximum? When is the resultant a minimum? When is the angle $\theta_R$ a maximum? Repeat for A = 100 lb and B = 50 lb.

**Fig. P2-99**

**C2-100** Two forces **A** and **B** are applied to an eye bolt using ropes as shown in Fig. P2-100. The resultant of the two forces has a magnitude R = 4 kN and makes an angle of 30° with the force **A** as shown. If both of the forces pull on the eye bolt as shown (ropes cannot push on the eye bolt), what is the range of angles ($\theta_{min} < \theta_B < \theta_{max}$) for which this problem has a solution? Calculate and plot the required magnitudes A and B as functions of the angle $\theta_B$ ($\theta_{min} < \theta_B < \theta_{max}$). Why is the magnitude of **B** a minimum when $\theta_B = 90°$?

**Fig. P2-100**

*A wind surfer moving at a constant velocity is in equilibrium under the action of wind, water, and gravity forces.*

# Chapter 3

# STATICS OF PARTICLES

**3-1 INTRODUCTION**         **3-3 EQUILIBRIUM OF A PARTICLE**

**3-2 FREE-BODY DIAGRAMS**         **S U M M A R Y**

## 3-1  INTRODUCTION

 Statics was defined in Chapter 1 as the branch of rigid-body mechanics concerned with bodies that are acted on by a balanced system of forces (the resultant of all forces acting on the body is zero) and hence are at rest or moving with a constant velocity in a straight line.

A body with negligible dimensions is commonly referred to as a particle. In mechanics, either large bodies or small bodies can be referred to as particles when the size and shape of the body have no effect on the response of the body to a system of forces. Under these conditions, the mass of the body can be assumed to be concentrated at a point. For example, the earth can be modeled as a particle for orbital motion studies since the size of the earth is insignificant when compared to the size of its orbit and the shape of the earth does not influence the description of its position or the action of forces applied to it.

Since it is assumed that the mass of a particle is concentrated at a point and that the size and shape of a particle can be neglected, a particle can be subjected only to a system of concurrent forces. Newton's first law of motion states that "in the absence of external forces ($\mathbf{R} = \mathbf{0}$), a particle originally at rest or moving with a constant velocity (in equilibrium) will remain at rest or continue to move with a constant velocity along a straight line." Thus, a necessary condition for equilibrium of a particle is

$$\mathbf{R} = \Sigma \mathbf{F} = \mathbf{0} \qquad (3\text{-}1)$$

A particle in equilibrium must also satisfy Newton's second law of motion, which can be expressed in equation form (Eq. 1-1) as

$$\mathbf{R} = \Sigma \mathbf{F} = m\mathbf{a} \qquad (1\text{-}1)$$

In order to satisfy both Eqs. 1-1 and 3-1

$$m\mathbf{a} = \mathbf{0}$$

Since the mass of the particle is not zero, the acceleration of a particle in equilibrium is zero ($a = 0$). Thus, a particle initially at rest will remain at rest, and a particle moving with a constant velocity will maintain that velocity if $R = 0$. Therefore, Eq. 3-1 is both a necessary condition and a sufficient condition for equilibrium.

The particle assumption is valid for many practical applications and thus provides a means for introducing the student to some interesting engineering problems early in a statics course. For this reason, this short chapter on statics (equilibrium) of a particle has been introduced before the more difficult problems associated with equilibrium of a rigid body (which involves the concepts of moments and distributed loads) are considered.

The force system acting on a body in a typical statics problem consists of known forces and unknown forces. Both must be clearly identified before a solution to a specific problem is attempted. A method commonly used to identify all forces acting on a body in a given situation is described in the following section.

## 3-2 FREE-BODY DIAGRAMS

Problems in engineering mechanics (statics and dynamics) are concerned with the external effects produced when a system of forces acts on a physical body. The external effects typically include motion of the body in dynamics problems or development of support reactions to resist motion of the body in statics problems. Solving an engineering mechanics problem usually requires identification of all external forces acting on a "body of interest." A carefully prepared drawing or sketch that shows a "body of interest" separated from all interacting bodies is known as a free-body diagram (FBD). Once the body of interest is selected, the forces exerted by all other bodies on the one being considered must be determined and shown on the diagram. It is important that "all" forces acting "on" the body of interest be shown. Recall also that "a force cannot exist unless there is a body to exert the force." Frequently, the student will overlook and omit a force from the free-body diagram or show a force on the free-body diagram when there is no body present to exert the force.

The actual procedure for drawing a free-body diagram consists of two essential steps:

1. Make a decision regarding what body is to be isolated and analyzed. Prepare a sketch of the external boundary of the body selected.
2. Represent all forces, known and unknown, that are applied by other bodies to the isolated body with vectors in their correct positions.

The number of forces on a free-body diagram is determined by noting the number of bodies that exert forces on the body of interest. These forces may be either forces of contact or body forces. An important body force is the earth-pull on (or weight of) a body.

Each known force should be shown on a free-body diagram with its correct magnitude and direction. Letter symbols are used for the magnitudes of unknown forces. If a force has a known line of action but an unknown magnitude and sense, the sense of the force can be assumed. The correct sense will become apparent after solving for the

# CONCEPTUAL EXAMPLE 3-1: FREE-BODY DIAGRAMS

*A worker is using a hoist and cable to remove an engine from a car as shown in Fig. CE3-1a. Draw a free-body diagram for the part of this system contained within the dotted box shown on the figure.*

(a)

(b)

(c)

**Fig. CE3-1**

## SOLUTION

A concept that is fundamental to the solution of engineering problems is the free-body diagram. A carefully prepared sketch that shows a body or a group of bodies separated from all interacting bodies and with all external forces applied is known as a free-body diagram.

    The system within the dotted box of Fig. CE3-1a is shown isolated from its surroundings in Fig. CE3-1b. The system consists of the engine, short cables used to fasten the engine to the cable ring, a part of the cable between the ring and the hoist, and a part of the cable between the ring and the worker. A careful examination of this isolated system indicates that three bodies were removed during the isolating process; namely, the earth, the hoist, and the worker. The actions of these bodies on the isolated system are shown as forces $T_1$, $T_2$, and $W$ on the free-body diagram of Fig. CE3-1c. Tensions $T_1$ and $T_2$ represent forces that the cables have the ability to transmit from the hoist and worker to the ring, respectively. The force $W$ represents the force exerted on the engine by the Earth's gravitational field (the weight of the engine).

    Most engineers consider a complete and correct free-body diagram to be the single most important tool for the solution of mechanics problems.

Cylinder supported by
smooth surfaces

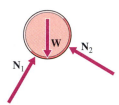

Free-body diagram

**Figure 3-1**   Free-body diagram for a
cylinder supported by two smooth
surfaces.

Block held on a smooth inclined
surface with a flexible cable

Free-body diagram

**Figure 3-2**   Free-body diagram for a
block held on a smooth inclined surface with a flexible cable.

unknown magnitude. By definition, the magnitude of a force is always positive; therefore, if the solution yields a negative magnitude,
the minus sign indicates that the sense of the force is opposite to that
assumed on the free-body diagram.

If both magnitude and direction of a force acting on the body of
interest is unknown (such as a pin reaction in a pin-connected structure),
it is frequently convenient to show the two rectangular components of
the force on the free-body diagram instead of the actual force. In this
way, one deals with two forces of unknown magnitude but known direction. After the two rectangular components of the force are determined, the magnitude and direction of the actual force can easily be
found. However, do not show *both* the force of unknown magnitude
and its rectangular components on the same diagram.

The word *free* in the name "free-body diagram" emphasizes the
idea that all bodies exerting forces on the body of interest are removed
or withdrawn and are replaced by the forces they exert. Do not show
both the bodies removed and the forces exerted by them on the free-
body diagram. Sometimes it may be convenient to indicate, by lightweight dotted lines, the faint outlines of the bodies removed, in order
to visualize the geometry and specify dimensions required for solution
of the problem.

In drawing a free-body diagram of a given body, certain assumptions are made regarding the nature of the forces (reactions) exerted by
other bodies on the body of interest. Two common assumptions are the
following:

1. If a surface of contact at which a force is applied by one body to
   another body has only a small degree of roughness, it may be
   assumed to be smooth (frictionless), and hence the action (or reaction) of the one body on the other is directed normal to the
   surface of contact as shown in Fig. 3-1.

2. A body that possesses only a small degree of bending stiffness
   (resistance to bending), such as a cord, rope, chain, or the like, may
   be considered to be perfectly flexible, and hence the pull of such
   a body on any other body is directed along the axis of the flexible
   body as shown in Fig. 3-2.

A large number of additional assumptions will be discussed in Section
6-2.2 in the chapter on equilibrium of rigid bodies, where the topic of
idealization of supports and connections is considered.

The term *body of interest* used in the definition of a free-body diagram may mean any definite part of a structure or machine such as an
eye bar in a bridge truss or a connecting rod in an automobile engine.
The body of interest may also be taken as a group of physical bodies
joined together (considered as one body), such as an entire bridge or
a complete engine.

The importance of drawing a free-body diagram before attempting
to solve a mechanics problem cannot be overemphasized. A procedure
that can be followed to construct a complete and correct free-body diagram contains the following four steps:

**Constructing a Free-Body Diagram**

*Step 1.* Decide which body or combination of bodies is to be shown on the free-body diagram.
*Step 2.* Prepare a drawing or sketch of the outline of this isolated or free body.
*Step 3.* Carefully trace around the boundary of the free body and identify all the forces exerted by contacting or attracting bodies that were removed during the isolation process.
*Step 4.* Choose the set of coordinate axes to be used in solving the problem and indicate their directions on the free-body diagram. Place any dimensions required for solution of the problem on the diagram.

Application of these four steps to any statics or dynamics problem should produce a complete and correct free-body diagram, which is an essential first step for the solution of any problem.

The free-body diagram is the "road map" for writing the equations of equilibrium. Every equation of equilibrium must be supported by a properly drawn, complete, free-body diagram. The symbols used in the equations of equilibrium must match the symbols used on the free-body diagram. For example, use $A \cos 30°$ rather than $A_x$ if $A$ is used to represent a force of known direction (30° with respect to the $x$-axis) on the free-body diagram.

Most engineers consider a free-body diagram to be the single most important tool for the solution of mechanics problems. The free-body diagram clearly identifies the body to which the principles of mechanics are to be applied. It also clearly identifies all forces that act on the body and where they are applied. The use of a complete and proper free-body diagram can reduce many of the errors commonly made in solving mechanics problems.

## 3-3 EQUILIBRIUM OF A PARTICLE

In Section 3-1 it was noted that the term *particle* is used in statics to describe a body when the size and shape of the body will not significantly affect the solution of the problem being considered and when the mass of the body can be assumed to be concentrated at a point. As a result, we observed that a particle can be subjected only to a system of concurrent forces and that the necessary and sufficient conditions for equilibrium can be expressed mathematically as

$$\mathbf{R} = \Sigma\mathbf{F} = 0 \tag{3-1}$$

where $\Sigma\mathbf{F}$ is the vector sum of all forces acting on the particle.

### 3-3.1 Two-Dimensional Problems

For a system of coplanar (say the $xy$-plane), concurrent forces, Eq. 3-1 can be written as

$$R = R_x + R_y = R_n + R_t$$
$$= R_x\mathbf{i} + R_y\mathbf{j} = R_n\mathbf{e}_n + R_t\mathbf{e}_t$$
$$= \Sigma F_x\mathbf{i} + \Sigma F_y\mathbf{j} = \Sigma F_n\mathbf{e}_n + \Sigma F_t\mathbf{e}_t = 0 \qquad \text{(3-2)}$$

Equation 3-2 is satisfied only if

$$R_x = R_x\mathbf{i} = \Sigma F_x\mathbf{i} = 0$$
$$R_y = R_y\mathbf{j} = \Sigma F_y\mathbf{j} = 0$$
$$R_n = R_n\mathbf{e}_n = \Sigma F_n\mathbf{e}_n = 0$$
$$R_t = R_t\mathbf{e}_t = \Sigma F_t\mathbf{e}_t = 0$$

In scalar form, these equations become

$$R_x = \Sigma F_x = 0$$
$$R_y = \Sigma F_y = 0 \qquad \text{(3-3a)}$$

or

$$R_n = \Sigma F_n = 0$$
$$R_t = \Sigma F_t = 0 \qquad \text{(3-3b)}$$

That is, the sum of the rectangular components in *any* direction must be zero. Though this would appear to give an infinite number of equations, no more than two of the equations are independent. The remaining equations could be obtained from combinations of the two independent equations. However, it is sometimes convenient to use $\Sigma F_x = 0$ and $\Sigma F_n = 0$ as the two independent equations rather than $\Sigma F_x = 0$ and $\Sigma F_y = 0$ (see Example 3-3).

Equations 3-3 can be used to determine two unknown quantities (two magnitudes, two slopes, or a magnitude and a slope). The procedure is illustrated in the following examples.

## CONCEPTUAL EXAMPLE 3-2: EQUILIBRIUM OF A PARTICLE

*Four forces act on a small airplane in flight, as shown in Fig. CE3-2: its weight, the thrust provided by the engine, the lift provided by the wings, and the drag resulting from its motion through the air. Is it possible for the plane to be in equilibrium under the action of these four forces?*

Fig. CE3-2

### SOLUTION

A moving body may be in equilibrium if it moves with a constant velocity (acceleration $\mathbf{a} = 0$) along a straight line since if $\mathbf{a} = 0$, Newton's second law states that $\Sigma\mathbf{F} = m\mathbf{a} = 0$. The four forces acting on the airplane can be represented as a concurrent force system; therefore, two scalar equilibrium equa-

tions are available to solve for any unkown forces. Summing forces in the x'-
and y'-directions eliminates the need to solve simultaneous equations. Thus,

$$+\searrow\Sigma F_{x'} = -F_T + F_D \cos 10° + W \sin 10°$$
$$= -F_T + 3 \cos 10° + 25 \sin 10° = 0$$

from which
$$F_T = 7.296 \text{ kN} \cong 7.30 \text{ kN}$$

$$+\nearrow\Sigma F_{y'} = F_L + F_D \sin 10° - W \cos 10°$$
$$= F_L + 3 \sin 10° - 25 \cos 10° = 0$$

from which
$$F_L = 24.099 \text{ kN} \cong 24.1 \text{ kN}$$

A fundamental characteristic of an object in equilibrium is a zero acceleration.

## EXAMPLE PROBLEM 3-1

Two flexible cables $A$ and $B$ are used to support a 220-lb traffic light as shown
in Fig. 3-3a. Determine the tension in each of the cables.

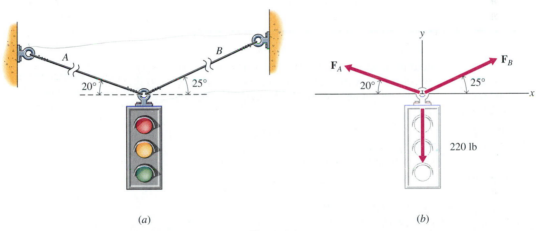

(a)                          (b)

**Fig. 3-3**

### SOLUTION

The light is subjected to a system of coplanar, concurrent forces. The necessary
and sufficient conditions for equilibrium are given by Eqs. 3-3 as $\Sigma F_x = 0$ and
$\Sigma F_y = 0$. Applying these equations, by using the free-body diagram for the
traffic light shown in Fig. 3-3b, yields

> A flexible cable always exerts a
> tensile force with a line of action
> along the axis of the cable.

$$+\rightarrow\Sigma F_x = F_B \cos 25° - F_A \cos 20° = 0 \qquad F_B = 1.0368 \, F_A$$
$$+\uparrow\Sigma F_y = F_B \sin 25° + F_A \sin 20° - 220$$
$$= 1.0368 \, F_A \sin 25° + F_A \sin 20° - 220 = 0$$

Solving yields

$$F_A = 281.98 \text{ lb} \cong 282 \text{ lb} \qquad\qquad \text{Ans.}$$
$$F_B = 292.36 \text{ lb} \cong 292 \text{ lb} \qquad\qquad \text{Ans.}$$

A homogeneous sphere with a mass of 50 kg rests against two smooth planes that form a V-shaped trough as shown in Fig. 3-4a. Determine the forces exerted on the sphere by the planes at contact points $A$ and $B$.

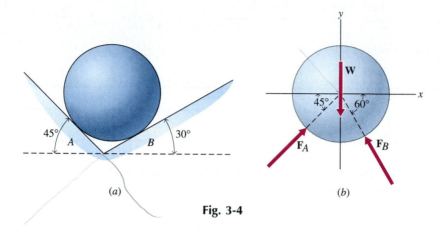

**Fig. 3-4**

## SOLUTION

The sphere is subjected to a system of coplanar, concurrent forces. The necessary and sufficient conditions for equilibrium are given by Eqs. 3-3 as $\Sigma F_x = 0$ and $\Sigma F_y = 0$. Applying these equations, by using the free-body diagram for the sphere shown in Fig. 3-4b, yields

$$+\rightarrow \Sigma F_x = F_A \cos 45° - F_B \cos 60° = 0 \qquad F_B = 1.4142\, F_A$$
$$+\uparrow \Sigma F_y = F_A \sin 45° + F_B \sin 60° - mg$$
$$= F_A \sin 45° + 1.4142\, F_A \sin 60° - 50(9.81) = 0$$

Solving yields

$$F_A = 253.90 \text{ lb} \cong 254 \text{ N} \qquad\qquad \text{Ans.}$$
$$F_B = 359.07 \text{ lb} \cong 359 \text{ N} \qquad\qquad \text{Ans.}$$

A smooth surface exerts a compressive force with a line of action normal to the surface at the point of contact.

A free-body diagram of a particle subjected to the action of four forces is shown in Fig. 3-5a. Determine the magnitudes of forces $F_1$ and $F_2$ so that the particle is in equilibrium.

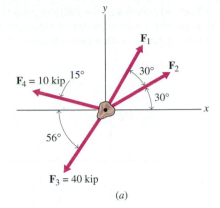

(a)

## SOLUTION

The particle is subjected to a system of coplanar, concurrent forces. The necessary and sufficient conditions for equilibrium are given by Eqs. 3-3 as $\Sigma F_x = 0$ and $\Sigma F_y = 0$. Applying these equations, by using the free-body diagram shown in Fig. 3-5a, yields

$$+\rightarrow \Sigma F_x = F_{1x} + F_{2x} + F_{3x} + F_{4x}$$
$$= F_1 \cos 60° + F_2 \cos 30° - 40 \cos 56° - 10 \cos 15°$$
$$= 0.5000F_1 + 0.8660F_2 - 22.37 - 9.659 = 0$$

from which

$$F_1 + 1.732F_2 = 64.06 \qquad (a)$$

$$+\uparrow \Sigma F_y = F_{1y} + F_{2y} + F_{3y} + F_{4y}$$
$$= F_1 \sin 60° + F_2 \sin 30° - 40 \sin 56° + 10 \sin 15°$$
$$= 0.8660F_1 + 0.5000F_2 - 33.16 + 2.588 = 0$$

from which

$$F_1 + 0.5774F_2 = 35.30 \qquad (b)$$

Solving Eqs. $a$ and $b$ simultaneously yields

$$F_1 = 20.9 \text{ kip} \qquad \text{Ans.}$$
$$F_2 = 24.9 \text{ kip} \qquad \text{Ans.}$$

Alternatively, summing forces in a direction $n$ perpendicular to the line of action of force $F_2$, as shown in Fig. 3-5b, yields

$$+\nwarrow \Sigma F_n = F_{1n} + F_{3n} + F_{4n}$$
$$= F_1 \sin 30° - 40 \sin 26° + 10 \sin 45° = 0$$

from which

$$F_1 = 20.93 = 20.9 \text{ kip} \qquad \text{Ans.}$$

Once $F_1$ is known, forces can be summed in any other direction to obtain $F_2$. Thus, summing forces in the $x$-direction yields

$$+\rightarrow \Sigma F_x = F_{1x} + F_{2x} + F_{3x} + F_{4x}$$
$$= 20.93 \cos 60° + F_2 \cos 30° - 40 \cos 56° - 10 \cos 15° = 0$$

from which

$$F_2 = 24.90 = 24.9 \text{ kip} \qquad \text{Ans.}$$

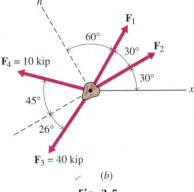

(b)

**Fig. 3-5**

The $x$- and $y$-components of the resultant of any number of forces in a plane are the algebraic sums of the $x$- and $y$-components of the individual forces.

Summing forces in a direction perpendicular to one of the unknown forces eliminates the need to solve simultaneous equations.

A free-body diagram of a particle subjected to the action of four forces is shown in Fig. 3-6a. Determine the magnitude and direction (angle $\theta$) of force $\mathbf{F}_4$ so that the particle is in equilibrium.

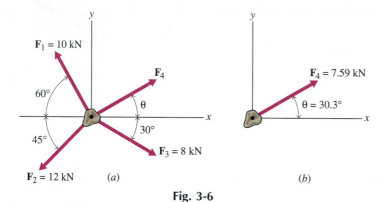

**Fig. 3-6**

## SOLUTION

The particle is subjected to a system of coplanar, concurrent forces. The necessary and sufficient conditions for equilibrium are given by Eqs. 3-3 as $\Sigma F_x = 0$ and $\Sigma F_y = 0$. Applying these equations, by using the free-body diagram shown in Fig. 3-6a, yields

$$+\rightarrow\Sigma F_x = F_{1x} + F_{2x} + F_{3x} + F_{4x}$$
$$= -10 \cos 60° - 12 \cos 45° + 8 \cos 30° + F_4 \cos \theta = 0$$

from which

$$F_{4x} = F_4 \cos \theta = 6.557 \text{ kN} \qquad (a)$$

$$+\uparrow\Sigma F_y = F_{1y} + F_{2y} + F_{3y} + F_{4y}$$
$$= 10 \sin 60° - 12 \sin 45° - 8 \sin 30° + F_4 \sin \theta = 0$$

from which

$$F_{4y} = F_4 \sin \theta = 3.825 \text{ kN} \qquad (b)$$

Once the components $F_{4x}$ and $F_{4y}$ are known, Eqs. 2-1 are used to determine $F_4$ and $\theta$:

$$F_4 = \sqrt{F_{4x}^2 + F_{4y}^2} = \sqrt{(6.557)^2 + (3.825)^2} = 7.59 \text{ kN} \qquad \text{Ans.}$$
$$\theta = \tan^{-1}\frac{F_{4y}}{F_{4x}} = \tan^{-1}\frac{3.825}{6.557} = 30.3° \qquad \text{Ans.}$$

The results are shown in Fig. 3-6b.

> The $x$- and $y$-components of the resultant of any number of forces in a plane are the algebraic sums of the $x$- and $y$-components of the individual forces.

A homogeneous steel cylinder weighing 500 lb is supported by a flexible cable and a smooth inclined plane as shown in Fig. 3-7a. Determine the tension $T$ in the cable and the force $N$ exerted by the inclined plane on the cylinder.

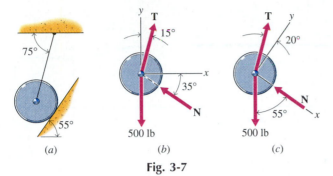

**Fig. 3-7**

## SOLUTION

The tension $T$ in the cable and the force $N$ exerted by the inclined plane on the cylinder can be determined by considering the free-body diagram of the cylinder shown in Fig. 3-7b. From this free-body diagram it can be seen that the cylinder is subjected to a system of coplanar, concurrent forces. Thus, the necessary and sufficient conditions for equilibrium are given by Eqs. 3-3 as $\Sigma F_x = 0$ and $\Sigma F_y = 0$. Applying these equations yields

$$+\rightarrow\Sigma F_x = T \sin 15° - N \cos 35° = 0.2588T - 0.8192N = 0$$

from which

$$T = 3.165N \qquad\qquad\qquad (a)$$

$$+\uparrow\Sigma F_y = T \cos 15° + N \sin 35° - 500 = 0.9659T + 0.5736N - 500 = 0$$

from which

$$N = 871.7 - 1.684T \qquad\qquad\qquad (b)$$

Solving Eqs. $a$ and $b$ simultaneously yields

$$T = 436 \text{ lb} \quad \measuredangle 75° \qquad\qquad \text{Ans.}$$
$$N = 137.7 \text{ lb} \quad 35° \searrow \qquad\qquad \text{Ans.}$$

The solution to a problem can frequently be simplified by proper selection of the coordinate system. For example, if the $x$- and $y$-axes are chosen parallel and perpendicular to the inclined surface as shown on the free-body diagram of Fig. 3-7c, the need to solve simultaneous equations is eliminated. Thus,

$$+\nearrow\Sigma F_y = T \cos 20° - 500 \sin 55° = 0.9397T - 409.6 = 0$$

from which

$$T = 436 \text{ lb} \quad \measuredangle 75° \qquad\qquad \text{Ans.}$$

Once the force $T$ is known, $\Sigma F_x = 0$ yields $N$ directly. Thus,

$$+\searrow\Sigma F_x = -N - T \sin 20° + 500 \cos 55° = 0$$

from which

$$N = 286.8 - 0.3420T$$
$$= 286.8 - 0.3420(436) = 137.7 \text{ lb} \quad 35° \searrow \qquad \text{Ans.}$$

A flexible cable always exerts a tensile force with a line of action along the axis of the cable.

A smooth surface exerts a compressive force with a line of action normal to the surface at the point of contact.

The cable system shown in Fig. 3-6a is being used to lift body A. The cable system is in equilibrium at the cable positions shown in the figure when a 500-N force is applied at joint 1. Determine the tensions in cables A, B, C, and D and the mass of a body A that is being lifted.

(a)

## SOLUTION

The tensions in the cables can be determined by using the free-body diagrams of joints 1 and 2 as shown in Figs. 3-8b and 3-8c, respectively. Each joint is subjected to a system of coplanar, concurrent forces. Thus, the necessary and sufficient conditions for equilibrium are given by Eqs. 3-3 as $\Sigma F_x = 0$ and $\Sigma F_y = 0$.

From the free-body diagram for joint 1 (see Fig. 3-8b),

$$+\uparrow \Sigma F_y = T_C \sin 10° - 500 = 0$$

from which

$$T_C = \frac{500}{\sin 10°} = 2879 \text{ N}$$

$$+\rightarrow \Sigma F_x = T_A - T_C \cos 10° = 0$$

from which

$$T_A = T_C \cos 10° = 2879 \cos 10° = 2835 \text{ N}$$

From the free-body diagram for joint 2 (see Fig. 3-8c),

$$+\rightarrow \Sigma F_x = T_B \sin 20° - T_A = 0$$

from which

$$T_B = \frac{T_A}{\sin 20°} = \frac{2835}{\sin 20°} = 8289 \text{ N}$$

$$+\uparrow \Sigma F_y = T_B \cos 20° - T_D = 0$$

from which

$$T_D = T_B \cos 20° = 8289 \cos 20° = 7789 \text{ N}$$

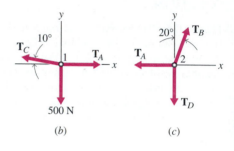

(b)                     (c)

The cable tensions rounded to three significant figures are

$$T_A = 2840 \text{ N} \qquad \text{Ans.}$$
$$T_B = 8290 \text{ N} \qquad \text{Ans.}$$
$$T_C = 2880 \text{ N} \qquad \text{Ans.}$$
$$T_D = 7790 \text{ N} \qquad \text{Ans.}$$

(d)

**Fig. 3-8**

Finally, from the free-body diagram for body A (see Fig. 3-8d),

$$+\uparrow \Sigma F_y = T_D - W_A = 0$$

from which

$$W_A = T_D = 7789 \text{ N}$$

Thus,

$$m_A = \frac{W_A}{g} = \frac{7789}{9.81} = 794 \text{ kg} \qquad \text{Ans.}$$

A flexible cable always exerts a tensile force with a line of action along the axis of the cable.

## 3-3.2 Three-Dimensional Problems

For a three-dimensional system of concurrent forces, Eq. 3-1 can be written as

$$\begin{aligned}
\mathbf{R} &= \Sigma \mathbf{F} \\
&= \mathbf{R}_x + \mathbf{R}_y + \mathbf{R}_z \\
&= R_x \mathbf{i} + R_y \mathbf{j} + R_z \mathbf{k} \\
&= \Sigma F_x \mathbf{i} + \Sigma F_y \mathbf{j} + \Sigma F_z \mathbf{k} = \mathbf{0}
\end{aligned} \tag{3-4}$$

Equation 3-4 is satisfied only if

$$\begin{aligned}
\mathbf{R}_x &= R_x \mathbf{i} = \Sigma F_x \mathbf{i} = \mathbf{0} \\
\mathbf{R}_y &= R_y \mathbf{j} = \Sigma F_y \mathbf{j} = \mathbf{0} \\
\mathbf{R}_z &= R_z \mathbf{k} = \Sigma F_z \mathbf{k} = \mathbf{0}
\end{aligned} \tag{3-5}$$

In scalar form, these equations become

$$\begin{aligned}
R_x &= \Sigma F_x = 0 \\
R_y &= \Sigma F_y = 0 \\
R_z &= \Sigma F_z = 0
\end{aligned} \tag{3-6}$$

Equations 3-5 and 3-6 can be used to determine three unknown quantities (three magnitudes, three slopes, or any combination of three magnitudes and slopes). The procedure is illustrated in the following examples. Example 3-7 illustrates the scalar method of solution for a three-dimensional problem. Example 3-8 illustrates the vector method of solution for a similar problem.

## EXAMPLE PROBLEM 3-7

A free-body diagram of a particle subjected to the action of four forces is shown in Fig. 3-9. Determine the magnitude and the coordinate direction angles of the unknown force $\mathbf{F}_4$ so that the particle is in equilibrium.

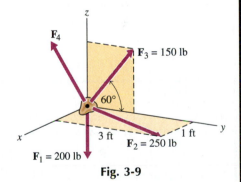

**Fig. 3-9**

### SOLUTION

The necessary and sufficient conditions for equilibrium of a particle subjected to a three-dimensional system of concurrent forces are given by Eqs. 3-6 as

$$R_x = \Sigma F_x = 0 \qquad R_y = \Sigma F_y = 0 \qquad R_z = \Sigma F_z = 0$$

The scalar components for each of the forces shown on the free-body diagram are as follows:

For $\mathbf{F}_1$:

$$\begin{aligned}
F_{1x} &= 0 \\
F_{1y} &= 0 \\
F_{1z} &= -200 \text{ lb}
\end{aligned}$$

For $\mathbf{F}_2$:

$$\begin{aligned}
F_{2x} &= F_2 \cos \theta_{2x} = 250(1/\sqrt{10}) = 79.06 \text{ lb} \\
F_{2y} &= F_2 \cos \theta_{2y} = 250(3/\sqrt{10}) = 237.17 \text{ lb} \\
F_{2z} &= 0
\end{aligned}$$

For $\mathbf{F}_3$:

$$F_{3x} = 0$$
$$F_{3y} = F_3 \cos \theta_{3y} = 150 \cos 60° = 75.00 \text{ lb}$$
$$F_{3z} = F_3 \cos \theta_{3z} = 150 \cos 30° = 129.90 \text{ lb}$$

Thus,

$$R_x = \Sigma F_x = F_{1x} + F_{2x} + F_{3x} + F_{4x}$$
$$= 0 + 79.06 + 0 + F_{4x} = 0 \qquad F_{4x} = -79.06 \text{ lb}$$
$$R_y = \Sigma F_y = F_{1y} + F_{2y} + F_{3y} + F_{4y}$$
$$= 0 + 237.17 + 75.00 + F_{4y} = 0 \qquad F_{4y} = -312.17 \text{ lb}$$
$$R_z = \Sigma F_z = F_{1z} + F_{2z} + F_{3z} + F_{4z}$$
$$= -200 + 0 + 129.90 + F_{4z} = 0 \qquad F_{4z} = 70.10 \text{ lb}$$

> The $x$-, $y$-, and $z$-components of the resultant of any number of forces in a three-dimensional system are the algebraic sums of the $x$-, $y$-, and $z$-components of the individual forces.

Once the rectangular components of the force $\mathbf{F}_4$ are known, Eqs. 2-6 can be used to determine its magnitude and coordinate direction angles. Thus

$$F_4 = \sqrt{F_{4x}^2 + F_{4y}^2 + F_{4z}^2}$$
$$= \sqrt{(-79.06)^2 + (-312.17)^2 + (70.10)^2} = 329.57 \cong 330 \text{ lb} \qquad \text{Ans.}$$

$$\theta_x = \cos^{-1} \frac{F_{4x}}{F_4} = \cos^{-1} \frac{-79.06}{329.57} = 103.88° \cong 103.9° \qquad \text{Ans.}$$

$$\theta_y = \cos^{-1} \frac{F_{4y}}{F_4} = \cos^{-1} \frac{-312.17}{329.57} = 161.24° \cong 161.2° \qquad \text{Ans.}$$

$$\theta_z = \cos^{-1} \frac{F_{4z}}{F_4} = \cos^{-1} \frac{70.10}{329.57} = 77.72° \cong 77.7° \qquad \text{Ans.}$$

## EXAMPLE PROBLEM 3-8

A free-body diagram of a particle subjected to the action of four forces is shown in Fig. 3-10. Determine the magnitude and the coordinate direction angles of the unknown force $\mathbf{F}_D$ so that the particle is in equilibrium.

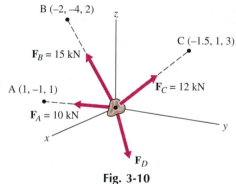

**Fig. 3-10**

### SOLUTION

The necessary and sufficient conditions for equilibrium of a particle subjected to a three-dimensional system of concurrent forces are given by Eq. 3-4 as

$$\mathbf{R} = \Sigma \mathbf{F} = \Sigma F_x \mathbf{i} + \Sigma F_y \mathbf{j} + \Sigma F_z \mathbf{k} = 0$$

Since the coordinates of points $A$, $B$, and $C$ on the lines of action of forces $F_A$, $F_B$, and $F_C$ are given, vector methods will be used for the solution. The four forces can be written in Cartesian vector form as

$$\mathbf{F}_A = F_A \mathbf{e}_A = 10 \left[ \frac{1\mathbf{i} - 1\mathbf{j} + 1\mathbf{k}}{\sqrt{(1)^2 + (-1)^2 + (1)^2}} \right] = 5.774\mathbf{i} - 5.774\mathbf{j} + 5.774\mathbf{k} \text{ kN}$$

$$\mathbf{F}_B = F_B \mathbf{e}_B = 15 \left[ \frac{-2\mathbf{i} - 4\mathbf{j} + 2\mathbf{k}}{\sqrt{(-2)^2 + (-4)^2 + (2)^2}} \right] = -6.124\mathbf{i} - 12.247\mathbf{j} + 6.124\mathbf{k} \text{ kN}$$

$$\mathbf{F}_C = F_C \mathbf{e}_C = 12 \left[ \frac{-1.5\mathbf{i} + 1\mathbf{j} + 3\mathbf{k}}{\sqrt{(-1.5)^2 + (1)^2 + (3)^2}} \right] = -5.143\mathbf{i} + 3.429\mathbf{j} + 10.286\mathbf{k} \text{ kN}$$

$$\mathbf{F}_D = F_D \mathbf{e}_D = F_{Dx}\mathbf{i} + F_{Dy}\mathbf{j} + F_{Dz}\mathbf{k}$$

> A force $\mathbf{F}$ can be written in Cartesian vector form by multiplying the magnitude of the force (a scalar) by a unit vector along the line of action of the force.

Substituting the values for the four forces into Eq. 3-4 yields

$$\mathbf{R} = \mathbf{F}_A + \mathbf{F}_B + \mathbf{F}_C + \mathbf{F}_D = \Sigma F_x \mathbf{i} + \Sigma F_y \mathbf{j} + \Sigma F_z \mathbf{k}$$
$$= (F_{Dx} + 5.774 - 6.124 - 5.123)\mathbf{i} + (F_{Dy} - 5.774 - 12.247 + 3.429)\mathbf{j}$$
$$+ (F_{Dz} + 5.774 + 6.124 + 10.286)\mathbf{k} = 0$$

Therefore,

$$\mathbf{F}_D = F_{Dx}\mathbf{i} + F_{Dy}\mathbf{j} + F_{Dz}\mathbf{k} = 5.473\mathbf{i} + 14.592\mathbf{j} - 22.184\mathbf{k} \text{ kN}$$

Once the rectangular components of the force $\mathbf{F}_4$ are known, Eqs. 2-6 can be used to determine its magnitude and coordinate direction angles. Thus

$$F_D = \sqrt{F_{Dx}^2 + F_{Dy}^2 + F_{Dz}^2}$$
$$= \sqrt{(5.473)^2 + (14.592)^2 + (-22.184)^2} = 27.111 \text{ kN} \cong 27.1 \text{ kN} \quad \text{Ans.}$$

$$\theta_{Dx} = \cos^{-1}\frac{F_{Dx}}{F_D} = \cos^{-1}\frac{5.473}{27.111} = 78.35° \cong 78.4° \quad \text{Ans.}$$

$$\theta_{Dy} = \cos^{-1}\frac{F_{Dy}}{F_D} = \cos^{-1}\frac{14.592}{27.111} = 57.44° \cong 57.4° \quad \text{Ans.}$$

$$\theta_{Dz} = \cos^{-1}\frac{F_{Dz}}{F_D} = \cos^{-1}\frac{-22.184}{27.111} = 144.91° \cong 144.9° \quad \text{Ans.}$$

## EXAMPLE PROBLEM 3-9

A 3000-lb cylinder is supported by a system of cables as shown in Fig. 3-11a. Determine the tensions in cables $A$, $B$, and $C$.

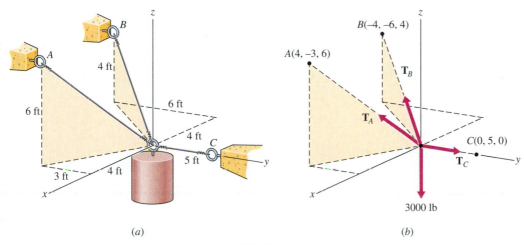

(a)          (b)

**Fig. 3-11**

## SOLUTION

A free-body diagram for the cylinder is shown in Fig. 3-11b. The cable forces and the weight of the cylinder form a three-dimensional system of concurrent forces with three unknown cable tensions $T_A$, $T_B$, and $T_C$. The coordinates of the support points for each of the cables is shown on the free-body diagram in $(x, y, z)$ format as an aid in writing vector equations for the cable tensions. The necessary and sufficient conditions for equilibrium of the cylinder are given by Eq. 3-4 as

$$\mathbf{R} = \Sigma \mathbf{F} = \mathbf{T}_A + \mathbf{T}_B + \mathbf{T}_C + \mathbf{W} = 0$$

The cable tensions and the weight of the cylinder can be expressed in Cartesian vector form as

$$\mathbf{T}_A = T_A\left[\frac{4\mathbf{i} - 3\mathbf{j} + 6\mathbf{k}}{\sqrt{(4)^2 + (-3)^2 + (6)^2}}\right] = 0.5121T_A\mathbf{i} - 0.3841T_A\mathbf{j} + 0.7682T_A\mathbf{k}\ \text{lb}$$

$$\mathbf{T}_B = T_B\left[\frac{-4\mathbf{i} - 6\mathbf{j} + 4\mathbf{k}}{\sqrt{(-4)^2 + (-6)^2 + (4)^2}}\right] = -0.4851T_B\mathbf{i} - 0.7276T_B\mathbf{j} + 0.4851T_B\mathbf{k}\ \text{lb}$$

$$\mathbf{T}_C = T_C\mathbf{j}\ \text{lb}$$
$$\mathbf{W} = -3000\mathbf{k}\ \text{lb}$$

A force **F** can be written in Cartesian vector form by multiplying the magnitude of the force (a scalar) by a unit vector along the line of action of the force.

Substituting the values for the four forces into Eq. 3-4 yields

$$\mathbf{R} = (0.5121T_A - 0.4851T_B)\mathbf{i} + (-0.3841T_A - 0.7276T_B + T_C)\mathbf{j} + (0.7682T_A + 0.4851T_B - 3000)\mathbf{k} = 0$$

The three scalar equations obtained from this vector equation are

$$\Sigma F_x = 0.5121T_A - 0.4851T_B = 0 \qquad\qquad (a)$$
$$\Sigma F_y = -0.3841T_A - 0.7276T_B + T_C = 0 \qquad (b)$$
$$\Sigma F_z = 0.7682T_A + 0.4851T_B - 3000 = 0 \qquad (c)$$

From Eq. *a*:

$$T_B = 1.0557T_A$$

From Eq. *c*.

$$0.7682T_A + 0.4851T_B = 3000$$
$$0.7682T_A + 0.4851(1.0557T_A) = 3000$$

Solving for $T_A$ yields

$$T_A = 2343\ \text{lb} \cong 2.34\ \text{kip} \qquad\qquad \text{Ans.}$$

Also,

$$T_B = 1.0557T_A$$
$$= 1.0557(2343) = 2474\ \text{lb} \cong 2.47\ \text{kip} \qquad \text{Ans.}$$

Finally, from Eq. *c*:

$$T_C = 0.3841T_A + 0.7276T_B$$
$$= 0.3841(2343) + 0.7276(2474)$$
$$= 2700\ \text{lb} \cong 2.70\ \text{kip} \qquad\qquad \text{Ans.}$$

Most pocket calculators have programs for solving simple systems of linear equations.

A block is supported by a system of cables as shown in Fig. 3-12a. The weight of the block is 500 N. Determine the tensions in cables A, B, C, and D.

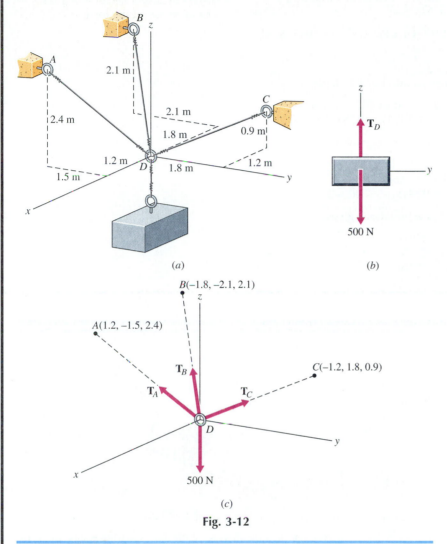

(a)

(b)

(c)

**Fig. 3-12**

## SOLUTION

A free-body diagram for the block is shown in Fig. 3-12b. From the equilibrium equation $\Sigma F_z = 0$,

$$+\uparrow \Sigma F_z = T_D - 500 = 0 \qquad T_D = 500 \text{ N}\uparrow \qquad \text{Ans.}$$

A free-body diagram for joint $D$ of the cable system is shown in Fig. 3-12c. This diagram shows that joint $D$ is subjected to a three-dimensional system of concurrent forces consisting of the three unknown cable tensions $T_A$, $T_B$, $T_C$ and the known cable tension $T_D = 500$ N. The coordinates of the support points $A$, $B$, and $C$ for cables $AD$, $BD$, and $CD$, respectively, are shown on the free-body diagram in $(x, y, z)$ format as an aid in writing vector equations for the cable tensions. The necessary and sufficient conditions for equilibrium of the joint are given by Eq. 3-4 as

$$\mathbf{R} = \Sigma \mathbf{F} = \mathbf{T}_A + \mathbf{T}_B + \mathbf{T}_C + \mathbf{T}_D = \mathbf{0}$$

The cable tensions can be expressed in Cartesian vector form as

$$\mathbf{T}_A = T_A\left[\frac{1.2\mathbf{i} - 1.5\mathbf{j} + 2.4\mathbf{k}}{\sqrt{9.450}}\right] = 0.3904T_A\mathbf{i} - 0.4880T_A\mathbf{j} + 0.7807T_A\mathbf{k}\ \text{N}$$

$$\mathbf{T}_B = T_B\left[\frac{-1.8\mathbf{i} - 2.1\mathbf{j} + 2.1\mathbf{k}}{\sqrt{12.060}}\right] \doteq -0.5183T_B\mathbf{i} - 0.6047T_B\mathbf{j} + 0.6047T_B\mathbf{k}\ \text{N}$$

$$\mathbf{T}_C = T_C\left[\frac{-1.2\mathbf{i} + 1.8\mathbf{j} + 0.9\mathbf{k}}{\sqrt{5.490}}\right] = -0.5121T_C\mathbf{i} + 0.7682T_C\mathbf{j} + 0.3841T_C\mathbf{k}\ \text{N}$$

$$\mathbf{T}_D = -500\mathbf{k}\ \text{N}$$

Substituting these values for the cable tensions into Eq. 3-4 yields

$$\mathbf{R} = (0.3904T_A - 0.5183T_B - 0.5121T_C)\mathbf{i} + (-0.4880T_A - 0.6047T_B + 0.7682T_C)\mathbf{j} + (0.7807T_A + 0.6047T_B + 0.3841T_C - 500)\mathbf{k} = \mathbf{0}$$

The three scalar equations obtained from this vector equation are

$$\Sigma F_x = 0.3904T_A - 0.5183T_B - 0.5121T_C = 0$$
$$\Sigma F_y = -0.4880T_A - 0.6047T_B + 0.7682T_C = 0$$
$$\Sigma F_z = 0.7807T_A + 0.6047T_B + 0.3841T_C - 500 = 0$$

The simultaneous solution of these three linear equations gives

$$T_A = 459.23\ \text{N} \cong 459\ \text{N} \qquad \text{Ans.}$$
$$T_B = 32.44\ \text{N} \cong 32.4\ \text{N} \qquad \text{Ans.}$$
$$T_C = 317.26\ \text{N} \cong 317\ \text{N} \qquad \text{Ans.}$$

# PROBLEMS

## Introductory Problems

**3-1\*** Block *A* of Fig. P3-1 rests on a smooth (frictionless) surface. If the block weighs 25 lb, determine the force exerted on the block by the surface and the force **P** required to prevent motion of the block.

**Fig. P3-1**

**3-2\*** Block *A* of Fig. P3-2 rests on a smooth (frictionless) surface. If the mass of the block is 100 kg, determine the force exerted on the block by the surface and the force **P** required to prevent motion of the block.

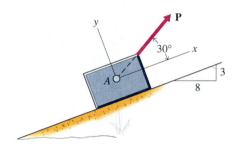

**Fig. P3-2**

**3-3** A homogeneous cylinder weighing 500 lb rests against two smooth planes that form a trough as shown in Fig. P3-3. Determine the forces exerted on the cylinder by the planes at contact points *A* and *B*.

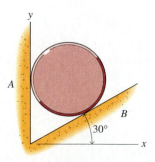

**Fig. P3-3**

**3-4** A homogeneous cylinder with a mass of 250 kg is supported against a smooth surface by a cable as shown in Fig. P3-4. Determine the forces exerted on the cylinder by the cable and by the smooth surface at contact point C.

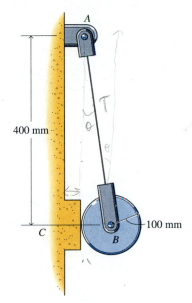

400 mm

100 mm

**Fig. P3-4**

**3-5\*** A 1000-lb block is supported by a strut and two cables as shown in Fig. P3-5. Determine the forces exerted on the pin at C by the strut and by the two cables.

25°

60°

**Fig. P3-5**

**3-6\*** A body with a mass of 750 kg is supported by the flexible cable system shown in Fig. P3-6. Determine the tensions in cables AC, BC, and CD.

60°

30°

**Fig. P3-6**

**3-7** The collar A shown in Fig. P3-7 is free to slide on the smooth rod BC. Determine the forces exerted on the collar by the cable and by the rod when the force **F** = 900 lb is applied to the collar.

30°

20°

Cable

**F** = 900 lb

**Fig. P3-7**

**3-8\*** An automobile stuck in a muddy field is being moved by using a cable fastened to a tree as shown in Fig. P3-8. When the cable is in the position shown and force **P** = 500 N, determine the x- and y-components of the cable force being applied to the automobile.

P

Cable

5°

5°

Tree

**Fig. P3-8**

**3-9** Determine the magnitudes of forces **F**$_2$ and **F**$_3$ so that the particle shown in Fig. P3-9 is in equilibrium.

y

**F**$_2$

60°

**F**$_1$ = 500 lb

45°

x

**F**$_3$

**Fig. P3-9**

**3-10\*** Determine the magnitude and direction angle $\theta$ of force $F_4$ so that the particle shown in Fig. P3-10 is in equilibrium.

**Fig. P3-10**

**3-11** Determine the forces in cables $A$ and $B$ if block $W$ of Fig. P3-11 weighs 350 lb.

**Fig. P3-11**

**3-12** A body with a mass of 300 kg is supported by the flexible cable system shown in Fig. P3-12. Determine the tensions in cables $A$, $B$, $C$, $D$, and $E$.

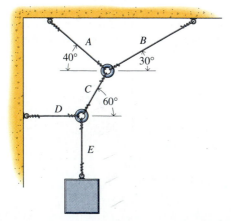

**Fig. P3-12**

100

## Intermediate Problems

**3-13\*** Two flower pots are supported with cables as shown in Fig. P3-13. If pot $A$ weighs 10 lb and pot $B$ weighs 8 lb, determine the tension in each of the cables and the slope of cable $BC$.

**Fig. P3-13**

**3-14\*** Three smooth homogeneous cylinders $A$, $B$, and $C$ are stacked in a V-shaped trough as shown in Fig. P3-14. Each cylinder has a diameter of 500 mm and a mass of 100 kg. Determine the forces exerted on cylinder $A$ by the inclined surfaces.

**Fig. P3-14**

**3-15** Three smooth homogeneous cylinders $A$, $B$, and $C$ are stacked in a V-shaped trough as shown in Fig. P3-15. Cylinder $A$ weighs 100 lb; cylinders $B$ and $C$ each weigh 200 lb. All cylinders have a 5-in. diameter. Determine the minimum angle $\theta$ for equilibrium.

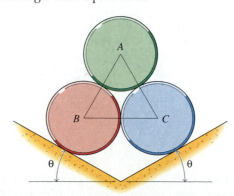

**Fig. P3-15**

**3-16\*** The masses of cylinders $A$ and $B$ of Fig. P3-16 are 40 kg and 75 kg, respectively. Determine the forces exerted on the cylinders by the inclined surfaces and the magnitude and direction of the force exerted by cylinder $A$ on cylinder $B$ when the cylinders are in equilibrium. Assume that all surfaces are smooth.

**Fig. P3-16**

**3-17** A continuous cable is used to support blocks $A$ and $B$ as shown in Fig. P3-17. Block $A$ is supported by a small wheel that is free to roll on the cable. Determine the displacement $y$ of block $A$ for equilibrium if the weights of blocks $A$ and $B$ are 50 lb and 75 lb, respectively.

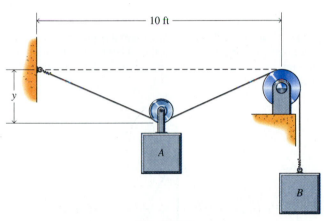

**Fig. P3-17**

**3-18** Two bodies $A$ and $B$ weighing 800 N and 200 N, respectively, are held in equilibrium on perpendicular surfaces by a connecting flexible cable that makes an angle $\theta$ with the horizontal as shown in Fig. P3-18. Find the reactions of the surfaces on the bodies, the tension in the cable, and the angle $\theta$. Assume all surfaces to be smooth.

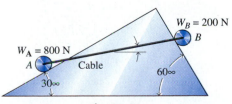

**Fig. P3-18**

## Challenging Problems

**3-19\*** The particle shown in Fig. P3-19 is in equilibrium under the action of the four forces shown on the free-body diagram. Determine the magnitude and the coordinate direction angles of the unknown force $\mathbf{F}_4$.

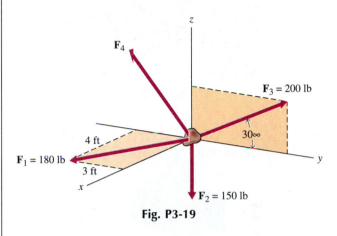

**Fig. P3-19**

**3-20\*** The particle shown in Fig. P3-20 is in equilibrium under the action of the four forces shown on the free-body diagram. Determine the magnitude and the coordinate direction angles of the unknown force $\mathbf{F}_4$.

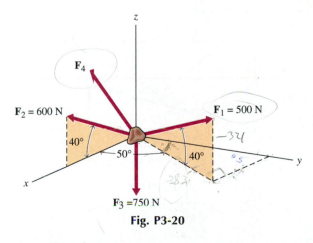

**Fig. P3-20**

20, 23, 35

Wed 5-7

**101**

**3-21** The block shown in Fig. P3-21 weighs 500 lb. Determine the tensions in cables $AB$, $AC$, and $AD$.

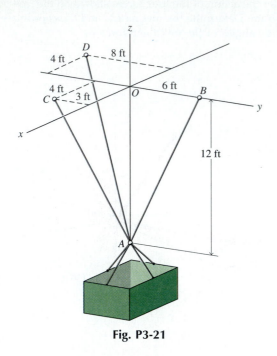

Fig. P3-21

**3-22** Struts $AB$ and $AC$ of Fig. P3-22 can transmit only axial tensile or compressive forces. Determine the forces in struts $AB$ and $AC$ and the tension in cable $AD$ when force $F = 1250$ N.

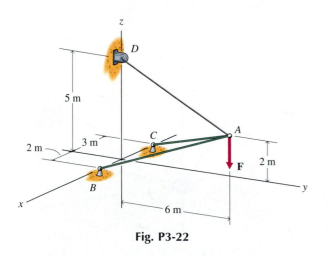

Fig. P3-22

**3-23*** Determine the forces in legs $AB$, $AC$, and $AD$ of the tripod shown in Fig. P3-23 when force $\mathbf{F} = 75$ lb. The legs can transmit only axial forces.

Fig. P3-23

**3-24*** Column $AB$ of Fig. P3-24 can support a maximum axial compressive force of 7500 N. Determine the tensions in each of the cables when this level of load exists in the column.

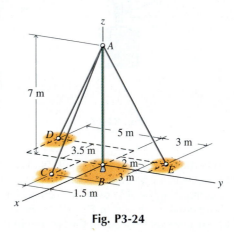

Fig. P3-24

**3-25** The circular disk shown in Fig. P3-25 weighs 750 lb. Determine the tensions in cables $A$, $B$, and $C$.

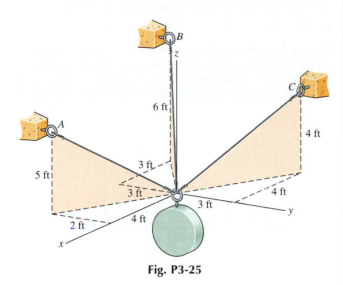

**Fig. P3-25**

**3-26** Two forces are applied in a horizontal plane to a loading ring at the top of a post as shown in Fig. P3-26. The post can transmit only an axial compressive force. Two guy wires $AC$ and $BC$ are used to hold the loading ring in equilibrium. Determine the force transmitted by the post and the forces in the two guy wires.

**Fig. P3-26**

## Computer Problems

**C3-27** A 75-lb stoplight is suspended between two poles as shown in Fig. P3-27. Neglect the weight of the flexible cables and plot the tension in both cables as a function of the sag distance $d$ ($0 \le d \le 8$ ft). Determine the minimum sag $d$ for which both tensions are less than

**a.** 100 lb.     **b.** 250 lb.     **c.** 500 lb.

**Fig. P3-27**

**C3-28** A 50-kg load is suspended from a pulley as shown in Fig. P3-28. The tension in the flexible cable does not change as it passes around the small frictionless pulleys, and the weight of the cable may be neglected. Plot the force $P$ required for equilibrium as a function of the sag distance $d$ ($0 \le d \le 1$ m). Determine the minimum sag $d$ for which $P$ is less than

**a.** Twice the weight of the load.
**b.** Four times the weight of the load.
**c.** Eight times the weight of the load.

**Fig. P3-28**

**C3-29** Three identical steel disks are stacked in a box as shown in Fig. P3-29. The weight and diameter of the smooth disks are 50 lb and 12 in., respectively. Plot the three forces exerted on disk C (by disk A, by the side wall, and by the floor) as a function of the distance b between the walls of the box (24 in. ≤ b ≤ 36 in.). Determine the range of b for which

a. The force at the floor is larger than the other two forces.
b. None of the three forces exceeds 100 lb.
c. None of the three forces exceeds 200 lb.

**C3-30** Two small wheels are connected by a lightweight rigid rod as shown in Fig. P3-30. Plot the angle $\theta$ (between the rod and the horizontal) as a function of the weight $W_1$ ($W_1 \le 10 W_2$). Determine the weight $W_1$ for which

a. $\theta = -50°$.  b. $\theta = 10°$.  c. $\theta = 25°$.

Do you think the equilibrium positions of parts a, b, and c are stable? (That is, if the wheels were disturbed slightly, do you think they would return to the equilibrium position or do you think they would slide off of the triangular surface?)

Fig. P3-30

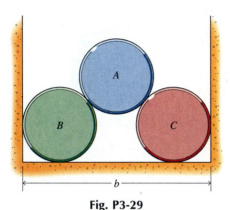

Fig. P3-29

---

# SUMMARY

The term *particle* is used in statics to describe a body when the size and shape of the body will not significantly affect the solution of the problem being considered and when the mass of the body can be assumed to be concentrated at a point. As a result, a particle can be subjected only to a system of concurrent forces and the necessary and sufficient conditions for equilibrium can be expressed mathematically as

$$\mathbf{R} = \Sigma \mathbf{F} = 0 \tag{3-1}$$

The force system acting on a body in a typical problem consists of known forces and unknown forces. Both must be clearly identified before a solution to a specific problem is attempted. A carefully prepared drawing that shows a "body of interest" separated from all other interacting bodies and with all external forces applied is known as a free-body diagram (FBD). The importance of drawing a free-body diagram before attempting to solve a mechanics problem cannot be overemphasized. A procedure that can be followed to construct a complete and correct free-body diagram contains the following four steps:

1. Decide which body or combination of bodies is to be shown on the free-body diagram.
2. Prepare a drawing or sketch of the outline of this isolated or free body.

3. Carefully trace around the boundary of the free body and identify all of the forces exerted by contacting or attracting bodies that were removed during the isolation process.

4. Choose the set of coordinate axes to be used in solving the problem and indicate their directions on the free-body diagram. Place any dimensions required for solution of the problem on the diagram.

Each known force should be shown on a free-body diagram with its correct magnitude, slope, and sense. Letter symbols can be used for the magnitudes of unknown forces. If a force has a known line of action but an unknown magnitude and sense, the sense of the force can be assumed. By definition, the magnitude of a force is always positive; therefore, if the solution yields a negative magnitude, the minus sign indicates that the sense of the force is opposite to that assumed on the free-body diagram. Most engineers consider an appropriate free-body diagram to be the single most important tool for the solution of mechanics problems.

For a three-dimensional system of concurrent forces, Eq. 3-1 can be written as

$$\begin{aligned}
\mathbf{R} = \Sigma\mathbf{F} &= \mathbf{R}_x + \mathbf{R}_y + \mathbf{R}_z \\
&= R_x\mathbf{i} + R_y\mathbf{j} + R_z\mathbf{k} \\
&= \Sigma F_x\mathbf{i} + \Sigma F_y\mathbf{j} + \Sigma F_z\mathbf{k} = 0
\end{aligned} \qquad (3\text{-}4)$$

Equation 3-4 is satisfied only if

$$\Sigma F_x = 0 \qquad \Sigma F_y = 0 \qquad \Sigma F_z = 0 \qquad (3\text{-}6)$$

Equations 3-6 can be used to determine three unknown quantities (three magnitudes, three slopes, or any combination of three magnitudes and slopes).

# REVIEW PROBLEMS

**3-31\*** Determine the magnitude and direction angle $\theta$ of force $F_4$ so that the particle shown in Fig. P3-31 is in equilibrium.

**Fig. P3-31**

**3-32\*** Determine the magnitude and direction angle $\theta$ of force $F_4$ so that the particle shown in Fig. P3-32 is in equilibrium.

**Fig. P3-32**

**3-33** Two 10-in.-diameter pipes and a 6-in.-diameter pipe are supported in a pipe rack as shown in Fig. P3-33. The 10-in.-diameter pipes each weigh 350 lb and the 6-in.-diameter pipe weighs 225 lb. Determine the forces exerted on the pipes by the supports at contact surfaces A, B, and C. Assume all surfaces to be smooth.

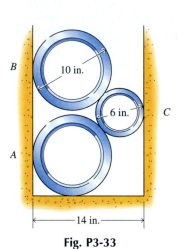

**Fig. P3-33**

**3-34** The mass of block A in Fig. P3-34 is 250 kg. Block A is supported by a small wheel that is free to roll on the continuous cable between supports B and C. The length of the cable is 42 m. Determine the distance x and the tension T in the cable when the system is in equilibrium.

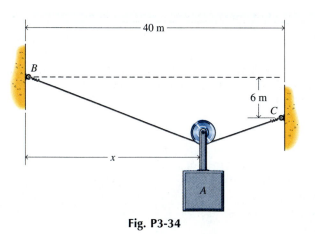

**Fig. P3-34**

**3-35\*** Two bodies weighing 150 lb and 200 lb, respectively, rest on a cylinder and are connected by a rope as shown in Fig. P3-35. Find the reactions of the cylinder on the bodies, the tension in the rope, and the angle $\theta$. Assume all surfaces to be smooth.

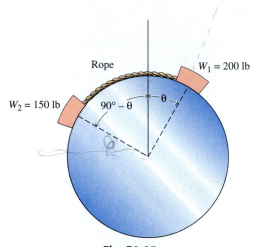

**Fig. P3-35**

**3-36\*** The particle shown in Fig. P3-36 is in equilibrium under the action of the four forces shown on the free-body diagram. Determine the tensions in cables A, B, and C.

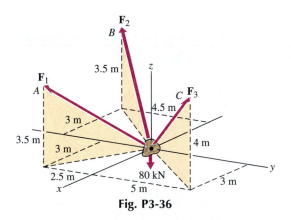

**Fig. P3-36**

**3-37** The hot-air balloon shown in Fig. P3-37 is tethered with three mooring cables. If the net lift of the balloon is 900 lb, determine the force exerted on the balloon by each of the cables.

**Fig. P3-37**

**3-38** The traffic light shown in Fig. P3-38 is supported by a system of cables. Determine the tensions in cables $A$, $B$, and $C$ if the traffic light has a mass of 100 kg.

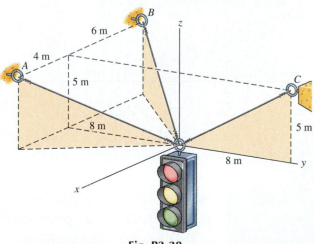

**Fig. P3-38**

## Computer Problems

**C3-39** A 1000-lb load is securely fastened to a hoisting rope as shown in Fig. P3-39. The tension in the flexible cable does not change as it passes around the small frictionless pulley at the right support. The weight of the cable may be neglected. Plot the force $P$ required for equilibrium as a function of the sag distance $d$ $(0 \leq d \leq 10 \text{ ft})$. Determine the minimum sag $d$ for which $P$ is less than

**a.** Twice the weight of the load.
**b.** Four times the weight of the load.
**c.** Eight times the weight of the load.

Plot both the force $P$ and the tension $T_{AB}$ in segment $AB$ of the cable on the same graph.

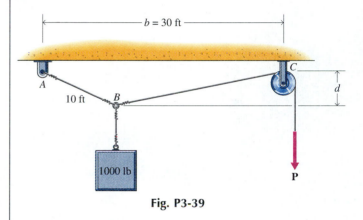

**Fig. P3-39**

**C3-40** A pair of steel disks are stacked in a box as shown in Fig. P3-40. The masses and diameters of the smooth disks are $m_A = 5$ kg, $m_B = 25$ kg, $d_A = 100$ mm, and $d_B = 200$ mm. Plot the two forces exerted on disk $A$ (by disk $B$ and by the side wall) as a function of the distance $b$ between the walls of the box $(200 \text{ mm} \leq b \leq 300 \text{ mm})$. Determine the range of $b$ for which

**a.** The force at the side wall is less than $W_A$, the weight of disk $A$.
**b.** Neither of the two forces exceeds $2W_A$.
**c.** Neither of the two forces exceeds $4W_A$.

**Fig. P3-40**

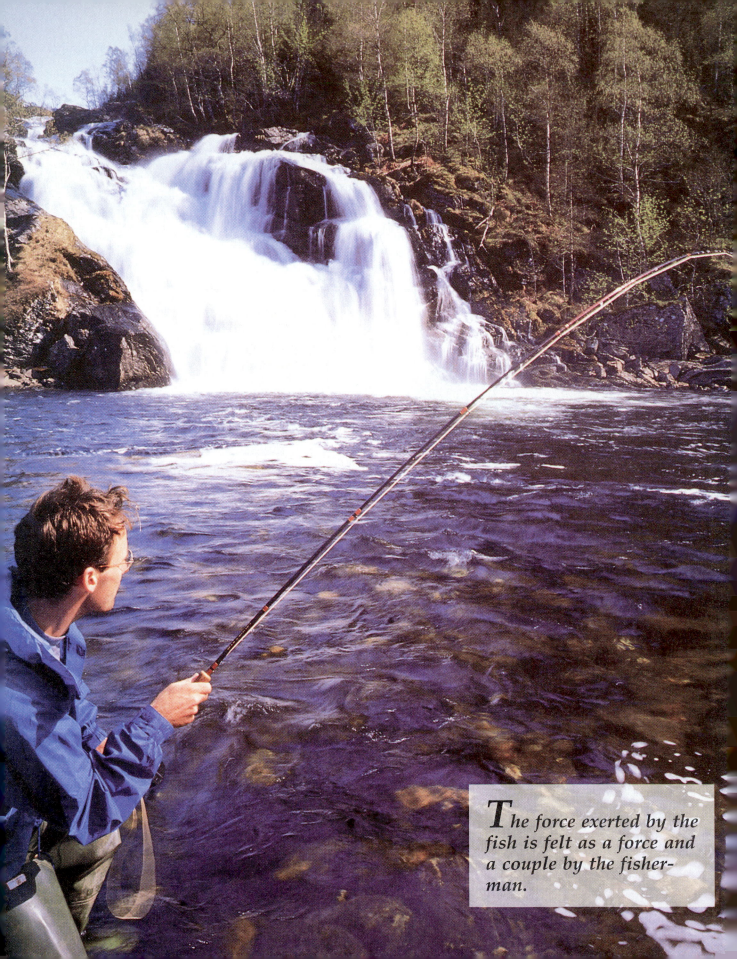

*The force exerted by the fish is felt as a force and a couple by the fisherman.*

# Chapter 4

# RIGID BODIES: EQUIVALENT FORCE/ MOMENT SYSTEMS

## 4-1    INTRODUCTION

 The resultant force **R** of a system of two or more concurrent forces $\mathbf{F}_1, \mathbf{F}_2, \ldots, \mathbf{F}_n$ was defined in Chapter 2 as the single force that will produce the same effect on a body as the original system of forces. When the resultant force **R** of a concurrent force system is zero, the body on which the system of forces acts is in equilibrium, and the force system is said to be balanced. Methods to determine resultants of coplanar and noncoplanar concurrent force systems were discussed in Chapter 2 and applied to the equilibrium of a particle in Chapter 3.

For the case of a three-dimensional body that has a definite size and shape, the particle idealization discussed in Chapter 3 is no longer valid, in general, since the forces acting on the body are usually not concurrent. For these more general force systems, the condition **R** = 0 is a necessary but not a sufficient condition for equilibrium of the body. A second restriction related to the tendency of a force to produce rotation of a body must also be satisfied and gives rise to the concept of a moment. In this chapter, the moment of a force about a point and the moment of a force about a line (axis) will be defined and methods will be developed for finding the resultant forces and the resultant moments (couples) for four other force systems that can be applied to a rigid body: namely,

1. Coplanar, parallel force systems.
2. Coplanar, nonparallel, nonconcurrent force systems.
3. Noncoplanar, parallel force systems.
4. Noncoplanar, nonparallel, nonconcurrent force systems.

The resultants of these force systems will be used in later chapters dealing with equilibrium (Chapter 6) and motion (Chapter 16) of rigid bodies.

## 4-2   MOMENTS AND THEIR CHARACTERISTICS

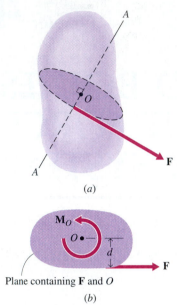

(a)

(b)

Plane containing **F** and $O$

**Figure 4-1**   Moment of a force **F** about point $O$.

The moment of a force about a point or axis is a measure of the tendency of the force to rotate a body about that point or axis. For example, the moment of force **F** about point $O$ in Fig. 4-1$a$ is a measure of the tendency of the force to rotate the body about line $AA$. Line $AA$ is perpendicular to the plane containing force **F** and point $O$.

A moment has both a magnitude and a direction, and adds according to the parallelogram law of addition; therefore, it is a vector quantity. The magnitude of a moment $|M|$ is defined as the product of the magnitude of a force $|F|$ and the perpendicular distance $d$ from the line of action of the force to the axis. Thus, in Fig. 4-1$b$, the magnitude of the moment of the force **F** about point $O$ (actually about axis $AA$, which is perpendicular to the page and passes through point $O$) is

$$M_O = |\mathbf{M}_O| = |\mathbf{F}|d \qquad (4\text{-}1)$$

Point $O$ is called the moment center, distance $d$ is called the moment arm, and line $AA$ is called the axis of the moment.

The direction (sense) of a moment in a two-dimensional problem can be specified by using a small curved arrow about the point as shown in Fig. 4-1$b$. If the force tends to produce a counterclockwise rotation, the moment is defined to be positive. In a similar manner, if the force tends to produce a clockwise rotation, the moment is defined to be negative.

Since the magnitude of a moment of a force is the product of a force and a length, the dimensional expression for a moment is $FL$. In the U.S. Customary System, the units commonly used for moments are lb · ft and lb · in. or in. · lb, ft · lb, and ft · kip. In the SI System, the units commonly used for moments are N · m, kN · m, and so on. It is immaterial whether the unit of force or the unit of length is stated first.

The procedure for determining moments in simple two-dimensional problems (coplanar forces in, say, the $xy$-plane) is illustrated in the following examples.

# CONCEPTUAL EXAMPLE 4-1: MOMENT OF A FORCE

*Experience indicates that it is easier to turn a wheel if the force is applied near the rim of the wheel rather than near the axle. Explain why the moment of a force depends on the magnitude of the force, the direction of the force, and where the force is applied relative to the axis of rotation.*

## SOLUTION

The concept of a moment of a force can be illustrated by considering the door shown in Fig. CE4-1. The force **F** of Fig. CE4-1a is perpendicular to the door and tends to rotate the door about a line through the axes of the hinges. Greater force is required to rotate the door when the force is applied near the hinge line rather than at the knob. In Fig. CE4-1a, the moment of force **F** about the hinge line is defined as $M = Fd$, where $M$ is the moment of the force, $F$ is the magnitude of the force, and $d$ is the perpendicular distance from the hinge line to the line of action of the force. The distance $d$ is called the moment arm of the force.

If the force **F** is not perpendicular to the door (see Fig. CE4-1b), a larger force **F** is required to rotate the door. The tendency of an inclined force **F** to rotate the door can be seen by resolving force **F** into components as shown in Fig. 4-1c. Component $\mathbf{F}_y$ does not cause rotation because its line of action passes through the hinge line ($d = 0$). Component $\mathbf{F}_z$ does not cause rotation because its line of action is parallel to the hinge line. Only component $\mathbf{F}_x$, which is perpendicular to the plane of the door and at a distance $d$ from the hinge line, tends to rotate the door.

(a)                    (b)                    (c)

**Fig. CE4-1**

Three forces are applied to the bar shown in Fig. 4-2. Determine

a. The moment of force $\mathbf{F}_A$ about point $E$.
b. The moment of force $\mathbf{F}_E$ about point $A$.
c. The moment of force $\mathbf{F}_D$ about point $B$.

**Fig. 4-2**

## SOLUTION

The moment of a force $\mathbf{F}$ about an arbitrary point $O$ is defined as the product of the magnitude of the force $|\mathbf{F}|$ and the perpendicular distance $d$ from the line of action of the force to the point. Thus,

$$M_O = |\mathbf{M}_O| = |\mathbf{F}|d$$

a. $M_E = |\mathbf{F}_A|d_{E/A} = 100(10) = 1000 \text{ in. } \cdot \text{lb}$
$\qquad \mathbf{M}_E = 1000 \text{ in. } \cdot \text{lb} \curvearrowleft$ 　　　　　　　　 Ans.

b. $M_A = |\mathbf{F}_E|d_{A/E} = 200(12) = 2400 \text{ in. } \cdot \text{lb}$
$\qquad \mathbf{M}_A = 2400 \text{ in. } \cdot \text{lb} \curvearrowleft$ 　　　　　　　 Ans.

c. $M_B = |\mathbf{F}_D|d_{B/D} = 300(14) = 4200 \text{ in. } \cdot \text{lb}$
$\qquad \mathbf{M}_B = 4200 \text{ in. } \cdot \text{lb} \curvearrowright$ 　　　　　　　 Ans.

Perpendicular distances from the lines of action of the forces to the points are easily determined; therefore, magnitudes of the moments are determined using the definition of a moment ($M = Fd$).

The direction (sense) of a moment in simple two-dimensional problems can be specified by using a small curved arrow about the point. A positive moment tends to produce a counterclockwise rotation.

## EXAMPLE PROBLEM 4-2

Three forces are applied to a bracket as shown in Fig. 4-3. Determine

a. The moment of force $\mathbf{F}_C$ about point $B$.
b. The moment of force $\mathbf{F}_D$ about point $A$.
c. The moment of force $\mathbf{F}_B$ about point $C$.

**Fig. 4-3**

---

### SOLUTION

The moment of a force $\mathbf{F}$ about an arbitrary point $O$ is defined as the product of the magnitude of the force $|\mathbf{F}|$ and the perpendicular distance $d$ from the line of action of the force to the point. Thus,

$$M_O = |\mathbf{M}_O| = |F|d$$

a. $M_B = |F_C|d_{B/C} = 90(160 + 120)$
$$= 25,200 \text{ N} \cdot \text{mm} = 25.2 \text{ N} \cdot \text{m}$$
$$\mathbf{M}_B = 25.2 \text{ N} \cdot \text{m}\curvearrowleft \qquad \text{Ans.}$$

b. $M_A = |F_D|d_{A/D} = 120(350 + 200)$
$$= 66,000 \text{ N} \cdot \text{mm} = 66.0 \text{ N} \cdot \text{m}$$
$$\mathbf{M}_A = 66.0 \text{ N} \cdot \text{m}\curvearrowleft \qquad \text{Ans.}$$

c. $M_C = |F_B|d_{C/B} = 150[(160 + 120) \cos 30°]$
$$= 36,373 \text{ N} \cdot \text{mm} \cong 36.4 \text{ N} \cdot \text{m}$$
$$\mathbf{M}_C = 36.4 \text{ N} \cdot \text{m}\curvearrowright \qquad \text{Ans.}$$

Perpendicular distances from the lines of action of the forces to the points are easily determined; therefore, magnitudes of the moments are determined using the definition of a moment ($M = Fd$).

The direction (sense) of a moment in simple two-dimensional problems can be specified by using a small curved arrow about the point. A positive moment tends to produce a counterclockwise rotation.

Three forces are applied to a triangular plate as shown in Fig. 4-4a. Determine

a. The moment of force $\mathbf{F}_B$ about point $C$.
b. The moment of force $\mathbf{F}_C$ about point $B$.
c. The moment of force $\mathbf{F}_D$ about point $B$.

(a)                    (b)

**Fig. 4-4**

## SOLUTION

The moment of a force $\mathbf{F}$ about an arbitrary point $O$ is defined as the product of the magnitude of the force $|\mathbf{F}|$ and the perpendicular distance $d$ from the line of action of the force to the point. Thus,

$$M_O = |\mathbf{M}_O| = |\mathbf{F}|d$$

In this example, the problem is slightly more involved because of the geometry. As shown in Fig. 4-4b, the perpendicular distances from the forces to the points of interest are determined as follows

$$\phi = \tan^{-1}\frac{20}{15} = 53.13°$$

$$L_{AC} = \sqrt{20^2 + 15^2} = 25 \text{ in.}$$
$$d_{C/B} = L_{AC} - 10 = 25 - 10 = 15 \text{ in.}$$
$$d_{B/C} = 15 \sin 53.13° = 12.00 \text{ in.}$$
$$d_{B/D} = 15 - 15 \sin 53.13° = 3.00 \text{ in.}$$

a.  $M_C = |F_B|d_{C/B} = 400(15) = 6000 \text{ in.} \cdot \text{lb}$
$$\mathbf{M}_C = 6000 \text{ in.} \cdot \text{lb} \circlearrowright \qquad\qquad \text{Ans.}$$

b.  $M_B = |F_C|d_{B/C} = 300(12) = 3600 \text{ in.} \cdot \text{lb}$
$$\mathbf{M}_B = 3600 \text{ in.} \cdot \text{lb} \circlearrowleft \qquad\qquad \text{Ans.}$$

c.  $M_B = |F_D|d_{B/D} = 250(3) = 750 \text{ in.} \cdot \text{lb}$
$$\mathbf{M}_B = 750 \text{ in.} \cdot \text{lb} \circlearrowleft \qquad\qquad \text{Ans.}$$

Trigonometry is used with the diagram shown in Fig. 4-4b to determine moment arms. Then, magnitudes of the moments are determined using the definition of a moment ($M = Fd$).

The direction (sense of a moment in simple two-dimensional problems can be specified by using a small curved arrow about the point. A positive moment tends to produce a counterclockwise rotation.

Four forces are applied to a plate as shown in Fig. 4-5a. Determine

a. The moment of force $\mathbf{F}_B$ about point $A$.
b. The moment of force $\mathbf{F}_C$ about point $B$.
c. The moment of force $\mathbf{F}_C$ about point $A$.

**Fig. 4-5**

## SOLUTION

The moment of a force $\mathbf{F}$ about an arbitrary point $O$ is defined as the product of the magnitude of the force $|\mathbf{F}|$ and the perpendicular distance $d$ from the line of action of the force to the point. Thus,

$$M_O = |\mathbf{M}_O| = |\mathbf{F}|d$$

As shown in Fig. 4-5b, the perpendicular distances from the forces to the points of interest are as follows

$$d_{A/B} = 250 \sin 45° = 176.78 \text{ mm}$$
$$d_{B/C} = 400 \cos 30° = 346.41 \text{ mm}$$
$$d_{A/C} = 400 \cos 30° - 250 \sin 30° = 221.41 \text{ mm}$$

a. $M_A = |\mathbf{F}_B|d_{A/B} = 15(10^3)(176.78)(10^{-3}) = 2.652(10^3) \text{ N} \cdot \text{m} \cong 2.65 \text{ kN} \cdot \text{m}$
   $\mathbf{M}_A \cong 2.65 \text{ kN} \cdot \text{m} \curvearrowleft$        Ans.

b. $M_B = |\mathbf{F}_C|d_{B/C} = 20(10^3)(346.41)(10^{-3}) = 6.928(10^3) \text{ N} \cdot \text{m} \cong 6.93 \text{ kN} \cdot \text{m}$
   $\mathbf{M}_B \cong 6.93 \text{ kN} \cdot \text{m} \curvearrowleft$        Ans.

c. $M_A = |\mathbf{F}_C|d_{A/C} = 20(10^3)(221.41)(10^{-3}) = 4.428(10^3) \text{ N} \cdot \text{m} \cong 4.43 \text{ kN} \cdot \text{m}$
   $\mathbf{M}_A \cong 4.43 \text{ kN} \cdot \text{m} \curvearrowleft$        Ans.

Trigonometry is used with the diagram shown in Fig. 4-5b to determine moment arms. Then, magnitudes of the moments are determined using the definition of a moment ($M = Fd$).

The direction (sense) of a moment in simple two-dimensional problems can be specified by using a small curved arrow about the point. A positive moment tends to produce a counterclockwise rotation.

# PROBLEMS

### Introductory Problems

**4-1\*** Determine the moment of the 250-lb force shown in Fig. P4-1 about point $A$.

**Fig. P4-1**

**4-2\*** Determine the moments of the 225-N force shown in Fig. P4-2 about points $A$, $B$, and $C$.

**Fig. P4-2**

**4-3** Two forces are applied at a point in a plane as shown in Fig. P4-3. Determine

**a.** The moments of force $F_1$ about points $A$ and $B$.
**b.** The moments of force $F_2$ about points $B$ and $C$.

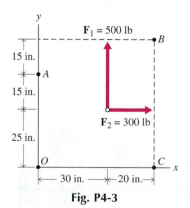

**Fig. P4-3**

**4-4** Two forces are applied at a point in a plane as shown in Fig. P4-4. Determine

**a.** The moments of force $F_1$ about points $B$ and $C$.
**b.** The moments of force $F_2$ about points $A$ and $C$.

**Fig. P4-4**

**4-5\*** Two forces are applied to a bracket as shown in Fig. P4-5. Determine

**a.** The moment of force $F_1$ about point $A$.
**b.** The moment of force $F_2$ about point $B$.

**Fig. P4-5**

**4-6\*** Two forces are applied to a bracket as shown in Fig. P4-6. Determine

**a.** The moment of force $F_1$ about point $A$.
**b.** The moment of force $F_2$ about point $B$.

**Fig. P4-6**

**4-7** Two forces are applied to a beam as shown in Fig. P4-7. Determine the moments of forces $F_1$ and $F_2$ about point $A$.

**Fig. P4-7**

**4-8** Two forces are applied to a beam as shown in Fig. P4-8. Determine the moments of forces $F_1$ and $F_2$ about point $A$.

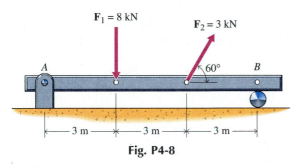

**Fig. P4-8**

## Intermediate Problems

**4-9\*** Three forces are applied to a circular plate as shown in Fig. P4-9. Determine

a. The moment of force $F_1$ about point $O$.
b. The moment of force $F_3$ about point $O$.
c. The moment of force $F_2$ about point $A$.

**Fig. P4-9**

**4-10\*** Determine the moment of the 500-N force shown in Fig. P4-10 about points $A$ and $B$.

**Fig. P4-10**

**4-11** Determine the moment of the 350-lb force shown in Fig. P4-11 about points $A$ and $B$.

**Fig. P4-11**

**4-12** Determine the moment of the 750-N force shown in Fig. P4-12 about points $A$, $B$, and $C$.

**Fig. P4-12**

**117**

## Challenging Problems

**4-13*** Two forces are applied to an angle bracket as shown in Fig. P4-13. Determine the moments of forces $F_1$ and $F_2$ about points $A$ and $B$.

**Fig. P4-13**

**4-14*** Two forces are applied to an eye bracket as shown in Fig. P4-14. Determine the moments of forces $F_1$ and $F_2$ about points $A$ and $B$.

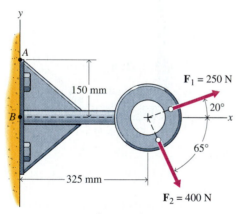

**Fig. P4-14**

**4-15** Two forces are applied to a bracket as shown in Fig. P4-15. Determine the moments of forces $F_1$ and $F_2$ about points $A$ and $B$.

**Fig. P4-15**

**4-16** Two forces are applied to a beam as shown in Fig. P4-16. Determine the moments of forces $F_1$ and $F_2$ about points $A$ and $B$.

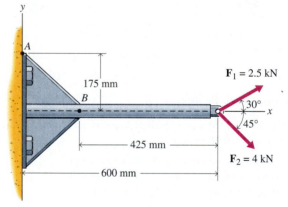

**Fig. P4-16**

### 4-2.1 Principle of Moments: Varignon's Theorem

A concept often used in solving mechanics (statics, dynamics, mechanics of materials) problems is the principle of moments. This principle, when applied to a system of forces, states that the moment **M** of the resultant **R** of a system of forces with respect to any axis or point is equal to the vector sum of the moments of the individual forces of the system with respect to the same axis or point. The resultant **R** of a system of forces $F_1$, $F_2$ ..., $F_n$ is

$$R = F_1 + F_2 + \cdots + F_n$$

Applying the principle of moments gives the moment of the system of forces about an arbitrary point $O$ as

$$M_O = Rd_R = F_1 d_1 + F_2 d_2 + \cdots + F_n d_n$$

where all $d$'s are distances from the point $O$ (moment axis) to a perpendicular intersection with the line of action of the respective force. Positive moments tend to produce a counterclockwise rotation about the point while negative moments tend to produce a clockwise rotation about the point.

Application of this principle to a pair of concurrent forces is known as Varignon's theorem. Varignon's theorem can be illustrated by using the concurrent force system, shown in Fig. 4-6, where $\mathbf{R}$ is the resultant of forces $\mathbf{A}$ and $\mathbf{B}$, which lie in the $xy$-plane. The point of concurrency $A$ and the moment center $O$ have been arbitrarily selected to lie on the $y$-axis. The distances $d$, $a$, and $b$ are the perpendicular distances from the moment center $O$ to the forces $\mathbf{R}$, $\mathbf{A}$, and $\mathbf{B}$, respectively. The angles $\gamma$, $\alpha$, and $\beta$ (measured from the $x$-axis) locate the forces $\mathbf{R}$, $\mathbf{A}$, and $\mathbf{B}$, respectively.

The magnitudes of the moments produced by the resultant $\mathbf{R}$ and by the two forces $\mathbf{A}$ and $\mathbf{B}$ with respect to point $O$ are

$$\begin{aligned} M_R &= Rd = R(h \cos \gamma) \\ M_A &= Aa = A(h \cos \alpha) \\ M_B &= Bb = B(h \cos \beta) \end{aligned} \qquad (a)$$

From Fig. 4-6 note also that

$$R \cos \gamma = A \cos \alpha + B \cos \beta \qquad (b)$$

Substituting Eqs. $a$ into Eq. $b$ and multiplying both sides of the equation by $h$ yields

$$M_R = M_A + M_B \qquad (4\text{-}2)$$

Equation 4-2 indicates that the moment of the resultant $\mathbf{R}$ with respect to a point $O$ is equal to the sum of the moments of the forces $\mathbf{A}$ and $\mathbf{B}$ with respect to the same point $O$.

The following examples illustrate the use of the principle of moments.

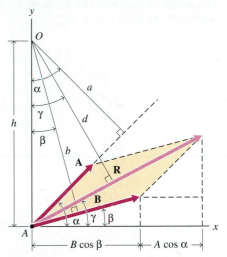

**Figure 4-6** Principle of moments for two concurrent forces.

Determine the moment about point $O$ (actually the $z$-axis through point $O$) of the 500-lb force shown in Fig. 4-7$a$.

**Fig. 4-7**

---

## SOLUTION

Inspection of Fig. 4-7$b$ shows that the perpendicular distance $d$ from the line of action of the force **F** to point $O$ is

$$d = 5 \cos 30° - 3 \sin 30° = 2.830 \text{ ft}$$

Thus,

$$\curvearrowright +M_O = -Fd = -500(2.830) = -1415 \text{ ft} \cdot \text{lb}$$
$$M_O = 1415 \text{ ft} \cdot \text{lb}\curvearrowright \qquad \text{Ans.}$$

Alternatively, the moment of force **F** about point $O$ also can be determined by using the principle of moments. The magnitudes of the rectangular components of the 500-lb force are

$$F_x = F \cos 30° = 500 \cos 30° = 433.0 \text{ lb}$$
$$F_y = F \sin 30° = 500 \sin 30° = 250.0 \text{ lb}$$

Once the forces $F_x$ and $F_y$ are known, the moment $M_O$ is

$$\curvearrowright +M_O = F_y(3) - F_x(5) = 250.0(3) - 433.0(5) = -1415 \text{ ft} \cdot \text{lb}$$
$$M_O = 1415 \text{ ft} \cdot \text{lb}\curvearrowright \qquad \text{Ans.}$$

> The principle of moments provides a more convenient method for analysis than direct application of the definition of a moment ($M = Fd$) since the moment arm for each component force is easier to establish than the perpendicular distance $d$.

Use the principle of moments to determine the moment about point $B$ of the 300-N force shown in Fig. 4-8a.

(a)

(b)

**Fig. 4-8**

## SOLUTION

The magnitudes of the rectangular components of the 300-N force are

$$F_x = F \cos 30° = 300 \cos 30° = 259.8 \text{ N}$$
$$F_y = F \sin 30° = 300 \sin 30° = 150.0 \text{ N}$$

Once the forces $F_x$ and $F_y$ are known, the moment $M_B$ is

$$\curvearrowright + M_B = -F_x(0.250) - F_y(0.200)$$
$$= -259.8(0.250) - 150.0(0.200) = -95.0 \text{ N} \cdot \text{m}$$
$$\mathbf{M}_B = 95.0 \text{ N} \cdot \text{m} \curvearrowright \qquad \text{Ans.}$$

The moment about point $B$ can also be determined by moving the force **F** along its line of action (principle of transmissibility) to points $C$ or $D$ as shown in Fig. 4-8b. For both of these points, one component of the force produces no moment about point $B$. Thus, for point $C$

$$d_2 = 250 + 200 \tan 30° = 365.5 \text{ mm}$$
$$\curvearrowright + M_B = -F_x d_2 = -259.8(0.3655) = -95.0 \text{ N} \cdot \text{m}$$
$$\mathbf{M}_B = 95.0 \text{ N} \cdot \text{m} \curvearrowright \qquad \text{Ans.}$$

For point $D$

$$d_3 = 200 + 250 \cot 30° = 633.0 \text{ mm}$$
$$\curvearrowright + M_B = -F_y d_3 = -150.0(0.633) = -95.0 \text{ N} \cdot \text{m}$$
$$\mathbf{M}_B = 95.0 \text{ N} \cdot \text{m} \curvearrowright \qquad \text{Ans.}$$

> Force **F** is a sliding vector; therefore, it produces the same moment about point $B$ when it is applied at any point along its line of action on or off the body.

Three forces $\mathbf{F}_1$, $\mathbf{F}_2$, and $\mathbf{F}_3$ are applied to a beam as shown in Fig. 4-9a. Determine the moments $\mathbf{M}_{A1}$, $\mathbf{M}_{A2}$, and $\mathbf{M}_{A3}$ about point $A$ produced by each of the forces.

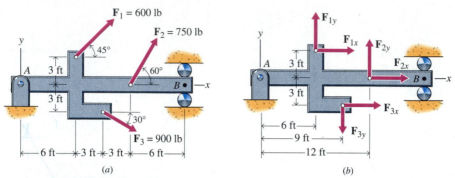

Fig. 4-9

## SOLUTION

The magnitudes of the rectangular components (see Fig. 4-9b) of the three forces are

$$F_{1x} = F_1 \cos 45° = 600 \cos 45° = 424.3 \text{ lb}$$
$$F_{1y} = F_1 \sin 45° = 600 \sin 45° = 424.3 \text{ lb}$$
$$F_{2x} = F_2 \cos 60° = 750 \cos 60° = 375.0 \text{ lb}$$
$$F_{2y} = F_2 \sin 60° = 750 \sin 60° = 649.5 \text{ lb}$$
$$F_{3x} = F_3 \cos 30° = 900 \cos 30° = 779.4 \text{ lb}$$
$$F_{3y} = F_3 \sin 30° = 900 \sin 30° = 450.0 \text{ lb}$$

The moments $\mathbf{M}_{A1}$, $\mathbf{M}_{A2}$, and $\mathbf{M}_{A3}$ are determined by using the principle of moments. Thus

$$\curvearrowright + M_{A1} = F_{1y}(6) - F_{1x}(3) = 424.3(6) - 424.3(3) = 1273 \text{ ft} \cdot \text{lb}$$
$$M_{A1} = 1273 \text{ ft} \cdot \text{lb} \curvearrowleft \qquad \text{Ans.}$$
$$\curvearrowright + M_{A2} = F_{2y}(12) - F_{2x}(0) = 649.5(12) - 375.0(0) = 7794 \text{ ft} \cdot \text{lb}$$
$$M_{A2} = 7794 \text{ ft} \cdot \text{lb} \curvearrowleft \qquad \text{Ans.}$$
$$\curvearrowright + M_{A3} = F_{3x}(3) - F_{3y}(9) = 779.4(3) - 450.0(9) = -1712 \text{ ft} \cdot \text{lb}$$
$$M_{A3} = 1712 \text{ ft} \cdot \text{lb} \curvearrowright \qquad \text{Ans.}$$

The principle of moments provides a more convenient method for analysis than direct application of the definition of a moment ($M = Fd$) since the moment arm for each component force is easier to establish than the perpendicular distance $d$.

# PROBLEMS

## Introductory Problems

**4-17*** Determine the moment of the 300-lb force shown in Fig. P4-17 about point $A$.

**Fig. P4-17**

**4-18*** Determine the moment of the 250-N force shown in Fig. P4-18 about point $A$.

**Fig. P4-18**

**4-19** Determine the moment of the 750-lb force shown in Fig. P4-19 about point $O$.

**Fig. P4-19**

**4-20** A 160-N force is applied to the handle of a door as shown in Fig. P4-20. Determine the moments of the force about hinges $A$ and $B$.

**Fig. P4-20**

**4-21*** Determine the moment of the 425-lb force shown in Fig. P4-21 about point $B$.

**Fig. P4-21**

**4-22** Determine the moments of the 16-kN force shown in Fig. P4-22 about points $A$ and $B$.

**Fig. P4-22**

123

## Intermediate Problems

**4-23*** A 50-lb force is applied to the handle of a lug wrench, which is being used to tighten the nuts on the rim of an automobile tire as shown in Fig. P4-23. The diameter of the bolt circle is $5\frac{1}{2}$ in. Determine the moments of the force about the axle for the wheel (point $O$) and about the point of contact of the wheel with the pavement (point $A$).

**Fig. P4-23**

**4-24*** Two forces $F_1$ and $F_2$ are applied to a beam as shown in Fig. P4-24. Determine

**a.** The moment of force $F_1$ about point $A$.
**b.** The moment of force $F_2$ about point $B$.

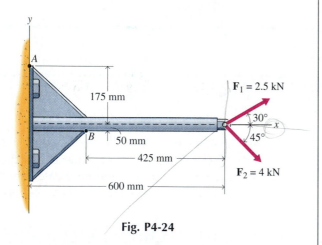

**Fig. P4-24**

**4-25** Three forces $F_A$, $F_B$, and $F_C$ are applied to a beam as shown in Fig. P4-25. Determine

**a.** The moments of forces $F_A$ and $F_C$ about point $O$.
**b.** The moment of force $F_B$ about point $D$.

**Fig. P4-25**

**4-26** Three forces $F_1$, $F_2$, and $F_3$ are applied to a bracket as shown in Fig. P4-26. Determine the moments of each of the forces about point $B$.

**Fig. P4-26**

## Challenging Problems

**4-27*** Determine the moments of the 50-lb force shown in Fig. P4-27 about points $A$ and $B$.

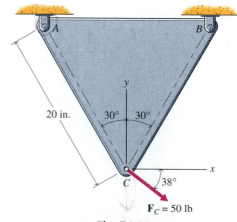

**Fig. P4-27**

124

**4-28\*** Determine the moments of the 450-N force shown in Fig. P4-28 about points $A$ and $B$.

**Fig. P4-28**

**4-29** Determine the moments of the 300-lb force shown in Fig. P4-29 about points $A$ and $B$.

**Fig. P4-29**

**4-30** Determine the moments of the 300-N force shown in Fig. P4-30 about points $A$ and $B$.

**Fig. P4-30**

## 4-3 VECTOR REPRESENTATION OF A MOMENT

For some two-dimensional problems and for most three-dimensional problems, use of Eq. 4-1 for moment determinations is not convenient owing to difficulties in determining the perpendicular distance $d$ between the line of action of the force and the moment center $O$. For these types of problems, a vector approach simplifies moment calculations.

In Fig. 4-1, the moment of the force $\mathbf{F}$ about point $O$ can be represented by the expression

$$\mathbf{M}_O = \mathbf{r} \times \mathbf{F} \qquad (4\text{-}3)$$

where $\mathbf{r}$ is a position vector from point $O$ to a point $A$ on the line of action of the force $\mathbf{F}$, as shown in Fig. 4-10. By definition, the cross product (see Appendix A) of the two intersecting vectors $\mathbf{r}$ and $\mathbf{F}$ is

$$\mathbf{M}_O = \mathbf{r} \times \mathbf{F} = |\mathbf{r}||\mathbf{F}| \sin \alpha\, \mathbf{e} \qquad (4\text{-}4)$$

where $\alpha$ is the angle $(0 \leq \alpha \leq 180°)$ between the two intersecting vectors $\mathbf{r}$ and $\mathbf{F}$, and $\mathbf{e}$ is a unit vector perpendicular to the plane contain-

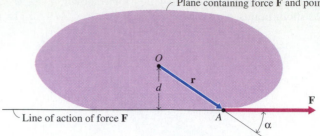

Figure 4-10   Cross-product definition of the moment of a force **F** about point $O$.

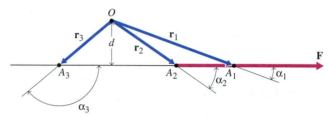

Figure 4-11   Different position vectors yield the same moment of force **F** about point $O$.

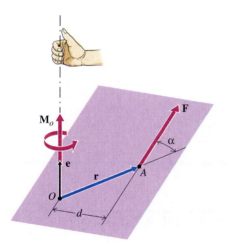

Figure 4-12   Right-hand rule for positive moments.

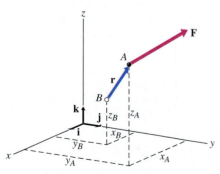

Figure 4-13   Cartesian vector form of position vector **r**.

ing vectors **r** and **F**. It is obvious from Fig. 4-10 that the term $|\mathbf{r}|\sin\alpha$ equals the perpendicular distance $d$ from the line of action of the force to the moment center $O$. Note also from Fig. 4-11 that the distance $d$ is independent of the position $A$ along the line of action of the force since

$$|\mathbf{r}_1|\sin\alpha_1 = |\mathbf{r}_2|\sin\alpha_2 = |\mathbf{r}_3|\sin\alpha_3 = d$$

Thus, Eq. 4-4 can be written

$$\mathbf{M}_O = |\mathbf{F}|d\mathbf{e} = Fd\mathbf{e} = M_O\mathbf{e} \qquad (4\text{-}5)$$

In Eq. 4-5 the direction of the unit vector **e** is determined (see Fig. 4-12) by using the right-hand rule (fingers of the right hand curl from positive **r** to positive **F** and the thumb points in the direction of positive $\mathbf{M}_O$). Thus, Eq. 4-5 yields both the magnitude $M_O$ and the direction **e** of the moment $\mathbf{M}_O$. It is important to note that the sequence $\mathbf{r} \times \mathbf{F}$ must be maintained in calculating moments since the sequence $\mathbf{F} \times \mathbf{r}$ will produce a moment with the opposite sense. The vector product is *not* a commutative operation.

### 4-3.1   Moment of a Force About a Point

The vector **r** from the point about which a moment is to be determined (say point $B$ of Fig. 4-13) to any point on the line of action of a force **F** (say point $A$ of Fig. 4-13) can be expressed in terms of unit vectors **i**, **j**, and **k**, and the coordinates $(x_A, y_A, z_A)$ and $(x_B, y_B, z_B)$ of points $A$ and $B$, respectively. The position vector $\mathbf{r}_A$ can be written in Cartesian vector form as

$$\mathbf{r}_A = x_A\mathbf{i} + y_A\mathbf{j} + z_A\mathbf{k}$$

Similarly, the location of point $B$ with respect to the origin of coordinates can be specified by using a position vector $\mathbf{r}_B$, which can be written in Cartesian vector form as

$$\mathbf{r}_B = x_B\mathbf{i} + y_B\mathbf{j} + z_B\mathbf{k}$$

Observe in Fig. 4-14 that

$$\mathbf{r}_A = \mathbf{r}_B + \mathbf{r}_{A/B}$$

Therefore,

$$
\begin{aligned}
\mathbf{r} = \mathbf{r}_{A/B} = \mathbf{r}_A - \mathbf{r}_B \\
= (x_A\mathbf{i} + y_A\mathbf{j} + z_A\mathbf{k}) - (x_B\mathbf{i} + y_B\mathbf{j} + z_B\mathbf{k}) \\
= (x_A - x_B)\mathbf{i} + (y_A - y_B)\mathbf{j} + (z_A - z_B)\mathbf{k} \quad \text{(4-6)}
\end{aligned}
$$

where the subscript $A/B$ indicates $A$ with respect to $B$.

Equation 4-3 is applicable for both the two-dimensional case (forces in, say, the $xy$-plane) and the three-dimensional case (forces with arbitrary space orientations).

**The Two-Dimensional Case** Consider first the moment $\mathbf{M}_O$ about the origin of coordinates (see Fig. 4-15$a$) produced by a force $\mathbf{F}$ in the $xy$-plane. The line of action of the force passes through point $A$. For this special case (see Fig. 4-15$b$)

$$\mathbf{F} = F_x\mathbf{i} + F_y\mathbf{j}$$

and the position vector $\mathbf{r}$ from the origin $O$ to point $A$ is

$$\mathbf{r} = r_x\mathbf{i} + r_y\mathbf{j}$$

From the vector definition of a moment as expressed by Eq. 4-3,

$$
\begin{aligned}
\mathbf{M}_O = \mathbf{r} \times \mathbf{F} = (r_x\mathbf{i} + r_y\mathbf{j}) \times (F_x\mathbf{i} + F_y\mathbf{j}) \\
= r_xF_x(\mathbf{i} \times \mathbf{i}) + r_xF_y(\mathbf{i} \times \mathbf{j}) + r_yF_x(\mathbf{j} \times \mathbf{i}) + r_yF_y(\mathbf{j} \times \mathbf{j}) \\
= r_xF_x(\mathbf{0}) + r_xF_y(\mathbf{k}) + r_yF_x(-\mathbf{k}) + r_yF_y(\mathbf{0}) \\
= (r_xF_y - r_yF_x)\,\mathbf{k} = M_z\mathbf{k} \quad \text{(4-7a)}
\end{aligned}
$$

The vector product $\mathbf{r} \times \mathbf{F}$ for this two-dimensional case can be written in determinant form as

$$\mathbf{M}_O = \mathbf{r} \times \mathbf{F} = \begin{vmatrix} \mathbf{i} & \mathbf{j} & \mathbf{k} \\ r_x & r_y & 0 \\ F_x & F_y & 0 \end{vmatrix} = (r_xF_y - r_yF_x)\mathbf{k} = M_z\mathbf{k} \quad \text{(4-7b)}$$

Thus, for the two-dimensional case, the moment $\mathbf{M}_O$ about point $O$ due to a force $\mathbf{F}$ in the $xy$-plane is perpendicular to the plane (directed along the $z$-axis). The moment is completely defined by the scalar quantity

$$M_O = M_z = r_xF_y - r_yF_x \quad \text{(4-8)}$$

since a positive value for $M_O$ indicates a tendency to rotate the body in a counterclockwise direction, which by the right-hand rule is along the positive $z$-axis. Similarly, a negative value indicates a tendency to rotate the body in a clockwise direction, which requires a moment in the negative $z$-direction.

The following examples illustrate the use of vector algebra for determining moments about a point in two-dimensional problems.

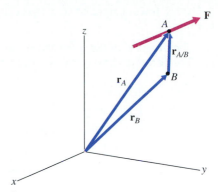

**Figure 4-14**  Position vector $\mathbf{r}_{A/B}$ in terms of vectors $\mathbf{r}_A$ and $\mathbf{r}_B$.

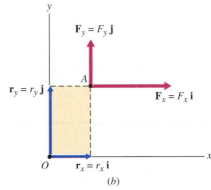

**Figure 4-15**  Moment of a force $\mathbf{F}$ in the $xy$-plane.

A 1000-N force is applied to a beam cross section as shown in Fig. 4-16. Determine

a. The moment of the force about point $O$.
b. The perpendicular distance $d$ from point $B$ to the line of action of the force.

**Fig. 4-16**

## SOLUTION

a. The force **F** and the position vector **r** from point $O$ to point $A$ can be expressed in Cartesian vector form as

$$\mathbf{F} = 1000\,(0.80\mathbf{i} + 0.60\mathbf{j}) = (800\mathbf{i} + 600\mathbf{j})\ \text{N}$$
$$\mathbf{r} = \mathbf{r}_{A/O} = (0.100\mathbf{i} + 0.200\mathbf{j})\ \text{m}$$

$$\mathbf{M}_O = \mathbf{r} \times \mathbf{F} = \begin{vmatrix} \mathbf{i} & \mathbf{j} & \mathbf{k} \\ r_x & r_y & 0 \\ F_x & F_y & 0 \end{vmatrix} = (r_xF_y - r_yF_x)\mathbf{k} = M_z\mathbf{k}$$

$$\mathbf{M}_O = (r_xF_y - r_yF_x)\mathbf{k} = [(0.100)(600) - (0.200)(800)]\mathbf{k}$$
$$= -100\mathbf{k}\ \text{N} \cdot \text{m} = 100\ \text{N} \cdot \text{m} \curvearrowleft \qquad \text{Ans.}$$

b. The position vector **r** from point $B$ to point $A$ is

$$\mathbf{r} = \mathbf{r}_{A/B} = (0.100\mathbf{i} + 0.350\mathbf{j})\ \text{m}$$
$$\mathbf{M}_B = (r_xF_y - r_yF_x)\mathbf{k} = [(0.100)(600) - (0.350)(800)]\mathbf{k}$$
$$= -220\mathbf{k}\ \text{N} \cdot \text{m} = 220\ \text{N} \cdot \text{m} \curvearrowleft$$

$$d = \frac{|\mathbf{M}_B|}{|\mathbf{F}|} = \frac{220}{1000} = 0.220\ \text{m} = 220\ \text{mm} \qquad \text{Ans.}$$

The moment of a force **F** about a point $O$ is given by the vector cross product $\mathbf{M}_O = \mathbf{r} \times \mathbf{F}$ where **r** is a position vector from point $O$ to any point on the line of action of force **F**. Cartesian vector analysis automatically gives the sign of the moment.

Four forces are applied to a square plate as shown in Fig. 4-17. Determine the moments produced by each of the forces about the origin $O$ of the $xy$-coordinate system.

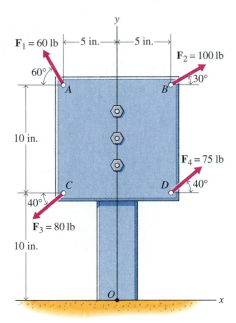

**Fig. 4-17**

---

## SOLUTION

The four forces and the four position vectors can be expressed in Cartesian vector form.

$$\begin{aligned}
\mathbf{F}_1 &= (-30.0\mathbf{i} + 52.0\mathbf{j}) \text{ lb} & \mathbf{r}_{A/O} &= (-5.00\mathbf{i} + 20.0\mathbf{j}) \text{ in.} \\
\mathbf{F}_2 &= (86.6\mathbf{i} + 50.0\mathbf{j}) \text{ lb} & \mathbf{r}_{B/O} &= (5.00\mathbf{i} + 20.0\mathbf{j}) \text{ in.} \\
\mathbf{F}_3 &= (-61.3\mathbf{i} - 51.4\mathbf{j}) \text{ lb} & \mathbf{r}_{C/O} &= (-5.00\mathbf{i} + 10.0\mathbf{j}) \text{ in.} \\
\mathbf{F}_4 &= (57.5\mathbf{i} + 48.2\mathbf{j}) \text{ lb} & \mathbf{r}_{D/O} &= (5.00\mathbf{i} + 10.0\mathbf{j}) \text{ in.}
\end{aligned}$$

$$\mathbf{M}_O = \mathbf{r} \times \mathbf{F} = \begin{vmatrix} \mathbf{i} & \mathbf{j} & \mathbf{k} \\ r_x & r_y & 0 \\ F_x & F_y & 0 \end{vmatrix} = (r_x F_y - r_y F_x)\mathbf{k} = M_z \mathbf{k}$$

$$\begin{aligned}
\mathbf{M}_{O1} &= (r_{1x}F_{1y} - r_{1y}F_{1x})\mathbf{k} = [(-5.00)(52.0) - (20.0)(-30.0)]\mathbf{k} \\
&= 340\mathbf{k} \text{ in} \cdot \text{lb} = 340 \text{ in} \cdot \text{lb} \curvearrowleft \qquad \text{Ans.} \\
\mathbf{M}_{O2} &= (r_{2x}F_{2y} - r_{2y}F_{2x})\mathbf{k} = [(5.00)(50.0) - (20.0)(86.6)]\mathbf{k} \\
&= -1482\mathbf{k} \text{ in} \cdot \text{lb} = 1482 \text{ in} \cdot \text{lb} \curvearrowright \qquad \text{Ans.} \\
\mathbf{M}_{O3} &= (r_{3x}F_{3y} - r_{3y}F_{3x})\mathbf{k} = [(-5.00)(-51.4) - (10.0)(-61.3)]\mathbf{k} \\
&= 870\mathbf{k} \text{ in} \cdot \text{lb} = 870 \text{ in} \cdot \text{lb} \curvearrowleft \qquad \text{Ans.} \\
\mathbf{M}_{O4} &= (r_{4x}F_{4y} - r_{4y}F_{4x})\mathbf{k} = [(5.00)(48.2) - (10.0)(57.5)]\mathbf{k} \\
&= -334\mathbf{k} \text{ in} \cdot \text{lb} = 334 \text{ in} \cdot \text{lb} \curvearrowright \qquad \text{Ans.}
\end{aligned}$$

The moment of a force $\mathbf{F}$ about a point $O$ is given by the vector cross product $\mathbf{M}_O = \mathbf{r} \times \mathbf{F}$ where $\mathbf{r}$ is a position vector from point $O$ to any point on the line of action of force $\mathbf{F}$. Cartesian vector analysis automatically gives the sign of the moment.

# PROBLEMS

Use the vector definition $\mathbf{M} = \mathbf{r} \times \mathbf{F}$ in the solution of the following problems.

## Introductory Problems

**4-31*** Determine the moment of the 375-lb force shown in Fig. P4-31 about point $O$.

**Fig. P4-31**

**4-32*** Determine the moment of the 675-N force shown in Fig. P4-32 about point $O$.

**Fig. P4-32**

**4-33** Determine the moment of the 760-lb force shown in Fig. P4-33 about point $A$.

**Fig. P4-33**

**4-34** Determine the moment of the 750-N force shown in Fig. P4-34 about point $B$.

**Fig. P4-34**

**4-35** A 250-lb force is applied to a beam as shown in Fig. P4-35. Determine the moment of the force about point $A$.

**Fig. P4-35**

**4-36** A 500-N force is applied to a beam as shown in Fig. P4-36. Determine

**a.** The moment of the force about point $B$.
**b.** The moment of the force about point $C$.

**Fig. P4-36**

130

## Intermediate Problems

**4-37\*** Two forces $F_1$ and $F_2$ are applied to a triangular plate as shown in Fig. P4-37. Determine

**a.** The moment of force $F_1$ about point $A$.
**b.** The moment of force $F_2$ about point $B$.

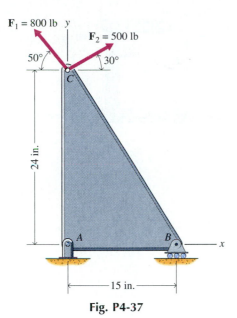

Fig. P4-37

**4-38\*** Two forces $F_1$ and $F_2$ are applied to a bracket as shown in Fig. P4-38. Determine

**a.** The moment of force $F_1$ about point $O$.
**b.** The moment of force $F_2$ about point $A$.

Fig. P4-38

**4-39** Two forces $F_1$ and $F_2$ are applied to a rectangular plate as shown in Fig. P4-39. Determine

**a.** The moment of force $F_1$ about point $B$.
**b.** The moment of force $F_2$ about point $A$.

Fig. P4-39

**4-40** A 450-N force is applied to a bracket as shown in Fig. P4-40. Determine the moment of the force

**a.** About point $B$.
**b.** About point $C$.

Fig. P4-40

## Challenging Problems

**4-41\*** Two forces $F_1$ and $F_2$ are applied to a gusset plate as shown in Fig. P4-41. Determine

**a.** The moment of force $F_1$ about point $A$.
**b.** The moment of force $F_2$ about point $B$.

Fig. P4-41

**131**

**4-42\*** Two forces are applied to an eyebar as shown in Fig. P4-42. Determine

a. The moment of force $F_1$ about point $A$.
b. The moment of force $F_2$ about point $B$.

**Fig. P4-42**

**4-44** A 650-N force is applied to a bracket as shown in Fig. P4-44. Determine the moment of the force

a. About point $D$.
b. About point $E$.

**Fig. P4-44**

**4-43** A 583-lb force is applied to a bracket as shown in Fig. P4-43. Determine the moment of the force

a. About point $D$.
b. About point $E$.

**Fig. P4-43**

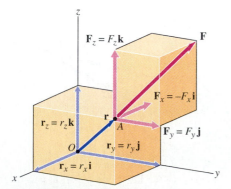

**Figure 4-18** Moment of a force **F** with a space orientation.

**The Three-Dimensional Case** The moment $M_O$ about the origin of coordinates $O$ produced by a force **F** with a space (three-dimensional) orientation can also be determined by using Eq. 4-3. For this general case (see Fig. 4-18), the force **F** can be expressed in Cartesian vector form as

$$\mathbf{F} = F_x\mathbf{i} + F_y\mathbf{j} + F_z\mathbf{k}$$

and the position vector **r** from the origin $O$ to an arbitrary point $A$ on the line of action of the force as

$$\mathbf{r} = r_x\mathbf{i} + r_y\mathbf{j} + r_z\mathbf{k}$$

From the vector definition of a moment as expressed by Eq. 4-3

$$
\begin{aligned}
\mathbf{M}_O = \mathbf{r} \times \mathbf{F} &= (r_x\mathbf{i} + r_y\mathbf{j} + r_z\mathbf{k}) \times (F_x\mathbf{i} + F_y\mathbf{j} + F_z\mathbf{k}) \\
&= r_xF_x(\mathbf{i}\times\mathbf{i}) + r_xF_y(\mathbf{i}\times\mathbf{j}) + r_xF_z(\mathbf{i}\times\mathbf{k}) \\
&\quad + r_yF_x(\mathbf{j}\times\mathbf{i}) + r_yF_y(\mathbf{j}\times\mathbf{j}) + r_yF_z(\mathbf{j}\times\mathbf{k}) \\
&\quad + r_zF_x(\mathbf{k}\times\mathbf{i}) + r_zF_y(\mathbf{k}\times\mathbf{j}) + r_zF_z(\mathbf{k}\times\mathbf{k}) \\
&= (r_yF_z - r_zF_y)\mathbf{i} + (r_zF_x - r_xF_z)\mathbf{j} + (r_xF_y - r_yF_x)\mathbf{k} \\
&= M_x\mathbf{i} + M_y\mathbf{j} + M_z\mathbf{k} \qquad\qquad\qquad \textbf{(4-9a)}
\end{aligned}
$$

The vector product $\mathbf{r} \times \mathbf{F}$ for this three-dimensional case can be written in determinant form as

$$
\begin{aligned}
\mathbf{M}_O = \mathbf{r} \times \mathbf{F} &= \begin{vmatrix} \mathbf{i} & \mathbf{j} & \mathbf{k} \\ r_x & r_y & r_z \\ F_x & F_y & F_z \end{vmatrix} \\[4pt]
&= (r_yF_z - r_zF_y)\mathbf{i} + (r_zF_x - r_xF_z)\mathbf{j} + (r_xF_y - r_yF_x)\mathbf{k} \\
&= M_x\mathbf{i} + M_y\mathbf{j} + M_z\mathbf{k} \qquad\qquad\qquad \textbf{(4-9b)}
\end{aligned}
$$

where

$$
\begin{aligned}
M_x &= r_yF_z - r_zF_y \\
M_y &= r_zF_x - r_xF_z \qquad\qquad \textbf{(4-10)} \\
M_z &= r_xF_y - r_yF_x
\end{aligned}
$$

are the three scalar components of the moment of force $\mathbf{F}$ about point $O$. The magnitude of the moment $|\mathbf{M}_O|$ (see Fig. 4-19) is

$$
|\mathbf{M}_O| = \sqrt{M_x^2 + M_y^2 + M_z^2} \qquad\qquad \textbf{(4-11)}
$$

Alternatively, the moment $\mathbf{M}_O$ can be written as

$$
\mathbf{M}_O = M_O\mathbf{e} \qquad\qquad \textbf{(4-12)}
$$

where

$$
\mathbf{e} = \cos\theta_x\,\mathbf{i} + \cos\theta_y\,\mathbf{j} + \cos\theta_z\,\mathbf{k} \qquad\qquad \textbf{(4-13)}
$$

The direction cosines associated with the unit vector $\mathbf{e}$ are

$$
\cos\theta_x = \frac{M_x}{|\mathbf{M}_O|} \qquad \cos\theta_y = \frac{M_y}{|\mathbf{M}_O|} \qquad \cos\theta_z = \frac{M_z}{|\mathbf{M}_O|} \qquad \textbf{(4-14)}
$$

A moment obeys all the rules of vector combination and can be considered a sliding vector with a line of action coinciding with the moment axis.

Any point on the line of action of force $\mathbf{F}$ can be used to determine the moment about an arbitrary point $O$. For the force shown in Fig. 4-20

$$
\begin{aligned}
\mathbf{M}_O &= \mathbf{r}_B \times \mathbf{F} \\
\mathbf{r}_B &= \mathbf{r}_A + \mathbf{r}_{B/A} \\
\mathbf{M}_O &= (\mathbf{r}_A + \mathbf{r}_{B/A}) \times \mathbf{F} = (\mathbf{r}_A \times \mathbf{F}) + (\mathbf{r}_{B/A} \times \mathbf{F})
\end{aligned}
$$

Since $\mathbf{r}_{B/A}$ is collinear with $\mathbf{F}$,

$$
\mathbf{r}_{B/A} \times \mathbf{F} = \mathbf{O}
$$

Therefore,

$$
\mathbf{M}_O = \mathbf{r}_A \times \mathbf{F}
$$

which further illustrates that the moment of a force about an arbitrary point $O$ can be determined by drawing the position vector $\mathbf{r}$ from point $O$ to any point on the line of action of the force.

**Figure 4-19** Direction angles associated with the moment vector $\mathbf{M}_o$.

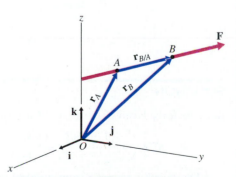

**Figure 4-20** Different position vectors yield the same moment of force $\mathbf{F}$ about point $O$.

The principle of moments discussed in Section 4-2.1 is not restricted to two concurrent forces but may be extended to any force system. The proof for an arbitrary number of forces follows from the distributive property of the vector product. Thus,

$$\mathbf{M}_O = \mathbf{r} \times \mathbf{R}$$

but

$$\mathbf{R} = \mathbf{F}_1 + \mathbf{F}_2 + \cdots + \mathbf{F}_n$$

therefore

$$\mathbf{M}_O = \mathbf{r} \times (\mathbf{F}_1 + \mathbf{F}_2 + \cdots + \mathbf{F}_n)$$
$$= (\mathbf{r} \times \mathbf{F}_1) + (\mathbf{r} \times \mathbf{F}_2) + \cdots + (\mathbf{r} \times \mathbf{F}_n)$$

Thus

$$\mathbf{M}_O = \mathbf{M}_R = \mathbf{M}_1 + \mathbf{M}_2 + \cdots + \mathbf{M}_n \tag{4-15}$$

Equation 4-15 indicates that the moment of the resultant of any number of forces is equal to the sum of the moments of the individual forces.

The following example illustrates the use of vector algebra for determining moments about a point in three-dimensional problems.

## EXAMPLE PROBLEM 4-10

A bar is bent and loaded as shown in Fig. 4-21. Determine

a.   The moment of force **F** about point $O$.
b.   The perpendicular distance $d$ from point $O$ to the line of action of the force.

**Fig. 4-21**

## SOLUTION

a.   The force **F** can be written in Cartesian vector form as

$$\mathbf{F} = 875\left[\frac{75\mathbf{i} + 150\mathbf{j} + 140\mathbf{k}}{\sqrt{(75)^2 + (150)^2 + (140)^2}}\right] = 300.4\mathbf{i} + 600.8\mathbf{j} + 560.7\mathbf{k} \text{ N}$$

The position vector **r** from point $O$ to point $A$ can be written in Cartesian vector form as

$$\mathbf{r}_{A/O} = 0.200\mathbf{i} + 0.250\mathbf{j} + 0.150\mathbf{k} \text{ m}$$

The moment of force $\mathbf{F}$ about point $O$ is then given by the expression

$$\mathbf{M}_O = \mathbf{r}_{A/O} \times \mathbf{F} = \begin{vmatrix} \mathbf{i} & \mathbf{j} & \mathbf{k} \\ r_x & r_y & r_z \\ F_x & F_y & F_z \end{vmatrix} = \begin{vmatrix} \mathbf{i} & \mathbf{j} & \mathbf{k} \\ 0.200 & 0.250 & 0.150 \\ 300.4 & 600.8 & 560.7 \end{vmatrix}$$

$$= 50.06\mathbf{i} - 67.08\mathbf{j} + 45.06\mathbf{k} \ \text{N} \cdot \text{m}$$
$$\cong 50.1\mathbf{i} - 67.1\mathbf{j} + 45.1\mathbf{k} \ \text{N} \cdot \text{m} \qquad \text{Ans.}$$

The moment of force $\mathbf{F}$ about point $O$ is given by the vector cross product $\mathbf{M}_O = \mathbf{r} \times \mathbf{F}$ where $\mathbf{r}$ is a position vector from point $O$ to any point on the line of action of force $\mathbf{F}$.

**b.** The magnitude of moment $\mathbf{M}_O$ is obtained by using the expression

$$M_O = |\mathbf{M}_O| = \sqrt{M_x^2 + M_y^2 + M_z^2} = \sqrt{(50.06)^2 + (-67.08)^2 + (45.06)^2}$$
$$= 95.06 \ \text{N} \cdot \text{m} \cong 95.1 \ \text{N} \cdot \text{m}$$

The distance $d$ is obtained by using the definition of a moment. Thus,

$$d = \frac{M}{F} = \frac{95.06}{875} = 0.10864 \ \text{m} \cong 108.6 \ \text{mm} \qquad \text{Ans.}$$

## EXAMPLE PROBLEM 4-11

An 800-lb force is applied to a lever-shaft assembly as shown in Fig. 4-22. Determine

**a.** The moment of force $\mathbf{F}$ about point $O$.
**b.** The perpendicular distance $d$ from point $O$ to the line of action of the force.

**Fig. 4-22**

## SOLUTION

**a.** The force $\mathbf{F}$ can be written in Cartesian vector form as

$$\mathbf{F} = 800 \left[ \frac{-5.0\mathbf{i} + 35.0\mathbf{j} - 40.0\mathbf{k}}{\sqrt{(-5.0)^2 + (35.0)^2 + (-40.0)^2}} \right] = -74.93\mathbf{i} + 524.5\mathbf{j} - 599.4\mathbf{k} \ \text{lb}$$

The position vector $\mathbf{r}$ from point $O$ to point $A$ can be written in Cartesian vector form as

$$\mathbf{r}_{A/O} = 27.5\mathbf{i} + 35.0\mathbf{j} \ \text{in.}$$

The moment of force $\mathbf{F}$ about point $O$ is then given by the expression

$$\mathbf{M}_O = \mathbf{r}_{A/O} \times \mathbf{F} = \begin{vmatrix} \mathbf{i} & \mathbf{j} & \mathbf{k} \\ r_x & r_y & r_z \\ F_x & F_y & F_z \end{vmatrix} = \begin{vmatrix} \mathbf{i} & \mathbf{j} & \mathbf{k} \\ 27.5 & 35.0 & 0 \\ -74.93 & 524.5 & -599.4 \end{vmatrix}$$

$$= -20{,}979\mathbf{i} + 16{,}484\mathbf{j} + 17{,}046\mathbf{k} \ \text{in.} \cdot \text{lb}$$
$$\cong -21.0\mathbf{i} + 16.48\mathbf{j} + 17.05\mathbf{k} \ \text{in.} \cdot \text{kip} \qquad \text{Ans.}$$

The moment of force $\mathbf{F}$ about point $O$ is given by the vector cross product $\mathbf{M}_O = \mathbf{r} \times \mathbf{F}$ where $\mathbf{r}$ is a position vector from point $O$ to any point on the line of action of force $\mathbf{F}$.

**b.** The magnitude of moment $\mathbf{M}_O$ is obtained by using the expression

$$M_O = |\mathbf{M}_O| = \sqrt{M_x^2 + M_y^2 + M_z^2} = \sqrt{(-20{,}979)^2 + (16{,}484)^2 + (17{,}046)^2}$$
$$= 31{,}661 \text{ in.} \cdot \text{lb} \cong 31.7 \text{ in.} \cdot \text{kip}$$

The distance $d$ is obtained by using the definition of a moment. Thus,

$$d = \frac{M}{F} = \frac{31{,}661}{800} = 39.58 \text{ in.} \cong 39.6 \text{ in.} \qquad \text{Ans.}$$

## EXAMPLE PROBLEM 4-12

A force with a magnitude of 840 N acts at a point in a body as shown in Fig. 4-23. Determine

**a.** The moment of the force about point $B$.
**b.** The angles $\theta_x$, $\theta_y$, and $\theta_z$ between the axis of the moment and the positive $x$-, $y$-, and $z$-coordinate axes.
**c.** The perpendicular distance $d$ from point $B$ to the line of action of the force.

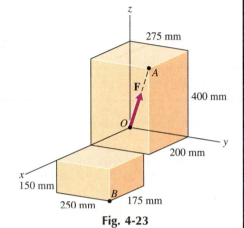

**Fig. 4-23**

### SOLUTION

**a.** The force $\mathbf{F}$ can be written in Cartesian vector form as

$$\mathbf{F} = 840 \left[ \frac{200\mathbf{i} + 275\mathbf{j} + 400\mathbf{k}}{\sqrt{(200)^2 + (275)^2 + (400)^2}} \right] = 320\mathbf{i} + 440\mathbf{j} + 640\mathbf{k} \text{ N}$$

The position vector $\mathbf{r}$ from point $B$ to point $O$ can be written in Cartesian vector form as

$$\mathbf{r} = -0.375\mathbf{i} - 0.250\mathbf{j} + 0.150\mathbf{k} \text{ m}$$

The moment of force $\mathbf{F}$ about point $B$ is then given by the expression

$$\mathbf{M}_B = \mathbf{r} \times \mathbf{F} = \begin{vmatrix} \mathbf{i} & \mathbf{j} & \mathbf{k} \\ r_x & r_y & r_z \\ F_x & F_y & F_z \end{vmatrix} = \begin{vmatrix} \mathbf{i} & \mathbf{j} & \mathbf{k} \\ -0.375 & -0.250 & 0.150 \\ 320 & 440 & 640 \end{vmatrix}$$

$$= -226\,\mathbf{i} + 288\,\mathbf{j} - 85\,\mathbf{k} \text{ N} \cdot \text{m} \qquad \text{Ans.}$$

> The moment of force $\mathbf{F}$ about point $B$ is given by the vector cross product $\mathbf{M}_B = \mathbf{r} \times \mathbf{F}$ where $\mathbf{r}$ is a position vector from point $B$ to any point on the line of action of force $\mathbf{F}$.

**b.** The magnitude of moment $\mathbf{M}_B$ is obtained by using the expression

$$|\mathbf{M}_B| = \sqrt{M_x^2 + M_y^2 + M_z^2} = \sqrt{(-226)^2 + (288)^2 + (-85)^2} = 375.8 \text{ N} \cdot \text{m}$$

The angles $\theta_x$, $\theta_y$, and $\theta_z$ between the axis of the moment and the positive $x$-, $y$-, and $z$-coordinate axes are

$$\theta_x = \cos^{-1} \frac{M_x}{|\mathbf{M}_O|} = \cos^{-1} \frac{-226}{375.8} = 126.97° \cong 127.0° \qquad \text{Ans.}$$

$$\theta_y = \cos^{-1} \frac{M_y}{|\mathbf{M}_O|} = \cos^{-1} \frac{288}{375.8} = 39.97° \cong 40.0° \qquad \text{Ans.}$$

$$\theta_z = \cos^{-1} \frac{M_z}{|\mathbf{M}_O|} = \cos^{-1} \frac{-85}{375.8} = 103.07° \cong 103.1° \qquad \text{Ans.}$$

**c.** The distance $d$ is obtained by using the definition of a moment. Thus,

$$d = \frac{M}{F} = \frac{375.8}{840} = 0.4474 \text{ m} \cong 447 \text{ mm} \qquad \text{Ans.}$$

# PROBLEMS

## Introductory Problems

**4-45*** A force with a magnitude of 970 lb acts at a point in a body as shown in Fig. P4-45. Determine the moment of the force about point C.

**Fig. P4-45**

**4-46** A force with a magnitude of 890 N acts at a point in a body as shown in Fig. P4-46. Determine the moment of the force about point C.

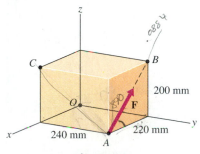

**Fig. P4-46**

**4-47*** A force with a magnitude of 928 lb acts at a point in a body as shown in Fig. P4-47. Determine the moment of the force about point O.

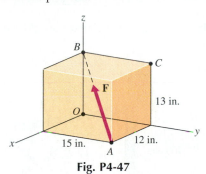

**Fig. P4-47**

**4-48** A force with a magnitude of 860 N acts at a point in a body as shown in Fig. P4-48. Determine

**a.** The moment of the force about point C.
**b.** The perpendicular distance from the line of action of the force to point C.

**Fig. P4-48**

**4-49*** A force with a magnitude of 650 lb acts at a point in a body as shown in Fig. P4-49. Determine

**a.** The moment of the force about point A.
**b.** The direction angles associated with the moment vector.

**Fig. P4-49**

**4-50*** A force with a magnitude of 1000 N acts at a point in a body as shown in Fig. P4-50. Determine

**a.** The moment of the force about point A.
**b** The direction angles associated with the moment vector.

**Fig. P4-50**

**4-51** A force with a magnitude of 400 lb acts at a point in a body as shown in Fig. P4-51. Determine

a. The moment of the force about point $O$.
b. The direction angles associated with the moment vector.

**Fig. P4-51**

**Intermediate Problems**

**4-52*** Determine the moment of the 760-N force shown in Fig. P4-52 about point $B$.

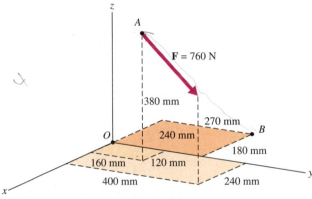

**Fig. P4-52**

**4-53*** Determine the moment of the 580-lb force shown in Fig. P4-53 about point $B$.

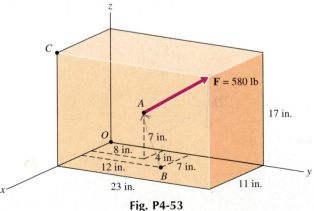

**Fig. P4-53**

**4-54** A 720-N force is applied to a T-bar as shown in Fig. P4-54. Determine

a. The moment of the force about point $O$.
b. The direction angles associated with the moment vector.

**Fig. P4-54**

**4-55** A 750-lb force is applied to a pipe bracket as shown in Fig. P4-55. Determine

a. The moment of the force about point $O$.
b. The direction angles associated with the moment vector.

**Fig. P4-55**

**4-56*** A 610-N force is applied to a lever attached to a post as shown in Fig. P4-56. Determine

a. The moment of the force about point $O$ at the base of the post.
b. The direction angles associated with the moment vector.

**Fig. P4-56**

## Challenging Problems

**4-57*** Determine the moment of the 1000-lb force shown in Fig. P4-57 about point $O$.

**Fig. P4-57**

**4-58*** Determine the moment of the 480-N force shown in Fig. P4-58 about point $O$.

**Fig. P4-58**

**4-59** A force with a magnitude of 580 lb acts at a point in a body as shown in Fig. P4-59. Determine

**a.** The moment of the force about point $B$.
**b.** The direction angles associated with the moment vector.

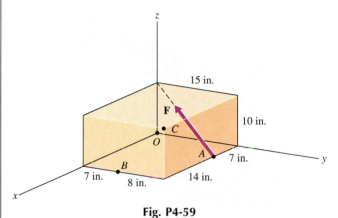

**Fig. P4-59**

**4-60** A force with a magnitude of 585 N acts at a point in a body as shown in Fig. P4-60. Determine

**a.** The moment of the force about point $C$.
**b.** The direction angles associated with the moment vector.

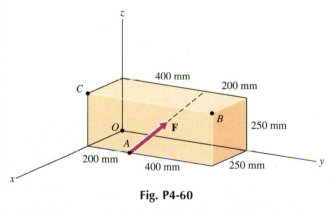

**Fig. P4-60**

## 4-3.2   Moment of a Force About a Line (Axis)

The moment $\mathbf{M}_O$ of a force $\mathbf{F}$ about a point $O$ was defined as the vector product

$$\mathbf{M}_O = \mathbf{r} \times \mathbf{F} \qquad (4\text{-}3)$$

In physical terms, the moment $\mathbf{M}_O$ represents the tendency of the force $\mathbf{F}$ to rotate a body about an axis through point $O$ that is perpendicular to a plane containing force $\mathbf{F}$ and position vector $\mathbf{r}$, as shown in Fig. 4-24. The sense of moment $\mathbf{M}_O$ is indicated by the right-hand rule for vector products.

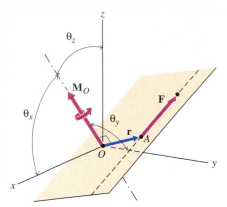

**Figure 4-24**   Direction angles associated with the moment vector $\mathbf{M}_o$.

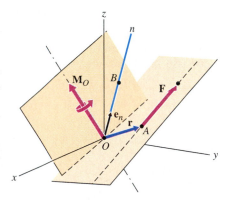

**Figure 4-25**   Unit vector $\mathbf{e}_n$ associated with an arbitrary line $OB$.

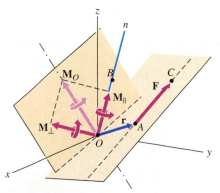

**Figure 4-26**   Rectangular components of moment $\mathbf{M}_o$ parallel and perpendicular to line $OB$.

In many engineering situations, a body is constrained to rotate about a given axis. If the force $\mathbf{F}$ acts in a plane perpendicular to the given axis, Eq. 4-3 can be used directly to determine the moment of the force about the axis. If the force does not act in a plane perpendicular to the axis, then use of Eq. 4-3 serves only as an intermediate step in a process that permits the moment of the force about the required axis to be determined.

The moment $\mathbf{M}_{OB}$ of a force with respect to a line (say, line $OB$ in Fig. 4-25) can be determined by first calculating the moment $\mathbf{M}_O$ about point $O$ on the line (or about any other point on the line). Then, the moment vector $\mathbf{M}_O$ can be resolved into components that are parallel $\mathbf{M}_\parallel$ and perpendicular $\mathbf{M}_\perp$ to the line $OB$, as shown in Fig. 4-26. If $\mathbf{e}_n$ is a unit vector in the $n$-direction along line $OB$, as shown in Fig. 4-25, then

$$\mathbf{M}_{OB} = \mathbf{M}_\parallel = (\mathbf{M}_O \cdot \mathbf{e}_n)\, \mathbf{e}_n$$
$$= [(\mathbf{r} \times \mathbf{F}) \cdot \mathbf{e}_n]\, \mathbf{e}_n = M_{OB}\, \mathbf{e}_n \qquad (4\text{-}16)$$

These two operations, the vector product $\mathbf{r} \times \mathbf{F}$ of the position vector $\mathbf{r}$ and the force $\mathbf{F}$ to obtain the moment $\mathbf{M}_O$ about point $O$, followed by the scalar product $\mathbf{M}_O \cdot \mathbf{e}_n$ of the moment $\mathbf{M}_O$ about point $O$ and the unit vector $\mathbf{e}_n$ along the desired moment axis, to obtain the moment $M_{OB}$ can be performed in sequence or combined into one operation. The quantity inside the brackets is called the triple scalar product. The triple scalar product can be written in determinant form as

$$M_{OB} = \mathbf{M}_O \cdot \mathbf{e}_n = (\mathbf{r} \times \mathbf{F}) \cdot \mathbf{e}_n = \begin{vmatrix} \mathbf{i} & \mathbf{j} & \mathbf{k} \\ r_x & r_y & r_z \\ F_x & F_y & F_z \end{vmatrix} \cdot \mathbf{e}_n \qquad (4\text{-}17)$$

or alternatively as

$$M_{OB} = \mathbf{M}_O \cdot \mathbf{e}_n = (\mathbf{r} \times \mathbf{F}) \cdot \mathbf{e}_n = \begin{vmatrix} e_{nx} & e_{ny} & e_{nz} \\ r_x & r_y & r_z \\ F_x & F_y & F_z \end{vmatrix} \qquad (4\text{-}18)$$

where $e_{nx}$, $e_{ny}$, and $e_{nz}$ are the Cartesian components (direction cosines) of the unit vector $\mathbf{e}_n$. The unit vector $\mathbf{e}_n$ is usually selected in the direction from $O$ toward $B$. A positive coefficient of $\mathbf{e}_n$ in the expression $\mathbf{M}_{OB}$ means that the moment vector has the same sense as that selected for $\mathbf{e}_n$, whereas a negative sign indicates that $\mathbf{M}_{OB}$ is opposite to the sense of $\mathbf{e}_n$.

In the process of determining the moment of a force $\mathbf{F}$ about an arbitrary line $OB$, the position vector $\mathbf{r}$ can be taken from any point on the line $OB$ (the $n$-axis of Fig. 4-26) to any point on the line of action of force $\mathbf{F}$. For example, the position vector $\mathbf{r}_{C/B}$ can be used instead of $\mathbf{r}_{A/O}$ if it is easier to write or more convenient for some other reason. The use of several different position vectors is illustrated in Example 4-15.

The following examples illustrate the use of vector algebra for determining moments about lines in three-dimensional problems.

The force **F** in Fig. 4-27 can be expressed in Cartesian vector form as

$$\mathbf{F} = 60\mathbf{i} + 100\mathbf{j} + 120\mathbf{k} \text{ lb}$$

Determine the moment of force **F** about line $BC$.

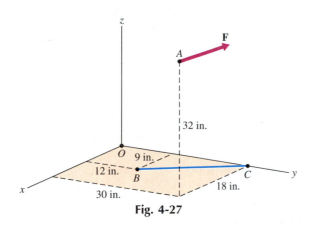

**Fig. 4-27**

### SOLUTION

The moment of force **F** about line $BC$ will be determined by using a two-step process. First, the moment of force **F** about point $B$ will be determined. The position vector $\mathbf{r}_{A/B}$ from point $B$ to point $A$ can be written in Cartesian vector form as

$$\mathbf{r}_{A/B} = (18 - 9)\mathbf{i} + (30 - 12)\mathbf{j} + (32 - 0)\mathbf{k} = 9\mathbf{i} + 18\mathbf{j} + 32\mathbf{k} \text{ in.}$$

The moment of force **F** about point $B$ is then given by the expression

$$\mathbf{M}_B = \mathbf{r}_{A/B} \times \mathbf{F} = \begin{vmatrix} \mathbf{i} & \mathbf{j} & \mathbf{k} \\ 9 & 18 & 32 \\ 60 & 100 & 120 \end{vmatrix} = -1040\mathbf{i} + 840\mathbf{j} - 180\mathbf{k} \text{ in.} \cdot \text{lb}$$

> The moment of force **F** about point $B$ is given by the vector cross product $\mathbf{M}_B = \mathbf{r} \times \mathbf{F}$ where **r** is a position vector from point $B$ to any point on the line of action of force **F**.

A unit vector $\mathbf{e}_{BC}$ along line $BC$ with a sense from $B$ to $C$ is

$$\mathbf{e}_{BC} = \frac{-9\mathbf{i} + 18\mathbf{j}}{\sqrt{405}} = -0.4472\mathbf{i} + 0.8944\mathbf{j}$$

The scalar component of moment $\mathbf{M}_B$ along line $BC$ is then determined by using the dot product

$$M_{BC} = \mathbf{M}_B \cdot \mathbf{e}_{BC} = (-1040\mathbf{i} + 840\mathbf{j} - 180\mathbf{k}) \cdot (-0.4472\mathbf{i} + 0.8944\mathbf{j})$$
$$= 1216.38 \text{ in.} \cdot \text{lb} \cong 1216 \text{ in.} \cdot \text{lb} \qquad \text{Ans.}$$

> The dot product $\mathbf{M}_B \cdot \mathbf{e}_{BC}$ gives the projection or scalar component of moment $\mathbf{M}_B$ along line $BC$.

Finally, the moment of force **F** about line $BC$ can be expressed in Cartesian vector form as

$$\mathbf{M}_{BC} = M_{BC}\mathbf{e}_{BC} = 1216.38(-0.4472\mathbf{i} + 0.8944\mathbf{j})$$
$$= -543.97\mathbf{i} + 1087.93\mathbf{j} \text{ in.} \cdot \text{lb} \cong -544\mathbf{i} + 1088\mathbf{j} \text{ in.} \cdot \text{lb} \qquad \text{Ans.}$$

> Moment $\mathbf{M}_{BC}$ can be written in Cartesian vector form by multiplying the magnitude $M_{BC}$ of the moment (a scalar) by a unit vector $\mathbf{e}_{BC}$ along line $BC$.

The magnitude of force **F** in Fig. 4-28 is 500 N. Determine the scalar component of the moment at point $O$ about line $OC$.

**Fig. 4-28**

## SOLUTION

The moment of force **F** about line $OC$ can be determined by using a two-step process. The $z$-coordinate of point $A$ that will be needed to write a vector equation for force **F** is $z_A = 800 \tan 20° = 291.2$ mm. The force **F** can be written in Cartesian vector form as

$$\mathbf{F} = 500 \left[ \frac{-850\mathbf{i} - 800\mathbf{j} + 458.8\mathbf{k}}{\sqrt{(-850)^2 + (-800)^2 + (458.8)^2}} \right] = -338.86\mathbf{i} - 318.93\mathbf{j} + 182.91\mathbf{k} \text{ N}$$

The position vector $\mathbf{r}_{B/O}$ can be written in Cartesian vector form as

$$\mathbf{r}_{B/O} = -0.850\mathbf{i} + 0.750\mathbf{k} \text{ m}$$

The moment of force **F** about point $O$ is then given by the expression

$$\mathbf{M}_O = \mathbf{r}_{B/O} \times \mathbf{F} = \begin{vmatrix} \mathbf{i} & \mathbf{j} & \mathbf{k} \\ -0.850 & 0 & 0.750 \\ -338.86 & -318.93 & 182.91 \end{vmatrix}$$

$$= 239.20\mathbf{i} - 98.67\mathbf{j} + 271.09\mathbf{k} \text{ N} \cdot \text{m}$$

A unit vector $\mathbf{e}_{OC}$ along line $OC$ with a sense from $O$ to $C$ is

$$\mathbf{e}_{OC} = -1.000\mathbf{i}$$

The scalar component of moment $\mathbf{M}_O$ along line $OC$ is then determined by using the dot product

$$M_{OC} = \mathbf{M}_O \cdot \mathbf{e}_{OC} = (239.20\mathbf{i} - 409.62\mathbf{j}) + (403.93\mathbf{k}) \cdot (-1.00\mathbf{i})$$
$$= -239.20 \text{ N} \cdot \text{m} \cong -239 \text{ N} \cdot \text{m} \qquad \text{Ans.}$$

Alternatively, the scalar moment component $M_{OC}$ can be determined in a single step by using the expression

$$M_{OC} = \begin{vmatrix} e_{nx} & e_{ny} & e_{nz} \\ r_x & r_y & r_z \\ F_x & F_y & F_z \end{vmatrix} = \begin{vmatrix} -1.000 & 0 & 0 \\ -0.850 & 0 & 0.750 \\ -338.86 & -318.93 & 182.91 \end{vmatrix}$$

$$= -239.20 \text{ N} \cdot \text{m} \cong -239 \text{ N} \cdot \text{m} \qquad \text{Ans.}$$

> The moment of force **F** about point $O$ is given by the vector cross product $\mathbf{M}_O = \mathbf{r} \times \mathbf{F}$ where **r** is a position vector from point $O$ to any point on the line of action of force **F**.

> The dot product $\mathbf{M}_O \cdot \mathbf{e}_{OC}$ gives the projection or scalar component of moment $\mathbf{M}_O$ along line $OC$.

The force **F** in Fig. 4-29 has a magnitude of 440 lb. Determine

a.  The moment $\mathbf{M}_B$ of the force about point $B$.
b.  The component of moment $\mathbf{M}_B$ parallel to line $BC$.
c.  The component of moment $\mathbf{M}_B$ perpendicular to line $BC$.
d.  The unit vector associated with the component of moment $\mathbf{M}_B$ perpendicular to line $BC$.

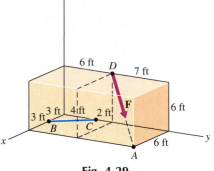

**Fig. 4-29**

## SOLUTION

a.  The force **F** can be written in Cartesian vector form as

$$\mathbf{F} = 440 \left[ \frac{6\mathbf{i} + 7\mathbf{j} - 6\mathbf{k}}{\sqrt{(6)^2 + (7)^2 + (-6)^2}} \right] = 240\mathbf{i} + 280\mathbf{j} - 240\mathbf{k} \text{ lb}$$

The position vector $\mathbf{r}_{A/B}$ can be written in Cartesian vector form as

$$\mathbf{r}_{A/B} = 3\mathbf{i} + 13\mathbf{j} \text{ ft}$$

The moment of force **F** about point $B$ is then given by the expression

$$\mathbf{M}_B = \mathbf{r}_{A/B} \times \mathbf{F} = \begin{vmatrix} \mathbf{i} & \mathbf{j} & \mathbf{k} \\ r_x & r_y & r_z \\ F_x & F_y & F_z \end{vmatrix} = \begin{vmatrix} \mathbf{i} & \mathbf{j} & \mathbf{k} \\ 3 & 13 & 0 \\ 240 & 280 & -240 \end{vmatrix}$$

$$= -3120\mathbf{i} + 720\mathbf{j} - 2280\mathbf{k} \text{ ft} \cdot \text{lb} \qquad \text{Ans.}$$

> The moment of force **F** about point $B$ is given by the vector cross product $\mathbf{M}_B = \mathbf{r} \times \mathbf{F}$ where **r** is a position vector from point $B$ to any point on the line of action of force **F**.

b.  The unit vector $\mathbf{e}_{BC}$ associated with line $BC$ is

$$\mathbf{e}_{BC} = -\frac{3}{5}\mathbf{i} + \frac{4}{5}\mathbf{j} = -0.60\mathbf{i} + 0.80\mathbf{j}$$

The component of moment $\mathbf{M}_B$ parallel to line $BC$ is

$$\mathbf{M}_{BC} = \mathbf{M}_{\parallel} = (\mathbf{M}_B \cdot \mathbf{e}_{BC})\mathbf{e}_{BC} = M_{BC}\mathbf{e}_{BC}$$
$$M_{BC} = \mathbf{M}_B \cdot \mathbf{e}_{BC} = (-3120\mathbf{i} + 720\mathbf{j} - 2280\mathbf{k}) \cdot (-0.60\mathbf{i} + 0.80\mathbf{j})$$
$$= -3120(-0.60) + 720(0.80) = 2448 \text{ ft} \cdot \text{lb}$$
$$\mathbf{M}_{BC} = M_{BC}\mathbf{e}_{BC} = 2448(-0.60\mathbf{i} + 0.80\mathbf{j}) = -1469\mathbf{i} + 1958\mathbf{j} \text{ ft} \cdot \text{lb} \qquad \text{Ans.}$$

> The dot product $\mathbf{M}_B \cdot \mathbf{e}_{BC}$ gives the projection or scalar component of moment $\mathbf{M}_B$ along line $BC$.

> Moment $\mathbf{M}_{BC}$ can be written in Cartesian vector form by multiplying the magnitude $M_{BC}$ of the moment (a scalar) by a unit vector $\mathbf{e}_{BC}$ along line $BC$.

Alternatively, the moment $M_{BC}$ can be determined in a single operation. A different point on the line can also be used. For example, consider point $C$ and the position vector $\mathbf{r}_{D/C}$.

$$r_{D/C} = 2\mathbf{j} + 6\mathbf{k} \text{ ft}$$

In this case,

$$M_{BC} = \begin{vmatrix} e_{nx} & e_{ny} & e_{nz} \\ r_x & r_y & r_z \\ F_x & F_y & F_z \end{vmatrix} = \begin{vmatrix} -0.60 & 0.80 & 0 \\ 0 & 2 & 6 \\ 240 & 280 & -240 \end{vmatrix}$$

$$(-0.60)(-2160) - (0.80)(-1440) = 2448 \text{ ft} \cdot \text{lb}$$

c.  The moment $\mathbf{M}_{\perp}$ is obtained as the difference between $\mathbf{M}_B$ and $\mathbf{M}_{\parallel}$ since $\mathbf{M}_{\parallel}$ and $\mathbf{M}_{\perp}$ are the two rectangular components of $\mathbf{M}_B$. Thus,

$$\mathbf{M}_{\perp} = \mathbf{M}_B - \mathbf{M}_{\parallel} = \mathbf{M}_B - \mathbf{M}_{BC}$$
$$= (-3120\mathbf{i} + 720\mathbf{j} - 2280\mathbf{k}) - (-1469\mathbf{i} + 1958\mathbf{j})$$
$$= -1651\mathbf{i} - 1238\mathbf{j} - 2280\mathbf{k} \text{ ft} \cdot \text{lb} \qquad \text{Ans.}$$

> The component of $\mathbf{M}_B$ perpendicular to line $BC$ is the difference between $\mathbf{M}_B$ and $\mathbf{M}_{BC}$ since $\mathbf{M}_{BC}$ and $\mathbf{M}_{\perp}$ are rectangular components of $\mathbf{M}_B$.

**d.** The magnitude of moment $\mathbf{M}_\perp$ is

$$|\mathbf{M}_\perp| = \sqrt{(-1651)^2 + (-1238)^2 + (-2280)^2} = 3075 \text{ ft} \cdot \text{lb}$$

Therefore

$$\mathbf{e}_\perp = \frac{-1651}{3075}\mathbf{i} + \frac{-1238}{3075}\mathbf{j} + \frac{-2280}{3075}\mathbf{k} = -0.537\mathbf{i} - 0.403\mathbf{j} - 0.741\mathbf{k} \quad \text{Ans.}$$

As a check:

$$\mathbf{e}_\| \cdot \mathbf{e}_\perp = (-0.60\mathbf{i} + 0.80\mathbf{j}) \cdot (-0.5369\mathbf{i} - 0.4026\mathbf{j} - 0.7415\mathbf{k})$$
$$= 0.00006 \cong 0$$

which verifies, except for round-off in the expression for $\mathbf{e}_\perp$, that the moment components $\mathbf{M}_\|$ and $\mathbf{M}_\perp$ are perpendicular.

# PROBLEMS

### Introductory Problems

**4-61\*** The magnitude of the force $\mathbf{F}$ in Fig. P4-61 is 450 lb. Determine the scalar component of the moment at point $B$ about line $BC$.

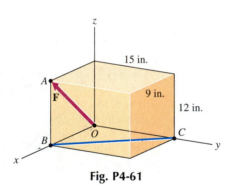

**Fig. P4-61**

**4-62\*** The magnitude of the force $\mathbf{F}$ in Fig. P4-62 is 595 N. Determine the scalar component of the moment at point $O$ about line $OC$.

**Fig. P4-62**

**4-63** The magnitude of the force $\mathbf{F}$ in Fig. P4-63 is 680 lb. Determine the scalar component of the moment at point $O$ about line $OC$.

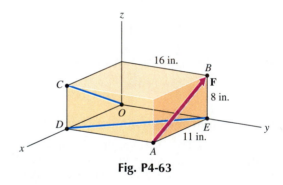

**Fig. P4-63**

**4-64** The magnitude of the force $\mathbf{F}$ in Fig. P4-64 is 635 N. Determine the scalar component of the moment at point $O$ about line $OC$.

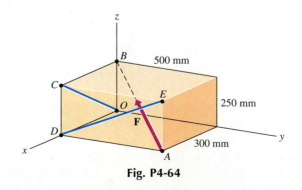

**Fig. P4-64**

144

**4-65\*** The magnitude of the force **F** in Fig. P4-65 is 680 lb. Determine the scalar component of the moment at point $D$ about line $DE$.

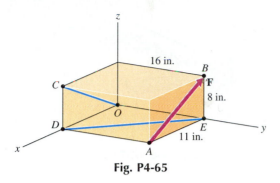

**Fig. P4-65**

**4-66\*** The magnitude of the force **F** in Fig. P4-66 is 635 N. Determine the scalar component of the moment at point $D$ about line $DE$.

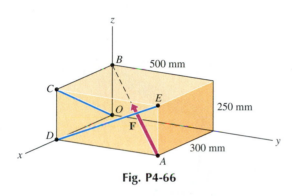

**Fig. P4-66**

**4-67** Determine the scalar component of the moment of the 800-lb force shown in Fig. P4-67 about line $OA$.

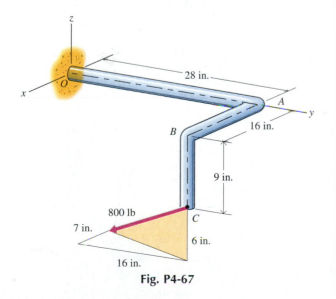

**Fig. P4-67**

**4-68** Determine the scalar component of the moment of the 750-N force shown in Fig. P4-68 about line $OC$.

**Fig. P4-68**

**4-69\*** Determine the scalar component of the moment of the 500-lb force shown in Fig. P4-69 about line $AC$.

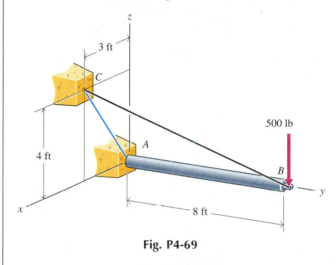

**Fig. P4-69**

**4-70\*** Determine the scalar component of the moment of the 750-N force shown in Fig. P4-70 about the axis of shaft $AB$.

**Fig. P4-70**

**4-71** Determine the scalar component of the moment of the 100-lb force shown in Fig. P4-71 about the axis of the hinges (line *AB*).

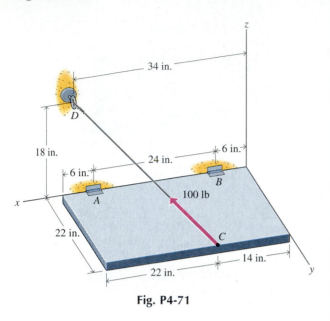

**Fig. P4-71**

## Intermediate Problems

**4-72\*** Determine the moment of the 610-N force shown in Fig. P4-72 about line *CD*. Express the result in Cartesian vector form.

**Fig. P4-72**

**4-73\*** A 200-lb force **F** is applied to a lever-shaft assembly as shown in Fig. P4-73. Determine the moment of the force about line *OC*. Express the result in Cartesian vector form.

**Fig. P4-73**

**4-74** A bracket is subjected to an 825-N force as shown in Fig. P4-74. Determine the moment of the force about line *OB*. Express the result in Cartesian vector form.

**Fig. P4-74**

**4-75** A pipe bracket is subjected to a 200-lb force as shown in Fig. P4-75. Determine the moment of the force about line *BC*. Express the result in Cartesian vector form.

**Fig. P4-75**

**4-76*** A 534-N force **F** is applied to a lever-shaft assembly as shown in Fig. P4-76. Determine the moment of the force about line *OB*. Express the result in Cartesian vector form.

**Fig. P4-76**

## Challenging Problems

**4-77*** A curved bar is subjected to a 660-lb force as shown in Fig. P4-77. Determine the moment of the force about line *BC*. Express the result in Cartesian vector form.

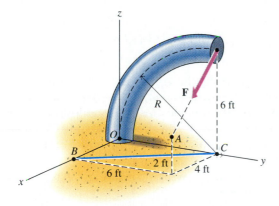

**Fig. P4-77**

**4-78*** A bracket is subjected to a 384-N force as shown in Fig. P4-78. Determine the moment of the force about line *OC*. Express the result in Cartesian vector form.

**Fig. P4-78**

**4-79** The magnitude of force **F** in Fig. P4-79 is 781 lb. Determine

**a.** The component of the moment at point *C* parallel to line *CD*.

**b.** The component of the moment at point *C* perpendicular to line *CD* and the direction angles associated with this moment vector.

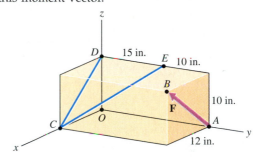

**Fig. P4-79**

**4-80** The magnitude of force **F** in Fig. P4-80 is 976 N. Determine

**a.** The component of the moment at point *C* parallel to line *CE*.

**b.** The component of the moment at point *C* perpendicular to line *CE* and the direction angles associated with this moment vector.

**Fig. P4-80**

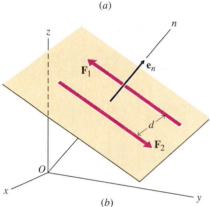

**Figure 4-30**   Moment of two equal, noncollinear, parallel forces of opposite sense.

## 4-4   COUPLES

A couple is a system of forces whose resultant force is zero but whose resultant moment about a point is not zero. A couple can be represented by two equal, noncollinear, parallel forces of opposite sense as shown in Fig. 4-30a. Since the two forces are equal, parallel, and of opposite sense, the sum of the forces in any direction is zero. Therefore, a couple tends only to rotate the body on which it acts. The moment of a couple is defined as the sum of the moments of the pair of forces that comprise the couple. For the two forces $\mathbf{F}_1$ and $\mathbf{F}_2$ shown in Fig. 4-30a, the moments of the couple about points $A$ and $B$ in the plane of the couple are

$$M_A = |\mathbf{F}_2|d \qquad M_B = |\mathbf{F}_1|d$$

However,

$$|\mathbf{F}_1| = |\mathbf{F}_2| = F$$

Therefore,

$$M_A = M_B = Fd$$

which indicates that the magnitude of the moment of a couple about a point in the plane of the couple is equal to the magnitude of one of the forces multiplied by the perpendicular distance between the forces.

Other characteristics of a couple can be determined by considering two parallel forces in space such as those shown in Fig. 4-30a. The sum of the moments of the two forces about any point $O$ is

$$\mathbf{M}_O = \mathbf{r}_1 \times \mathbf{F}_1 + \mathbf{r}_2 \times \mathbf{F}_2$$

or since $\mathbf{F}_2$ equals $-\mathbf{F}_1$

$$\mathbf{M}_O = \mathbf{r}_1 \times \mathbf{F}_1 + \mathbf{r}_2 \times (-\mathbf{F}_1)$$
$$= (\mathbf{r}_1 - \mathbf{r}_2) \times \mathbf{F}_1 = \mathbf{r}_{A/B} \times \mathbf{F}_1$$

where $\mathbf{r}_{A/B}$ is the position vector from any point $B$ on $\mathbf{F}_2$ to any point $A$ on $\mathbf{F}_1$. Therefore, from the definition of the vector cross product

$$\mathbf{M}_O = \mathbf{r}_{A/B} \times \mathbf{F}_1$$
$$= |\mathbf{r}_{A/B}||\mathbf{F}_1|\sin \alpha \, \mathbf{e}_n = F_1 d \, \mathbf{e}_n \tag{4-19}$$

where $d$ is the perpendicular distance between the forces of the couple and $\mathbf{e}_n$ is a unit vector perpendicular to the plane of the couple (see Fig. 4-30b) with its sense in the direction specified for moments by the right-hand rule. Equation 4-19 indicates that the moment of a couple does not depend on the location of the moment center $O$. Thus, the moment of a couple about any point is the same, which indicates that a couple is a free vector.

The characteristics of a couple, which control its "external effect" on a rigid body, are as follows:

1.   The magnitude of the moment of the couple.
2.   The sense (direction of rotation) of the couple.
3.   The orientation of the moment of the couple (axis about which rotation is induced). As indicated in Fig. 4-30b, the orientation of the moment of the couple is specified by providing the aspect of the plane of the couple, that is, the direction or slope of the plane (not its location) as defined by a normal $n$ to the plane.

# CONCEPTUAL EXAMPLE 4-2: MOMENT OF A COUPLE

*A lug wrench is being used to tighten a lug nut on an automobile wheel, as shown in Fig. CE4-2. Two equal, parallel forces of opposite sense are applied to the wrench. Describe the effects of the forces on the wheel.*

## SOLUTION

Two equal, parallel forces of opposite sense are called a couple. Since the sum of the forces in any direction is zero, a couple has no tendency to translate a body in any direction but tends only to rotate the body on which it acts. For example, for the two forces ($F_1 = F_2 = F$) shown in Fig. CE4-2, whose lines of action are located a distance $d/2$ from the axis of the wrench, the moments about points $A$, $B$, $C$, and $D$ are

$$M_A = F_2 d = Fd \downarrow$$
$$M_B = F_1(d/2) + F_2(d/2) = (F_1 + F_2)(d/2) = 2F(d/2) = Fd \downarrow$$
$$M_C = F_1 d = Fd \downarrow$$
$$M_D = F_1(d + a) - F_2 a = F_1 d + (F_1 - F_2)a = F_1 d = Fd \downarrow$$

**Fig. CE4-2**

Clearly, in these four cases, the moment of the couple is the same for the different points about which moments have been determined (axes through the points that are perpendicular to the plane of the couple). A unique feature of a couple is that it has the same moment with respect to every point in space. The moment of a couple is a vector (a free vector) whose direction is normal to the plane of the couple and whose sense is determined in accordance with the right-hand rule by using the twist of the couple to indicate the direction of rotation. The magnitude of the couple moment equals the product of the magnitude of one of the couple forces and the perpendicular distance between the forces ($M = Fd$).

Equation 4-19 indicates that several transformations of a couple can be made without changing any of the external effects of the couple on the body. For example,

1. A couple can be translated to a parallel position in its plane (see Figs. 4-31a and 4-31b) or to any parallel plane (since position vectors $\mathbf{r}_1$ and $\mathbf{r}_2$ do not appear in Eq. 4-19).
2. A couple can be rotated in its plane (see Figs. 4-31a and 4-31c).
3. The magnitude of the two forces of a couple and the distance between them can be changed provided the product $Fd$ remains constant (see Figs. 4-31a and 4-31d).

For two-dimensional problems, a couple is frequently represented by a curved arrow on a sketch of the body as shown in Fig. 4-31e. The

**Figure 4-31**   Equivalent representations of the same couple.

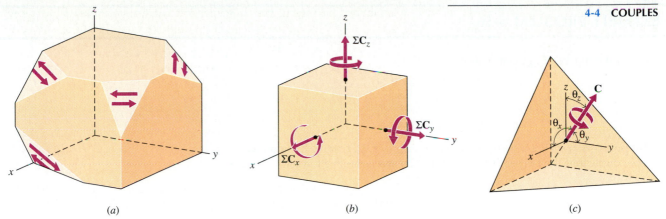

*(a)*      *(b)*      *(c)*

**Figure 4-32** Combining several couples to obtain a resultant couple **C**.

magnitude of the moment of the couple $|\mathbf{M}| = M = Fd$ is provided, and
the curved arrow indicates the sense of the couple. A couple can also
be represented formally as a vector as indicated in Fig. 4-31*f*.

Any number of couples $\mathbf{C}_1, \mathbf{C}_2, \ldots, \mathbf{C}_n$ in a plane can be combined
to yield a resultant couple **C** equal to the algebraic sum of the indi-
vidual couples. A system of couples in space (see Fig. 4-32*a*) can be
combined into a single resultant couple **C** by representing each couple
of the system (since a couple is a free vector) by a vector, drawn for
convenience from the origin of a set of rectangular axes. Each couple
can then be resolved into components $\mathbf{C}_x$, $\mathbf{C}_y$, and $\mathbf{C}_z$ along the coordinate axes. These vector components represent couples lying in the
$yz$-, $zx$-, and $xy$-planes, respectively. The $x$-, $y$-, and $z$-components of the
resultant couple **C** are obtained as $\Sigma\mathbf{C}_x$, $\Sigma\mathbf{C}_y$, and $\Sigma\mathbf{C}_z$ as shown in Fig.
4-32*b*. The original system of couples is thus reduced to three couples
lying in the coordinate planes. The resultant couple **C** for the system
(see Fig. 4-32*c*) can be written in vector form as

$$\mathbf{C} = \Sigma\mathbf{C}_x + \Sigma\mathbf{C}_y + \Sigma\mathbf{C}_z = \Sigma C_x\mathbf{i} + \Sigma C_y\mathbf{j} + \Sigma C_z\mathbf{k} \qquad \textbf{(4-20)}$$

The magnitude of the couple **C** is

$$|\mathbf{C}| = \sqrt{(\Sigma C_x)^2 + (\Sigma C_y)^2 + (\Sigma C_z)^2} \qquad \textbf{(4-21)}$$

Alternatively, the couple **C** can be written as

$$\mathbf{C} = C\mathbf{e} \qquad \textbf{(4-22)}$$

where

$$\mathbf{e} = \cos\theta_x\,\mathbf{i} + \cos\theta_y\,\mathbf{j} + \cos\theta_z\,\mathbf{k}$$

The direction cosines associated with the unit vector **e** are

$$\theta_x = \cos^{-1}\frac{\Sigma C_x}{|\mathbf{C}|} \qquad \theta_y = \cos^{-1}\frac{\Sigma C_y}{|\mathbf{C}|} \qquad \theta_z = \cos^{-1}\frac{\Sigma C_z}{|\mathbf{C}|} \qquad \textbf{(4-23)}$$

Two parallel forces of opposite sense

$$\mathbf{F}_1 = -75\mathbf{i} + 100\mathbf{j} - 200\mathbf{k} \text{ N}$$

and

$$\mathbf{F}_2 = 75\mathbf{i} - 100\mathbf{j} + 200\mathbf{k} \text{ N}$$

act at points $A$ and $O$ of a body as shown in Fig. 4-33. Determine the moment of the couple and the perpendicular distance between the two forces.

**Fig. 4-33**

## SOLUTION

The position vector $\mathbf{r}_{A/O}$ can be written in Cartesian vector form as

$$\mathbf{r}_{A/O} = 0.075\mathbf{i} + 0.090\mathbf{j} + 0.150\mathbf{k} \text{ m}$$

The moment of the couple is obtained from the expression

$$\mathbf{M}_O = \mathbf{r}_{A/O} \times \mathbf{F}_1 = \begin{vmatrix} \mathbf{i} & \mathbf{j} & \mathbf{k} \\ 0.075 & 0.090 & 0.150 \\ -75 & 100 & -200 \end{vmatrix}$$

$$= -33.00\mathbf{i} + 3.75\mathbf{j} + 14.25\mathbf{k} \text{ N} \cdot \text{m} \qquad \text{Ans.}$$

> The moment of force $\mathbf{F}_1$ about point $O$ is given by the vector cross product $\mathbf{M}_O = \mathbf{r} \times \mathbf{F}_1$ where $\mathbf{r}$ is a position vector from point $O$ to any point on the line of action of force $\mathbf{F}_1$.

The magnitude of moment $\mathbf{M}_O$ is

$$M_O = |\mathbf{M}_O| = \sqrt{M_{Ox}^2 + M_{Oy}^2 + M_{Oz}^2}$$
$$= \sqrt{(-33.00)^2 + (3.75)^2 + (14.25)^2} = 36.14 \text{ N} \cdot \text{m}$$

The magnitude of force $\mathbf{F}_1$ is

$$F_1 = |\mathbf{F}_1| = \sqrt{F_{1x}^2 + F_{1y}^2 + F_{1z}^2} = \sqrt{(-75)^2 + (100)^2 + (200)^2} = 235.85 \text{ N}$$

The distance $d$ is obtained by using the definition of a moment. Thus,

$$d = \frac{M}{F} = \frac{36.14}{235.85} = 0.15323 \text{ m} \cong 153.2 \text{ mm} \qquad \text{Ans.}$$

A beam is loaded with a system of forces as shown in Fig. 4-34. Express the resultant of the force system in Cartesian vector form.

**Fig. 4-34**

---

## SOLUTION

An examination of Fig. 4-34 indicates that the force system consists of a system of three couples in the same plane. A scalar analysis yields, for forces $A$ and $D$:

$$M_1 = F_A d_1 = 500(12) = 6000 \text{ in.} \cdot \text{lb} \downarrow = -6000 \text{ in.} \cdot \text{lb}$$

For forces $B$ and $C$, the perpendicular distance between forces $B$ and $C$ is not obvious from Fig. 4-34; therefore, components of the forces will be used.

$$F_{Bx} = F_B \cos 60° = 750 \cos 60° = 375 \text{ lb}$$
$$F_{By} = F_B \sin 60° = 750 \sin 60° = 649.5 \text{ lb}$$
$$M_2 = F_{Bx} d_2 = 375(10) = 3750 \text{ in.} \cdot \text{lb} \downarrow = -3750 \text{ in.} \cdot \text{lb}$$
$$M_3 = F_{By} d_3 = 649.5(12) = 7794 \text{ in.} \cdot \text{lb} \uparrow = +7794 \text{ in.} \cdot \text{lb}$$

For forces $E$ and $F$:

$$M_4 = F_E d_4 = 800(12) = 9600 \text{ in.} \cdot \text{lb} \downarrow = -9600 \text{ in.} \cdot \text{lb}$$

For any number of couples in a plane, the resultant couple **C** is equal to the algebraic sum of the individual couples. Thus,

$$C = \Sigma M = -6000 + 7794 - 3750 - 9600 = -11{,}556 \text{ in.} \cdot \text{lb} = -963 \text{ ft} \cdot \text{lb}$$
$$\mathbf{C} = -963\mathbf{k} \text{ ft} \cdot \text{lb} \qquad \text{Ans.}$$

Alternatively, a vector analysis yields:

$$\mathbf{C} = (\mathbf{r}_{D/A} \times \mathbf{F}_D) + (\mathbf{r}_{B/C} \times \mathbf{F}_B) + (\mathbf{r}_{E/F} \times \mathbf{F}_E)$$
$$= [(12\mathbf{i} - 10\mathbf{j}) \times (-500\mathbf{j})] + [(12\mathbf{i} + 10\mathbf{j}) \times (375\mathbf{i} + 649.5\mathbf{j})]$$
$$+ [(12\mathbf{j}) \times (800\mathbf{i})] = -11{,}556\mathbf{k} \text{ in} \cdot \text{lb} = -963\mathbf{k} \text{ ft} \cdot \text{lb} \qquad \text{Ans.}$$

## EXAMPLE PROBLEM 4-18

The magnitudes of the four couples applied to the block shown in Fig. 4-35 are $|C_1| = 75$ N · m, $|C_2| = 50$ N · m, $|C_3| = 60$ N · m, and $|C_4| = 90$ N · m. Determine the magnitude of the resultant couple **C** and the direction angles associated with the unit vector **e** used to describe the normal to the plane of the resultant couple **C**.

**Fig. 4-35**

### SOLUTION

The $x$-, $y$-, and $z$-components of the resultant couple **C** are

$$\Sigma C_x = C_1 = 75 \text{ N} \cdot \text{m}$$
$$\Sigma C_y = C_2 + C_4 \cos 30° = 50 + 90 \cos 30° = 127.9 \text{ N} \cdot \text{m}$$
$$\Sigma C_z = C_3 + C_4 \sin 30° = 60 + 90 \sin 30° = 105.0 \text{ N} \cdot \text{m}$$

The resultant couple **C** for the system can be written in vector form as

$$\mathbf{C} = \Sigma\mathbf{C}_x + \Sigma\mathbf{C}_y + \Sigma\mathbf{C}_z = 75\mathbf{i} + 127.9\mathbf{j} + 105.0\mathbf{k} \text{ N} \cdot \text{m}$$

The magnitude of the couple **C** is

$$|\mathbf{C}| = \sqrt{(\Sigma C_x)^2 + (\Sigma C_y)^2 + (\Sigma C_z)^2}$$
$$= \sqrt{(75.0)^2 + (127.9)^2 + (105.0)^2} = 181.7 \text{ N} \cdot \text{m} \qquad \text{Ans.}$$

The direction angles are

$$\theta_x = \cos^{-1}\frac{\Sigma C_x}{|\mathbf{C}|} = \cos^{-1}\frac{75.0}{181.7} = 65.6° \qquad \text{Ans.}$$

$$\theta_y = \cos^{-1}\frac{\Sigma C_y}{|\mathbf{C}|} = \cos^{-1}\frac{127.9}{181.7} = 45.3° \qquad \text{Ans.}$$

$$\theta_z = \cos^{-1}\frac{\Sigma C_z}{|\mathbf{C}|} = \cos^{-1}\frac{105.0}{181.7} = 54.7° \qquad \text{Ans.}$$

> A system of couples in space can be combined into a single resultant couple **C**. The $x$-, $y$-, and $z$-components of the resultant couple are the algebraic sums of the $x$-, $y$-, and $z$-components of the individual couples.

---

# PROBLEMS

### Introductory Problems

**4-81\*** Determine the moment of the couple shown in Fig. P4-81 and the perpendicular distance between the two forces.

**Fig. P4-81**

**4-82\*** Determine the moment of the couple shown in Fig. P4-82 and the perpendicular distance between the two forces.

**Fig. P4-82**

**4-83** Two parallel forces of opposite sense, $F_1 = (-70i - 120j - 80k)$ lb and $F_2 = (70i + 120j + 80k)$ lb, act at points $B$ and $A$ of a body as shown in Fig. P4-83. Determine the moment of the couple and the perpendicular distance between the two forces.

**Fig. P4-83**

**4-84** Two parallel forces of opposite sense, $F_1 = (125i + 200j + 250k)$ N and $F_2 = (-125i - 200j - 250k)$ N, act at points $A$ and $B$ of a body as shown in Fig. P4-84. Determine the moment of the couple and the perpendicular distance between the two forces.

**Fig. P4-84**

**4-85*** A bracket is loaded with a system of forces as shown in Fig. P4-85. Express the resultant of the force system in Cartesian vector form.

**Fig. P4-85**

**4-86*** A plate is loaded with a system of forces as shown in Fig. P4-86. Express the resultant of the force system in Cartesian vector form.

**Fig. P4-86**

**4-87** A bracket is loaded with a system of forces as shown in Fig. P4-87. Express the resultant of the force system in Cartesian vector form.

**Fig. P4-87**

**4-88** A plate is loaded with a system of forces as shown in Fig. P4-88. Express the resultant of the force system in Cartesian vector form.

**Fig. P4-88**

**4-89*** Determine the total moment of the two couples shown in Fig. P4-89 and the direction angles associated with the total moment vector.

**Fig. P4-89**

**4-90*** Three couples are applied to a rectangular block as shown in Fig. P4-90. Determine the magnitude of the resultant couple **C** and the direction angles associated with the resultant couple vector.

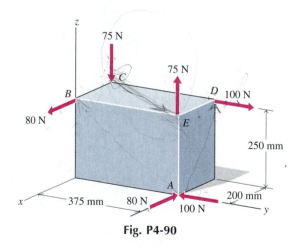

**Fig. P4-90**

**4-91** Three couples are applied to a bent bar as shown in Fig. P4-91. Determine the magnitude of the resultant couple **C** and the direction angles associated with the resultant couple vector.

**Fig. P4-91**

**4-92** The input and output torques from a gear box are shown in Fig. P4-92. Determine the magnitude and direction of the resultant torque **T**.

**Fig. P4-92**

## Challenging Problems

**4-93*** Three couples are applied to a bent bar as shown in Fig. P4-93. Determine

a. The magnitude of the resultant couple **C** and the direction angles associated with the resultant couple vector.

b. The scalar component of the resultant couple **C** about line *OA*.

Fig. P4-93

**4-94*** Three couples are applied to a bent bar as shown in Fig. P4-94. Determine

**a.** The magnitude of the resultant couple **C** and the direction angles associated with the resultant couple vector.

**b.** The scalar component of the resultant couple **C** about line OA.

Fig. P4-94

**4-95** Three couples are applied to a rectangular block as shown in Fig. P4-95. Determine

**a.** The magnitude of the resultant couple **C** and the direction angles associated with the resultant couple vector.

**b.** The scalar component of the resultant couple **C** about line OA.

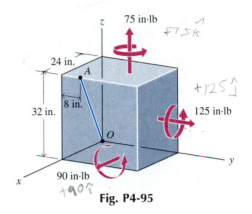

Fig. P4-95

**4-96** Two couples are applied to a rectangular block as shown in Fig. P4-96. Determine

**a.** The magnitude of the resultant couple **C** and the direction angles associated with the resultant couple vector.

**b.** The scalar component of the resultant couple **C** about line AB.

Fig. P4-96

## 4-5 RESOLUTION OF A FORCE INTO A FORCE AND A COUPLE

In many problems in mechanics it is convenient to resolve a force **F** into a parallel force **F** and a couple **C**. Thus, in Fig. 4-36a, let **F** represent a force acting on a body at point A. An arbitrary point O in the body and the plane containing both force **F** and point O are shown shaded in Fig. 4-36a. The aspect of the shaded plane in the body can

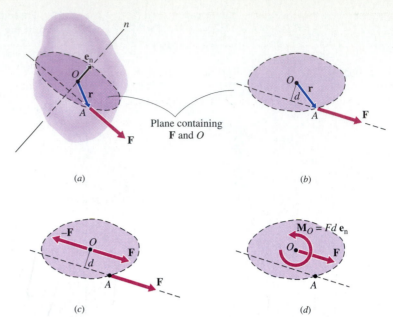

**Figure 4-36**   Resolving a force **F** into a parallel force **F** and a couple **C**.

be described by using its outer normal $n$ and a unit vector **e** along the outer normal. A two-dimensional representation of the shaded plane is shown in Fig. 4-36b. If two equal and opposite collinear forces **F**, parallel to the original force, are introduced at point $O$, as shown in Fig. 4-36c, the three forces have the same external effect on the body as the original force **F**, since the effects of the two equal, opposite, and collinear forces cancel. The new three-force system can be considered to be a force **F** acting at point $O$ (parallel to the original force and of the same magnitude and sense), and a couple **C** that has the same moment as the moment of the original force about point $O$, as shown in Fig. 4-36d. The magnitudes and action lines of the forces of this couple, however, may be changed in accordance with the transformations of a couple discussed in Section 4-4.

Since a force can be resolved into a force and a couple lying in the same plane, it follows, conversely, that a force and a couple lying in the same plane can be combined into a single force in the plane by reversing the procedure illustrated in Fig. 4-36. Thus, the sole external effect of combining a couple with a force is to move the action line of the force into a parallel position. The magnitude and sense of the force remain unchanged.

The procedure for transforming a force into a force **F** and a couple **C** is illustrated in the following examples.

*A lug wrench is being used to tighten a lug nut on an automobile wheel, as shown in Fig. CE4-3a. Replace the force F applied by the mechanic's hand to the lug wrench by an equivalent force/couple system at the lug nut.*

(c)

(a)    (b)

**Fig. CE4-3**

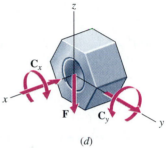

(d)

---

## SOLUTION

Two force systems are equivalent if they exert the same push or pull on a body and produce the same moment about any point. Often it is necessary in the solution of a mechanics problem to move the line of action of a force to a parallel position. This can be accomplished by using the theory of couples to yield an equivalent force/couple system.

The line of action of the force applied to the wrench by the mechanic's hand can be transferred to a parallel line that passes through the lug nut by introducing two equal, opposite, collinear forces ($\mathbf{F}$ and $-\mathbf{F}$) at the nut, as shown in Fig. CE4-3b. These two forces at the nut have the same magnitude as the original force and lines of action parallel to it. The couple formed by the force $-\mathbf{F}$ at the nut and the force $\mathbf{F}$ applied by the mechanic's hand can be translated to a parallel position and represented as a couple $\mathbf{C}$ about an axis $n$ through the nut (see Fig. CE4-3b) that is perpendicular to the plane of the three forces. Thus, as shown in Fig. CE4-3c, the effect of the force $\mathbf{F}$ exerted on the lug wrench by the mechanic's hand is to produce a tendency of the nut to translate (caused by force $\mathbf{F}$) and a tendency to rotate (caused by couple $\mathbf{C}$) about an axis $n$ through point $O$ that is not the axis of the nut. Couple $\mathbf{C}$ is the moment of force $\mathbf{F}$ of Fig. CE4-2a about point $O$. Couple $\mathbf{C}$ is a vector quantity; therefore, it can be resolved (see Fig. CE4-3d) into a component $\mathbf{C}_x$ that tends to rotate the nut and a component $\mathbf{C}_y$ that tends to deform the nut if forces become excessive. Couples $\mathbf{C}_x$ and $\mathbf{C}_y$ are rectangular components of couple $\mathbf{C}$ about lines $Ox$ and $Oy$, respectively.

A 300-lb force $\mathbf{F}_A$ is applied to a bracket at point $A$ as shown in Fig. 4-37$a$. Replace the force $\mathbf{F}_A$ by a force $\mathbf{F}_O$ and a couple $\mathbf{C}$ at point $O$.

Fig. 4-37

## SOLUTION

Force $\mathbf{F}_O$ has the same magnitude and sense as force $\mathbf{F}_A$. Couple $\mathbf{C}$ has the same magnitude and sense as the moment of force $\mathbf{F}_A$ about point $O$. Thus,

$$F_O = F_A = 300 \text{ lb}$$

The magnitude and sense of the couple is obtained by using the definition of a moment. Thus, as shown in Fig. 4-37$b$, the perpendicular distance $d_{A/O}$ from point $O$ to the line of action of the force $\mathbf{F}$ is

$$d_{A/O} = 20 \sin 25° + 30 \cos 25° = 35.64 \text{ in.}$$

The couple $C$ is then determined as

$$C = |\mathbf{F}_A|\, d_{A/O} = 300(35.64) = 10,692 \text{ in.} \cdot \text{lb} \cong 10,690 \text{ in.} \cdot \text{lb}$$

Therefore, as shown in Fig. 4-28$c$,

$$\mathbf{F}_O = 300 \text{ lb} \searrow 25° \qquad \text{Ans.}$$
$$\mathbf{C} = 10,690 \text{ in.} \cdot \text{lb} \uparrow \qquad \text{Ans.}$$

Alternatively, the force $\mathbf{F}_O$ and couple $\mathbf{C}$ can be expressed in Cartesian vector form as

$$\mathbf{F}_O = \mathbf{F}_A = F_A \mathbf{e}_A = 300(-\cos 25°\mathbf{i} + \sin 25°\mathbf{j})$$
$$= -271.89\mathbf{i} + 126.79\mathbf{j} \text{ lb}$$
$$\cong -272\mathbf{i} + 126.8\mathbf{j} \text{ lb} \qquad \text{Ans.}$$
$$\mathbf{C} = \mathbf{r}_{A/O} \times \mathbf{F}_A = (20\mathbf{i} + 30\mathbf{j}) \times (-271.89\mathbf{i} + 126.8\mathbf{j})$$
$$= 10,690\mathbf{k} \text{ in.} \cdot \text{lb} \qquad \text{Ans.}$$

> Any force $\mathbf{F}$ can be moved to a parallel position through an arbitrary point $O$ if a couple $\mathbf{C}$ is added that has a moment equal to the moment of the original force $\mathbf{F}$ about point $O$.

The force **F** shown in Fig. 4-38a has a magnitude of 763 N. Replace the force **F** by a force **F**$_O$ at point $O$ and a couple **C**.

a. Express the force **F**$_O$ and the couple **C** in Cartesian vector form.
b. Determine the direction angles $\theta_x$, $\theta_y$, and $\theta_z$ associated with the couple vector.

(a)                                    (b)

**Fig. 4-38**

## SOLUTION

a. The force **F**$_A$ that is equal to force **F** shown in Fig. 4-38a can be expressed in Cartesian vector form as

$$\mathbf{F}_A = \mathbf{F} = Fe_F = 763\left[\frac{100-50}{\sqrt{23,300}}\mathbf{i} + \frac{200-80}{\sqrt{23,300}}\mathbf{j} + \frac{0-80}{\sqrt{23,300}}\mathbf{k}\right]$$
$$= 250\mathbf{i} + 600\mathbf{j} - 400\mathbf{k}\ \text{N} \qquad \text{Ans.}$$

> A force **F** can be written in Cartesian vector form by multiplying the magnitude of the force (a scalar) by a unit vector along the line of action of the force.

The moment of the force **F** about point $O$ is

$$\mathbf{C} = \mathbf{r}_{A/O} \times \mathbf{F} = \begin{vmatrix} \mathbf{i} & \mathbf{j} & \mathbf{k} \\ r_x & r_y & r_z \\ F_x & F_y & F_z \end{vmatrix} = \begin{vmatrix} \mathbf{i} & \mathbf{j} & \mathbf{k} \\ 0.05 & 0.08 & 0.08 \\ 250 & 600 & -400 \end{vmatrix}$$

$$= -80.0\mathbf{i} + 40.0\mathbf{j} + 10.0\mathbf{k}\ \text{N} \cdot \text{m} \qquad \text{Ans.}$$

> The moment of force **F** about point $O$ is given by the vector cross product $\mathbf{M}_O = \mathbf{r} \times \mathbf{F}$ where **r** is a position vector from point $O$ to any point on the line of action of force **F**.

b. The magnitude of the couple **C** is

$$|\mathbf{C}| = \sqrt{C_x^2 + C_y^2 + C_z^2} = \sqrt{(-80.0)^2 + (40.0)^2 + (10.0)^2} = 90.0\ \text{N} \cdot \text{m}$$

The direction angles are

$$\theta_x = \cos^{-1}\frac{M_x}{|\mathbf{C}|} = \cos^{-1}\frac{-80.0}{90.0} = 152.7° \qquad \text{Ans.}$$

$$\theta_y = \cos^{-1}\frac{M_y}{|\mathbf{C}|} = \cos^{-1}\frac{40.0}{90.0} = 63.6° \qquad \text{Ans.}$$

$$\theta_z = \cos^{-1}\frac{M_z}{|\mathbf{C}|} = \cos^{-1}\frac{10.0}{90.0} = 83.6° \qquad \text{Ans.}$$

The force **F** and the couple **C** are shown in Fig. 4-38b.

# PROBLEMS

## Introductory Problems

**4-97\*** Replace the 50-lb force shown in Fig. P4-97 by a force at point $A$ and a couple. Express your answer in Cartesian vector form.

**Fig. P4-97**

**4-98\*** Replace the 300-N force shown in Fig. P4-98 by a force at point $B$ and a couple. Express your answer in Cartesian vector form.

**Fig. P4-98**

**4-99** Replace the 275-lb force shown in Fig. P4-99 by a force at point $A$ and a couple. Express your answer in Cartesian vector form.

**Fig. P4-99**

**4-100\*** Replace the 675-N force shown in Fig. P4-100 by a force at point $B$ and a couple. Express your answer in Cartesian vector form.

**Fig. P4-100**

**4-101** Replace the 300-lb force shown in Fig. P4-101 by a force at point $B$ and a couple. Express your answer in Cartesian vector form.

**Fig. P4-101**

**4-102** Replace the 600-N force shown in Fig. P4-102 by a force at point $A$ and a couple. Express your answer in Cartesian vector form.

**Fig. P4-102**

**4-103*** Replace the 900-lb force shown in Fig. P4-103 by a force at point B and a couple. Express your answer in Cartesian vector form.

**Fig. P4-103**

**4-104** Replace the 350-N force shown in Fig. P4-104 by a force at point B and a couple. Express your answer in Cartesian vector form.

**Fig. P4-104**

## Intermediate Problems

**4-105*** Replace the 50-lb force shown in Fig. P4-105 by a force at point B and a couple. Express your answer in Cartesian vector form.

**Fig. P4-105**

**4-106*** Replace the 300-N force shown in Fig. P4-106 by a force at point B and a couple. Express your answer in Cartesian vector form.

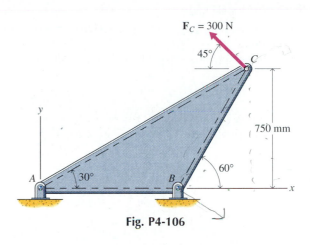

**Fig. P4-106**

**4-107** Replace the 300-lb force shown in Fig. P4-107 by a force at point A and a couple. Express your answer in Cartesian vector form.

**Fig. P4-107**

**4-108** Replace the 450-N force shown in Fig. P4-108 by a force at point A and a couple. Express your answer in Cartesian vector form.

**Fig. P4-108**

## Challenging Problems

**4-109\*** The force **F** shown in Fig. P4-109 has a magnitude of 580 lb. Replace the force **F** by a force $\mathbf{F}_B$ at point $B$ and a couple **C**.

**a.** Express the force $\mathbf{F}_B$ and the couple **C** in Cartesian vector form.

**b.** Determine the direction angles associated with the couple vector.

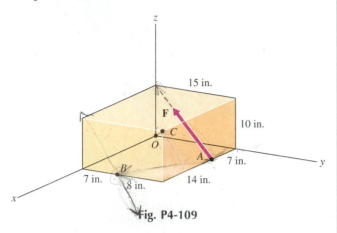

**Fig. P4-109**

**4-110\*** The force **F** shown in Fig. P4-110 has a magnitude of 494 N. Replace the force **F** by a force $\mathbf{F}_C$ at point $C$ and a couple **C**.

**a.** Express the force $\mathbf{F}_C$ and the couple **C** in Cartesian vector form.

**b.** Determine the direction angles associated with the couple vector.

**Fig. P4-110**

**4-111** Replace the 660-lb force shown in Fig. P4-111 by a force at point $B$ and a couple. Express your answer in Cartesian vector form.

**Fig. P4-111**

**4-112** Replace the 500-N force shown in Fig. P4-112 by a force at hinge $C$ and a couple. Express your answer in Cartesian vector form.

**Fig. P4-112**

## 4-6 SIMPLIFICATION OF A FORCE SYSTEM: RESULTANTS

Two force systems are equivalent if they produce the same external effect when applied to a rigid body. The "resultant" of any force system is the simplest equivalent system to which the given system will reduce. For some systems, the resultant is a single force. For other systems, the simplest equivalent system is a couple. Still other force systems reduce to a force and a couple as the simplest equivalent system.

## 4-6.1 Coplanar Force Systems

The resultant of a system of coplanar forces $\mathbf{F}_1, \mathbf{F}_2, \mathbf{F}_3, \ldots, \mathbf{F}_n$ can be determined by using the rectangular components of the forces in any two convenient perpendicular directions. Thus, for a system of forces in the $xy$-plane, such as the one shown in Fig. 4-39$a$, the resultant force $\mathbf{R}$ can be expressed as

$$\mathbf{R} = \mathbf{R}_x + \mathbf{R}_y = R_x\mathbf{i} + R_y\mathbf{j} = R\mathbf{e}$$

where

$$R_x = \Sigma F_x \qquad R_y = \Sigma F_y$$
$$R = |\mathbf{R}| = \sqrt{(\Sigma F_x)^2 + (\Sigma F_y)^2} \qquad \text{(4-24)}$$
$$\mathbf{e} = \cos\theta_x\,\mathbf{i} + \cos\theta_y\,\mathbf{j}$$
$$\cos\theta_x = \frac{\Sigma F_x}{R} \qquad \cos\theta_y = \frac{\Sigma F_y}{R}$$

If force components other than rectangular components are used, the law of cosines and the law of sines must be used to determine the magnitude and direction of the resultant since Eqs. 4-24 apply only to rectangular components.

The location of the line of action of $\mathbf{R}$ with respect to an arbitrary point $O$ (say the origin of the $xy$-coordinate system) can be computed by using the principle of moments. The moment of $\mathbf{R}$ about point $O$ must equal the sum of the moments of the forces $\mathbf{F}_1, \mathbf{F}_2, \mathbf{F}_3, \ldots, \mathbf{F}_n$ of the original system about the same point $O$, as shown in Fig. 4-39$b$. Thus,

$$Rd_R = F_1d_1 + F_2d_2 + F_3d_3 + \cdots + F_nd_n = \Sigma M_O$$

Therefore

$$d_R = \frac{\Sigma M_O}{R} \qquad \text{(4-25)}$$

where $\Sigma M_O$ stands for the algebraic sum of the moments of the forces of the original system about point $O$. The direction of $d_R$ is selected so that the product $Rd_R$ produces a moment about point $O$ with the same sense (clockwise or counterclockwise) as the algebraic sum of the moments of the forces of the original system about point $O$.

The location of the line of action of $\mathbf{R}$ with respect to point $O$ can also be specified by determining the intercept of the line of action of the force with one of the coordinate axes. For example in Fig. 4-39$c$, the intercept $x_R$ is determined from the equation

$$x_R = \frac{\Sigma M_O}{R_y} \qquad \text{(4-26)}$$

since the component $R_x$ of the resultant force $\mathbf{R}$ does not produce a moment about point $O$. The special case for a system of coplanar parallel forces is illustrated in Fig. 4-40.

In the event that the resultant force $\mathbf{R}$ of a system of coplanar forces $\mathbf{F}_1, \mathbf{F}_2, \ldots, \mathbf{F}_n$ is zero but the moment $\Sigma M_O$ is not zero, the resultant is a couple $\mathbf{C}$ whose vector is perpendicular to the plane containing the forces (the $xy$-plane for the case being discussed). Thus, the resultant of a coplanar system of forces may be either a force $\mathbf{R}$ or a couple $\mathbf{C}$.

(a)

(b)

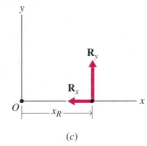

(c)

**Figure 4-39** Resultant $\mathbf{R}$ for a coplanar system of forces.

(a)

(b)

**Figure 4-40** Resultant $\mathbf{R}$ for a coplanar system of parallel forces.

Four forces are applied to a rectangular plate as shown in Fig. 4-41a. Determine the magnitude and direction of the resultant of the four forces.

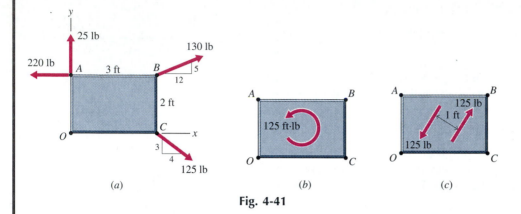

Fig. 4-41

## SOLUTION

The resultant of a system of coplanar forces is either a force **R** or a couple **C**. The resultant force is

$$R_x = \Sigma F_x = -220 + (12/13)(130) + (4/5)(125) = 0$$
$$R_y = \Sigma F_y = 25 + (5/13)(130) - (3/5)(125) = 0$$
$$\mathbf{R} = R_x\mathbf{i} + R_y\mathbf{j} = \mathbf{0} \qquad \text{Ans.}$$

> The $x$- and $y$-components of the resultant of any number of coplanar forces are the algebraic sums of the $x$- and $y$-components of the individual forces.

The couple **C,** if it exists, can be determined by summing moments about any convenient point in the plate. An examination of the force system indicates that the determination can be simplified by summing moments about point $B$. Thus,

$$\begin{aligned}
\curvearrowright +\Sigma M_B &= -F_{Ay}d_A + F_{Cx}d_C \\
&= -25(3) + (4/5)(125)(2) = 125 \text{ ft} \cdot \text{lb} \qquad \text{Ans.}
\end{aligned}$$

Alternatively, the moments can be summed about point $O$. Thus,

$$\begin{aligned}
\curvearrowright +\Sigma M_O &= F_{Ax}d_{Ax} - F_{Bx}d_{Bx} + F_{By}d_{By} - F_{Cy}d_{Cy} \\
&= 220(2) - (12/13)(130)(2) + (5/13)(130)(3) - (3/5)(125)(3) \\
&= 125 \text{ ft} \cdot \text{lb} \qquad \text{Ans.}
\end{aligned}$$

Since the resultant of the force system is a couple, it can be represented by a curved arrow at any point on the plate, as illustrated in Fig. 4-41b, or by a pair of equal but opposite parallel forces, as illustrated in Fig. 4-41c.

> The principle of moments provides a more convenient method for analysis than direct application of the definition of a moment ($M = Fd$) since the moment arm for each component force is easier to establish.

## EXAMPLE PROBLEM 4-22

Three forces and a couple are applied to a bracket as shown in Fig. 4-42a. Determine

a. The magnitude and direction of the resultant.
b. The perpendicular distance $d_R$ from point $O$ to the line of action of the resultant.
c. The distance $x_R$ from point $O$ to the intercept of the line of action of the resultant with the $x$-axis.

(a)

### SOLUTION

a. The resultant of a system of coplanar forces is either a force **R** or a couple **C**. The resultant force is

$$R_x = \Sigma F_x = 500 \cos 60° + 300 = 550 \text{ N}$$
$$R_y = \Sigma F_y = 500 \sin 60° + 200 = 633 \text{ N}$$
$$R = |\mathbf{R}| = \sqrt{(\Sigma F_x)^2 + (\Sigma F_y)^2}$$
$$= \sqrt{(550)^2 + (633)^2} = 838.6 \text{ N} = 839 \text{ N} \qquad \text{Ans.}$$
$$\theta_x = \cos^{-1} \frac{\Sigma F_x}{R} = \cos^{-1} \frac{550}{838.6} = 49.0° \qquad \text{Ans.}$$

b. The line of action of the resultant force **R** with respect to point $O$ can be located by summing moments about point $O$. Thus,

$$\curvearrowleft + \Sigma M_O = -300(0.400) - 500 \cos 60° \, (0.700) - 200(0.450) + 600 = 215 \text{ N} \cdot \text{m}$$

From the definition of a moment,

$$\Sigma M_O = R d_R$$
$$d_R = \frac{\Sigma M_O}{R} = \frac{215}{838.6} = 0.2564 \text{ m} \cong 256 \text{ mm} \qquad \text{Ans.}$$

c. At the intercept of the line of action of resultant **R** with the $x$-axis, the rectangular component $R_x$ produces no moment about point $O$. Thus, from the definition of a moment,

$$\Sigma M_O = R_y x_R$$
$$x_R = \frac{\Sigma M_O}{R_y} = \frac{215}{633} = 0.3397 \text{ m} \cong 340 \text{ mm} \qquad \text{Ans.}$$

The results are illustrated in Fig. 4-42b.

(b)

**Fig. 4-42**

The $x$- and $y$-components of the resultant of any number of coplanar forces are the algebraic sums of the $x$- and $y$-components of the individual forces.

The principle of moments provides a more convenient method for analysis than direct application of the definition of a moment ($M = Fd$) since the moment arm for each component force is easier to establish.

# PROBLEMS

## Introductory Problems

**4-113*** Determine the magnitude and direction of the resultant of the two forces shown in Fig. P4-113 and the perpendicular distance $d_R$ from point $A$ to the line of action of the resultant.

**Fig. P4-113**

**4-114*** Determine the magnitude and direction of the resultant of the two forces shown in Fig. P4-114 and the perpendicular distance $d_R$ from point $A$ to the line of action of the resultant.

**Fig. P4-114**

**4-115** Replace the three forces shown in Fig. P4-115 by an equivalent force-couple system at point $B$.

**Fig. P4-115**

**4-116** Replace the three forces shown in Fig. P4-116 by an equivalent force-couple system at point $O$.

**Fig. P4-116**

**4-117*** Three forces are applied to a beam as shown in Fig. P4-117. Determine the resultant **R** of the three forces and the location of its line of action with respect to support $A$.

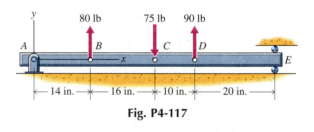

**Fig. P4-117**

**4-118*** Four forces are applied to a truss as shown in Fig. P4-118. Determine the magnitude and direction of the resultant of the four forces and the perpendicular distance $d_R$ from point $A$ to the line of action of the resultant.

**Fig. P4-118**

## Intermediate Problems

**4-119\*** Replace the three forces shown in Fig. P4-119 by an equivalent force-couple system at point *B*.

**Fig. P4-119**

**4-120\*** Replace the three forces shown in Fig. P4-120 by an equivalent force-couple system at point *A*.

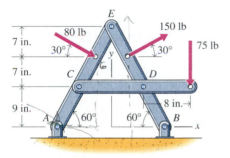

**Fig. P4-120**

**4-121** Replace the three forces shown in Fig. P4-121 by an equivalent force-couple system at point *A*.

**Fig. P4-121**

**4-122** Replace the three forces shown in Fig. P4-122 by an equivalent force-couple system at point *C*.

**Fig. P4-122**

## Challenging Problems

**4-123\*** Replace the three forces shown in Fig. P4-123 by an equivalent force-couple system at point *O*.

**Fig. P4-123**

**4-124\*** Four forces are applied to a post as shown in Fig. P4-124. Determine the resultant **R** of the four forces and the location of its line of action with respect to support *E*.

**Fig. P4-124**

**4-125** Four forces and a couple are applied to a rectangular plate as shown in Fig. P4-125. Determine the magnitude and direction of the resultant of the force-couple system and the distance $x_R$ from point $O$ to the intercept of the line of action of the resultant with the $x$-axis.

Fig. P4-125

**4-126** Four forces and a couple are applied to a frame as shown in Fig. P4-126. Determine the magnitude and direction of the resultant and the perpendicular distance $d_R$ from point $A$ to the line of action of the resultant.

Fig. P4-126

## 4-6.2 Noncoplanar, Parallel Force Systems

If all forces of a three-dimensional system are parallel, the resultant force **R** is the algebraic sum of the forces of the system. The line of action of the resultant is determined by using the principle of moments. Thus, as shown in Fig. 4-43a, for a system of forces $\mathbf{F}_1$, $\mathbf{F}_2$, ..., $\mathbf{F}_n$, perpendicular to the $xy$-plane,

$$\mathbf{R} = \mathbf{F}_1 + \mathbf{F}_2 + \cdots + \mathbf{F}_n = R\mathbf{k} = \Sigma F\,\mathbf{k} \tag{4-27}$$
$$\mathbf{M}_O = \mathbf{r} \times \mathbf{R} = \mathbf{r}_1 \times \mathbf{F}_1 + \mathbf{r}_2 \times \mathbf{F}_2 + \mathbf{r}_3 \times \mathbf{F}_3 + \cdots + \mathbf{r}_n \times \mathbf{F}_n \tag{4-28}$$

The intersection of the line of action of the resultant force **R** with the $xy$-plane (see Fig. 4-43b) is located by equating the moments of **R** about the $x$- and $y$-axes to the sums of the moments of the forces of the system $\mathbf{F}_1$, $\mathbf{F}_2$, . . . , $\mathbf{F}_n$ about the $x$- and $y$-axes. It is important to note from Eqs. 4-10 that $M_x = F_z r_y$ while $M_y = -F_z r_x$. Therefore,

$$Rx_R = F_1 x_1 + F_2 x_2 + \cdots + F_n x_n = -\Sigma M_y$$
$$Ry_R = F_1 y_1 + F_2 y_2 + \cdots + F_n y_n = \Sigma M_x$$

which gives

$$x_R = -\frac{\Sigma M_y}{R} \qquad y_R = \frac{\Sigma M_x}{R} \tag{4-29}$$

The signs (+ and −) for $M_x$ and $M_y$ were determined by using the right-hand rule. In the event that the resultant force **R** of the system of parallel forces is zero but the moments $\Sigma M_x$ and $\Sigma M_y$ are not zero, the resultant is a couple **C** whose vector lies in a plane perpendicular to the forces (in the $xy$-plane for the system of forces illustrated in Fig. 4-43). Thus, the resultant of a noncoplanar system of parallel forces may be either a force **R** or a couple **C**. The following example illustrates the procedure for determining resultants of noncoplanar, parallel force systems.

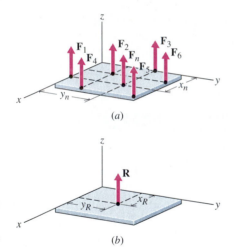

**Figure 4-43** Resultant **R** for a noncoplanar system of parallel forces.

Determine the resultant of the parallel force system shown in Fig. 4-44a and locate the intersection of the line of action of the resultant with the $xy$-plane.

(a)                                     (b)

**Fig. 4-44**

## SOLUTION

The resultant **R** of the force system is

$$\mathbf{R} = \Sigma F\, \mathbf{k}$$
$$= (75 + 30 - 80 - 50 + 120)\mathbf{k} = 95\mathbf{k}\ \text{lb} \qquad \text{Ans.}$$

> The resultant of a number of noncoplanar parallel forces is the algebraic sum of the individual forces.

The intersection $(x_R, y_R)$ of the line of action of the resultant with the $xy$-plane is determined by summing moments about the $x$- and $y$-axes. With positive moments defined by using the right-hand rule

$$\Sigma M_y = -75(4) - 30(6) + 80(2) + 50(4) - 120(2) = -360\ \text{ft} \cdot \text{lb}$$
$$\Sigma M_x = 75(0) + 30(2) - 80(2) - 50(4) + 120(6) = 420\ \text{ft} \cdot \text{lb}$$

Therefore, as shown in Fig. 4-44b,

$$x_R = -\frac{\Sigma M_y}{R} = -\frac{-360}{95} = 3.79\ \text{ft} \qquad \text{Ans.}$$

$$y_R = \frac{\Sigma M_x}{R} = \frac{420}{95} = 4.42\ \text{ft} \qquad \text{Ans.}$$

The coordinates $(x_R, y_R)$ can also be determined by using a vector analysis:

$$\mathbf{M}_O = \mathbf{r} \times \mathbf{R} = \mathbf{r}_1 \times \mathbf{F}_1 + \mathbf{r}_2 \times \mathbf{F}_2 + \mathbf{r}_3 \times \mathbf{F}_3 + \mathbf{r}_4 \times \mathbf{F}_4 + \mathbf{r}_5 \times \mathbf{F}_5$$
$$= [(x_R\mathbf{i} + y_R\mathbf{j}) \times (95\mathbf{k})]$$
$$= [(4\mathbf{i}) \times (75\mathbf{k})] + [(6\mathbf{i} + 2\mathbf{j}) \times (30\mathbf{k})] + [(2\mathbf{i} + 2\mathbf{j}) \times (-80\mathbf{k})]$$
$$\quad + [(4\mathbf{i} + 4\mathbf{j}) \times (-50\mathbf{k})] + [(2\mathbf{i} + 6\mathbf{j}) \times (120\mathbf{k})]$$
$$= -95x_R\mathbf{j} + 95y_R\mathbf{i} = 420\mathbf{i} - 360\mathbf{j}$$

> The moment of the resultant **R** about any axis equals the algebraic sum of the moments of the individual forces about the same axis.

Solving for $x_R$ and $y_R$ yields

$$x_R = \frac{360}{95} = 3.79\ \text{ft} \qquad \text{Ans.}$$

$$y_R = \frac{420}{95} = 4.42\ \text{ft} \qquad \text{Ans.}$$

# PROBLEMS

## Introductory Problems

**4-127*** Determine the resultant of the parallel force system shown in Fig. P4-127 and locate the intersection of the line of action of the resultant with the *xy*-plane.

**Fig. P4-127**

**4-128*** Determine the resultant of the parallel force system shown in Fig. P4-128 and locate the intersection of the line of action of the resultant with the *xy*-plane.

**Fig. P4-128**

**4-129** Determine the resultant of the parallel force system shown in Fig. P4-129 and locate the intersection of the line of action of the resultant with the *xy*-plane.

**Fig. P4-129**

**4-130** Determine the resultant of the parallel force system shown in Fig. P4-130 and locate the intersection of the line of action of the resultant with the *xy*-plane.

**Fig. P4-130**

**4-131*** Determine the resultant of the parallel force system shown in Fig. P4-131 and locate the intersection of the line of action of the resultant with the *yz*-plane.

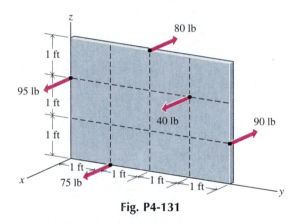

**Fig. P4-131**

**4-132** Determine the resultant of the parallel force system shown in Fig. P4-132 and locate the intersection of the line of action of the resultant with the *xz*-plane.

**Fig. P4-132**

## Challenging Problems

**4-133\*** The resultant of the parallel force system shown in Fig. P4-133 is a couple, which can be expressed in Cartesian vector form as

$$C = (-1160j + 2250k) \text{ in.} \cdot \text{lb}$$

Determine the magnitudes of forces $F_1$, $F_2$, and $F_3$.

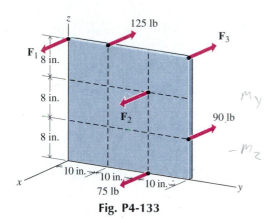

**Fig. P4-133**

**4-134\*** The resultant of the parallel force system shown in Fig. P4-134 is a couple, which can be expressed in Cartesian vector form as

$$C = (-180i + 435j) \text{ N} \cdot \text{m}$$

Determine the magnitudes of forces $F_1$, $F_2$, and $F_3$.

**Fig. P4-134**

**4-135** The resultant of the parallel force system shown in Fig. P4-135 is a couple, which can be expressed in Cartesian vector form as

$$C = (850j - 950k) \text{ in.} \cdot \text{lb}$$

Determine the magnitude and sense of each of the unknown forces ($F_1$, $F_2$, and $F_3$).

**Fig. P4-135**

**4-136** The resultant of the parallel force system shown in Fig. P4-136 is a couple, which can be expressed in Cartesian vector form as

$$C = (200i + 350j) \text{ N} \cdot \text{m}$$

Determine the magnitude and sense of each of the unknown forces ($F_1$, $F_2$, and $F_3$).

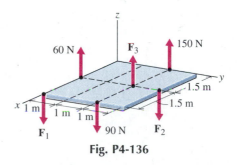

**Fig. P4-136**

## 4-6.3 General Force Systems

The resultant **R** of a general, three-dimensional system of forces $F_1$, $F_2$, $F_3$, . . . , $F_n$, such as the one shown in Fig. 4-45a, can be determined by resolving each force of the system into an equal parallel force through any point (taken for convenience as the origin $O$ of a system of coordinate axes) and a couple as shown in Fig. 4-45b. Thus, as shown in Fig. 4-45c, the given system is replaced by two systems:

(a)

(b)

(c)

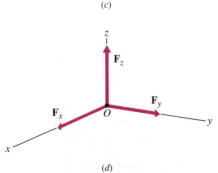

(d)

1. A system of noncoplanar, concurrent forces through the origin $O$ that have the same magnitudes and directions as the forces of the original system.
2. A system of noncoplanar couples.

Each force and couple of the two systems can be resolved into components along the coordinate axes as shown in Figs. 4-45d and 4-45e. The resultant of the concurrent force system is a force through the origin $O$ that can be expressed as

$$\mathbf{R} = \mathbf{R}_x + \mathbf{R}_y + \mathbf{R}_z = R_x\mathbf{i} + R_y\mathbf{j} + R_z\mathbf{k} = R\mathbf{e} \qquad \text{(4-30)}$$

where

$$R_x = \Sigma F_x \qquad R_y = \Sigma F_y \qquad R_z = \Sigma F_z$$
$$R = |\mathbf{R}| = \sqrt{(\Sigma F_x)^2 + (\Sigma F_y)^2 + (\Sigma F_z)^2}$$
$$\mathbf{e} = \cos\theta_x\,\mathbf{i} + \cos\theta_y\,\mathbf{j} + \cos\theta_z\,\mathbf{k}$$
$$\cos\theta_x = \frac{\Sigma F_x}{|\mathbf{R}|} \qquad \cos\theta_y = \frac{\Sigma F_y}{|\mathbf{R}|} \qquad \cos\theta_z = \frac{\Sigma F_z}{|\mathbf{R}|}$$

The resultant of the system of noncoplanar couples is a couple $\mathbf{C}$ that can be expressed as

$$\mathbf{C} = \Sigma\mathbf{C}_x + \Sigma\mathbf{C}_y + \Sigma\mathbf{C}_z = \Sigma C_x\mathbf{i} + \Sigma C_y\mathbf{j} + \Sigma C_z\mathbf{k} = C\mathbf{e} \qquad \text{(4-31)}$$

where

$$C = |\mathbf{C}| = \sqrt{(\Sigma C_x)^2 + (\Sigma C_y)^2 + (\Sigma C_z)^2}$$
$$\mathbf{e} = \cos\theta_x\,\mathbf{i} + \cos\theta_y\,\mathbf{j} + \cos\theta_z\,\mathbf{k}$$
$$\cos\theta_x = \frac{\Sigma C_x}{|\mathbf{C}|} \qquad \cos\theta_y = \frac{\Sigma C_y}{|\mathbf{C}|} \qquad \cos\theta_z = \frac{\Sigma C_z}{|\mathbf{C}|}$$

The resultant force $\mathbf{R}$ and the resultant couple $\mathbf{C}$ shown in Fig. 4-45f together constitute the resultant of the system with respect to point $O$. In special cases, the resultant couple $\mathbf{C}$ may vanish, leaving the force $\mathbf{R}$ as the resultant of the system. Again in special cases, the resultant force $\mathbf{R}$ may vanish, leaving the couple $\mathbf{C}$ as the resultant of the system. If the resultant force $\mathbf{R}$ and the resultant couple $\mathbf{C}$ both vanish, the resultant of the system is zero and the system is in equilibrium. Thus, the resultant of a general force system may be a force $\mathbf{R}$, a couple $\mathbf{C}$, or a force $\mathbf{R}$ and a couple $\mathbf{C}$.

When the couple $\mathbf{C}$ is perpendicular to the resultant force $\mathbf{R}$, as

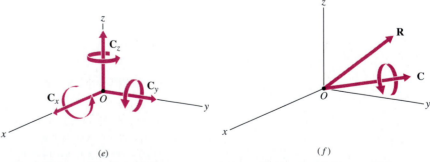

(e)

(f)

**Figure 4-45**   Resultant $\mathbf{R}$ for a general, three-dimensional system of forces.

**Figure 4-46**  Replacing a resultant force **R** and a resultant couple **C** with a parallel force **R**.

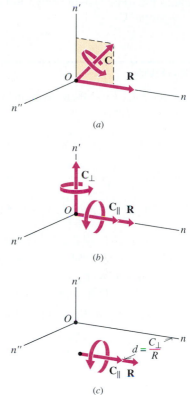

(a)

(b)

(c)

**Figure 4-47**  Replacing a resultant force **R** and a resultant couple **C** with a positive wrench.

shown in Fig. 4-46, the two can be combined to form a single force **R** whose line of action is a distance $d = C/R$ from point $O$ in a direction that makes the direction of the moment of $R$ about $O$ the same as that of $C$.

Another transformation of the resultant (**R** and **C**) of a general force system is illustrated in Figs. 4-47 and 4-48. In this case, the couple **C** is resolved into components parallel and perpendicular to the resultant force **R**, as shown in Fig. 4-47b. The resultant force **R** and the perpendicular component of the couple $C_\perp$ can be combined as illustrated in Fig. 4-46. In addition, the parallel component of the couple $C_\parallel$ can be translated to coincide with the line of action of the resultant force **R**, as shown in Fig. 4-47c. The combination of couple $C_\parallel$ and resultant force **R** is known as a wrench. The action may be described as a push (or pull) and a twist about an axis parallel to the push (or pull). When the force and moment vectors have the same sense, as shown in Fig. 4-47c, the wrench is positive. When the vectors have the opposite sense, as shown in Fig. 4-48c, the wrench is negative.

The analysis of general, three-dimensional force systems is illustrated in the following examples.

(a)

(b)

(c)

**Figure 4-48**  Replacing a resultant force **R** and a resultant couple **C** with a negative wrench.

Replace the force system shown in Fig. 4-49*a* with a force **R** through point *O* and a couple **C**.

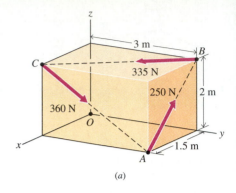

(*a*)

## SOLUTION

The three forces and their positions with respect to point *O* can be written in Cartesian vector form as

$$\mathbf{F}_A = 250\left[-\frac{1.5}{2.5}\mathbf{i} + \frac{2.0}{2.5}\mathbf{k}\right] = (-150\mathbf{i} + 200\mathbf{k}) \text{ N} \qquad \mathbf{r}_{A/O} = (1.5\mathbf{i} + 3.0\mathbf{j}) \text{ m}$$

$$\mathbf{F}_B = 335\left[\frac{1.5}{3.354}\mathbf{i} - \frac{3.0}{3.354}\mathbf{j}\right] = (150\mathbf{i} - 300\mathbf{j}) \text{ N} \qquad \mathbf{r}_{B/O} = (3.0\mathbf{j} + 2.0\mathbf{k}) \text{ m}$$

$$\mathbf{F}_C = 360\left[\frac{3.0}{3.606}\mathbf{j} - \frac{2.0}{3.606}\mathbf{k}\right] = (300\mathbf{j} - 200\mathbf{k}) \text{ N} \qquad \mathbf{r}_{C/O} = (1.5\mathbf{i} + 2.0\mathbf{k}) \text{ m}$$

Each of the three forces can be replaced by an equal force through point *O* and a couple. The vector sum of the concurrent forces is

$$\mathbf{R} = \mathbf{F}_A + \mathbf{F}_B + \mathbf{F}_C$$
$$= -150\mathbf{i} + 200\mathbf{k} + 150\mathbf{i} - 300\mathbf{j} + 300\mathbf{j} - 200\mathbf{k} = \mathbf{0} \qquad \text{Ans.}$$

The moment $\mathbf{M}_O$ of the resultant couple is

$$\mathbf{C} = \mathbf{M}_O = \mathbf{r}_{A/O} \times \mathbf{F}_A + \mathbf{r}_{B/O} \times \mathbf{F}_B + \mathbf{r}_{C/O} \times \mathbf{F}_C$$

$$= \begin{vmatrix} \mathbf{i} & \mathbf{j} & \mathbf{k} \\ 1.5 & 3.0 & 0 \\ -150 & 0 & 200 \end{vmatrix} + \begin{vmatrix} \mathbf{i} & \mathbf{j} & \mathbf{k} \\ 0 & 3.0 & 2.0 \\ 150 & -300 & 0 \end{vmatrix} + \begin{vmatrix} \mathbf{i} & \mathbf{j} & \mathbf{k} \\ 1.5 & 0 & 2.0 \\ 0 & 300 & -200 \end{vmatrix}$$

$$= 600\mathbf{i} + 300\mathbf{j} + 450\mathbf{k} \text{ N} \cdot \text{m} \qquad \text{Ans.}$$

The magnitude of the resultant couple is

$$|\mathbf{C}| = \sqrt{C_x^2 + C_y^2 + C_z^2}$$
$$= \sqrt{(600)^2 + (300)^2 + (450)^2} = 807.8 \cong 808 \text{ N} \cdot \text{m} \qquad \text{Ans.}$$

Finally, the direction angles that locate the axis of the couple are

$$\theta_x = \cos^{-1}\frac{C_x}{|\mathbf{C}|} = \cos^{-1}\frac{600}{807.8} = 42.0° \qquad \text{Ans.}$$

$$\theta_y = \cos^{-1}\frac{C_y}{|\mathbf{C}|} = \cos^{-1}\frac{300}{807.8} = 68.2° \qquad \text{Ans.}$$

$$\theta_z = \cos^{-1}\frac{C_z}{|\mathbf{C}|} = \cos^{-1}\frac{450}{807.8} = 56.1° \qquad \text{Ans.}$$

The resultant of the force system is the couple shown in Fig. 4-49*b*.

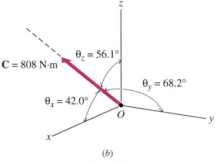

(*b*)

**Fig. 4-49**

A force **F** can be written in Cartesian vector form by multiplying the magnitude of the force (a scalar) by a unit vector along the line of action of the force.

Any force **F** can be moved to a parallel position through an arbitrary point *A* if a couple **C** is added that has a moment equal to the moment of the original force **F** about point *A*.

A system of couples in space can be combined into a single resultant couple **C**. The *x*-, *y*-, and *z*-components of the resultant couple are the algebraic sums of the *x*-, *y*-, and *z*-components of the individual couples.

Three forces are applied to a rigid body as shown in Fig. 4-50a.

a. Reduce the forces to a wrench.
b. Determine the intersection of the wrench with the $xy$-plane.

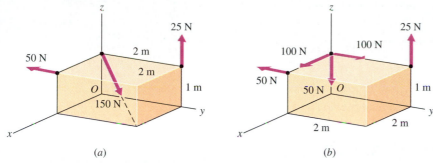

**Fig. 4-50**

---

## SOLUTION

a. The rectangular components of the three forces are shown in Fig. 4-50b. From these data, the equivalent force **R** and the couple **C** at point $O$ can be written in Cartesian vector form as

$$\mathbf{R} = 100\mathbf{i} + 50\mathbf{j} - 25\mathbf{k} \text{ N}$$
$$\mathbf{C} = 100\mathbf{j} - 100\mathbf{k} \text{ N} \cdot \text{m}$$

The magnitude of the resultant force is

$$|\mathbf{R}| = \sqrt{R_x^2 + R_y^2 + R_z^2}$$
$$= \sqrt{(100)^2 + (50)^2 + (-25)^2} = 114.56 \cong 114.6 \text{ N} \cdot \text{m} \qquad \text{Ans.}$$

The direction angles that locate the line of action of the resultant force are

$$\theta_x = \cos^{-1} \frac{R_x}{|\mathbf{R}|} = \cos^{-1} \frac{100}{114.56} = 29.2° \qquad \text{Ans.}$$

$$\theta_y = \cos^{-1} \frac{R_y}{|\mathbf{R}|} = \cos^{-1} \frac{50}{114.56} = 64.1° \qquad \text{Ans.}$$

$$\theta_z = \cos^{-1} \frac{R_z}{|\mathbf{R}|} = \cos^{-1} \frac{-25}{114.56} = 102.6° \qquad \text{Ans.}$$

A unit vector $\mathbf{e}_R$ along the line of action of the resultant force **R** is

$$\mathbf{e} = \cos\theta_x \mathbf{i} + \cos\theta_y \mathbf{j} + \cos\theta_z \mathbf{k}$$
$$= 0.8729\mathbf{i} + 0.4365\mathbf{j} - 0.2182\mathbf{k}$$

The components of the couple parallel and perpendicular to the line of action of the resultant force **R** are

$$|\mathbf{C}_{\parallel}| = \mathbf{C} \cdot \mathbf{e}_R = (100\mathbf{j} - 100\mathbf{k}) \cdot (0.8729\mathbf{i} + 0.4365\mathbf{j} - 0.2182\mathbf{k})$$
$$= 65.46 \cong 65.5 \text{ N} \cdot \text{m} \qquad \text{Ans.}$$
$$\mathbf{C}_{\parallel} = |\mathbf{C}_{\parallel}| \mathbf{e}_R = 65.46(0.8729\mathbf{i} + 0.4365\mathbf{j} - 0.2182\mathbf{k})$$
$$= 57.14\mathbf{i} + 28.57\mathbf{j} - 14.28\mathbf{k} \text{ N} \cdot \text{m}$$
$$\mathbf{C}_{\perp} = \mathbf{C} - \mathbf{C}_{\parallel} = -57.14\mathbf{i} + 71.43\mathbf{j} - 85.72\mathbf{k} \text{ N} \cdot \text{m}$$

> The dot product $\mathbf{C} \cdot \mathbf{e}_R$ gives the projection or scalar component of the couple **C** along the line of action of resultant **R**.

Thus, the force **R** and the couple **C** that comprise the wrench are

$$\mathbf{R} = 100\mathbf{i} + 50\mathbf{j} - 25\mathbf{k} \text{ N} \qquad \text{Ans.}$$
$$\mathbf{C}_{\parallel} = 57.1\mathbf{i} + 28.6\mathbf{j} - 14.28\mathbf{k} \text{ N} \cdot \text{m} \qquad \text{Ans.}$$

(a)                 (b)              (c)

**Fig. 4-51**

**b.** The resultant force **R** and the couple $\mathbf{C}_\perp$ can be combined by translating force **R** such that $\mathbf{r} \times \mathbf{R} = \mathbf{C}_\perp$ as illustrated schematically in Fig. 4-51. Thus,

$$\begin{vmatrix} \mathbf{i} & \mathbf{j} & \mathbf{k} \\ x_R & y_R & 0 \\ 100 & 50 & -25 \end{vmatrix} = -57.14\mathbf{i} + 71.43\mathbf{j} - 85.72\mathbf{k}$$

$$-25y_R\mathbf{i} + 25x_R\mathbf{j} + (50x_R - 100y_R)\mathbf{k} = -57.14\mathbf{i} + 71.43\mathbf{j} - 85.72\mathbf{k}$$

From which

$$-25y_R = -57.14 \qquad y_R = 2.29 \text{ m} \qquad \text{Ans.}$$
$$25x_R = 71.43 \qquad x_R = 2.86 \text{ m} \qquad \text{Ans.}$$

> The component of couple **C** perpendicular to the line of action of resultant **R** is the difference between couple **C** and $\mathbf{C}_\parallel$ since $\mathbf{C}_\parallel$ and $\mathbf{C}_\perp$ are rectangular components of couple **C**.

## EXAMPLE PROBLEM 4-26

Forces are applied at points $A$, $B$, and $C$ of the bar shown in Fig. 4-52a. Replace this system of forces with a force **R** through point $O$ and a couple **C**.

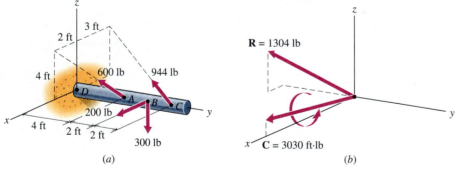

(a)                                 (b)

**Fig. 4-52**

### SOLUTION

The forces and their positions with respect to point $O$ are

$$\mathbf{F}_A = 600\left[\frac{2}{6}\mathbf{i} - \frac{4}{6}\mathbf{j} + \frac{4}{6}\mathbf{k}\right] = 200\mathbf{i} - 400\mathbf{j} + 400\mathbf{k} \text{ lb}$$

$$\mathbf{F}_B = 200\mathbf{i} - 300\mathbf{k} \text{ lb}$$

$$\mathbf{F}_C = 944\left[-\frac{3}{9.434}\mathbf{i} - \frac{8}{9.434}\mathbf{j} + \frac{4}{9.434}\mathbf{k}\right] = -300\mathbf{i} - 800\mathbf{j} + 400\mathbf{k} \text{ lb}$$

$$\mathbf{r}_{A/O} = 4\mathbf{j} \text{ ft} \qquad \mathbf{r}_{B/O} = 6\mathbf{j} \text{ ft} \qquad \mathbf{r}_{C/O} = 8\mathbf{j} \text{ ft}$$

> A force **F** can be written in Cartesian vector form by multiplying the magnitude of the force (a scalar) by a unit vector along the line of action of the force.

Each of the three forces can be replaced by an equal force through point $O$ and a couple. The vector sum of these concurrent forces is the resultant force $\mathbf{R}$.

$$\mathbf{R} = \mathbf{F}_A + \mathbf{F}_B + \mathbf{F}_C$$
$$= 200\mathbf{i} - 400\mathbf{j} + 400\mathbf{k} + 200\mathbf{i} - 300\mathbf{k} - 300\mathbf{i} - 800\mathbf{j} + 400\mathbf{k}$$
$$= 100\mathbf{i} - 1200\mathbf{j} + 500\mathbf{k} \text{ lb} \qquad \text{Ans.}$$

The magnitude of the resultant force is

$$|\mathbf{R}| = \sqrt{R_x^2 + R_y^2 + R_z^2} = \sqrt{(100)^2 + (-1200)^2 + (500)^2} = 1304 \text{ lb} \quad \text{Ans.}$$

The direction angles that locate the line of action of the resultant force are

$$\theta_x = \cos^{-1}\frac{R_x}{|\mathbf{R}|} = \cos^{-1}\frac{100}{1304} = 85.6° \qquad \text{Ans.}$$
$$\theta_y = \cos^{-1}\frac{R_y}{|\mathbf{R}|} = \cos^{-1}\frac{-1200}{1304} = 157.0° \qquad \text{Ans.}$$
$$\theta_z = \cos^{-1}\frac{R_z}{|\mathbf{R}|} = \cos^{-1}\frac{500}{1304} = 67.5° \qquad \text{Ans.}$$

The moment $\mathbf{M}_O$ of the resultant couple is

$$\mathbf{C} = \mathbf{M}_O = \mathbf{r}_{A/O} \times \mathbf{F}_A + \mathbf{r}_{B/O} \times \mathbf{F}_B + \mathbf{r}_{C/O} \times \mathbf{F}_C$$
$$= 1600\mathbf{i} - 800\mathbf{k} - 1800\mathbf{i} - 1200\mathbf{k} + 3200\mathbf{i} + 2400\mathbf{k}$$
$$= 3000\mathbf{i} + 400\mathbf{k} \text{ ft} \cdot \text{lb} \qquad \text{Ans.}$$

The magnitude of the resultant couple is

$$|\mathbf{C}| = \sqrt{C_x^2 + C_y^2 + C_z^2} = \sqrt{(3000)^2 + (400)^2} = 3027 \cong 3030 \text{ ft} \cdot \text{lb} \quad \text{Ans.}$$

The direction angles that locate the axis of the couple are

$$\theta_x = \cos^{-1}\frac{C_x}{|\mathbf{C}|} = \cos^{-1}\frac{3000}{3027} = 7.66° \qquad \text{Ans.}$$
$$\theta_y = \cos^{-1}\frac{C_y}{|\mathbf{C}|} = \cos^{-1}\frac{0}{3027} = 90.0° \qquad \text{Ans.}$$
$$\theta_z = \cos^{-1}\frac{C_z}{|\mathbf{C}|} = \cos^{-1}\frac{400}{3027} = 82.4° \qquad \text{Ans.}$$

The resultant of the force system is the force $\mathbf{R}$ and the couple $\mathbf{C}$ shown in Fig. 4-52b.

> A system of forces in space can be combined into a single resultant force $\mathbf{R}$. The $x$-, $y$-, and $z$-components of the resultant force are the algebraic sums of the $x$-, $y$-, and $z$-components of the individual forces.

> Any force $\mathbf{F}$ can be moved to a parallel position through an arbitrary point $O$ if a couple $\mathbf{C}$ is added that has a moment equal to the moment of the original force $\mathbf{F}$ about point $O$.

> A system of couples in space can be combined into a single resultant couple $\mathbf{C}$. The $x$-, $y$-, and $z$-components of the resultant couple are the algebraic sums of the $x$-, $y$-, and $z$-components of the individual couples.

# PROBLEMS

### Introductory Problems

**4-137\*** Replace the force system shown in Fig. P4-137 with a force $\mathbf{R}$ through point $D$ and a couple $\mathbf{C}$.

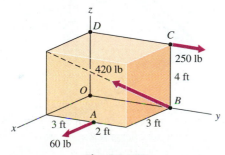

**Fig. P4-137**

**4-138** Replace the force system shown in Fig. P4-138 with a force $\mathbf{R}$ through point $D$ and a couple $\mathbf{C}$.

**Fig. P4-138**

**4-139\*** Forces are applied at points A, B, and C of the bar shown in Fig. P4-139. Replace this system of forces with a force **R** through point O and a couple **C**.

Fig. P4-139

**4-140\*** Forces are applied at points A, B, and C of the bar shown in Fig. P4-140. Replace this system of forces with a force **R** through point O and a couple **C**.

Fig. P4-140

**4-141** Forces are applied at points B, C, and D of the bar shown in Fig. P4-141. Replace this system of forces with a force **R** through point A and a couple **C**.

Fig. P4-141

**4-142** Forces are applied at points A, B, and C of the bar shown in Fig. P4-142. Replace this system of forces with a force **R** through point O and a couple **C**.

Fig. P4-142

## Intermediate Problems

**4-143\*** Forces are applied at points A, B, and C of the bar shown in Fig. P4-143. Replace this system of forces with a force **R** through point O and a couple **C**.

Fig. P4-143

**4-144\*** Forces are applied at points B and C of the bar shown in Fig. P4-144. Replace this system of forces with a force **R** through point O and a couple **C**.

Fig. P4-144

**4-145** Forces are applied at points $B$ and $D$ of the bar shown in Fig. P4-145. Replace this system of forces with a force $\mathbf{R}$ through point $A$ and a couple $\mathbf{C}$.

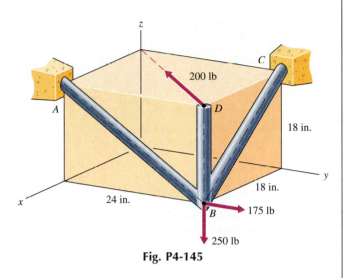

Fig. P4-145

**4-147*** Forces are applied at points $A$, $B$, and $C$ of the bar shown in Fig. P4-147. Replace this system of forces with a force $\mathbf{R}$ through point $O$ and a couple $\mathbf{C}$.

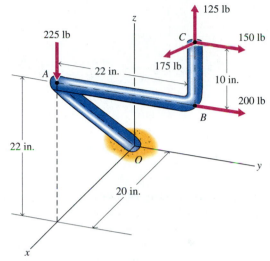

Fig. P4-147

**4-146** Forces are applied at points $A$ and $B$ of the bar shown in Fig. P4-146. Replace this system of forces with a force $\mathbf{R}$ through point $O$ and a couple $\mathbf{C}$.

Fig. P4-146

**4-148** Forces are applied at points $A$ and $B$ of the bar shown in Fig. P4-148. Replace this system of forces with a force $\mathbf{R}$ through point $O$ and a couple $\mathbf{C}$.

Fig. P4-148

## Challenging Problems

**4-149*** Reduce the forces shown in Fig. P4-149 to a wrench and locate the intersection of the wrench with the xz-plane.

**Fig. P4-149**

**4-150*** Reduce the forces shown in Fig. P4-150 to a wrench and locate the intersection of the wrench with the xz-plane.

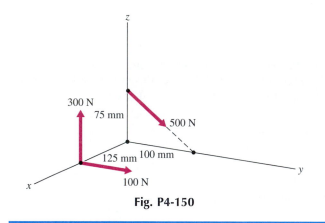

**Fig. P4-150**

**4-151** Reduce the forces shown in Fig. P4-151 to a wrench and locate the intersection of the wrench with the xy-plane.

**Fig. P4-151**

**4-152** Reduce the forces shown in Fig. P4-152 to a wrench and locate the intersection of the wrench with the xy-plane.

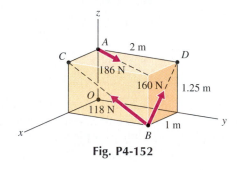

**Fig. P4-152**

# SUMMARY

For a three-dimensional body, the particle idealization discussed in Chapter 3 is not valid, in general, since the forces acting on the body are usually not concurrent. For a general force system, $\mathbf{R} = 0$ is a necessary but not a sufficient condition for equilibrium of the body. A second restriction related to the tendency of a force to produce rotation of a body must also be satisfied and gives rise to the concept of a moment. In this chapter, the moment of a force about a point and the moment of a force about a line (axis) were defined and methods were developed for finding the resultant forces and the resultant moments (couples) for any general force system that may be applied to the body.

A moment is a vector quantity since it has both a magnitude and a direction, and adds according to the parallelogram law of addition. The moment of a force **F** about a point $O$ can be represented by the vector cross product:

$$\mathbf{M}_O = \mathbf{r} \times \mathbf{F} \tag{4-3}$$

where **r** is a position vector from point $O$ to any point on the line of action of the force **F**. The cross product of the two intersecting vectors **r** and **F** is

$$\mathbf{M}_O = \mathbf{r} \times \mathbf{F} = M_x\mathbf{i} + M_y\mathbf{j} + M_z\mathbf{k} = M_O\mathbf{e} \tag{4-9}$$

where

$$M_x = r_yF_z - r_zF_y \qquad M_y = r_zF_x - r_xF_z \qquad M_z = r_xF_y - r_yF_x$$

are the three scalar components of the moment. The magnitude of the moment $\mathbf{M}_O$ is

$$M_O = |\mathbf{M}_O| = \sqrt{M_x^2 + M_y^2 + M_z^2} \tag{4-11}$$

The direction cosines associated with the unit vector **e** are

$$\cos \theta_x = \frac{M_x}{|\mathbf{M}_O|} \qquad \cos \theta_y = \frac{M_y}{|\mathbf{M}_O|} \qquad \cos \theta_z = \frac{M_z}{|\mathbf{M}_O|} \tag{4-14}$$

The moment $\mathbf{M}_{OB}$ of a force **F** about a line $OB$ in a direction $n$ specified by the unit vector $\mathbf{e}_n$ is

$$\mathbf{M}_{OB} = (\mathbf{M}_O \cdot \mathbf{e}_n)\mathbf{e}_n = [(\mathbf{r} \times \mathbf{F}) \cdot \mathbf{e}_n]\mathbf{e}_n = M_{OB}\mathbf{e}_n \tag{4-16}$$

Two equal parallel forces of opposite sense are called a couple. A couple has no tendency to translate a body in any direction but tends only to rotate the body on which it acts. Several transformations of a couple can be made without changing any of the external effects of the couple on the body. A couple (1) can be translated to a parallel position in its plane or to any parallel plane, and (2) can be rotated in its plane. Also, (3) the magnitude $F$ of the two forces of a couple and the distance $d$ between them can be changed provided the product $Fd$ remains constant. Any system of couples in a plane or in space can be combined into a single resultant couple **C**.

Any force **F** can be resolved into a parallel force **F** and a couple **C**. Alternatively, a force and a couple in the same plane can be combined into a single force. The sole effect of combining a couple with a force is to move the action line of the force into a parallel position. The magnitude and sense of the force remain unchanged.

The resultant of a force system acting on a rigid body is the simplest force system that can replace the original system without altering the external effect of the system on the body. For a coplanar system of forces, the resultant is either a force **R** or a couple **C**. For a three-dimensional system of forces, the resultant may be a force **R**, a couple **C**, or both a force **R** and a couple **C**. The combination of a force **R** and a couple **C** (a push or pull together with a twist about the line of action of the force) is known as a wrench.

# REVIEW PROBLEMS

**4-153*** A 2500-lb jet engine is suspended from the wing of an airplane as shown in Fig. P4-153. Determine the moment produced by the engine at point $A$ in the wing when the plane is

a. On the ground with the engine not operating.
b. In flight with the engine developing a thrust **T** of 15,000 lb.

**Fig. P4-153**

**4-154** Determine the moment of the 1650-N force shown in Fig. P4-154 about point $O$.

**Fig. P4-154**

**4-155*** The driving wheel of a truck is subjected to the force-couple system shown in Fig. P4-155. Replace this system by an equivalent single force and determine the point of application of the force along the vertical diameter of the wheel.

**Fig. P4-155**

**4-156*** A bracket is subjected to the force-couple system shown in Fig. P4-156. Determine

a. The magnitude and direction of the resultant force **R.**
b. The perpendicular distance from support $O$ to the line of action of the resultant.

**Fig. P4-156**

**4-157**   Determine the resultant of the parallel force system shown in Fig. P4-157 and locate the intersection of its line of action with the xy-plane.

**Fig. P4-157**

**4-158**   A 200-N force is applied at corner B of a rectangular plate as shown in Fig. P4-158. Determine

**a.**   The moment of the force about point O.
**b.**   The moment of the force about line OD.

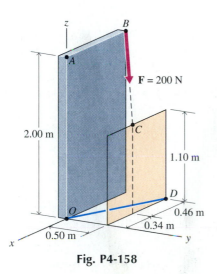

**Fig. P4-158**

**4-159\***   A Z-section is loaded with a system of forces as shown in Fig. P4-159. Express the resultant of this force system in Cartesian vector form.

**Fig. P4-159**

**4-160\***   A bent rod supports two forces as shown in Fig. P4-160. Determine

**a.**   The moment of the two forces about point O.
**b.**   The moment of the two forces about line OA.

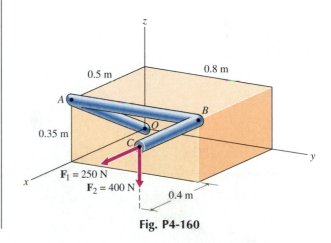

**Fig. P4-160**

**4-161** The force **F** in Fig. P4-161 can be expressed in Cartesian vector form as $\mathbf{F} = 60\mathbf{i} + 100\mathbf{j} + 120\mathbf{k}$ lb. Determine the scalar component of the moment at point $B$ about line $BC$.

**Fig. P4-161**

**4-162** A bent rod supports a 450-N force as shown in Fig. P4-162.

a. Replace the 450-N force with a force **R** through point $O$ and a couple **C**.

b. Determine the twisting moments produced by force **F** in the three different segments of the rod.

**Fig. P4-162**

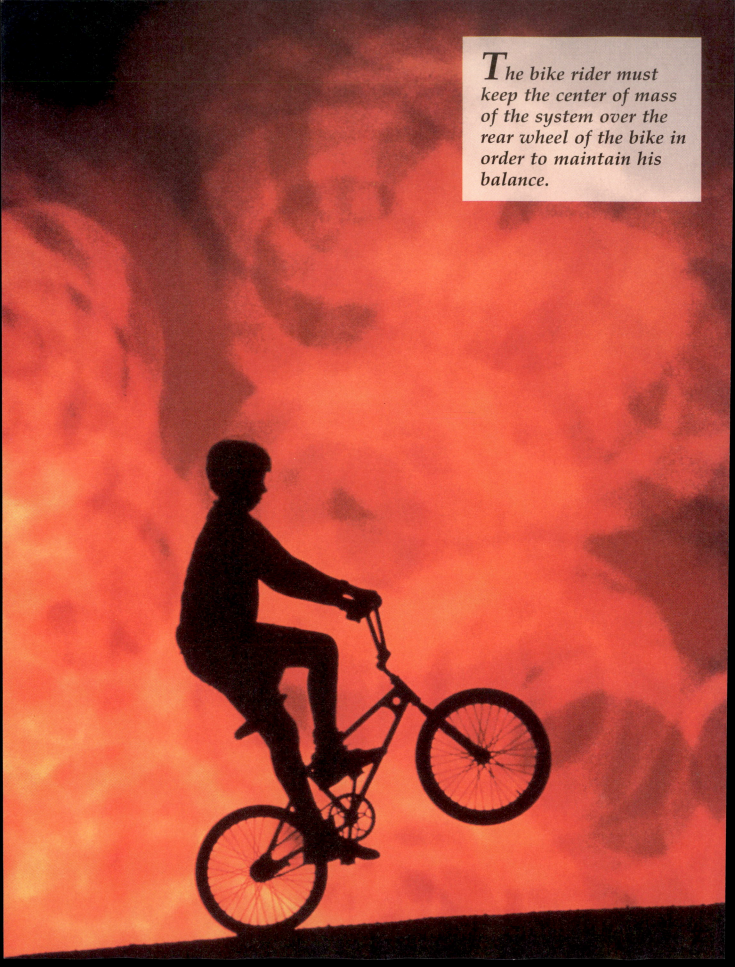

*The bike rider must keep the center of mass of the system over the rear wheel of the bike in order to maintain his balance.*

# Chapter 5

# DISTRIBUTED FORCES: CENTROIDS AND CENTER OF GRAVITY

## 5-1 INTRODUCTION

In previous chapters, we have dealt primarily with concentrated forces each of which could be represented as a simple vector quantity that has a magnitude, a direction, a line of action, and in some instances a point of application. In many instances, the loads are not concentrated at a point but are distributed along a line (such as the contact line of a roller bearing illustrated in Fig. 5-1a) or over a surface (such as the wind load on a sign illustrated in Fig. 5-1b). In these cases, the loads are known as distributed loads. The distribution can be uniform or nonuniform. Other forces, known as body forces, which result from gravitational, electrical, and magnetic effects, are distributed over the mass of the body. When the areas over which the loads are applied become significant with respect to the size of the body, the assumption of a concentrated force is no longer valid, and the actual distribution of the load must be included in the analysis.

A distributed force at any point is characterized by its intensity and its direction. A force distributed over an area and acting normal to the surface (usually due to the action of a liquid or gas) is known as a pressure. Internal distributed forces in solids (known as stresses) may or may not act normal to the surface of interest. The units for both pressure and stress are force per unit area (lb/in.$^2$ or N/m$^2$). Forces distributed over the volume of a body (body forces) are measured in units of force per unit volume (lb/in.$^3$ or N/m$^3$).

In the preceding chapters, moments of forces about points or axes have been considered. In the analysis of many problems in engineering, expressions are encountered that represent moments of masses, forces, volumes, areas, or lines with respect to axes or planes. For ex-

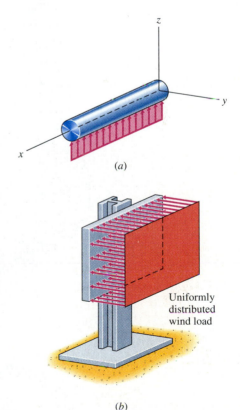

**Figure 5-1** Examples of distributed loads.

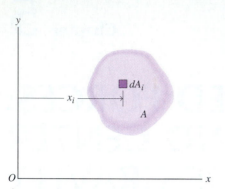

**Figure 5-2**   Moment of the element of area $dA_i$ about the $y$-axis.

ample, consider the moment of an area $A$ (in the $xy$-plane) about the $y$-axis as shown in Fig. 5-2. Since an area is a distributed quantity, its moment about an axis cannot be defined as the product of the area and the perpendicular distance of the area from the axis (similar to the manner in which the moment of a concentrated force was defined) because the different parts of the area are at different distances from the axis; hence, the distance of the area from the axis is indefinite and meaningless. The area can, however, be considered to be composed of a large number of very small (differential) elements of area $dA$, and the moment $dM$ of an element about an axis can then be defined as the product of the area of the element and the distance of the element from the axis. Thus,

$$dM_i = x_i \, dA_i \qquad (5\text{-}1)$$

where the subscript $i$ denotes the $i$th element. The moment of the area $A$ about the $y$-axis is then defined as the algebraic sum of the moments of the $n$ elements of area about the $y$-axis. Thus,

$$M_y = \sum_{i=1}^{n} x_i \, dA_i \qquad (5\text{-}2)$$

or in integral form

$$M_y = \int_A x \, dA \qquad (5\text{-}3)$$

The moment of a mass, force, volume, area, or line with respect to a line or a plane can be defined in a similar way.

The moment thus defined is called the first moment of the quantity being considered since the first power of the distance ($x$ in the above case) is used in the expression. Later, in Chapter 10, integrals of the form $\int_A x^2 \, dA$ will be introduced. Such integrals are known as second moments since the second power of the distance appears in the expression.

The sign of the moment of an element about an axis may be positive or negative since the coordinate of the element may be positive or negative whereas the masses, forces, volumes, areas, and lengths are always positive. Likewise, the moment of the quantity (mass, force, volume, area, or length) about an axis or plane may be positive, negative, or zero since the sum of the positive moments of the elements may be larger than, smaller than, or equal to the sum of the negative moments of the elements.

The dimensional expression for the moment of a line is length squared ($L^2$); therefore, the moment of a line about an axis or plane will have the units in.$^2$, ft$^2$, mm$^2$, m$^2$, and so on. Similarly, the dimensions of the moments of an area and of a volume are length cubed ($L^3$) and length to the fourth power ($L^4$), respectively.

Concepts required for the treatment of distributed forces include center of mass, center of gravity, and centroid. These topics are developed in the next two sections of this chapter.

## 5-2 CENTER OF GRAVITY AND CENTER OF MASS

### 5-2.1 Center of Gravity

An immediate application of equivalent force-couples is the replacement of distributed forces such as gravitational forces with a single equivalent force. For example, consider the set of $n$ particles shown in Fig. 5-3$a$. Approximating the weights of the $n$ particles as a parallel force system (Fig. 5-3$b$),[1] the external effect of the $n$ weights is the same as that of a single equivalent force **W** acting at point $G$ as shown in Fig. 5-3$c$. The equivalent force **W** is called the weight of the body, and the point $G$ through which the force acts is called the **center of gravity.** As in Section 4-6.2, the direction of the weight force **W** is parallel to all of the individual weights $\mathbf{W}_i$, its magnitude is the sum of the magnitudes of the individual weights (so that the sum of forces is the same in Figs. 5-3$b$ and 5-3$c$),

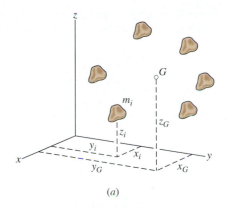

(a)

$$W = \sum_{i=1}^{n} W_i \tag{5-4a}$$

and the location of $G$ (so that the sum of moments is the same on Figs. 5-3$b$ and 5-3$c$)[2] is

$$M_{yz} = Wx_G = \sum_{i=1}^{n} W_i x_i \quad \text{or} \quad x_G = \frac{1}{W} \sum_{i=1}^{n} W_i x_i \tag{5-4b}$$

$$M_{zx} = Wy_G = \sum_{i=1}^{n} W_i y_i \quad \text{or} \quad y_G = \frac{1}{W} \sum_{i=1}^{n} W_i y_i \tag{5-4c}$$

$$M_{xy} = Wz_G = \sum_{i=1}^{n} W_i z_i \quad \text{or} \quad z_G = \frac{1}{W} \sum_{i=1}^{n} W_i z_i \tag{5-4d}$$

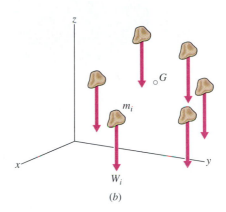

(b)

where $x_i$, $y_i$, $z_i$, and $x_G$, $y_G$, $z_G$ are shown in Fig. 5-3$a$.

If the particles form a continuous body, as shown in Fig. 5-4$a$, the summations must be replaced by integrals over the volume of the body, giving

$$W = \int dW \tag{5-5a}$$

$$M_{yz} = Wx_G = \int x \, dW \quad \text{or} \quad x_G = \frac{1}{W} \int x \, dW \tag{5-5b}$$

$$M_{zx} = Wy_G = \int y \, dW \quad \text{or} \quad y_G = \frac{1}{W} \int y \, dW \tag{5-5c}$$

$$M_{xy} = Wz_G = \int z \, dW \quad \text{or} \quad z_G = \frac{1}{W} \int z \, dW \tag{5-5d}$$

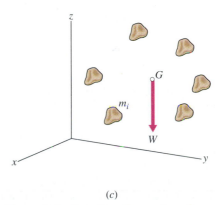

(c)

Equations 5-5$b$, 5-5$c$, and 5-5$d$ can be combined into a single vector equation by multiplying the first, second, and third equations by **i, j,**

**Figure 5-3** Locating the center of gravity $G$ of a system of particles.

---

[1]The weight forces are actually concurrent at the center of the Earth. However, most engineering structures are small compared to the radius of the Earth and the error introduced by assuming the forces are parallel is very small.

[2]Equation 5-4$d$ is not directly obtainable from Figs. 5-4$b$ and 5-4$c$. By the principle of transmissibility, the force **W** can act anywhere along the vertical line through $G$ without affecting its moment. Equation 5-4$d$ is obtained by rotating the figures, both the weights and the coordinate system, so that gravity acts along the $y$-axis instead of the $z$-axis. Then the sum of moments gives Eq. 5-4$d$.

*Show how the concept of a center of gravity could be used to locate the geographical center of the contiguous 48 states of the United States.*

(a)          (b)          (c)

**Fig. CE5-1**

### SOLUTION

When a body is made of a homogeneous material, its center of gravity is located at the geometric center of the body. Figure CE5-1 shows a map of the United States that has been mounted on a homogeneous sheet of cardboard. When the map is suspended by a cord at point $A$ (see Fig. CE5-1a), the center of gravity of the map is located on the vertical line coinciding with the axis of the cord. This results from the fact that two forces hold the map in equilibrium: the tension in the cord and the weight of the map. These two forces must be equal, opposite, and collinear, and, by definition, the weight of the map acts through the center of gravity of the map. When the map is suspended from a second point $B$ (see Fig. CE5-1b), the center of gravity lies on the vertical line through point $B$ and hence lies at the point of intersection of this line and the first. If the map is now suspended from a third point $C$ (see Fig. CE5-1c), a vertical line through $C$ will pass through the point of intersection of the first two lines.

As indicated on Fig. CE5-1, the center of gravity (geographical center of the contiguous 48 states) is located near Lebanon, Kansas, which is midway across Kansas and approximately 12 miles south of the Kansas/Nebraska border.

Mathematical methods, developed in the following sections of the book, can be used to provide center of gravity (and center of mass) locations for many shapes of interest to engineers.

and **k,** respectively, and adding. Thus,

$$Wx_G\mathbf{i} + Wy_G\mathbf{j} + Wz_G\mathbf{k} = \sum_{i=1}^{n} W_i x_i \mathbf{i} + \sum_{i=1}^{n} W_i y_i \mathbf{j} + \sum_{i=1}^{n} W_i z_i \mathbf{k}$$

from which

$$W(x_G\mathbf{i} + y_G\mathbf{j} + z_G\mathbf{k}) = \sum_{i=1}^{n} W_i(x_i\mathbf{i} + y_i\mathbf{j} + z_i\mathbf{k})$$

which reduces to

$$\mathbf{M}_O = W\mathbf{r}_G = \sum_{i=1}^{n} W_i\mathbf{r}_i \quad \text{or} \quad \mathbf{r}_G = \frac{1}{W}\sum_{i=1}^{n} W_i\mathbf{r}_i \qquad \text{(5-6)}$$

where $\mathbf{r}_i = x_i\mathbf{i} + y_i\mathbf{j} + z_i\mathbf{k}$ is the position vector from the origin $O$ to the
$i$th particle (Fig. 5-4$b$) and $\mathbf{r}_G = x_G\mathbf{i} + y_G\mathbf{j} + z_G\mathbf{k}$ is the position vector
from the origin to the center of gravity. Similarly for Eqs. 5-5, if the par-
ticles form a continuous body,

$$W\mathbf{r}_G = \int \mathbf{r}\, dW \quad \text{or} \quad \mathbf{r}_G = \frac{1}{W}\int \mathbf{r}\, dW \qquad \text{(5-7)}$$

The center of gravity $G$ located by Eqs. 5-4 through 5-7 represents
the point at which all of the weight of the body could be concentrated
without changing the external effects on the body.

### 5-2.2 Center of Mass

For practical engineering work where the size of a body is small in
comparison to the size of the Earth, all of the particles that make up
the body can be assumed to be at the same distance from the center of
the Earth; therefore, they experience the same gravitational acceleration $g$. Dividing Eqs. 5-5 by the gravitational constant gives

$$m = \int dm \qquad \text{(5-8a)}$$

$$m x_G = \int x\, dm \quad \text{or} \quad x_G = \frac{1}{m}\int x\, dm \qquad \text{(5-8b)}$$

$$m y_G = \int y\, dm \quad \text{or} \quad y_G = \frac{1}{m}\int y\, dm \qquad \text{(5-8c)}$$

$$m z_G = \int z\, dm \quad \text{or} \quad z_G = \frac{1}{m}\int z\, dm \qquad \text{(5-8d)}$$

The location of the center of mass $(x_G, y_G, z_G)$ is the same as the
center of gravity, and both points will be labeled $G$. However, the center of mass is defined equally well in the weightlessness of space as it
is on the surface of the Earth.

The moments $M_{yz}$, $M_{zx}$, and $M_{xy}$ are called first moments of the
weight forces relative to the $yz$-, $zx$-, and $xy$-planes, respectively. The
integrand in each case is the first power of the distance to the respec-
tive plane. Similarly, the integrals $\int x\, dm$, $\int y\, dm$, and $\int z\, dm$ are called the
first moments of the mass (and integrals of the form $\int x\, dA$ are called
first moments of area) although they are not strictly moments since $dm$
(and $dA$) are not actually forces. Later, integrals of the form $\int x^2\, dm$ (and
$\int x^2\, dA$) will be introduced. Such integrals are called second moments
of mass (and second moments of area) since the second power of the
distance appears in these expressions.

(a)

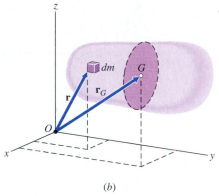

(b)

**Figure 5-4**  Representing a continuous body as a system of particles.

Locate the center of gravity for the six particles shown in Fig. 5-5 if $W_A = 50$ lb, $W_B = 25$ lb, $W_C = 30$ lb, $W_D = 35$ lb, $W_E = 20$ lb, and $W_F = 40$ lb.

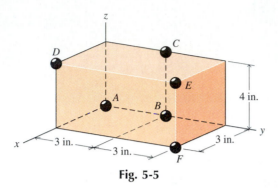

**Fig. 5-5**

## SOLUTION

The center of gravity for the system of particles is located by using the principle of moments for a parallel system of gravitational forces $\mathbf{W}_i$ and their resultant $\mathbf{W}$. The total weight $W$ for the six particles is

$$W = W_A + W_B + W_C + W_D + W_E + W_F = 50 + 25 + 30 + 35 + 20 + 40 = 200 \text{ lb}$$

Summing moments for the individual particles yields

$$M_{yz} = \Sigma W_i x_i = W_A x_A + W_B x_B + W_C x_C + W_D x_D + W_E x_E + W_F x_F$$
$$= 50(0) + 25(0) + 30(0) + 35(3) + 20(3) + 40(3) = 285 \text{ in.} \cdot \text{lb}$$
$$M_{zx} = \Sigma W_i y_i = W_A y_A + W_B y_B + W_C y_C + W_D x_D + W_E y_E + W_F y_F$$
$$= 50(0) + 25(3) + 30(3) + 35(0) + 20(6) + 40(6) = 525 \text{ in.} \cdot \text{lb}$$
$$M_{xy} = \Sigma W_i z_i = W_A z_A + W_B z_B + W_C z_C + W_D x_D + W_E z_E + W_F z_F$$
$$= 50(0) + 25(0) + 30(4) + 35(4) + 20(4) + 40(0) = 340 \text{ in.} \cdot \text{lb}$$

The principle of moments states that the sum of the moments of the weights $W_i$ equals the product of the total weight $W$ and the corresponding coordinate of the center of gravity $G$. Thus,

$$W x_G = \Sigma W_i x_i \qquad x_G = \frac{\Sigma W_i x_i}{W} = \frac{285}{200} = 1.425 \text{ in.} \qquad \text{Ans.}$$

$$W y_G = \Sigma W_i y_i \qquad y_G = \frac{\Sigma W_i y_i}{W} = \frac{525}{200} = 2.625 \text{ in.} \qquad \text{Ans.}$$

$$W z_G = \Sigma W_i z_i \qquad z_G = \frac{\Sigma W_i z_i}{W} = \frac{340}{200} = 1.700 \text{ in.} \qquad \text{Ans.}$$

The resultant weight $\mathbf{W}$ of a system of particles is the sum of the weights of the individual particles; furthermore, the first moment of the resultant is the algebraic sum of the first moments of the individual particles about any axis or plane.

# PROBLEMS

## Introductory Problems

**5-1\*** Locate the center of gravity for the three particles shown in Fig. P5-1 if $W_A = 25$ lb, $W_B = 30$ lb, and $W_C = 45$ lb.

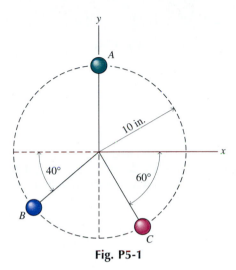

**Fig. P5-1**

**5-2\*** Locate the center of mass for the three particles shown in Fig. P5-2 if $m_A = 26$ kg, $m_B = 21$ kg, and $m_C = 36$ kg.

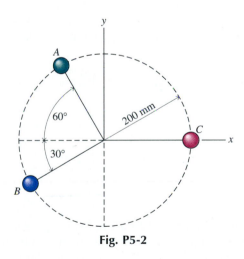

**Fig. P5-2**

**5-3** Locate the center of gravity for the four particles shown in Fig. P5-3 if $W_A = 20$ lb, $W_B = 25$ lb, $W_C = 30$ lb, and $W_D = 40$ lb.

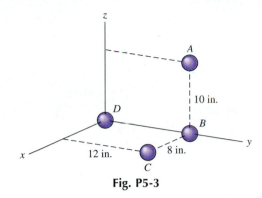

**Fig. P5-3**

**5-4** Locate the center of mass for the four particles shown in Fig. P5-4 if $m_A = 16$ kg, $m_B = 24$ kg, $m_C = 14$ kg, and $m_D = 36$ kg.

**Fig. P5-4**

## Intermediate Problems

**5-5\*** Locate the center of gravity for the five particles shown in Fig. 5-5 if $W_A = 25$ lb, $W_B = 35$ lb, $W_C = 15$ lb, $W_D = 28$ lb, and $W_E = 16$ lb.

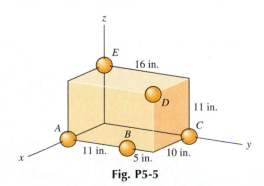

**Fig. P5-5**

**195**

**5-6\*** Locate the center of mass for the five particles shown in Fig. 5-6 if $m_A = 2$ kg, $m_B = 3$ kg, $m_C = 4$ kg, $m_D = 3$ kg, and $m_E = 2$ kg.

Fig. P5-5

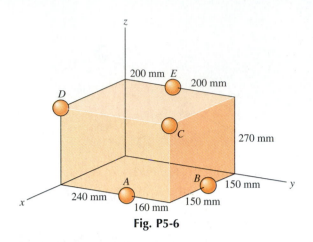

Fig. P5-6

**5-7** Locate the center of gravity for the five particles shown in Fig. 5-5 if $W_A = 15$ lb, $W_B = 24$ lb, $W_C = 35$ lb, $W_D = 18$ lb, and $W_E = 26$ lb.

**5-8** Locate the center of mass for the five particles shown in Fig. 5-6 if $m_A = 6$ kg, $m_B = 9$ kg, $m_C = 5$ kg, $m_D = 7$ kg, and $m_E = 4$ kg.

**5-9\*** Three bodies with masses of 2, 4, and 6 slugs are located at points (2, 3, 4), (3, −4, 5), and (−3, 4, 6), respectively. Locate the mass center of the system if the distances are measured in feet.

**5-10** Three bodies with masses of 3, 6, and 7 kg are located at points (4, −3, 1), (−1, 3, 2), and (2, 2, −4), respectively. Locate the mass center of the system if the distances are measured in meters.

## 5-3 CENTROIDS OF VOLUMES, AREAS, AND LINES

### 5-3.1 Centroids of Volumes

If the density $\rho$ (mass per unit volume) of the material that makes up a body is constant (the same for every particle which makes up the body), then $m = \rho V$, $dm = \rho\, dV$, and Eqs. 5-8 can be divided by the density $\rho$ to get

$$V = \int dV \qquad (5\text{-}9a)$$

$$Vx_C = \int x\, dV \quad \text{or} \quad x_C = \frac{1}{V}\int x\, dV \qquad (5\text{-}9b)$$

$$Vy_C = \int y\, dV \quad \text{or} \quad y_C = \frac{1}{V}\int y\, dV \qquad (5\text{-}9c)$$

$$Vz_C = \int z\, dV \quad \text{or} \quad z_C = \frac{1}{V}\int z\, dV \qquad (5\text{-}9d)$$

where $V$ is the volume of the body. The coordinates $x_C$, $y_C$, and $z_C$, defined by Eqs. 5-9, depend only on the geometry of the body and are independent of the physical properties. The point located by such a set of coordinates is known as the centroid $C$ of the volume of the body. The term *centroid* is usually used in connection with geometrical figures (volumes, areas, and lines) whereas the terms *center of mass* and *center of gravity* are used in connection with physical bodies.

Note that the centroid $C$ of a body can be defined whether the body is homogeneous (made of a uniform material; constant density)

**196**

or not. If the body is homogeneous, the centroid will have the same position as the center of mass and the center of gravity. If the density of the material varies from point to point within the body, the center of gravity of the body and the centroid of the volume occupied by the body will usually be at different points, as indicated in Fig. 5-6. Since the density of the lower portion of the cone in Fig. 5-6 is greater than the density of the upper portion of the cone, the center of gravity G (which depends on the weights of the two parts) will be below the centroid C (which depends only on the volume of the two parts).

## 5-3.2 Centroids of Areas

If the body consists of a homogeneous, thin plate of uniform thickness $t$ and surface area $A$, then $m = \rho t A$, $dm = \rho t\, dA$, and Eqs. 5-8 can be divided by $\rho t$ to get

$$A = \int dA \tag{5-10a}$$

$$Ax_C = \int x\, dA \quad \text{or} \quad x_C = \frac{1}{A}\int x\, dA \tag{5-10b}$$

$$Ay_C = \int y\, dA \quad \text{or} \quad y_C = \frac{1}{A}\int y\, dA \tag{5-10c}$$

$$Az_C = \int z\, dA \quad \text{or} \quad z_C = \frac{1}{A}\int z\, dA \tag{5-10d}$$

For a thin three-dimensional shell (such as the dome covering a football stadium), three coordinates $x_C$, $y_C$, and $z_C$ are required to specify the location of the centroid C of the shell. The location of the centroid in this case often does not lie on the shell. For example, the centroid of a basketball is not located on the rubber shell of the ball but is located at the center of the hollow cavity inside the shell. For a flat plate in the $xy$-plane, only two coordinates in the plane of the plate $x_C$ and $y_C$ are required to specify the location of the centroid C of the plate. Even in this case, the centroid may lie inside or outside of the plane area.

## 5-3.3 Centroids of Lines

If the body consists of a homogeneous curved wire with a small uniform cross-sectional area $A$ and length $L$, then $m = \rho A L$, $dm = \rho A\, dL$, and Eqs. 5-8 can be divided by $\rho A$ to get

$$L = \int dL \tag{5-11a}$$

$$Lx_C = \int x\, dL \quad \text{or} \quad x_C = \frac{1}{L}\int x\, dL \tag{5-11b}$$

$$Ly_C = \int y\, dL \quad \text{or} \quad y_C = \frac{1}{L}\int y\, dL \tag{5-11c}$$

$$Lz_C = \int z\, dL \quad \text{or} \quad z_C = \frac{1}{L}\int z\, dL \tag{5-11d}$$

Two or three coordinates, depending on the shape, are required to specify the location of the centroid of the line defining the shape of the wire. Unless the line is straight, the centroid will usually not lie on the line.

**Figure 5-6** Centroid and center of gravity locations for a body with a specific weight that varies.

## 5-3.4   Centroid, Center of Mass, or Center of Gravity by Integration

The procedure involved in the determination, by integration, of the coordinates of the centroid, center of mass, or center of gravity of a body can be summarized as follows:

1. Prepare a sketch of the body approximately to scale.
2. Establish a coordinate system. Rectangular coordinates are used with most shapes that have flat planes for boundaries. Polar coordinates are usually used for shapes with circular boundaries. Whenever a line or plane of symmetry exists in a body, a coordinate axis or plane should be chosen to coincide with this line or plane. The centroid, center of mass, or center of gravity will always lie on such a line or plane since the moments of symmetrically located pairs of elements (one with a positive coordinate and the other with an equal negative coordinate) will always cancel.
3. Select an element of volume, area, or length. For center of mass or center of gravity determinations, determine the mass or weight of the element by using the appropriate expression (constant or variable) for the density or specific weight. The element can frequently be selected so that only single integration is required for the complete body or for the several parts into which the body can be divided. Sometimes, however, it may be necessary to use double integration or perhaps triple integration for some shapes. If possible, the element should be chosen so that all parts are the same distance from the reference axis or plane. This distance will be the moment arm for first moment determinations. When the parts of the element are at different distances from the reference axis or plane, the location of the centroid, center of mass, or center of gravity of the element must be known in order to establish the moment arm for moment calculations. Integrate the expression to determine the volume, area, length, mass, or weight of the body.
4. Write an expression for the first moment of the element with respect to one of the reference axes or planes. Integrate the expression to determine the first moment with respect to the reference axis or plane.
5. Use the appropriate equation (Eqs. 5-4, 5-5, etc.) to obtain the coordinate of the centroid, center of mass, or center of gravity with respect to the axis or plane.
6. Repeat steps 3 to 5, using different reference axes or planes for the other coordinates of the centroid, center of mass, or center of gravity.
7. Locate the centroid, center of mass, or center of gravity on the sketch. Gross errors are often detected by using this last step.

Locate the centroid of the rectangular area shown in Fig. 5-7a.

**Fig. 5-7**

## SOLUTION

Symmetry considerations require that the centroid of a rectangular area be located at the center of the rectangle. Thus, for the rectangular area shown in Fig. 5-7a, $x_C = b/2$ and $y_C = h/2$. These results are established by integration in the following fashion. For the differential element of area shown in Fig. 5-7b, $dA = b\,dy$. The element $dA$ is located a distance $y$ from the $x$-axis; therefore, the moment of the area about the $x$-axis is

$$M_x = \int_A y\,dA = \int_0^h y\,(b\,dy) = b\left[\frac{y^2}{2}\right]_0^h = \frac{bh^2}{2}$$

> If an area has an axis of symmetry, the centroid is located on that axis; and, if the area has two axes of symmetry, the centroid is located at the point of intersection of the two axes.

The principle of moments states that

$$Ay_C = \int y\,dA = M_x$$

Therefore,

$$y_C = \frac{M_x}{A} = \frac{bh^2/2}{bh} = \frac{h}{2} \qquad \text{Ans.}$$

In a similar manner, using an element of area $dA = h\,dx$, the moment of the area about the $y$-axis is

$$M_y = \int_A x\,dA = \int_0^b x\,(h\,dx) = h\left[\frac{x^2}{2}\right]_0^b = \frac{hb^2}{2}$$

The principle of moments states that

$$Ax_C = \int x\,dA = M_y$$

Therefore,

$$x_C = \frac{M_y}{A} = \frac{hb^2/2}{bh} = \frac{b}{2} \qquad \text{Ans.}$$

The element of area $dA = b\,dy$, used to calculate $M_x$, was not used to calculate $M_y$ since all parts of the horizontal strip are located at different distances $x$ from the $y$-axis. As a result of this example, it is now known that $x_C = b/2$ for the element of area $dA = b\,dy$ shown in Fig. 5-7b. This result will be used frequently in later examples to simplify the integrations.

## EXAMPLE PROBLEM 5-3

Locate the $y$-coordinate of the centroid of the area of the quarter circle shown in Fig. 5-8a.

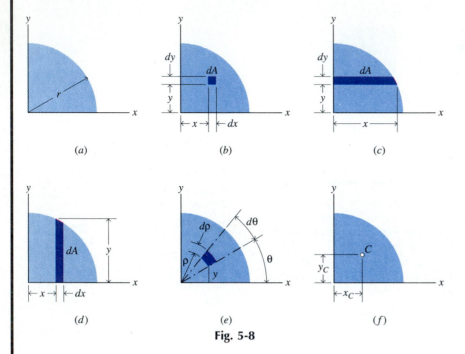

Fig. 5-8

---

## SOLUTION

Four different elements will be used to solve this problem.

### METHOD 1:   Double integral in rectangular coordinates

For the differential element shown in Fig. 5-8b, $dA = dy\, dx$. The element $dA$ is at a distance $y$ from the $x$-axis; therefore, the moment of the area about the $x$-axis is

$$M_x = \int_A y\, dA = \int_0^r \int_0^{\sqrt{r^2 - x^2}} y\, dy\, dx$$

$$= \int_0^r \left[ \frac{y^2}{2} \right]_0^{\sqrt{r^2 - x^2}} dx = \int_0^r \frac{r^2 - x^2}{2}\, dx = \left[ \frac{r^2 x}{2} - \frac{x^3}{6} \right]_0^r = \frac{r^3}{3}$$

The principle of moments states that

$$Ay_C = \int y\, dA = M_x$$

Therefore,

$$y_C = \frac{M_x}{A} = \frac{r^3/3}{\pi r^2/4} = \frac{4r}{3\pi} \qquad \text{Ans.}$$

> Double integration is required to solve the problem if a differential element is chosen with sides $dx$ and $dy$ at distances $x$ and $y$ from the reference axes.

## METHOD 2:  Single integral using a horizontal strip

Alternatively, the element of area can be selected as shown in Fig. 5-8c. For this element, which is located a distance $y$ from the $x$-axis, $dA = x\,dy = \sqrt{r^2 - y^2}\,dy$. Therefore, the moment of the area about the $x$-axis is

$$M_x = \int_A y\,dA = \int_0^r y\,\sqrt{r^2 - y^2}\,dy = \left[-\frac{(r^2 - y^2)^{3/2}}{3}\right]_0^r = \frac{r^3}{3}$$

The principle of moments states that

$$Ay_C = \int y\,dA = M_x$$

Therefore,

$$y_C = \frac{M_x}{A} = \frac{r^3/3}{\pi r^2/4} = \frac{4r}{3\pi} \qquad \text{Ans.}$$

The problem can be solved with a single integration if a differential element is chosen as a strip with width $x$ that varies as a function of $y$.

## METHOD 3:  Single integral using a vertical strip

The element of area could also be selected as shown in Fig. 5-8d. For this element, $dA = y\,dx = \sqrt{r^2 - x^2}\,dx$; however, all parts of the element are at different distances $y$ from the $x$-axis. For this type of element, the results of Example Problem 5-2 can be used to compute a moment $dM_x$, which can be integrated to yield moment $M_x$. Thus,

$$dM_x = \frac{y}{2}\,dA = \frac{y}{2}\,y\,dx = \frac{y^2}{2}\,dx = \frac{r^2 - x^2}{2}\,dx$$

$$M_x = \int_A dM_x = \int_0^r \frac{r^2 - x^2}{2}\,dx = \left[\frac{r^2 x}{2} - \frac{x^3}{6}\right]_0^r = \frac{r^3}{3}$$

The principle of moments states that

$$Ay_C = \int y\,dA = M_x$$

Therefore,

$$y_C = \frac{M_x}{A} = \frac{r^3/3}{\pi r^2/4} = \frac{4r}{3\pi} \qquad \text{Ans.}$$

Similarly, the problem can be solved with a single integration if a differential element is chosen as a strip with height $y$ that varies as a function of $x$.

## METHOD 4:  Double integral using polar coordinates

Finally, polar coordinates can be used to locate the centroid of the quarter circle. With polar coordinates, the element of area is $dA = \rho\,d\theta\,d\rho$ and the distance from the $x$-axis to the element is $y = \rho \sin \theta$ as shown in Fig. 5-8e. Thus,

$$M_x = \int_A y\,dA = \int_0^r \int_0^{\pi/2} \rho^2 \sin \theta\,d\theta\,d\rho$$

$$= \int_0^r \rho^2 \left[-\cos \theta\right]_0^{\pi/2} d\rho = \int_0^r \rho^2\,d\rho = \left[\frac{\rho^3}{3}\right]_0^r = \frac{r^3}{3}$$

The principle of moments states that

$$Ay_C = \int y\,dA = M_x$$

Therefore,

$$y_C = \frac{M_x}{A} = \frac{r^3/3}{\pi r^2/4} = \frac{4r}{3\pi} \qquad \text{Ans.}$$

Frequently, polar coordinates are more efficient when circular boundaries are involved.

In a completely similar manner, the $x$-coordinate of the centroid is obtained as

$$x_C = \frac{M_y}{A} = \frac{r^3/3}{\pi r^2/4} = \frac{4r}{3\pi}$$

The results are illustrated in Fig. 5-8f.

## EXAMPLE PROBLEM 5-4

Locate the centroid of the triangular area shown in Fig. 5-9a.

(a)

### SOLUTION

If the horizontal differential element of area shown in Fig. 5-9b is selected for the integration, the same element can be used for the complete area. From similar triangles in Figure 5-9b,

$$\frac{h-y}{w} = \frac{h}{b} \quad \text{or} \quad w = \frac{b}{h}(h-y)$$

Therefore, for this element, $dA = w\,dy = (b/h)(h-y)\,dy$ and the distance from the $x$-axis to the element is $y$. Thus, the moment of the area about the $x$-axis is

$$M_x = \int_A y\,dA = \int_0^h \frac{b}{h}(hy - y^2)\,dy = \frac{b}{h}\left[\frac{hy^2}{2} - \frac{y^3}{3}\right]_0^h = \frac{bh^2}{6}$$

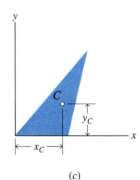

(b)

The principle of moments states that

$$Ay_C = \int y\,dA = M_x$$

Therefore,

$$y_C = \frac{M_x}{A} = \frac{bh^2/6}{bh/2} = \frac{h}{3} \qquad \text{Ans.}$$

The distance $x$ from the $y$-axis to the centroid of the element of area $dA$ is

$$x = \frac{ay}{h} + \frac{w}{2} = \frac{b}{2} + \frac{(2a-b)y}{2h}$$

Thus, the moment of the area about the $y$-axis is

$$M_y = \int_A x\,dA = \int_0^h \left[\frac{b}{2} + \frac{(2a-b)y}{2h}\right]\left[\frac{b}{h}(h-y)\right]dy$$

$$= \frac{1}{2h^2}\left[b^2h^2y - \frac{b^2hy^2}{2} + \frac{2abhy^2}{2} - \frac{2aby^3}{3} - \frac{b^2hy^2}{2} + \frac{b^2y^3}{3}\right]_0^h$$

$$= \frac{bh(a+b)}{6}$$

(c)

**Fig. 5-9**

The principle of moments states that

$$Ax_C = \int x\,dA = M_y$$

Therefore,

$$x_C = \frac{M_y}{A} = \frac{bh(a+b)/6}{bh/2} = \frac{a+b}{3} \qquad \text{Ans.}$$

The results are shown in Fig. 5-9c.

> The width of the element can be expressed as a function of $y$ once it is observed that the entire triangle is similar to the triangle that lies above the shaded element.

A circular arc of thin homogeneous wire is shown in Fig. 5-10a.

**a.** Locate the $x$- and $y$-coordinates of the mass center.
**b.** Use the results of part $a$ to determine the coordinates of the mass center for a half circle.

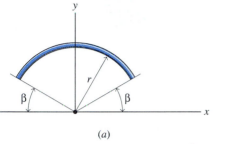

(a)                                   (b)

**Fig. 5-10**

## SOLUTION

**a.** The wire can be assumed to consist of a large number of differential elements of length $dL$ as shown in Fig. 5-10b. The mass of each of these elements is

$$dm = \rho \, dV = \rho A \, dL = \rho A (r \, d\theta)$$

Therefore, the total mass of the wire is

$$m = \int dm = \int_{\beta}^{\pi-\beta} \rho A r \, d\theta = \rho A r \int_{\beta}^{\pi-\beta} d\theta = \rho A r (\pi - 2\beta)$$

The distance $y$ from the $x$-axis to the element $dm$ is $y = r \sin \theta$. Thus,

$$my_C = \int y \, dm = \int_{\beta}^{\pi-\beta} (r \sin \theta)(\rho A r \, d\theta)$$

$$= \rho A r^2 \int_{\beta}^{\pi-\beta} \sin \theta \, d\theta = \rho A r^2 (2 \cos \beta)$$

Therefore,

$$y_C = \frac{2\rho A r^2 \cos \beta}{\rho A r (\pi - 2\beta)} = \frac{2r \cos \beta}{\pi - 2\beta} \qquad \text{Ans.}$$

Since the length of wire is symmetric about the $y$-axis,

$$x_C = 0 \qquad \text{Ans.}$$

**b.** For the half circle, $\beta = 0$

$$y_C = \frac{2r}{\pi} \qquad \text{Ans.}$$

$$x_C = 0 \qquad \text{Ans.}$$

> Polar coordinates are more efficient for this determination since the shape of the wire is a circular arc.

Locate the center of gravity $G$ of the homogeneous right circular cone shown in Fig. 5-11a, which has an altitude $h$ and radius $r$ and is made of a material with a specific weight $\gamma$.

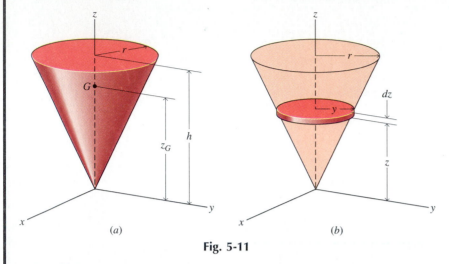

**Fig. 5-11**

## SOLUTION

From symmetry, it is obvious that $x_G = y_G = 0$. The coordinate $z_G$ of the center of gravity $G$ of the cone can be located by using the differential element of volume shown in Fig. 5-11b. The weight $dW$ of the differential element is

$$dW = \gamma\, dV = \gamma\, (\pi y^2)\, dz = \gamma\pi \left(\frac{rz}{h}\right)^2 dz = \frac{\gamma\pi r^2}{h^2}\, z^2\, dz$$

The principle of moments states that

$$Wz_G = \int z\, dW = M_{xy}$$

Therefore,

$$z_G = \frac{\int z\, dW}{\int dW} = \frac{\int_V z\, (\gamma\, dV)}{\int_V \gamma\, dV}$$

Thus,

$$M_{xy} = Wz_G = \int z\, dW = \int_0^h \frac{\gamma\pi r^2}{h^2}\, z^3\, dz = \frac{1}{4}\, \gamma\pi r^2 h^2$$

The weight of the cone is

$$W = \int_V \gamma\, dV = \int_0^h \frac{\gamma\pi r^2}{h^2}\, z^2\, dz = \frac{\gamma\pi r^2}{h^2} \left[\frac{z^3}{3}\right]_0^h = \frac{1}{3}\, \gamma\pi r^2 h$$

Therefore,

$$z_G = \frac{M_{xy}}{W} = \frac{\gamma\pi r^2 h^2/4}{\gamma\pi r^2 h/3} = \frac{3h}{4} \qquad \text{Ans.}$$

Since the $xz$-plane and the $yz$-plane are planes of symmetry,

$$x_G = y_G = 0 \qquad \text{Ans.}$$

This problem can be solved with a single integration if a thin circular disk element is used.

Locate the centroid of the volume of the hemisphere shown in Fig. 5-12a.

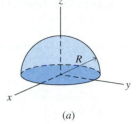

(a)

## SOLUTION

From symmetry, it is obvious that $x_C = y_C = 0$. The coordinate $z_C$ of the centroid $C$ of the hemisphere will be located by using two different methods.

**METHOD 1:   Single integral using rectangular coordinates**

For the differential element of volume shown in Fig. 5-12b,

$$dV = \pi y^2\, dz = \pi(R^2 - z^2)\, dz$$
$$V = \int_V dV = \int_0^R \pi(R^2 - z^2)\, dz = \pi\left[R^2 z - \frac{z^3}{3}\right]_0^R = \frac{2}{3}\pi R^3$$

> This problem can be solved with a single integration if a thin circular disk element is used.

and

$$M_{xy} = \int_V z\, dV = \int_0^R \pi(R^2 z - z^3)\, dz = \left[\frac{\pi R^2 z^2}{2} - \frac{\pi z^4}{2}\right]_0^R = \frac{\pi R^4}{4}$$

The principle of moments states that

$$V z_C = \int z\, dV = M_{xy}$$

Therefore,

$$z_C = \frac{M_{xy}}{V} = \frac{\pi R^4/4}{2\pi R^3/3} = \frac{3}{8}R \qquad \text{Ans.}$$

**METHOD 2:   Triple integral using spherical coordinates**

For the element of volume shown in Fig. 5-12c,

$$z = r \cos\phi$$
$$dV = (r\, d\phi)(r \sin\phi\, d\theta)\, dr = r^2 \sin\phi\, dr\, d\phi\, d\theta$$

> Triple integration is required to solve this problem if spherical coordinates are used.

Thus,

$$M_{xy} = \int_V z\, dV = \int_0^{2\pi}\int_0^{\pi/2}\int_0^R r^3 \sin\phi \cos\phi\, dr\, d\phi\, d\theta$$
$$= \frac{R^4}{4}\int_0^{2\pi}\int_0^{\pi/2} \sin\phi \cos\phi\, d\phi\, d\theta$$
$$= \frac{R^4}{8}\int_0^{2\pi}\left[\sin^2\phi\right]_0^{\pi/2} d\theta = \frac{R^4}{8}\int_0^{2\pi} d\theta = \frac{\pi R^4}{4}$$

The principle of moments states that

$$V z_C = \int z\, dV = M_{xy}$$

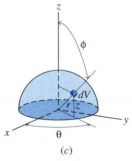

(c)

**Fig. 5-12**

Therefore,

$$z_C = \frac{M_{xy}}{V} = \frac{\pi R^4/4}{2\pi R^3/3} = \frac{3}{8}R \qquad \text{Ans.}$$

Locate the centroid of the arc of thin wire shown in Fig. 5-13a.

## SOLUTION

The wire can be assumed to consist of a large number of differential elements of length $dL$ as shown in Fig. 5-13b. The length $dL$ of each of these elements is given by the expression

$$dL = \sqrt{(dx)^2 + (dy)^2} = \sqrt{(dx/dy)^2 + 1}\, dy$$

For the parabolic arc $y^2 = 12x$:

$$dL = \sqrt{(y/6)^2 + 1}\, dy = \frac{1}{6}\sqrt{y^2 + 36}\, dy$$

$$L = \frac{1}{6}\int_0^{12} \sqrt{y^2 + 36}\, dy$$

$$= \frac{1}{6}\left[\frac{y}{2}\sqrt{y^2 + 36} + 18\ln\left(y + \sqrt{y^2 + 36}\right)\right]_0^{12} = 17.747 \text{ in.}$$

The element $dL$ is located at distances $x$ and $y$ from the coordinate $y$- and $x$-axes, respectively. Thus,

$$M_x = \int_L y\, dL = \frac{1}{6}\int_0^{12} y\sqrt{y^2 + 36}\, dy = \frac{1}{6}\left[\frac{1}{3}\sqrt{(y^2 + 36)^3}\right]_0^{12} = 122.16 \text{ in.}^2$$

$$M_y = \int_L x\, dL = \frac{1}{72}\int_0^{12} y^2\sqrt{y^2 + 36}\, dy$$

$$= \frac{1}{72}\left[\frac{y}{4}\sqrt{(y^2 + 36)^3} - \frac{36y}{8}\sqrt{(y^2 + 36)} - 162\ln\left(y + \sqrt{y^2 + 36}\right)\right]_0^{12}$$

$$= 87.31 \text{ in.}^2$$

The principle of moments states that

$$Ly_C = \int y\, dL = M_x$$

$$Lx_C = \int x\, dL = M_y$$

Therefore,

$$y_C = \frac{M_x}{L} = \frac{122.16}{17.747} = 6.88 \text{ in.} \qquad\qquad \text{Ans.}$$

$$x_C = \frac{M_y}{A} = \frac{87.312}{17.747} = 4.92 \text{ in.} \qquad\qquad \text{Ans.}$$

**Fig. 5-13**

The length $dL$ of the element is given by the Pythagorean theorem as

$$dL = \sqrt{(dx)^2 + (dy)^2}$$

# PROBLEMS

## Introductory Problems

Locate the centroid of the shaded area shown in each of the following figures.

**5-11\*** Fig. P5-11 if $b = 12$ in. and $h = 8$ in.

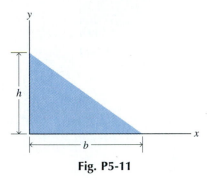

**Fig. P5-11**

**5-12\*** Fig. P5-12 if $b = 200$ mm and $h = 300$ mm.

**Fig. P5-12**

**5-13** Fig. P5-13.

$$\frac{x^2}{a^2} + \frac{y^2}{b^2} = 1$$

**Fig. P5-13**

**5-14** Fig. P5-14.

$$y = \frac{x^2}{2b}$$

**Fig. P5-14**

**5-15\*** Fig. P5-15.

$$y = \sqrt{x}$$

**Fig. P5-15**

**5-16** Fig. P5-16.

$$25y = 50x - x^2$$

**Fig. P5-16**

**5-17\*** Fig. P5-17.

$$y = b$$

$$y^2 = \frac{b^2 x}{a}$$

**Fig. P5-17**

**5-18\*** Fig. P5-18.

**Fig. P5-18**

**5-19** Fig. P5-19.

**Fig. P5-19**

**5-20** Fig. P5-20.

**Fig. P5-20**

## Intermediate Problems

**5-21\*** Fig. P5-21.

**Fig. P5-21**

**5-22\*** Fig. P5-22.

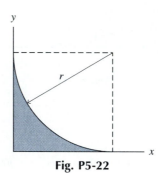

**Fig. P5-22**

**5-23** Locate the centroid of the volume obtained by revolving the shaded area shown in Fig. P5-23 about the $x$-axis.

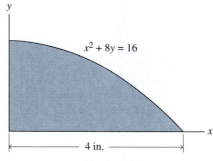

**Fig. P5-23**

**5-24** Locate the centroid of the volume obtained by revolving the shaded area shown in Fig. P5-24 about the $x$-axis.

**Fig. P5-24**

## Challenging Problems

Locate the centroid of the curved slender rod shown in each of the following figures.

**5-25\*** Fig. P5-25.

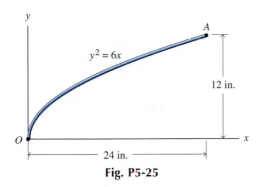

**Fig. P5-25**

**5-26** Fig. P5-26 if $b = 50$ mm.

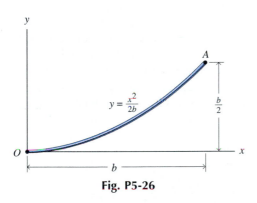

**Fig. P5-26**

**5-27\*** Locate the centroid of the volume of the portion of a right circular cone shown in Fig. P5-27.

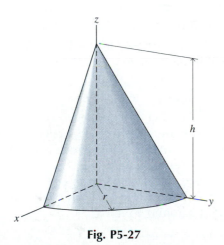

**Fig. P5-27**

**5-28\*** Locate the mass center of the hemisphere shown in Fig. P5-28 if the density $\rho$ at any point $P$ is proportional to the distance from the $xy$-plane to point $P$.

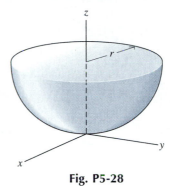

**Fig. P5-28**

**5-29** Locate the mass center of the right circular cone shown in Fig. P5-29 if the density $\rho$ at any point $P$ is proportional to the distance from the $xy$-plane to point $P$.

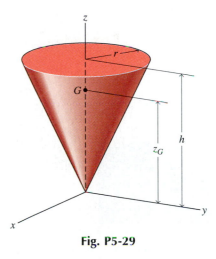

**Fig. P5-29**

**5-30** Locate the centroid of the volume of the tetrahedron shown in Fig. P5-30.

**Fig. P5-30**

*(a)*        *(b)*        *(c)*

**Figure 5-14** Centroid locations for several common shapes.

## 5-4    CENTROIDS OF COMPOSITE BODIES

The centroid of an area is a property of the shape of the area and is independent of the coordinate system used to compute it. That is, no matter what coordinate system is used to compute the location of the centroid of a rectangle, it will always be found to be halfway between the opposite sides as shown in Fig. 5-14*a*.

Similarly, the centroid of a quarter circle will always be $\frac{4r}{3\pi}$ from the radial lines bounding the quarter circle (Fig. 5-14*b*) and the centroid of a triangle will always be one-third of the way (measured perpendicularly to the side) from any side of the triangle to the opposite vertex (Fig. 5-14*c*). Once the centroids of these common shapes are known, they need not be continually recalculated. In fact, the known locations of the centroids of common simple shapes can be used to locate the centroids of more complicated shapes.

For example, the area shown in Fig. 5-15 consists of two parts—a triangle and a rectangle. Integrating to find the area and the centroid gives

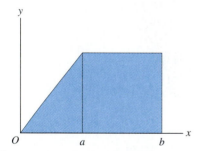

**Figure 5-15** Composite body consisting of a rectangle and a triangle.

$$A = \int_{x=0}^{x=a} dA + \int_{x=a}^{x=b} dA = A_{\text{tri}} + A_{\text{rect}}$$

$$M_y = x_C A = \int_{x=0}^{x=a} x \, dA + \int_{x=a}^{x=b} x \, dA$$

$$= (x_C A)_{\text{tri}} + (x_C A)_{\text{rect}} \qquad \textbf{(5-12)}$$

The location of the centroid of the composite area consisting of the triangle and rectangle is the sum of the first moments of the two component parts divided by the total area of the two component parts. In general, if an area is composed of $n$ simple shapes whose centroid locations are known (or can be looked up easily), the centroid of the larger area can be found from

$$x_C A = \sum_{i=1}^{n} x_{Ci} A_i \qquad A = \sum_{i=1}^{n} A_i \qquad \textbf{(5-13)}$$

Composite areas with holes are also handled easily by using Eq. 5-13. Consider the composite area of Fig. 5-16*b*, which consists of a circular hole cut out of a rectangle and triangle.

Denoting the triangle by $t$, the rectangle by $r$, the circular hole by $h$, and the composite body by $c$, the centroid location of the solid triangle-rectangle area (Fig. 5-16*a*) is found from

$$(x_C A)_t + (x_C A)_r = (x_C A)_c + (x_C A)_h$$

**Figure 5-16**   A hole in a composite area is treated as a negative area.

The first, second, and last terms are easily computed using known properties of simple shapes. Therefore, the centroid of the composite body can be found from

$$(x_C A)_c = (x_C A)_t + (x_C A)_r - (x_C A)_h$$
$$A_C = A_t + A_r - A_h$$

But these equations are identical to Eq. 5-13 if the area of the hole is considered to be a negative quantity.

Similar equations can be developed for composite lines, volumes, masses, and weights. The final results would show the $A$'s of Eqs. 5-13 replaced with $L$'s, $V$'s, $m$'s, and $W$'s, respectively. Tables 5-1 and 5-2 contain a listing of centroid locations for some common shapes.

## TABLE 5-1 CENTROID LOCATIONS FOR A FEW COMMON LINE SEGMENTS AND AREAS

**Circular arc**

$$L = 2r\alpha$$

$$x_C = \frac{r \sin \alpha}{\alpha}$$

$$x_C = 0$$

**Circular sector**

$$A = r^2\alpha$$

$$x_C = \frac{2r \sin \alpha}{3\alpha}$$

$$y_C = 0$$

**Quarter circular arc**

$$L = \frac{\pi r}{2}$$

$$x_C = \frac{2r}{\pi}$$

$$y_C = \frac{2r}{\pi}$$

**Quadrant of a circle**

$$A = \frac{\pi r^2}{4}$$

$$x_C = \frac{4r}{3\pi}$$

$$y_C = \frac{4r}{3\pi}$$

**Semicircular arc**

$$L = \pi r$$

$$x_C = r$$

$$y_C = \frac{2r}{\pi}$$

**Semicircular area**

$$A = \frac{\pi r^2}{2}$$

$$x_C = r$$

$$y_C = \frac{4r}{3\pi}$$

**Rectangular area**

$$A = bh$$

$$x_C = \frac{b}{2}$$

$$y_C = \frac{h}{2}$$

**Quadrant of an ellipse**

$$A = \frac{\pi ab}{4}$$

$$x_C = \frac{4a}{3\pi}$$

$$y_C = \frac{4b}{3\pi}$$

**Triangular area**

$$A = \frac{bh}{2}$$

$$x_C = \frac{2b}{3}$$

$$y_C = \frac{h}{3}$$

**Parabolic spandrel**

$$A = \frac{bh}{3}$$

$$x_C = \frac{3b}{4}$$

$$y_C = \frac{3h}{10}$$

**Triangular area**

$$A = \frac{bh}{2}$$

$$x_C = \frac{a + b}{3}$$

$$y_C = \frac{h}{3}$$

**Quadrant of a parabola**

$$A = \frac{2bh}{3}$$

$$x_C = \frac{5b}{8}$$

$$y_C = \frac{2h}{5}$$

## TABLE 5-2   CENTROID LOCATIONS FOR A FEW COMMON VOLUMES

### Rectangular parallelepiped

$$V = abc$$

$$x_C = \frac{a}{2}$$

$$y_C = \frac{b}{2}$$

$$z_C = \frac{c}{2}$$

### Rectangular tetrahedron

$$V = \frac{abc}{6}$$

$$x_C = \frac{a}{4}$$

$$y_C = \frac{b}{4}$$

$$z_C = \frac{c}{4}$$

### Circular cylinder

$$V = \pi r^2 L$$

$$x_C = 0$$

$$y_C = \frac{L}{2}$$

$$z_C = 0$$

### Semicylinder

$$V = \frac{\pi r^2 L}{2}$$

$$x_C = 0$$

$$y_C = \frac{L}{2}$$

$$z_C = \frac{4r}{3\pi}$$

### Hemisphere

$$V = \frac{2\pi r^3}{3}$$

$$x_C = 0$$

$$y_C = 0$$

$$z_C = \frac{3r}{8}$$

### Paraboloid

$$V = \frac{\pi r^2 h}{2}$$

$$x_C = 0$$

$$y_C = \frac{2h}{3}$$

$$z_C = 0$$

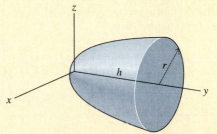

### Right circular cone

$$V = \frac{\pi r^2 h}{3}$$

$$x_C = 0$$

$$y_C = \frac{3h}{4}$$

$$z_C = 0$$

### Half cone

$$V = \frac{\pi r^2 h}{6}$$

$$x_C = 0$$

$$y_C = \frac{3h}{4}$$

$$z_C = \frac{r}{\pi}$$

Locate the centroid of the slender rod shown in Fig. 5-17a.

(a)                                      (b)

**Fig. 5-17**

## SOLUTION

The rod can be divided into five parts as shown in Fig. 5-17b. The centroid for the composite rod is determined by listing the length, the centroid location, and the first moment for the individual parts in a table and applying the principle of moments. Thus,

$$L_3 = \pi r = \pi(8) = 25.13 \text{ in.}$$

$$z_{C3} = \frac{2r}{\pi} = \frac{2(8)}{\pi} = 5.09 \text{ in.}$$

> The principle of moments can be used to locate the centroid of any composite body if first moments of the individual parts are known or can be determined.

| Part | $L_i$ (in.) | $x_{Ci}$ (in.) | $M_{yz}$ (in.²) | $y_{Ci}$ (in.) | $M_{zx}$ (in.²) | $z_{Ci}$ (in.) | $M_{xy}$ (in.²) |
|------|------|------|------|------|------|------|------|
| 1 | 9 | 10 | 90 | 4.5 | 40.5 | 0 | 0 |
| 2 | 10 | 5 | 50 | 0 | 0 | 0 | 0 |
| 3 | 25.13 | 0 | 0 | 8 | 201.0 | 5.093 | 128.0 |
| 4 | 10 | 5 | 50 | 16 | 160 | 0 | 0 |
| 5 | 9 | 10 | 90 | 16 | 144 | 4.5 | 40.5 |
| Σ | 63.13 | | 280 | | 545.5 | | 168.5 |

The principle of moments states that

$$Lx_C = \int x \, dL = M_{yz} \qquad x_C = \frac{M_{yz}}{L} = \frac{280}{63.13} = 4.44 \text{ in.} \qquad \text{Ans.}$$

$$Ly_C = \int y \, dL = M_{zx} \qquad y_C = \frac{M_{zx}}{L} = \frac{545.5}{63.13} = 8.64 \text{ in.} \qquad \text{Ans.}$$

$$Lz_C = \int z \, dL = M_{xy} \qquad z_C = \frac{M_{xy}}{L} = \frac{168.5}{63.13} = 2.67 \text{ in.} \qquad \text{Ans.}$$

Locate the centroid of the composite area shown in Fig. 5-18a.

## SOLUTION

The composite area can be divided into four parts, a rectangle, a triangle, a quarter circle, and a circle, as shown in Fig. 5-18b. Recall that the area of the hole is negative since it must be subtracted from the area of the rectangle. The centroid locations for each of these parts can be determined from the relationships listed in Table 5-1.

For the triangle:

(a)

$$x_{C2} = \frac{2b}{3} = \frac{50}{3} = 16.67 \text{ mm}$$

$$y_{C2} = 50 + \frac{h}{3} = 50 + \frac{50}{3} = 66.67 \text{ mm}$$

For the quarter circle:

$$x_{C3} = \frac{4r}{3\pi} = -\frac{4(50)}{3\pi} = -21.22 \text{ mm}$$

$$y_{C3} = 50 + \frac{4r}{3\pi} = 50 + \frac{4(50)}{3\pi} = 71.22 \text{ mm}$$

The centroid for the composite area is determined by listing the area, centroid location, and first moments for the individual parts in a table and applying the principle of moments. Thus,

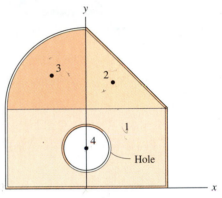

(b)

**Fig. 5-18**

| Part | $A_i$ (mm²) | $x_{Ci}$ (mm) | $M_y$ (mm³) | $y_{Ci}$ (mm) | $M_x$ (mm³) |
|------|-------------|---------------|-------------|---------------|-------------|
| 1 | 5000 | 0 | 0 | 25 | 125,000 |
| 2 | 1250 | 16.67 | 20,838 | 66.67 | 83,338 |
| 3 | 1963 | −21.22 | −41,665 | 71.22 | 139,805 |
| 4 | −707 | 0 | 0 | 25 | −17,675 |
| Σ | 7506 | | −20,827 | | 330,468 |

The principle of moments states that

$$x_C A = \Sigma x_{Ci} A_i = M_y \qquad x_C = \frac{M_y}{A} = \frac{-20,827}{7506} = -2.77 \text{ mm} \qquad \text{Ans.}$$

$$y_C A = \Sigma y_{Ci} A_i = M_x \qquad y_C = \frac{M_x}{A} = \frac{330,468}{7506} = 44.0 \text{ mm} \qquad \text{Ans.}$$

The principle of moments can be used to locate the centroid of any composite area if first moments of the individual parts are known or can be determined.

# PROBLEMS

## Introductory Problems

**5-31*** Locate the centroid of the slender rod shown in Fig. P5-31.

**Fig. P5-31**

**5-32** Locate the centroid of the slender rod shown in Fig. P5-32.

**Fig. P5-32**

Locate the centroid of the shaded area shown in each of the following figures.

**5-33*** Fig. P5-33.

**Fig. P5-33**

**5-34*** Fig. P5-34.

**Fig. P5-34**

**5-35** Fig. P5-35.

**Fig. P5-35**

**5-36**  Fig. P5-36.

80 mm

80 mm

120 mm

80 mm | 80 mm | 80 mm

**Fig. P5-36**

**5-37***  Fig. P5-37.

3 in. | 3 in.

2 in.

6 in.        2 in.

2 in.

6 in.        6 in.

**Fig. P5-37**

**5-38**  Fig. P5-38.

50 mm

25 mm

50 mm

25 mm

25 mm

50 mm | 50 mm

**Fig. P5-38**

**5-39***  Fig. P5-39.

7 in.        7 in.

7 in.

6 in.

7 in.

**Fig. P5-39**

**5-40**  Fig. P5-40.

75 mm

75 mm

50 mm

75 mm        75 mm

**Fig. P5-40**

## Intermediate Problems

**5-41***  Locate the centroid of the slender rod shown in Fig. P5-41.

5 in.

12 in.

10 in.

**Fig. P5-41**

**5-42** Locate the centroid of the slender rod shown in Fig. P5-42.

**Fig. P5-42**

Locate the centroid of the shaded area shown in each of the following figures.

**5-43\*** Fig. P5-43.

**Fig. P5-43**

**5-44\*** Fig. P5-44.

**Fig. P5-44**

**5-45** Fig. P5-45.

**Fig. P5-45**

**5-46** Fig. P5-46.

**Fig. P5-46**

## Challenging Problems

**5-47\*** Locate the centroid of the slender rod shown in Fig. P5-47.

**Fig. P5-47**

**5-48**   Locate the centroid of the slender rod shown in Fig. P5-48.

**Fig. P5-48**

**5-49\***   A bracket is made of brass ($\gamma = 0.316$ lb/in.³) and aluminum ($\gamma = 0.100$ lb/in.³) plates as shown in Fig. P5-49.

**a.**   Locate the centroid of the bracket.
**b.**   Locate the center of gravity of the bracket.

**Fig. P5-49**

**5-50\***   A cylinder with a hemispherical cavity and a conical cap is shown in Fig. P5-50.

**a.**   Locate the centroid of the composite volume if $R = 140$ mm, $L = 250$ mm, and $h = 300$ mm.

**Fig. P5-50**

**b.**   Locate the center of mass of the composite volume if the cylinder is made of steel ($\rho = 7870$ kg/m³) and the cap is made of aluminum ($\rho = 2770$ kg/m³).

**5-51**   A bracket is made of steel ($\gamma = 0.284$ lb/in.³) and aluminum ($\gamma = 0.100$ lb/in.³) plates as shown in Fig. P5-51.

**a.**   Locate the centroid of the bracket.
**b.**   Locate the center of gravity of the bracket.

**Fig. P5-51**

**5-52**   A cylinder with a conical cavity and a hemispherical cap is shown in Fig. P5-52.

**a.**   Locate the centroid of the composite volume if $R = 200$ mm, and $h = 250$ mm.
**b.**   Locate the center of mass of the composite volume if the cylinder is made of brass ($\rho = 8750$ kg/m³) and the cap is made of aluminum ($\rho = 2770$ kg/m³).

**Fig. P5-52**

**5-53\*** Locate the center of gravity of the bracket shown in Fig. P5-53 if the holes have 6-in. diameters.

**5-54** Locate the center of mass of the machine component shown in Fig. P5-54. The brass ($\rho = 8750$ kg/m³) disk $C$ is mounted on the steel ($\rho = 7870$ kg/m³) shaft $B$.

**Fig. P5-53**

**Fig. P5-54**

## 5-5 THEOREMS OF PAPPUS AND GULDINUS

Two theorems stated by Pappus[3] and Guldinus[4] before the development of calculus can be used to determine the surface area generated by revolving a plane curve or the volume generated by revolving an area about an axis that does not intersect any part of the plane curve or area. Applications of the theorems require use of the equations previously developed for locating the centroids of lines and areas.

> **Theorem 1** The area $A$ of a surface of revolution generated by revolving a plane curve of length $L$ about any nonintersecting axis in its plane is equal to the product of the length of the curve and the length of the path traveled by the centroid of the curve.

If the curve $AB$ of Fig. 5-19 is revolved about the $y$-axis, the increment of surface area $dA$ generated by the element of length $dL$ of line $AB$ is

$$dA = 2\pi z \, dL$$

Thus, the total surface area $A$ generated by revolving the line $AB$ about the $y$-axis is

$$A = 2\pi \int_L z \, dL$$

The coordinate $z_{CL}$ of the centroid $C$ of a line in the $yz$-plane is given by Eq. 5-11 as

$$z_{CL} = \frac{1}{L} \int_0^L z \, dL$$

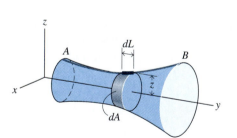

**Figure 5-19** Determining the surface area of a body of revolution.

[3]Pappus of Alexandria (about A.D. 380), a Greek geometer.
[4]Paul Guldin (1577–1643), a Swiss mathematician.

Thus, the surface area $A$ generated by revolving the line $AB$ about the $y$-axis is

$$A = 2\pi z_{CL} L \qquad (5\text{-}14)$$

where $2\pi z_{CL}$ is the distance traveled by the centroid $C$ of the line $L$ in generating the surface area $A$. The theorem is also valid if the line $AB$ is rotated through an angle $\theta$ other than $2\pi$ radians. Thus, for any angle of rotation $\theta$ ($0 \le \theta \le 2\pi$), the surface area $A$ generated is

$$A = \theta z_{CL} L \qquad (5\text{-}15)$$

Surface areas of a wide variety of shapes can be determined by using Theorem 1. Examples include the surface area of the cylinder, cone, sphere, torus, ellipsoid, and any other surface of revolution for which the generating line can be described and the location of its centroid determined.

> Theorem 2 The volume $V$ of the solid of revolution generated by revolving a plane area $A$ about any nonintersecting axis (the axis may be a boundary of area $A$ but may not pass through the area) in its plane is equal to the product of the area and the length of the path traveled by the centroid of the area.

If the plane area $A$ of Fig. 5-20 is revolved about the $y$-axis, the increment of volume $dV$ generated by the element of area $dA$ is

$$dV = 2\pi z \, dA$$

Thus, the total volume $V$ generated by revolving the area $A$ about the $y$-axis is

$$V = 2\pi \int_A z \, dA$$

**Figure 5-20** Determining the volume of a body of revolution.

The coordinate $z_{CA}$ of the centroid $C$ of an area in the $yz$-plane is given by Eq. 5-10 as

$$z_{CA} = \frac{1}{A} \int_A z \, dA$$

Thus, the volume $V$ generated by revolving the area $A$ about the $y$-axis is

$$V = 2\pi z_{CA} A \qquad (5\text{-}16)$$

where $2\pi z_{CA}$ is the distance traveled by the centroid $C$ of the area $A$ in generating the volume $V$. The theorem is also valid if the area $A$ is rotated through an angle $\theta$ other than $2\pi$ radians. Thus, for any angle of rotation $\theta$ ($0 \le \theta \le 2\pi$), the volume $V$ generated is

$$V = \theta z_{CA} A \qquad (5\text{-}17)$$

The following examples illustrate the procedure for determining surface areas and volumes for solids of revolution by using the theorems of Pappus and Guldinus.

Determine the surface area $A$ and volume $V$ of the solid (torus) generated by revolving the circle shown in Fig 5-21 through an angle of 360° about the $y$-axis.

**Fig. 5-21**

### SOLUTION

The circumference $L_c$ and area $A_c$ of the circle shown in Fig. 5-21 are

$$L_c = 2\pi r \qquad A_c = \pi r^2$$

The centroid $C_L$ of the boundary line $L_c$ and the centroid $C_A$ of the area $A_c$ of the circle shown in Fig. 5-21 are both located at the position $x = R$. Thus,

$$x_{CL} = x_{CA} = R$$

Applying the theorems of Pappus and Guldinus for areas and volumes yields the surface area and volume for the torus generated by revolving the circle about the $y$-axis.

$$A = 2\pi x_{CL} L_c = 2\pi(R)(2\pi r) = 4\pi^2 R r \qquad \text{Ans.}$$
$$V = 2\pi x_{CA} A_c = 2\pi(R)(\pi r^2) = 2\pi^2 R r^2 \qquad \text{Ans.}$$

The plane curve or area being used to generate a surface area or volume must not intersect the axis about which the curve or area is being revolved.

Determine the surface area $A$ and volume $V$ of the solid generated by revolving the shaded area shown in Fig. 5-22$a$ through an angle of 360° about the $y$-axis.

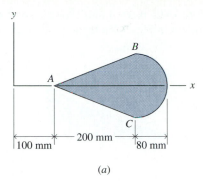

(a)

## SOLUTION

For cases where the centroid $C_L$ of the perimeter and $C_A$ of the area are not available by inspection or from tables, the theorems of Pappus and Guldinus can be used by summing results for the individual parts.

### Area Calculations

The lengths and centroid locations of segments $BAC$ and $BC$ of Fig. 5-22$b$ are

$$L_{ABC} = 2\sqrt{(200)^2 + (80)^2} = 430.8 \text{ mm}$$
$$L_{BC} = \pi R = \pi(80) = 251.3 \text{ mm}$$
$$(x_{CL})_{ABC} = 100 + \frac{200}{2} = 200 \text{ mm}$$
$$(x_{CL})_{BC} = 300 + \frac{2(80)}{\pi} = 350.9 \text{ mm}$$

> The plane curve or area being used to generate a surface area or volume must not intersect the axis about which the curve or area is being revolved.

Applying Theorem 1 for areas gives

$$
\begin{aligned}
A = A_{ABC} + A_{BC} &= 2\pi L_{ABC}(x_{CL})_{ABC} + 2\pi L_{AB}(x_{CL})_{BC} \\
&= 2\pi(430.8)(200) + 2\pi(251.3)(350.9) \\
&= 1.0954(10^6) \text{ mm}^2 \cong 1.095 \text{ m}^2 \qquad \text{Ans.}
\end{aligned}
$$

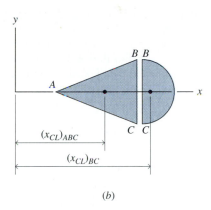

(b)

### Volume Calculations

The areas and centroid locations of segments $BAC$ and $BC$ of Fig. 5-22$c$ are

$$A_{ABC} = \frac{1}{2}bh = \frac{1}{2}(160)(200) = 16{,}000 \text{ mm}^2$$
$$A_{BC} = \frac{1}{2}\pi R^2 = \frac{1}{2}\pi(80)^2 = 10{,}053 \text{ mm}^2$$
$$(x_{CA})_{ABC} = 100 + \frac{2}{3}(200) = 233.3 \text{ mm}$$
$$(x_{CA})_{BC} = 300 + \frac{4(80)}{3\pi} = 334.0 \text{ mm}$$

Applying Theorem 2 for volumes gives

$$
\begin{aligned}
V = V_{ABC} + V_{BC} &= 2\pi A_{ABC}(x_{CA})_{ABC} + 2\pi A_{AB}(x_{CA})_{BC} \\
&= 2\pi(16{,}000)(233.3) + 2\pi(10{,}053)(334.0) \\
&= 44.55(10^6) \text{ mm}^3 \cong 0.0446 \text{ m}^3 \qquad \text{Ans.}
\end{aligned}
$$

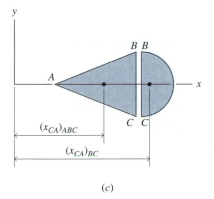

(c)

**Fig. 5-22**

## EXAMPLE PROBLEM 5-13

Determine the volume $V$ of the solid generated by revolving the shaded area shown in Fig. 5-23$a$ through an angle of 360° about the $x$-axis.

(a)                                                    (b)

**Fig. 5-23**

### SOLUTION

The theorems of Pappus and Guldinus can be extended further, to cases where the centroid $C_A$ of an area or $C_L$ of a line is not available by inspection or from tables, by applying the theorems to infinitesimal elements of area or length and integrating to find the total volume or surface area. Use of this procedure eliminates the need to determine centroid locations for the finite lengths or areas.

A differential element from the shaded area of Fig. 5-23$a$ is shown in Fig. 5-23$b$. The volume generated by revolving this element about the $x$-axis is

$$dV = 2\pi(y_{CA})_{dA}(dA) = 2\pi\left(\frac{y}{2}\right)(y\ dx) = \pi y^2\ dx$$

Integrating to obtain the total volume yields

$$V = \int_V dV = \int_0^b \pi y^2\ dx$$
$$= \int_0^b \pi\left(\frac{a^2}{b^4}\right)x^4\ dx = \left[\frac{\pi a^2 x^5}{5b^4}\right]_0^b = \frac{\pi a^2 b}{5} \qquad \text{Ans.}$$

The axis about which an area is being revolved to generate a volume may be a boundary of the area but may not intersect the area.

# PROBLEMS

Use the theorems of Pappus and Guldinus to solve the following problems.

### Introductory Problems

Determine the surface area $A$ and volume $V$ of the solid body generated by revolving

**5-55\*** The shaded area of Fig. P5-55 through an angle of 360° about the $y$-axis.

**Fig. P5-55**

**5-56\*** The shaded area of Fig. P5-56 through an angle of 360° about the $y$-axis.

**Fig. P5-56**

**5-57** The shaded area of Fig. P5-57 through an angle of 360° about the $x$-axis.

**Fig. P5-57**

**5-58** The shaded area of Fig. P5-58 through an angle of 360° about the $y$-axis.

**Fig. P5-58**

**5-59\*** The shaded area of Fig. P5-59 through an angle of 360° about the $y$-axis.

**Fig. P5-59**

**5-60** The shaded area of Fig. P5-60 through an angle of 360° about the $y$-axis.

**Fig. P5-60**

## Intermediate Problems

Determine the volume $V$ of the solid body generated by revolving

**5-61*** The shaded area of Fig. P5-61 through an angle of 360° about the x-axis.

**Fig. P5-61**

**5-62*** The shaded area of Fig. P5-62 through an angle of 360° about the x-axis.

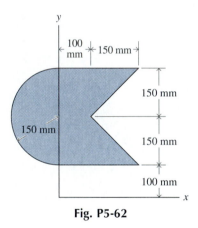

**Fig. P5-62**

**5-63** The shaded area of Fig. P5-63 through an angle of 180° about the y-axis.

**Fig. P5-63**

**5-64** The shaded area of Fig. P5-64 through an angle of 270° about the y-axis.

**Fig. P5-64**

**5-65*** The shaded area of Fig. P5-65 through an angle of 360° about the x-axis.

**Fig. P5-65**

**5-66** The shaded area of Fig. P5-66 through an angle of 360° about the y-axis.

**Fig. P5-66**

**226**

**5-67\*** The shaded area of Fig. P5-67 through an angle of 360° about the y-axis.

**Fig. P5-67**

**5-68** The shaded area of Fig. P5-68 through an angle of 360° about the y-axis.

**Fig. P5-68**

## 5-6 DISTRIBUTED LOADS ON BEAMS

When a load is applied to a rigid body, it is often distributed along a line or over an area $A$. In many instances, it is convenient to replace this distributed load with a resultant force that is equivalent to the distributed load. Thus, the quantities to be determined are the magnitude of the force and the location of its line of action.

Consider the beam shown in Fig. 5-24a. Since distributed loads on beams usually do not vary across the width of the cross section of the beam, the actual load intensity on the beam can be multiplied by the width of the beam to yield a distributed line load $w$ (N/m or lb/ft) whose magnitude varies only with position $x$ along the beam. The distributed load $w$ versus position $x$ curve is known as a load diagram. The magnitude of the differential force $d\mathbf{R}$ exerted on the beam by the distributed load $w$ in an increment of length $dx$ (see Fig. 5-24b) is

$$dR = w\,dx$$

Therefore, the magnitude of the single concentrated resultant force $\mathbf{R}$ that is equivalent to the distributed load $w$ is

$$R = \int_L w\,dx \qquad\qquad \textbf{(5-18)}$$

Thus, the magnitude of the resultant force $\mathbf{R}$ is equal to the area $A$ under the load diagram, as shown in Fig. 5-24c.

The location of the line of action of the resultant force $\mathbf{R}$ can be determined by equating the moment of the resultant force $\mathbf{R}$ about an arbitrary point $O$ to the moment of the distributed load about the same point $O$. The moment produced by the force $dR$ about point $O$ of Fig. 5-24b is

$$dM_O = x\,dR$$

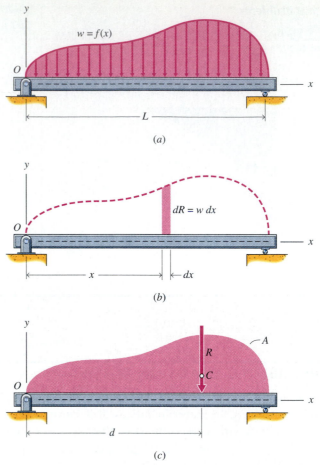

**Figure 5-24**   Replacing a distributed load on a beam with a resultant force **R**.

where $dR = w\,dx$. Therefore, the total moment produced by the distributed load $w$ about point $O$ is

$$M_O = \int dM_O = \int x\,dR \qquad (a)$$

The moment produced by the resultant force **R** about point $O$ is

$$M_O = Rd \qquad (b)$$

where $d$ is the distance along the beam from point $O$ to the line of action of the resultant force **R**. Thus, from Eqs. $a$ and $b$

$$M_O = Rd = \int x\,dR = x_C R \qquad \text{or} \qquad d = x_C = \frac{M_O}{R} \qquad (5\text{-}19)$$

Equation 5-19 indicates that the line of action of the resultant force **R** passes through the centroid of the area $A$ under the load diagram, as shown in Fig. 5-24c.

The following examples illustrate the procedure for determining the resultant of a distributed load $w$ and the location of its line of action by using the procedures previously developed for locating the centroids of areas.

## CONCEPTUAL EXAMPLE 5-2: DISTRIBUTED LOADS

*The lift force provided by the surface area of an airplane wing can be represented as a distributed line load w = f(x) along the main spar of the wing, as shown in Fig. CE5-2. Determine the resultant L of this distributed load and locate its line of action with respect to the vertical centerline of the plane's fuselage.*

**Fig. CE5-2**

### SOLUTION

For purposes of determining external effects such as support reactions for the wing, it is convenient to replace the distributed load $w = f(x)$ with a resultant force **L** at some point along the length of the beam. The magnitude of the lift force is obtained from the expression

$$L = \int w\, dx$$

which indicates that $L$ equals the area under the load diagram. The line of action of the load is obtained from the first-moment expression

$$Ld = \int xw\, dx$$

which indicates that the line of action of the resultant force passes through the centroid of the area under the load diagram. For distributions such as the trapezoidal distribution shown in Fig. CE5-2, known values for areas and centroid locations can be used to determine $R$ and $d$. In instances where $w$ is a more complicated function of position, integration methods or numerical procedures may be required to determine $L$ and $d$.

A beam is subjected to a system of loads that can be represented by the load diagram shown in Fig. 5-25a. Determine the resultant of this system of distributed loads and locate its line of action with respect to the left support of the beam.

Fig. 5-25

## SOLUTION

The magnitude of the resultant **R** of the distributed load shown in Fig. 5-25a is equal to the area under the load diagram. The load diagram can be divided into two triangles and a rectangle. Thus, from Table 5-1,

> When the intensity of a distributed load varies in a single direction $[w = f(x)]$, the resultant **R** is represented by the area under the load-distribution diagram and its line of action passes through the centroid of the area of the diagram.

$$F_1 = A_1 = \frac{1}{2}b_1h_1 = \frac{1}{2}(2)(300) = 300 \text{ N}$$

$$x_{C1} = \frac{2}{3}b_1 = \frac{2}{3}(2) = 1.333 \text{ m}$$

$$F_2 = A_2 = b_2h_2 = (6)(300) = 1800 \text{ N}$$

$$x_{C2} = 2 + \frac{1}{2}b_2 = 2 + \frac{1}{2}(6) = 5.00 \text{ m}$$

$$F_3 = A_3 = \frac{1}{2}b_3h_3 = \frac{1}{2}(4)(300) = 600 \text{ N}$$

$$x_{C3} = 8 + \frac{1}{3}b_3 = 8 + \frac{1}{3}(4) = 9.33 \text{ m}$$

The equivalent forces for the three different areas and the locations of their lines of action are shown in Fig. 5-25b. Thus,

$$R = F_1 + F_2 + F_3 = 300 + 1800 + 600 = 2700 \text{ N} \qquad \text{Ans.}$$

The line of action of the resultant with respect to the left support is located by summing moments about point $A$. Thus,

$$M_A = Rd = F_1x_{C1} + F_2x_{C2} + F_3x_{C3}$$
$$= 300(1.333) + 1800(5.00) + 600(9.33) = 15,000 \text{ N} \cdot \text{m}$$

Finally,

$$d = x_C = \frac{M_A}{R} = \frac{15,000}{2700} = 5.56 \text{ m} \qquad \text{Ans.}$$

The resultant force **F** and its line of action are shown in Fig. 5-25c.

A beam is subjected to a system of loads that can be represented by the load diagram shown in Fig. 5-26a. Determine the resultant of this system of loads and locate its line of action with respect to the left support of the beam.

**Fig. 5-26**

## SOLUTION

The magnitude of the resultant **R** of the distributed load shown in Fig. 5-26a is equal to the area under the load diagram, and the line of action of the resultant passes through the centroid of the area.

The area under the load diagram is determined by using the element of area shown in Fig. 5-26b. Thus,

$$A = \int_A w \, dx = \int_0^L w_{max} \sin \frac{\pi x}{2L} \, dx = \frac{2Lw_{max}}{\pi} \left[ -\cos \frac{\pi x}{2L} \right]_0^L = \frac{2Lw_{max}}{\pi}$$

Thus,

$$R = A = \frac{2Lw_{max}}{\pi} = 0.637 w_{max} L \qquad \text{Ans.}$$

The moment of the area about support $A$ is

$$M_A = \int_A x \, (w \, dx) = \int_0^L w_{max} \, x \sin \frac{\pi x}{2L} \, dx$$

$$= w_{max} \left[ \frac{4L^2}{\pi^2} \sin \frac{\pi x}{2L} - \frac{2L}{\pi} x \cos \frac{\pi x}{2L} \right]_0^L = \frac{4L^2 w_{max}}{\pi^2}$$

Using the principle of moments

$$d = x_C = \frac{M_A}{A} = \frac{4L^2 w_{max}/\pi^2}{2Lw_{max}/\pi} = \frac{2L}{\pi} = 0.637L \qquad \text{Ans.}$$

The results are shown in Fig. 5-26c.

# PROBLEMS

Determine the resultant **R** of the system of distributed loads and locate its line of action with respect to the left support for

## Introductory Problems

**5-69*** The beam shown in Fig. P5-69.

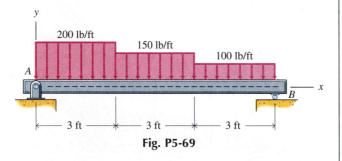

Fig. P5-69

**5-70*** The beam shown in Fig. P5-70.

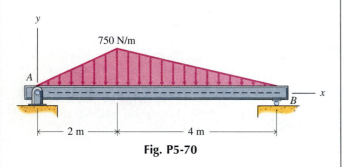

Fig. P5-70

**5-71** The beam shown in Fig. P5-71.

Fig. P5-71

**5-72** The beam shown in Fig. P5-72.

Fig. P5-72

## Intermediate Problems

**5-73*** The beam shown in Fig. P5-73.

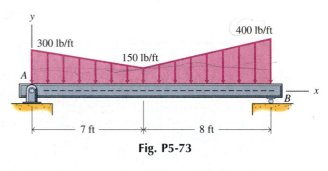

Fig. P5-73

**5-74*** The beam shown in Fig. P5-74.

Fig. P5-74

**5-75**   The beam shown in Fig. P5-75.

**Fig. P5-75**

**5-76**   The beam shown in Fig. P5-76.

**Fig. P5-76**

**Challenging Problems**

**5-77*****   The beam shown in Fig. P5-77.

**Fig. P5-77**

**5-78*****   The beam shown in Fig. P5-78.

$x_e = \dfrac{MY}{A}$

$A = \int 266 x^-$

**Fig. P5-78**

**5-79**   The beam shown in Fig. P5-79.

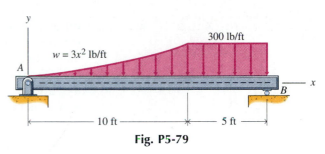

**Fig. P5-79**

**5-80**   The beam shown in Fig. P5-80.

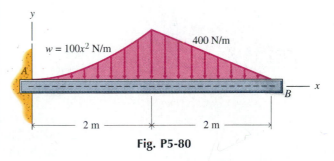

**Fig. P5-80**

## 5-7   FORCES ON SUBMERGED SURFACES

A fluid (either a liquid or a gas) at rest can, by definition, transmit compressive forces but not tensile or shear forces. Since a shear force acts tangent to a surface, a fluid at rest can exert only a compressive normal force (known as a pressure) on a submerged surface. The pressure, called a hydrostatic pressure (equal in all directions), is due to the weight of the fluid above any point on the submerged surface; therefore, fluid pressures vary linearly with depth in fluids with a constant specific weight. The absolute pressure $p_A$ at a depth $d$ is

$$p_A = p_0 + \gamma d = p_0 + \rho g d \qquad (5\text{-}20)$$

where

$p_0$ = atmospheric pressure at the surface of the fluid
$\gamma$ = specific weight of the fluid
$\rho$ = density of the fluid
$g$ = gravitational acceleration

In the U.S. Customary system of units, the specific weight $\gamma$ of fresh water is 62.4 lb/ft³. In the SI system of units, the density $\rho$ of fresh water is 1000 kg/m³. The gravitational acceleration $g$ is 32.2 ft/s² in the U.S. Customary system and 9.81 m/s² in the SI system.

In general, pressure-measuring instruments record pressures above atmospheric pressure. Such pressures are known as gage pressures, and it is obvious from Eq. 5-20 that the gage pressure $p_g$ is

$$p_g = p_A - p_0 = \gamma d = \rho g d \qquad \text{(5-21)}$$

For the analysis of many engineering problems involving fluid forces, it is necessary to determine the resultant force **R** due to the distribution of pressure on a submerged surface and the location of the intersection of the line of action of the resultant force with the submerged surface. The point $P$ on the submerged surface where the line of action of the resultant force **R** intersects the submerged surface is known as the center of pressure.

## 5-7.1   Forces on Submerged Plane Surfaces

For the case of fluid pressures on submerged plane surfaces, the load diagram (area) introduced in Section 5-6 for a distributed load along a line becomes a pressure solid (volume), as shown in Fig. 5-27a, since the intensity of a distributed load (pressure) on the submerged surface varies over an area instead of a length. When the distributed pressure $p$ is applied to an area in the $xy$-plane, the ordinate $p(x, y)$ along the $z$-axis represents the intensity of the force (force per unit area). The magnitude of the increment of force $d\mathbf{R}$ on an element of area $dA$ is

$$dR = p\,dA = dV_{ps}$$

where $dV_{ps}$ is an element of volume of the pressure solid, as shown in Fig. 5-27a. The magnitude of the resultant force **R** acting on the submerged surface is

$$R = \int_A p\,dA = \int_V dV_{ps} = V_{ps} \qquad \text{(5-22)}$$

where $V_{ps}$ is the volume of the pressure solid.

The line of action of the resultant force **R** with respect to the $x$- and $y$-axes (called the center of pressure) can be located by using the principle of moments.

For moments about the $y$-axis:

$$Rd_x = \int x\,dR = \int_A x\,p\,dA = \int_V x\,dV_{ps} = x_{Cps}V_{ps} \qquad \text{(5-23a)}$$

For moments about the $x$-axis:

$$Rd_y = \int y\,dR = \int_A y\,p\,dA = \int_V y\,dV_{ps} = y_{Cps}V_{ps} \qquad \text{(5-23b)}$$

(a)

(b)

**Figure 5-27**   Replacing a distributed fluid pressure load on a submerged plane surface with a resultant force **R**.

Figure 5-27 indicates that the line of action of the resultant force **R** passes through the centroid $C_V$ of the volume of the pressure solid. If the pressure is uniformly distributed over the area, the center of pressure $P$ will coincide with the centroid $C_A$ of the area. If the pressure is not uniformly distributed over the area, the center of pressure $P$ and the centroid $C_A$ of the area will be at different points, as shown in Fig. 5-27b.

## 5-7.2  Forces on Submerged Curved Surfaces

Equations 5-22 and 5-23 apply only to submerged plane surfaces; however, many surfaces of interest in engineering applications—such as those associated with pipes, dams, and tanks—are curved. For such problems, the resultant force **R** and the intersection of its line of action with the curved surface can be determined by integration for each individual problem, but general formulas applicable to a broad class of problems cannot be developed. To overcome this difficulty, the procedure illustrated in Fig. 5-28 has been developed.

In Fig. 5-28a, a cylindrical gate with a radius $a$ and a length $L$ is being used to close an opening in the wall of a tank containing a fluid. The pressure distribution on the gate is shown in Fig. 5-28b. From such a distribution, horizontal and vertical components of the resultant force can be determined by integration and combined to yield the resultant force **R**. The pressure-solid approach can also be used to determine the resultant force **R** if horizontal and vertical planes are used to isolate the gate and a volume of fluid in contact with the gate, as shown in Fig. 5-28c. The force exerted on the horizontal fluid surface by the fluid pressure is

$$F_{1v} = p_1 A_h = \gamma(d - a)(aL)$$

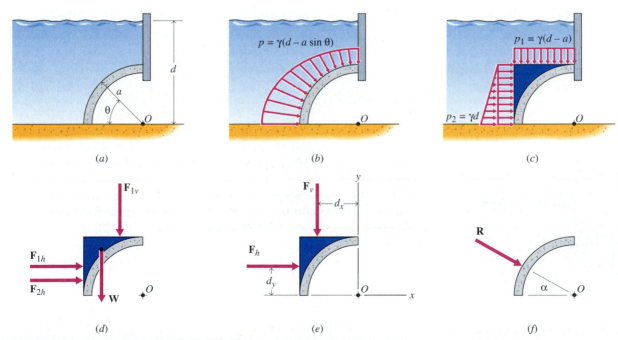

**Figure 5-28** Replacing a distributed fluid pressure load on a submerged curved surface with a resultant force **R**.

Similarly on the vertical surface,

$$F_{1h} = p_1 A_v = \gamma(d-a)(aL)$$
$$F_{2h} = (p_2 - p_1)A_v = \gamma a(aL)$$

The volume of fluid $V_f$ has a weight $W$, which is given by the expression

$$W = \gamma V_f = \gamma(a^2 - \frac{1}{4}\pi a^2)L$$

The four forces $F_{1v}$, $F_{1h}$, $F_{2h}$, and $W$ together with their lines of action are shown in Fig. 5-28d. The two vertical forces and the two horizontal forces can be combined to give

$$F_v = F_{1v} + W$$
$$F_h = F_{1h} + F_{2h}$$

where $F_v$ and $F_h$ are the rectangular components of a resultant force **R**. That is, **R** is the resultant of $F_{1v}$, $F_{1h}$, $F_{2h}$, and $W$, which are the forces exerted by the adjoining water and the earth on the volume of water in contact with the gate. This force is the same as the force exerted by the water on the gate because the volume of water in contact with the gate is in equilibrium and the force exerted on the water by the gate is equal in magnitude and opposite in direction to the force exerted on the gate by the water. The magnitude of the resultant is

$$R = \sqrt{(F_h)^2 + (F_v)^2}$$

The slope of the line of action of the resultant is given by the expression

$$\alpha = \tan^{-1}\frac{F_v}{F_h}$$

Finally, the location of the line of action of the resultant with respect to an arbitrary point can be determined by summing moments about the point. For point $O$ shown in Fig. 5-28e,

$$Rd = F_v d_x - F_h d_y$$

For the cylindrical gate, the line of action of the resultant passes through point $O$, as shown in Fig 5-28f. This results from the fact that the pressure always acts normal to the surface; therefore, for the cylindrical gate, the line of action of each increment $d\mathbf{R}$ of the resultant passes through point $O$. In other words, the increments form a system of concurrent forces.

The following examples illustrate the procedure for determining the resultant force and locating the center of pressure for submerged surfaces by using pressure-solid approaches.

## EXAMPLE PROBLEM 5-16

The water behind a dam is 100 m deep as shown in Fig. 5-29a. Determine

**a.** The magnitude of the resultant force **R** exerted on a 30-m length of the dam by the water pressure.
**b.** The distance from the water surface to the center of pressure.

(a)

(b)

(c)

**Fig. 5-29**

### SOLUTION

A cross section through the pressure solid is shown in Fig. 5-29b. At the base of the dam, the pressure is

$$p = \gamma d = \rho g d = 1000(9.81)(100)$$
$$= 981(10^3) \ N/m^2 = 981 \ kN/m^2$$

**a.** Thus, for the 30-m length of dam, the volume of the pressure solid is

$$V_{ps} = \frac{1}{2}pA = \frac{1}{2}(981)(10^3)(100)(30) = 1471.5(10^6) \ N \cong 1472 \ MN$$

$$R = V_{ps} \cong 1472 \ MN \qquad \text{Ans.}$$

**b.** Since the width of the pressure solid is constant and the cross section is a triangle, the distance from the water surface to the centroid of the solid is

$$d_P = \frac{2}{3}d = \frac{2}{3}(100) = 66.67 \ m \cong 66.7 \ m \qquad \text{Ans.}$$

The results are shown in Fig. 5-29c.

When a fluid is at rest, it is capable of transmitting contact forces only in a direction perpendicular to the surface of contact. The resultant **R** is represented by the volume of the pressure solid and its line of action passes through the centroid of the volume.

The width of the rectangular gate shown in Fig. 5-30 is 8 ft. Determine the magnitude of the resultant force **R** exerted on the gate by the water pressure and the location of the center of pressure $d_p$ with respect to the hinge at the top of the gate.

(a)                                                    (b)

**Fig. 5-30**

## SOLUTION

A cross section through the pressure solid is shown in Fig. 5-30b. At the top of the gate, the pressure is

$$p_T = \gamma d_T = 62.4(10) = 624 \text{ lb/ft}^2$$

At the bottom of the gate the pressure is

$$p_B = \gamma d_B = 62.4(25) = 1560 \text{ lb/ft}^2$$

The volume of the pressure solid can be separated into two parts.

For part 1:

$$V_{ps1} = p_T hw = 624(15)(8) = 74{,}880 \text{ lb}$$

For part 2:

$$V_{ps2} = \frac{1}{2}(p_B - p_T)hw = \frac{1}{2}(1560 - 624)(15)(8) = 56{,}160 \text{ lb}$$

Therefore,

$$R = R_1 + R_2 = V_{ps1} + V_{ps2} = 74{,}880 + 56{,}160$$
$$= 131{,}040 \text{ lb} \cong 131.0 \text{ kip} \qquad \text{Ans.}$$

Summing moments about point $A$ yields

$$M_A = R_1 y_{C1} + R_2 y_{C2}$$
$$= 74{,}880(7.5) + 56{,}160(10) = 1{,}123{,}200 \text{ ft} \cdot \text{lb}$$

The principle of moments states that

$$R d_P = R_1 y_{C1} + R_2 y_{C2} = M_A$$

Therefore,

$$d_P = \frac{M_A}{R} = \frac{1{,}123{,}200}{131{,}040} = 8.571 \text{ ft} \cong 8.57 \text{ ft} \qquad \text{Ans.}$$

> When a fluid is at rest, it is capable of transmitting contact forces only in a direction perpendicular to the surface of contact. Dividing the pressure distribution into two parts simplifies the calculation.

The width of the rectangular gate shown in Fig. 5-31a is 4 m. Determine the magnitude of the resultant force **R** exerted on the gate by the water pressure and the location of the center of pressure $d_P$ with respect to the hinge at the bottom of the gate.

(a)                                         (b)

**Fig. 5-31**

## SOLUTION

A cross section through the pressure solid is shown in Fig. 5-31b. At the top of the gate the pressure is

$$p_T = \rho g d_T = 1000(9.81)(7.5) = 73{,}575 \text{ N/m}^2$$

At the bottom of the gate the pressure is

$$p_B = \rho g d_B = 1000(9.81)(12.5) = 122{,}625 \text{ N/m}^2$$

The volume of the pressure solid can be separated into two parts.

For part 1:

$$V_{ps1} = p_T L w = 73{,}575(10)(4) = 2.943(10^6) \text{ N}$$

For part 2:

$$V_{ps2} = \frac{1}{2}(p_B - p_A)Lw$$

$$= \frac{1}{2}(122{,}625 - 73{,}575)(10)(4) = 0.981(10^6) \text{ N}$$

Therefore,

$$R = R_1 + R_2 = V_{ps1} + V_{ps2} = 2.943(10^6) + 0.981(10^6)$$
$$= 3.924(10^6) \text{ N} \cong 3.92 \text{ MN} \qquad \text{Ans.}$$

Summing moments about point B yields

$$M_B = V_{ps1}d_{C1} + V_{ps2}d_{C2} = 2.943(10^6)(5.00) + 0.981(10^6)(3.333)$$
$$= 17.985(10^6) \text{ N} \cdot \text{m} \cong 17.99(10^6) \text{ N} \cdot \text{m}$$

The principle of moments states that

$$Rd_P = R_1 d_{C1} + R_2 d_{C2} = M_B$$

Therefore,

$$d_P = \frac{M_B}{R} = \frac{17.985(10^6)}{3.924(10^6)} = 4.583 \text{ m} \cong 4.58 \text{ m} \qquad \text{Ans.}$$

When a fluid is at rest, it is capable of transmitting contact forces only in a direction perpendicular to the surface of contact. Dividing the pressure distribution into two parts simplifies the calculation.

# PROBLEMS

## Introductory Problems

**5-81*** If the dam shown in Fig. P5-81 is 200 ft wide, determine the magnitude of the resultant force **R** exerted on the dam by the water ($\gamma = 62.4$ lb/ft$^3$) pressure.

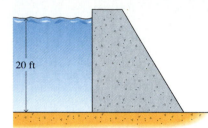

**Fig. P5-81**

**5-82*** If the dam shown in Fig. P5-82 is 50 m wide, determine the magnitude of the resultant force **R** exerted on the dam by the water ($\rho = 1000$ kg/m$^3$) pressure.

**Fig. P5-82**

**5-83** A glass-walled fish tank is 2 ft wide by 6 ft long by 3 ft deep. When the water ($\gamma = 62.4$ lb/ft$^3$) in the tank is 2.5 ft deep, determine the magnitude of the resultant force **R** exerted on a 2- by 3-ft end plate by the water pressure and the distance from the water surface to the center of pressure.

**5-84** A flat steel plate is used to seal an opening 1 m wide by 2 m high in the vertical wall of a large water ($\rho = 1000$ kg/m$^3$) tank. When the water level in the tank is 15 m above the top of the opening, determine the magnitude of the resultant force **R** exerted on the plate by the water pressure and the distance from the centroid of the area of the plate to the center of pressure.

**5-85*** The width of the rectangular gate shown in Fig. P5-85 is 8 ft. Determine the magnitude of the resultant force **R** exerted on the gate by the water ($\gamma = 62.4$ lb/ft$^3$) pressure and the location of the center of pressure with respect to the hinge at the bottom of the gate.

**Fig. P5-85**

**5-86** The width of the rectangular gate shown in Fig. P5-86 is 2 m. Determine the magnitude of the resultant force **R** exerted on the gate by the water ($\rho = 1000$ kg/m$^3$) pressure and the location of the center of pressure with respect to the hinge at the top of the gate.

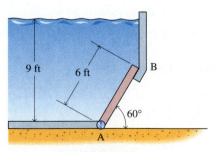

**Fig. P5-86**

## Intermediate Problems

**5-87*** The width of the rectangular gate shown in Fig. P5-87 is 4 ft. Determine the magnitude of the resultant force **R** exerted on the gate by the water ($\gamma = 62.4$ lb/ft$^3$) pressure and the location of the center of pressure with respect to the hinge at the bottom of the gate.

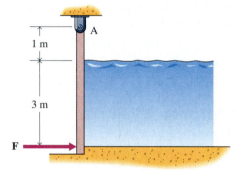

**Fig. P5-87**

**5-88\*** The width of the rectangular gate shown in Fig. P5-88 is 4 m. Determine the magnitude of the resultant force **R** exerted on the gate by the water ($\rho = 1000$ kg/m³) pressure and the location of the center of pressure with respect to the hinge at the bottom of the gate.

Fig. P5-88

**5-89** The width of the rectangular gate shown in Fig. P5-89 is 3 ft. Determine the magnitude of the resultant force **R** exerted on the gate by the water ($\gamma = 62.4$ lb/ft³) pressure and the location of the center of pressure with respect to the hinge at the bottom of the gate.

Fig. P5-89

**5-90** The width of the cylindrical gate shown in Fig. P5-90 is 4 ft. Determine the magnitude and the slope of the line of action of the resultant force **R** exerted on the gate by the water ($\rho = 1000$ kg/m³) pressure if $d = 10$ ft and $a = 4$ ft.

Fig. P5-90

**5-91** Determine the magnitude and locate the line of action of the resultant force **R** exerted by the water ($\gamma = 62.4$ lb/ft³) pressure on a 5-ft length of the dam shown in Fig. P5-91.

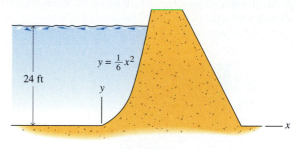

Fig. P5-91

## Challenging Problems

**5-92\*** A flat plate is used to seal an opening in a large water ($\rho = 1000$ kg/m³) tank as shown in Fig. P5-92. If the opening has the cross section shown in Fig. P5-92b, determine the magnitude of the resultant force **R** exerted on the plate by the water pressure and the location of the center of pressure with respect to the bottom of the opening if $d = 5$ m, $h = 2$ m, and $b = 2$ m.

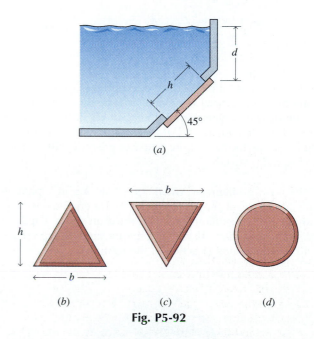

Fig. P5-92

**5-93\*** A flat plate is used to seal an opening in a large water ($\gamma = 62.4$ lb/ft³) tank as shown in Fig. P5-92. If the opening has the cross section shown in Fig. P5-92c, determine the magnitude of the resultant force **R** exerted on the plate by the water pressure and the location of the center of pressure with respect to the bottom of the opening if $d = 15$ ft, $h = 5$ ft, and $b = 5$ ft.

**5-94** A flat plate is used to seal an opening in a large water ($\rho = 1000$ kg/m³) tank as shown in Fig. P5-92. If the opening has the cross section shown in Fig. P5-92d, determine the magnitude of the resultant force **R** exerted on the plate by the water pressure and the location of the center of pressure with respect to the bottom of the opening if $d = 10$ m and $h = 1$ m.

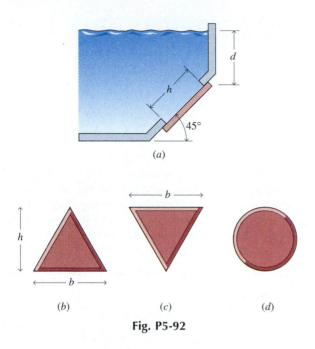

(a)

(b)      (c)      (d)

**Fig. P5-92**

**5-95** A water ($\gamma = 62.4$ lb/ft³) trough 6 ft wide by 10 ft long by 3 ft deep has semicircular ends. When the trough is full of water, determine the magnitude of the resultant force **R** exerted on an end of the trough by the water pressure and the location of the center of pressure with respect to the water surface.

**5-96** A 4-m-diameter water ($\rho = 1000$ kg/m³) pipe is filled to the level shown in Fig. P5-96. Determine the magnitude of the resultant force **R** exerted on a 2-m length of the curved section $AB$ of the pipe by the water pressure and locate the line of action of the resultant.

**Fig. P5-96**

## Computer Problems

**C5-97** A rectangular gate holds back water ($\gamma = 62.4$ lb/ft³) as shown in Fig. P5-97. The gate is 10 ft high and 8 ft wide and is pivoted at a point 4.5 ft from its bottom edge. When the water level is sufficiently low, the gate rests against the stop at $C$ and does not touch the stop at $A$. When the water level is sufficiently high, the gate presses against the stop at $A$ and does not touch the stop at $C$.

**a.** Plot $F_A$, $F_{Bx}$, and $F_C$, the horizontal components of the forces at the two stops and the pivot, as a function of the water depth $h$ ($5 \leq h \leq 35$ ft).

**b.** Plot $d$, the location of the center of pressure relative to the pivot $B$ as a function of the water depth $h$ ($5 \leq h \leq 35$ ft).

**c.** If the stop at $A$ were removed, at what depth of water would the gate rotate and allow the water to drain out? Where is the location of the center of pressure for this depth?

**Fig. P5-97**

**C5-98** A water tank at the aquarium has a circular window 2 m in diameter in a vertical wall as shown in Fig. P5-98.

**a.** Plot $R$, the resultant of the water ($\rho = 1000$ kg/m³) pressure on the glass window, as a function of the water depth $h$ ($0.5 \leq h \leq 5$ m).

**Fig. P5-98**

**b.** Plot $d$, the location of the center of pressure relative to the center of the circular window, as a function of the water depth $h$ ($0.5 \le h \le 5$ m).

**C5-99** A wine barrel is modeled as a circular cylinder 4 ft in diameter and 6 ft long on its side as shown in Fig. P5-99.

**a.** Plot $R$, the resultant of the wine ($\gamma = 56$ lb/ft$^3$) pressure on the end of the barrel, as a function of the wine depth $h$ ($0 \le h \le 4$ ft).

**b.** Plot $d$, the location of the center of pressure relative to the center of the circle, as a function of the wine depth $h$ ($0 \le h \le 4$ ft).

**Fig. P5-99**

**C5-100** The tank shown in Fig. P5-100 is basically rectangular (1 m square by 2 m long). It is being raised from its side to its end. If the tank is one-fourth full of oil ($\rho = 850$ kg/m$^3$),

**a.** Plot $h$, the depth of oil in the tank, as a function of the angle $\theta$ ($0° \le \theta \le 90°$).

**b.** Plot $R$, the resultant of the oil pressure on the end of the tank, as a function of $\theta$ ($0° \le \theta \le 90°$).

**c.** Plot $d$, the location of the center of pressure relative to the center of the square, as a function of the angle $\theta$ ($0° \le \theta \le 90°$).

**Fig. P5-100**

# SUMMARY

Previously, we dealt with concentrated forces, which we represented with a vector having a magnitude, a direction, a line of action, and in some cases a point of application. In many instances, surface loads on a body are not concentrated at a point but are distributed along a length or over an area. Other forces, known as body forces, are distributed over the volume of the body. A distributed force at any point is characterized by its intensity and its direction. A force distributed over an area and acting normal to the area is known as a pressure.

Previously, moments of forces about points or axes were considered. In engineering analysis, equations are also encountered that represent moments of masses, forces, volumes, areas, or lines with respect to axes or planes. Such moments are called first moments of the quantity being considered since the first power of a distance is used in the expression.

The term *center of mass* is used to denote the point in a physical body where the mass can be conceived to be concentrated so that the moment of the concentrated mass with respect to an axis or plane equals the moment of the distributed mass with respect to the same axis or plane. The term *center of gravity* is used to denote the point in the body through which the weight of the body acts, regardless of the position (or orientation) of the body. The location of the center of mass in a body is determined by using equations of the form

$$M_{yz} = mx_G = \int_m x \, dm \quad \text{or} \quad x_G = \frac{1}{m} \int_m x \, dm$$

243

If the density $\rho$ of a body is constant,

$$x_C = \frac{1}{V} \int_V x \, dV \qquad y_C = \frac{1}{V} \int_V y \, dV \qquad z_C = \frac{1}{V} \int_V z \, dV \qquad \text{(5-9)}$$

Equations 5-9 indicate that the coordinates $x_C$, $y_C$, and $z_C$ depend only on the geometry of the body and are independent of the physical properties. The point located by such a set of coordinates is known as the centroid of the volume of the body. The term *centroid* is usually used in connection with geometrical figures (volumes, areas, and lines), whereas the terms *center of mass* and *center of gravity* are used in connection with physical bodies. The centroid of a volume has the same position as the center of gravity of the body if the body is homogeneous. If the specific weight is variable, the center of gravity of the body and the centroid of the volume will be at different points.

When a load applied to a rigid body is distributed along a line or over an area $A$, it is often convenient for purposes of static analysis to replace this distributed load with a resultant force **R** that is equivalent to the distributed load $w$. For a beam with a distributed load along its length, the magnitude of the resultant force is determined from the expression

$$R = \int_L w \, dx \qquad \text{(5-18)}$$

which indicates that the magnitude of the resultant force is equal to the area under the load diagram used to represent the distributed load. The line of action of the resultant force passes through the centroid of the area under the load diagram. For the case of fluid pressures on submerged plane surfaces, the load diagram (area) for a distributed load along a line becomes a pressure solid (volume) since the intensity of the pressure on the submerged surface varies over an area instead of a length. The magnitude of the resultant force acting on the submerged surface is

$$R = \int_A p \, dA = \int_V dV_{ps} = V_{ps} \qquad \text{(5-22)}$$

which indicates that the magnitude of the resultant force is equal to the volume of the pressure solid used to represent the distributed load. The line of action of the resultant force passes through the centroid of the volume of the pressure solid. If the pressure is uniformly distributed over the area, the center of pressure (point of intersection of the line of action of the resultant force with the area) will coincide with the centroid of the area. If the pressure is not uniformly distributed over the area, the center of pressure and the centroid of the area will be at different points.

# REVIEW PROBLEMS

**5-101\*** Locate the centroid of the shaded area shown in Fig. P5-101.

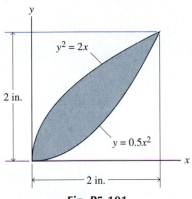

**Fig. P5-101**

**5-102** Locate the centroid of the shaded area shown in Fig. P5-102.

**Fig. P5-102**

**5-103** Locate the centroid of the volume shown in Fig. P5-103 if $R = 10$ in. and $h = 32$ in.

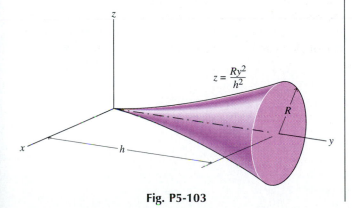

**Fig. P5-103**

**5-104\*** Two channel sections and a plate are used to form the cross section shown in Fig. P5-104. Each of the channels has a cross-sectional area of 2605 mm². Locate the $y$-coordinate of the centroid of the composite section with respect to the top surface of the plate.

**Fig. P5-104**

**5-105** Determine the volume of material removed when the groove is cut in the circular shaft shown in Fig. P5-105.

**Fig. P5-105**

**5-106*** Locate the centroid and the mass center of the volume shown in Fig. P5-106, which consists of an aluminum cylinder ($\rho = 2770$ kg/m³) and a steel ($\rho = 7870$ kg/m³) cylinder and sphere.

**Fig. P5-106**

**5-107*** Determine the resultant **R** of the system of distributed loads on the beam of Fig. P5-107 and locate its line of action with respect to the left support of the beam.

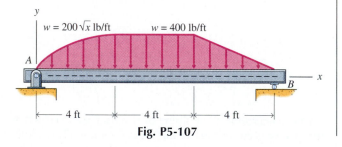

**Fig. P5-107**

**5-108** Determine the resultant **R** of the system of distributed loads on the beam of Fig. P5-108 and locate its line of action with respect to the left support of the beam.

**Fig. P5-108**

**5-109** A square viewing window is located in a large water tank as shown in Fig. P5-109. Determine the resultant force **R** exerted by the water pressure on the window and locate its line of action with respect to point $A$ at the top of the window if $d = 35$ ft and $h = 25$ ft. Use 62.4 lb/ft³ for the specific weight $\gamma$ of water.

**Fig. P5-109**

*R*igid body principles that involve both force and moment equilibrium are needed to determine forces in the joints and hydraulic cylinders of the crane lifting a bundle of logs.

# EQUILIBRIUM OF RIGID BODIES

## 6-1 INTRODUCTION

 The concept of equilibrium was introduced in Chapter 3 and applied to a system of forces acting on a particle. Since any system of forces acting on a particle is a concurrent force system, a particle is in equilibrium if the resultant **R** of the force system acting on the particle is zero. For the case of a rigid body, it was shown in Chapter 4 that the most general force system can be expressed in terms of a resultant force **R** and a resultant couple **C**.

Later in Dynamics it will be shown that all motion can be represented as a sum of a translation and a rotation. Any tendency of the body to translate is produced by the resultant forces **R** and any tendency of the body to rotate is produced by the resultant couple **C**. Therefore, for a rigid body to be in equilibrium, both the resultant force **R** and the resultant couple **C** must vanish. These two conditions can be expressed by the two vector equations

$$\mathbf{R} = \Sigma F_x\, \mathbf{i} + \Sigma F_y\, \mathbf{j} + \Sigma F_z\, \mathbf{k} = \mathbf{0}$$
$$\mathbf{C} = \Sigma M_x\, \mathbf{i} + \Sigma M_y\, \mathbf{j} + \Sigma M_z\, \mathbf{k} = \mathbf{0}$$

(6-1)

Equations 6-1 can be expressed in scalar form as

$$\Sigma F_x = 0 \qquad \Sigma F_y = 0 \qquad \Sigma F_z = 0$$
$$\Sigma M_x = 0 \qquad \Sigma M_y = 0 \qquad \Sigma M_z = 0$$

(6-2)

Equations 6-2 are the necessary conditions for equilibrium of a rigid body. They are also the sufficient conditions for equilibrium if all the forces acting on the body can be determined from these equations.

The forces and moments that act on a rigid body are either external or internal. Forces applied to a rigid body by another body or by the earth are external forces. Fluid pressure on the wall of a tank or the force applied by a truck wheel to a road surface are examples of external forces. The weight of a body is another example of an external force. Internal forces hold the particles forming the rigid body together. If the body of interest is composed of several parts, the forces holding the parts together are also defined as internal forces.

External forces can be divided into applied forces and reaction forces. Forces exerted on the body by external sources are the applied forces. Forces exerted on the body by supports or connections are the reaction forces. Our concern in this chapter is only with external forces and the moments these external forces produce. Since internal forces occur as equal and opposite pairs, they have no effect on equilibrium of the overall rigid body. Internal forces will be considered in Chapter 8.

The best way to identify all forces acting on a body of interest is to use the free-body diagram approach introduced in Chapter 3 when equilibrium of a particle was considered. This free-body diagram of the body of interest must show all the applied forces and all the reaction forces exerted on the body at the supports. In Section 6-2 free-body diagrams for rigid bodies are considered and the reaction forces exerted on bodies by a number of different types of supports are specified.

## 6-2 FREE-BODY DIAGRAMS

The concept of a free-body diagram was introduced in Chapter 3 and used to solve equilibrium problems involving bodies that could be idealized as a particle. In all these problems, the external forces (applied forces and reaction forces) could be represented as a concurrent force system. In the more general case of a rigid body, systems of forces other than concurrent force systems are encountered and the free-body diagrams become much more complicated. The basic procedure for drawing the diagram, however, remains the same and consists essentially of the following four steps:

*Step 1.* Decide which body or combination of bodies is to be isolated or cut free from its surroundings.

*Step 2.* Prepare a drawing or sketch of the outline of this isolated or free body.

*Step 3.* Carefully trace around the boundary of the free body and identify and label all the forces and moments exerted by contacting or interacting bodies that were removed during the isolation process.

*Step 4.* Choose the set of coordinate axes to be used in solving the problem and indicate these directions on the free-body diagram.

Application of these four steps to any statics or dynamics problem will produce a complete and correct free-body diagram, which is *an essential first step for the solution of the problem.*

Forces that are known should be added to the diagram and labeled with their proper magnitudes and directions. Letter symbols can be used to represent magnitudes of forces that are unknown. If the correct sense of an unknown force is not obvious, the sense can be arbitrarily assumed. The algebraic sign of the calculated value of the unknown force will indicate the sense of the force. A plus sign indicates that the force is in the direction assumed, and a minus sign indicates that the force is in a direction opposite from that assumed.

When connections or supports are removed from the isolated body, the actions of these connections or supports must be represented by forces and/or moments on the free-body diagram. The forces and moments used to represent the actions of common connections and supports used with bodies subjected to two-dimensional force systems are identified and discussed in Section 6-2.1. A similar discussion for connections and supports used with bodies subjected to three-dimensional force systems is presented in Section 6-2.2.

### 6-2.1  Idealization of Two-Dimensional Supports and Connections

Common types of supports and connections used with rigid bodies subjected to two-dimensional force systems, together with the forces and moments used to represent the actions of these supports and connections on a free body, are listed in Table 6-1.

### 6-2.2  Idealization of Three-Dimensional Supports and Connections

Common types of supports and connections used with rigid bodies subjected to three-dimensional force systems, together with the forces and moments used to represent the actions of these supports and connections on a free-body diagram, are listed in Table 6-2.

Free-body diagrams for a number of structures and machine components are presented in the following examples.

TABLE 6-1   TWO-DIMENSIONAL REACTIONS AT SUPPORTS AND CONNECTIONS

1. Gravitational attraction

**Fig. 6-1**

The gravitational attraction of the Earth on a body (see Fig. 6-1) is the weight **W** of the body. The line of action of the force **W** passes through the center of gravity of the body and is directed toward the center of the Earth.

2. Flexible cord, rope, chain, or cable

**Fig. 6-2**

A flexible cord, rope, chain, or cable (see Fig. 6-2) always exerts a tensile force **R** on the body. The line of action of the force **R** is known; it is tangent to the cord, rope, chain, or cable at the point of attachment.

3. Rigid link

**Fig. 6-3**

A rigid link (see Fig. 6-3) can exert either a tensile or a compressive force **R** on the body. The line of action of the force **R** is known; it must be directed along the axis of the link (see Section 6-3.1 for proof).

4. Ball, roller, or rocker

**Fig. 6-4**

A ball, roller, or rocker (see Fig. 6-4) can exert a compressive force **R** on the body. The line of action of the force **R** is perpendicular to the surface supporting the ball, roller, or rocker.

**TABLE 6-1** **(continued)**

### 5. Smooth surface

**Fig. 6-5**

A smooth surface, either flat or curved (see Fig. 6-5), can exert a compressive force **R** on the body. The line of action of the force **R** is perpendicular to the smooth surface at the point of contact between the body and the smooth surface.

### 6. Smooth pin

OR

**Fig. 6-6**

A smooth pin (see Fig. 6-6) can exert a force **R** of unknown magnitude $R$ and direction $\theta$ on the body. As a result, the force **R** is usually represented on a free-body diagram by its rectangular components $R_x$ and $R_y$.

### 7. Rough surface

DUE TO FRICTION

**Fig. 6-7**

Rough surfaces (see Fig. 6-7) are capable of supporting a tangential frictional force $R_t$ as well as a compressive normal force $R_n$. As a result, the force **R** exerted on a body by a rough surface is a compressive force **R** at an unknown angle $\theta$. The force **R** is usually represented on a free-body diagram by its rectangular components $R_n$ and $R_t$.

### 8. Pin in a smooth guide

**Fig. 6-8**

A pin in a smooth guide (see Fig. 6-8) can transmit only a force **R** perpendicular to the surfaces of the guide. The sense of **R** is assumed on the figure and may be either downward to the left or upward to the right.

**TABLE 6-1 (continued)**

### 9. Collar on a smooth shaft

**Fig. 6-9**

A collar on a smooth shaft (see Fig. 6-9) that is pin-connected to a body can transmit only a force **R** perpendicular to the axis of the shaft. When the connection between the collar and the body is fixed (see Fig. 6-10), the collar can transmit both a force **R** and a moment **M** perpendicular to the axis of the shaft. If the shaft is not smooth, a tangential frictional force $R_t$ as well as a normal force $R_n$ can be transmitted.

**Fig. 6-10**

### 10. Fixed support

**Fig. 6-11**

A fixed support (see Fig. 6-11) can exert both a force **R** and a couple **C** on the body. The magnitude $R$ and the direction $\theta$ of the force **R** are not known. Therefore, the force **R** is usually represented on a free-body diagram by its rectangular components $R_x$ and $R_y$ and the couple **C** by its moment **M**.

### 11. Linear elastic spring

**Fig. 6-12**

The force **R** exerted on a body by a linear elastic spring (see Fig. 6-12) is proportional to the change in length of the spring. The spring will exert a tensile force if lengthened and a compressive force if shortened. The line of action of the force is along the axis of the spring.

### 12. Ideal pulley

**Fig. 6-13**

Pulleys (see Fig. 6-13) are used to change the direction of a rope or cable. The pin connecting an ideal pulley to a member can exert a force **R** of unknown magnitude $R$ and direction $\theta$ on the body. The force **R** is usually represented on a free-body diagram by its rectangular components $R_x$ and $R_y$. Also, since the pin is smooth (frictionless), the tension **T** in the cable must remain constant to satisfy moment equilibrium about the axis of the pulley.

**TABLE 6-2   THREE-DIMENSIONAL REACTIONS AT SUPPORTS AND CONNECTIONS**

**1. Ball and socket**

**Fig. 6-14**

A ball-and-socket joint (see Fig. 6-14) can transmit a force **R** but no moment. The force **R** is usually represented on a free-body diagram by its three rectangular components $R_x$, $R_y$, and $R_z$.

**2. Hinge**

**Fig. 6-15**

A hinge (see Fig. 6-15) is normally designed to transmit a force **R** in a direction perpendicular to the axis of the hinge pin. The design may also permit a force component to be transmitted along the axis of the pin. Individual hinges have the ability to transmit small moments about axes perpendicular to the axis of the pin. However, properly aligned pairs of hinges transmit only forces under normal conditions of use. Thus, the action of a hinge is represented on a free-body diagram by the force components $R_x$, $R_y$, and $R_z$ and the moments $M_x$ and $M_z$ when the axis of the pin is in the $y$-direction.

**3. Ball bearing**

**Fig. 6-16**

Ideal (smooth) ball bearings (see Fig. 6-16) are designed to transmit a force **R** in a direction perpendicular to the axis of the bearing. The action of the bearing is represented on a free-body diagram by the force components $R_x$ and $R_z$ when the axis of the bearing is in the $y$-direction.

**4. Journal bearing**

**Fig. 6-17**

Journal bearings (see Fig. 6-17) are designed to transmit a force **R** in a direction perpendicular to the axis of the bearing. Individual journal bearings (also known as bushings) have the ability to transmit small moments about axes perpendicular to the axis of the shaft. However, properly aligned pairs of bearings transmit only forces perpendicular to the axis of the shaft under normal conditions of use. Therefore, the action of a journal bearing is represented on a free-body diagram by the force components $R_x$ and $R_z$ and the couple moments $M_x$ and $M_z$ when the axis of the bearing is in the $y$-direction.

**TABLE 6-2    (continued)**

5. Thrust bearing

A thrust bearing (see Fig. 6-18), as the name implies, is designed to transmit force components both perpendicular and parallel (thrust) to the axis of the bearing. Individual thrust bearings have the ability to transmit small moments about axes perpendicular to the axis of the shaft. However, properly aligned pairs of bearings transmit only forces under normal conditions of use. Therefore, the action of a thrust bearing is represented on a free-body diagram by the force components $R_x$, $R_y$, and $R_z$ and the couple moments $M_x$ and $M_z$ when the axis of the bearing is in the $y$-direction.

**Fig. 6-18**

6. Smooth pin and bracket

A pin and bracket (see Fig. 6-19) is designed to transmit a force **R** in a direction perpendicular to the axis of the pin but may also transmit a force component along the axis of the pin. The unit also has the ability to transmit small moments about axes perpendicular to the axis of the pin. Therefore, the action of a smooth pin and bracket is represented on a free-body diagram by the force components $R_x$, $R_y$, and $R_z$ and the couple moments $M_x$ and $M_z$ when the axis of the pin is in the $y$-direction.

**Fig. 6-19**

7. Fixed support

A fixed support (see Fig. 6-20) can resist both a force **R** and a couple **C**. The magnitudes and directions of the force and couple are not known. Thus, the action of a fixed support is represented on a free-body diagram by the force components $R_x$, $R_y$, and $R_z$ and the moments $M_x$, $M_y$, and $M_z$.

**Fig. 6-20**

Draw the free-body diagram for the beam shown in Fig. 6-21a.

(a)

(b)

**Fig. 6-21**

## SOLUTION

Two concentrated forces $P_1$ and $P_2$ are applied to the beam. The weight of the beam is represented by the force **W**, which has a line of action that passes through the center of gravity $G$ of the beam. The beam is supported at the left end with a smooth pin and bracket and at the right end with a roller. The action of the left support is represented by the forces $A_x$ and $A_y$. The action of the roller is represented by the force $B_y$, which acts normal to the surface of the bearing plate. A complete free-body diagram for the beam is shown in Fig. 6-21b.

A free-body diagram for the beam should show the beam completely isolated from its surroundings. Bodies removed during the isolation process are the pin support at $A$, the roller support at $B$, and the Earth. All applied loads plus the actions on the beam of each body removed in the isolating process should be represented with appropriate vectors on the free-body diagram.

Draw the free-body diagram for the beam shown in Fig. 6-22a. Neglect the weight of the beam.

(a)

(b)

**Fig. 6-22**

## SOLUTION

A couple **M** and a concentrated load **P** are applied to the beam. Also, the beam supports a uniformly distributed load over a length $a$ at the left end of the beam and a body $D$ through a pulley and cable system at the right end of the beam. The support at the left end of the beam is fixed, and its action on the beam is represented by the forces $\mathbf{A}_x$ and $\mathbf{A}_y$ and the couple $\mathbf{M}_A$. The action of the cable at the right end of the beam is represented by a tensile force **T** with a line of action in the direction of the cable. A complete free-body diagram for the beam is shown in Fig. 6-22b. Note that the pulley and body $D$ are not a part of the free-body diagram for the beam.

Bodies removed during the isolation process are the fixed support at $A$ and the cable at $B$. All applied loads plus the actions on the beam of each body removed in the isolating process should be represented with appropriate vectors on the free-body diagram.

Draw a free-body diagram for

**a.** The cylinder shown in Fig. 6-23.
**b.** The bar shown in Fig. 6-23.

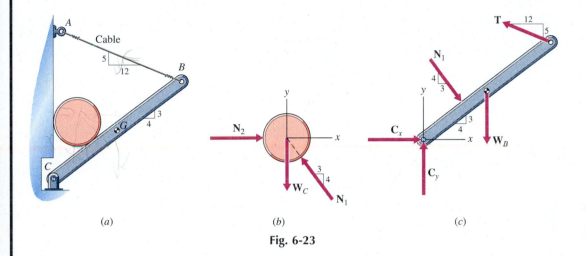

(a)          (b)          (c)

**Fig. 6-23**

## SOLUTION

**a.** Forces $N_1$ and $N_2$ act normal to the surface of the cylinder at points of contact with the bar and the wall. The weight $W_C$ of the cylinder acts through the center of gravity $G$ of the cylinder.

**b.** The action of the pin at support $C$ of the bar is represented by force components $C_x$ and $C_y$. Force $N_1$ acts normal to the surface of the bar at the point where the cylinder contacts the bar. The cable at $B$ exerts a force $T$ on the bar that is tangent to the cable at the point of attachment. The weight $W_B$ of the bar acts through the center of gravity $G$ of the bar.

Bodies removed during the isolation process for the cylinder are the wall, the bar, and the Earth. The actions on the cylinder of each body removed in the isolating process should be represented with appropriate vectors on the free-body diagram.

Bodies removed during the isolation process for the bar are the pin support at $C$, the cylinder, the cable at $B$, and the Earth. The actions on the bar of each body removed in the isolating process should be represented with appropriate vectors on the free-body diagram.

Draw a free-body diagram for

**a.** Bar *AD* shown in Fig. 6-24.
**b.** Bar *CF* shown in Fig. 6-24.

Neglect the weights of the bars.

**Fig. 6-24**

---

## SOLUTION

**a.** The action of the pin at support *A* is represented by force components $\mathbf{A}_x$ and $\mathbf{A}_y$. The cable at *B* exerts a force $\mathbf{T}$ on bar *AD* that is tangent to the cable at the point of attachment. The pin at *C* exerts a force $\mathbf{C}$ on bar *AD* that is normal to the surface of the slot in the bar.
**b.** The pin at *C* exerts a force $\mathbf{C}$ on bar *CF* that is normal to the surface of the slot in bar *AD*. The cable at *E* exerts a force $\mathbf{T}$ on bar *AD* that is tangent to the cable at the point of attachment. The action of the pin at support *F* is represented by force components $\mathbf{F}_x$ and $\mathbf{F}_y$.

Bodies removed during the isolation process for bar *AD* are the pin support at *A*, the cable at *B*, and the pin at *C*. The applied load $\mathbf{P}_1$ plus the actions on the bar of each body removed in the isolating process should be represented with appropriate vectors on the free-body diagram.

Bodies removed during the isolation process for bar *CF* are the pin at *C*, the cable at *E*, and the pin support at *F*. The applied load $\mathbf{P}_2$ plus the actions on the bar of each body removed in the isolating process should be represented with appropriate vectors on the free-body diagram.

A cylinder is supported on a smooth inclined surface by a two-bar frame as shown in Fig. 6-25a. Draw the free-body diagram **(a)** for the cylinder, **(b)** for the two-bar frame, and **(c)** for the pin at C.

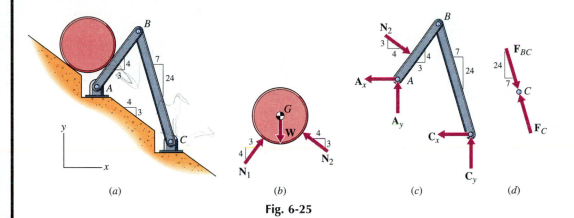

**Fig. 6-25**

## SOLUTION

**a.** The free-body diagram for the cylinder is shown in Fig. 6-25b. The weight **W** of the cylinder acts through the center of gravity G. The forces $\mathbf{N}_1$ and $\mathbf{N}_2$ act normal to the smooth surfaces at the points of contact.

**b.** The free-body diagram for the two-bar frame is shown in Fig. 6-25c. The action of the smooth pin and bracket supports at points A and C are represented by forces $\mathbf{A}_x$ and $\mathbf{A}_y$ and $\mathbf{C}_x$ and $\mathbf{C}_y$, respectively. Since bar BC is a link, the resultant $\mathbf{F}_C$ of forces $\mathbf{C}_x$ and $\mathbf{C}_y$ must have a line of action along the axis of the link. As a result, the free-body diagram for pin C can be drawn as shown in Fig. 6-25d. Note that the pin forces at B are internal and do not appear on the free-body diagram.

Bodies removed during the isolation process for the cylinder are the inclined surface, the frame, and the Earth. The actions on the cylinder of each body removed in the isolating process should be represented with appropriate vectors on the free-body diagram.

Bodies removed during the isolation process for the frame are the pin support at A, the cylinder, and the pin support at C. The actions on the frame of each body removed in the isolating process should be represented with appropriate vectors on the free-body diagram.

Draw free-body diagrams (*a*) for the pulley, (*b*) for the post *AB*, and (*c*) for the beam *CD* shown in Fig. 6-26*a*.

(*a*)                    (*b*)                    (*c*)

**Fig. 6-26**

## SOLUTION

a.  The free-body diagram for the pulley with the cable removed is shown in Fig. 6-26*b*. The action of the cable on the pulley is represented by a distributed force as shown in the first part of the figure. The action of the post on the pulley (through the pin at *B*) is a force **R,** which is equal in magnitude but opposite in sense to the resultant of the distributed force exerted by the cable. For the solution of problems, a free-body diagram consisting of the pulley and the contacting portion of the cable, as shown in the last part of Fig. 6-26*b*, is used so that the distributed load becomes an internal force, which does not appear in the analysis.

b.  The post is fixed at the base and loaded by the pulley and beam through pins at points *B* and *C*, respectively. The free-body diagram is shown in Fig. 6-26*c*. Note that the forces $B_x$ and $B_y$ exerted on the post by the pulley are equal in magnitude but opposite in direction to the forces $B_x$ and $B_y$ exerted on the pulley by the post.

c.  The free-body diagram for the beam is also shown in Fig. 6-26*c*. The masses $m_1$ and $m_2$ must be converted to forces ($W = m\mathbf{g}$) before their actions can be represented on the free-body diagram.

> Bodies removed during the isolation process for the pulley are the cable and the pin at *B*. The actions on the pulley of each body removed in the isolating process should be represented with appropriate vectors on the free-body diagram.

> Bodies removed during the isolation process for post *AB* are the fixed support at *A*, the pin at *B*, and the pin at *C*. The actions on the post of each body removed in the isolating process should be represented with appropriate vectors on the free-body diagram.

Draw a free-body diagram for the door shown in Fig. 6-27. Assume that the hinges at supports $C$ and $D$ are properly aligned so that they are not required to transmit moment components.

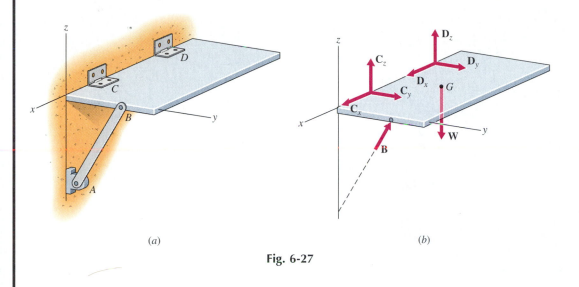

(a)                                                   (b)

**Fig. 6-27**

---

## SOLUTION

The link at $B$ exerts a force $\mathbf{B}$ on the door that is directed along the axis of the link. Since the hinges at $C$ and $D$ are properly aligned, moment components are not transmitted by individual hinges; therefore, their actions can be represented by force components $\mathbf{C}_x$, $\mathbf{C}_y$, and $\mathbf{C}_z$ at hinge $C$ and force components $\mathbf{D}_x$, $\mathbf{D}_y$, and $\mathbf{D}_z$ at hinge $D$. The weight $\mathbf{W}$ of the door acts through the center of gravity $G$ of the door.

Bodies removed during the isolation process for the door are the link at $B$, the hinge at $C$, the hinge at $D$, and the Earth. The actions on the door of each body removed in the isolating process should be represented with appropriate vectors on the free-body diagram.

Draw the free-body diagram for the curved bar $AC$ shown in Fig. 6-28a, which is supported by a ball-and-socket joint at $A$, a flexible cable at $B$, and a pin and bracket at $C$. Neglect the weight of the bar.

(a)

(b)

**Fig. 6-28**

## SOLUTION

The action of the ball-and-socket joint at support $A$ is represented by three rectangular force components $A_x$, $A_y$, and $A_z$. The action of the pin and bracket at support $C$ can be represented by force components $C_x$, $C_y$, and $C_z$ and moment components $M_x$ and $M_z$. The action of the cable is represented by the cable tension $T$. A complete free-body diagram for bar $AC$ is shown in Fig. 6-28b.

Bodies removed during the isolation process for the bar are the ball-and-socket joint at $A$, the cable at $B$, and the pin and bracket at $C$. The applied loads $P_1$ and $P_2$ plus actions on the bar of each body removed in the isolating process should be represented with appropriate vectors on the free-body diagram.

# PROBLEMS

Draw complete free-body diagrams for the bodies specified in the following problems. Neglect the weight of the member except where the problem statement indicates that the weight of the member is to be considered. Assume that all surfaces are smooth unless indicated otherwise.

## Introductory Problems

**6-1** The cantilever beam shown in Fig. P6-1, which has a weight **W.**

**Fig. P6-1**

**6-2** The beam shown in Fig. P6-2, which has a mass *m.*

**Fig. P6-2**

**6-3** The cylinder shown in Fig. P6-3, which has a weight **W.**

**Fig. P6-3**

**6-4** The curved bar shown in Fig. P6-4.

**Fig. P6-4**

**6-5** The curved bar shown in Fig. P6-5.

**Fig. P6-5**

**6-6** The angle bracket shown in Fig. P6-6.

**Fig. P6-6**

**6-7** The curved bar shown in Fig. P6-7.

**Fig. P6-7**

**6-8** The beam shown in Fig. P6-8.

**Fig. P6-8**

**6-9** The sled shown in Fig. P6-9.

**Fig. P6-9**

**6-10** The diving board shown in Fig. P6-10.

**Fig. P6-10**

**6-11** The cart shown in Fig. P6-11, which has a weight **W.**

**Fig. P6-11**

**6-12** The lawn mower shown in Fig. P6-12, which has a weight **W** and is resting on a rough surface.

**Fig. P6-12**

**Intermediate Problems**

**6-13** **a.** The cylinder, which has a weight **W**, **b.** bar *ACE*, and **c.** bar *BCD* shown in Fig. P6-13.

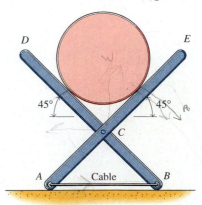

**Fig. P6-13**

**6-14**  **a.** The cylinder, which has a mass $m$, and **b.** the frame shown in Fig. P6-14.

**Fig. P6-14**

**6-15**  **a.** Bracket $AB$ shown in Fig. P6-15.

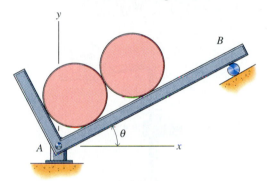

**Fig. P6-15**

**6-16**  **a.** Bar $AC$ and **b.** bar $DE$ shown in Fig. P6-16.

**Fig. P6-16**

**6-17**  **a.** The cylinder, which has a weight **W**, **b.** bar $AC$, and **c.** bar $BCD$ shown in Fig. P6-17.

**Fig. P6-17**

**6-18**  **a.** Bar $AB$ and **b.** bar $CB$ shown in Fig. P6-18.

**Fig. P6-18**

**6-19**  **a.** Bar $BE$ and **b.** bar $DF$ shown in Fig. P6-19.

**Fig. P6-19**

**6-20**  **a.** Bar $AC$ and **b.** bar $DF$ shown in Fig. P6-20.

**Fig. P6-20**

**6-21** The bent bar shown in Fig. P6-21, which is fixed at the wall at $A$.

**Fig. P6-21**

**6-22** The bent bar shown in Fig. P6-22, which is fixed at the wall at $A$.

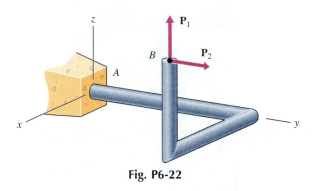

**Fig. P6-22**

**6-23** The shaft shown in Fig. P6-23. The bearing at $A$ is a thrust bearing, and the bearing at $D$ is a ball bearing. Neglect the weights of the shaft and the levers.

**Fig. P6-23**

**6-24** The block shown in Fig. P6-24, which has a mass $m$. The support at $A$ is a ball-and-socket joint. The support at $B$ is a pin and bracket.

**Fig. P6-24**

**6-25** The bent bar shown in Fig. P6-25. The support at $A$ is a journal bearing and the supports at $B$ and $C$ are ball bearings.

**Fig. P6-25**

**6-26** The bar bracket shown in Fig. P6-26. The support at $B$ is a ball-and-socket joint. The ends of the bars at $A$ and $C$ rest against smooth surfaces.

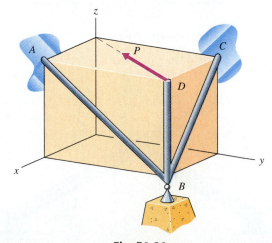

**Fig. P6-26**

**6-27**  The bar shown in Fig. P6-27. The bar rests against a smooth vertical wall at end $D$. The support at $A$ is a ball-and-socket joint. The cable is not continuous at $B$.

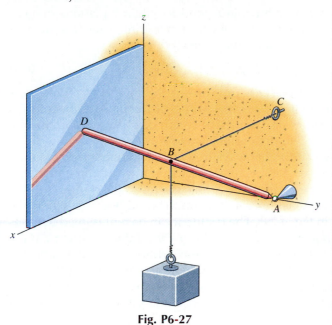

**Fig. P6-27**

**6-28**  The door shown in Fig. P6-28, which has a weight $W$.

**Fig. P6-28**

**6-29**  The bent bar shown in Fig. P6-29. The support at $A$ is a ball-and-socket joint, the supports at $B$ are a cable and a link, and the support at $C$ is a link.

**Fig. P6-29**

**6-30**  The bent bar shown in Fig. P6-30. Supports at $B$ and $C$ are ball bearings. The horizontal and vertical surfaces at $A$ are smooth.

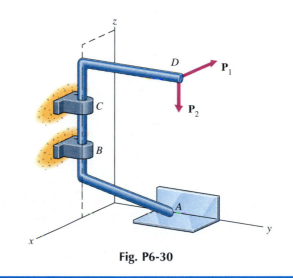

**Fig. P6-30**

## 6-3  EQUILIBRIUM IN TWO DIMENSIONS

The term *two dimensional* is frequently used to describe problems in which the forces involved are contained in a plane (say the $xy$-plane) and the axes of all couples are perpendicular to the plane containing the forces. For two-dimensional problems, since a force in the $xy$-plane has no $z$-component and produces no moments about the $x$- or $y$-axes, Eqs. 6-1 reduce to

$$\mathbf{R} = \Sigma F_x\, \mathbf{i} + \Sigma F_y\, \mathbf{j} = 0$$
$$\mathbf{C} = \Sigma M_z\, \mathbf{k} = 0$$

(6-3)

Thus, three of the six independent scalar equations of equilibrium (Eqs.

6-2) are automatically satisfied; namely,

$$\Sigma F_z = 0 \qquad \Sigma M_x = 0 \qquad \Sigma M_y = 0$$

Therefore, there are only three independent scalar equations of equilibrium for a rigid body subjected to a two-dimensional system of forces. The three equations can be expressed as

$$\Sigma F_x = 0 \qquad \Sigma F_y = 0 \qquad \Sigma M_A = 0 \qquad \textbf{(6-4)}$$

The third equation represents the sum of the moments of all forces about a $z$-axis through any point $A$ on or off the body. Equations 6-4 are both the necessary and sufficient conditions for equilibrium of a body subjected to a two-dimensional system of forces.

There are two additional ways in which the equations of equilibrium can be expressed for a body subjected to a two-dimensional system of forces. The resultant force $\textbf{R}$ and the resultant couple $\textbf{C}$ of the general two-dimensional force system on the rigid body shown in Fig. 6-29a are shown in Fig. 6-29b. The resultant can be represented in terms of its scalar components as shown in Fig. 6-29c. If the condition $\Sigma M_A = 0$ is satisfied, $\textbf{C} = \textbf{0}$. If, in addition, the condition $\Sigma F_x = 0$ is satisfied, $\textbf{R} = \Sigma F_y\,\textbf{j}$. For any point $B$ on or off the body that does not lie on the $y$-axis, the equation $\Sigma M_B = 0$ can be satisfied only if $\Sigma F_y = 0$. Thus, an alternative set of scalar equilibrium equations for two-dimensional problems is

$$\Sigma F_x = 0 \qquad \Sigma M_A = 0 \qquad \Sigma M_B = 0 \qquad \textbf{(6-5)}$$

where points $A$ and $B$ must have different $x$-coordinates.

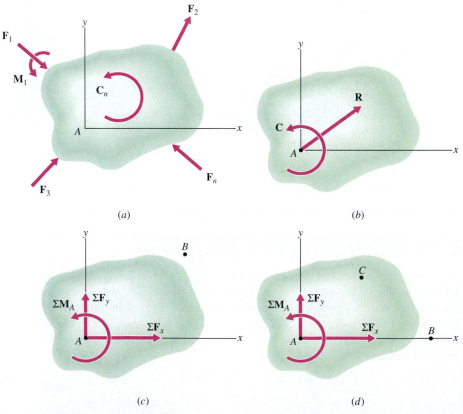

(a)

(b)

(c)

(d)

**Figure 6-29**   Representations of resultant force $\textbf{R}$ and resultant couple $\textbf{C}$ at point $A$ for a two-dimensional system of forces.

The conditions of equilibrium for a two-dimensional force system can also be expressed by using three moment equations. Again, if the condition $\Sigma M_A = 0$ is satisfied, $\mathbf{C} = \mathbf{0}$. In addition, for a point $B$ (see Fig. 6-29d) on the x-axis on or off the body (except at point $A$), the equation $\Sigma M_B = 0$ can be satisfied, once $\Sigma M_A = 0$, only if $\Sigma F_y = 0$. Thus, $\mathbf{R} = \Sigma F_x \mathbf{i}$. Finally, for any point $C$ (see Fig. 6-29d) on or off the body that does not lie on the x-axis, the equation $\Sigma M_C = 0$ can be satisfied only if $\Sigma F_x = 0$. Thus, a second set of alternative scalar equilibrium equations for two-dimensional problems is

$$\Sigma M_A = 0 \qquad \Sigma M_B = 0 \qquad \Sigma M_C = 0 \qquad \textbf{(6-6)}$$

where $A$, $B$, and $C$ are any three points not on the same straight line.

## 6-3.1   The Two-Force Body (Two-Force Members)

Equilibrium of a body under the action of two forces occurs with sufficient frequency to warrant special attention. For example, consider the link with negligible weight shown in Fig. 6-30a. Any forces exerted on the link by the frictionless pins at $A$ and $B$ can be resolved into components along and perpendicular to the line connecting points $A$ and $B$ as shown in Fig 6-30b. From the equilibrium equations,

$$\Sigma F_x = 0 \qquad A_x - B_x = 0 \qquad A_x = B_x$$
$$\Sigma F_y = 0 \qquad A_y - B_y = 0 \qquad A_y = B_y$$

Forces $A_y$ and $B_y$, however, form a couple that must be zero when the link is in equilibrium; therefore, $A_y = B_y = 0$. Thus, for two-force members, equilibrium requires that

1.  The forces acting on the member must be equal, opposite, and collinear as shown in Fig. 6-30c. The line of action of the force will pass through the frictionless pin connections at $A$ and $B$.
2.  The weights of the members must be negligible.
3.  The connections to the member must not support a moment about the connection. This condition was used in the previous discussion when it was assumed that the pins used to apply loads to the link were frictionless.

Finally, the shape of the member, as shown in Fig. 6-30d, has no effect on the three conditions for equilibrium. The loads carried by the member and the type of connection provide the requirements for equilibrium independent of the shape of the member.

## 6-3.2   The Three-Force Body (Three-Force Members)

Equilibrium of a body under the action of three forces (see Fig. 6-31) also represents a special situation. If a body is in equilibrium under the action of three forces, then the lines of action of the three forces must be concurrent (i.e., meet at a common point); otherwise, the noncurrent force would exert a moment about the point of concurrency of the other two forces. A body subjected to three parallel forces represents a special case of the three-force body. For this case, the point of concurrency is assumed to be at infinity.

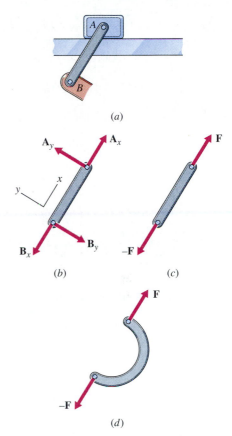

**Figure 6-30**   Example of and free-body diagram for a two-force body.

**Figure 6-31**   Example of and free-body diagram for a three-force body.

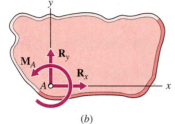

**Figure 6-32**   Resultant force **R** and resultant couple **C** at point $A$ for a two-dimensional system of forces.

## 6-3.3   Statically Indeterminate Reactions and Partial Constraints

Consider a body subjected to a system of coplanar forces $\mathbf{F}_1$, $\mathbf{F}_2$, $\mathbf{F}_3$, $\mathbf{F}_4$,..., $\mathbf{F}_n$, as shown in Fig. 6-32a. This system of forces can be replaced by an equivalent force-couple system at an arbitrary point $A$ as shown in Fig. 6-32b. The resultant force has been represented by its rectangular components $\mathbf{R}_x$ and $\mathbf{R}_y$ and the couple by $\mathbf{M}_A$. In order for the body to be in equilibrium, the supports must be capable of exerting an equal and opposite force-couple system on the body. As an example, consider the supports shown in Fig. 6-33a. The pin support at $A$, as shown in Fig. 6-33b, can exert forces in the $x$- and $y$-directions to prevent a translation of the body but no moment to prevent a rotation about an axis through $A$. The link at $B$, which exerts a force in the $y$-direction, produces the moment about point $A$ required to prevent rotation. Thus, all motion of the body is prevented, and the body is in equilibrium under the action of the three forces $\mathbf{A}_x$, $\mathbf{A}_y$, and $\mathbf{B}_y$. The forces exerted on the body by the supports are known as constraints (restrictions to movement). When the equations of equilibrium are sufficient to determine the unknown forces at the supports, as in Fig. 6-33, the body is said to be statically determinate with adequate (proper) constraints.

Three support reactions for a body subjected to a coplanar system of forces does not always guarantee that the body is statically determinate with adequate constraints. For example, consider the body and supports shown in Fig. 6-34a. The pin support at $A$ (see Fig. 6-34b) can exert forces in the $x$- and $y$-directions to prevent a translation of the

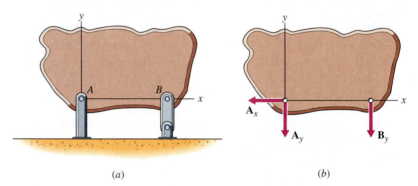

**Figure 6-33**   Body with adequate constraints.

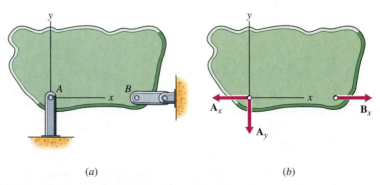

**Figure 6-34**   Body with partial constraints.

body, but since the line of action of force $\mathbf{B}_x$ passes through point $A$, it does not exert the moment required to prevent rotation of the body about point $A$. Similarly, the three links shown in Fig. 6-35a can prevent rotation of the body about any point (see Fig. 6-35b) and translation of the body in the $y$-direction but not translation of the body in the $x$-direction. The bodies in Figs. 6-34 and 6-35 are partially (improperly) constrained, and the equations of equilibrium are not sufficient to determine all the unknown reactions. A body with an adequate number of reactions is improperly constrained when the constraints are arranged in such a way that the support forces are either concurrent or parallel. Partially constrained bodies can be in equilibrium for specific systems of forces. For example, the reactions $\mathbf{R}_A$ and $\mathbf{R}_B$ for the beam shown in Fig. 6-36a can be determined by using the equilibrium equations $\Sigma F_y = 0$ and $\Sigma M_A = 0$. The beam is improperly constrained, however, since motion in the $x$-direction would occur if any of the applied loads had a small $x$-component.

Finally, if the link at $B$ in Fig. 6-33a is replaced with a pin support, as shown in Fig. 6-37a, an additional reaction $\mathbf{B}_x$ (see Fig. 6-37b) is obtained that is not required to prevent movement of the body. Obviously, the three independent equations of equilibrium will not provide sufficient information to determine the four unknowns. Constrained bodies with extra supports are statically indeterminate, since relations involving physical properties of the body, in addition to the equations of equilibrium, are required to determine some of the unknown reactions. Statically indeterminate problems are studied in "Mechanics of Materials" courses. The supports not required to maintain equilibrium of the body are called redundant supports. Typical examples of redundant supports for beams are shown in Fig. 6-38. The roller at support $B$ of the cantilever beam shown in Fig. 6-38a can be removed without affecting equilibrium of the beam. Similarly, the roller support at either $B$ or $C$ could be removed from the beam of Fig. 6-38b without disturbing equilibrium of the beam.

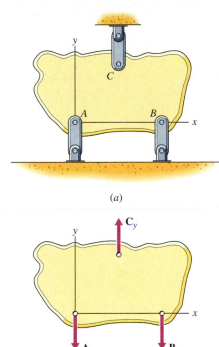

(a)

(b)

**Figure 6-35** Body with partial constraints.

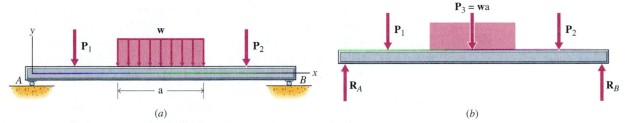

(a)

(b)

**Figure 6-36** Beam with partial constraints.

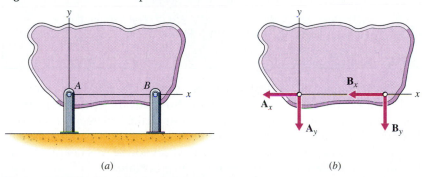

(a)

(b)

**Figure 6-37** Body with a redundant support.

**Figure 6-38** Beams with redundant supports.

### 6-3.4 Problem Solving

In Section 1-6, a procedure was outlined for solving engineering-type problems. The procedure consisted of three phases:

1. Problem definition and identification
2. Model development and simplification
3. Mathematical solution and interpretation of results

Application of this procedure to equilibrium-type problems yields the following steps.

---

**Steps for Analyzing and Solving Problems**

1. Read the problem carefully. Many student difficulties arise from failure to observe this preliminary step.
2. Identify the result requested.
3. Prepare a scaled sketch and tabulate the information provided.
4. Identify the equilibrium equations to be used to obtain the result.
5. Draw the appropriate free-body diagram. Carefully draw and label all applied forces and support reactions. Establish a convenient set of coordinate axes. Use a right-handed system in case vector cross products must be employed. Compare the number of unknowns on the free-body diagram with the number of independent equations of equilibrium. Draw additional diagrams if needed.
6. Apply the appropriate force and moment equations.
7. Report the answer with the appropriate number of significant figures and the appropriate units.
8. Study the answer and determine if it is reasonable. As a check, write some other equilibrium equations and see if they are satisfied by the solution.

---

Bodies subjected to coplanar force systems are not very complex, and a scalar solution is usually suitable for analysis. This results from the fact that moments can be expressed as scalars instead of vectors. For the more general case of rigid bodies subjected to three-dimensional force systems (discussed in the next section of this chapter), vector analysis methods are usually more appropriate.

The following examples illustrate the solution of coplanar force problems.

*A beam is loaded and supported as shown in Fig. CE6-1. Determine the force
in member BC and the reaction at support A.*

(a)

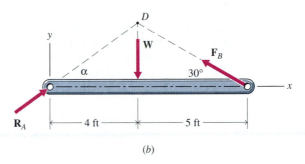

(b)

## SOLUTION

A free-body diagram for the beam is shown in Fig. CE6-1*b*. The distributed
load on the beam can be replaced with a resultant force **W** when external effects (reactions) are being determined. The magnitude of force **W** is equal to
the area under the load diagram, and the line of action of force **W** passes
through the centroid of the same area. Thus,

$$W = \frac{1}{2}(500)(3) = 750 \text{ lb} \qquad d_x = 3 + \frac{1}{3}(3) = 4 \text{ ft}$$

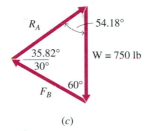

(c)

**Fig. CE6-1**

Member *BC* is a two-force member; therefore, force $\mathbf{F}_B$ lies along the line
between points *B* and *C*. In order for the beam to be in equilibrium, the line
of action of the reaction at support *A* must pass through point *D* (the point of
concurrence of forces **W** and $\mathbf{F}_B$); otherwise, force $\mathbf{R}_A$ would produce a moment about point *D*. Once it is observed that the beam is a three-force mem-
ber, angle $\alpha$ can be determined from the equation

$$\alpha = \tan^{-1} \frac{5 \tan 30°}{4} = 35.82°$$

The force $\mathbf{F}_B$ and the reaction $\mathbf{R}_A$ can then be determined either by using
the law of sines together with the force polygon shown in Fig. CE6-1*c* or by
summing moments about points *A* and *B*. Solving using the force polygon
yields

$$F_B = \frac{W \sin 54.18°}{\sin 65.82°} = \frac{750 \sin 54.18°}{\sin 65.82°} = 667 \text{ lb}$$

$$R_A = \frac{W \sin 60°}{\sin 65.82°} = \frac{750 \sin 60°}{\sin 65.82°} = 712 \text{ lb}$$

Alternatively, summing moments about points *A* and *B* yields

$$+\text{\large$\circlearrowleft$}\Sigma M_A = F_B(9 \sin 30°) - 750(4) = 0 \qquad F_B = 667 \text{ lb}$$
$$+\text{\large$\circlearrowleft$}\Sigma M_B = -R_A(9 \sin 35.82°) + 750(5) = 0 \qquad R_A = 712 \text{ lb}$$

## CONCEPTUAL EXAMPLE 6-2: EQUILIBRIUM OF A RIGID BODY

*The mass of the man shown in Fig. CE6-2a is 80 kg. How much force must he exert on the rope of the chair lift in order to support himself?*

### SOLUTION

Free-body diagrams for the man and for pulley $B$ of the lift system are shown in Fig. CE6-2b. Summing moments about the axis of the pulley verifies that the tension force in a continuous rope which passes over a frictionless pulley has a constant magnitude. Summing forces in a vertical direction then yields

$$+\uparrow\Sigma F_y = T_1 + T_1 - T_2 = 0 \qquad \text{from which} \qquad T_2 = 2T_1$$

Turning now to the free-body diagram for the man and summing forces in a vertical direction yields

$$\begin{aligned}
+\uparrow\Sigma F_y &= T_1 + T_2 - W \\
&= T_1 + 2T_1 - mg \\
&= 3T_1 - 80(9.81) = 0 \qquad \text{from which} \qquad T_1 = 263 \text{ N}
\end{aligned}$$

(a)

(b)

**Fig. CE6-2**

A beam is loaded and supported as shown in Fig. 6-39a. Determine the reactions at supports $A$ and $B$. Neglect the weight of the beam.

(a)

(b)

**Fig. 6-39**

## SOLUTION

A free-body diagram of the beam is shown in Fig. 6-39b. The reaction at support $A$ is represented by force components $\mathbf{A}_x$ and $\mathbf{A}_y$. The reaction at $B$ is a vertical force $\mathbf{B}$.

The reaction at support $B$ can be determined by summing moments about point $A$, since force $\mathbf{B}$ will be the only unknown appearing in the equation. Thus,

$$+\curvearrowleft \Sigma M_A = -500(3) - 800(6) - 700(9) - 400(12) + B(15) = 0$$
$$B = 1160 \text{ lb}$$
$$\mathbf{B} = 1160\mathbf{j} \text{ lb} = 1160 \text{ lb}\uparrow \qquad\qquad \text{Ans.}$$

Similarly, the reaction component $A_y$ at support $A$ is determined by summing moments about point $B$. Note that force $A_x$, if it exists, does not produce a moment about point $B$. Thus,

$$+\curvearrowleft \Sigma M_B = -A_y(15) + 500(12) + 800(9) + 700(6) + 400(3) = 0$$
$$A_y = 1240 \text{ lb}$$

Finally, from the equilibrium equation $\Sigma F_x = 0$,

$$+\rightarrow \Sigma F_x = A_x = 0$$

Therefore,

$$\mathbf{A} = 1240\mathbf{j} \text{ lb} = 1240 \text{ lb}\uparrow \qquad\qquad \text{Ans.}$$

The equilibrium equation $\Sigma F_y = 0$ can be used to provide a check of the results.

Scalar algebra provides a simpler solution than the vector cross product for moment calculations in this problem.

Summing moments about point $B$ yields a value for force $A_y$ that is independent of the previous determination for force $B$.

A pipe strut $BC$ is loaded and supported as shown in Fig. 6-40$a$. The strut has a uniform cross section and a mass of 50 kg. Determine the tension in the cable and the reaction at support $C$.

(a)                          (b)

**Fig. 6-40**

## SOLUTION

A free-body diagram of the strut is shown in Fig. 6-40$b$. The reaction at support $C$ is represented by force components $\mathbf{C}_x$ and $\mathbf{C}_y$. The cable at $B$ exerts a horizontal tensile force $\mathbf{T}$. The weight of the bar is

$$W = mg = 50(9.81) = 490.5 \text{ N}$$

The cable tension $\mathbf{T}$ can be determined by summing moments about point $C$.

$$+\circlearrowleft \Sigma M_C = T(1000) - 750(800) - 490.5(800) = 0$$
$$T = 992.4 \text{ N} \cong 992 \text{ N}$$
$$\mathbf{T} = -992 \, \mathbf{i} \text{ N} = 992 \text{ N} \leftarrow \qquad\qquad \text{Ans.}$$

Once cable tension $\mathbf{T}$ has been determined, the equilibrium equations $\Sigma F_x = 0$ and $\Sigma F_y = 0$ can be used to determine the reaction components $C_x$ and $C_y$ at support $C$.

$$+\rightarrow \Sigma F_x = C_x - T = C_x - 992.4 = 0$$
$$C_x = 992.4 \text{ N} \cong 992 \text{ N} \rightarrow$$
$$+\uparrow \Sigma F_y = C_y - 750 - 490.5 = 0$$
$$C_y = 1240.5 \text{ N} \cong 1241 \text{ N} \uparrow$$
$$C = \sqrt{C_x^2 + C_y^2} = \sqrt{(992.4)^2 + (1240.5)^2} = 1588.6 \text{ N} \cong 1589$$
$$\theta = \tan^{-1}\frac{C_y}{C_x} = \tan^{-1}\frac{1240.5}{992.4} = 51.3°$$
$$\mathbf{C} = 992\mathbf{i} + 1241\mathbf{j} \text{ N} = 1589 \text{ N} \measuredangle 51.3° \qquad\qquad \text{Ans.}$$

The equilibrium equation $\Sigma M_B = 0$ can be used to provide a check of results.

> Choosing a convenient axis about which to sum moments can simplify the solution of many problems by eliminating the need to solve simultaneous equations.

A 75-lb load is supported by an angle bracket, pulley, and cable as shown in Fig. 6-41a. Determine

a.  The force exerted on the bracket by the pin at *C*.
b.  The reactions at supports *A* and *B* of the bracket.

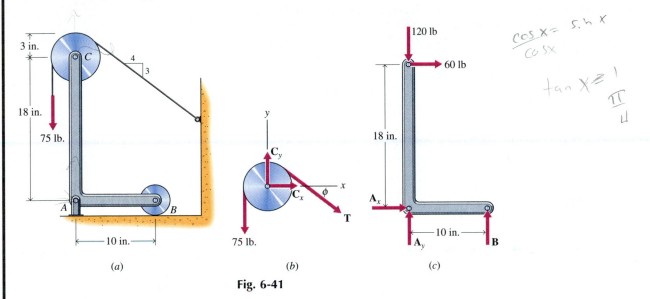

Fig. 6-41

---

## SOLUTION

a.  The force exerted on the bracket by the pulley is determined by using the free-body diagram of the pulley shown in Fig. 6-41b. The pin reaction on the pulley at *C* is shown as forces $C_x$ and $C_y$. The force in the cable between the pulley and the wall is shown as a tension **T**. The angle between the tension *T* and the *x*-axis is

$$\phi = \tan^{-1}\frac{3}{4} = 36.87°$$

The cable tension **T** can be determined by summing moments about point C.

$$+\!\!\!\downarrow\!\Sigma M_C = 75(3) - T(3) = 0 \qquad T = 75 \text{ lb}$$

This result indicates that the tension in a cable must remain constant to satisfy moment equilibrium about the axis of the pulley. This result is based on the ideal pulley assumption: namely, that the pin supporting the pulley is smooth (frictionless).

Once the cable tension **T** has been determined, the equilibrium equations $\Sigma F_x = 0$ and $\Sigma F_y = 0$ can be used to determine the reaction components $C_x$ and $C_y$.

$$+\!\rightarrow\!\Sigma F_x = C_x + 75\cos 36.87° = 0$$
$$C_x = -60 \text{ lb} = 60 \text{ lb}\leftarrow$$
$$+\!\uparrow\!\Sigma F_y = C_y - 75 - 75\sin 36.87° = 0$$
$$C_y = 120 \text{ lb} = 120 \text{ lb}\uparrow$$

Choosing a convenient axis about which to sum moments can simplify the solution of many problems by eliminating the need to solve simultaneous equations.

Therefore, the force exerted on the bracket by pin $C$ is

$$C_x = 60 \text{ lb} \rightarrow \qquad C_y = 120 \text{ lb} \downarrow$$

$$C = \sqrt{C_x^2 + C_y^2} = \sqrt{(60.0)^2 + (120.0)^2} = 134.2 \text{ lb}$$

$$\theta = \tan^{-1}\frac{C_y}{C_x} = \tan^{-1}\frac{120.0}{60.0} = 63.4°$$

$$\mathbf{C} = 60.0\mathbf{i} - 120\mathbf{j} \text{ lb} = 134.2 \text{ lb} \searrow 63.4°$$

b. The reactions at supports $A$ and $B$ of the bracket are determined by using the free-body diagram of the bracket shown in Fig. 6-41c. The reaction at support $A$ is represented by force components $\mathbf{A}_x$ and $\mathbf{A}_y$. The reaction at $B$ is a vertical force $\mathbf{B}$. The forces at $C$ are equal and opposite to those acting on the pulley.

The reaction at support $B$ can be determined by summing moments about point $A$, since force $\mathbf{B}$ will be the only unknown appearing in the equation. Thus,

$$+ \downarrow \Sigma M_A = B(10) - 60(18) = 0$$
$$B = 108 \text{ lb}$$
$$+ \rightarrow \Sigma F_x = A_x + 60 = 0$$
$$A_x = -60 \text{ lb} = 60 \text{ lb} \leftarrow$$
$$+ \uparrow \Sigma F_y = A_y + B - 120$$
$$= A_y + 108 - 120 = 0$$
$$A_y = 12 \text{ lb} = 12 \text{ lb} \uparrow$$
$$A = \sqrt{A_x^2 + A_y^2} = \sqrt{(60)^2 + (12)^2} = 61.2 \text{ lb}$$
$$\theta = \tan^{-1}\frac{A_y}{A_x} = \tan^{-1}\frac{12}{60} = 11.31°$$
$$\mathbf{A} = -60.0\mathbf{i} + 12.00\mathbf{j} \text{ lb} = 61.2 \text{ lb} \nwarrow 11.31° \qquad \text{Ans.}$$
$$\mathbf{B} = 108.0\mathbf{j} \text{ lb} = 108.0 \uparrow \qquad \text{Ans.}$$

Alternatively, summing moments about point $B$ yields a value for force $A_y$ that is independent of the previous determination for force $B$.

A rope and pulley system is used to support a body W as shown in Fig. 6-42a. Each pulley is free to rotate. One rope is continuous over pulleys A and B; the other is continuous over pulley C. Determine the tension **T** in the rope over pulleys A and B required to hold body W in equilibrium if the mass of body W is 175 kg.

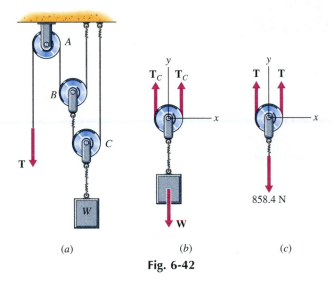

(a)          (b)          (c)

**Fig. 6-42**

## SOLUTION

The tension in the cable supporting pulley C is obtained from the equilibrium equation $\Sigma F_y = 0$ and the free-body diagram for pulley C shown in Fig. 6-42b.

$$+\uparrow\Sigma F_y = 2T_C - W = 0$$

$$T_C = \frac{1}{2}W = \frac{1}{2}mg = \frac{1}{2}(175)(9.81) = 858.4 \text{ N}$$

From the free-body diagram for pulley B shown in Fig. 6-42c.

$$+\uparrow\Sigma F_y = 2T - 858.4 = 0$$
$$T = 429.2 \text{ N} \cong 429 \text{ N} \qquad\qquad \text{Ans.}$$

> The tension is constant in a continuous cable over a system of frictionless pulleys (see Example Problem 6-11).

A beam is loaded and supported as shown in Fig. 6-43a. Determine the components of the reactions at supports $A$ and $B$.

(a)

(b)

**Fig. 6-43**

## SOLUTION

A free-body diagram of the beam is shown in Fig. 6-43b. The action of the pin support at $A$ is represented by force components $\mathbf{A}_x$ and $\mathbf{A}_y$. The action of the roller support at $B$ is represented by the force $\mathbf{B}_y$, which acts perpendicular to the horizontal surface at $B$. The 200 lb/ft distributed load can be represented on the free-body diagram by a fictitious resultant $\mathbf{R}$ with a line of action at a distance $x_C$ from the left support. Thus,

$$R = A = 200(6) = 1200 \text{ lb} \qquad x_C = 19 + \frac{1}{2}(6) = 22 \text{ ft}$$

> The resultant $\mathbf{R}$ of a distributed load is given by the area under the load diagram and its line of action passes through the centroid of the area of the diagram.

The resultant $\mathbf{R}$ can be used *only* for calculating external effects (reactions). For determinations involving internal surfaces (such as those presented later, in Chapter 8), the actual distributed load must be used. The beam is subjected to a coplanar system of parallel forces in the $y$-direction; therefore, $\mathbf{A}_x = \mathbf{0}$. The two remaining equilibrium equations are available to solve for $\mathbf{A}_y$ and $\mathbf{B}_y$. Determination of $\mathbf{B}_y$:

$$+\circlearrowleft \Sigma M_A = 0: \qquad B_y(25) + 750 - 800(13) - 1200(22) = 0$$
$$B_y = +1442 \text{ lb} \qquad \mathbf{B}_y = +1442 \text{ lb} \uparrow \qquad \text{Ans.}$$

Determination of $\mathbf{A}_y$:

$$+\circlearrowleft \Sigma M_B = 0: \qquad -A_y(25) + 750 + 800(12) + 1200(3) = 0$$
$$A_y = +558 \text{ lb} \qquad \mathbf{A}_y = 558 \text{ lb} \uparrow \qquad \text{Ans.}$$

> Summing moments about point $B$ yields a value for force $A_y$ that is independent of the previous determination for force $B$.

Alternatively (or as a check):

$$+\uparrow \Sigma F_y = 0: \qquad A_y - 800 - 1200 + 1442 = 0$$
$$A_y = +558 \text{ lb} \qquad \mathbf{A}_y = 558 \text{ lb} \uparrow$$

A beam is loaded and supported as shown in Fig. 6-44a. Determine the components of the reactions at supports A and B.

**Fig. 6-44**

## SOLUTION

A free-body diagram of the beam is shown in Fig. 6-44b. The action of the pin support at A is represented by force components $A_x$ and $A_y$. The action of the roller support at B is represented by the force $B_y$, which acts perpendicular to the horizontal surface at B. The distributed loads can be represented temporarily on the free-body diagram by fictitious resultants $R_1$, $R_2$, and $R_3$ with lines of action at distances $x_{C1}$, $x_{C2}$, $x_{C3}$, respectively, from the left support. Thus,

$$R_1 = A_1 = \frac{1}{2}(6)(3) = 9.00 \text{ kN} \qquad x_{C1} = \frac{2}{3}(3) = 2.0 \text{ m}$$

$$R_2 = A_2 = \frac{1}{2}(3)(3) = 4.50 \text{ kN} \qquad x_{C2} = 3 + \frac{1}{3}(3) = 4.0 \text{ m}$$

$$R_3 = A_3 = 3(3) = 9.00 \text{ kN} \qquad x_{C3} = 3 + \frac{1}{2}(3) = 4.5 \text{ m}$$

The beam is subjected to a coplanar system of parallel forces in the $y$-direction; therefore, $A_x = \mathbf{0}$. The two remaining equilibrium equations are available to solve for $A_y$ and $B_y$.
Determination of $B_y$:

$$+\lcurvearrowright \Sigma M_A = 0: \qquad B_y(6) - 9.00(2.0) - 4.50(4.0) - 9.00(4.5) = 0$$
$$B_y = +12.75 \text{ kN} \qquad \mathbf{B}_y = 12.75 \text{ kN}\uparrow \qquad \text{Ans.}$$

Determination of $A_y$:

$$+\lcurvearrowright \Sigma M_B = 0: \qquad -A_y(6) + 9.00(4.0) + 4.5(2.0) + 9.00(1.5) = 0$$
$$A_y = +9.75 \text{ kN} \qquad \mathbf{A}_y = 9.75 \text{ kN}\uparrow \qquad \text{Ans.}$$

Alternatively: (or as a check)

$$+\uparrow \Sigma F_y = 0: \qquad A_y - 9.00 - 4.50 - 9.00 + 12.75 = 0$$
$$A_y = +9.75 \text{ kN} \qquad \mathbf{A}_y = 9.75 \text{ kN}\uparrow$$

> The resultant **R** of a distributed load is given by the area under the load diagram and its line of action passes through the centroid of the area of the diagram.

> The distributed load on the right half of the beam is divided into a rectangle and a triangle for which areas and centroid locations are known.

> Summing moments about point B yields a value for force $A_y$ that is independent of the previous determination for force $B_y$.

A pin-connected three-bar frame is loaded and supported as shown in Fig. 6-45a. Determine the reactions at supports $A$ and $B$.

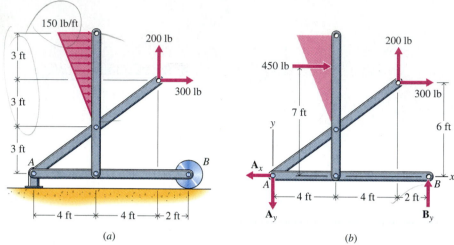

(a)　　　　　　　　(b)

**Fig. 6-45**

## SOLUTION

A free-body diagram of the frame is shown in Fig. 6-45b. The action of the pin support at $A$ is represented by force components $A_x$ and $A_y$. The action of the roller support at $B$ is represented by the force $B_y$, which acts perpendicular to the horizontal surface at $B$. The distributed load can be represented temporarily on the free-body diagram by a fictitious resultant $R$, with a line of action at a distance $y_C$ above support $A$. Thus,

$$R = A = \frac{1}{2}(150)(6) = 450 \text{ lb} \qquad y_C = 3 + \frac{2}{3}(6) = 7.00 \text{ ft}$$

Since the frame is subjected to a general coplanar force system, three equilibrium equations are available to solve for $A_x$, $A_y$, and $B_y$.

Determination of $B_y$:

$$+\!\!\downarrow\!\Sigma M_A = 0: \qquad B_y(10) - 450(7) + 200(8) - 300(6) = 0$$
$$B_y = +335 \text{ lb} \qquad \mathbf{B} = \mathbf{B}_y = 335 \text{ lb}\uparrow \qquad \text{Ans.}$$

Determination of $A_x$:

$$+\!\rightarrow\!\Sigma F_x = 0: \qquad -A_x + 450 + 300 = 0$$
$$A_x = +750 \text{ lb} \qquad \mathbf{A}_x = 750 \text{ lb}\leftarrow \qquad \text{Ans.}$$

Determination of $A_y$:

$$+\!\!\downarrow\!\Sigma M_B = 0: \qquad A_y(10) - 450(7) - 200(2) - 300(6) = 0$$
$$A_y = +535 \text{ lb} \qquad \mathbf{A}_y = 535 \text{ lb}\downarrow$$

The reaction at support $A$ is

$$A = \sqrt{(A_x)^2 + (A_y)^2} = \sqrt{(750)^2 + (535)^2} = 921 \text{ lb}$$
$$\theta = \tan^{-1}\frac{A_y}{A_x} = \tan^{-1}\frac{535}{750} = 35.5° \qquad \mathbf{A} = 921 \text{ lb}\!\!\nearrow 35.5° \qquad \text{Ans.}$$

> The resultant $R$ of a distributed load is given by the area under the load diagram and its line of action passes through the centroid of the area of the diagram.

> Summing moments about point $A$ yields force $B_y$ directly.

> Summing moments about point $B$ yields a value for force $A_y$ that is independent of the previous determination for force $B_y$.

A pin-connected two-bar frame is loaded and supported as shown in Fig. 6-46a. Determine the reactions at supports $A$ and $B$.

Fig. 6-46

---

## SOLUTION

A free-body diagram of the frame is shown in Fig. 6-46b. The action of the pin support at $A$ is represented by the single force $A$ at a known angle (since bar $AC$ is a two-force member) $\theta_A = \tan^{-1}(200/300) = 33.69°$. The action of the pin support at $B$ is represented by force components $B_x$ and $B_y$. Since the frame is subjected to a general coplanar force system, three equilibrium equations are available to solve for the unknown magnitudes of forces $A$, $B_x$, and $B_y$.
Determination of $A$:

$$+\circlearrowleft\Sigma M_B = 0: \quad -A \sin 33.69°(0.6) + 400(0.16) + 600(0.1) = 0$$
$$A = +372.6 \text{ N} \quad \mathbf{A} = 373 \text{ N} \angle 33.7° \quad \text{Ans.}$$

Determination of $B_x$:

$$+\rightarrow\Sigma F_x = 0: \quad A \cos 33.69° + B_x - 600 = 0$$
$$372.6 \cos 33.69° + B_x - 600 = 0$$
$$B_x = +290 \text{ N} \quad \mathbf{B}_x = 290 \text{ N} \rightarrow$$

Determination of $B_y$:

$$+\circlearrowleft\Sigma M_A = 0: \quad B_y(0.6) - 400(0.44) + 600(0.1) = 0$$
$$B_y = +193.3 \text{ N} \quad \mathbf{B}_y = 193.3 \text{ N}\uparrow$$

The reaction at support $B$ is

$$B = \sqrt{(B_x)^2 + (B_y)^2} = \sqrt{(290)^2 + (193.3)^2} = 348.5 \text{ N}$$
$$\theta_B = \tan^{-1}\frac{B_y}{B_x} = \tan^{-1}\frac{193.3}{290} = 33.69° \quad \mathbf{B} = 349 \text{ N}\angle 33.7° \quad \text{Ans.}$$

The results are shown in Fig. 6-46c.

---

Since bar $AC$ is a two-force member, the line of action of force $A$ must pass through pins $A$ and $C$.

Summing moments about point $B$ yields force $A$ directly.

Summing moments about point $A$ yields a value for force $B_y$ that is independent of the previous determination for force $A$.

A bar weighing 250 lb is supported by a post and cable as shown in Fig. 6-47a. Assume that all surfaces are smooth. Determine the tension in the cable and the forces on the bar at the contacting surfaces.

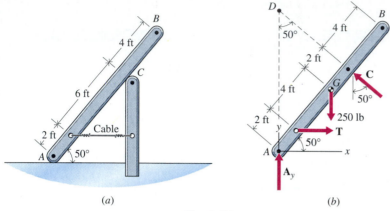

(a)                                    (b)

**Fig. 6-47**

## SOLUTION

A free-body diagram of the bar is shown in Fig. 6-47b. All surfaces are smooth; therefore, the reaction at $A$ is a vertical force $\mathbf{A}$ and the reaction at $C$ is a force $\mathbf{C}$ perpendicular to the bar. The cable exerts a tension $\mathbf{T}$ on the bar in the direction of the cable. Since the bar is subjected to a general coplanar force system, three equilibrium equations are available to solve for the unknown magnitudes of forces $\mathbf{A}$, $\mathbf{C}$, and $\mathbf{T}$.

Determination of $\mathbf{T}$:

The determination of the cable tension $\mathbf{T}$ can be simplified by taking moments about the point of concurrence (off the bar) of forces $\mathbf{A}$ and $\mathbf{C}$. Thus,

$$+\!\!\downarrow\!\!\Sigma M_D = 0: \quad T[(8/\sin 50°) - 2\sin 50°] - 250(6\cos 50°) = 0$$
$$T = +108.2 \text{ lb} \qquad \mathbf{T} = 108.2 \text{ lb}\rightarrow \qquad\qquad \text{Ans.}$$

Determination of $\mathbf{A}$:

$$+\!\!\downarrow\!\!\Sigma M_C = 0: \quad -A(8\cos 50°) + T(6\sin 50°) + 250(2\cos 50°) = 0$$
$$-A(8\cos 50°) + 108.2(6\sin 50°) + 250(2\cos 50°) = 0$$
$$A = +159.2 \text{ lb} \qquad \mathbf{A} = 159.2 \text{ lb}\uparrow \qquad\qquad \text{Ans.}$$

Determination of $\mathbf{C}$:

$$+\!\!\downarrow\!\!\Sigma M_A = 0: \quad -T(2\sin 50°) - 250(6\cos 50°) + C(8) = 0$$
$$-(108.2)(2\sin 50°) - 250(6\cos 50°) + C(8) = 0$$
$$C = +141.2 \text{ lb} \qquad \mathbf{C} = 141.2 \text{ lb} \angle 40° \qquad\qquad \text{Ans.}$$

> Choosing a convenient axis about which to sum moments can simplify the solution of many problems by eliminating the need to solve simultaneous equations.

A cylinder with a mass of 50 kg is supported on an inclined surface by a pin-connected two-bar frame as shown in Fig. 6-48a. Assume that all surfaces are smooth. Determine

a. The forces exerted on the cylinder by the contacting surfaces.
b. The reactions at supports $A$ and $C$ of the two-bar frame.

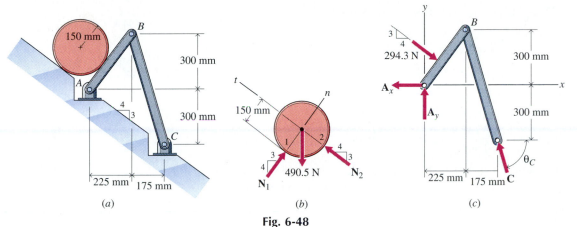

Fig. 6-48

## SOLUTION

a. A free-body diagram of the cylinder is shown in Fig. 6-48b. The actions of the inclined surface and frame are represented by forces $N_1$ and $N_2$, respectively. The action of the Earth (weight) is

$$W = mg = 50(9.81) = 490.5 \text{ N}$$

Determination of $N_1$:

$$+\nearrow\Sigma F_n = 0: \quad N_1 - \frac{4}{5}(490.5) = 0$$

$$N_1 = 392.4 \text{ N} \quad N_1 = 392 \text{ N} \angle 53.1° \quad \text{Ans.}$$

Determination of $N_2$:

$$+\nwarrow\Sigma F_t = 0: \quad N_2 - \frac{3}{5}(490.5) = 0$$

$$N_2 = 294.3 \text{ N} \quad N_2 = 294 \text{ N} \searrow 36.9° \quad \text{Ans.}$$

b. A free-body diagram of the frame is shown in Fig. 6-48c. The action of the pin support at $C$ is represented by the single force $\mathbf{C}$ at a known angle $\theta_C$, since bar $BC$ is a two-force member:

$$\theta_C = \tan^{-1}\frac{600}{175} = 73.74°$$

The action of the pin support at $A$ is represented by force components $\mathbf{A}_x$ and $\mathbf{A}_y$. Since the frame is subjected to a general coplanar force system, three equilibrium equations are available to solve for the unknown magnitudes of the three forces $\mathbf{C}$, $\mathbf{A}_x$, and $\mathbf{A}_y$.

Determination of $\mathbf{C}$:

$$+\langle\Sigma M_A = 0: \quad C\sin 73.74°(0.4) - C\cos 73.74°(0.3) - 294.3(0.15) = 0$$

$$C = +147.15 \text{ N} \quad \mathbf{C} = 147.2 \text{ N} \searrow 73.7° \quad \text{Ans.}$$

Since the cylinder is subjected to a system of three forces ($N_1$, $N_2$, and $W$), the forces must be concurrent in order for the cylinder to be in equilibrium.

Selecting axes which coincide with the lines of action of forces $N_1$ and $N_2$ eliminates the need to solve simultaneous equations.

Determination of $A_x$:

$$+\rightarrow\Sigma F_x = 0: \qquad -A_x - C\cos\theta_C + \frac{4}{5}(294.3) = 0$$

$$-A_x - 147.15\cos 73.74° + \frac{4}{5}(294.3) = 0$$

$$A_x = +194.2 \text{ N} \qquad \mathbf{A}_x = 194.2 \text{ N}\leftarrow$$

Determination of $A_y$:

$$+\uparrow\Sigma F_y = 0: \qquad A_y + C\sin\theta_C - \frac{3}{5}(294.3) = 0$$

$$A_y + 147.15\sin 73.74° - \frac{3}{5}(294.3) = 0$$

$$A_y = +35.32 \text{ N} \qquad \mathbf{A}_y = 35.3 \text{ N}\uparrow$$

The reaction at support $A$ is

$$A = \sqrt{(A_x)^2 + (A_y)^2} = \sqrt{(194.2)^2 + (35.32)^2} = 197.4 \text{ N}$$

$$\theta_A = \tan^{-1}\frac{A_y}{A_x} = \tan^{-1}\frac{35.32}{194.2} = 10.31° \qquad \mathbf{A} = 197.4 \text{ N}\measuredangle10.31° \qquad \text{Ans.}$$

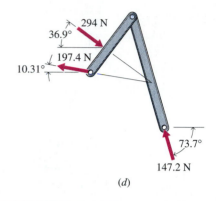

(d)

The results are shown in Fig. 6-48d. Since the frame is a three-force body, the forces $\mathbf{N}_2$, $\mathbf{A}$, and $\mathbf{C}$ must be concurrent, as illustrated in the figure.

A rigid crank arm is loaded and supported as shown in Fig. 6-49a. Determine the components of the reaction at support $A$ and the force exerted on the arm by the link at $C$.

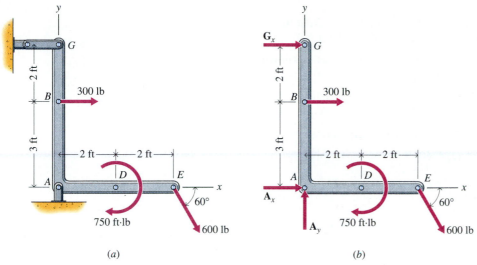

**Fig. 6-49**

## SOLUTION

A free-body diagram of the crank arm is shown in Fig. 6-49b. The action of the pin support at $A$ is represented by force components $\mathbf{A}_x$ and $\mathbf{A}_y$. The action of the link at $G$ is represented by a force $\mathbf{G}_x$ that acts along the axis of the link. Since the crank arm is subjected to a general coplanar force system, three equilibrium equations are available to solve for the unknown magnitudes of the three forces $\mathbf{A}_x$, $\mathbf{A}_y$, and $\mathbf{G}_x$.

### SOLUTION BY SCALAR ANALYSIS

The force $\mathbf{G}_x$ exerted on the arm by the link can be determined by summing moments about point $A$. Thus,

$$+\!\downarrow\!\Sigma M_A = -G_x(5) - 300(3) - 750 - 600 \sin 60° \, (4) = 0$$
$$G_x = -745.7 \text{ lb} \quad \mathbf{G}_x \cong 746 \text{ lb}\leftarrow \qquad \text{Ans.}$$

Once the force $\mathbf{G}_x$ has been determined, the equilibrium equations $\Sigma F_x = 0$ and $\Sigma F_y = 0$ can be used to determine the reaction components $A_x$ and $A_y$.

$$+\rightarrow\Sigma F_x = A_x + G_x + 300 + 600 \cos 60°$$
$$= A_x + (-745.7) + 300 + 600 \cos 60° = 0$$
$$A_x = 145.7 \text{ lb} \quad \mathbf{A}_x = 145.7 \text{ lb}\rightarrow \qquad \text{Ans.}$$
$$+\!\uparrow\!\Sigma F_y = A_y - 600 \sin 60° = 0$$
$$A_y = 519.6 \text{ lb} \quad \mathbf{A}_y \cong 520 \text{ lb}\uparrow \qquad \text{Ans.}$$

### SOLUTION BY VECTOR ANALYSIS

The equations of equilibrium for a general coplanar system of forces can be expressed in Cartesian vector form as

$$\mathbf{R} = \Sigma F_x \mathbf{i} + \Sigma F_y \mathbf{j} = 0 \qquad \mathbf{C} = \Sigma M_z \mathbf{k} = 0$$

Scalar algebra provides a simpler solution than the vector cross product for moment calculations in most two-dimensional problems.

Alternatively, summing moments about point $G$ yields a value for force $A_x$ that is independent of the previous determination for force $G_x$.

The applied loads can be written in Cartesian vector form as

$$\mathbf{B} = 300\mathbf{i} \text{ lb} \qquad \mathbf{E} = 600(\cos 60°\mathbf{i} - \sin 60°\mathbf{j})$$
$$\mathbf{C}_D = -750\mathbf{k} \text{ ft} \cdot \text{lb} \qquad = 300\mathbf{i} - 519.6\mathbf{j} \text{ lb}$$

The reactions can be written as

$$\mathbf{A} = A_x\mathbf{i} + A_y\mathbf{j} \qquad \mathbf{G} = G_x\mathbf{i}$$

Therefore,

$$\mathbf{R} = \Sigma F_x\mathbf{i} + \Sigma F_y\mathbf{j} = (A_x + G_x + 300 + 300)\mathbf{i} + (A_y - 519.6)\mathbf{j}$$
$$= (A_x + G_x + 600)\mathbf{i} + (A_y - 519.6)\mathbf{j} = \mathbf{0}$$

Summing moments about point $A$ to minimize the number of unknowns

$$\mathbf{C} = \Sigma M_A\mathbf{k} = \mathbf{r}_B \times \mathbf{F}_B + \mathbf{r}_G \times \mathbf{F}_G + \mathbf{r}_E \times \mathbf{F}_E + \mathbf{C}_D$$
$$= [3\mathbf{j} \times 300\mathbf{i}] + [5\mathbf{j} \times G_x\mathbf{i}] + [4\mathbf{i} \times 300\mathbf{i} - 519.6\mathbf{j}] - 750\mathbf{k}$$
$$= -900\mathbf{k} - 5G_x\mathbf{k} - 2078.4\mathbf{k} - 750\mathbf{k} = (-3728.4 - 5G_x)\mathbf{k} = \mathbf{0}$$

The scalar equations obtained from these two vector equations are

$$A_x + G_x + 600 = 0$$
$$A_y - 519.6 = 0$$
$$5G_x + 3728.4 = 0$$

Solving yields

$$A_x = 145.7 \text{ lb} \qquad A_y = 519.6 \text{ lb} \qquad G_x = -745.7 \text{ lb}$$

Therefore,

$$\mathbf{A} \cong 145.7\mathbf{i} + 520\mathbf{j} \text{ lb} \qquad \mathbf{G} \cong -746\mathbf{i} \text{ lb} \qquad\qquad \text{Ans.}$$

These results are identical to those obtained using a scalar analysis. For two-dimensional problems, scalar methods are usually preferred.

A pin-connected truss is loaded and supported as shown in Fig. 6-50a. The body $W$ has a mass of 100 kg. Determine the components of the reactions at supports $A$ and $B$.

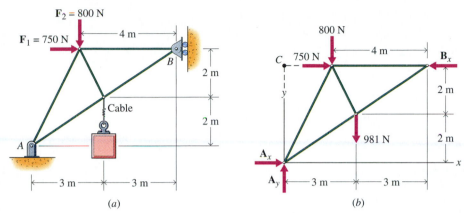

Fig. 6-50

## SOLUTION

The tension $T$ in the cable between the body $W$ and the truss is

$$T = mg = 100(9.81) = 981 \text{ N}$$

A free-body diagram of the truss is shown in Fig. 6-50b. The action of the pin support at $A$ is represented by force components $\mathbf{A}_x$ and $\mathbf{A}_y$. The action of the roller support at $B$ is represented by the force $\mathbf{B}_x$, which acts perpendicular to the vertical surface at $B$. Since the truss is subjected to a general coplanar force system, three equilibrium equations are available to solve for the unknown magnitudes of the three forces $\mathbf{A}_x$, $\mathbf{A}_y$, and $\mathbf{B}_x$.
Determination of $\mathbf{B}_x$:

$$+\!\!\downarrow\!\Sigma M_A = 0 \qquad B_x(4) - 800(2) - 750(4) - 981(3) = 0$$
$$B_x = +1885.8 \text{ N} \qquad \mathbf{B}_x = 1886 \text{ N}\leftarrow \qquad \text{Ans.}$$

Determination of $\mathbf{A}_x$:

$$+\!\!\rightarrow\!\Sigma F_x = 0: \qquad A_x + 750 - B_x = A_x + 750 - 1885.8 = 0$$
$$A_x = +1135.8 \text{ N} \qquad \mathbf{A}_x = 1136 \text{ N}\rightarrow \qquad \text{Ans.}$$

Determination of $\mathbf{A}_y$:

$$+\!\!\uparrow\!\Sigma F_y = 0: \qquad A_y - 800 - 981 = 0$$
$$A_y = +1781 \text{ N} \qquad \mathbf{A}_y = 1781 \text{ N}\uparrow \qquad \text{Ans.}$$

The component $\mathbf{A}_x$ can also be determined directly by summing moments about point $C$.

$$+\!\!\downarrow\!\Sigma M_C = 0: \qquad A_x(4) - 800(2) - 981(3) = 0$$
$$A_x = +1135.8 \text{ N} \qquad \mathbf{A}_x = 1136 \text{ N}\rightarrow$$

Choosing a convenient axis about which to sum moments can usually simplify the solution of most problems.

## SOLUTION BY VECTOR ANALYSIS

The equations of equilibrium for a general coplanar system of forces can be expressed in Cartesian vector form as

$$\mathbf{R} = \Sigma F_x\, \mathbf{i} + \Sigma F_y\, \mathbf{j} = \mathbf{0} \qquad \mathbf{C} = \Sigma M_z\, \mathbf{k} = \mathbf{0}$$

The applied loads can be written in Cartesian vector form as

$$\mathbf{F}_1 = 750\mathbf{i}\ \text{N} \qquad \mathbf{F}_2 = -800\mathbf{j}\ \text{N} \qquad \mathbf{F}_3 = -981\mathbf{j}\ \text{N}$$

The reactions can be written as

$$\mathbf{A} = A_x\mathbf{i} + A_y\mathbf{j} \qquad \mathbf{B} = -B_x\mathbf{i}$$

Therefore,

$$\mathbf{R} = \Sigma F_x\, \mathbf{i} + \Sigma F_y\, \mathbf{j}$$
$$= (A_x - B_x + 750)\mathbf{i} + (A_y - 800 - 981)\mathbf{j} = \mathbf{0}$$

Summing moments about point $A$ to minimize the number of unknowns, we have

$$\mathbf{C} = \Sigma M_A\mathbf{k}$$
$$= \mathbf{r}_1 \times \mathbf{F}_1 + \mathbf{r}_2 \times \mathbf{F}_2 + \mathbf{r}_3 \times \mathbf{F}_3 + \mathbf{r}_B \times \mathbf{B}$$
$$= [(2\mathbf{i} + 4\mathbf{j}) \times 750\mathbf{i}] + [(2\mathbf{i} + 4\mathbf{j}) \times (-800\mathbf{j})]$$
$$\quad + [(3\mathbf{i} + 2\mathbf{j}) \times (-981\mathbf{j})] + [(6\mathbf{i} + 4\mathbf{j}) \times (-B_x\mathbf{i})]$$
$$= -3000\mathbf{k} - 1600\mathbf{k} - 2943\mathbf{k} + 4B_x\mathbf{k} = (-7543 + 4B_x)\mathbf{k} = \mathbf{0}$$

The scalar equations obtained from these two vector equations are

$$A_x - B_x + 750 = 0$$
$$A_y - 800 - 981 = 0$$
$$-7543 + 4B_x = 0$$

Solving yields

$$A_x = 1136\ \text{N} \qquad A_y = 1781\ \text{N} \qquad B_x = 1886\ \text{N}$$

Therefore

$$\mathbf{A} = 1136\mathbf{i} + 1781\mathbf{j}\ \text{N} \qquad \mathbf{B} = -1886\mathbf{i}\ \text{N} \qquad\qquad \text{Ans.}$$

These results are obviously identical to those obtained using a scalar analysis. For two-dimensional problems, scalar methods are usually preferred.

# PROBLEMS

## Introductory Problems

**6-31\*** A curved slender bar is loaded and supported as shown in Fig. P6-31. Determine the reaction at support A.

**Fig. P6-31**

**6-32\*** A curved slender bar is loaded and supported as shown in Fig. P6-32. Determine the reactions at supports A and B.

**Fig. P6-32**

**6-33** A beam is loaded and supported as shown in Fig. P6-33. The beam has a uniform cross section and weighs 250 lb. Determine the reactions at supports A and B.

**Fig. P6-33**

**6-34** A beam is loaded and supported as shown in Fig. P6-34. The beam has a uniform cross section and a mass of 180 kg. Determine the reaction at support A.

**Fig. P6-34**

**6-35\*** A beam is loaded and supported as shown in Fig. P6-35. The beam has a uniform cross section and weighs 425 lb. Determine the reactions.

**Fig. P6-35**

**6-36\*** A beam is loaded and supported as shown in Fig. P6-36. The beam has a uniform cross section and a mass of 120 kg. Determine the reactions at supports A and B.

**Fig. P6-36**

**6-37** A structural member is loaded and supported as shown in Fig. P6-37. The member has a uniform cross section and weighs 208 lb. Determine the reactions at supports A and B.

**Fig. P6-37**

**6-38** A beam is loaded and supported as shown in Fig. P6-38. The beam has a uniform cross section and a mass of 20 kg. Determine the reaction at support A and the tension T in the cable.

Fig. P6-41

**6-42** An angle bracket is loaded and supported as shown in Fig. P6-42. Determine the force exerted by the cable at *A* and the reaction at support *B*.

Fig. P6-42

and supported as shown in Fig.
form cross section and weighs
s at supports *A* and *E*.

400 lb

D
E

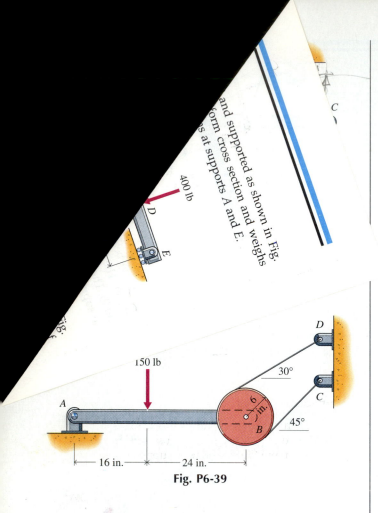

150 lb

A

6 in.

B

30°

C

45°

D

16 in. 24 in.

Fig. P6-39

**6-40*** An angle bracket is loaded and supported as shown in Fig. P6-40. Determine the reactions at supports *A* and *B*.

B

100 mm

100 mm

500 N

A

C

x

220 mm

350 N

Fig. P6-40

**6-41** A 30-lb force **P** is applied to the brake pedal of an automobile as shown in Fig. P6-41. Determine the force **Q** applied to the brake cylinder and the reaction at support *A*.

**6-43*** Determine the reactions at supports *A* and *B* of the curved bar shown in Fig. P6-43.

Fig. P6-43

**6-44\*** Determine the force exerted by the cable at $B$ and the reaction at support $A$ of the curved bar shown in Fig. P6-44.

**Fig. P6-44**

**6-45** A rope and pulley system is used to support a body as shown in Fig. P6-45. Each pulley is free to rotate, and the ropes are continuous over the pulleys. Determine the force **P** required to hold the body in equilibrium if the weight **W** of the body is 400 lb.

**Fig. P6-45**

**6-46** A rope and pulley system is used to support a body as shown in Fig. P6-46. Each pulley is free to rotate, and the ropes are continuous over the pulleys. Determine the force **P** required to hold the body in equilibrium if the mass $m$ of the body is 250 kg.

**Fig. P6-46**

**Intermediate Problems**

**6-47\*** A beam is loaded and supported as shown in Fig. P6-47. The beam has a uniform cross section and weighs 975 lb. Determine the reactions at supports $A$ and $B$.

**Fig. P6-47**

**6-48\*** A beam is loaded and supported as shown in Fig. P6-48. The beam has a uniform cross section and a mass of 275 kg. Determine the reaction at support $A$.

**Fig. P6-48**

**6-49** Determine the force **P** required to push the 300-lb cylinder over the small block shown in Fig. P6-49.

**Fig. P6-49**

**6-50** The mass of the cylinder shown in Fig. P6-50 is 100 kg. Determine the reactions at contact points $A$ and $B$. All surfaces are smooth.

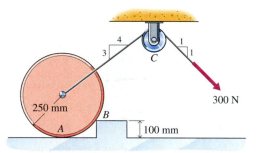

**Fig. P6-50**

**6-51\*** A lever is loaded and supported as shown in Fig. P6-51. Determine the force exerted on the lever by link $CD$ and the reaction at support $A$.

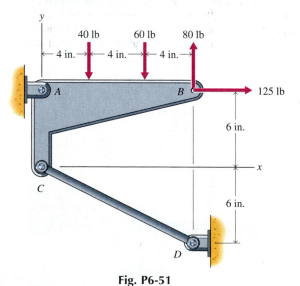

**Fig. P6-51**

**6-52\*** Determine the reactions at supports $A$ and $B$ of the curved bar shown in Fig. P6-52.

**Fig. P6-52**

**6-53** Determine the force exerted by the cable at $B$ and the reaction at support $A$ of the bar shown in Fig. P6-53.

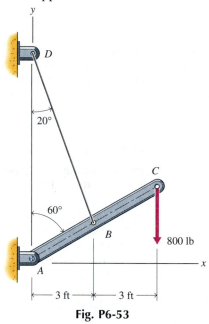

**Fig. P6-53**

**6-54** The man shown in Fig. P6-54 has a mass of 75 kg; the beam has a mass of 40 kg. The beam is in equilibrium with the man standing at the end and pulling on the cable. Determine the force exerted on the cable by the man and the reaction at support $C$.

**Fig. P6-54**

**296**

## Challenging Problems

**6-55\*** A rope and pulley system is used to support a body as shown in Fig. P6-55. Each pulley is free to rotate, and the ropes are continuous over the pulleys. Determine the force **P** required to hold the body in equilibrium if the weight **W** of the body is 250 lb.

**Fig. P6-55**

**6-56\*** Pulleys 1 and 2 of the rope and pulley system shown in Fig. P6-56 are connected and rotate as a unit. The radii of the pulleys are 100 mm and 300 mm, respectively. Rope A is fastened to pulley 1 at point A'. Rope B is fastened to pulley 2 at point B'. Rope C is continuous over pulleys 3 and 4. Determine the tension **T** in rope C required to hold body W in equilibrium if the mass of body W is 225 kg.

**Fig. P6-56**

**6-57** Pulleys A and B of the chain hoist shown in Fig. P6-57 are connected and rotate as a unit. The chain is continuous, and each of the pulleys contains slots that prevent the chain from slipping. Determine the force **F** required to hold a 1000-lb block W in equilibrium if the radii of pulleys A and B are 3.5 and 4.0 in., respectively.

**Fig. P6-57**

**6-58** A wagon with a mass of 3500 kg is held in equilibrium on an inclined surface by using a cable as shown in Fig. P6-58. Determine the force in the cable and the forces exerted on the wheels at A and B by the inclined surface.

**Fig. P6-58**

**6-59\*** The lawn mower shown in Fig. P6-59 weighs 35 lb. Determine the force **P** required to move the mower at a constant velocity and the forces exerted on the front and rear wheels by the inclined surface.

Fig. P6-59

Fig. 6-62

**6-60\*** The coal wagon shown in Fig. P6-60 is used to haul coal from a mine. If the mass of the coal and wagon is 2000 kg, determine the force **P** required to move the wagon at a constant velocity and the forces exerted on the front and rear wheels by the inclined surface.

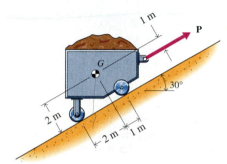

Fig. P6-60

**6-61** Bar *AB* is supported in a horizontal position by two cables as shown in Fig. P6-61. Determine the magnitude of force **P** and the force in each of the cables.

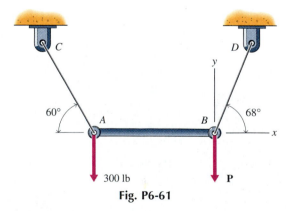

Fig. P6-61

**6-62\*** A frame is loaded and supported as shown in Fig. P6-62. Determine the reactions at supports *A* and *E*.

**6-63** The wrecker truck of Fig. P6-63 has a weight of 15,000 lb and a center of gravity at *G*. The force exerted on the rear (drive) wheels by the ground consists of both a normal component $B_y$ and a tangential component $B_x$, while the force exerted on the front wheel consists of a normal force $A_y$ only. Determine the maximum pull *P* the wrecker can exert when $\theta = 30°$ if $B_x$ cannot exceed $0.8B_y$ (because of friction considerations) and the wrecker does not tip over backward (the front wheels remain in contact with the ground).

Fig. P6-63

**6-64** Bar *AB* of Fig. P6-64 has a uniform cross section, a mass of 25 kg, and a length of 1 m. Determine the angle $\theta$ for equilibrium.

Fig. P6-64

**298**

**6-65*** The crane and boom shown in Fig. P6-65 weigh 12,000 lb and 600 lb, respectively. When the boom is in the position shown, determine

a. The maximum load that can be lifted by the crane.
b. The tension in the cable used to raise and lower the boom when the load being lifted is 3600 lb.
c. The pin reaction at boom support A when the load being lifted is 3600 lb.

**Fig. P6-65**

**6-66** The garage door ABCE shown in Fig. P6-66 is being raised by a cable DE. The one-piece door is a homogeneous rectangular slab weighing 225 lb. Frictionless rollers B and C run in tracks at each side of the door as shown. Determine the tension T in the cable and the forces B and C on the frictionless rollers when d = 75 in.

**Fig. P6-66**

## Computer Problems

**C6-67** The lever shown in Fig. P6-67 is formed in a quarter circular arc of radius 20 in. Plot A and B, the magnitude of the pin force at A and the force on the smooth support B, as a function of $\theta$ ($0° \le \theta \le 85°$), the angle at which the support is located.

**Fig. P6-67**

**C6-68** A lightweight pipe is slipped over the handle of a wrench to give extra leverage (Fig. P6-68). If the inside of the pipe is smooth and fits loosely about the handle of the wrench, plot the forces exerted on the wrench at A and B as a function of the overlap distance d ($0 \le d \le 200$ mm).

**Fig. P6-68**

**C6-69** The crane and boom shown in Fig. P6-69 weigh 12,000 lb and 600 lb, respectively. The pulleys at D and E are small, and the cables attached to them are essentially parallel.

a. Plot d, the location of the resultant force relative to point C, as a function of the boom angle $\theta$ ($0° \le \theta \le 80°$), when the crane is lifting a 3600-lb load.
b. Plot A, the magnitude of the reaction force on the pin at A, as a function of $\theta$ ($0° \le \theta \le 80°$), when the crane is lifting a 3600-lb load.

**Fig. P6-69**

**c.** It is desired that the resultant force on the tread always be at least 1 ft behind $C$ to ensure that the crane is never in danger of tipping over. Plot $W_{max}$, the maximum load that may be lifted, as a function of $\theta$ $(0° \leq \theta \leq 80°)$.

**C6-70** The tower crane shown in Fig. P6-70 is used to lift construction materials. The counterweight $C$ weighs 31,000 N; the motor $M$ weighs 4500 N; the weight of the boom $AB$ is 36,000 N and can be considered as acting at point $G$; and the weight of the tower is 23,000 N and can be considered as acting at its midpoint.

**a.** If the tower is lifting a 9000-N load, calculate and plot the reaction forces on the feet at $D$ and $E$ as a function of the distance $x$ at which the weight is being lifted.

**b.** It is desired that the reaction forces at the feet $D$ and $E$ always be greater than 4500 N to ensure that the tower is never in danger of tipping over. Plot $W_{max}$, the maximum load that can be lifted as a function of the distance $x$ $(0 \leq x \leq 36 \text{ m})$.

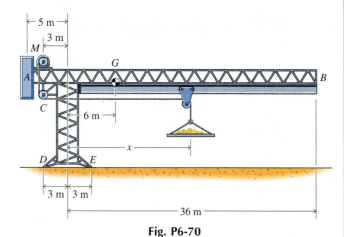

**Fig. P6-70**

**C6-71** The hydraulic cylinder $BC$ is used to tip the box of the dump truck shown in Fig. P6-71. If the combined weight of the box and the load is 22,000 lb and acts through the center of gravity $G$,

**a.** Plot $C$, the force in the hydraulic cylinder, as a function of the angle $\theta$ $(0 \leq \theta \leq 80°)$.

**b.** Plot $A$, the magnitude of the reaction force on the pin at $A$, as a function of $\theta$ $(0° \leq \theta \leq 80°)$.

**Fig. P6-71**

**C6-72** An extension ladder is being raised into position by a hydraulic cylinder as shown in Fig. P6-72. If the ladder has a mass of 500 kg and center of gravity at $G$,

**a.** Plot $B$, the force in the hydraulic cylinder, as a function of the angle $\theta$ $(10° \leq \theta \leq 90°)$.

**b.** Plot $A$, the magnitude of the reaction force on the pin at $A$, as a function of $\theta$ $(10° \leq \theta \leq 90°)$.

**Fig. P6-72**

## 6-4   EQUILIBRIUM IN THREE DIMENSIONS

A general three-dimensional system of forces $\mathbf{F}_1, \mathbf{F}_2, ..., \mathbf{F}_n$ and couples $\mathbf{C}_1, \mathbf{C}_2, ..., \mathbf{C}_n$ can be replaced by an equivalent system consisting of a system of noncoplanar, concurrent forces and a system of noncoplanar couples. The resultant $\mathbf{R}$ of the concurrent force system can be expressed as

$$\mathbf{R} = \Sigma F_x\, \mathbf{i} + \Sigma F_y\, \mathbf{j} + \Sigma F_z\, \mathbf{k} \qquad \text{(4-30)}$$

The resultant $\mathbf{C}$ of the noncoplanar system of couples can be expressed as

$$\mathbf{C} = \Sigma M_x\, \mathbf{i} + \Sigma M_y\, \mathbf{j} + \Sigma M_z\, \mathbf{k} \qquad \text{(4-31)}$$

The resultant force $\mathbf{R}$ and the resultant couple $\mathbf{C}$ together constitute the resultant of the general three-dimensional force system. Equations 4-30 and 4-31 indicate that the resultant of the force system may be a force $\mathbf{R}$, a couple $\mathbf{C}$, or both a force $\mathbf{R}$ and a couple $\mathbf{C}$. Thus, a rigid body subjected to a general three-dimensional system of forces will be in equilibrium if $R = C = 0$, which requires that

$$\Sigma F_x = 0 \qquad \Sigma F_y = 0 \qquad \Sigma F_z = 0$$

and

$$\qquad \text{(6-2)}$$

$$\Sigma M_x = 0 \qquad \Sigma M_y = 0 \qquad \Sigma M_z = 0$$

Thus, there are six independent scalar equations of equilibrium for a rigid body subjected to a general three-dimensional system of forces. The first three equations indicate that the $x$-, $y$-, and $z$-components of the resultant force $\mathbf{R}$ acting on a body in equilibrium are zero. The second three equations express the further equilibrium requirement that there are no couple components acting on the body about any of the coordinate axes or about axes parallel to the coordinate axes. These six equations are the necessary conditions for equilibrium of a rigid body. They are also the sufficient conditions for equilibrium if all the forces acting on the body can be determined from these equations. The six scalar equations (Eq. 6-2) are independent since each can be satisfied independently of the others.

The following examples illustrate the solution of three-dimensional problems.

A homogeneous flat plate that weighs 500 lb is supported by a shaft $AB$ and a cable $C$ as shown in Fig. 6-51$a$. The bearing at $A$ is a ball bearing, and the bearing at $B$ is a thrust bearing. The bearings are properly aligned; therefore, they transmit only force components. When the three forces shown in Fig. 6-51$a$ are being applied to the plate, determine the reactions at bearings $A$ and $B$ and the tension in cable $C$. Use a scalar analysis for the solution but express the final results in Cartesian vector form.

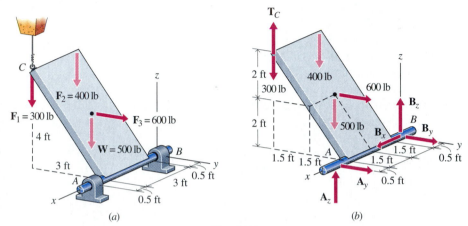

Fig. 6-51

## SOLUTION

A free-body diagram for the plate is shown in Fig. 6-51$b$. The bearing at $A$ is a ball bearing; therefore, this reaction is represented by two force components $A_y$ and $A_z$. The bearing at $B$ is a thrust bearing; therefore, this reaction is represented by three force components $B_x$, $B_y$, and $B_z$. The force in the cable is represented by tension $T_C$. The plate is subjected to a general, three-dimensional system of forces; therefore, six equilibrium equations (Eqs. 6-2) are available for determining the six unknown force components indicated on the free-body diagram. Summing forces in the positive $x$-, $y$-, and $z$-directions and summing moments about the $x$-, $y$-, and $z$-axes in accordance with the right-hand rule yields

Scalar analysis is often used to solve three-dimensional problems if moment arms for the forces are easy to determine.

$$\Sigma F_x = B_x = 0 \qquad\qquad B_x = 0 \qquad (a)$$

$$\Sigma F_y = A_y + B_y + F_3 = A_y + B_y + 600 = 0$$
$$A_y + B_y = -600 \text{ lb} \qquad (b)$$

$$\Sigma F_z = A_z + B_z + T_C - F_1 - F_2 - W$$
$$= A_z + B_z + T_C - 300 - 400 - 500 = 0$$
$$A_z + B_z + T_C = 1200 \text{ lb} \qquad (c)$$

$$\Sigma M_x = F_1(3) + F_2(3) - F_3(2) + W(1.5) - T_C(3)$$
$$= 300(3) + 400(3) - 600(2) + 500(1.5) - T_C(3) = 0$$
$$T_C = 550 \text{ lb} \qquad (d)$$

The problem would be statically indeterminate without the assumption of zero moments about lines perpendicular to the axis of the thrust bearing.

$$\Sigma M_y = F_1(3.5) + F_2(0.5) + W(2) - T_C(3.5) - A_z(4.0)$$
$$= 300(3.5) + 400(0.5) + 500(2) - T_C(3.5) - A_z(4.0) = 0$$
$$4.0A_z + 3.5T_C = 2250 \text{ ft} \cdot \text{lb} \qquad (e)$$

$$\Sigma M_z = A_y(4) + F_3(2)$$
$$= A_y(4) + 600(2) = 0 \qquad\qquad A_y = -300 \text{ lb} \qquad (f)$$

From Eqs. *b* and *f*:

$$A_y + B_y = -300 + B_y = -600 \text{ lb}$$
$$B_y = -300 \text{ lb}$$

From Eqs. *d* and *e*:

$$4.0A_z + 3.5T_C = 4.0A_z + 3.5(550) = 2250 \text{ ft} \cdot \text{lb}$$
$$A_z = 81.3 \text{ lb}$$

From Eq. *d*:

$$A_z + B_z + T_C = 81.3 + B_z + 550 = 1200 \text{ lb}$$
$$B_z = 568.7 \text{ lb}$$

The results from the scalar analysis can be expressed in Cartesian vector form as

$$\mathbf{A} = -300\mathbf{j} + 81.3\mathbf{k} \text{ lb} \qquad\qquad \text{Ans.}$$
$$\mathbf{B} = -300\mathbf{j} + 569\mathbf{k} \text{ lb} \qquad\qquad \text{Ans.}$$
$$\mathbf{T}_C = 550\mathbf{k} \text{ lb} \qquad\qquad \text{Ans.}$$

The negative signs associated with force components $A_y$ and $B_y$ indicate that they act in opposite directions to those shown on the free-body diagram.

The masses of cartons 1, 2, and 3, which rest on the platform shown in Fig. 6-52a, are 300 kg, 100 kg, and 200 kg, respectively. The mass of the platform is 500 kg. Determine the tensions in the three cables A, B, and C that support the platform.

(a)

**Fig. 6-52**

(b)

## SOLUTION

A free-body diagram for the platform is shown in Fig. 6-52b. The platform is subjected to a system of parallel forces in the z-direction; therefore, the equilibrium equations $\Sigma F_x = 0$, $\Sigma F_y = 0$, and $\Sigma M_z = 0$ are automatically satisfied. One force equation ($\Sigma F_z = 0$) and two moment equations ($\Sigma M_x = 0$ and $\Sigma M_y = 0$) are available to solve for the unknown cable tensions.

> Scalar analysis is often used to solve three-dimensional problems if moment arms for the forces are easy to determine.

The weight of the platform and the forces exerted on the platform by the three cartons are

$$F_1 = m_1 g = 300(9.81) = 2943 \text{ N} \qquad F_3 = m_3 g = 200(9.81) = 1962 \text{ N}$$
$$F_2 = m_2 g = 100(9.81) = 981 \text{ N} \qquad W = m_p g = 500(9.81) = 4905 \text{ N}$$

From the equilibrium equation $\Sigma F_z = 0$,

$$+\uparrow \Sigma F_z = T_A + T_B + T_C - F_1 - F_2 - F_3 - W$$
$$= T_A + T_B + T_C - 2943 - 981 - 1962 - 4905 = 0$$

From which

$$T_A + T_B + T_C = 10{,}791 \text{ N} \qquad\qquad (a)$$

From the equilibrium equations $\Sigma M_x = 0$ and $\Sigma M_y = 0$,

$$+\downdownarrows \Sigma M_x = T_A y_A + T_B y_B + T_C y_C - F_1 y_1 - F_2 y_2 - F_3 y_3 - W y_W$$
$$= T_A(0) + T_B(1) + T_C(4) - 2943(1) - 981(1) - 1962(3) - 4905(2) = 0$$

from which

$$T_B + 4T_C = 19{,}620 \text{ N} \cdot \text{m} \qquad\qquad (b)$$

$$+\downdownarrows \Sigma M_y = T_A x_A - T_B x_B - T_C x_C + F_1 x_1 + F_2 x_2 + F_3 x_3 + W x_W$$
$$= T_A(0) - T_B(3) - T_C(1) + 2943(1) + 981(2) + 1962(1) + 4905(1.5) = 0$$

from which

$$3T_B + T_C = 14{,}224.5 \text{ N} \cdot \text{m} \qquad\qquad (c)$$

Solving Eqs. a, b, and c simultaneously yields

$$T_A = 3344 \text{ N} \cong 3.34 \text{ kN}\uparrow \qquad\qquad \text{Ans.}$$
$$T_B = 3389 \text{ N} \cong 3.39 \text{ kN}\uparrow \qquad\qquad \text{Ans.}$$
$$T_C = 4058 \text{ N} \cong 4.06 \text{ kN}\uparrow \qquad\qquad \text{Ans.}$$

A post and bracket is used to support a pulley as shown in Fig. 6-53a. A cable passing over the pulley transmits a 500-lb load as shown in the figure. Determine the reaction at support $A$ of the post.

## SOLUTION

A free-body diagram of the post and bracket with the pulley removed is shown in Fig. 6-53b. The applied load at $B$ and the position vector from $A$ to $B$ can be written in Cartesian vector form as

$$\mathbf{B} = -500\mathbf{i} - 500\mathbf{k} \text{ lb} \qquad \mathbf{r}_{B/A} = 12\mathbf{j} + 21\mathbf{k} \text{ in.}$$

The fixed support at $A$ can resist both a force and a couple. Therefore, the reaction at $A$ can be written in Cartesian vector form as

$$\mathbf{A} = A_x\mathbf{i} + A_y\mathbf{j} + A_z\mathbf{k} \qquad \mathbf{C}_A = M_{Ax}\mathbf{i} + M_{Ay}\mathbf{j} + M_{Az}\mathbf{k}$$

The conditions for equilibrium are

$$\mathbf{R} = \Sigma F_x\,\mathbf{i} + \Sigma F_y\,\mathbf{j} + \Sigma F_z\,\mathbf{k} = (A_x - 500)\mathbf{i} + A_y\mathbf{j} + (A_z - 500)\mathbf{k} = 0 \quad (a)$$

Summing moments about $A$ to eliminate the maximum number of unknowns

$$\mathbf{C} = \Sigma M_x\,\mathbf{i} + \Sigma M_y\,\mathbf{j} + \Sigma M_z\,\mathbf{k}$$
$$= M_{Ax}\mathbf{i} + M_{Ay}\mathbf{j} + M_{Az}\mathbf{k} + (\mathbf{r}_{B/A} \times \mathbf{B})$$

$$= M_{Ax}\mathbf{i} + M_{Ay}\mathbf{j} + M_{Az}\mathbf{k} + \begin{vmatrix} \mathbf{i} & \mathbf{j} & \mathbf{k} \\ 0 & 12 & 21 \\ -500 & 0 & -500 \end{vmatrix}$$

$$= (M_{Ax} - 6000)\mathbf{i} + (M_{Ay} - 10{,}500)\mathbf{j} + (M_{Az} + 6000)\mathbf{k} = 0 \quad (b)$$

Equating the coefficients of $\mathbf{i}$, $\mathbf{j}$, and $\mathbf{k}$ to zero in Eqs. $a$ and $b$ and solving yields

$$A_x = 500 \text{ lb} \qquad A_y = 0 \text{ lb} \qquad A_z = 500 \text{ lb}$$
$$M_{Ax} = 6000 \text{ in.} \cdot \text{lb} \qquad M_{Ay} = 10{,}500 \text{ in.} \cdot \text{lb} \qquad M_{Az} = -6000 \text{ in.} \cdot \text{lb}$$

The reaction at support $A$ can be expressed in Cartesian vector form as

$$\mathbf{A} = 500\mathbf{i} + 500\mathbf{k} \text{ lb} \qquad \mathbf{C}_A = 6000\mathbf{i} + 10{,}500\mathbf{j} - 6000\mathbf{k} \text{ in.} \cdot \text{lb} \qquad \text{Ans.}$$

The magnitude of the force at support $A$ is

$$|\mathbf{A}| = \sqrt{A_x^2 + A_y^2 + A_z^2} = \sqrt{(500)^2 + (0)^2 + (500)^2} = 707 \text{ lb}$$

The magnitude of the moment at support $A$ is

$$|\mathbf{C}_A| = \sqrt{M_{Ax}^2 + M_{Ay}^2 + M_{Az}^2}$$
$$= \sqrt{(6000)^2 + (10{,}500)^2 + (-6000)^2} = 13{,}500 \text{ in.} \cdot \text{lb}$$

The direction angles associated with the couple $\mathbf{C}_A$ are

$$\theta_x = \cos^{-1}\frac{M_{Ax}}{|\mathbf{C}_A|} = \cos^{-1}\frac{6000}{13{,}500} = 63.6° \qquad \text{Ans.}$$

$$\theta_y = \cos^{-1}\frac{M_{Ay}}{|\mathbf{C}_A|} = \cos^{-1}\frac{10{,}500}{13{,}500} = 38.9° \qquad \text{Ans.}$$

$$\theta_z = \cos^{-1}\frac{M_{Az}}{|\mathbf{C}_A|} = \cos^{-1}\frac{-6000}{13{,}500} = 116.4° \qquad \text{Ans.}$$

The components of the force $\mathbf{A}$ and the couple $\mathbf{C}_A$ are shown in Fig. 6-53c.

Choosing a convenient point about which to sum moments can usually simplify the solution of most three-dimensional problems.

(a)

(b)

(c)

**Fig. 6-53**

The door shown in Fig. 6-54a has a mass of 25 kg and is supported in a horizontal position by two hinges and a bar. The hinges have been properly aligned; therefore, they exert only force reactions on the door. Assume that the hinge at B resists any force along the axis of the hinge pins. Determine the reactions at supports A, B, and D.

Fig. 6-54

## SOLUTION

A free-body diagram of the door is shown in Fig. 6-54b. The weight, the hinge reactions at A and B, and the bar reaction at D can be written as

$$\mathbf{W} = -mg\mathbf{k} = -25(9.81)\mathbf{k} = -245.3\mathbf{k} \text{ N}$$
$$\mathbf{A} = A_y\mathbf{j} + A_z\mathbf{k} \text{ N}$$
$$\mathbf{B} = B_x\mathbf{i} + B_y\mathbf{j} + B_z\mathbf{k} \text{ N}$$
$$\mathbf{D} = 0.5647D\mathbf{j} + 0.8253D\mathbf{k} \text{ N}$$

For the door to be in equilibrium

$$\mathbf{R} = \Sigma F_x\,\mathbf{i} + \Sigma F_y\mathbf{j} + \Sigma F_z\,\mathbf{k}$$
$$= (B_x)\mathbf{i} + (A_y + B_y + 0.5647D)\mathbf{j} + (A_z + B_z + 0.8253D - 245.3)\mathbf{k} = \mathbf{0} \quad (a)$$

Summing moments about B to eliminate the maximum number of unknowns

$$\mathbf{C} = \Sigma C_x\,\mathbf{i} + \Sigma C_y\,\mathbf{j} + \Sigma C_z\,\mathbf{k}$$
$$= (\mathbf{r}_{A/B} \times \mathbf{A}) + (\mathbf{r}_{W/B} \times \mathbf{W}) + (\mathbf{r}_{D/B} \times \mathbf{D})$$

$$= \begin{vmatrix} \mathbf{i} & \mathbf{j} & \mathbf{k} \\ 1.2 & 0 & 0 \\ 0 & A_y & A_z \end{vmatrix} + \begin{vmatrix} \mathbf{i} & \mathbf{j} & \mathbf{k} \\ 0.6 & 0.5 & 0 \\ 0 & 0 & -245.3 \end{vmatrix} + \begin{vmatrix} \mathbf{i} & \mathbf{j} & \mathbf{k} \\ 1.6 & 0.65 & 0 \\ 0 & 0.5647D & 0.8253D \end{vmatrix}$$

$$\mathbf{C} = (-122.65 + 0.5364D)\mathbf{i} + (-1.2A_z + 147.18 - 1.3205D)\mathbf{j}$$
$$+ (1.2A_y + 0.9035D)\mathbf{k} = \mathbf{0} \quad (b)$$

The six scalar equations obtained from vector Eqs. a and b are

$$B_x = 0$$
$$A_y + B_y + 0.5647D = 0$$
$$A_z + B_z + 0.8253D - 245.3 = 0$$
$$-122.65 + 0.5364D = 0$$
$$-1.2A_z + 147.18 - 1.3205D = 0$$
$$1.2A_y + 0.9035D = 0$$

Solving these equations simultaneously yields

$$A_z = -128.96 \text{ N}$$
$$A_y = -172.15 \text{ N}$$
$$B_x = 0$$
$$B_y = 43.03 \text{ N}$$
$$B_z = 185.56 \text{ N}$$
$$D = 228.65 \text{ N}$$

The reactions at hinges $A$ and $B$ and the force exerted by the bar at $D$ can be expressed in Cartesian vector form as

| | | |
|---|---|---|
| $\mathbf{A} = -172.2\mathbf{j} - 129.0\mathbf{k}$ N | $\|\mathbf{A}\| = 215$ N | Ans. |
| $\mathbf{B} = 43.0\mathbf{j} + 185.6\mathbf{k}$ N | $\|\mathbf{B}\| = 190.5$ N | Ans. |
| $\mathbf{D} = 129.1\mathbf{j} + 188.7\mathbf{k}$ N | $\|\mathbf{D}\| = 229$ N | Ans. |

A scalar analysis can frequently be used in three-dimensional problems when a single unknown is the only quantity required. For example, the force exerted by the bar at support $D$ can be determined by summing moments about the axis of the hinges. Thus,

$$+ \downharpoonleft \Sigma M_x = 0 \qquad D(0.95 \sin \theta_D) - W(0.50) = 0$$
$$D(0.95 \sin 34.38°) - 245.3(0.50) = 0$$
$$D = 228.6 = 229 \text{ N}$$

> The problem would be statically indeterminate without the assumption of zero moments about lines perpendicular to the axis of the hinges.

# PROBLEMS

## Introductory Problems

**6-73\*** Determine the reaction at support $A$ of the pipe system shown in Fig. P6-73 when the force applied to the pipe wrench is 50 lb.

**Fig. P6-73**

**6-74\*** The rectangular plate of uniform thickness shown in Fig. P6-74 has a mass of 500 kg. Determine the tensions in the three cables supporting the plate.

**Fig. P6-74**

**6-75** A 500-lb homogeneous circular plate is supported by three cables as shown in Fig. P6-75. Determine the tensions in the three cables.

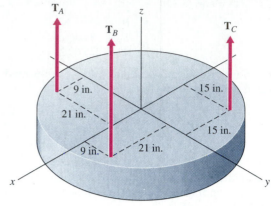

**Fig. P6-75**

**6-76** The triangular plate of uniform thickness shown in Fig. P6-76 has a mass of 400 kg. Determine the tensions in the two cables supporting the plate and the reaction at the ball support.

**Fig. P6-76**

**6-77\*** The bent bar shown in Fig. P6-77 is supported with three brackets that exert only force reactions on the bar. Determine the reactions at supports $A$, $B$, and $C$.

**Fig. P6-77**

**6-78\*** The bent bar shown in Fig. P6-78 is supported with two brackets that exert only force reactions on the bar. End C of the bar rests against smooth horizontal and vertical surfaces. Determine the reactions at supports A, B, and C.

**Fig. P6-78**

**6-80** Bar AC of Fig. P6-80 rests against a smooth surface at end C and is supported at end A with a ball-and-socket joint. The cable at B is attached midway between the ends of the bar. Determine the reactions at supports A and C and the tension in the cable at B.

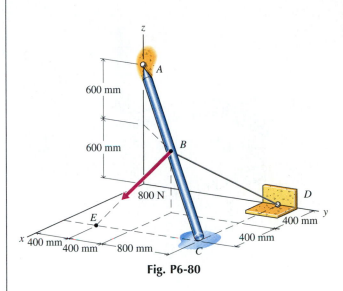

**Fig. P6-80**

### Intermediate Problems

**6-81\*** A shaft is loaded through a pulley and a lever (see Fig. P6-81) that are fixed to the shaft. Friction between the belt and pulley prevents slipping of the belt. Determine the force **P** required for equilibrium and the reactions at supports A and B. The support at A is a ball bearing, and the support at B is a thrust bearing. The bearings exert only force reactions on the shaft.

**6-79** Bar AB is used to support an 850-lb load as shown in Fig. P6-79. End A of the bar is supported with a ball-and-socket joint. End B of the bar is supported with two cables. Determine the components of the reaction at support A and the tensions in the two cables.

**Fig. P6-79**

**Fig. P6-81**

**6-82\*** The shaft with two levers shown in Fig. P6-82 is used to change the direction of a force. Determine the force **P** required for equilibrium and the reactions at supports $A$ and $B$. The support at $A$ is a ball bearing, and the support at $B$ is a thrust bearing. The bearings exert only force reactions on the shaft.

750 N

200 mm

200 mm

$A$

200 mm    250 mm

200 mm

$B$

**P**

**Fig. P6-82**

**6-83** The plate shown in Fig. P6-83 is supported in a horizontal position by two hinges and a cable. The hinges have been properly aligned; therefore, they exert only force reactions on the plate. Assume that the hinge at $B$ resists any force along the axis of the hinge pins. Determine the reactions at supports $A$ and $B$ and the tension in the cable.

$D$

5 in.

20 in.

5 in.

$x$

10 in.

100 lb

$B$    30°

$A$    $G$

10 in.    $C$

15 in.

15 in.    $y$

300 lb

**Fig. P6-83**

**6-84** The block $W$ shown in Fig. P6-84 has a mass of 250 kg. Bar $AB$ rests against a smooth vertical wall at end $B$ and is supported at end $A$ with a ball-and-socket joint. The two cables are attached to a point on the bar midway between the ends. Determine the reactions at supports $A$ and $B$ and the tensions in the two cables.

$z$

400 mm

700 mm

$C$    200 mm

$B$

300 mm

$D$

600 mm    300 mm    600 mm

200 mm

$A$    $y$

500 mm

$x$

$W$

**Fig. P6-84**

**6-85\*** The plate shown in Fig. P6-85 weighs 200 lb and is supported in a horizontal position by a hinge and a cable. Determine the reactions at the hinge and the tension in the cable.

12 in.

$z$

$C$

20 in.    13 in.    13 in.

2 in.    $A$    2 in.

$x$    $B$    28 in.    $y$

**Fig. P6-85**

**6-86** Beam $CD$ of Fig. P6-86 is supported at the left end $C$ by a smooth pin and bracket and at the right end $D$ by a continuous cable that passes around a frictionless pulley. The lines of action of the force in the cable pass through point $D$. Determine the components of the reaction at support $C$ and the force in the cable when a 5-kN load $W$ is being supported by the beam.

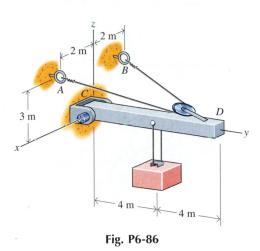

**Fig. P6-86**

## Challenging Problems

**6-87*** The plate shown in Fig. P6-87 weighs 150 lb and is supported in a horizontal position by two hinges and a cable. The hinges have been properly aligned; therefore, they exert only force reactions on the plate. Assume that the hinge at $B$ resists any force along the axis of the hinge pins. Determine the reactions at supports $A$ and $B$ and the tension in the cable.

**Fig. P6-87**

**6-88*** A bar is supported by a ball-and-socket joint, link, and cable as shown in Fig. P6-88. Determine the reactions at supports $A$ (ball-and-socket joint) and $B$ (link) and the tension in the cable.

**Fig. P6-88**

**6-89** A beam is supported by a ball-and-socket joint and two cables as shown in Fig. P6-89. Determine the reaction at support $A$ (the ball-and-socket joint) and the tensions in the two cables.

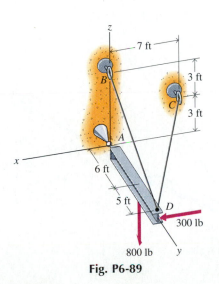

**Fig. P6-89**

**6-90** The plate shown in Fig. P6-90 has a mass of 75 kg. The brackets at supports $A$ and $B$ exert only force reactions on the plate. Each of the brackets can resist a force along the axis of pins in one direction only. Determine the reactions at supports $A$ and $B$ and the tension in the cable.

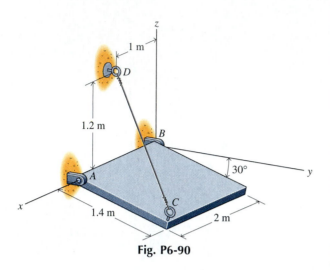

**Fig. P6-90**

**6-91\*** A bar is supported by a ball-and-socket joint and two cables as shown in Fig. P6-91. Determine the reaction at support $A$ (the ball-and-socket joint) and the tensions in the two cables.

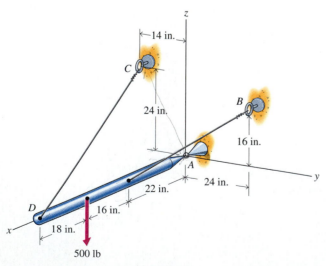

**Fig. P6-91**

**6-92** The plate shown in Fig. P6-92 has a mass of 100 kg. The hinges at supports $A$ and $B$ exert only force reactions on the plate. Assume that the hinge at $B$ resists any force along the axis of the hinge pins. Determine the reactions at supports $A$ and $B$ and the tension in the cable.

**Fig. P6-92**

## Computer Problems

**C6-93** An I-beam is supported by a ball-and-socket joint at $A$, by a rope $BC$, and by a horizontal force $\mathbf{P}$ as shown in Fig. P6-93. In addition, a 560-lb load is suspended from a movable support at $D$. If the uniform beam is 80 in. long and weighs 225 lb,

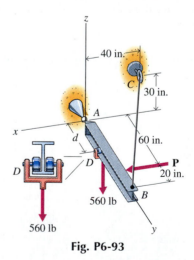

**Fig. P6-93**

**a.** Plot the magnitude of the force $\mathbf{P}$ required to keep the beam aligned with the $y$-axis as a function of the distance $d$ ($0 \le d \le 80$ in.).
**b.** On the same graph plot $T_{BC}$, the tension in the rope, and $A$, the magnitude of the force exerted on the ball-and-socket joint.

**C6-94** A hand winch is used to raise a 75-kg load as shown in Fig. P6-94. If the force **P** is always perpendicular to both the handle *DE* and the arm *CD*, plot *A* and *B*, the magnitudes of the bearing forces, as a function of the angle $\theta$ ($0 \le \theta \le 360°$).

End view

**Fig. P6-94**

# SUMMARY

Any system of forces acting on a rigid body can be expressed in terms of a resultant force **R** and a resultant couple **C**. Therefore, for a rigid body to be in equilibrium, both the resultant force **R** and the resultant couple **C** must vanish. These two conditions can be expressed by the two vector equations

$$\mathbf{R} = \Sigma F_x \, \mathbf{i} + \Sigma F_y \, \mathbf{j} + \Sigma F_2 \, \mathbf{k} = \mathbf{0}$$
$$\mathbf{C} = \Sigma M_x \, \mathbf{i} + \Sigma M_y \, \mathbf{j} + \Sigma M_z \, \mathbf{k} = \mathbf{0}$$

(6-1)

Equation 6-1 can be expressed in scalar form as

$$\Sigma F_x = 0 \qquad \Sigma F_y = 0 \qquad \Sigma F_z = 0$$
$$\Sigma M_x = 0 \qquad \Sigma M_y = 0 \qquad \Sigma M_z = 0$$

(6-2)

Equations 6-2 are the necessary conditions for equilibrium of a rigid body. They are also the sufficient conditions for equilibrium if all the forces acting on the body can be determined from these equations.

In order to study the force system acting on a body, it is necessary to identify all forces, both known and unknown, that act on the body. The best way to identify all forces acting on a body is to use the free-body diagram approach. Special care must be exercised when representing the actions of connections and supports on the free-body diagram.

The term *two-dimensional* is used to describe problems in which the forces involved are contained in a plane (say the $xy$-plane) and the axes of all couples are perpendicular to the plane containing the forces. For two-dimensional problems, the equations of equilibrium reduce to

$$\mathbf{R} = \Sigma F_x \, \mathbf{i} + \Sigma F_y \, \mathbf{j} = \mathbf{0}$$
$$\mathbf{C} = \Sigma M_z \, \mathbf{k} = \mathbf{0}$$

(6-3)

or in scalar form to

$$\Sigma F_x = 0 \qquad \Sigma F_y = 0 \qquad \Sigma M_A = 0$$

(6-4)

The third equation represents the sum of the moments of all forces about a $z$-axis through any point $A$ on or off the body. Equations 6-4 are both the necessary and sufficient conditions for equilibrium of a body subjected to a two-dimensional system of forces. Alternative forms of Eqs. 6-4 are

$$\Sigma F_x = 0 \qquad \Sigma M_A = 0 \qquad \Sigma M_B = 0$$

(6-5)

where points $A$ and $B$ must have different $x$-coordinates, and

$$\Sigma M_A = 0 \qquad \Sigma M_B = 0 \qquad \Sigma M_C = 0$$

(6-6)

where $A$, $B$, and $C$ are any three points not on the same straight line.

# REVIEW PROBLEMS

**6-95*** A curved slender bar is loaded and supported as shown in Fig. P6-95. Determine the reactions at supports $A$ and $B$ and the tension **T** in the cable.

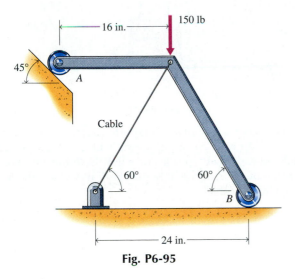

**Fig. P6-95**

**6-97** A cylinder is supported by a bar and cable as shown in Fig. P6-97. The weight of the cylinder is 150 lb and the weight of the bar is 20 lb. If all surfaces are smooth, determine the reaction at support $C$ of the bar and the tension **T** in the cable.

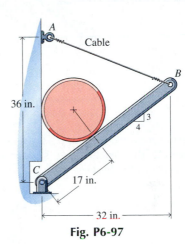

**Fig. P6-97**

**6-98*** A cylinder is supported by a bar and cable as shown in Fig. P6-98. The mass of the 200-mm diameter cylinder is 75 kg. Determine the reaction at support $A$ and the tension **T** in the cable.

**6-96*** Determine the reactions at supports $A$ and $B$ of the truss shown in Fig. P6-96.

**Fig. P6-96**

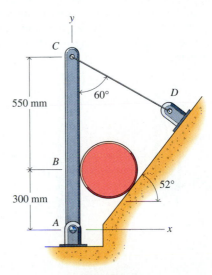

**Fig. P6-98**

**315**

**6-99** Two beams are loaded and supported as shown in Fig. P6-99. Determine the reactions at supports $A$, $B$, and $C$. Neglect the weights of the beams.

**Fig. P6-99**

**6-100** A bracket is loaded and supported as shown in Fig. P6-100. Determine the reactions at supports $A$ and $B$. Neglect the weight of the bracket.

**Fig. P6-100**

**6-101\*** The shaft shown in Fig. P6-101 is part of a drive system in a factory. Friction between the belts and pulleys prevents slipping of the belts. Determine the torque **T** required for equilibrium and the reactions at supports $A$ and $B$. The support at $A$ is a journal bearing, and the support at $B$ is a thrust bearing. The bearings exert only force reactions on the shaft.

**Fig. P6-101**

**6-102\*** A beam is loaded and supported as shown in Fig. P6-102. Determine the reactions at supports $A$ and $B$ when $m_1 = 75$ kg and $m_2 = 225$ kg.

**Fig. P6-102**

**6-103** A bar is loaded and supported as shown in Fig. P6-103. End $A$ of the bar is supported with a ball-and-socket joint and end $B$ with a cable and a link. Determine the components of the reaction at supports $A$ and $B$ and the tension **T** in the cable.

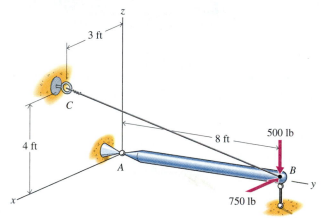

**Fig. P6-103**

**6-104** A bar is loaded and supported as shown in Fig. P6-104. End $A$ of the bar is supported with a ball-and-socket joint. Determine the components of the reaction at support $A$ and the tensions $T_C$ and $T_D$ in the two cables.

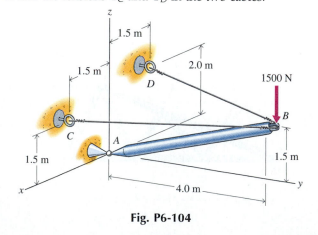

**Fig. P6-104**

**6-105\*** The 200-lb plate shown in Fig. P6-105 is supported in a horizontal position by two hinges and a continuous cable. The hinges have been properly aligned; therefore, they exert only force reactions on the plate. Assume that the hinge at *B* resists any force along the axis of the hinge pins. Determine the reactions at supports *A* and *B* and the tension **T** in the cable.

**6-106** A bent bar is loaded and supported as shown in Fig. P6-106. End *A* of the bar is supported with a ball-and-socket joint, point *B* with a cable and link, and point *C* with a link. Determine the components of the reaction at support *A* and the forces in the cable and links.

**Fig. P6-105**

**Fig. P6-106**

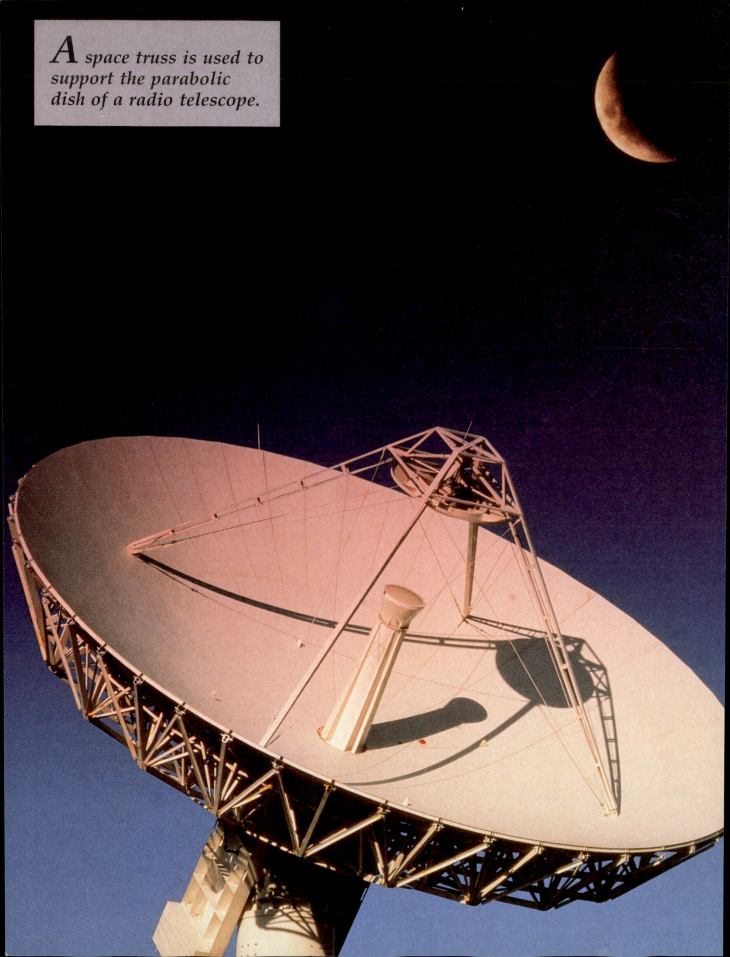

*A space truss is used to support the parabolic dish of a radio telescope.*

# TRUSSES, FRAMES, AND MACHINES

## 7-1 INTRODUCTION

In Chapter 6 the equations of equilibrium were used to determine the external support reactions on a rigid body. Determination of the support reactions, however, is only the first step in the analysis of engineering structures and machines. In this chapter the equations of equilibrium are used to determine the joint forces in structures composed of pin-connected members. Determination of these joint forces is a necessary first step in choosing the fasteners (type, size, material, etc.) used to hold the structure together. In Chapter 8 this analysis will be extended to show how to determine the resultant of the internal forces in a member. Determination of the internal forces is needed to design the members that make up the structure.

Like all forces (internal, external, applied, or reactive), the joint forces always occur in equal-magnitude, opposite-direction pairs. If not separated by means of a free-body diagram from the rest of the structure (or environment), the pair of forces need not be considered when writing the equations of equilibrium. Therefore, in order to determine the joint forces, the structure must be divided into two or more parts. Then the joint forces at the separation points become external forces on each free-body diagram and will enter the equations of equilibrium. Application of the equations of equilibrium to the various parts of a simple truss, frame, or machine allows the determination of all the forces acting at the connections if the problem is statically determinate.

Although there are many different types of engineering structures, this chapter will focus on two of the more important and common types:

(a)

(b)

(c)

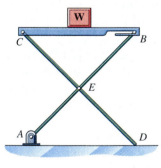

(d)

**Figure 7-1**  Typical examples of engineering structures known as trusses.

**Trusses,** which are structures composed entirely of two force members. Trusses generally consist of triangular subelements and are constructed and supported so as to prevent any motion. Bridge supports such as those shown in Fig. 7-1 are trusses. Their airy structure can carry a great load for a relatively small structural weight.

**Frames,** which are structures that always contain at least one member acted on by forces at three or more points. The table structure of Fig. 7-2 is a frame. Frames are also constructed and supported so as to prevent any motion. Framelike structures that are not fully constrained are often called **machines** or **mechanisms.**

## 7-2  PLANE TRUSSES

A truss is a structure composed of usually straight members joined together at their end points and loaded only at the joints (Fig. 7-1). The airy structure of a truss provides greater strength over large spans than would more solid types of structures. Trusses are commonly seen supporting the roofs of buildings as well as large railroad and highway bridges. Although not commonly seen, trusses also form the skeletal structure of many large buildings.

**Planar trusses** lie in a single plane, and all applied loads must lie in the same plane. Planar trusses are often used in pairs to support bridges, as shown in Fig. 7-3. All members of the truss *ABCDEF* lie in the same vertical plane. Loads on the floor of the bridge are carried by means of the floor construction to the joints *ABCD*. The loads thus transmitted to the joints act in the same vertical plane as the truss.

**Space trusses** are structures that are not contained in a single plane and/or are loaded out of the plane of the structure. Classic examples

**Figure 7-2**  Typical example of a frame.

**Figure 7-3**  Stringers and floor beams transmit loads to the joints of a bridge truss.

are the truss structures supporting large antennas and windmills. Space trusses will be discussed in Section 7-3.

There are four main assumptions made in the analysis of trusses. One result of the assumptions is that all members of the idealized structure are two-force members. Such structures are much easier to analyze than more general structures having the same number of members. Although the assumptions are idealizations of actual structures, real trusses behave according to the idealizations to a high degree of approximation. The resulting error is usually small enough to justify the assumptions.

Truss members are connected together at their ends only.

The first assumption means that the truss of Figs. 7-1a and 7-3 should be drawn as in Fig. 7-4. In actual practice, the main top and bottom chords frequently consist of members that span several joints rather than a series of shorter members between joints. The members of a truss are usually long and slender, however, and can support little lateral load or bending moment. Hence, the noncontinuous member assumption is usually acceptable. Because members are not continuous through a joint, no confusion should result from drawing trusses using the simple line representations of Fig. 7-1.

Truss members are connected together by frictionless pins.

In real trusses, the members are usually bolted, welded, or riveted to a gusset plate, as shown in Fig. 7-5a, rather than connected by an idealized frictionless pin as shown in Fig. 7-5b. However, experience has shown the frictionless pin to be an acceptable idealization as long as the axes of the members all intersect at a single point.

The truss structure is loaded only at the joints.

As stated earlier, the members that make up a truss are usually long and slender. Like cables, such members can withstand large tensile (axial) load but cannot withstand moments or large lateral loads. Loads must either be applied directly to the joints as indicated in the diagrams of Fig. 7-1 or must be carried to the joints by a floor structure such as shown in Fig. 7-3.

The weights of the members may be neglected.

In the analysis of trusses, the weight of the members is frequently neglected. While this may be acceptable for small trusses, it may not be acceptable for a large bridge truss. Again, because the members can withstand little bending moment or lateral load, experience has shown that little error results from assuming that the load acts at the joints of the truss. Common practice is to assume that half of the weight of each member acts at the two joints that it connects.

The result of these four assumptions is that forces act only at the ends of the members. Also, because the pins are assumed frictionless, there is no moment applied to the ends of the members. Therefore, by the analysis of Section 6-3, each member is a two-force member sup-

**Figure 7-4** Idealized truss consisting of pin-connected, two-force members with all loads applied at the joints.

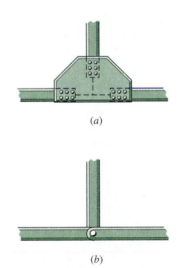

**Figure 7-5** Actual riveted truss joint idealized as a frictionless pin connection.

**Figure 7-7**   Free-body diagrams for the individual members and pins of a simple three-member truss.

**Figure 7-6**   Truss member idealized as a slender two-force member.

porting only an axial force, as shown in Fig. 7-6. In its simplest form, a truss (such as that shown in Fig. 7-7*a*) consists of a collection of two-force members held together by frictionless pins, as shown in Fig. 7-7*b*.

For general two-force members, the forces act along the line joining the points where the forces are applied. Since truss members are usually straight, however, the forces will act along the axis of the member as shown in Figs. 7-6 and 7-7[1]. Forces that pull on the end of a member (as in Figs. 7-6 and 7-7) tend to elongate the member. Such forces are called tensile forces, and the member is said to be in tension. Forces that push on the end of the member tend to shorten the member. Such forces are called compressive forces, and the member is said to be in compression. It may be noted from Fig. 7-6 that when a joint exerts forces that pull on the end of a member, the member exerts forces that also pull on the joint.

It is important to distinguish which members of a truss are in compression and which are in tension. The long, slender members that make up a truss are very strong in tension but tend to bend or buckle under large compressive loads. Truss members in compression either must be made thicker than the other truss members or must be braced to prevent buckling.

One end of large bridge trusses is usually allowed to "float" on a rocker or roller support as shown in Fig. 7-1. Aside from the mathematical requirement (in a planar equilibrium problem, only three support reactions can be determined), such a support is needed to allow for the expansion or contraction of the structure due to temperature variations.

The basic building block of all trusses is a triangle. In order to retain its shape and support the large loads applied to them, trusses must

---

[1]For curved two-force members, however, the line joining the ends is not the axis of the member. All the trusses considered in this chapter will contain only straight two-force members. For a brief discussion of the effect of curved two-force members, see Section 7-2.4.

be rigid structures. The simplest structure that is rigid (independent of how it is supported) is a triangle. Of course, the word *rigid* does not mean that a truss will not deform at all under loading. It will undergo very small deformations, but will very nearly retain its original shape.

"Rigid" is often interpreted also to mean that the truss will retain its shape when removed from its supports or when one of the supports is free to slide. In this sense, the truss of Fig. 7-8 is rigid, whereas the truss of Fig. 7-9 is not. The truss of Fig. 7-9 is called a **compound truss,** and the lack of internal rigidity is made up for by an extra external support reaction.

The basic building block of all trusses is a triangle. Large trusses are constructed by attaching several triangles together. One method of construction starts with a basic triangular element such as triangle *ABC* of Fig. 7-10. Additional triangular elements are added one at a time by attaching one new joint (for example, *D*) to the truss, using two new members (for example, *BD* and *CD*). A truss that can be constructed in this fashion is called a **simple truss.** Although it might appear that all trusses composed of triangles are simple trusses, such is not the case. For example, neither of the trusses of Figs. 7-8 or 7-9 is a simple truss.

The importance of a simple truss is that it allows a simple way to check the rigidity and solvability of a truss. Clearly, since a simple truss is constructed solely of triangular elements, it is always rigid. Also, since each new joint brings two new members with it, a simple relationship exists between the number of joints *j* and the number of members *m* in a simple plane truss:

$$m = 2j - 3 \qquad\qquad \textbf{(7-1)}$$

Equation 7-1 expresses the condition necessary to guarantee that the number of equations to be solved ($2j$) is the same as the number of unknowns to be solved for ($m$ member forces and 3 support reactions). For example, consider the simple truss shown in Fig. 7-11a, which has 8 joints, 13 members, and 3 support reactions. Application of Eq. 7-1 yields

$$m = 2j - 3 \qquad \text{or} \qquad 13 = 2(8) - 3 = 13$$

**Figure 7-8** Plane truss with internal rigidity.

**Figure 7-9** Plane truss that lacks internal rigidity.

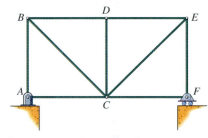

**Figure 7-10** A simple planar truss constructed using triangular elements.

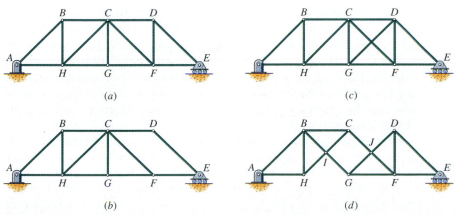

(a)

(b)

(c)

(d)

**Figure 7-11** Examples of simple, unstable, and statically indeterminate trusses.

(b)

(c)

(d)

**Fig. 7-11**

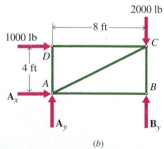

(a)

(b)

**Figure 7-12** Simple truss with five
members and four joints. Free-body
diagram for the complete truss.

In the truss of Fig. 7-11*b*, member *DF* of the truss of Fig. 7-11*a* has been
removed. In this case the truss is unstable and will collapse under load.
This condition is expressed by the equation

$$m < 2j - 3$$

Similarly, in the truss of Fig. 7-11*c*, an additional member *DG* has been
added to the truss of Fig. 7-11*a*. In this case, there are more unknowns
than there are independent equations and the truss is statically indeterminate because of the presence of the redundant member. This condition is expressed by the equation

$$m > 2j - 3$$

Although Eq. 7-1 ensures that a simple plane truss is rigid and
solvable, it is neither sufficient nor necessary to ensure that a nonsimple plane truss is rigid and solvable. For example, the nonsimple plane
trusses of Figs. 7-8 and 7-9 are both rigid (at least while attached to
their supports) and solvable, although one (Fig. 7-8) satisfies Eq. 7-1
and the other (Fig. 7-9) does not. A tempting generalization of Eq.
7-1 is

$$m = 2j - r \tag{7-2}$$

where *r* is the number of support reactions. Both trusses of Figs. 7-8
and 7-9 satisfy Eq. 7-2, as do all simple trusses that have the customary three support reactions. However, constructions can be envisioned
(see Fig. 7-11*d*) for which even Eq. 7-2 is not a proper test of the solvability of a truss.

### 7-2.1    Method of Joints

Consider the truss of Fig. 7-12*a*, whose free-body diagram is shown in
Fig. 7-12*b*. Since the entire truss is a rigid body in equilibrium, each
part must also be in equilibrium. The method of joints consists of taking the truss apart, drawing separate free-body diagrams of each part—
each member and each pin—as in Fig. 7-13, and applying the equations of equilibrium to each part of the truss in turn.

The free-body diagrams of the members in Fig. 7-13 have only axial forces applied to their ends because of the assumptions about how
a truss is constructed and loaded. The symbol $T_{BC}$ is used to represent
the unknown force in member *BC*. (No significance is attached to the
order of the subscripts; that is, $T_{BC} = T_{CB}$.) Since the lines of action of
the member forces are all known, the force in each member is completely specified by giving its magnitude and sense; that is, whether
the force points away from the member as in Fig. 7-13 or toward the
member. Thus the force (a vector) is represented by the scalar symbol
$T_{BC}$. The sense of the force will be taken from the sign of $T_{BC}$; positive
indicates the direction drawn on the free-body diagram, negative indicates the opposite direction.

Forces that point away from a member as in Fig. 7-13 tend to stretch
the member and are called tensile. Forces that point toward a member
tend to compress the member and are called compressive. Whether a
member is in compression or tension is usually not known ahead of
time. Although some people try to guess and draw some of the forces
in tension and others in compression, it is not necessary to do so. In
this book, the free-body diagrams will all be drawn as though all mem-

bers are in tension. A negative value for a force in the solution will indicate that the member was really in compression. This can be reported
either by saying that $T_{BC} = -2500$ lb or by saying that $T_{BC} = 2500$ lb
(C). The latter is preferable since it does not depend on whether member $BC$ was assumed to be in tension or in compression in the free-
body diagram.

According to Newton's third law (of action and reaction), the force
exerted on a member by a pin and the force exerted on a pin by a member are equal and opposite. Therefore the same symbol $T_{AB}$ is used for
the force exerted by the member $AB$ on pin $B$ and for the force exerted
by pin $B$ on member $AB$. Having drawn the free-body diagrams of the
members as two-force members ensures that the members are in equilibrium. Then there is no further information to be gained from the
free-body diagrams of the two-force members and they may be discarded for the remainder of the analysis. The analysis of the truss reduces to considering the equilibrium of the joints that make up the
truss—hence the name *method of joints*.

Equilibrium of the joints that make up the truss is expressed by
drawing a separate free-body diagram for each joint and writing the
equilibrium equation

$$\Sigma \mathbf{F} = 0 \tag{7-3}$$

for each joint. Since each joint consists of concurrent forces in a plane,
moment equilibrium gives no useful information and Eq. 7-1 has only
two independent components. Therefore, for a plane truss containing
$j$ pins, there will be a total of $2j$ independent scalar equations available.
But according to Eq. 7-1, this is precisely the number of independent
equations needed to solve for the $m$ member forces and 3 support reactions of a simple truss.

Solution of the $2j$ equations is significantly simplified if a joint can
be found on which only two unknown forces and one or more known
forces act (for example joint $D$ of Fig. 7-13). In this case, the two equa-

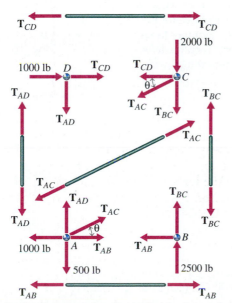

**Figure 7-13** Free-body diagrams for the individual members and pins of
the truss shown in Fig. 7-12.

A simple truss is used to support a 4-kN load as shown in Fig. CE7-1a. Determine the support reactions and the force in each member of the truss.

Fig. CE7-1

## SOLUTION

The support reactions are obtained by using the free-body diagram for the complete truss shown in Fig. CE7-1b. Three equilibrium equations are available to solve for the three unknown reaction components. Thus,

$$+\circlearrowright\Sigma M_B = A_x(2) - 4(4) = 0 \quad \text{from which} \quad A_x = 8 \text{ kN}\rightarrow$$
$$+\circlearrowright\Sigma M_A = B_x(2) - 4(4) = 0 \quad\quad\quad\quad B_x = 8 \text{ kN}\leftarrow$$
$$+\uparrow\Sigma F_y = B_y - 4 = 0 \quad\quad\quad\quad\quad B_y = 4 \text{ kN}\uparrow$$

An exploded set of free-body diagrams for the three joints of the truss is shown in Fig. CE7-1c. Two equilibrium equations are available to solve for unknown forces at each of these joints. An examination of this diagram shows that each joint has only two unknown forces after the support reactions have been determined. Since the unknown forces at joint C can be determined without using the support reactions, joint C will be used as a starting point for the analysis of the truss.

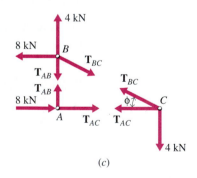

$$\phi = \tan^{-1}\frac{2}{4} = 26.57°$$

Applying the equilibrium equations $\Sigma F_x = 0$ and $\Sigma F_y = 0$ at joint C yields

$$+\uparrow\Sigma F_y = T_{BC}\sin 26.57° - 4 = 0 \quad\quad T_{BC} = 8.943 \text{ kN} \cong 8.94 \text{ kN } (T)$$
$$+\rightarrow\Sigma F_x = -T_{BC}\cos 26.57° - T_{AC}$$
$$= -8.943\cos 26.57° - T_{AC} = 0 \quad T_{AC} = -7.999 \text{ kN} \cong 8.00 \text{ kN } (C)$$

Applying the equilibrium equation $\Sigma F_y = 0$ at joint A yields

$$+\uparrow\Sigma F_y = T_{AB} = 0 \quad\quad T_{AB} = 0$$

The remaining equation $\Sigma F_x = 0$ at joint A and the two equilibrium equations at joint B could be used to determine the support reactions without using the overall equilibrium equations for the truss.

tions for this joint can be solved independent of the rest of the equations. If such a joint is not readily available, one can usually be created by solving the equilibrium equations for the entire truss first. Once two of the unknown forces have been determined, they can be treated as known forces on the free-body diagrams of the other joints. The joints are solved sequentially in this fashion until all forces are known.

As mentioned earlier, a negative value for a force (such as $T_{DC}$) indicates that the member is in compression rather than in tension. It is unnecessary to go back to the free-body diagram and change the direction of the arrow. In fact, doing so is likely only to cause confusion. The free-body diagrams should all be drawn consistently. A negative value for a symbol on one free-body diagram translates to the same negative value for the same symbol on another free-body diagram.

Once all the forces have been determined, a summary should be made listing the magnitude of the force in each member and whether the member is in tension or compression (see the example problems).

Finally, it must be noted that the equations of equilibrium for the entire truss are contained in the equations of equilibrium of the joints (see Example Problems 7-1 and 7-2). That is, if all the joints are in equilibrium and all the members are in equilibrium, then the entire truss is also in equilibrium. A consequence of this is that the three support reactions can be determined along with the $m$ member forces from the $2j$ equations of equilibrium of the joints. Overall equilibrium in this case may be used as a check of the solution. However, if overall equilibrium is used first to determine the support reactions and help start the method of joints, then three of the $2j$ joint equations of equilibrium will be redundant and may be used as a check of the solution.

Use the method of joints to determine the force in each member of the truss shown in Fig. 7-14a. State whether each member is in tension or compression.

**Fig. 7-14**

## SOLUTION

The truss of Fig. 7-14a is a simple truss with $m = 5$ members and $j = 4$ joints. Therefore, the eight equations obtained from equilibrium of the four joints (two from each joint) can be solved for the forces in the five members and the three support reactions. An exploded set of free-body diagrams for the four joints of the truss is shown in Fig. 7-14b. An examination of this diagram shows that joint A has only two unknown forces; therefore, joint A will be used as a starting point for the analysis of the truss.

Applying the equilibrium equations $\Sigma F_x = 0$ and $\Sigma F_y = 0$ at joint A yields

$+\uparrow \Sigma F_y = T_{AB} \sin 30° - 500 = 0 \qquad T_{AB} = 1000 \text{ lb} = 1000 \text{ lb } (T)$

$+\rightarrow \Sigma F_x = T_{AD} + T_{AB} \cos 30°$

$\qquad = T_{AD} + 1000 \cos 30° = 0 \qquad T_{AD} = -866.0 \text{ lb} = 866 \text{ lb } (C)$

Once forces $T_{AB}$ and $T_{AD}$ are known, joint B can be used to solve for two additional unknown member forces $T_{BC}$ and $T_{BD}$. Applying the equilibrium equations $\Sigma F_{x'} = 0$ and $\Sigma F_{y'} = 0$ at joint B yields

$+\nwarrow \Sigma F_{y'} = -T_{BD} \sin 60° - 1000 \sin 60° = 0 \qquad T_{BD} = -1000 \text{ lb} = 1000 \text{ lb } (C)$

$+\nearrow \Sigma F_{x'} = T_{BC} - T_{AB} + T_{BD} \cos 60° - 1000 \cos 60°$

$\qquad = T_{BC} - 1000 + (-1000) \cos 60°$

$\qquad - 1000 \cos 60° = 0 \qquad T_{BC} = 2000 \text{ lb} = 2000 \text{ lb } (T)$

All member forces are shown on the free-body diagrams as tensions. As a result, in the final solutions, positive member forces indicate tension and negative member forces indicate compression.

Joints must be selected in such an order that there are no more than two unknown forces at the joint.

Selecting $x'$ and $y'$ axes parallel and perpendicular to member AB at joint B eliminates the need to solve simultaneous equations.

Once forces $T_{BC}$ and $T_{BD}$ are known, two unknown forces act at joint $D$ and three unknown forces act at joint $C$. Therefore, applying the equilibrium equations $\Sigma F_x = 0$ and $\Sigma F_y = 0$ at joint $D$ yields

$$+\uparrow\Sigma F_y = T_{CD} + T_{BD}\sin 30°$$
$$= T_{CD} + (-1000)\sin 30° = 0 \qquad T_{CD} = 500 \text{ lb} = 500 \text{ lb } (T)$$
$$+\rightarrow\Sigma F_x = -D_x - T_{AD} - T_{BD}\cos 30°$$
$$= -D_x - (-866) - (-1000)\cos 30° = 0 \qquad D_x = 1732 \text{ lb} = 1732 \text{ lb}\leftarrow$$

Finally, applying the equilibrium equations $\Sigma F_x = 0$ and $\Sigma F_y = 0$ at joint $C$ yields the two remaining support reactions as

$$+\uparrow\Sigma F_y = C_y - T_{CD} - T_{BC}\cos 60°$$
$$= C_y - 500 - 2000\cos 60° = 0 \qquad C_y = 1500 \text{ lb} = 1500 \text{ lb}\uparrow$$
$$+\rightarrow\Sigma F_x = C_x - T_{BC}\sin 60°$$
$$= C_x - 2000\sin 60° = 0 \qquad C_x = 1732 \text{ lb} = 1732 \text{ lb}\rightarrow$$

In the preceding analysis, the support reactions were determined from the equilibrium equations of the joints. These results can be checked by considering equilibrium of the complete truss. A free-body diagram of the truss is shown in Fig. 7-14c. The reaction at support $D$ can be determined by summing moments about point $C$. Thus,

$$+\!\!\downarrow\Sigma M_C = -D_x(8\tan 30°) - 1000(4) - 500(8) = 0 \qquad D_x = 400 \text{ lb}\leftarrow$$

The reaction components at $C$ are determined by applying the equilibrium equations $\Sigma F_x = 0$ and $\Sigma F_y = 0$. Thus,

$$+\rightarrow\Sigma F_x = C_x - D_x = C_x - 1732 = 0 \qquad C_x = 1732 \text{ lb}\rightarrow$$
$$+\uparrow\Sigma F_y = C_y - 1000 - 500 = 0 \qquad C_y = 1500 \text{ lb}\uparrow$$

Therefore, the desired answers are

$$T_{AB} = 1000 \text{ lb} = 1000 \text{ lb } (T) \qquad \text{Ans.}$$
$$T_{AD} = -866.0 \text{ lb} = 866 \text{ lb } (C) \qquad \text{Ans.}$$
$$T_{BC} = 2000 \text{ lb} = 2000 \text{ lb } (T) \qquad \text{Ans.}$$
$$T_{BD} = -1000 \text{ lb} = 1000 \text{ lb } (C) \qquad \text{Ans.}$$
$$T_{CD} = 500 \text{ lb} = 500 \text{ lb } (T) \qquad \text{Ans.}$$

Use the method of joints to determine the force in each member of the truss shown in Fig. 7-15a. State whether each member is in tension or compression.

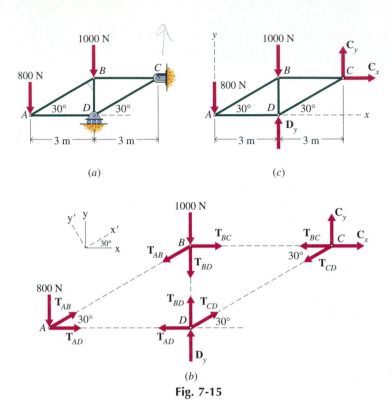

(a)

(c)

(b)

**Fig. 7-15**

## SOLUTION

The truss of Fig. 7-15a is a simple truss with $m = 5$ members and $j = 4$ joints. Therefore, the eight equations obtained from equilibrium of the four joints (two from each joint) can be solved for the forces in the five members and the three support reactions. An exploded set of free-body diagrams for the four joints of the truss is shown in Fig. 7-15b. An examination of this diagram shows that joint $A$ has only two unknown forces; therefore, joint $A$ will be used as a starting point for the analysis of the truss.

Applying the equilibrium equations $\Sigma F_x = 0$ and $\Sigma F_y = 0$ at joint $A$ yields

$$+\uparrow\Sigma F_y = T_{AB} \sin 30° - 800 = 0 \qquad T_{AB} = 1600 \text{ N} = 1600 \text{ N } (T)$$
$$+\rightarrow\Sigma F_x = T_{AD} + T_{AB} \cos 30°$$
$$= T_{AD} + 1600 \cos 30° = 0 \qquad T_{AD} = -1385.6 \text{ N} \cong 1386 \text{ N } (C)$$

Once forces $T_{AB}$ and $T_{AD}$ are known, joint $B$ can be used to solve for two additional unknown member forces $T_{BC}$ and $T_{BD}$. Applying the equilibrium equations $\Sigma F_x = 0$ and $\Sigma F_y = 0$ at joint $B$ yields

$$+\rightarrow\Sigma F_x = T_{BC} - T_{AB} \cos 30°$$
$$= T_{BC} - 1600 \cos 30° = 0 \qquad T_{BC} = 1385.6 \text{ N} \cong 1386 \text{ N } (T)$$
$$+\uparrow\Sigma F_y = -T_{BD} - T_{AB} \sin 30° - 1000$$
$$= -T_{BD} - 1600 \sin 30° - 1000 = 0 \qquad T_{BD} = -1800 \text{ N} = 1800 \text{ N } (C)$$

Once forces $T_{BC}$ and $T_{BD}$ are known, two unknown forces act at joint $D$ and three unknown forces act at joint $C$. Therefore, applying the equilibrium equations $\Sigma F_x = 0$ and $\Sigma F_y = 0$ at joint $D$ yields

$$+\rightarrow \Sigma F_x = T_{CD} \cos 30° - T_{AD}$$
$$= T_{CD} \cos 30° - (-1385.6) = 0 \qquad T_{CD} = -1600 \text{ N} = 1600 \text{ N}(C)$$
$$+\uparrow \Sigma F_y = D_y + T_{BD} + T_{CD} \sin 30°$$
$$= D_y + (-1800) + (-1600) \sin 30° = 0 \qquad D_y = 2600 \text{ N} = 2600 \text{ N}\uparrow$$

Finally, applying the equilibrium equations $\Sigma F_x = 0$ and $\Sigma F_y = 0$ at joint $C$ yields the two remaining support reactions as

$$+\uparrow \Sigma F_y = C_y - T_{CD} \sin 30°$$
$$= C_y - (-1600) \sin 30° = 0 \qquad C_y = -800 \text{ N} = 800 \text{ lb}\downarrow$$
$$+\rightarrow \Sigma F_x = C_x - T_{BC} - T_{CD} \cos 30°$$
$$= C_x - 1385.6 - (-1600) \cos 30° = 0 \qquad C_x = 0$$

In the preceding analysis, the support reactions were determined from the equilibrium equations of the joints. These results can be checked by considering the equilibrium of the complete truss. A free-body diagram of the truss is shown in Fig. 7-15c. The reaction at support $D$ can be determined by summing moments about point $C$. Thus,

$$+\curvearrowleft \Sigma M_C = -D_y(3) - 1000(3) - 800(6) = 0 \qquad D_y = 2600 \text{ N}\uparrow$$

The reaction components at $C$ are determined by applying the equilibrium equations $\Sigma F_x = 0$ and $\Sigma F_y = 0$. Thus,

$$+\rightarrow \Sigma F_x = C_x = 0 \qquad C_x = 0$$
$$+\uparrow \Sigma F_y = C_y + D_y - 1000 - 800$$
$$= C_y + 2600 - 1000 - 800 = 0 \qquad C_y = -800 \text{ N} = 800 \text{ N}\downarrow$$

Therefore, the desired answers are

$$T_{AB} = 1600 \text{ N} = 1600 \text{ N} \ (T) \qquad \text{Ans.}$$
$$T_{AD} = -1385.6 \text{ N} \cong 1386 \text{ N} \ (C) \qquad \text{Ans.}$$
$$T_{BC} = 1385.6 \text{ N} \cong 1386 \text{ N} \ (T) \qquad \text{Ans.}$$
$$T_{BD} = -1800 \text{ N} = 1800 \text{ N} \ (C) \qquad \text{Ans.}$$
$$T_{CD} = -1600 \text{ N} = 1600 \text{ N} \ (C) \qquad \text{Ans.}$$

All member forces were shown as tensions on the free-body diagrams; therefore, the compressive forces in members $BD$ and $CD$ must be entered as negative numbers in the equilibrium equations for joint $C$.

Use the method of joints to determine the force in each member of the truss shown in Fig. 7-16a. State whether each member is in tension or compression.

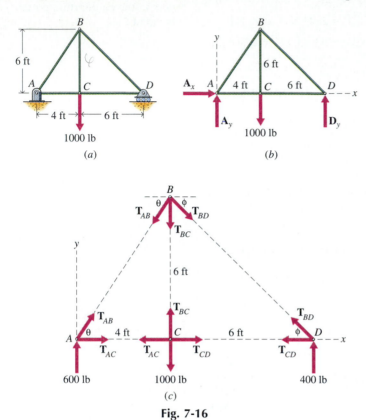

Fig. 7-16

---

## SOLUTION

An examination of Fig. 7-16a indicates that no joint is available with only two unknown forces (support reactions included) that can be used to initiate the analysis; therefore, the reactions at supports $A$ and $D$ will be determined first. A free-body diagram of the complete truss is shown in Fig. 7-16b. The reaction at support $A$ is represented by force components $\mathbf{A}_x$ and $\mathbf{A}_y$. The reaction at $D$ is a vertical force $\mathbf{D}_y$.

The reaction at support $D$ can be determined by summing moments about point $A$. Thus,

$$+\circlearrowleft\Sigma M_A = D_y(10) - 1000(4) = 0 \qquad D_y = 400 \text{ lb}$$

Similarly, the reaction component $A_y$ at support $A$ is determined by summing moments about point $D$. Thus,

$$+\circlearrowleft\Sigma M_D = -A_y(10) + 1000(6) = 0 \qquad A_y = 600 \text{ lb}$$

Finally, from the equilibrium equation $\Sigma F_x = 0$,

$$+\rightarrow\Sigma F_x = A_x = 0 \qquad A_x = 0$$

The equilibrium equation $\Sigma F_y = 0$ can be used to provide a check of these results. Thus,

$$+\uparrow\Sigma F_y = A_y + D_y - 1000 = 600 + 400 - 1000 = 0$$

> Summing moments about point $D$ yields a value for force $A_y$ that is independent of the previous determination for force $D_y$.

Two independent equilibrium equations are available at each joint for use in solving for unknown member forces. An examination of the exploded set of free-body diagrams for the four joints shown in Fig. 7-16c indicates that the analysis for member forces can be started at either joint $A$ or joint $D$ since both contain only two unknown forces.

Applying the equilibrium equations $\Sigma F_x = 0$ and $\Sigma F_y = 0$ at joint $A$ after determining the angle $\theta$ yields

$$\theta = \tan^{-1} \frac{6}{4} = 56.31°$$

$+\uparrow\Sigma F_y = T_{AB} \sin 56.31° + 600 = 0 \qquad T_{AB} = -721.11 \text{ lb} \cong 721 \text{ lb (C)}$

$+\rightarrow\Sigma F_x = T_{AC} + T_{AB} \cos 56.31°$

$\qquad = T_{AC} + (-721.11) \cos 56.31° = 0 \qquad T_{AC} = 400.0 \text{ lb} = 400 \text{ lb (T)}$

Once forces $T_{AB}$ and $T_{AC}$ are known, either joint $B$ or joint $C$ can be used to solve for additional member forces. From a free-body diagram for joint $C$,

$+\rightarrow\Sigma F_x = T_{CD} - T_{AC}$

$\qquad = T_{CD} - 400.0 = 0 \qquad T_{CD} = 400.0 \text{ lb} = 400 \text{ lb (T)}$

$+\uparrow\Sigma F_y = T_{BC} - 1000 = 0 \qquad T_{BC} = 1000 \text{ lb} = 1000 \text{ lb (T)}$

Finally, applying the equilibrium equation $\Sigma F_y = 0$ at joint $D$ after determining the angle $\phi$ yields

$$\phi = \tan^{-1} \frac{6}{6} = 45.0°$$

$+\uparrow\Sigma F_y = T_{BD} \sin 45.0° + 400 = 0 \qquad T_{BD} = -565.69 \text{ lb} \cong 566 \text{ lb (C)}$

A check on the accuracy of the previous results is provided by summing forces in the $x$-direction at joint $D$. Thus,

$+\rightarrow\Sigma F_x = -T_{CD} - T_{BD} \cos 45.0°$

$\qquad = -400.0 - (-565.69) \cos 45.0° = -0.003 \cong 0$

A further check on the accuracy of the results is made by checking the equilibrium of joint $B$. Thus,

$+\rightarrow\Sigma F_x = T_{BD} \cos \phi - T_{AB} \cos \theta$

$\qquad = -565.69 \cos 45.0° - (-721.11) \cos 56.31° = -0.004 \cong 0$

$+\uparrow\Sigma F_x = -T_{BD} \sin \phi - T_{AB} \sin \theta - T_{BC}$

$\qquad = -(-565.69) \sin 45.0° - (-721.11) \sin 56.31° - 1000 = 0.003 \cong 0$

Therefore, the desired answers are

$$T_{AB} = -721.11 \text{ lb} \cong 721 \text{ lb (C)} \qquad \text{Ans.}$$
$$T_{AC} = 400.0 \text{ lb} = 400 \text{ lb (T)} \qquad \text{Ans.}$$
$$T_{BC} = 1000 \text{ lb} = 1000 \text{ lb (T)} \qquad \text{Ans.}$$
$$T_{BD} = -565.69 \text{ lb} \cong 566 \text{ lb (C)} \qquad \text{Ans.}$$
$$T_{CD} = 400.0 \text{ lb} = 400 \text{ lb (T)} \qquad \text{Ans.}$$

Use the method of joints to determine the force in each member of the truss shown in Fig. 7-17a. State whether each member is in tension or compression.

(a)

(b)

(c)

**Fig. 7-17**

## SOLUTION

An examination of Fig. 7-17a indicates that no joint is available with only two unknown forces (support reactions included) that can be used to initiate the analysis; therefore, the reactions at supports A and D will be determined first. A free-body diagram of the complete truss is shown in Fig. 7-17b. The reaction at support A is represented by force components $A_x$ and $A_y$. The reaction at D is a horizontal force $D_x$.

The reaction at support D can be determined by summing moments about point A. Thus,

$$+ \circlearrowleft \Sigma M_A = D_x(3 \cos 30°) - 500\left(\frac{1}{2}\right)(3 + 3 \sin 30°) = 0 \qquad D_x = 433.0 \text{ kN}$$

The reaction components $A_x$ and $A_y$ at support A are determined by summing forces in the x- and y-directions. Thus,

$$+ \rightarrow \Sigma F_x = A_x - D_x = A_x - 433.0 = 0 \qquad A_x = 433.0 \text{ kN}$$
$$+ \uparrow \Sigma F_y = A_y - 500 = 0 \qquad A_y = 500 \text{ kN}$$

Two independent equilibrium equations are available at each joint for use in solving for unknown member forces. An examination of the exploded set of free-body diagrams for the four joints, shown in Fig. 7-17c, indicates that the analysis for member forces can be started at either joint A or joint D since both contain only two unknown forces.

Applying the equilibrium equations $\Sigma F_{x'} = 0$ and $\Sigma F_{y'} = 0$ at joint $A$ yields

$$+\nwarrow \Sigma F_{y'} = T_{AB} \sin 30° - 433.0 \sin 30° + 500 \sin 60° = 0$$
$$T_{AB} = -433.0 \text{ kN} = 433 \text{ kN(C)}$$
$$+\nearrow \Sigma F_{x'} = T_{AC} + T_{AB} \cos 30° + 433.0 \cos 30° + 500 \cos 60°$$
$$= T_{AC} + (-433.0) \cos 30° + 433.0 \cos 30° + 500 \cos 60° = 0$$
$$T_{AC} = -250.0 \text{ kN} = 250 \text{ kN } (C)$$

Once forces $T_{AB}$ and $T_{AC}$ are known, either joint $B$ or joint $C$ can be used to solve for additional member forces. From a free-body diagram for joint $C$,

$$+\nwarrow \Sigma F_{y'} = T_{BC} - 500 \sin 60° = 0 \qquad T_{BC} = 433.0 \text{ kN} = 433 \text{ kN } (T)$$
$$+\nearrow \Sigma F_{x'} = T_{CD} - T_{AC} - 500 \cos 60°$$
$$= T_{CD} - (-250.0) - 500 \cos 60° = 0 \qquad T_{CD} = 0$$

Finally, applying the equilibrium equation $\Sigma F_x = 0$ at joint $D$ yields

$$+\rightarrow \Sigma F_x = -T_{BD} - T_{CD} \cos 30° - 433.0$$
$$= -T_{BD} - (0) \cos 30° - 433.0 = 0 \qquad T_{BD} = -433.0 \text{ kN} = 433 \text{ kN } (C)$$

A check on the accuracy of the results is made by checking the equilibrium of joint $B$. Thus,

$$+\rightarrow \Sigma F_x = T_{BD} + T_{BC} \cos 60° - T_{AB} \cos 60°$$
$$= -433.0 + 433 \cos 60° - (-433.0) \cos 60° = 0$$
$$+\uparrow \Sigma F_x = -T_{AB} \sin 60° - T_{BC} \sin 60°$$
$$= -(-433.0) \sin 60° - (433.0) \sin 60° = 0$$

Therefore, the desired answers are

$$
\begin{aligned}
T_{AB} &= -433.0 \text{ kN} = 433 \text{ kN } (C) && \text{Ans.}\\
T_{AC} &= -250.0 \text{ kN} = 250 \text{ kN } (C) && \text{Ans.}\\
T_{BC} &= \phantom{-}433.0 \text{ kN} = 433 \text{ kN } (T) && \text{Ans.}\\
T_{BD} &= -433.0 \text{ kN} = 433 \text{ kN } (C) && \text{Ans.}\\
T_{CD} &= 0 && \text{Ans.}
\end{aligned}
$$

Use the method of joints to find the force in each member of the truss of Fig. 7-18a.

(a)                                    (b)

**Fig. 7-18**

## SOLUTION

Truss *ABCD* is a simple truss with $m = 5$ members and $j = 4$ joints. Therefore, the eight equations obtained from the equilibrium of the four joints can be solved for the three support reactions as well as the forces in all five members.

The first step is to draw a free-body diagram of the entire truss (Fig. 7-18b) and write the equilibrium equations

$$+\rightarrow \Sigma F_x = 1000 + A_x = 0$$
$$+\uparrow \Sigma F_y = -2000 + A_y + B_y = 0$$
$$\downarrow + \Sigma M_A = -(4)(1000) - (8)(2000) + 8B_y = 0$$

These equations can be solved to get

$$A_x = -1000 \text{ lb} \qquad B_y = 2500 \text{ lb} \qquad A_y = -500 \text{ lb}$$

Next, draw a free-body diagram of pin *D* and solve the equilibrium equations

$$+\rightarrow \Sigma F_x = 1000 + T_{CD} = 0$$
$$+\uparrow \Sigma F_y = -T_{AD} = 0$$

to get

$$T_{CD} = -1000 \text{ lb} \qquad T_{AD} = 0 \text{ lb}$$

Next draw a free-body diagram of pin *C* and solve the equilibrium equations

$$+\rightarrow \Sigma F_x = -T_{CD} - T_{AC} \cos \theta = 0$$
$$+\uparrow \Sigma F_y = -2000 - T_{BC} - T_{AC} \sin \theta = 0$$

where

$$\sin \theta = \frac{AD}{AC} = \frac{4}{\sqrt{4^2 + 8^2}} = 0.4472$$

and

$$\cos \theta = \frac{CD}{AC} = \frac{8}{\sqrt{4^2 + 8^2}} = 0.8944$$

But $T_{CD} = -1000$ lb, so

$$T_{AC} = -\frac{T_{CD}}{\cos \theta} = -\frac{-1000}{0.8944} = 1118 \text{ lb}$$

Joints must be selected in such an order that there are no more than two unknown forces at the joint. Once the support reactions are known, the analysis can be started at either joint *B* or joint *D* since both contain only two unknown forces.

**Fig. 7-19**

and

$$T_{BC} = -2000 - (1118)(0.4472) = -2500 \text{ lb}$$

Next, draw a free-body diagram of pin $B$ and write the equilibrium equations

$$+\rightarrow \Sigma F_x = -T_{AB} = 0$$
$$+\uparrow \Sigma F_y = T_{BC} + 2500 = 0$$

The first of these equations can be solved to get

$$T_{AB} = 0 \text{ lb}$$

The second equation contains no unknowns, however, since the value of $T_{BC}$ has already been found. The second equation can be used to check the consistency of the answers:

$$(-2500) + 2500 = 0 \qquad \text{(check)}$$

Finally, draw a free-body diagram of pin $A$ and write

$$+\rightarrow \Sigma F_x = T_{AB} + T_{AC} \cos \theta - 1000 = 0$$
$$+\uparrow \Sigma F_y = T_{AD} + T_{AC} \sin \theta - 500 = 0$$

Again, there are no unknowns in these equations since the values of $T_{AB}$, $T_{AC}$, and $T_{AD}$ have already been found. These two equations again reduce to a check of the consistency of the solution:

$$0 + (1118)(0.8944) - 1000 = -0.06$$
$$0 + (1118)(0.4472) - 500 = -0.03$$

The difference is less than the rounding performed on $T_{AB}$, $T_{AC}$, and $T_{AD}$ and so the solution checks.

The desired answers then are

$$T_{AB} - T_{AD} = 0 \text{ lb} \qquad \text{Ans.}$$
$$T_{AC} = 1118 \text{ lb } (T) \qquad \text{Ans.}$$
$$T_{BC} = 2500 \text{ lb } (C) \qquad \text{Ans.}$$
$$T_{CD} = 1000 \text{ lb } (C) \qquad \text{Ans.}$$

(The fact that $T_{AB}$ and $T_{AD}$ both came out zero is a peculiarity of the loading and does not mean that members $AB$ and $AD$ should be eliminated from the truss. For a slightly different loading situation, the forces in these members will not be zero. Even for the given loading condition, members $AB$ and $AD$ are necessary to ensure the rigidity of the truss. Without member $AB$, for example, the truss would collapse if the roller support at $B$ were disturbed slightly to the right or left.)

The truss shown in Fig. 7-20a supports one side of a bridge; an identical truss supports the other side. Floor beams carry vehicle loads to the truss joints. A 2000-kg car is stopped on the bridge. Calculate the force in each member of the truss using the method of joints.

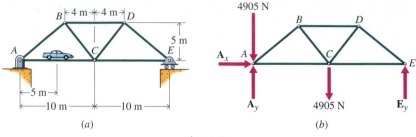

(a)

(b)

**Fig. 7-20**

## SOLUTION

The truss of Fig. 7-20a is a simple truss with $m = 7$ members and $j = 5$ joints. Therefore, the ten equations obtained from the equilibrium of the five joints can be solved for the three support reactions as well as the forces in all seven members.

Half of the car's weight $(1000)(9.81) = 9810$ newtons (N) is carried by the truss shown and the other half is carried by the truss on the other side of the bridge. Since the car is midway between joints $A$ and $C$, 4905 N will be applied to joint $A$ and 4905 N will be applied to joint $C$.

The next step is to draw a free-body diagram of the entire truss (Fig. 7-20b) and write the equilibrium equations:

$$+\rightarrow \Sigma F_x = A_x = 0$$
$$+\uparrow \Sigma F_y = A_y - 4905 - 4905 + E_y = 0$$
$$\downarrow + \Sigma M_A = 20E_y - (10)(4905) = 0$$

These equations can be solved immediately to get

$$A_x = 0 \text{ N} \qquad E_y = 2453 \text{ N} \qquad A_y = 7357 \text{ N}$$

Now the free-body diagram (Fig. 7-21a) of pin $A$ is drawn and the equilibrium equations written:

$$+\rightarrow \Sigma F_x = A_x + T_{AC} + T_{AB} \cos \theta = 0$$
$$+\uparrow \Sigma F_y = A_y - 4905 + T_{AB} \sin \theta = 0$$

where $A_x = 0$ N, $A_y = 7357$ N, and $\theta = \tan^{-1}(5/6) = 39.81°$. This gives

$$T_{AB} = -3830 \text{ N} \qquad T_{AC} = 2942 \text{ N}$$

Since one of the three forces applied at pin $B$ is now known, the free-body diagram of pin $B$ (Fig. 7-21b) is drawn next. The equilibrium equations for this pin are:

$$+\rightarrow \Sigma F_x = -T_{AB} \cos \theta + T_{BD} + T_{BC} \cos \phi = 0$$
$$+\uparrow \Sigma F_y = -T_{AB} \sin \theta - T_{BC} \sin \phi = 0$$

where $T_{AB} = -3830$ N and $\phi = \tan^{-1}(5/4) = 51.34°$. They are solved to get

$$T_{BC} = 3140 \text{ N} \qquad T_{BD} = -4904 \text{ N}$$

At this point, pin $C$, pin $D$, or pin $E$ could be solved next, since each of these pins has only two forces whose values have not yet been found applied to them. For this example the free-body diagram of pin $D$ (Fig. 7-21d) will be

The weight of the car is supported by a bridge floor structure that transmits loads to truss members only at the joints.

Joints must be selected in such an order that there are no more than two unknown forces at the joint. Once the support reactions are known, the analysis can be started at either joint $A$ or joint $E$ since both contain only two unknown forces.

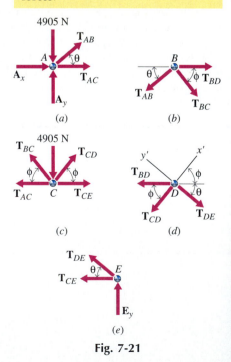

**Fig. 7-21**

considered next. Writing the standard horizontal and vertical components of the equilibrium equations gives

$$+\rightarrow\Sigma F_x = -(-4904) - T_{CD} \cos \phi + T_{DE} \cos \theta = 0$$
$$+\uparrow\Sigma F_y = -T_{CD} \sin \phi - T_{DE} \sin \theta = 0$$

Both of these equations contain both of the unknown forces $T_{CD}$ and $T_{DE}$. Although the solution of this pair of equations is not particularly difficult, the calculations can be simplified if the equilibrium equations are written in terms of components that are along and perpendicular to member $CD$. This gives

$$+\nearrow\Sigma F_{x'} = -(-4904) \cos \phi - T_{CD} + T_{DE} \cos (\theta + \phi) = 0$$
$$+\nwarrow\Sigma F_{y'} = (-4904) \sin \phi - T_{DE} \sin (\theta + \phi) = 0$$

The second of these equations can be solved immediately to get

$$T_{DE} = -3830 \text{ N}$$

Then

$$T_{CD} = 3140 \text{ N}$$

Moving to pin $C$, the free-body diagram (Fig. 7-21$c$) is drawn and the equilibrium equations written:

$$+\rightarrow\Sigma F_x = -(2942) - (3140) \cos \phi + (3140) \cos \phi + T_{CE} = 0$$
$$+\uparrow\Sigma F_y = (3140) \sin \phi - (4905) + (3140) \sin \phi = 0$$

The first of these equations gives

$$T_{CE} = 2942 \text{ N}$$

Since the values of all the forces in the second equation have already been found, this equation reduces to a check of the consistency of the results:

$$(3140) \sin 51.34° - (4905) + (3140) \sin 51.34° = -1.16$$

(The small number $-1.16$ is due to rounding all the intermediate answers to four significant figures. Keeping more accuracy in the intermediate values would reduce the residual, so the solution checks.)

Finally, draw the free-body diagram of pin $E$ (Fig. 7-21$c$) and write the equilibrium equations:

$$+\rightarrow\Sigma F_x = -T_{CE} - T_{DE} \cos \theta = 0$$
$$+\uparrow\Sigma F_y = T_{DE} \sin \theta + E_y = 0$$

Again, there are no unknowns left to be solved for. These equations are used simply as a check:

$$-(2942) - (-3830) \cos 39.81° = -0.10$$
$$(-3830) \sin 39.81° + (2453) = 0.87$$

and again the solution checks.

The required answers are

| | | |
|---|---|---|
| $T_{AB} = T_{DE} = 3830 \text{ N } (C)$ | | Ans. |
| $T_{AC} = T_{CE} = 2942 \text{ N } (T)$ | | Ans. |
| $T_{BC} = T_{CD} = 3140 \text{ N } (T)$ | | Ans. |
| $T_{BD} = 4904 \text{ N } (C)$ | | Ans. |

The truss shown (Fig. 7-22a) supports one end of a 40-ft-wide, 24-ft-high outdoor movie screen, which weighs 7000 lb. An identical truss supports the other end of the screen. A 20-mph wind blowing directly at the screen creates a wind loading of 1.042 lb/ft² on the screen. Calculate the maximum tensile and compressive forces in the members of the truss and indicate in which members they occur.

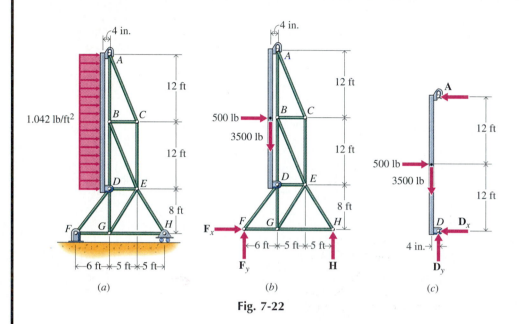

**Fig. 7-22**

## SOLUTION

Truss ABCDEFGH is a simple truss with $m = 13$ members and $j = 8$ joints. Therefore, the 16 equations obtained from the equilibrium of the eight joints can be solved for the three support reactions as well as the forces in all 13 members.

The first step is to draw a free-body diagram of the entire truss (Fig. 7-22b) to find the support reactions. The distributed wind loading will be replaced with its equivalent force of

$$(1.042 \text{ lb/ft}^2)(960 \text{ ft}^2) = 1000 \text{ lb}$$

applied at the center of the screen. Only half of the weight and distributed force are carried by the truss at each end and are shown on the free-body diagram of Fig. 7-22b. The equilibrium equations are

$$+\rightarrow \Sigma F_x = 500 + F_x = 0$$
$$+\uparrow \Sigma F_y = F_y + H - 3500 = 0$$
$$\downarrow + \Sigma M_F = -(6 - 4/12)(3500) - (20)(500) + 16H = 0$$

Solution of these equations gives the support reactions as

$$F_x = -500 \text{ lb} \qquad F_y = 1635 \text{ lb} \qquad H = 1865 \text{ lb}$$

The next step is to draw a free-body diagram of the screen (Fig. 7-22c) to calculate the forces it exerts at the joints of the truss. Again the distributed wind loading is replaced with its equivalent force applied at the center of the

The wind load and weight of the screen are supported by a structure with connections that transmit only a horizontal load to truss members at joint A and both horizontal and vertical loads to truss members at joint D.

screen. The equations of equilibrium are

$$+\rightarrow \Sigma F_x = 500 - A - D_x = 0$$
$$+\uparrow \Sigma F_y = D_y - 3500 = 0$$
$$\curvearrowleft+\Sigma M_D = (4/12)(3500) + 24A - (12)(500) = 0$$

Solution of these equations gives

$$A = 201.4 \text{ lb} \qquad D_x = 298.6 \text{ lb} \qquad D_y = 3500 \text{ lb}$$

Now the joints will be solved in order starting from the top. First draw the free-body diagram of pin $A$ (Fig. 7-23$a$) and write the equilibrium equations:

$$+\rightarrow \Sigma F_x = A + T_{AC} \sin \theta = 0$$
$$+\uparrow \Sigma F_y = -T_{AB} - T_{AC} \cos \theta = 0$$

where $A = 201.4$ lb and $\theta = \tan^{-1}(5/12) = 22.62°$. These are solved to get

$$T_{AB} = 483.3 \text{ lb} \qquad \text{and} \qquad T_{AC} = -523.6 \text{ lb}$$

At this point, the free-body diagram of pin $B$ involves three new unknowns and cannot yet be solved. Instead, draw the free-body diagram of pin $C$ (Fig. 7-23$c$) and solve the equilibrium equations:

$$+\rightarrow \Sigma F_x = -T_{BC} - (-523.6) \sin \theta = 0 \qquad T_{BC} = 201.4 \text{ lb}$$
$$+\uparrow \Sigma F_y = (-523.6) \cos \theta - T_{CE} = 0 \qquad T_{CE} = -483.3 \text{ lb}$$

Now draw the free-body diagram of pin $B$ (Fig. 7-23$b$) and solve the equilibrium equations:

$$+\rightarrow \Sigma F_x = 201.4 + T_{BE} \sin \theta = 0 \qquad T_{BD} = 966.6 \text{ lb}$$
$$+\uparrow \Sigma F_y = 483.3 - T_{BD} - T_{BE} \cos \theta = 0 \qquad T_{BE} = -523.6 \text{ lb}$$

At this point, the free-body diagrams of pin $D$ (Fig. 7-23$d$) and pin $E$ (Fig. 7-23$e$) each contain three unknowns and are not directly solvable. Instead, draw the free-body diagram of pin $F$ (Fig. 7-23$f$) and solve the equilibrium equations:

$$+\rightarrow \Sigma F_x = T_{DF}(3/5) + T_{FG} + (-500) = 0 \qquad T_{DF} = -2044 \text{ lb}$$
$$+\uparrow \Sigma F_y = T_{DF}(4/5) + 1635 = 0 \qquad T_{FG} = 1726 \text{ lb}$$

Now the equilibrium equations for pin $D$ (Fig. 7-23$d$)

$$+\rightarrow \Sigma F_x = 298.6 - (-2044)(3/5) + T_{DE} = 0$$
$$+\uparrow \Sigma F_y = -3500 + 966.6 - (-2044)(4/5) - T_{DG} = 0$$

contain only two member forces that have not yet been found, and these equations can be solved to get

$$T_{DE} = -1525 \text{ lb} \qquad \text{and} \qquad T_{DG} = -898.2 \text{ lb}$$

Next, draw the free-body diagram of pin $E$ (Fig. 7-23$e$) and write the equilibrium equations:

$$+\rightarrow \Sigma F_x = -(-1525) - (-523.6) \sin \theta - T_{EG} \cos \phi + T_{EH} \cos \phi = 0$$
$$+\uparrow \Sigma F_y = (-483.3) + (-523.6) \cos \theta - T_{EG} \sin \phi - T_{EH} \sin \phi = 0$$

where $\phi = \tan^{-1}(8/5) = 58.0°$. These are solved to get

$$T_{EG} = 1059 \text{ lb} \qquad \text{and} \qquad T_{EH} = -2199 \text{ lb}$$

Then draw the free-body diagram of pin $G$ (Fig. 7-23$g$) and write the equilibrium equations:

$$+\rightarrow \Sigma F_x = T_{GH} + 1059 \cos \phi - 1726 = 0$$
$$+\uparrow \Sigma F_y = (-898.2) + 1059 \sin \phi = 0$$

The first equation gives

$$T_{GH} = 1165 \text{ lb}$$

Joints must be selected in such an order that there are no more than two unknown forces at the joint. Once the support reactions are known, the analysis can be started at joint $A$, joint $F$, or joint $H$ since each contains only two unknown forces.

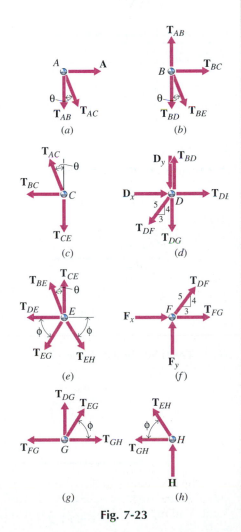

Fig. 7-23

There are no unknowns left in the second equation, however, and it is used to check the consistency of the results:

$$(-898.2) + (1059) \sin 58.0° = -0.12$$

The remainder is due to the roundoff of values above.

Finally, draw the free-body diagram of pin $H$ (Fig. 7-23$h$) and write the equilibrium equations

$$+\rightarrow \Sigma F_x = -T_{GH} - T_{EH} \cos \phi = 0$$
$$+\uparrow \Sigma F_y = T_{EH} \sin \phi + H = 0$$

There are no unknowns left in these equations, and they reduce to a check of the results:

$$-(1165) - (-2199) \cos 58.0° = 0.29 \qquad \text{(check)}$$
$$(-2199) \sin 58.0° + 1865 = 0.14 \qquad \text{(check)}$$

A search through the member forces found above gives that the maximum tensile force occurs in member $FG$ whereas the maximum compressive force occurs in member $EH$:

$$\text{max tensile} = T_{FG} = 1726 \text{ lb } (T) \qquad \text{Ans.}$$
$$\text{max compressive} = T_{EH} = 2200 \text{ lb } (C) \qquad \text{Ans.}$$

(h)

# PROBLEMS

Use the method of joints to determine the force in each member of the truss shown in the following problems. State whether each member is in tension or compression.

## Introductory Problems

**7-1*** The truss shown in Fig. P7-1.

**Fig. P7-1**

**7-2*** The truss shown in Fig. P7-2.

**Fig. P7-2**

**7-3** The truss shown in Fig. P7-3.

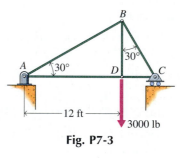

**Fig. P7-3**

**7-4** The truss shown in Fig. P7-4.

**Fig. P7-4**

**7-5\*** The truss shown in Fig. P7-5.

Fig. P7-5

**7-6\*** The truss shown in Fig. P7-6.

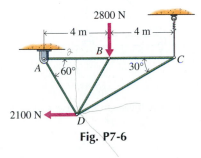

Fig. P7-6

**7-7** The truss shown in Fig. P7-7.

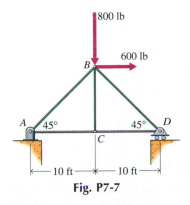

Fig. P7-7

**7-8** The truss shown in Fig. P7-8.

Fig. P7-8

**7-9\*** The truss shown in Fig. P7-9.

Fig. P7-9

**7-10\*** The truss shown in Fig. P7-10.

Fig. P7-10

**7-11** The truss shown in Fig. P7-11.

Fig. P7-11

**7-12** The truss shown in Fig. P7-12.

first
3 in chpt 7

**Fig. P7-12**

**7-13\*** Determine the forces in members *AB*, *AE*, and *EF* of the truss shown in Fig. P7-13.

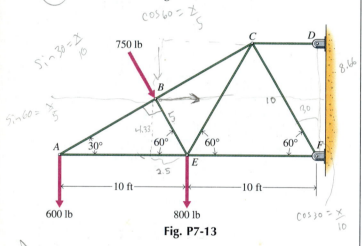

**Fig. P7-13**

**7-14\*** Determine the forces in members *CD*, *DE*, and *DF* of the truss shown in Fig. P7-14.

**Fig. P7-14**

**7-15** Determine the forces in members *AD*, *CD*, and *DE* of the truss shown in Fig. P7-15.

**Fig. P7-15**

**7-16** Determine the forces in members *BC*, *CD*, and *DF* of the truss shown in Fig. P7-16.

**Fig. P7-16**

### Intermediate Problems

**7-17\*** A crate weighing 4000 lb is attached by light, inextensible cables to the truss of Fig. P7-17. Determine the force in each member of the truss.

**Fig. P7-17**

**7-18** Determine the forces in members $BC$, $BE$, and $CE$ of the truss shown in Fig. P7-18. The length of all members is 3 m.

**Fig. P7-18**

**7-19\*** The Gambrel truss shown in Fig. P7-19 supports one side of a bridge; an identical truss supports the other side. Floor beams carry vehicle loads to the truss joints. Calculate the forces in members $BC$, $BG$, and $CG$ when a truck weighing 7500 lb is stopped in the middle of the bridge as shown. The center of gravity of the truck is midway between the front and rear wheels.

**Fig. P7-19**

**7-20\*** The Gambrel truss shown in Fig. P7-20 supports one side of a bridge; an identical truss supports the other side. Floor beams carry vehicle loads to the truss joints. Calculate the forces in members $CD$, $CG$, and $DG$ when a truck

**Fig. P7-20**

having a mass of 3500 kg is stopped in the middle of the bridge as shown. The center of gravity of the truck is 1 m in front of the rear wheels.

**7-21** Determine the forces in members $CD$, $DG$, and $GF$ of the truss shown in Fig. P7-21.

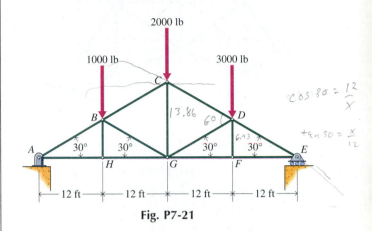

**Fig. P7-21**

**7-22** Determine the forces in members $CD$, $DI$, and $HI$ of the truss shown in Fig. P7-22.

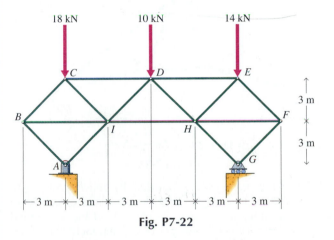

**Fig. P7-22**

## Challenging Problems

**7-23\*** Determine the force in each member of the pair of trusses shown in Fig. P7-23.

**Fig. P7-23**

**7-24*** Determine the forces in members *DE*, *EI*, and *HI* of the truss shown in Fig. P7-24.

**Fig. P7-24**

**7-25** Determine the forces in members *BC*, *CF*, and *FG* of the truss shown in Fig. P7-25.

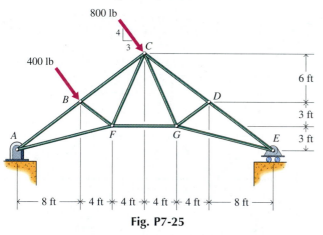

**Fig. P7-25**

**7-26** Snow on a roof supported by the Pratt truss of Fig. P7-26 can be approximated as a distributed load of 250 N/m (measured along the roof). Treat the distributed load as you would the weight of the members; that is, replace the total load on each of the upper members as a vertical force, half applied to the joint at each end of the member. Determine the forces in members *BC*, *CG*, *CH*, and *GH*.

**Fig. P7-26**

**7-27*** Snow on a roof supported by the Howe truss of Fig. P7-27 can be approximated as a distributed load of 20 lb/ft (measured along the roof). Treat the distributed load as you would the weight of the members; that is, replace the total load on each of the upper members as a vertical force, half applied to the joint at each end of the member. Determine the forces in members *BC*, *BG*, *CG*, and *GH*.

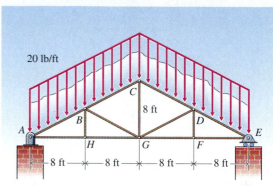

**Fig. P7-27**

**7-28*** Determine the forces in members *CF*, *CG*, and *FG* of the Warren bridge truss shown in Fig. P7-28. Each member is made of steel having a mass per unit of length of 6 kg/m. Assume that the weight of each member can be represented as a vertical force, half of which is applied at each end of the member.

**Fig. P7-28**

**7-29** Determine the forces in members *CD*, *CF*, and *FG* of the inverted Mansard truss shown in Fig. P7-29. Each member is made of timber having a weight per unit length of 6 lb/ft. Assume that the weight of each member can be represented as a vertical force, half of which is applied at each end of the member.

**Fig. P7-29**

**346**

**7-30** For the simple truss of Fig. P7-30 show that overall equilibrium of the truss is a consequence of equilibrium of all of the pins; hence, the equations of overall equilibrium give no new information. (Hint: Write the equations for equilibrium of each of the pins and eliminate the unknown member forces from these equations.)

**Fig. P7-30**

## Computer Problems

**C7-31** A three-bar truss is loaded and supported as shown in Fig. P7-31. Plot the force in each member of the truss as a function of angle $\theta$ ($0 \le \theta \le 90°$).

**Fig. P7-31**

**C7-32** A truss is loaded and supported as shown in Fig. P7-32. Plot the forces in members $BC$, $BG$, and $GH$ of the truss as a function of angle $\theta$ ($0 \le \theta \le 90°$).

**Fig. P7-32**

**C7-33** An overhead crane consists of an I-beam supported by a simple truss as shown in Fig. P7-33. If the uniform I-beam weighs 400 lb, plot the force in members $BC$, $CF$, and $EF$ as a function of the position $d$ ($0 \le d \le 8$ ft).

**Fig. P7-33**

**C7-34** The simple truss shown in Fig. P7-34 supports one side of a bridge: an identical truss supports the other side. A 2000-kg car is stopped on the bridge at location $c$. Floor beams carry the vehicle loads to the joints. Treat the car as a point mass at location $c$, and plot the force in members $AB$, $AE$, $BC$, and $BE$ as a function of the car's position $c$ ($0 \le c \le 48$ m).

**Fig. P7-34**

*(a)*

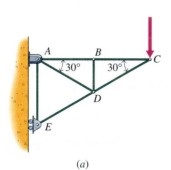

*(b)*

**Figure 7-24**   Example of a truss with
zero-force members.

*(a)*

*(b)*

*(c)*

**Figure 7-25**   Zero-force members at
joints with no external forces applied.

## 7-2.2   Zero-Force Members

It is often the case that certain members of a given truss carry no load.
Zero-force members in a truss usually arise in one of two general ways.
The first is this:

> When only two members form a noncollinear truss joint and no external
> load or support reaction is applied to the joint, then the members must
> be zero-force members.

The truss of Fig. 7-24*a* is an example of this condition. The free-body
diagram of pin *C* is drawn in Fig. 7-24*b*. The equations of equilibrium
for this joint

$$+\rightarrow \Sigma F_x = -T_{BC} - T_{CD} \cos 30° = 0$$
$$+\uparrow \Sigma F_y = -T_{CD} \sin 30° = 0$$

are trivially solved to get

$$T_{CD} = 0 \quad \text{and} \quad T_{BC} = 0$$

That is, for this particular truss and for this particular loading, the two
members *BC* and *CD* could be removed without affecting the solution
or even (in this particular case) the stability of the truss.

The second way in which zero-force members normally arise in a
truss is this:

> When three members form a truss joint for which two of the members are
> collinear and the third forms an angle with the first two, then the non-
> collinear member is a zero-force member provided no external force or
> support reaction is applied to that joint. The two collinear members carry
> equal loads (either both tension or both compression).

Such a condition arises, for example, when the load of Fig. 7-24*a* is
moved from pin *B* to pin *C* as in Fig. 7-25*a*. The free-body diagram of
pin *B* is drawn in Fig. 7-25*b*. The equations of equilibrium for this joint
are

$$+\rightarrow \Sigma F_x = -T_{AB} + T_{BC} = 0$$
$$+\uparrow \Sigma F_y = -T_{BD} = 0$$

Thus, since joint *B* is now unloaded, the force in member *BD* vanishes
and the forces in members *AB* and *BC* are equal in magnitude—either
both tension (both positive) or both compression (both negative).

Once it is known that *BD* is a zero-force member, the same rea-
soning can then be used to show that member *AD* carries no load. The
free-body diagram of pin *D* is drawn in Fig. 7-25*c*. In order to simplify
the calculations, coordinate axes are chosen along and normal to the
collinear members *CD* and *DE*. The equations of equilibrium are then

$$+\nearrow \Sigma F_x = -T_{DE} - T_{AD} \cos 60° + T_{BD} \cos 60° + T_{CD} = 0$$
$$+\nwarrow \Sigma F_y = T_{AD} \sin 60° + T_{BD} \sin 60° = 0$$

But since $T_{BD} = 0$ (*BD* is already known to be a zero-force member),
then

$$T_{AD} = 0 \quad \text{and} \quad T_{DE} = T_{CD}$$

Thus, for the loading of Fig. 7-25*a*, both members *AD* and *BD* are zero-
force members.

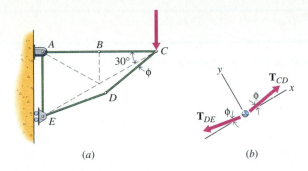

(a)                    (b)

**Figure 7-26**   Zero-force members are needed to guarantee stability of the truss.

These zero-force members cannot simply be removed from the truss and discarded, however. They are needed to guarantee the stability of the truss. If members $AD$ and $BD$ were removed, there would be nothing to prevent some small disturbance from moving pin $D$ slightly out of alignment as in Fig. 7-26$a$. Then the free-body diagram of pin $D$ would look like Fig. 7-26$b$. Again, choosing axes along and normal to the line $CE$ gives the equilibrium equations

$$+\nearrow \Sigma F_x = -T_{DE} \cos \phi + T_{CD} \cos \phi = 0$$
$$+\nwarrow \Sigma F_y = T_{DE} \sin \phi + T_{CD} \sin \phi = 0$$

The first of these equations requires that $T_{CD} = T_{DE}$, whereas the second requires that $T_{CD} = -T_{DE}$. The only way both of these equations can be satisfied is if both forces equal zero. But equilibrium of pin $C$ requires that $T_{CD}$ not be zero. What has happened, of course, is that the truss is no longer in static equilibrium. Pin $D$ will continue to buckle outward and the truss will collapse.

A seemingly trivial solution to the stability problem would be to replace the two members $CD$ and $DE$ with a single member $CE$ and to replace the two members $AB$ and $BC$ with a single member $AC$. Though this solution would satisfy the statics part of the problem, it would not take care of the tendency for long slender members to buckle when subjected to large compressive loads. Therefore, long members such as member $CE$ of Fig. 7-25 are usually replaced by a pair of shorter members and the midjoint braced if analysis of the truss indicates that the member is likely to be in compression for some expected loading. Long members such as member $AC$ of Fig. 7-24 must also be replaced by a pair of shorter members and the midjoint braced if it is ever desired to load the truss at some point along the long member.

Thus one must not be too quick to discard truss members just because they carry no load for a given configuration. These members are often needed to carry part of the load when the applied loading changes, and they are almost always needed to guarantee the stability of the truss.

Although recognizing these and other special joint-loading conditions can simplify the analysis of a truss, such recognition is not required to solve the truss. If one does not recognize that a member is a zero-force member, drawing the free-body diagram and writing the equilibrium equations will immediately show that it is a zero-force member. Also, these shortcuts should be applied with care. If there is any doubt about whether or not a member is a zero-force member, the prudent choice is to draw the free-body diagram and solve for the member force.

Find all the zero force members in the simple Fink truss of Fig. 7-27*a* for the loading shown.

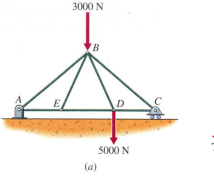

**Fig. 7-27**

---

## SOLUTION

Joint $E$ connects three members, two of which ($AE$ and $DE$) are collinear. The joint has no loads applied to it. Therefore the force in the noncollinear member ($BE$) must be zero. This is easily checked by drawing the free-body diagram for pin $E$ (Fig. 7-27*b*) and writing the equation of equilibrium in the direction perpendicular to the collinear members $AE$ and $DE$

> Zero-force members can be identified by inspection and verified by application of an equilibrium equation in a direction perpendicular to the collinear members.

$$+\uparrow\Sigma F_y = T_{BE}\sin\theta = 0$$

which gives immediately that

$$T_{BE} = 0$$

Although joint $D$ looks similar to joint $E$, there is a load applied at this joint, and the foregoing argument does not apply (as is easily checked by drawing the free-body diagram and writing the equations of equilibrium for joint $D$).

The only zero-force member for the given loading condition is member $BE$.                                                                              Ans.

Identify all the zero-force members in the scissors truss of Fig. 7-28a for the loading shown.

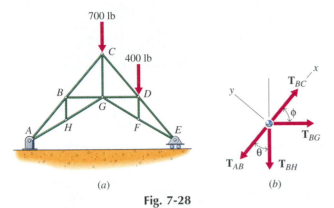

Fig. 7-28

## SOLUTION

Joint $F$ connects three members, two of which ($EF$ and $FG$) are collinear. The joint is unloaded, so the force in the noncollinear member ($DF$) must be zero. Similarly, at joint $H$ members $AH$ and $GH$ are collinear and the joint is unloaded; hence, the force in member $BH$ is also zero.

Joint $B$ connects two collinear members ($AB$ and $BC$) and two noncollinear members ($BG$ and $BH$). The joint is unloaded, and the force in one of the noncollinear members ($BH$) is known to be zero. Therefore, the force in the other noncollinear member ($BG$) must also be zero. Again, this can be easily checked by drawing the free-body diagram for pin $B$ (Fig. 7-28b) and writing the equation of equilibrium in the direction perpendicular to the collinear members $AB$ and $BC$:

$$+\nwarrow\Sigma F_y = -T_{BH}\sin\theta - T_{BG}\sin\phi = 0$$

which gives

$$T_{BG} = 0$$

Note that the argument applied to joint $B$ does not apply to joint $D$ since joint $D$ has a load applied to it.

Therefore, the zero-force members for the given loading are $BG$, $BH$, and $DF$.                                                                Ans.

> Zero-force members can be identified by inspection and verified by application of an equilibrium equation in a direction perpendicular to the collinear members.

# PROBLEMS

Identify the zero-force members present when the trusses shown in the following problems are subjected to the loadings indicated.

**7-35** The truss shown in Fig. P7-35.

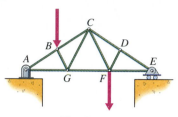

**Fig. P7-35**

**7-36** The truss shown in Fig. P7-36.

**Fig. P7-36**

**7-37** The truss shown in Fig. P7-37.

**Fig. P7-37**

**7-38** The truss shown in Fig. P7-38.

**Fig. P7-38**

**7-39** The truss shown in Fig. P7-39.

**Fig. P7-39**

**7-40** The truss shown in Fig. P7-40.

**Fig. P7-40**

**7-41** The truss shown in Fig. P7-41.

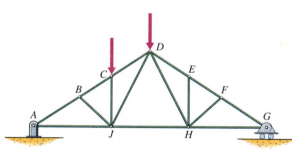

**Fig. P7-41**

**7-42** The truss shown in Fig. P7-42.

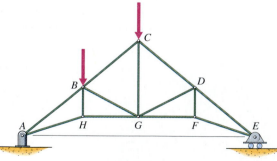

**Fig. P7-42**

## 7-2.3 Method of Sections

As stated in the method of joints, if an entire truss is in equilibrium, then each and every part of the truss is also in equilibrium. That does not mean, however, that the truss must be broken up into its most elemental parts—individual members and pins. In the method of sections, the truss will be divided up into just two pieces. Each of these pieces is also a rigid body in equilibrium.

For example, the truss of Fig. 7-29a can be divided into two parts by passing an imaginary section *aa* through some of its members. Of course, the section must pass entirely through the truss so that complete free-body diagrams can be drawn for each of the two pieces. Since the whole truss is in equilibrium, the part of the truss to the left of section *aa* and the part of the truss to the right of section *aa* are both in equilibrium also.

The free-body diagrams of the two parts are drawn in Figs. 7-29a and 7-29c, and include the forces on each cut member that was exerted by the other part of the member, which was cut away. Since the members are all straight two-force members, the forces in these members must act along the members as shown. Forces in members that have not been cut are internal to the rigid bodies and are not shown on the free-body diagrams. Thus, in order to determine the force in member *CF*, the section must cut through that member.

As is the case with any rigid body in plane equilibrium, three independent equations of equilibrium can be written for each rigid body. The resulting six equations are sufficient to solve for the six unknowns—the forces in the three cut members and the three support reactions. As with the method of joints, the solution of the equations can be simplified if the support reactions are determined from overall equilibrium before the truss is sectioned. Then the equations for either part will yield the remaining three unknown forces. In this case, the equilibrium equations for the remaining portion of the truss give no new information; they merely repeat the other equilibrium equations (see Problem 7-69).

If a section cuts through four or more members whose forces are unknown, then the method of sections will not generate enough equations of equilibrium to solve for all the unknown forces. Though it still might be possible to obtain values for one or two of the forces (see Example Problem 7-13), it is usually best to use a section that cuts through no more than three members whose forces are unknown.

It will often happen that a section that cuts no more than three members and that passes through a given member of interest cannot

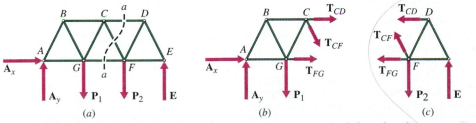

(a)                    (b)                    (c)

**Figure 7-29**   Free-body diagrams for the parts of a truss to the left and right of an imaginary plane.

*Determine the force in member FG of the truss shown in Fig. CE7-2a.*

Fig. CE7-2

(a)

(b)

## SOLUTION

The force in member $FG$ can be determined by using the method of joints; however, a more efficient procedure for determining a single force is known as the method of sections. The method of sections makes use of the fact that any and all parts of a truss are in equilibrium if the complete truss is in equilibrium. For example, consider a part of the truss of Fig. CE7-2a to the left of an imaginary plane passed through members $CD$, $DG$, and $FG$. A free-body diagram for this part of the truss is shown in Fig. CE7-2b. The loads acting on this part of the truss are the applied external loads and the three forces exerted on cut members $CD$, $DG$, and $FG$ by the part of the truss which has been removed.

The reaction components at support $A$ are $A_x = 0$ and $A_y = 2000$ lb. Since the lines of action of forces $T_{CD}$ and $T_{DG}$ pass through joint $D$, force $T_{FG}$ can be determined by summing moments about point $D$. Thus,

$$+\!\downarrow\!\Sigma M_D = T_{FG}(8) + 2000(12) + 1000(24) - 2000(36) = 0$$

from which

$$T_{FG} = 3000 \text{ lb} = 3000 \text{ lb } (T)$$

In a similar way, forces $T_{CD}$ and $T_{DG}$ can be determined by summing moments about points $G$ and $I$ (off the truss), respectively.

Note in this analysis that each equilibrium equation involves only one unknown force. Whenever possible, it is desirable to use equations that do not require use of previously determined results in order to prevent errors in previous calculations from affecting subsequent results. The method of sections has the advantage that the force in a member near the center of a truss usually can be determined without first obtaining the forces in all other members of the truss to one side of the member, and as a result, the force is independent of any errors in internal forces previously calculated.

be found. In such a case it may be necessary to draw a section through a nearby member and solve for the forces in it first. Then the method of joints can be used to find the forces in the members next to the cut section, or the truss can be further sectioned to find the force in the member of interest (see Example Problem 7-14). The method of joints is a series calculation in that each successive result depends on previous results. The method of sections is a parallel calculation in that the results can be made independent of previous results.

One of the principle advantages of using the method of section is that the force in a member near the center of a large truss usually can be determined without first obtaining the forces in the rest of the truss. As a result, the calculation of the force is independent of any errors in other internal forces previously calculated. To find the same force using the method of joints, however, would require that the force in a large number of other members be determined first. Any errors made in the determination of one member force will cause all subsequent forces to be in error as well.

Finally, the method of sections may be used as a spot check when the method of joints or a computer program is used to solve a large truss. Although it is unlikely that a computer will make an error in its computation, it is quite likely that the input data may be in error. Most often these errors occur when an operator incorrectly enters the coordinates of a joint, incorrectly specifies how the joints are connected, or incorrectly applies a load to the truss. In such cases, the method of sections can be used to independently check the forces in one or two interior members.

Each truss member in Fig. 7-30*a* has a length of 4 m. Use the method of sections to find the forces in members *BC*, *CG*, and *FG*.

**Fig. 7-30**

## SOLUTION

The forces in members *BC*, *CG*, and *FG* can be determined by passing an imaginary plane (see Fig. 7-30*b*) through the three members and drawing a free-body diagram of the part of the truss either to the left or to the right of the imaginary plane. Support reactions are needed for both of these diagrams so they will be determined first.

A free-body diagram of the complete truss is shown in Fig. 7-30*b*. The reaction at support *A* is represented by force components $A_x$ and $A_y$ and the reaction at *E* by a vertical force $\mathbf{E}_y$. The reaction at *E* can be determined by summing moments about point *A*. Thus,

$$+\circlearrowleft\Sigma M_A = E_y (12) - 30(2) - 18(6) - 24(10) = 0 \qquad E_y = 34 \text{ kN}$$

The reaction components $A_x$ and $A_y$ at support *A* are determined by summing forces in the *x*- and *y*-directions. Thus,

$$+\rightarrow\Sigma F_x = A_x = 0 \qquad\qquad A_x = 0$$
$$+\uparrow\Sigma F_y = A_y + E_y - 30 - 18 - 24$$
$$= A_y + 34 - 30 - 18 - 24 = 0 \qquad A_y = 38 \text{ kN}$$

Free-body diagrams to the left and right of an imaginary plane through members *BC*, *CG*, and *FG* are shown in Fig. 7-30*c*. The free-body diagram to the left of the plane will be used for the force determinations in the members since it has fewer applied loads. Since the lines of action of forces $T_{CG}$ and $T_{FG}$ pass through joint *G*, the force $T_{BC}$ is determined by summing moments about point *G*. Thus,

$$+\circlearrowleft\Sigma M_G = -T_{BC}(4 \sin 60°) + 30(2) - 38(4) = 0$$
$$T_{BC} = -26.56 \text{ kN} \cong 26.6 \text{ kN } (C)$$

Similarly, since the lines of action of forces $T_{CG}$ and $T_{BC}$ pass through joint *C*, the force $T_{FG}$ is determined by summing moments about point *C*. Thus,

$$+\circlearrowleft\Sigma M_C = T_{FG} (4 \sin 60°) + 30(4) - 38(6) = 0 \qquad T_{FG} = 31.18 \text{ kN} \cong 31.2 \text{ kN } (T)$$

Finally, since forces $T_{BC}$ and $T_{FG}$ are horizontal forces, $T_{CG}$ can be determined by summing forces in the vertical direction. Thus,

$$+\uparrow\Sigma F_y = T_{CG} \sin 60° + 38 - 30 = 0 \qquad T_{CG} = -9.238 \text{ kN} \cong 9.24 \text{ kN } (C)$$

Therefore, the desired answers are

$$T_{BC} = -26.56 \text{ kN} \cong 26.6 \text{ kN } (C) \qquad\qquad \text{Ans.}$$
$$T_{CG} = -9.238 \text{ kN} \cong 9.24 \text{ kN } (C) \qquad\qquad \text{Ans.}$$
$$T_{FG} = 31.18 \text{ kN} \cong 31.2 \text{ kN } (T) \qquad\qquad \text{Ans.}$$

Alternatively, summing moments about point *E* would yield a value for force $A_y$ that is independent of the previous determination for force $E_y$.

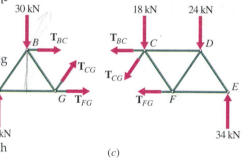

*(c)*

If a complete truss is in equilibrium, any part of the truss isolated as a free-body will also be in equilibrium if the internal forces exposed during the isolation process are properly applied.

Choosing a convenient axis about which to sum moments can simplify the solution of many problems by eliminating the need to solve simultaneous equations.

Use the method of sections to find the forces in members $AB$, $AF$, and $FG$ of the truss shown in Fig. 7-31$a$.

(a)

(b)

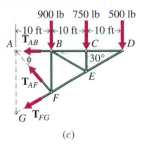
(c)

**Fig. 7-31**

## SOLUTION

The forces in members $AB$, $AF$, and $FG$ can be determined by passing an imaginary plane (see Fig. 7-31$b$) through the three members and drawing a free-body diagram of the part of the truss either to the left or to the right of the imaginary plane. Since support reactions are not needed for the free-body diagram to the right of the imaginary plane, it will be used for the force determinations in the members. Since the lines of action of forces $T_{AF}$ and $T_{FG}$ pass through joint $F$, the force $T_{AB}$ is determined by summing moments about point $F$. Thus,

$$+\circlearrowleft\Sigma M_F = T_{AB}(20 \tan 30°) - 750(10) - 500(20) = 0$$
$$T_{AB} = 1515.5 \text{ lb} \cong 1516 \text{ lb } (T)$$

Similarly, since the lines of action of forces $T_{AB}$ and $T_{FG}$ pass through joint $D$, the force $T_{AF}$ is determined by summing moments about point $C$. Thus,

$$\phi = \tan^{-1} \frac{20 \tan 30°}{10} = 49.11°$$

$$+\circlearrowleft\Sigma M_D = -T_{AF}(30 \sin 49.11°) + 900(20) + 750(10) = 0$$
$$T_{AF} = 1124.4 \text{ lb} \cong 1124 \text{ lb } (T)$$

Finally, since the lines of action of forces $T_{AB}$ and $T_{AF}$ pass through joint $A$, the force $T_{FG}$ is determined by summing moments about point $A$. Thus,

$$+\circlearrowleft\Sigma M_A = -T_{FG}(30 \sin 30°) - 900(10) - 750(20) - 500(30) = 0$$
$$T_{FG} = -2600 \text{ lb} = 2600 \text{ lb } (C)$$

Therefore, the desired answers are

$$T_{AB} = 1515.5 \text{ lb} \cong 1516 \text{ lb } (T) \qquad \text{Ans.}$$
$$T_{AF} = 1124.4 \text{ lb} \cong 1124 \text{ lb } (T) \qquad \text{Ans.}$$
$$T_{FG} = -2600 \text{ lb} = 2600 \text{ lb } (C) \qquad \text{Ans.}$$

If a complete truss is in equilibrium, any part of the truss isolated as a free-body will also be in equilibrium if the internal forces exposed during the isolation process are properly applied.

Choosing a convenient point about which to sum moments can usually simplify the solution of most problems when the method of sections is being used.

Alternatively, horizontal and vertical components of force $T_{AF}$ could be used in the equation $\Sigma M_D = 0$.

Use the method of sections to find the forces in members *EF*, *JK*, and *HJ* of the Baltimore truss drawn in Fig. 7-32*a*.

(a)

## SOLUTION

First, draw the free-body diagram of the entire truss (Fig. 7-32*b*) to solve for the support reactions at *A* and *D*. Summing moments about *A* gives

$$\downarrow + \Sigma M_A = 24D - (4)(2) - (8)(3) - (12)(5) - (16)(4) - (20)(3) = 0$$

which can be solved for *D* to get

$$D = 9 \text{ kN}$$

Then, force equilibrium gives

$$+\rightarrow \Sigma F_x = A_x = 0 \qquad\qquad A_x = 0 \text{ kN}$$
$$+\uparrow \Sigma F_y = A_y - 2 - 3 - 5 - 4 - 3 + D = 0 \qquad A_y = 8 \text{ kN}$$

(b)

Next, pass a section through members *EF*, *JK*, and *HJ* as shown on Fig. 7-32*b* and draw the free-body diagrams of the resulting two parts (Figs. 7-32*c* and 7-32*d*). The force in member *JK* can be found by summing forces in the vertical direction for either free-body diagram. For example, from Fig. 7-32*c*

$$+\uparrow \Sigma F_y = 8 - 2 - 3 - 5 + T_{JK} \sin 45° = 0 \qquad T_{JK} = 2.828 \text{ kN}$$

while from Fig. 7-32*d*

$$+\uparrow \Sigma F_y = 9 - 3 - 4 - T_{JK} \sin 45° = 0 \qquad T_{JK} = 2.828 \text{ kN}$$

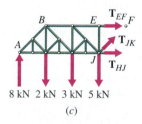

(c)

The force in member *HJ* can be found by summing moments about point *F* in Fig. 7-32*d*:

$$\downarrow + \Sigma M_F = (8)(9) - (4)(3) - 4T_{HJ} = 0$$

which gives

$$T_{HJ} = 15 \text{ kN}$$

Note that this is the same result as is obtained by summing moments about the point *F* in Fig. 7-32*c*:

$$\downarrow + \Sigma M_F = (12)(2) + (8)(3) + (4)(5) - (16)(8) + 4T_{HJ} = 0$$

which again gives

$$T_{HJ} = 15 \text{ kN}$$

(d)

**Fig. 7-32**

Finally, the force in member *EF* can be found by summing forces in the horizontal direction or by summing moments about *J*. Choosing the latter method and using Fig. 7-32*d* gives

$$\downarrow + \Sigma M_J = (12)(9) + 4T_{EF} - (8)(3) - (4)(4) = 0$$

or

$$T_{EF} = -17 \text{ kN}$$

The desired answers are

$$T_{EF} = 17 \text{ kN (C)} \qquad T_{HJ} = 15 \text{ kN (T)} \qquad T_{JK} = 2.83 \text{ kN (T)} \qquad \text{Ans.}$$

(Since overall equilibrium was first used to find the support reactions, equilibrium of either Fig. 7-32*c* or Fig. 7-32*d* can be used to solve for the member forces. Usually the part with the fewest forces acting on it will result in the simplest equations of equilibrium.)

Alternatively, summing moments about point *D* would yield a value for force $A_y$ that is independent of the previous determination for force *D*.

If a complete truss is in equilibrium, any part of the truss isolated as a free-body will also be in equilibrium if the internal forces exposed during the isolation process are properly applied.

Use the method of sections to find the forces in members $CD$ and $FG$ of the truss in Fig. 7-33a.

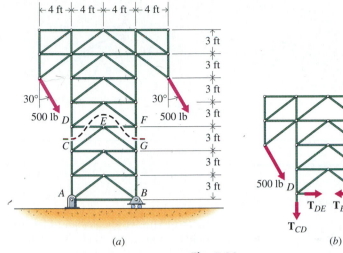

(a)                                    (b)

**Fig. 7-33**

## SOLUTION

Cut a section through members $CD$, $DE$, $EF$, and $FG$ as shown in Fig. 7-33a, and draw the free-body diagram of the upper part of the truss (Fig. 7-33b). Summing moments about $D$,

$$\downarrow + \Sigma M_D = (4)(500 \cos 30°) - (6)(500 \sin 30°) - 8T_{FG}$$
$$- (12)(500 \cos 30°) - (6)(500 \sin 30°) = 0$$

gives

$$T_{FG} = -808.0 \text{ lb}$$

Then, summing moments about $F$,

$$\downarrow + \Sigma M_F = (12)(500 \cos 30°) - (6)(500 \sin 30°) + 8T_{CD}$$
$$- (4)(500 \cos 30°) - (6)(500 \sin 30°) = 0$$

gives

$$T_{CD} = -58.01 \text{ lb}$$

The consistency of these answers can be checked using the vertical component of force equilibrium:

$$+ \uparrow \Sigma F_y = -2(500 \cos 30°) - (-808.0) - (-58.01) = -0.02$$

which is within the accuracy of the answers above.
The desired answers are

$$T_{CD} = 58.0 \text{ lb (C)} \qquad \qquad \text{Ans.}$$
$$T_{FG} = 808 \text{ lb (C)} \qquad \qquad \text{Ans.}$$

(Note that it was not necessary in this problem first to find the support reactions using overall equilibrium. Also note that neither $T_{DE}$ nor $T_{EF}$ can be found from this section. Either additional sections or the method of joints would be needed to find these forces if they needed to be found.)

> Horizontal and vertical components of the 500-lb applied loads are used in the equation $\Sigma M_D = 0$ since the moment arms for these components are easy to determine.

> Summing moments about point $F$ yields a value for force $T_{CD}$ that is independent of the previous determination for force $T_{FG}$.

Find the forces in members $BC$ and $BG$ in the Fink truss of Fig. 7-34a. All triangles are either equilateral or 30°–60°–90° right triangles, and the loads are all perpendicular to side $ABCD$.

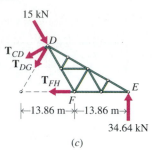

(a)          (b)          (c)

**Fig. 7-34**

## SOLUTION

First find the support reactions by drawing a free-body diagram of the entire truss (Fig. 7-34b) and writing the equilibrium equations

$$+\rightarrow \Sigma F_x = A_x + (15 + 30 + 30 + 30 + 15)(\sin 30°) = 0$$
$$+\uparrow \Sigma F_y = A_y - (15 + 30 + 30 + 30 + 15)(\cos 30°) + E = 0$$
$$\langle+\Sigma M_A = 41.57E - (6)(30) - (12)(30) - (18)(30) - (24)(15) = 0$$

which are solved to get

$$A_x = -60.00 \text{ kN} \qquad A_y = 69.28 \text{ kN} \qquad E = 34.64 \text{ kN}$$

A section cut through $BC$ or $BG$ would involve four unknown internal forces. The equilibrium equations cannot be solved until one or more of the forces are determined by some other means. A combination of the method of sections and the method of joints will be used to find the desired forces.

A section will be cut through the middle of the truss near to the members in which the forces are desired. First, cut a section $aa$ (Fig. 7-34a) through members $CD$, $DG$, and $FH$ and draw the free-body diagram of the right-hand part of the truss (Fig. 7-34c). Summing moments about point $H$,

$$\langle+\Sigma M_H = (27.72)(34.64) - (13.86 \cos 30°)(15) + (13.86)(T_{CD} \sin 30°) = 0$$

gives

$$T_{CD} = -112.58 \text{ kN}$$

Next, draw a free-body diagram of pin $C$ (Fig. 7-34d) and write the equilibrium equations along and perpendicular to members $BC$ and $CD$:

$$\nearrow \Sigma F_x = T_{CD} - T_{BC} = 0 \qquad \searrow \Sigma F_y = -30 - T_{CG} = 0$$

which are immediately solved to get

$$T_{BC} = T_{CD} = -112.58 \text{ kN} \qquad \text{and} \qquad T_{CG} = -30.00 \text{ kN}$$

Finally, cut a section $bb$ (Fig. 7-34a) through members $BC$, $BG$, $GH$, and $FH$ and draw the free-body diagram of the left-hand part of the truss (Fig. 7-34e). Again summing moments about point $H$,

$$\langle+\Sigma M_H = -(13.86)(69.28) + (12)(15) + (6)(30) - (13.86 \sin 30°)T_{BC} - 6T_{BG} = 0$$

gives

$$T_{BG} = 29.99 \text{ kN}$$

Thus, the desired answers are

$$T_{BC} = 112.6 \text{ kN } (C) \qquad\qquad \text{Ans.}$$
$$T_{BG} = 30.0 \text{ kN } (T) \qquad\qquad \text{Ans.}$$

> If possible, only three members with unknown forces should be cut when a part of the truss is isolated since there are only three independent equilibrium equations for a plane truss.

> Selecting axes which coincide with the lines of action of forces $T_{CD}$ and $T_{CG}$ eliminates the need to solve simultaneous equations.

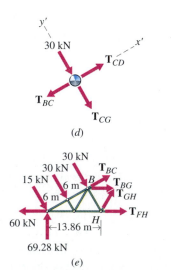

(d)

(e)

# PROBLEMS

Solve the following problems by the method of sections. State whether the members are in tension or compression.

## Introductory Problems

**7-43\*** The length of each truss member in Fig. P7-43 is 5 ft. Find the forces in members *CD*, *DE*, and *EF*.

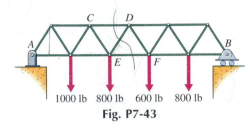

**Fig. P7-43**

**7-44\*** Determine the forces in members *CD*, *DE*, and *EF* of the truss shown in Fig. P7-44.

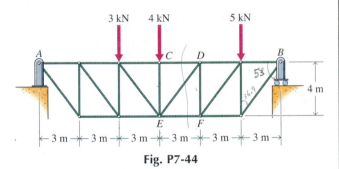

**Fig. P7-44**

**7-45** Determine the forces in members *CD*, *CF*, and *FG* of the bridge truss shown in Fig. P7-45.

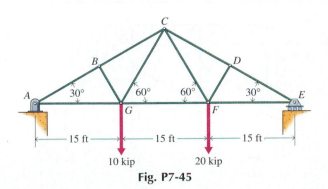

**Fig. P7-45**

**7-46** Determine the forces in members *BC*, *BF*, and *EF* of the truss shown in Fig. P7-46.

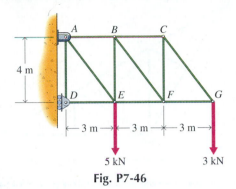

**Fig. P7-46**

**7-47\*** Determine the forces in members *BC*, *CF*, and *EF* of the stairs truss shown in Fig. P7-47.

**Fig. P7-47**

**7-48\*** Determine the forces in members *CD*, *CG*, and *GH* of the stairs truss shown in Fig. P7-48.

**Fig. P7-48**

**7-49** Determine the forces in members BC, CF, and EF of the truss shown in Fig. P7-49.

Fig. P7-49

**7-50** Determine the forces in members BC, CD, and DE of the truss shown in Fig. P7-50.

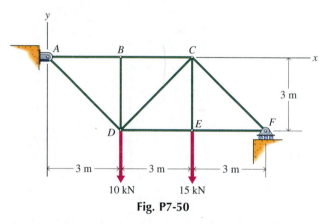

Fig. P7-50

**7-51\*** The truss shown in Fig. P7-51 supports a sign that weighs 3000 lb. The sign is connected to the truss at joints E, G, and H, and the connecting links are adjusted so that each joint carries one-third of the load. Determine the forces in members CD, CF, and FG.

Fig. P7-51

**7-52\*** Determine the forces in members CD, CE, and EF of the truss shown in Fig. P7-52.

Fig. P7-52

**7-53** Determine the forces in members CD and FG of the K-truss shown in Fig. P7-53. (Hint: Use section aa.)

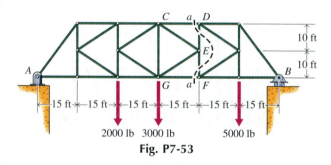

Fig. P7-53

**7-54** Determine the forces in members AB and FG of the truss shown in Fig. P7-54. (Hint: Use section aa.)

Fig. P7-54

362

**7-55*** Find the maximum load $P$ that can be supported by the truss of Fig. P7-55 without producing a force of more than 2500 lb in member $CD$.

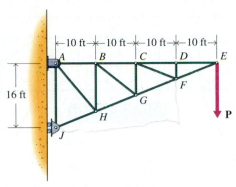

**Fig. P7-55**

**7-56*** Find the maximum load $P$ that can be supported by the truss of Fig. P7-56 without producing a force of more than 3 kN in member $EJ$.

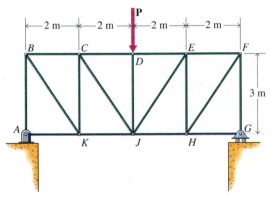

**Fig. P7-56**

**7-57** Determine the forces in members $BC$, $CF$, and $EF$ of the truss shown in Fig. P7-57.

**Fig. P7-57**

**7-58** Determine the forces in members $EF$, $EJ$, and $JK$ of the truss shown in Fig. P7-58.

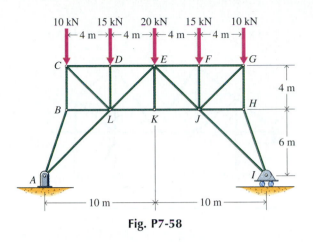

**Fig. P7-58**

## Intermediate Problems

**7-59*** Determine the forces in members $CD$, $CK$, and $LK$ of the Howe roof truss shown in Fig. P7-59.

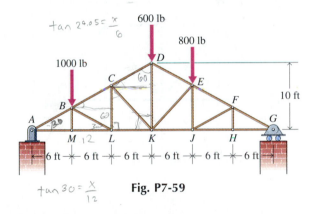

**Fig. P7-59**

**7-60*** Determine the forces in members $DE$, $EJ$, and $HJ$ of the roof truss shown in Fig. P7-60.

**Fig. P7-60**

**363**

**7-61** Determine the forces in members $CD$, $DF$, and $EF$ of the bridge truss shown in Fig. P7-61.

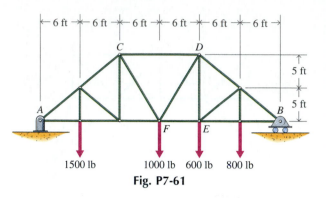

Fig. P7-61

**7-62** Determine the forces in members $BC$, $BG$, and $GH$ of the bridge truss shown in Fig. P7-62.

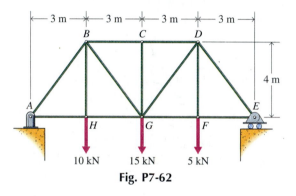

Fig. P7-62

**7-63\*** Determine the forces in members $DE$, $DJ$, and $JK$ of the truss shown in Fig. P7-63.

Fig. P7-63

**7-64** The homogeneous 3-kN sign is attached by means of short links to the signboard truss of Fig. P7-64. Determine the forces in members $CD$, $CE$, $EG$, and $FG$.

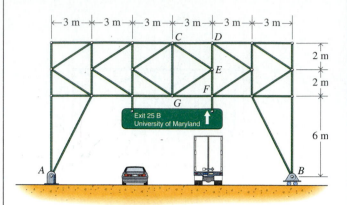

Fig. P7-64

## Challenging Problems

**7-65\*** The roof truss of Fig. P7-65 is composed of 30°-60°-90° right triangles and is loaded as shown. Determine the forces in members $CD$, $CE$, and $EF$.

Fig. P7-65

**7-66** Determine the forces in members $CD$, $DE$, and $DF$ of the roof truss shown in Fig. P7-66. Triangle $CDF$ is an equilateral triangle, and joints $E$ and $G$ are at the midpoints of their respective sides.

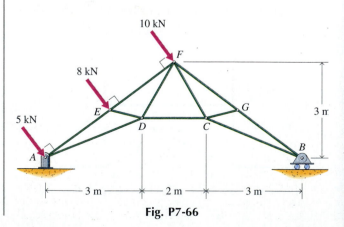

Fig. P7-66

**7-67\*** Determine the forces in members *DE*, *EG*, *FG*, and *FH* of the truss shown in Fig. P7-67.

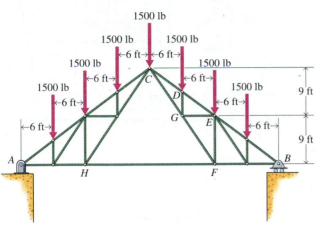

Fig. P7-67

**7-68** Determine the forces in members *CD*, *DF*, *EF*, and *FG* of the transmission line truss shown in Fig. P7-68.

Fig. P7-68

**7-69** Show that overall equilibrium of a truss is a consequence of equilibrium of the two separate parts generated by the method of sections. That is, section the bridge truss of Fig. P7-69 down the middle as indicated and write the equilibrium equations for each piece. Eliminate the member forces from the resulting six equations, and show that the result is equivalent to equilibrium of the whole truss.

Fig. P7-69

## Computer Problems

**C7-70** A truss is loaded and supported as shown in Fig. P7-70. Plot the forces in members *AF*, *BD*, and *BF* of the truss as a function of angle $\theta$ ($0 \le \theta \le 90°$).

Fig. P7-70

**C7-71** A truss is loaded and supported as shown in Fig. P7-71. Plot the forces in members *BC*, *CF*, and *FG* of the truss as a function of angle $\theta$ ($0 \le \theta \le 90°$).

Fig. P7-71

**C7-72** The Gambrel truss shown in Fig. P7-72 supports one side of a bridge; an identical truss supports the other side. A 3400-kg truck is stopped on the bridge at the location shown. Floor beams carry vehicle loads to the truss joints. If the center of gravity of the truck is located 1.5 m in front of the rear wheels, plot the force in members *AB*, *BC*, *BG*, and *GH* as a function of the truck's location *d* (0 ≤ *d* ≤ 20 m).

Fig. P7-72

---

**Figure 7-35** Straight two-force member subjected to a tensile load.

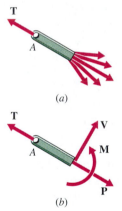

(a)

(b)

**Figure 7-36** Free-body diagrams for part of a straight two-force member.

**Figure 7-37** Curved two-force member subjected to a tensile load.

## 7-2.4  Forces in Straight and Curved Two-Force Members

All the truss members used in this section have been straight. As a result, the resultants of the forces at the ends of the members have always acted along the axis of the members. When the forces are directed away from the ends of the member as in Fig. 7-35, the member is said to be in tension and the force **T** is referred to as the tension force *in the member* and not just the tension force *on the end of the member*. That this description of the force is correct is easily shown by considering the equilibrium of a portion of the truss member.

Suppose that the truss member of Fig. 7-35 is cut normal to its axis at section *aa*. In general, there will be some complex force distribution acting on the cut surface as shown on the free-body diagram of Fig. 7-36a. However, according to the discussion in Section 4-6, this system of forces can be replaced with an equivalent force-couple as shown on the free-body diagram of Fig. 7-36b. Then, equilibrium of forces in the direction perpendicular to the axis of the member requires that **V**, the shear component of the equivalent force-couple, must be zero. Similarly, equilibrium of forces in the direction along the axis of the member requires that **P**, the axial component of the equivalent force-couple, must be equal in magnitude and opposite in direction to the force applied to the end of the member. Finally, moment equilibrium requires that **M**, the couple component of the equivalent force-couple, must be zero. That is, *if the forces at the ends of a straight two-force member are pulling on the member, then the forces on any cut section of the member also represent an axial force pulling on the cut section of the member*, regardless of where the member is cut. Therefore, it is proper to talk about the force **T** as the tension *in the member*.

When a two-force member is curved, however, the forces at the ends of the member do not act along the axis of the member. Instead, the forces act along the line joining the points where the forces are applied as shown in Fig. 7-37. If the member is cut normal to its axis at section *aa*, there will again be a complex force distribution acting on the cut surface, as shown in Fig. 7-38a. Of course, this system of forces can still be replaced with an equivalent force-couple as shown on the free-body diagram of Fig. 7-38b. However, force equilibrium now requires that the resultant **R** of the axial **P** and shear **V** components of the equivalent force-couple be equal in magnitude and opposite in direction to the force **T** at the end of the member, as shown in Fig. 7-38c. Since the forces **R** and **T** are not collinear, moment equilibrium now requires that $M = Td \neq 0$.

(a)          (b)

**Figure 7-38** Free-body diagrams for part of a curved two-force member.

It should be noted at this point that neither of the free-body diagrams shown in Fig. 7-38 represents a two-force body even though they both represent a part of a two-force body. The free-body diagram of Fig. 7-38*a* is not a two-force body, since the forces are distributed over the entire cut surface and hence do not all act at a single point. The free-body diagram of Fig. 7-38*b* is not a two-force body, since a couple acts on the body.

Therefore, *the design of straight two-force members need only consider axial forces,* but *curved two-force members must be designed to withstand shearing forces* **V** *and bending moments* **M** *as well as axial forces* **P**. To further complicate the problem, the strengths of the shear forces, bending moments, and axial forces depend on where the member is cut, as shown in Example Problem 7-15.

An arch support shown in Fig. 7-39a consists of two quarter circle members. Determine the axial force $P$, shear force $V$, and bending moment $M$ in member $AB$ as a function of the angle $\theta$.

Fig. 7-39

---

## SOLUTION

First draw the free-body diagram of pin $B$ (Fig. 7-39b). Since members $AB$ and $BC$ are two-force members, the forces $\mathbf{F}_{AB}$ and $\mathbf{F}_{BC}$ act at a 45° angle to the vertical. Then the equilibrium equations

$$+\rightarrow \Sigma F_x = F_{AB} \sin 45° - F_{BC} \sin 45° = 0$$
$$+\uparrow \Sigma F_y = F_{AB} \cos 45° + F_{BC} \cos 45° - 700 = 0$$

give the forces in the members $AB$ and $BC$

$$F_{AB} = F_{BC} = 495.0 \text{ lb}$$

Next, draw a free-body diagram of a section of member $AB$ (Fig. 7-39c). Summing forces horizontally and vertically and summing moments about $D$ gives

$$+\rightarrow \Sigma F_x = F_{AB} \sin 45° + P \sin \theta - V \cos \theta = 0$$
$$+\uparrow \Sigma F_y = F_{AB} \cos 45° + P \cos \theta + V \sin \theta = 0$$
$$\underset{\downarrow}{+} \Sigma M_D = M + (8 \sin \theta)(F_{AB} \cos 45°)$$
$$- (8 - 8 \cos \theta)(F_{BC} \sin 45°) = 0$$

or

$$P = -350(\cos \theta + \sin \theta) \text{ lb} \qquad \text{Ans.}$$
$$V = 350(\cos \theta - \sin \theta) \text{ lb} \qquad \text{Ans.}$$
$$M = 2800(1 - \sin \theta - \cos \theta) \text{ ft} \cdot \text{lb} \qquad \text{Ans.}$$

These results are plotted in Fig. 7-39d.

> Members $AB$ and $BC$ are two-force members.

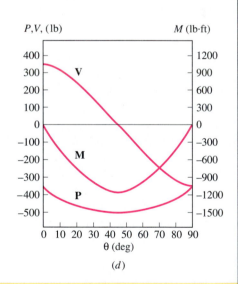

(d)

# 7-3 SPACE TRUSSES

A truss whose joints do not all lie in a plane and/or that is loaded and supported out of its plane is called a **space truss.** As with a planar truss, the members of a space truss may be treated as two-force members provided that no member is continuous through a joint, the external loading is applied only at the joints, and the joints consist of frictionless ball-and-socket connections. As with plane trusses, when the weight of the members is to be included, half of the weight of each member is applied to the joints at each end of the member.

The three-dimensional equivalent of a triangle is a tetrahedron, as shown in Fig. 7-40a. Space trusses are constructed from tetrahedral subunits. By analogy with a simple plane truss, a **simple space truss** is formed by adding tetrahedron units to the truss as shown in Fig. 7-40b. Since now each new joint brings three new members with it, the relationship between the number of joints $j$ and the number of members $m$ in a simple space truss is given by

$$m = 3j - 6 \qquad \text{(7-4)}$$

Clearly, simple space trusses, like simple plane trusses, are always rigid.

As in the case of plane trusses, space trusses can be analyzed using either the method of joints or the method of sections. For the method of joints, the truss is divided up into all its elemental members and joints. Also, as in the case of plane trusses, equilibrium of the members is guaranteed by the assumption that they are two-force members. Therefore, the members can again be discarded, and only the free-body diagrams of the pins need be considered. Equilibrium of the pins is expressed by writing the force equilibrium equation

$$\Sigma \mathbf{F} = 0 \qquad \text{(7-3)}$$

for each joint. Since each joint consists of a system of concurrent forces in three dimensions, moment equilibrium gives no useful information, but now Eq. 7-3 has three independent components. But according to Eq. 7-4, this is precisely the number of independent equations needed to solve for the $m$ member forces and six support reactions of a simple space truss. Aside from the larger number of equations and unknowns, the solution procedure is identical to that of a planar truss.

The procedure for the method of sections is also essentially identical to that of a planar truss. The space truss is divided in two by passing a section completely through the truss and drawing free-body diagrams of each part. Application of the equations of equilibrium

$$\Sigma \mathbf{F} = 0 \quad \text{and} \quad \Sigma \mathbf{M} = 0 \qquad \text{(7-5)}$$

to these two parts yields a total of 12 equations—6 for each part. These 12 equations are sufficient to determine the 6 support reactions and 6 internal member forces. However, it is usually difficult to pass a section through a typical space truss that cuts no more than 6 members. For this reason and because of the complexity of solving the large numbers of equations, the method of sections is not often used for space trusses.

(a)

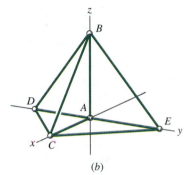

(b)

**Figure 7-40** Simple space truss constructed using tetrahedral subunits.

## EXAMPLE PROBLEM 7-16

The simple space truss of Fig. 7-41a is supported by a ball-and-socket joint at $E$ and by short links at $A$, $D$, and $C$. Determine the forces in each of the members.

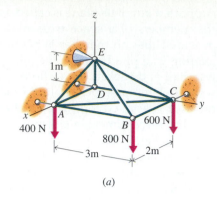

### SOLUTION

As a first step, draw the free-body diagram of the entire truss (Fig. 7-41b) and solve for the support reactions. Moment equilibrium about point $E$ gives

$$\Sigma M_E = (2i - k) \times (Aj - 400 k) + (2i + 3j - k) \times (-800k)$$
$$+ (3j - k) \times (Ci - 600k) + (-k) \times (Dj)$$
$$= 2Ak + 800j + Ai + 1600j - 2400i - 3Ck - 1800i - Cj + Di$$
$$= 0$$

The $x$-, $y$-, and $z$-components of this equation give

| | | |
|---|---|---|
| **j:** | $C = 2400$ | $C = 2400$ N |
| **k:** | $2A - 3C = 0$ | $A = 3600$ N |
| **i:** | $A + D = 4200$ | $D = 600$ N |

Force equilibrium in the $x$-, $y$-, and $z$-directions gives

| | |
|---|---|
| $\Sigma F_x = E_x + C = 0$ | $E_x = -2400$ N |
| $\Sigma F_y = E_y + A + D = 0$ | $E_y = -4200$ N |
| $\Sigma F_z = E_z - 400 - 600 - 800 = 0$ | $E_z = 1800$ N |

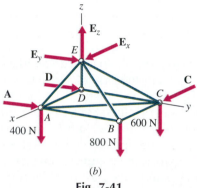

(b)

**Fig. 7-41**

The free-body diagram for pin $A$ (Fig. 7-42a) still has too many unknowns to solve, so pin $B$ will be considered first. Forces are shown on the free-body diagram (Fig. 7-42b) with an arrow to show the direction and a value for the magnitude. The directions of the forces are along the members; therefore, the force in member $BE$ applied at pin $B$ is

$$\mathbf{F}_{BE} = T_{BE} \frac{-2i - 3j + k}{\sqrt{2^2 + 3^2 + 1^2}} = -0.5345T_{BE}i - 0.8018T_{BE}j + 0.2673T_{BE}k$$

Similarly, $\mathbf{F}_{AB}$, $\mathbf{F}_{BC,}$ and the applied load $\mathbf{P}$ are

$$\mathbf{F}_{AB} = -T_{AB}j \qquad \mathbf{F}_{BC} = -T_{BC}i \qquad \mathbf{P} = -800k \text{ N}$$

The $x$-, $y$-, and $z$-components of the equilibrium equation $\Sigma \mathbf{F} = 0$ for this joint then become

| | | |
|---|---|---|
| **k:** | $0.2673T_{BE} = 800$ | $T_{BE} = 2993$ N |
| **i:** | $-T_{BC} - 0.5345T_{BE} = 0$ | $T_{BC} = -1600$ N |
| **j:** | $-T_{AB} - 0.8018T_{BE} = 0$ | $T_{AB} = -2400$ N |

Now the member forces acting on pin $A$ (Fig. 7-42a) are

$$\mathbf{F}_{AB} = T_{AB}j = -2400j \text{ N}$$

$$\mathbf{F}_{AC} = T_{AC} \frac{-2i + 3j}{\sqrt{2^2 + 3^2}} = -0.5547T_{AC}i + 0.8321T_{AC}j$$

$$\mathbf{F}_{AD} = -T_{AD}i$$

$$\mathbf{F}_{AE} = T_{AE} \frac{-2i + k}{\sqrt{2^2 + 1^2}} = -0.8944T_{AE}i + 0.4472T_{AE}k$$

The support reaction and applied load are

$$\mathbf{A} = 3600j \text{ N} \qquad \mathbf{P} = -400k \text{ N}$$

Putting these into the force equilibrium equation gives

$$
\begin{aligned}
\textbf{j:} && 0.8321T_{AC} &= -1200 & T_{AC} &= -1442.2 \text{ N} \\
\textbf{k:} && 0.4472T_{AE} &= 400 & T_{AE} &= 894.5 \text{ N} \\
\textbf{i:} && -0.5547T_{AC} - T_{AD} - 0.8944T_{AE} &= 0 & T_{AD} &= 0 \text{ N}
\end{aligned}
$$

(a)

Next, draw the free-body diagram of pin C (Fig. 7-42c). The force in member $AC$ applied at pin $C$, $\mathbf{F}_{CA}$, is equal in magnitude but opposite in direction to the force in the same member applied at pin $A$, $\mathbf{F}_{AC}$; therefore,

$$\mathbf{F}_{CA} = -\mathbf{F}_{AC} = 0.5547T_{AC}\mathbf{i} - 0.8321T_{AC}\mathbf{j} = -800\mathbf{i} + 1200\mathbf{j} \text{ N}$$

The rest of the member forces are again along the members,

$$\mathbf{F}_{CB} = T_{BC}\mathbf{i} = -1600\mathbf{i} \text{ N} \qquad \mathbf{F}_{CD} = -T_{CD}\mathbf{j}$$

$$\mathbf{F}_{CE} = T_{CE}\frac{-3\mathbf{j} + \mathbf{k}}{\sqrt{3^2 + 1^2}} = -0.9487T_{CE}\mathbf{j} + 0.3162T_{CE}\mathbf{k}$$

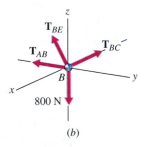

(b)

while the support reaction and applied load are

$$\mathbf{C} = C\mathbf{i} = 2400\mathbf{i} \text{ N} \qquad \mathbf{P} = -600\mathbf{k} \text{ N}$$

Substitution of these forces in the equilibrium equation yields

$$
\begin{aligned}
\textbf{i:} && -1600 - 800 + 2400 &= 0 \\
\textbf{j:} && -T_{CD} - 0.9487T_{CE} &= -1200 \\
\textbf{k:} && 0.3162T_{CE} &= 600
\end{aligned}
$$

The first equation has no unknowns and is just used as a check of the consistency of the answers. The remaining two equations give

$$T_{CD} = -600.2 \text{ N} \qquad T_{CE} = 1897.5 \text{ N}$$

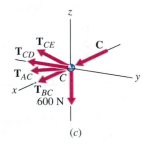

(c)

Next, consider pin D whose free-body diagram is drawn in Fig. 7-42d. The forces applied to this pin all act along the coordinate axes, so force equilibrium is easily written

$$T_{AD} = 0 \text{ N} \qquad T_{CD} = -600 \text{ N} \qquad T_{DE} = 0 \text{ N}$$

The first two equations just repeat information already known. The third says that member $DE$ is a zero-force member.

Finally, draw the free-body diagram of pin $E$ (Fig. 7-42e). The member forces and support reaction acting on this pin are

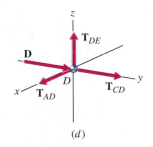

(d)

$$
\begin{aligned}
\mathbf{F}_{EA} &= -\mathbf{F}_{AE} = 0.8944T_{AE}\mathbf{i} - 0.4472T_{AE}\mathbf{k} = 800\mathbf{i} - 400\mathbf{k} \text{ N} \\
\mathbf{F}_{EB} &= -\mathbf{F}_{BE} = 0.5345T_{BE}\mathbf{i} + 0.8018T_{BE}\mathbf{j} - 0.2673T_{BE}\mathbf{k} = 1600\mathbf{i} + 2400\mathbf{j} - 800\mathbf{k} \text{ N} \\
\mathbf{F}_{EC} &= -\mathbf{F}_{CE} = 0.9487T_{CE}\mathbf{j} - 0.3162T_{CE}\mathbf{k} = 1800\mathbf{j} - 600\mathbf{k} \text{ N} \\
\mathbf{F}_{ED} &= -T_{DE}\mathbf{k} = 0 \text{ N} \\
\mathbf{E} &= -2400\mathbf{i} - 4200\mathbf{j} + 1800\mathbf{k} \text{ N}
\end{aligned}
$$

There are no unknowns left in these forces, and substitution into the force equilibrium equation just serves to check the consistency of the results:

$$
\begin{aligned}
\textbf{i:} && 800 + 1600 - 2400 &= 0 \\
\textbf{j:} && 2400 + 1800 - 4200 &= 0 \\
\textbf{k:} && -400 - 800 - 600 + 1800 &= 0
\end{aligned}
$$

The required answers are then

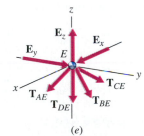

(e)

**Fig. 7-42**

| | | |
|---|---|---|
| $T_{AB} = 2400 \text{ N } (C)$ | $T_{AC} = 1442 \text{ N } (C)$ | Ans. |
| $T_{AD} = T_{DE} = 0 \text{ N}$ | $T_{AE} = 894 \text{ N } (T)$ | Ans. |
| $T_{BC} = 1600 \text{ N } (C)$ | $T_{BE} = 2993 \text{ N } (T)$ | Ans. |
| $T_{CD} = 600 \text{ N } (C)$ | $T_{CE} = 1898 \text{ N } (T)$ | Ans. |

The space truss of Fig. 7-43 is supported by a ball-and-socket joint at $A$ and by short links at $B$ and $C$. A 125-lb force acts in the $yz$-plane at $D$. Determine the support reactions and forces in each of the members.

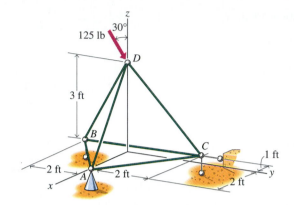

**Fig. 7-43**

## SOLUTION

The solution will be obtained using the method of joints. Since this truss is so simple, the support reactions will be found as part of the solution by joints rather than determined ahead of time using overall equilibrium.

The solution can start at pin $D$ since there is a known force and no more than three unknown forces applied to this pin (Fig. 7-44$d$). The forces are shown on the free-body diagram with an arrow to show the direction and a value for the magnitude. The directions of the forces are along the members; hence, the force in member $CD$ applied at pin $D$ is

$$\mathbf{F}_{DC} = T_{CD} \frac{-\mathbf{i} + 2\mathbf{j} - 3\mathbf{k}}{\sqrt{1^2 + 2^2 + 3^2}}$$

$$= -0.2673 T_{CD}\mathbf{i} + 0.5345 T_{CD}\mathbf{j} - 0.8018 T_{CD}\mathbf{k}$$

Similarly for the other forces,

$$\mathbf{F}_{DB} = T_{BD} \frac{-\mathbf{i} - 2\mathbf{j} - 3\mathbf{k}}{\sqrt{1^2 + 2^2 + 3^2}}$$

$$= -0.2673 T_{BD}\mathbf{i} - 0.5345 T_{BD}\mathbf{j} - 0.8018 T_{BD}\mathbf{k}$$

$$\mathbf{F}_{DA} = T_{AD} \frac{-2\mathbf{i} - 3\mathbf{k}}{\sqrt{2^2 + 3^2}} = 0.5547 T_{AD}\mathbf{i} - 0.8321 T_{AD}\mathbf{k}$$

Finally the applied load is

$$\mathbf{P} = 125.0 \sin 30°\mathbf{j} - 125.0 \cos 30°\mathbf{k} \text{ lb}$$
$$= 62.50\mathbf{j} - 108.25\mathbf{k} \text{ lb}$$

The $x$-, $y$-, and $z$-components of the equilibrium equation $\Sigma\mathbf{F} = \mathbf{0}$ for this joint then become

$$\mathbf{i:} \qquad -0.2673 T_{CD} - 0.2673 T_{BD} + 0.5547 T_{AD} = 0$$
$$\mathbf{j:} \qquad 0.5345 T_{CD} - 0.5345 T_{BD} + 62.50 = 0$$
$$\mathbf{k:} \quad -0.8018 T_{CD} - 0.8018 T_{BD} - 0.8321 T_{AD} - 108.25 = 0$$

> Joints must be selected in such an order that there are no more than three unknown forces at the joint.

respectively. These equations are easily solved to get

$$T_{AD} = -43.37 \text{ lb} \qquad T_{BD} = 13.466 \text{ lb} \qquad T_{CD} = -103.47 \text{ lb}$$

Next draw the free-body diagram of pin $B$ (Fig. 7-44$b$). The force in member $BD$ applied at pin $B$, $\mathbf{F}_{BD}$, is equal in magnitude but opposite in direction to the force in the same member applied at pin $D$, $\mathbf{F}_{DB}$; hence,

$$\begin{aligned}
\mathbf{F}_{BD} &= 0.2673T_{BD}\mathbf{i} + 0.5345T_{BD}\mathbf{j} + 0.8018T_{BD}\mathbf{k} \\
&= 3.599\mathbf{i} + 7.198\mathbf{j} + 10.797\mathbf{k} \text{ lb}
\end{aligned}$$

(a)

The other forces are again along the members; hence,

$$\mathbf{F}_{BC} = T_{BC}\mathbf{j}$$

$$\mathbf{F}_{BA} = T_{AB}\frac{3\mathbf{i} + 2\mathbf{j}}{\sqrt{3^2 + 2^2}} = 0.8321T_{AB}\mathbf{i} + 0.5547T_{AB}\mathbf{j}$$

$$\mathbf{B} = B\mathbf{k}$$

Thus, the equilibrium equations for this joint are

$$\begin{aligned}
\mathbf{i}: &\qquad 3{,}599 + 0.8321T_{AB} = 0 \\
\mathbf{j}: &\qquad 7.198 + T_{BC} + 0.5547T_{AB} = 0 \\
\mathbf{k}: &\qquad 10.797 + B = 0
\end{aligned}$$

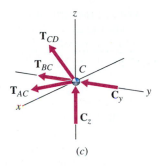

(b)

which are solved to get

$$T_{BC} = -4.799 \text{ lb} \qquad T_{AB} = -4.325 \text{ lb} \qquad B = -10.797 \text{ lb}$$

Next draw the free-body diagram of pin $C$ (Fig. 7-44$c$). The forces applied at pin $C$ are

$$\begin{aligned}
\mathbf{F}_{CD} &= 0.2673T_{CD}\mathbf{i} - 0.5345T_{CD}\mathbf{j} + 0.8018T_{CD}\mathbf{k} \\
&= -27.66\mathbf{i} + 55.30\mathbf{j} - 82.96\mathbf{k} \text{ lb} \\
\mathbf{F}_{CB} &= -T_{BC}\mathbf{j} = 4.799\mathbf{j} \text{ lb} \\
\mathbf{F}_{CA} &= T_{AC}\frac{3\mathbf{i} - 2\mathbf{j}}{\sqrt{3^2 + 2^2}} = 0.8321T_{AC}\mathbf{i} - 0.5547T_{AC}\mathbf{j} \\
&= 0.8321T_{AC}\mathbf{i} - 0.5547T_{AC}\mathbf{j} \\
\mathbf{C} &= -C_y\mathbf{j} + C_z\mathbf{k}
\end{aligned}$$

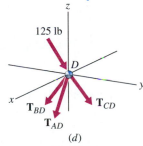

(c)

Putting these forces in the equilibrium equation produces

$$\begin{aligned}
\mathbf{i}: &\qquad -27.66 + 0.8321T_{AC} = 0 \\
\mathbf{j}: &\qquad 55.30 + 4.799 - 0.5547T_{AC} - C_y = 0 \\
\mathbf{k}: &\qquad -82.96 + C_z = 0
\end{aligned}$$

which gives

$$C_y = 41.66 \text{ lb} \qquad C_z = 82.96 \text{ lb} \qquad T_{AC} = 33.24 \text{ lb}$$

Finally, draw the free-body diagram of pin $A$ (Fig. 7-44$a$). The forces acting at this joint are

$$\begin{aligned}
\mathbf{F}_{AB} &= -0.8321T_{AB}\mathbf{i} - 0.5547T_{AB}\mathbf{j} \\
&= 3.599\mathbf{i} + 2.399\mathbf{j} \text{ lb} \\
\mathbf{F}_{AD} &= -0.5547T_{AD}\mathbf{i} + 0.8321T_{AD}\mathbf{k} \\
&= 24.06\mathbf{i} - 36.09\mathbf{k} \text{ lb} \\
\mathbf{F}_{AC} &= -0.8321T_{AC}\mathbf{i} + 0.5547T_{AC}\mathbf{j} \\
&= -27.66\mathbf{i} + 18.44\mathbf{j} \text{ lb} \\
\mathbf{A} &= -A_x\mathbf{i} + A_y\mathbf{j} + A_z\mathbf{k}
\end{aligned}$$

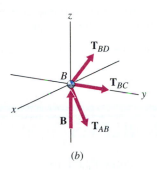

(d)

**Fig. 7-44**

Then, the equilibrium equation becomes

$$\begin{aligned}
\mathbf{i}: &\qquad 3.599 + 24.06 - 27.66 - A_x = 0 \\
\mathbf{j}: &\qquad 2.399 + 18.44 + A_y = 0 \\
\mathbf{k}: &\qquad -36.09 + A_z = 0
\end{aligned}$$

which gives

$$A_x = 0.00 \text{ lb} \qquad A_y = -20.84 \text{ lb} \qquad A_z = 36.09 \text{ lb}$$

The desired answers are then

$$\mathbf{A} = -20.84\mathbf{j} + 36.1\mathbf{k} \text{ lb} \qquad \text{Ans.}$$
$$\mathbf{B} = -10.80\mathbf{k} \text{ lb} \qquad \text{Ans.}$$
$$\mathbf{C} = -41.7\mathbf{j} + 83.0\mathbf{k} \text{ lb} \qquad \text{Ans.}$$

| | | |
|---|---|---|
| $T_{AB} = 4.32$ lb (C) | $T_{AC} = 33.2$ lb (T) | Ans. |
| $T_{AD} = 43.4$ lb (C) | $T_{BC} = 4.80$ lb (C) | Ans. |
| $T_{BD} = 13.47$ lb (T) | $T_{CD} = 103.5$ lb (C) | Ans. |

(The consistency of these answers can be checked by considering the overall equilibrium of the entire truss.)

# PROBLEMS

### Introductory Problems

**7-73*** The awning structure of Fig. P7-73 is supported by ball-and-socket joints at $A$ and $C$ and by a short link at $B$. The 50-lb force is parallel to the $y$-axis, and the 150-lb force is parallel to the $z$-axis. Determine the support reactions and the force in each member of this space truss.

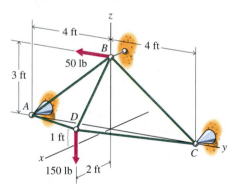

**Fig. P7-73**

**7-74*** The awning structure of Fig. P7-74 is supported by ball-and-socket joints at $A$ and $C$ and by a short link at $B$. The 750-N force is parallel to the $z$-axis. Determine the support reactions and the force in each member of this space truss.

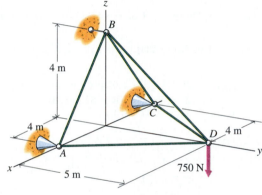

**Fig. P7-74**

**7-75** A wire is stretched tight between two pylons, one of which is shown in Fig. P7-75. The 250-lb force is parallel to

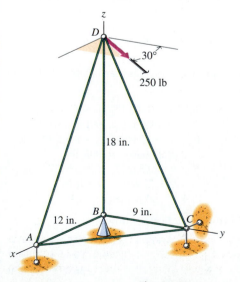

**Fig. P7-75**

the xy-plane and makes an angle of 30° with the y-axis. The supports are equivalent to a ball-and-socket joint at B and short links at A and C. Determine the support reactions and the force in each member of this space truss.

**7-76** A wire is stretched tight between two pylons, one of which is shown in Fig. P7-76. The 1.50-kN force is parallel to the xy-plane and makes an angle of 20° with the y-axis. The supports are equivalent to a ball-and-socket joint at B and short links at A and C. Determine the support reactions and the force in each member of this space truss.

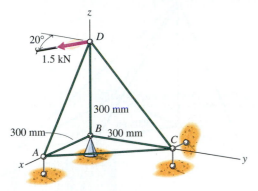

**Fig. P7-76**

## Intermediate Problems

**7-77*** The space truss shown in Fig. P7-77 is supported by a ball-and-socket joint at D and by short links at A and C. Determine the support reactions and the force in each member of this space truss.

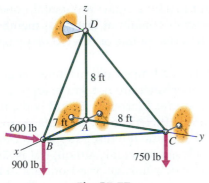

**Fig. P7-77**

**7-78*** The space truss shown in Fig. P7-78 is supported by a ball-and-socket joint at B and by short links at A and D. Determine the support reactions and the force in each member of this space truss.

**Fig. P7-78**

**7-79** The space truss shown in Fig. P7-79 is supported by a ball-and-socket joint at A and by short links at B, C, and D. Determine the support reactions and the forces in members BF, CE, CF, and DF of the truss.

**Fig. P7-79**

**7-80** The space truss shown in Fig. P7-80 is supported by a ball-and-socket joint at A and by short links at B and C. Determine the support reactions and the forces in members AB, BC, BD, and BE of the truss.

**Fig. P7-80**

## Challenging Problems

**7-81\*** A 120-lb sign hangs from joints $D$ and $E$ of the space truss of Fig. P7-81. The weight is supported equally by the two joints. In addition, a wind blowing straight at the sign exerts a force of 15 lb in the direction of the negative $y$-axis—also distributed evenly between the two joints. The truss supports are equivalent to a ball-and-socket joint at $B$ and short links at $A$, $C$, and $F$. Determine the support reactions and the force in each member of the truss.

**7-82** A sign which weighs 650 N hangs from joints $D$ and $E$ of the space truss of Fig. P7-82. The weight is supported equally by the two joints. In addition, a wind blowing straight at the sign exerts a force of 200 N in the direction of the negative $x$-axis—also distributed evenly between the two joints. The truss supports are equivalent to a ball-and-socket joint at $B$ and short links at $A$, $C$, and $F$. Determine the support reactions and the force in each member of the truss.

**Fig. P7-81**

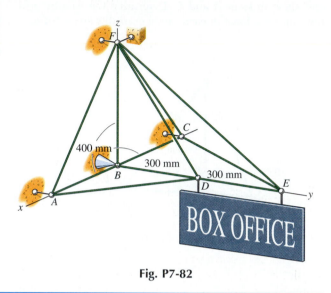

**Fig. P7-82**

## 7-4 FRAMES AND MACHINES

The first part of this chapter has dealt with very special types of structures, trusses, which are composed only of two-force members. Structures that contain other types of members are called frames or machines. Although frames and machines may also contain one or more two-force members, they always contain at least one member that is acted on by forces at more than two points or is acted on by both forces and moments.

The main distinction between frames and machines is that frames are rigid structures whereas machines are not. For example, the structure shown in Fig. 7-45a is a frame. Since it is a rigid body, three support reactions (Fig. 7-45b) are sufficient to fix it in place, and overall equilibrium is sufficient to determine the three support reactions.

The structure of Fig. 7-45c is a machine, although it is sometimes referred to as a nonrigid frame or a linkage. It is nonrigid in the sense that it depends on its supports to maintain its shape. The lack of internal rigidity is compensated for by an extra support reaction (Fig. 7-45d). In this case, overall equilibrium is not sufficient to determine all four support reactions. The structure must be taken apart and analyzed even if the only information desired is the support reactions.

In a more specific sense, the term *machine* is usually used to describe devices such as pliers, clamps, nutcrackers, and the like that are used to magnify the effect of forces. In each case, a force (input) is ap-

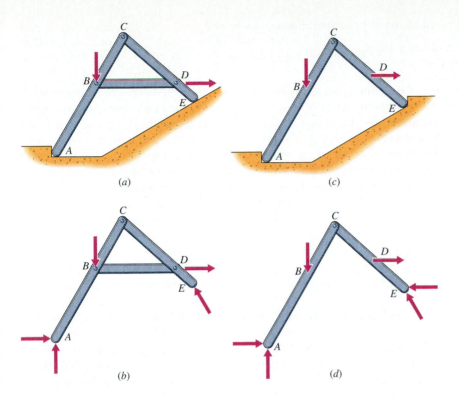

(a)

(c)

(b)

(d)

**Figure 7-45** Examples of rigid frames and nonrigid frames (machines).

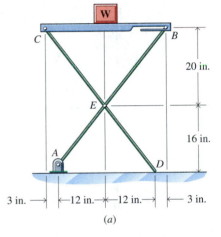

(a)

plied to the handle of the device and a much larger force (output) is applied by the device somewhere else. Like nonrigid frames, these machines must be taken apart and analyzed even if the only information desired is the relationship between the input and output forces.

As with the analysis of trusses, the method of solution for frames and machines consists of taking the structures apart, drawing free-body diagrams of each of the components, and writing the equations of equilibrium for each of the free-body diagrams. In the case of trusses, the direction of the force in all members was known and the method of joints reduced to solving a series of particle equilibrium problems. Since some of the members of frames and machines are not two-force members, however, the directions of the forces in these members is not known. The analysis of frames and machines will consist of solving for the equilibrium of a system of rigid bodies rather than a system of particles.

### 7-4.1  Frames

The method of analysis for frames can be demonstrated using the table of Fig. 7-2, which is reproduced here as Fig. 7-46a. None of the members that make up the table are two-force members, so the structure is definitely not a truss. Although the table can be folded up by unhooking the top from the leg, in normal use the table is a stable, rigid structure. Therefore, the table is a frame.

The analysis will be started by first drawing the free-body diagram of the entire table (Fig. 7-46b), for which the equations of equilibrium

$$+\rightarrow \Sigma F_x = A_x = 0$$
$$+\uparrow \Sigma F_y = A_y + D_y - W = 0$$
$$\downarrow + \Sigma M_A = 24D_y - 12W = 0$$

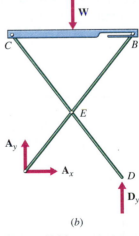

(b)

**Figure 7-46**  Table-type structure and free-body diagram for the complete structure.

**377**

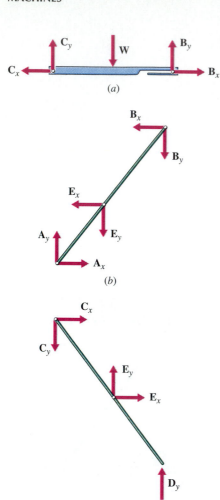

**Figure 7-47**   Free-body diagrams for
the individual members of the structure of Fig. 7-46.

yield the support reactions

$$A_x = 0 \qquad A_y = \frac{W}{2} \qquad D_y = \frac{W}{2}$$

Next, the table is taken apart and the free-body diagrams of each of its parts drawn (Fig. 7-47). Since none of the members is a two-force member, the directions of the forces at the joints $B$, $C$, and $E$ are not known—*they are not directed along the members!* Although the forces may be represented in terms of any convenient components, the free-body diagrams must take into account Newton's third law of action and re-action. That is, when drawing free-body diagrams, the forces exerted by one member on a second must be equal in magnitude and opposite in direction to the forces exerted by the second member on the first. For Fig. 7-47 this is effected by showing the components of the force exerted by member $AB$ on member $CD$ at joint $E$ to have equal mag-nitude and opposite direction to the components of the force exerted by member $CD$ on member $AE$ at joint $E$ and similarly for the other joints.

The floor at $D$ can exert only an upward force on the leg, and the force should be shown as such on the free-body diagrams. Similarly, the horizontal component of force exerted by the slot on the leg $AB$ can act only to the left and should be shown as such. If the values of these forces come out to be negative, either the solution is in error or the table is not in equilibrium. This suggests the general rule; consider the implications of a solution. Does the solution make sense? Does the so-lution fit the initial assumptions used to reach a solution?

The proper direction of the other force components are not all as clear. Though it is easy to guess that the vertical force components $\mathbf{B}_y$ and $\mathbf{C}_y$ act upward on the table top $BC$ and downward on the legs, it is not as easy to decide whether to draw $\mathbf{E}_y$ acting upward or down-ward on leg $AB$. At this point it does not matter, since the frictionless pin connections can support a force in either direction. As in the case of trusses, the direction that a force is shown on one member is unim-portant as long as the force is represented consistent with Newton's third law on each part of the structure. If a force component is assumed in the wrong direction, its symbol will simply end up having a nega-tive value. This can be accounted for in the report of final answers as shown in the example problems.

Although not all the members of a frame can be two-force mem-bers, it is possible and quite likely that one or more of the members will be two-force members. Take advantage of any such members and show that force as acting in its known direction. But, *be sure that all forces are not directed along the members.* Perhaps one of the most com-mon mistakes that students make is to treat frames like trusses; that is, to draw all forces as acting along the members and trying to apply the method of joints.

Unlike the analysis of trusses, the free-body diagrams of the pins of a frame are usually not drawn and analyzed separately. In the case of the truss, equilibrium of the members was assured by the two-force member assumption, the direction of the force exerted by each mem-ber on a pin was known, and equilibrium of the pins contained all the useful information of the problem. None of these statements apply to the analysis of frames, however, and it is seldom useful to analyze the equilibrium of the pins separately.

In most cases, it does not matter to which member a pin is attached when the structure is taken apart. There are, however, a few special situations in which it does matter:

> When a pin connects a support and two or more members, the pin must be assigned to one of the members. The support reactions are applied to the pin on this member.

> When a pin connects two or more members and a load is applied to the pin, the pin must be assigned to one of the members. The load is applied to the pin on this member.

Following these simple rules will avoid confusion as to where the loads and support reactions should be applied.

Special care is also warranted when one or more of the members that meet at a joint is a two-force member:

> Pins should never be assigned to two-force members.

> When all the members meeting at a joint are two-force members, the pin should be removed and analyzed separately as in the method of joints for a truss. That is, assign the pin to one of the multiforce members at the joint, or use the method of joints if all the members connected at a joint are two-force members.

While these last two "rules" are not strictly necessary, following them will prevent confusion when dealing with two-force members in frames. See Example Problem 7-21 for additional discussion and a numerical illustration.

Finally, the equations of equilibrium are written for each part of the frame and are solved for the joint forces. There are three independent equations of equilibrium (two force and one moment) for each part, and hence for the table parts of Fig. 7-47 there will be nine equations to solve for the six remaining unknown forces ($B_x$, $B_y$, $C_x$, $C_y$, $E_x$, $E_y$). Prior solution of the overall equilibrium of the frame for the support reactions will have reduced three of these equations to a check of the consistency of the answers.

## 7-4.2 Machines and Mechanisms

The method described for frames is also used to analyze machines and other nonrigid structures. In each case, the structure is taken apart, free-body diagrams are drawn for each part, and the equations of equilibrium are applied to each free-body diagram. For machines and non-rigid structures, however, the structure must be taken apart and analyzed even if the only information desired is the support reactions or the relationship between the external forces acting on it.

The method of analysis for machines can be demonstrated using the simple garlic press shown in Fig. 7-48a. Forces $H_1$ and $H_2$ applied to the handles (the *input forces*) are converted into forces $G_1$ and $G_2$ applied to the garlic clove (the *output forces*). Equilibrium of the entire press only gives that $H_1 = H_2$; it gives no information about the relationship between the input forces and the output forces.

In order to determine the relationship between the input forces and the output forces, the machine must be taken apart and free-body diagrams drawn for each of its parts as shown in Fig. 7-48b. Then the

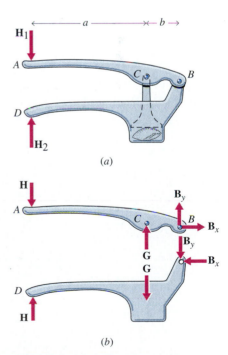

**Figure 7-48** Garlic press and free-body diagrams for the individual members of the press.

sum of moments about $B$ gives

$$(a + b)H = bG$$

or

$$G = \frac{a + b}{b}H$$

The ratio of the output and input forces is called the *mechanical advantage* (M.A.) of the machine

$$\text{mechanical advantage} = \frac{\text{output force}}{\text{input force}}$$

For the garlic press, the mechanical advantage is just

$$\text{M.A.} = \frac{a + b}{b}$$

A machine is a collection of members that produces a mechanical advantage. A mechanism is a collection of members that transfers and/or modifies motion, which may not produce a mechanical advantage.

The cord of Fig. 7-49a is wrapped around a frictionless pulley and supports a body with a mass $m$ of 100 kg. Determine the reaction components at supports $A$ and $E$ and the force exerted on bar $ABC$ by the pin at $B$.

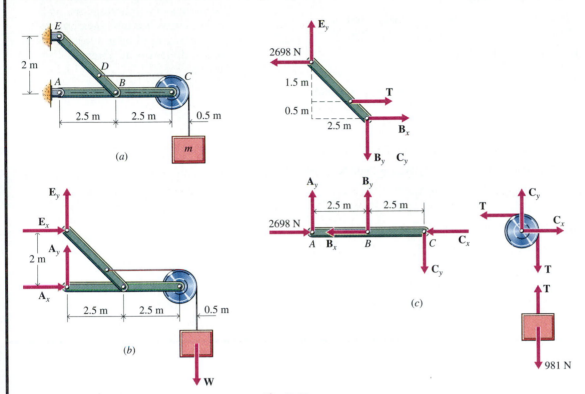

Fig. 7-49

## SOLUTION

An examination of the free-body diagram for the complete system shown in Fig. 7-49b shows four reaction components; therefore, all of the components cannot be determined by applying the three equilibrium equations available for this coplanar system of forces. The force $W$ shown on the diagram is

$$W = mg = 100(9.81) = 981 \text{ N}$$

The reaction component $E_x$ can be determined by summing moments about point $A$. Thus,

$$+\circlearrowleft\Sigma M_A = -E_x(2) - W(5.5)$$
$$= -E_x(2) - 981(5.5) = 0 \qquad E_x = -2698 \text{ N} \cong 2700 \text{ N} \leftarrow$$

The reaction component $A_x$ can then be determined by summing forces in the $x$-direction. Thus,

$$+\rightarrow\Sigma F_x = A_x + E_x = A_x + (-2698) = 0 \qquad A_x = 2698 \text{ N} \cong 2700 \text{ N} \rightarrow$$

Finally, summing forces in the $y$-direction yields

$$+\uparrow\Sigma F_y = A_y + E_y - 981 = 0 \qquad A_y + E_y = 981 \text{ N}$$

In order to determine forces $A_y$ and $E_y$, the system must be taken apart and analyzed. A complete set of free-body diagrams for all members of the system are shown in Fig. 7-49c. From the free-body diagram for the body with mass $m = 100$ kg or weight $W = 981$ N,

$$+\uparrow\Sigma F_y = T - 981 = 0 \qquad T = 981 \text{ N}$$

From a free-body diagram of the pulley,

$$+\rightarrow\Sigma F_x = C_x - T = C_x - 981 = 0 \qquad C_x = 981 \text{ N}$$
$$+\uparrow\Sigma F_y = C_y - T = C_y - 981 = 0 \qquad C_y = 981 \text{ N}$$

From the free-body diagram for bar $ABC$,

$$+\downarrow\Sigma M_A = B_y(2.5) - C_y(5.0)$$
$$= B_y(2.5) - 981(5.0) = 0 \qquad B_y = 1962 \text{ N} = 1962 \text{ N}\uparrow$$
$$+\downarrow\Sigma M_B = -A_y(2.5) - C_y(2.5)$$
$$= -A_y(2.5) - 981(2.5) = 0 \qquad A_y = -981 \text{ N} = 981 \text{ N}\downarrow$$
$$+\rightarrow\Sigma F_x = -B_x - C_x + 2698$$
$$= -B_x - 981 + 2698 = 0 \qquad B_x = 1717 \text{ N} = 1717 \text{ N}\leftarrow$$
$$B = \sqrt{(B_x)^2 + (B_y)^2} = \sqrt{(1717)^2 + (1962)^2} = 2607 \text{ N} \cong 2610 \text{ N}$$
$$\theta_x = \tan^{-1}\frac{B_y}{B_x} = \tan^{-1}\frac{1962}{1717} = \diagdown\ 48.81°$$

Finally, from the free-body diagram for bar $BDE$,

$$+\uparrow\Sigma F_y = E_y - B_y = E_y - 1962 = 0 \qquad E_y = 1962 \text{ N} = 1962 \text{ N}\uparrow$$

Therefore, the desired answers are

$$A_x = 2700 \text{ N}\rightarrow \qquad A_y = 981 \text{ N}\downarrow \qquad\qquad \text{Ans.}$$
$$E_x = 2700 \text{ N}\leftarrow \qquad E_y = 1962 \text{ N}\uparrow \qquad\qquad \text{Ans.}$$
$$\mathbf{B} = 2610 \text{ N} \diagdown\ 48.8° \qquad\qquad\qquad \text{Ans.}$$

Newton's third law (action and reaction) must be carefully observed when placing pin forces on mating members.

Only three independent scalar equations of equilibrium are available to solve for unknown quantities in any rigid body subjected to a two-dimensional system of forces.

Three bars are connected with smooth pins to form the A-frame shown in Fig. 7-50$a$. The weights of the bars are negligible. Determine the forces exerted by the pins at $B$ and $D$ on member $BD$ and the force exerted by the pin at $C$ on member $CDE$.

$\tan 58 = \dfrac{2}{x}$

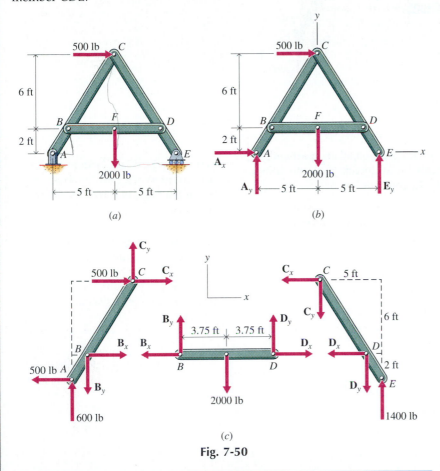

**Fig. 7-50**

---

## SOLUTION

An examination of the free-body diagram for the complete system shown in Fig. 7-50$b$ shows three reaction components; therefore, all of the components can be determined by applying the three equilibrium equations available for this coplanar system of forces. The reaction component $E_y$ is determined by summing moments about point $A$. Thus,

$$+\circlearrowleft \Sigma M_A = E_y(10) - 2000(5) - 500(8) = 0$$
$$E_y = 1400 \text{ lb} = 1400 \text{ lb}\uparrow$$

The reaction component $A_x$ is determined by summing forces in the $x$-direction. Thus,

$$+\rightarrow \Sigma F_x = A_x + 500 = 0 \qquad A_x = -500 \text{ lb} = 500 \text{ lb}\leftarrow$$

Finally, the reaction component $A_y$ is determined by summing moments about point $E$. Thus,

$$+\circlearrowleft \Sigma M_E = -A_y(10) - 2000(5) + 500(8) = 0$$
$$A_y = 600 \text{ lb} = 600 \text{ lb}\uparrow$$

(c)

**Fig. 7-50**

In order to determine forces **B** and **D,** the system must be taken apart and analyzed. A complete set of free-body diagrams for all members of the system are shown in Fig. 7-50c. From the free-body diagram for bar $BD$,

$$+\circlearrowleft\Sigma M_D = -B_y(7.5) + 2000(3.75) = 0 \qquad B_y = 1000 \text{ lb} = 1000 \text{ lb}\uparrow$$
$$+\circlearrowleft\Sigma M_B = D_y(7.5) - 2000(3.75) = 0 \qquad D_y = 1000 \text{ lb} = 1000 \text{ lb}\uparrow$$

From the free-body diagram for bar $CDE$,

$$+\circlearrowleft\Sigma M_C = -D_x(6) - D_y(3.75) + 1400(5)$$
$$= -D_x(6) - 1000(3.75) + 1400(5)$$
$$D_x = 541.7 \text{ lb} = 541.7 \text{ lb}\leftarrow$$
$$+\rightarrow\Sigma F_x = -C_x - D_x = -C_x - 541.7 = 0$$
$$C_x = -541.7 \text{ lb} = 541.7 \text{ lb}\rightarrow$$
$$+\uparrow\Sigma F_y = -C_y - D_y + 1400$$
$$= -C_y - 1000 + 1400 = 0$$
$$C_y = 400 \text{ lb} = 400 \text{ lb}\downarrow$$
$$C = \sqrt{(C_x)^2 + (C_y)^2} = \sqrt{(541.7)^2 + (400)^2} = 673.4 \text{ lb}$$
$$\theta_x = \tan^{-1}\frac{C_y}{C_x} = \tan^{-1}\frac{400}{541.7} = 36.44°$$

From the free-body diagram for bar $BD$,

$$+\rightarrow\Sigma F_x = D_x - B_x = 541.7 - B_x = 0$$
$$B_x = 541.7 \text{ lb} = 541.7 \text{ lb}\leftarrow$$
$$B = D = \sqrt{(D_x)^2 + (D_y)^2} = \sqrt{(541.7)^2 + (1000)^2} = 1137.3 \text{ lb}$$
$$\theta_x = \tan^{-1}\frac{D_y}{D_x} = \tan^{-1}\frac{1000}{541.7} = 61.56°$$

Therefore, the desired answers are

$$\mathbf{B} = 1137 \text{ lb} \searrow 61.6° \qquad\qquad \text{Ans.}$$
$$\mathbf{D} = 1137 \text{ lb} \nearrow 61.6° \qquad\qquad \text{Ans.}$$
$$\mathbf{C} = 673 \text{ lb} \nwarrow 36.4° \qquad\qquad \text{Ans.}$$

Newton's third law (action and reaction) must be carefully observed when placing pin forces on mating members.

Only three independent scalar equations of equilibrium are available to solve for unknown quantities in any rigid body subjected to a two-dimensional system of forces.

A bag of potatoes is resting on the chair of Fig. 7-51a. The force exerted by the potatoes on the frame at one side of the chair is equivalent to horizontal and vertical forces of 24 N and 84 N, respectively, at $E$ and a force of 28 N perpendicular to member $BH$ at $G$ (as shown in the free-body diagram of Fig. 7-51b). Find the forces acting on member $BH$.

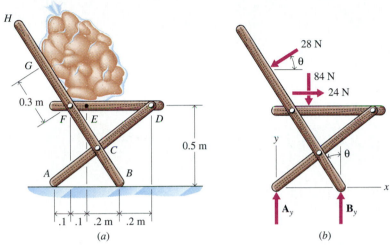

(a)     (b)

**Fig. 7-51**

## SOLUTION

An examination of the free-body diagram for the chair (see Fig. 7-51b) shows two vertical reaction components $\mathbf{A}_y$ and $\mathbf{B}_y$. Horizontal force components $\mathbf{A}_x$ and $\mathbf{B}_x$ do not exist since no horizontal forces are applied to the system consisting of the chair and the bag of potatoes (see Fig. 7-51a).

Before proceeding with an analysis of the chair, some dimensions needed in the analysis will be determined. From the sketch of the chair shown in Fig. 7-51c,

$$\theta = \tan^{-1} \frac{3}{5} = 30.96°$$

$$\phi = \tan^{-1} \frac{6}{5} = 50.19°$$

$$c = 0.3 + \frac{0.5}{\cos \theta} = 0.8831 \text{ m}$$

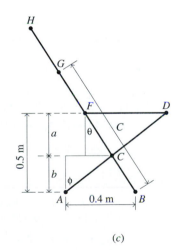

(c)

From Fig. 7-51c it can be seen that

$$a + b = 0.5$$
$$b \tan \phi = 0.1 + a \tan \theta$$
$$(0.5 - a)\left(\frac{6}{5}\right) = 0.1 + a\left(\frac{3}{5}\right)$$

From which

$$a = 0.2778 \text{ m} \qquad b = 0.2222 \text{ m}$$

The reaction $\mathbf{A}_y$ can be determined by summing moments about point $B$ of Fig. 7-51b. Thus,

$$+\!\!\bigcirc\!\!\downarrow \Sigma M_A = -A_y(0.4) - 24(0.5) + 84(0.2) + 28(0.8831) = 0$$
$$A_y = 73.81 \text{ N} \qquad \mathbf{A}_y = 73.81 \text{ N}\!\uparrow$$

**Fig. 7-52**

Summing forces in the $x$-direction,

$$+\rightarrow \Sigma F_x = 24 - 28 \cos \theta = 24 - 28 \cos 30.96° \cong 0$$

Summing forces in the $y$-direction,

$$+\uparrow \Sigma F_y = A_y + B_y - 84 - 28 \sin \theta$$
$$= 73.81 + B_y - 84 - 28 \sin 30.96° = 0$$
$$B_y = 24.59 \text{ N} \qquad \mathbf{B}_y = 24.59 \text{ N}\uparrow$$

Next, the chair is disassembled and free-body diagrams are drawn for each part (Fig. 7-52). An examination of member $BH$ shows that the member has four unknown force components; therefore, one of the components must be determined from equilibrium of a different member. For member $DF$, the equilibrium equation $\Sigma M_D = 0$ can be used to determine the force component $F_y$. Thus,

$$+\downarrow \Sigma M_D = 84(0.4) - F_y(0.5) = 0 \qquad F_y = 67.20 \text{ N}$$

Now the equations of equilibrium for member $BH$ can be used to determine the remaining unknown components. By summing moments about point $C$, two unknowns can be eliminated and force $F_x$ is determined as

$$+\downarrow \Sigma M_C = B_y(e) + F_y(d) - F_x(a) + 28\left(0.3 + \frac{a}{\cos \theta}\right)$$

$$= 24.59(b \tan \theta) + F_y(a \tan \theta) - F_x(a) + 28\left(0.3 + \frac{a}{\cos \theta}\right)$$

$$= 24.59(0.2222 \tan 30.96°) + 67.20(0.2778 \tan 30.96°)$$

$$- F_x(0.2778) + 28\left(0.3 + \frac{0.2778}{\cos 30.96°}\right) = 0 \qquad F_x = 115.00 \text{ N}$$

$$+\rightarrow \Sigma F_x = F_x + C_x - 28 \cos \theta$$
$$= 115.00 + C_x - 28 \cos 30.96° = 0 \qquad C_x = -90.98 \text{ N}$$
$$+\uparrow \Sigma F_y = B_y + C_y - F_y - 28 \sin \theta$$
$$= 24.59 + C_y - 67.20 - 28 \sin 30.96° = 0 \qquad C_y = 57.01 \text{ N}$$

The forces acting on member $BH$ are expressed in Cartesian vector form as

$$\mathbf{B} = 24.6 \mathbf{j} \text{ N} \qquad \text{Ans.}$$
$$\mathbf{C} = -91.0\mathbf{i} + 57.0\mathbf{j} \text{ N} \qquad \text{Ans.}$$
$$\mathbf{F} = 115.1\mathbf{i} + 67.2\mathbf{j} \text{ N} \qquad \text{Ans.}$$

plus the applied force of 28 N perpendicular to the bar at $G$. These forces are shown on the "report diagram" of Fig. 7-53.

Newton's third law (action and reaction) must be carefully observed when placing pin forces on mating members.

Only three independent scalar equations of equilibrium are available to solve for unknown quantities in any rigid body subjected to a two-dimensional system of forces.

**Fig. 7-53**

The weight of books on a shelf bracket is equivalent to a vertical force of 75 lb as shown on Fig. 7-54a. In addition, a vertical load of 50 lb is suspended from the middle of the lower brace $BC$. Find all forces acting on all three members of this frame.

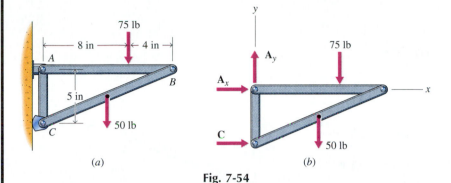

Fig. 7-54

## SOLUTION

First draw the free-body diagram of the entire shelf bracket as in Fig. 7-54b. The equations of overall equilibrium are

$$\mathord{\downarrow}+\Sigma M_A = 5C - (8)(75) - (6)(50) = 0$$
$$+\mathord{\rightarrow}\Sigma F_x = A_x + C = 0$$
$$+\mathord{\uparrow}\Sigma F_y = A_y - 75 - 50 = 0$$

which are solved to get the support reactions

$$A_x = -180 \text{ lb} \qquad A_y = 125 \text{ lb} \qquad C = 180 \text{ lb}$$

Next, dismember the bracket and draw separate free-body diagrams of each member (Fig. 7-55). Member $AC$ is a two-force member, and so pin $A$ will be assigned to member $AB$ and pin $C$ will be assigned to member $BC$. It does not matter to which member pin $B$ is assigned since no two-force members are attached to joint $B$ and joint $B$ is neither loaded nor attached to a support. Just to be definite, pin $B$ will be assigned to member $AB$ as shown in Fig. 7-55a. Then the equations of equilibrium for member $AB$ give

$$\mathord{\downarrow}\Sigma M_B = (4)(75) + 12T_{AC} - (12)(125) = 0 \qquad T_{AC} = 100 \text{ lb}$$
$$\mathord{\downarrow}+\Sigma M_A = 12B_y - (8)(75) = 0 \qquad B_y = 50 \text{ lb}$$
$$+\mathord{\rightarrow}\Sigma F_x = (-180) + B_x = 0 \qquad B_x = 180 \text{ lb}$$

It is easily verified that these values also satisfy the equations of equilibrium for the other free-body diagrams. These forces are all shown on the "report diagram" of Fig. 7-55.

> Newton's third law (action and reaction) must be carefully observed when placing pin forces on mating members.

> Pins should not normally be assigned to two-force members.

> Only three independent scalar equations of equilibrium are available to solve for unknown quantities in any rigid body subjected to a two-dimensional system of forces.

Fig. 7-55

387

**Fig. 7-56**

**Fig. 7-57**

## ALTERNATIVE SOLUTION

Pins $A$ and $C$ can be assigned to the two-force member $AC$ as shown on the free-body diagram of Fig. 7-57 if care is taken with the representation of the internal forces. The tension $T_{AC}$ shown on Fig. 7-55a represents the force exerted on member $AC$ by the pin $C$. Since the pin is now part of the member, the force $T_{AC}$ is an internal force and is not shown on the free-body diagram. The symbols $A_x$ and $A_y$ have already been used to represent the components of the support reaction, so $F_{Ax}$ and $F_{Ay}$ will be used for the components of the forces of action and reaction at pin $A$. Similarly, $F_{Cx}$ and $F_{Cy}$ will be used for the components of the forces of action and reaction at pin $C$ so as not to confuse these forces with the support reaction $C$.

As stated before, since no supports or two-force members are attached to pin $B$ and no loads are applied at pin $B$, it does not matter to which member pin $B$ is assigned. To illustrate this point, this time pin $B$ will be assigned to member $BC$.

Now the equations of equilibrium for member $AB$ give

$$\zeta+\Sigma M_A = 12B_y - (8)(75) = 0 \qquad B_y = 50 \text{ lb}$$
$$\zeta+\Sigma M_B = (4)(75) - 12F_{Ay} = 0 \qquad F_{Ay} = 25 \text{ lb}$$
$$+\rightarrow\Sigma F_x = F_{Ax} + B_x = 0 \qquad B_x = -F_{Ax}$$

Next, equilibrium of member $AC$ gives the equations

$$\zeta+\Sigma M_C = 5F_{Ax} + (5)(180) = 0 \qquad F_{Ax} = -180 \text{ lb}$$
$$\zeta+\Sigma M_A = (5)(180) - 5F_{Cx} = 0 \qquad F_{Cx} = 180 \text{ lb} = 125 - F_{Ay} - F_{Cy}$$
$$+\uparrow\Sigma F_y = 125 - 25 - F_{Cy} = 0 \qquad F_{Cy} = 100 \text{ lb}$$

Finally, returning to the free-body diagram of $AB$, the horizontal component of force equilibrium

$$+\rightarrow\Sigma F_x = F_{Ax} + B_x = (-180) + B_x = 0$$

> When pins are assigned to two-force members, the forces acting on the pins should be represented as forces of action and reaction between members and not as a pair of equal, opposite, collinear forces commonly used for two-force members.

**388**

**Fig. 7-58**

gives

$$B_x = 180 \text{ lb}$$

These forces are shown on the "report diagram" of Fig. 7-58. Although the forces at $B$ are clearly the same on Fig. 7-56 and Fig. 7-58, the forces at $A$ and $C$ appear to be different on the two diagrams. However, the resultant forces on the two diagrams are actually the same on both diagrams; they are only expressed using different components.

# PROBLEMS

### Introductory Problems

**7-83\*** Determine all forces acting on member $BCD$ of the linkage shown in Fig. P7-83.

**Fig. P7-83**

**7-84\*** Determine all forces acting on member $ABC$ of the linkage shown in Fig. P7-84.

**Fig. P7-84**

**7-85** A two-bar frame is loaded and supported as shown in Fig. F7-85. Determine the reactions at supports $A$ and $E$ and the force exerted on member $ABC$ by the pin at $C$.

**Fig. P7-85**

**7-86** Determine all forces acting on member *ABCD* of the frame shown in Fig. P7-86.

**Fig. P7-86**

**7-87\*** A pin-connected system of levers and bars is used as a toggle for a press as shown in Fig. P7-87. Determine the force **F** exerted on the can at *A* when a force **P** = 100 lb is applied to the lever at *G*.

**Fig. P7-87**

**7-88\*** A block with a mass of 150 kg is supported by a cable which passes over a 150-mm diameter pulley that is attached to a frame as shown in Fig. P7-88. Determine all forces acting on member *BCD* of the frame.

**Fig. P7-88**

**7-89** A pin-connected system of bars supports a 300-lb load as shown in Fig. P7-89. Determine the reactions at supports *A* and *B* and the force exerted by the pin at *C* on member *ACE*.

**Fig. P7-89**

**7-90** Determine all forces acting on member *ABE* of the frame shown in Fig. P7-90.

**Fig. P7-90**

**7-91\*** Determine all forces acting on member *ABCD* of the frame shown in Fig. P7-91.

**Fig. P7-91**

390

**7-92\*** The jaws and bolts of the wood clamp in Fig. P7-92 are parallel. The bolts pass through swivel mounts so that no moments act on them. The clamp exerts forces of 300 N on each side of the board. Treat the forces on the boards as uniformly distributed over the contact areas and determine the forces in each of the bolts. Show on a sketch all forces acting on the upper jaw of the clamp.

**Fig. P7-92**

**7-93** The spring clamp of Fig. P7-93 is used to hold the block $F$ against the floor. The force in the spring is $F = k(\ell - \ell_0)$, where $\ell$ is the present length of the spring, $\ell_0 = 3$ in. is the unstretched length of the spring, and $k = 240$ lb/ft is the spring constant. Determine all forces acting on member $ABC$ of the spring clamp and the force exerted by the spring clamp on the block $E$.

**Fig. P7-93**

**7-94** The spring clamp of Fig. P7-94 is used to hold block $E$ into the corner. The force in the spring is $F = k(\ell - \ell_0)$, where $\ell$ is the present length of the spring, $\ell_0 = 15$ mm is the unstretched length of the spring, and $k = 5000$ N/m is the spring constant. Determine all forces acting on member $ABC$ of the spring clamp and the force exerted by the spring clamp on the block $E$.

**Fig. P7-94**

**7-95\*** Forces of 25 lb are applied to the handles of the pipe pliers shown in Fig. P7-95. Determine the force exerted on the pipe at $D$ and the force exerted on handle $DAB$ by the pin at $A$.

**Fig. P7-95**

**7-96\*** Forces of 5 N are applied to the handles of the paper punch shown in Fig. P7-96. Determine the force exerted on the paper at $D$ and the force exerted on the pin at $B$ by handle $ABC$.

**Fig. P7-96**

**7-97** Determine the reactions at supports A and E and the force exerted by the pin at B on member ABC of the frame shown in Fig. P7-97.

**Fig. P7-97**

**7-98** Determine all forces acting on member ACE of the frame shown in Fig. P7-98. The diameter of the pulley at E is 120 mm. The mass of body W is 100 kg.

**Fig. P7-98**

## Intermediate Problems

**7-99\*** Determine all forces acting on member ACD of the frame shown in Fig. P7-99. The diameter of the pulley at D is 8 in. The weight of body W is 500 lb.

**Fig. P7-99**

**7-100\*** Determine all forces acting on member BCDE of the frame shown in Fig. P7-100. The diameters of the pulleys at C and E are 160 mm. The mass of body W is 200 kg.

**Fig. P7-100**

**7-101** Forces of 50 lb are applied to the handles of the bolt cutter of Fig. P7-101. Determine the force exerted on the bolt at E and all forces acting on the handle ABC.

**Fig. P7-101**

**7-102** A pair of vise grip pliers is shown in Fig. P7-102. Determine the force **F** exerted on the block by the jaws of the pliers when a force **P** = 100 N is applied to the handles.

**Fig. P7-102**

**7-103\*** The fold-down chair of Fig. P7-103 weighs 25 lb and has its center of gravity at G. Determine all forces acting on member ABC.

12 in.

9 in.

1 in.

B

A

D    G  C  E

9 in.

3 in. 3 in.

**Fig. P7-103**

**7-105** Determine all forces acting on member *DEF* of the frame shown in Fig. P7-105.

140 lb/ft

A    B    C

2 ft

75°    60°

F    E    D

2 ft    3 ft

**Fig. P7-105**

**7-106** Determine all forces acting on member *ABC* of the frame shown in Fig. P7-106.

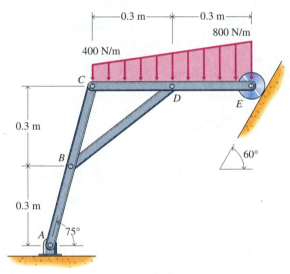

0.3 m    0.3 m

800 N/m

400 N/m

C

D    E

0.3 m

B    60°

0.3 m

A    75°

**Fig. P7-106**

**7-104\*** The tower crane of Fig. P7-104 is rigidly attached to the building at *F*. A cable is attached at *D* and passes over small frictionless pulleys at *A* and *E*. The object suspended from *C* weighs 1500 N. Determine all forces acting on member *ABCD*.

8 m    10 m    4.4 m

E

6 m

36.9

A    B    C    D

12 m

F

**Fig. P7-104**

**7-107\*** The hoist pulley structure of Fig. P7-107 is rigidly attached to the wall at *C*. A load of sand hangs from the cable that passes around the 1-ft-diameter frictionless pulley at *D*. The weight of the sand can be treated as a triangular distributed load with a maximum of 70 lb/ft. Determine all forces acting on member *ABC*.

2 ft    2 ft    3 ft

E

3 ft

D

A    B    C

70 lb/ft

2 ft

**Fig. P7-107**

**7-108** The sand on the tray of Fig. P7-108 can be treated as a triangular distributed load with a maximum of 800 N/m. The wheel at $C$ is frictionless. Determine all forces acting on member $ABC$.

**Fig. P7-108**

## Challenging Problems

**7-109*** Determine the reactions at supports $A$ and $B$ and the force exerted by the pin at $C$ on member $BCD$ of the frame shown in Fig. P7-109.

**Fig. P7-109**

**7-110*** A cylinder with a mass of 150 kg is supported by a two-bar frame as shown in Fig. P7-110. Determine all forces acting on member $ACE$.

**Fig. P7-110**

**7-111** A frame is loaded and supported as shown in Fig. P7-111. Determine all forces acting on member $ABCD$.

**Fig. P7-111**

**7-112** A frame is loaded and supported as shown in Fig. P7-112. Determine all forces acting on member $ADE$.

**Fig. P7-112**

**7-113*** A 200-lb cylinder is supported by a two-bar frame as shown in Fig. P7-113. Determine all forces acting on member $AEC$.

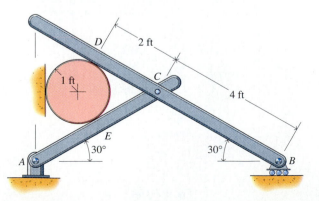

**Fig. P7-113**

**7-114** Figure P7-114 is a simplified sketch of the mechanism used to raise the bucket of a bulldozer. The bucket and its contents weigh 10 kN and have a center of gravity at $H$. Arm $ABCD$ has a weight of 2 kN and a center of gravity at $B$; arm $DEFG$ has a weight of 1 kN and a center of gravity at $E$. The weight of the hydraulic cylinders can be ignored. Calculate the force in the horizontal cylinders $CJ$ and $EI$ and all forces acting on arm $DEFG$ for the position shown.

**Fig. P7-114**

## Computer Problems

**C7-115** The mechanism shown in Fig. P7-115 is designed to keep its load level while raising it. A pin on the rim of the 4-ft-diameter pulley fits in a slot on arm $ABC$. Arms $ABC$ and $DE$ are each 4 ft long, and the package being lifted weighs 80 lb. The mechanism is raised by pulling on the rope, which is wrapped around the pulley.

**a.** Plot $P$, the force required to hold the platform as a function of the platform height $h$ ($0 \leq h \leq 5.5$ ft).
**b.** Plot $A$, $C$, and $E$—the magnitudes of the pin reaction forces at $A$, $C$, and $E$—as a function of $h$ ($0 \leq h \leq 5.5$ ft).

**Fig. P7-115**

**C7-116** Forces of $P = 100$ N are being applied to the handles of the vise grip pliers shown in Fig. P7-116. Plot the force applied on the block by the jaws as a function of the distance $d$ ($20 \leq d \leq 30$ mm).

**Fig. P7-116**

**C7-117** The door to an airplane hangar consists of two uniform sections, which are hinged at the middle as shown in Fig. P7-117. The door is raised by means of a cable, which is attached to a bar along the bottom edge of the door. Smooth rollers at the ends of the bar $C$ run in a smooth vertical channel. If the door is 30 ft wide, 15 ft tall, and weighs 1620 lb,

**a.** Plot $P$, the force required to hold the door open, as a function of the door opening height $h$ ($0.5 \leq h \leq 14.5$ ft).
**b.** Plot $A$ and $B$, the hinge forces, as a function of the height $h$ ($0.5 \leq h \leq 14.5$ ft).
**c.** What is the maximum height $h$ if the force in the hinges is not to exceed 5000 lb?

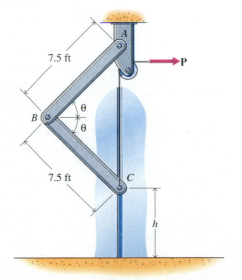

**Fig. P7-117**

**C7-118** A simplified sketch of the mechanism used to raise the bucket of a bulldozer is shown in Fig. P7-118. The bucket and its contents weigh 10 kN and have a center of gravity at $H$. Arm $ABCD$ has a weight of 2 kN and a center of gravity at $B$; arm $DEFG$ has a weight of 1 kN and a center of gravity at $E$. The weight of the hydraulic cylinders can be ignored. Determine the force in cylinders $CJ$ and $EI$ (which remain horizontal as angle $\theta$ is changed) and the support reaction at $A$ as a function of angle $\theta$ ($15° \leq \theta \leq 45°$).

**Fig. P7-118**

# SUMMARY

In previous chapters, equilibrium equations were used to determine rigid-body external support reactions. In this chapter, equilibrium equations are used to determine joint forces in structures comprised of pin-connected members. Like all forces (internal, external, applied, or reactive), the joint forces always occur in equal-magnitude, opposite-direction pairs. If not separated by means of a free-body diagram from the rest of the structure or environment, the pair of forces need not be considered when writing the equations of equilibrium. Therefore, in order to determine the joint forces, the structure must be divided into two or more parts.

Two broad categories of engineering structures were considered in this chapter: namely, trusses and frames. Trusses are rigid, fully constrained structures that are constructed using only pin-connected two-force members. Four main assumptions are made in the analysis of trusses: (1) Truss members are connected only at their ends; no member is continuous through a joint. (2) Members are connected by frictionless pins. (3) The truss structure is loaded only at the joints. (4) The weight of the members may be neglected. Because of these assumptions, truss members are modeled as two-force members with the forces acting at the ends of the member and directed along the axis of the member.

When the truss structure and the applied loads lie in the same plane, the structure is known as a planar truss. Space trusses are structures that are not contained in a single plane and/or are loaded out of the plane of the structure.

One method of analysis (the method of joints) for trusses is performed by drawing a free-body diagram for each pin (joint). Application of the vector equilibrium equation, $\Sigma \mathbf{F} = \mathbf{0}$, at each joint yields two algebraic equations, which can be solved for two unknowns. The pins are solved sequentially starting from a pin on which only two unknown forces and one or more known forces act. Once these forces are determined, their values can be applied to adjacent joints and treated as known quantities. This process is repeated until all unknown forces have been determined.

The method of joints is most often used when the forces in all of the members of a truss are to be determined.

A second method of analysis for trusses is the method of sections. When the method of sections is used, the truss is divided into two parts by passing an imaginary plane or curved section through the members of interest. Free-body diagrams may then be drawn for either or both parts of the truss. Since each part is a rigid body, three independent equations of equilibrium can be written for either part. Therefore, a section that cuts through no more than three members should be used.

It will often happen that a section that cuts no more than three members and that passes through a given member of interest cannot be found. In such a case it may be necessary to draw a section through a nearby member and solve for the forces in it first. Then the method of joints can be used to find the force in the member of interest.

One of the principle advantages of using the method of sections is that the force in a member near the center of a large truss usually can be determined without first obtaining the forces in the rest of the truss.

Structures that are not constructed entirely of two-force members are called frames or machines. Although frames and machines may also contain one or more two-force members, they always contain at least one member that is acted on by forces at more than two points or is acted on by both forces and moments. The main distinction between frames and machines is that frames are rigid structures, whereas machines are not.

As with the analysis of trusses, the method of solution for frames and machines consists of taking the structures apart, drawing free-body diagrams of each of the components, and writing the equations of equilibrium for each of the free-body diagrams. Since some of the members of frames and machines are not two-force members, however, the directions of the forces in these members is not known. The analysis of frames and machines consists of solving the equilibrium equations for a system of rigid bodies.

For machines and nonrigid structures, the structure must be taken apart and analyzed even if the only information desired is the support reactions or the relationship between the external forces (input and output forces) acting on it.

# REVIEW PROBLEMS

**7-119*** Determine the force in each member of the truss shown in Fig. P7-119.

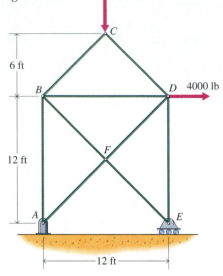

**Fig. P7-119**

**7-120*** A 500-kg mass is supported by a four-bar truss as shown in Fig. P7-120. Determine the force in each member of the truss.

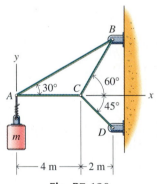

**Fig. P7-120**

**7-121** A block weighing 1000-lb is supported by a five-bar truss as shown in Fig. P7-121. Determine the force in each member of the truss.

**Fig. P7-121**

**7-122** The inverted bridge truss of Fig. P7-122 supports a roadway and vehicles giving the loading shown. Find the forces in members *CD*, *DG*, and *EG*.

**Fig. P7-122**

**7-123** Determine the force in each member of the space truss shown in Fig. P7-123. The support at *A* is a ball and socket. The supports at *D*, *E*, and *F* are links.

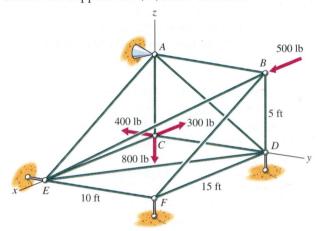

**Fig. P7-123**

**7-124** A scissors jack for an automobile is shown in Fig. P7-124. The screw threads exert a force **F** on the blocks at joints *A* and *B*. Determine the force **P** exerted on the automobile if $F = 800$ N and (a) $\theta = 15°$, (b) $\theta = 30°$, and (c) $\theta = 45°$.

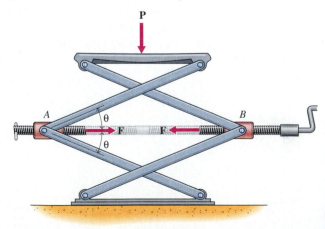

**Fig. P7-124**

**7-125*** In Fig. P7-125, a cable is attached to the structure at E, passes around the 2-ft-diameter, frictionless pulley at C, and then is attached to a 200-lb weight W. A second cable is attached between A and F. Determine all forces acting on member DEBFG.

**Fig. P7-125**

**7-126*** In Fig. P7-126, a cable is attached to the structure at E, passes around the 0.8-m-diameter frictionless pulley at A, and then is attached to a 1000-N weight W. Determine all forces acting on member ABCD.

**Fig. P7-126**

**7-127** A force of 20 lb is required to pull the stopper DE in Fig. P7-127. Determine all forces acting on member BCD.

**Fig. P7-127**

AB
AF

BCD

FCE

**7-128** A cord BD is used to keep the frame of Fig. P7-128 from collapsing under the distributed load w. Determine w when the tension in the cord is 600 N.

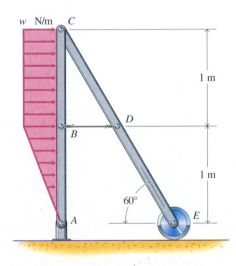

**Fig. P7-128**

399

**7-129\*** The mechanism of Fig. P7-129 is designed to keep a load level while raising it. A pin on the rim of the 4-ft-diameter pulley fits in a slot on arm *ABC*. Arms *ABC* and *DE* are each 4 ft long, and the package being lifted weighs 80 lb. The mechanism is raised by pulling on the rope, which is wrapped around the pulley. Determine the force **P** applied to the rope and all forces acting on the arm *ABC* when the package has been lifted 4 ft as shown.

**7-130** Three bars are connected with smooth pins to form the frame shown in Fig. P7-130. The weights of the bars are negligible. Determine

a. The force exerted by the pin at *D* on member *CDE*.
b. The reactions at supports *A* and *E*.

**Fig. P7-129**

**Fig. P7-130**

*Flexible cables are widely used for suspension bridges such as the Golden Gate Bridge in San Francisco.*

Chapter **8**

# INTERNAL FORCES IN STRUCTURAL MEMBERS

## 8-1 INTRODUCTION

When a structural member or machine component (cable, bar, shaft, beam, or column) is subjected to a system of external loads (applied loads and support reactions), a system of internal resisting forces develops within the member to balance the external forces. Consider a body subjected to a system of balanced external forces $F_1$, $F_2$, $F_3$, . . ., $F_n$, as shown in Fig. 8-1. These forces tend to either crush the body (compression) or pull it apart (tension). In either case, internal forces (resisting forces) develop within the body to resist the crushing or to hold the body together.

The resultant of the internal forces on a specific plane $aa$ within a body can be determined by assuming that the plane separates the body into two parts as shown in Fig. 8-2. Since the body is in equilibrium, each part of the body must also be in equilibrium under the action of the internal forces that develop on the plane separating the body into two parts. Therefore, the resultant of the internal forces acting on plane $aa$ can be determined by using either the left or right part of the body. In general, the distributed forces acting over the small elements of area $dA$ that make up the cross section $A$ are not uniformly distributed. The intensities of these internal forces (force per unit area) are called stresses. The problem of determining stress distributions in a specific body for a specific set of external loads is covered in textbooks dealing with the mechanics of deformable bodies.

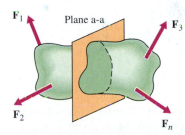

**Figure 8-1** Arbitrary body subjected to a balanced system of external forces.

**Figure 8-2** Free-body diagrams for the parts of a body to the left and right of an imaginary plane.

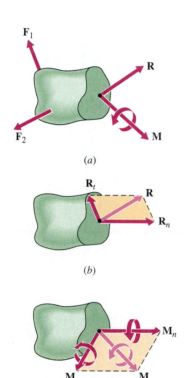

**Figure 8-3** Resultant force **R** and resultant moment **M** on an internal plane.

A free-body diagram for the left part of the body is shown in Fig. 8-3a. The internal force distribution on plane *aa* has been replaced by a resultant force **R** at a point on plane *aa* and a resultant moment **M**. The resultant force **R** can be resolved, as shown in Fig. 8-3b, into a component **R**$_n$ (normal force) perpendicular to plane *aa* and a component **R**$_t$ (shear force) tangent to plane *aa*. Similarly, the moment **M** can be resolved into a component **M**$_n$ (torsional or twisting moment) about an axis perpendicular to plane *aa* and a component **M**$_t$ (bending moment) about an axis tangent to plane *aa*, as shown in Fig. 8-3c. The components **R**$_t$ and **M**$_t$ can be resolved into rectangular components if an *xyz*-reference system is used.

The following procedure is suggested for determining the internal forces at a specific location in a member.

1. **Determine the Support Reactions.** Prepare a sketch of the body showing significant dimensions and all external loads (forces, bending moments, and torques) on the body in their exact locations. Support reactions should be determined before the plane of interest is exposed.

2. **Draw a Complete Free-Body Diagram.** Identify the plane of interest in the body. Prepare a free-body diagram for a portion of the body with the exposed plane of interest. Show all external loads on this part of the body and the force and moment resultants (or their components) on the exposed plane of interest.

3. **Apply the Equations of Equilibrium.** In the most general case, it will be possible to solve for six unknowns by using the equations

$$\Sigma F_x = 0 \qquad \Sigma F_y = 0 \qquad \Sigma F_z = 0$$
$$\Sigma M_x = 0 \qquad \Sigma M_y = 0 \qquad \Sigma M_z = 0$$

## 8-2 AXIAL FORCE AND TORQUE IN BARS AND SHAFTS

In Chapter 7, axially loaded two-force truss members and multiforce frame and machine elements were discussed in detail. Application of the equations of equilibrium to the various parts of the truss, frame, or machine allowed the determination of all forces acting at the smooth pin connections. In many other types of engineering applications, knowledge of the maximum axial force, maximum shear force, maximum twisting moment, or maximum bending moment transmitted by any cross section through the member is required to establish the ad-

*Two teams A and B are engaged in a tug of war. At the instant shown in Fig. CE8-1a, which team is winning and what is the magnitude of the force being transmitted by the rope segment between team members $A_3$ and $A_4$?*

(a)

(b)

**Fig. CE8-1**

## SOLUTION

Equilibrium of a system of collinear forces requires satisfaction of one force equation in the direction of the forces since all other equations are automatically satisfied. Thus, for the system shown in Fig. CE8-1a,

$$+\rightarrow \Sigma F = A_1 + A_2 + A_3 + A_4 + B_1 + B_2 + B_3 + B_4$$
$$= -100 - 80 - 75 - 90 + 80 + 70 + 90 + 105 = 0 \qquad \text{Ans.}$$

Therefore, the system is in equilibrium and neither team is winning.

Since the complete system consisting of teams $A$ and $B$ is in equilibrium, any part of the system isolated for consideration must also be in equilibrium under the action of the external loads applied to the part and the internal forces exposed during the isolating process. The force being transmitted by the rope segment between team members $A_3$ and $A_4$ can be determined by isolating team members $A_1$, $A_2$, and $A_3$ as a free-body diagram, as shown in Fig. CE8-1b. Force $F_{34}$ on this diagram represents the action of the part of the system removed during the isolating process. Summing forces in the direction of the forces yields

$$+\rightarrow \Sigma F = A_1 + A_2 + A_3 + F_{34}$$
$$= -100 - 80 - 75 + F_{34} = 0$$
$$F_{34} = 255 \text{ lb} = 255 \text{ lb} \rightarrow \qquad \text{Ans.}$$

In the event that the system is not in equilibrium, any unbalance of forces will produce an acceleration of the complete system in the direction of the unbalance in accordance with Newton's second law ($\mathbf{F} = m\mathbf{a}$). Such a situation represents a problem in dynamics.

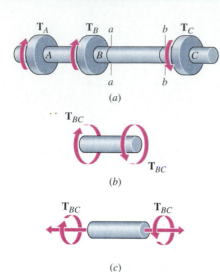

Figure 8-4  Torques transmitted by
internal planes of a shaft.

equacy of the member for its intended use. In this section, the equations of equilibrium are used to establish the variation of internal axial force along the length of an axially loaded member and the variation of resisting torque transmitted by transverse cross sections of a shaft. Axial force and torque diagrams are introduced as a way to visualize the distributions for the full lengths of the members.

An axial force diagram is a graph in which abscissas represent distances along the member and ordinates represent the internal axial forces at the corresponding cross sections. In plotting an axial force diagram, tensile forces are positive and compressive forces are negative. Example Problem 8-1 illustrates the computations required to construct an axial force diagram for a simple tension member subjected to four axial loads.

In a similar manner, a torque diagram is a graph in which abscissas represent distances along the member and ordinates represent the internal resisting torques at the corresponding cross sections. The sign convention used for torques is illustrated in Fig. 8-4. In the shaft shown in Fig. 8-4a, torque is applied to the shaft at gear $C$ and is removed at gears $A$ and $B$. Torques transmitted by cross sections $aa$ and $bb$ in the interval between gears $B$ and $C$ are shown in a pictorial fashion in Fig. 8-4b. A vector representation of these torques is shown in Fig. 8-4c. Positive torques point outward from the cross section when represented as a vector, according to the right-hand rule. Example Problem 8-2 illustrates the computations required to construct a torque diagram for a shaft subjected to four torques at different positions along the length of the shaft.

A steel bar with a rectangular cross section is used to transmit four axial loads as shown in Fig. 8-5a.

a. Determine the axial forces transmitted by cross sections in intervals *AB*, *BC*, and *CD* of the bar.
b. Draw an axial force diagram for the bar.

Equilibrium of a member subjected to a system of collinear forces requires only one force equation in the direction of the forces since all other equations are automatically satisfied.

## SOLUTION

a. The forces transmitted by cross sections in intervals *AB*, *BC*, and *CD* of the bar shown in Fig. 8-5a are obtained by using the three free-body diagrams shown in Figs. 8-5b, 8-5c, and 8-5d. Applying the force equilibrium equation $\Sigma F = 0$ along the axis of the bar yields

$$+\rightarrow \Sigma F = F_{AB} - 60 = 0 \qquad F_{AB} = +60 \text{ kN} \qquad \text{Ans.}$$
$$+\rightarrow \Sigma F = F_{BC} - 60 + 80 = 0 \qquad F_{BC} = -20 \text{ kN} \qquad \text{Ans.}$$
$$+\rightarrow \Sigma F = F_{CD} - 60 + 80 - 50 = 0 \qquad F_{CD} = +30 \text{ kN} \qquad \text{Ans.}$$

For all the above calculations, a free-body diagram of the part of the bar to the left of the imaginary cut has been used. A free-body diagram of the part of the bar to the right of the cut would have yielded identical results. In fact, for the determination of $F_{CD}$, the free-body diagram to the right of the cut would have been more efficient, since only the unknown force $F_{CD}$ and the 30-kN load would have appeared on the diagram.

b. The axial force diagram for the bar, constructed by using the results from part a, is shown in Fig. 8-5e. Note in the diagram that the abrupt changes in internal force are equal to the applied loads at pins A, B, C, and D. Thus, the axial force diagram could have been drawn directly below the sketch of the loaded bar of Fig. 8-5a, without the aid of the free-body diagrams shown in Fig. 8-5b, 8-5c, and 8-5d, by using the applied loads at pins A, B, C, and D.

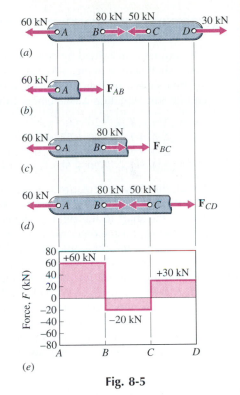

Fig. 8-5

A steel shaft is used to transmit torque from a motor to operating units in a factory. The torque is input at gear $B$ (see Fig. 8-6$a$) and is removed at gears $A$, $C$, $D$, and $E$.

**a.** Determine the torques transmitted by cross sections in intervals $AB$, $BC$, $CD$, and $DE$ of the shaft.
**b.** Draw a torque diagram for the shaft.

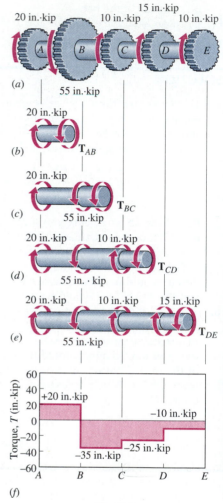

(a)

(b)

(c)

(d)

(e)

(f)

**Fig. 8-6**

## SOLUTION

**a.** The torques transmitted by cross sections in intervals $AB$, $BC$, $CD$, and $DE$ of the shaft shown in Fig. 8-6$a$ are obtained by using the four free-body diagrams shown in Figs. 8-6$b$, 8-6$c$, 8-6$d$, and 8-6$e$. Applying the moment equilibrium equation $\Sigma M = 0$ about the axis of the shaft yields,

$$+\downarrow\Sigma M = T_{AB} - 20 = 0 \qquad T_{AB} = +20 \text{ in.} \cdot \text{kip} \qquad \text{Ans.}$$
$$+\downarrow\Sigma M = T_{BC} - 20 + 55 = 0 \qquad T_{BC} = -35 \text{ in.} \cdot \text{kip} \qquad \text{Ans.}$$
$$+\downarrow\Sigma M = T_{CD} - 20 + 55 - 10 = 0 \qquad T_{CD} = -25 \text{ in.} \cdot \text{kip} \qquad \text{Ans.}$$
$$+\downarrow\Sigma M = T_{DE} - 20 + 55 - 10 - 15 = 0 \qquad T_{DE} = -10 \text{ in.} \cdot \text{kip} \qquad \text{Ans.}$$

For all the above calculations, a free-body diagram of the part of the shaft to the left of the imaginary cut has been used. A free-body diagram of the part of the shaft to the right of the cut would have yielded identical results. In fact, for the determination of $T_{CD}$ and $T_{DE}$, the free-body diagram to the right of the cut would have been more efficient, since fewer torques would have appeared on the diagram.

**b.** The torque diagram for the shaft, constructed by using the results from part $a$, is shown in Fig. 8-6$f$. Note in the diagram that the abrupt changes in torque are equal to the applied torques at gears $A$, $B$, $C$, $D$, and $E$. Thus, the torque diagram could have been drawn directly below the sketch of the shaft of Fig. 8-6$a$, without the aid of the free-body diagrams shown in Figs. 8-6$b$, 8-6$c$, 8-6$d$, and 8-6$e$, by using the applied torques at gears $A$, $B$, $C$, $D$, and $E$.

> Equilibrium of a shaft subjected to a system of torques requires only one moment equation about the axis of the shaft since all other equations are automatically satisfied.

# PROBLEMS

## Introductory Problems

**8-1*** A steel bar with a rectangular cross section is used to transmit four axial loads as shown in Fig. P8-1.

**a.** Determine the axial forces transmitted by transverse cross sections in intervals $AB$, $BC$, and $CD$ of the bar.

**b.** Draw an axial force diagram for the bar.

**Fig. P8-1**

**8-2*** A steel bar has three axial loads applied as shown in Fig. P8-2.

**a.** Determine the axial forces transmitted by transverse cross sections in intervals $AB$, $BC$, and $CD$ of the bar.

**b.** Draw an axial force diagram for the bar.

**Fig. P8-2**

**8-3** A bar is loaded and supported as shown in Fig. P8-3.

**a.** Determine the axial forces transmitted by transverse cross sections in intervals $AB$, $BC$, and $CD$ of the bar.

**b.** Draw an axial force diagram for the bar.

**Fig. P8-3**

**8-4** A bar is loaded and supported as shown in Fig. P8-4.

**a.** Determine the axial forces transmitted by transverse cross sections at points $A$, $B$, and $C$ of the bar.

**b.** Draw an axial force diagram for the bar.

**Fig. P8-4**

**8-5*** For the steel shaft shown in Fig. P8-5,

**a.** Determine the torque transmitted by transverse cross sections at points $A$, $B$, and $C$ of the shaft.

**b.** Draw a torque diagram for the shaft.

**Fig. P8-5**

**8-6** The motor shown in Fig. P8-6 supplies a torque of 500 N · m to shaft $BCDE$. The torques removed at gears $C$, $D$, and $E$ are 100 N · m, 150 N · m, and 250 N · m, respectively.

**a.** Determine the torques transmitted by transverse cross sections in intervals $BC$, $CD$, and $DE$ of the shaft.

**b.** Draw a torque diagram for the shaft.

**Fig. P8-6**

## Intermediate Problems

**8-7\*** For the steel shaft shown in Fig. P8-7,

**a.** Determine the maximum torque transmitted by any transverse cross section of the shaft.

**b.** Draw a torque diagram for the shaft.

**Fig. P8-7**

**8-8\*** For the steel shaft shown in Fig. P8-8,

**a.** Determine the maximum torque transmitted by any transverse cross section of the shaft.

**b.** Draw a torque diagram for the shaft.

**Fig. P8-8**

**8-9** For the steel shaft shown in Fig. P8-9,

**a.** Determine the maximum torque transmitted by any transverse cross section of the shaft.

**b.** Draw a torque diagram for the shaft.

**Fig. P8-9**

**8-10** For the steel shaft shown in Fig. P8-10,

**a.** Determine the maximum torque transmitted by any transverse cross section of the shaft.

**b.** Draw a torque diagram for the shaft.

**Fig. P8-10**

## Challenging Problems

**8-11\*** For the steel shaft shown in Fig. P8-11,

**a.** Determine the maximum torque transmitted by any transverse cross section of the shaft.

**b.** Draw a torque diagram for the shaft.

**Fig. P8-11**

**8-12\*** A bar is loaded and supported as shown in Fig. P8-12.

**a.** Determine the maximum axial load transmitted by any transverse cross section of the bar.

**b.** Draw an axial force diagram for the bar.

**Fig. P8-12**

**8-13** Four sets of flexible cables, spaced at 120° intervals, are used to stabilize a 400-ft communications tower. The

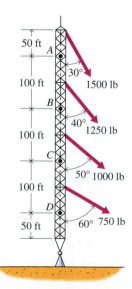

**Fig. P8-13**

tower and one cable from each set is shown in Fig. P8-13. The weight of each tower is 40 lb/ft, and the communications equipment at the top weighs 2000 lb. Determine the axial forces transmitted by transverse cross sections at points A, B, C, and D of the tower.

**8-14** Five 24-in.-diameter pulleys are keyed to a steel shaft as shown in Fig. P8-14. The pulleys carry belts that are used to drive machinery in a factory. Belt tensions for normal operating conditions are indicated on the figure. Determine the maximum torque transmitted by any transverse cross section of the shaft.

**Fig. P8-14**

## 8-3 AXIAL FORCE, SHEAR FORCE, AND BENDING MOMENTS IN MULTIFORCE MEMBERS

The frames and machines discussed in Chapter 7 all contained at least one multiforce member. Recall that the forces in these members were not directed along the axis of the member; therefore, a transverse cross section may be required to transmit axial forces, shear forces, and bending moments. In a general three-dimensional body, as discussed in Section 8-1, the internal forces on a specific plane are statically equivalent to a force-couple system, with the couple being dependent on the choice of location of the force. The point of reference to be used for all problems in this textbook is the centroid of the cross-sectional area. Many formulas derived later in "Mechanics of Materials" courses will require such a resolution of the internal forces.

For the case of the planar systems of external loads to be considered in this section, the internal force-couple system will consist of an axial force **P**, a shear force **V**, and a bending moment **M**. The procedure for determining these internal resisting forces and moments is illustrated in the following example.

Two bars are used to support a 500-lb load as shown in Fig. 8-7a. Determine the internal resisting forces and moment transmitted by section aa.

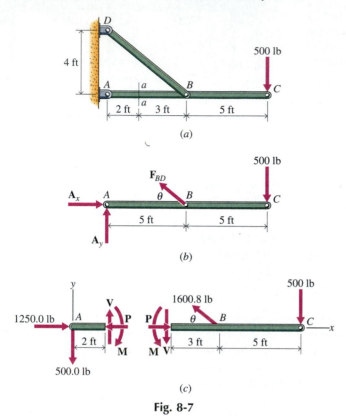

Fig. 8-7

## SOLUTION

The forces acting on bar $ABC$ are determined by using the free-body diagram shown in Fig. 8-7b. Bar $BD$ is a two-force member; therefore, the angle between the line of action of force $\mathbf{F}_{BD}$ and the axis of bar $ABC$ is

$$\theta = \tan^{-1}\frac{4}{5} = 38.66°$$

Force $\mathbf{F}_{BD}$ is determined by summing moments about point $A$ of Fig. 8-7b. Thus,

$$+\!\!\downarrow\!\Sigma M_A = F_{BD}\sin 38.66°\,(5) - 500(10) = 0 \qquad F_{BD} = 1600.8\text{ lb}$$

Once force $\mathbf{F}_{BD}$ is determined, forces $\mathbf{A}_x$ and $\mathbf{A}_y$ are determined by summing forces in the $x$- and $y$-directions. Thus,

$$+\!\!\rightarrow\!\Sigma F_x = A_x - F_{BD}\cos 38.66°$$
$$= A_x - 1600.8\cos 38.66° = 0 \qquad A_x = 1250.0\text{ lb}$$

$$+\!\!\uparrow\!\Sigma F_y = A_y + F_{BD}\sin 38.66° - 500$$
$$= A_x + 1600.8\sin 38.66° - 500 = 0 \qquad A_y = -500.0\text{ lb}$$

The internal resisting forces and moment transmitted by section aa are determined by passing an imaginary plane through the bar at section aa in order to expose the surface on which the internal forces act. Free-body diagrams for

> Only three independent scalar equations of equilibrium are available to solve for unknown quantities in any body or part of a body subjected to a two-dimensional system of forces.

the parts of the bar to the left and right of the imaginary plane are shown in Fig. 8-7c. Either part of the bar can be used for the determination of $P$, $V$, and $M$. The three equilibrium equations for the part of the bar to the left of the imaginary plane yield

$$+\rightarrow \Sigma F_x = 1250.0 - P = 0$$
$$P = 1250.0 \text{ lb} = 1250 \text{ lb} \leftarrow \qquad \text{Ans.}$$

$$+\uparrow \Sigma F_y = V - 500.0 = 0$$
$$V = 500.0 \text{ lb} = 500 \text{ lb} \uparrow \qquad \text{Ans.}$$

$$+\langle \Sigma M_A = V(2) - M = 500.0(2) - M = 0$$
$$M = 1000.0 \text{ ft} \cdot \text{lb} = 1000 \text{ ft} \cdot \text{lb} \downarrow \qquad \text{Ans.}$$

Similarly, the three equilibrium equations for the part of the bar to the right of the imaginary plane yield

$$+\rightarrow \Sigma F_x = P - 1600.8 \cos 38.66° = 0$$
$$P = 1250.0 \text{ lb} = 1250 \text{ lb} \rightarrow \qquad \text{Ans.}$$

$$+\uparrow \Sigma F_y = -V + 1600.8 \sin 38.66° - 500.0 = 0$$
$$V = 500.0 \text{ lb} = 500 \text{ lb} \downarrow \qquad \text{Ans.}$$

$$+\langle \Sigma M_C = V(8) - 1600.8 \sin 38.66° \, (5) + M$$
$$= 500.0(8) - 1600.8 \sin 38.66° \, (5) + M = 0$$
$$M = 1000.0 \text{ ft} \cdot \text{lb} = 1000 \text{ ft} \cdot \text{lb} \downarrow \qquad \text{Ans.}$$

The internal forces and moment acting on the part of the bar to the left of the imaginary plane are equal and opposite to those acting on the part of the bar to the right of the plane. Thus, as previously stated, the internal forces and moment have no effect on the overall equilibrium of the complete bar.

If bar ABC is in equilibrium, any part of the bar isolated as a free body will also be in equilibrium under the actions of the internal forces and moments exposed during the isolation process.

A bar and a cable are used to support a cylinder with a mass $m = 200$ kg as shown in Fig. 8-8a. Determine the internal resisting forces and moment transmitted by section *aa*.

(a)

$\mathbf{F}_D$

45°

$\mathbf{F}_B$

1962 N

(b)

**Fig. 8-8**

## SOLUTION

The force exerted on bar *ABC* by the cylinder is determined by using the free-body diagram of the cylinder shown in 8-8b. The weight of the cylinder is $W = mg = 200(9.81) = 1962$ N. Summing forces in the $y$-direction yields

$$+\uparrow\Sigma F_y = F_B \sin 45° - 1962 = 0 \qquad F_B = 2775 \text{ N}$$

The forces acting on bar *ABC* are determined by using the free-body diagram shown in Fig. 8-8c. For convenience, the $x$- and $y$-directions are selected parallel and perpendicular to the axis of the bar. Force $\mathbf{F}_C$ is determined by summing moments about point *A* of Fig. 8-8c. Thus,

$$+\curvearrowleft\Sigma M_A = F_C \sin 65° (4) - 2775(2) = 0 \qquad F_C = 1530.9 \text{ N}$$

Once force $\mathbf{F}_C$ is determined, forces $\mathbf{A}_x$ and $\mathbf{A}_y$ are determined by summing forces in the $x$- and $y$-directions. Thus,

$$+\nearrow\Sigma F_x = A_x - F_C \cos 65°$$
$$= A_x - 1530.9 \cos 65° = 0 \qquad A_x = 647.0 \text{ N}$$

$$+\nwarrow\Sigma F_y = A_y + F_C \sin 65° - 2775$$
$$= A_x + 1530.9 \sin 65° - 2775 = 0 \qquad A_y = 1387.5 \text{ N}$$

The internal resisting forces and moment transmitted by section *aa* are determined by passing an imaginary plane through the bar at section *aa* in order

(c)                                    (d)

to expose the surface on which the internal forces act. Free-body diagrams for
the parts of the bar to the left and right of the imaginary plane are shown in
Fig. 8-8d. Either part of the bar can be used for the determination of P, V, and
M. Using the three equilibrium equations for the part of the bar to the left of
the imaginary plane, because fewer forces are involved, yields

$$+\nearrow \Sigma F_x = 647.0 - P = 0$$
$$P = 647.0 \text{ lb} = 647 \text{ lb} \swarrow \qquad \text{Ans.}$$

$$+\nwarrow \Sigma F_y = 1387.5 - V = 0$$
$$V = 1387.5 \text{ N} \cong 1388 \text{ N} \searrow \qquad \text{Ans.}$$

$$+ \langle \Sigma M_A = -V(1) + M = -1387.5(1) + M = 0$$
$$M = 1387.5 \text{ N} \cdot \text{m} \cong 1388 \text{ N} \cdot \text{m} \rangle \qquad \text{Ans.}$$

Two bars and a cable are used to support a 500-lb load as shown in Fig. 8-9a. Determine the internal resisting forces and the moment transmitted by

**a.** Section *aa* in bar *BCD*.
**b.** Section *bb* in bar *EF*.

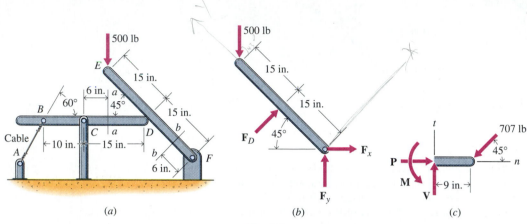

(a)    (b)    (c)

**Fig. 8-9**

---

### SOLUTION

The contact force $F_D$ at the point of contact between bar *BCD* and bar *EDF* (see Fig. 8-9a) can be determined by drawing a free-body diagram of bar *EDF* and summing moments about pin *F*, as shown in Fig. 8-9b. Since the line of action of force $F_D$ is perpendicular to the axis of bar *EF*,

$$+\downarrow\Sigma M_F = F_D(15) - 500(30) \cos 45° = 0 \qquad F_D = 707 \text{ lb}$$

> Only three independent scalar equations of equilibrium are available to solve for unknown quantities in any body or part of a body subjected to a two-dimensional system of forces.

**a.** Once the contact force $F_D$ is known, the internal resisting forces and moment on section *aa* can be determined by using the free-body diagram shown in Fig. 8-9c for the portion of the bar *BCD* to the right of section *aa*. Thus,

$$+\rightarrow\Sigma F_n = P - 707 \cos 45° = 0 \qquad P = 500 \text{ lb} \qquad \text{Ans.}$$
$$+\uparrow\Sigma F_t = V - 707 \sin 45° = 0 \qquad V = 500 \text{ lb} \qquad \text{Ans.}$$
$$+\downarrow\Sigma M_0 = M - 707(9) \sin 45° = 0 \qquad M = 4500 \text{ in.} \cdot \text{lb} \qquad \text{Ans.}$$

**b.** Similarly, the internal resisting forces and moment on section *bb* can be determined by using the free-body diagram shown in Fig. 8-9d for the portion of bar *EF* above section *bb*. Thus,

$$+\nwarrow\Sigma F_n = P - 500 \cos 45° = 0 \qquad P = 354 \text{ lb} \qquad \text{Ans.}$$
$$+\swarrow\Sigma F_t = V - 707 + 500 \sin 45° = 0 \qquad V = 354 \text{ lb} \qquad \text{Ans.}$$
$$+\downarrow\Sigma M_0 = M + 707(9) - 500(24) \cos 45° = 0 \qquad M = 2120 \text{ in.} \cdot \text{lb} \qquad \text{Ans.}$$

(d)

# PROBLEMS

## Introductory Problems

Determine the internal resisting forces and moment transmitted by section *aa* in

**8-15\*** The bracket shown in Fig. P8-15.

**Fig. P8-15**

**8-16\*** The bracket shown in Fig. P8-16.

**Fig. P8-16**

**8-17** The curved bar shown in Fig. P8-17.

**Fig. P8-17**

**8-18** The curved bar shown in Fig. P8-18.

**Fig. P8-18**

**8-19\*** The column shown in Fig. P8-19.

**Fig. P8-19**

**8-20\*** Bar *ABC* of the three-bar frame shown in Fig. P8-20.

**Fig. P8-20**

**417**

**8-21** The angle bracket shown in Fig. P8-21.

**Fig. P8-21**

**8-22** The angle bracket shown in Fig. P8-22.

**Fig. P8-22**

**8-23\*** The beam shown in Fig. P8-23.

**Fig. P8-23**

**8-24\*** The curved bar shown in Fig. P8-24.

**Fig. P8-24**

**8-25** The bar rack shown in Fig. P8-25 if each bar weighs 100 lb.

**Fig. P8-25**

**8-26** Bar *AB* of the sign support system shown in Fig. P8-26 if the mass of the sign is 25 kg. Section *aa* in bar *AB* is located midway between the points of attachment of the sign.

**Fig. P8-26**

## Intermediate Problems

**8-27\*** A three-bar frame is loaded and supported as shown in Fig. P8-27. Determine the internal resisting forces and moment transmitted by

**a.** Section *aa* in bar *BEF*.
**b.** Section *bb* in bar *ABCD*.

**Fig. P8-27**

**8-28\*** Two bars and a link are used to support a 750-N load as shown in Fig. P8-28. Determine the internal resisting forces and moment transmitted by section *aa* in bar *BCE*.

**Fig. P8-28**

**8-29** A three-bar frame is loaded and supported as shown in Fig. P8-29. Determine the internal resisting forces and moment transmitted by

**a.** Section *aa* in bar *BDF*.
**b.** Section *bb* in bar *ABC*.

**Fig. P8-29**

**8-30** A three-bar frame is loaded and supported as shown in Fig. P8-30. Determine the internal resisting forces and moment transmitted by

**Fig. P8-30**

**a.** Section *aa* in bar *BD*.
**b.** Section *bb* in bar *ABC*.

## Challenging Problems

**8-31\*** A three-bar frame is loaded and supported as shown in Fig. P8-31. Determine the internal resisting forces and moment transmitted by

**a.** Section *aa* in bar *ABC*.
**b.** Section *bb* in bar *CDE*.

**Fig. P8-31**

**8-32\*** A three-bar frame is loaded and supported as shown in Fig. P8-32. Determine the internal resisting forces

**Fig. P8-32**

and moment transmitted by

**a.** Section *aa* in bar *CGE*.
**b.** Section *bb* in bar *DEF*.

## Computer Problems

**C8-33** The hook shown in Fig. P8-33*a* supports a 10-kip load. Plot *P*, *V*, and *M*—the internal resisting forces and moment transmitted by a section of the hook—as a function

of the angle $\theta$ ($0 \le \theta \le 150°$). (Positive directions for the graphs are shown in Fig. P8-33*b*.)

10 kip

(*b*)

**C8-34** Forces of 100 N are being applied to the handles of the vise grip pliers shown in Fig. P8-34. Plot *P*, *V*, and *M*, the internal resisting forces and moment transmitted by section *aa* of the handle, as a function of the distance *d* ($20 \le d \le 30$ mm).

*R* = 10 in.

$\theta$

10 kip

(*a*)

**Fig. P8-33**

**Fig. P8-34**

## 8-4 SHEAR FORCES AND BENDING MOMENTS IN BEAMS

A structural member or machine component designed primarily to support forces acting perpendicular to the axis of the member is called a beam. The principal difference between beams and the axially loaded bars and torsionally loaded shafts of Section 8-2 is in the direction of the applied loads. In general, the length of a beam is large in comparison to the two cross-sectional dimensions. The line connecting the centroids of the cross sections of a beam is usually referred to as the centroidal or longitudinal axis of the beam. A beam may be straight or curved depending on the shape of this centroidal axis. Many beams are used in a horizontal position in building construction, but vertical and inclined beams are also found in other applications. The deformation of a beam is primarily one of bending. Some beams are loaded in pure bending, while others are subjected to bending loads in combination with axial, shear, and torsional loads. Most of the multiforce members of Section 8-3 were beams of this type. Slender members subjected primarily to axial compressive loads are known as columns. Slender members subjected to axial compressive loads in combination with loads that produce bending are usually referred to as beam columns. In this section, only long slender beams with transverse loads in a single plane (called the plane of bending) will be considered. Such loads

*A beam is loaded and supported as shown in Fig. CE8-2a. Determine the shear force $V$ and the bending moment $M$ on a transverse cross section at an arbitrary distance $x$ from the left support.*

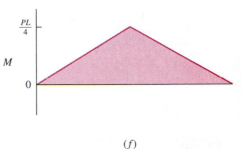

**Fig. CE8-2**

### SOLUTION

A free-body diagram for the complete beam is shown in Fig. CE8-2b. The reactions obtained by summing forces and moments are

$$A_x = 0 \qquad A_y = \frac{P}{2} \qquad B_y = \frac{P}{2}$$

The shear force $V$ and the bending moment $M$ in the interval $0 < x < L/2$ is determined by using the portion of the beam shown in Fig. CE8-2c. Both $V$ and $M$ on this free-body diagram are shown as positive quantities. Equilibrium of this portion of the beam requires that

$$+\uparrow \Sigma F_y = \frac{P}{2} - V = 0 \qquad V = \frac{P}{2}$$

$$+\downarrow \Sigma M_C = M - \frac{P}{2}x = 0 \qquad M = \frac{P}{2}x$$

The shear force $V$ and the bending moment $M$ in the interval $L/2 < x < L$ is determined by using the portion of the beam shown in Fig. CE8-2d. Again, the shear force $V$ and the bending moment $M$ are shown as positive quantities. Equilibrium of this portion of the beam requires that

$$+\uparrow \Sigma F_y = \frac{P}{2} - P - V = 0 \qquad V = -\frac{P}{2}$$

$$+\downarrow \Sigma M_D = M - \frac{P}{2}x + P\left(x - \frac{L}{2}\right) = 0 \qquad M = \frac{P}{2}(L - x)$$

Shear force and bending moment distributions for the complete beam can be obtained by plotting the results obtained in the previous determinations. Such graphs, shown in Figs. CE8-2e and CE8-2f, are known as shear and bending moment diagrams, respectively. Note in these diagrams that the plots of $V$ and $M$ are discontinuous at points where concentrated loads are applied. Note also that the maximum bending moment occurs where the shear force changes direction.

**Figure 8-10** Free-body diagrams for different beam supports.

**Figure 8-11** Types of loads typically applied to beams.

will cause only a shear force $V_r$ and a bending moment $M_r$ to be transmitted by an arbitrary cross section of the beam.

Before proceeding with an analysis of the shear force $V_r$ and the bending moment $M_r$ transmitted by a cross section of a beam, a few words will be said about beam supports, beam loadings, and the classification of beams.

The three types of beam supports commonly used in engineering practice include roller supports, pin supports, and fixed (or clamp) supports.

1. A roller support (see Fig. 8-10a) resists motion of the beam in a direction perpendicular to the axis of the beam. Hence, the reaction at a roller support for a horizontal beam is a vertical force. The beam is free to rotate at a roller support.

2. A pin support (see Fig. 8-10b) resists motion of the beam in any direction in the plane of loading. The reaction at a pin support for a horizontal beam is often represented by the horizontal and vertical components of the force. The beam is free to rotate at a pin support.

3. A fixed (or clamp) support (see Fig. 8-10c) prevents both rotation of the beam and motion of the beam in any direction in the plane of loading. The reaction at a fixed support can be represented by two force components and a moment.

Beams are classified according to the type of load they support. Beams may be subjected to concentrated loads, distributed loads, or couples (concentrated moments) that act alone or in any combination.

1. Loads that are applied to a very small portion of the length of a beam are called concentrated loads. A concentrated load (see Fig. 8-11a) can be idealized as a discrete force that acts at a specific point on the beam.

2. Loads that act over a finite length of the beam are called distributed loads. The distribution (see Fig. 8-11b) may be uniform or nonuniform. The weight of a beam is an example of a uniformly distributed load.

3. A concentrated moment (see Fig. 8-11c) is a couple produced by two equal and opposite forces applied to the beam at a particular section. The two ways used to represent the couple are shown on Fig. 8-11c.

Beams are also classified into types according to the kind of support used.

1. A beam supported by pins, rollers, or smooth surfaces at the ends (see Fig. 8-12a) is called a simply supported beam.

2. A simply supported beam that extends beyond its supports at either or both ends (see Fig. 8-12b) is called an overhanging beam.

3. A beam that is fixed at one end and free at the other end (see Fig. 8-12c) is called a cantilever beam.

4. A beam that is fixed at one end and simply supported at the other end (see Fig. 8-12d) is called a propped beam.

5. A beam with more than two simple supports (see Fig. 8-12e) is called a continuous beam.

6. A beam that is either fixed (no rotation) or restrained (limited rotation) at the ends (see Fig. 8-12f) is called a built-in beam.

**Figure 8-12** Beam classifications according to support conditions.

Beams can also be separated into statically determinate beams and statically indeterminate beams. When the support reactions can be obtained from the equations of statics alone, the beam is statically determinate. If the forces applied to a beam are limited to a plane (say the $xy$-plane), there are three equilibrium equations available for the determination of the support reactions. The equations are

$$\Sigma F_x = 0 \qquad \Sigma F_y = 0 \qquad \Sigma M_{zA} = 0$$

where $A$ is any point in the plane of loading (the $xy$-plane). Thus, at most, three reaction components can be determined. If all the applied forces and support reactions are perpendicular to the longitudinal axis of the beam, the equation $\Sigma F_x = 0$ is automatically satisfied. For such a beam to be statically determinate, only two unknown reactive forces can exist, since the number of equilibrium equations then available for the solution is reduced to two, namely

$$\Sigma F_y = 0 \qquad \text{and} \qquad \Sigma M_{zA} = 0$$

Examples of statically determinate beams are simple beams, overhanging beams, and cantilever beams.

The equations of equilibrium are not sufficient to determine the support reactions when the beam has more supports than are necessary to maintain equilibrium. Such beams are known as statically indeterminate beams, and the load-deformation properties of the beam, in addition to the equations of equilibrium, are required to determine the support reactions. Examples of statically indeterminate beams include the propped cantilever beam, the continuous beam, and the built-in beam. Only statically determinate beams will be treated in this textbook. Statically indeterminate beams will be treated in follow-on courses dealing with mechanics of materials.

Stresses and deflections in beams are functions of internal forces. The internal forces transmitted by a transverse cross section of a beam

**Figure 8-13**   Internal forces transmitted by an arbitrary transverse cross section of a beam.

are those forces and moments required to resist the external forces and maintain equilibrium. Consider the beam shown in Fig. 8-13a, which is subjected to a uniformly distributed load $w$, two concentrated loads $P_1$ and $P_2$, and support reactions at the ends of the beam. The support reactions can be determined by using the equilibrium equations and a free-body diagram of the entire beam. The internal forces transmitted by an arbitrary transverse cross section of the beam (say section $aa$) can be determined by passing a plane through the section so that the beam is separated into two parts. Since the entire beam is in equilibrium, each part of the beam separated by section $aa$ must also be in equilibrium under the action of the internal forces.

When the equilibrium equation $\Sigma F_y = 0$ is applied to the free-body diagram of Fig. 8-13b, the result can be written as

$$R - wx - P_1 = V_r$$

or

$$V_a = V_r$$

where $V_a$ is defined as the resultant of the external transverse forces acting on the part of a beam to either side of a section. This resultant force $V_a$ is called the transverse shear or just the shear at the section. As seen from Fig. 8-13b, the shear $V_a$ is equal in magnitude and opposite in sense to the resisting shear $V_r$. Since these shear forces are always equal in magnitude, they are frequently treated as though they were identical. For simplicity, the symbol $V$ will be used henceforth to represent both the transverse shear $V_a$ and the resisting shear $V_r$.

When the equilibrium equation $\Sigma M_{zA} = 0$ is applied to the free-body diagram of Fig. 8-13b, the result can be written as

$$Rx - \frac{wx^2}{2} - P_1(x - a) = M_r$$

or

$$M_a = M_r$$

where $M_a$ is defined as the algebraic sum of the moments of the external forces, acting on the part of the beam to either side of the section, with respect to an axis at the section perpendicular to the plane of bending (the $xy$-plane). The moment $M_a$ is called the bending moment or just the moment at the section. As seen from Fig. 8-13b, the bending moment $M_a$ is equal in magnitude and opposite in sense to the resisting moment $M_r$. Since these moments are always equal in magnitude, they are frequently treated as though they were identical. For simplicity, the symbol $M$ will be used henceforth to represent both the bending moment $M_a$ and the resisting moment $M_r$.

The bending moment $M_a$ and the transverse shear $V_a$ are not normally shown on a free-body diagram. The usual procedure is to show each external force individually as indicated in Fig. 8-13a. The variation of $V_r$ and $M_r$ along the beam can be expressed by means of equations or shown graphically by means of shear-force and bending-moment diagrams to be developed later, in Section 8-5.

A sign convention is necessary for the correct interpretation of results obtained from equations or diagrams for shear and moment. The

+ Shear      − Shear      + Moment      − Moment

(a)                  (b)

**Figure 8-14**  Sign conventions for shear and moment.

convention illustrated in Fig. 8-14 is widely used in the engineering profession. Note in Fig. 8-14b that a bending moment in a horizontal beam is positive at sections for which the top of the beam is in compression and the bottom is in tension. The signs of the terms in the preceding equations for $V_r$ and $M_r$ agree with this convention.

Since $M$ and $V$ vary with $x$, they are functions of $x$, and equations for $M$ and $V$ can be obtained from free-body diagrams of portions of the beam. The procedure is illustrated in Example Problem 8-6.

A beam is loaded and supported as shown in Fig. 8-15a. Write equations for the shear $V$ and bending moment $M$ for any section of the beam under the distributed load for the interval $2 < x < 8$.

(a)

(b)

**Fig. 8-15**

## SOLUTION

A free-body diagram of a portion of the beam from the left end to an arbitrary section under the distributed load is shown in Fig. 8-15b. Note that the resisting shear $V$ and the resisting moment $M$ are shown as positive values. In general, it is impossible to tell without calculation whether the shear and moment at a particular section are positive or negative. For this reason $V$ and $M$ should be represented as positive quantities on free-body diagrams so that the resulting equations will provide values with the correct algebraic sign.

The reaction at support $A$ is determined by using a free-body diagram for the entire beam and summing moments about support $B$. Thus,

$$R_A(10) - 2000(13) - 1000(6)(5) - 2000(2) = 0$$

from which

$$R_A = 6000 \text{ lb}$$

From the definition of $V$ or from the equilibrium equation $\Sigma F_y = 0$,

$$V = 6000 - 2000 - 1000(x - 2) = -1000x + 6000 \text{ lb} \qquad \text{Ans.}$$

From the definition of $M$, or from the equilibrium equation $\Sigma M_0 = 0$,

$$M = 6000x - 2000(x + 3) - 1000(x - 2)\left(\frac{x-2}{2}\right)$$

$$= -500x^2 + 6000x - 8000 \text{ ft} \cdot \text{lb} \qquad \text{Ans.}$$

The equations for $V$ and $M$ in the other intervals can be determined in a similar manner.

> The shear force $V$ at any cross section is the algebraic sum of the forces acting on the beam to the left of the cross section.

> The bending moment $M$ at any cross section is the moment with respect to an axis in the cross section of all forces and couples acting on the beam to the left of the cross section.

# PROBLEMS

## Introductory Problems

Using the coordinate axes shown, write equations for the shear $V$ and bending moment $M$ for any section

**8-35\*** In the interval $0 \le x \le 16$ ft of the beam shown in Fig. P8-35.

**Fig. P8-35**

**8-36\*** In the interval $0 \le x \le 6$ m of the beam shown in Fig. P8-36.

**Fig. P8-36**

**8-37** In the interval $0 \le x \le 10$ ft of the beam shown in Fig. P8-37.

**Fig. P8-37**

**8-38** In the interval $0 \le x \le 3$ m of the beam shown in Fig. P8-38.

**Fig. P8-38**

**8-39\*** In the interval $0 \le x \le 5$ ft of the beam shown in Fig. P8-39.

**Fig. P8-39**

**8-40** In the interval $0 \le x \le 4$ m of the beam shown in Fig. P8-40.

**Fig. P8-40**

## Intermediate Problems

**8-41\*** In the interval $4 \text{ ft} \le x \le 12$ ft of the beam shown in Fig. P8-41.

**Fig. P8-41**

**8-42\*** In the interval $3 \text{ m} \le x \le 8$ m of the beam shown in Fig. P8-42.

**Fig. P8-42**

**8-43** In the interval 10 ft ≤ x ≤ 20 ft of the beam shown in Fig. P8-43.

**Fig. P8-43**

**8-44** In the interval 0 ≤ x ≤ 4 m of the beam shown in Fig. P8-44.

**Fig. P8-44**

**8-45\*** In the interval 12 ft ≤ x ≤ 18 ft of the beam shown in Fig. P8-45.

**Fig. P8-45**

**8-46** In the interval 2 m ≤ x ≤ 6 m of the beam shown in Fig. P8-46.

**Fig. P8-46**

### Challenging Problems

**8-47\*** A beam is loaded and supported as shown in Fig. P8-47. Using the coordinate axes shown, write equations for the shear $V$ and bending moment $M$ for any section of the beam

**a.** In the interval $0 < x < 12$ ft.
**b.** In the interval $12$ ft $< x < 16$ ft.

**c.** Use the results of parts *a* and *b* to determine the magnitudes and locations of the maximum shear $V_{max}$ and the maximum bending moment $M_{max}$ in the portion of the beam between the supports.

**Fig. P8-47**

**8-48** A beam is loaded and supported as shown in Fig. P8-48. Using the coordinate axes shown, write equations for the shear $V$ and bending moment $M$ for any section of the beam

**a.** In the interval $0 < x < 2$ m.
**b.** In the interval $2$ m $< x < 4$ m.
**c.** In the interval $4$ m $< x < 8$ m.
**d.** Use the results of parts *a*, *b*, and *c* to determine the magnitudes and locations of the maximum shear $V_{max}$ and the maximum bending moment $M_{max}$ in the beam.

**Fig. P8-48**

### Computer Problems

**C8-49** A 500-lb load is supported by a roller on a beam as shown in Fig. P8-49.

**a.** Show that the maximum bending moment in the beam occurs at the roller.
**b.** Plot $|M|_{max}$, the maximum bending moment in the beam, as a function of the roller position $b$ $(0 ≤ b ≤ 25$ ft).

**Fig. P8-49**

**C8-50** A 8-kN load is supported by a cart that rolls along a beam as shown in Fig. P8-50.

**a.** Show that the maximum bending moment in the beam occurs at the wheel that is closer to the middle of the beam.

**b.** Plot $|M|_{max}$, the maximum bending moment in the beam, as a function of the cart's position $b$ ($1 \le b \le 6$ m).

**Fig. P8-50**

## 8-5 SHEAR-FORCE AND BENDING-MOMENT DIAGRAMS

Shear-force and bending-moment diagrams are used to provide a graphical representation of the variation of shear force $V$ and bending moment $M$ over the length of a beam. On a shear diagram, the transverse shear force $V$ is plotted as a function of position along the beam. On a moment diagram, the bending moment $M$ is plotted as a function of position along the beam. From such diagrams, maximum values of shear force and maximum values of bending moment and their locations are easily established.

Shear-force and bending-moment diagrams can be drawn by calculating the values of shear and moment at various sections along the beam and plotting enough points to obtain a smooth curve. Such a procedure is rather time-consuming since no single elementary expression can be written for the shear $V$ or the moment $M$, which applies to the entire length of the beam, unless the load is uniformly distributed or varies according to a known equation along the entire beam. Instead, it is necessary to divide the beam into intervals bounded by the abrupt changes in the loading.

Shear-force and bending-moment diagrams can also be established directly from a free-body diagram (load diagram) for the beam by using several simple mathematical relationships that exist between distributed loads and shears and between shears and moments. These relationships can be developed from a free-body diagram of an elemental length of a beam, as shown in Fig. 8-16. On this diagram, the upward direction is considered positive for the applied load $w$, and the shears and moments are shown as positive quantities according to the sign convention established in Section 8-4. Since the beam is in equilibrium, the element must also be in equilibrium. Applying the equilibrium equation $\Sigma F_y = 0$ to the element yields

$$\Sigma F_y = V + w\,dx - (V + dV) = 0$$

from which

**Figure 8-16** Free-body diagram for a differential length of beam supporting a distributed load.

$$dV = w\,dx \qquad \text{or} \qquad \frac{dV}{dx} = w \qquad (8\text{-}1)$$

The preceding equation indicates that, at any section in the beam, the slope of the shear diagram is equal to the intensity of loading. When

$w$ is known as a function of $x$, the equation can be integrated between definite limits to yield

$$\int_{x_1}^{x_2} x \, dw = \int_{V_1}^{V_2} dV = V_2 - V_1 \tag{8-2}$$

Thus, the change in shear between sections at $x_1$ and $x_2$ is equal to the area under the load diagram between the two sections, provided there are no concentrated forces in the interval $x_1 < x < x_2$.

Applying the equilibrium equation $\Sigma M_0 = 0$, to the element shown in Fig. 8-16 yields

$$\Sigma M_0 = M + V \, dx + w \, dx \, \frac{dx}{2} - (M + dM) = 0$$

from which

$$dM = V \, dx + w \, \frac{(dx)^2}{2}$$

Dividing by $dx$ and passing to the limit yields

$$\frac{dM}{dx} = V \qquad \text{or} \qquad dM = V \, dx \tag{8-3}$$

The preceding equation indicates that at any section in the beam the slope of the moment diagram is equal to the shear. The equation can be integrated between definite limits to give

$$\int_{x_1}^{x_2} V \, dx = \int_{M_1}^{M_2} dM = M_2 - M_1 \tag{8-4}$$

Thus, the change in moment between sections at $x_1$ and $x_2$ is equal to the area under the shear diagram between the two sections, provided there are no applied couples in the interval $x_1 < x < x_2$.

Note that the equations in this section were derived with the $x$-axis positive to the right, the applied loads positive upward, and the shear and moment signs as indicated in Fig. 8-14. If one or more of these assumptions are changed, the algebraic signs in the equation may need to be altered.

The relations just developed provide a second means for drawing shear and moment diagrams and for computing values of shear and moment at various sections along a beam. The method consists of drawing the shear diagram from the load diagram and the moment diagram from the shear diagram by means of Eqs. 8-1 through 8-4. This latter method, though it may not produce a precise curve, provides the shear and moment information usually required for designing beams and is less time-consuming than the first method.

Complete shear and moment diagrams should indicate values of shear and moment at each section where the load changes abruptly and at sections where they are maximum or minimum (negative maximum values). Sections where the shear and moment are zero should also be located. When all loads and reactions are known, the shear and moment at the ends of the beam can be determined by inspection. Both shear and moment are zero at the free end of a beam unless a force or a couple or both are applied there; in this case, the shear is the same as the force and the moment the same as the couple. At a simply supported or pinned end, the shear must equal the end reaction and the moment must be zero. At a built-in or fixed end, the reactions are the shear and moment values.

Once a starting point for the shear diagram is established, the diagram can be sketched immediately below and aligned with the load diagram by using the definition of shear and the fact that the slope of the shear diagram can be obtained from the load diagram. When positive directions are chosen as upward and to the right, a positive distributed load, one acting upward, will result in a positive slope on the shear diagram, and a negative load will give a negative slope. A concentrated force will produce an abrupt change in shear. The change in shear between any two sections is given by the area under the load diagram between the two sections. The change of shear at a concentrated force is equal to the concentrated force.

The moment diagram is drawn immediately below and aligned with the shear diagram in the same manner. The slope at any point on the moment diagram is given by the shear at the corresponding point on the shear diagram, a positive shear representing a positive slope, where upward and to the right are positive, and a negative shear representing a negative slope. The change in moment between any two sections is given by the area under the shear diagram between the two sections. A couple applied to a beam will cause the moment to change abruptly by an amount equal to the moment of the couple.

Example Problems 8-7, 8-8, and 8-9 illustrate the construction of shear and moment diagrams, directly from the load diagram, by using Eqs. 8-1 through 8-4.

A beam is loaded and supported as shown in Fig. 8-17a. Draw complete shear and moment diagrams for the beam.

(a)

(b)

(c)

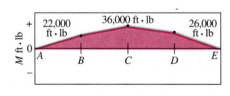

(d)

**Fig. 8-17**

## SOLUTION

A free-body diagram, or load diagram, for the beam is shown in Fig. 8-17b. The reaction at $E$ is determined by summing moments about point $A$.

$$+\downarrow\Sigma M_A = E_y(16) - 2000(4) - 6000(8) - 4000(12) = 0 \qquad E_y = 6500 \text{ lb}\uparrow$$

The reaction components at $A$ are determined by summing forces in the $x$- and $y$-directions. Thus,

$$+\uparrow\Sigma F_y = A_y - 2000 - 6000 - 4000 + E_y$$
$$= A_y - 2000 - 6000 - 4000 + 6500 = 0 \qquad A_y = 5500 \text{ lb}\uparrow$$
$$+\rightarrow F_x = A_x = 0 \qquad\qquad\qquad A_x = 0$$

The shear and moment diagrams for the beam can be drawn directly from the load diagram without writing the shear and moment equations. The shear diagram, shown in Fig. 8-17c, is drawn below the load diagram. The shear at the left end of the beam is zero. At support $A$, the shear changes abruptly by

At a concentrated load, the shear diagram changes abruptly by the magnitude of the load and in the direction of the load.

an amount equal to the magnitude of the reaction $A_y$ and in the direction of $A_y$. Thus,

$$V_A = 0 + 5500 \text{ lb} = 5500 \text{ lb}$$

The slope of the shear diagram is equal to the load $w$; therefore, the slope of the diagram between $A$ and $B$ is zero since there is no distributed load $w$ in this interval of the beam. The change in shear from $A$ to $B$ is equal to the area under the load diagram; thus,

$$V_B = V_A + \Delta V = 5500 + 0 = 5500 \text{ lb}$$

A concentrated load $P$ is applied at $B$; therefore, the shear changes abruptly by the magnitude of the load and in the direction of the load. Thus,

$$V_{B'} = V_B + P = 5500 - 2000 = 3500 \text{ lb}$$

The shear diagram for the remaining intervals of the beam are drawn in exactly the same fashion since there is no distributed load $w$ in any interval of the beam. Thus,

$$
\begin{aligned}
V_C &= V_{B'} + \Delta V = 3500 + 0 = 3500 \text{ lb} \\
V_{C'} &= V_C + \Delta V = 3500 - 6000 = -2500 \text{ lb} \\
V_D &= V_{C'} + \Delta V = -2500 + 0 = -2500 \text{ lb} \\
V_{D'} &= V_D + \Delta V = -2500 - 4000 = -6500 \text{ lb} \\
V_E &= V_{D'} + \Delta V = -6500 + 0 = -6500 \text{ lb} \\
V_{E'} &= V_E + \Delta V = -6500 + 6500 = 0
\end{aligned}
$$

The moment diagram is drawn below the shear diagram, as shown in Fig. 8-17d. The moment at the left support of the beam is zero. The slope of the moment diagram is equal to the shear $V$; therefore, the slope of the diagram is constant ($V = 5500$ lb) from $A$ to $B$. The change in moment from $A$ to $B$ is equal to the area under the shear diagram; thus,

$$M_B = M_A + \Delta M = 0 + 5500(4) = 22{,}000 \text{ ft} \cdot \text{lb}$$

Similarly, since the shear is constant (but different) in all intervals of the beam,

$$
\begin{aligned}
M_C &= M_B + \Delta M = 22{,}000 + 3500(4) = 36{,}000 \text{ ft} \cdot \text{lb} \\
M_D &= M_C + \Delta M = 36{,}000 + (-2500)(4) = 26{,}000 \text{ ft} \cdot \text{lb} \\
M_E &= M_D + \Delta M = 26{,}000 + (-6500)(4) = 0
\end{aligned}
$$

which verifies the absence of any moment at the pin support $E$. If the moment at $E$ had not worked out to be zero, it would indicate that an error had been made somewhere in the analysis of support reactions, shears, or moments.

When there is no load in a section of a beam, the shear diagram is a straight line with a zero slope ($dV/dx = w = 0$).

The change in shear between two points is equal to the area under the load diagram between the two points.

Closing of the shear diagram provides a check of the equilibrium equation $\Sigma F_y = 0$ for the complete beam.

When the shear is constant in a section of a beam, the moment diagram has a constant slope ($dM/dx = V$).

Closing of the moment diagram provides a check of the equilibrium equation $\Sigma M_E = 0$ at the right end of the beam.

A beam is loaded and supported as shown in Fig. 8-18a. Draw complete shear and moment diagrams for the beam.

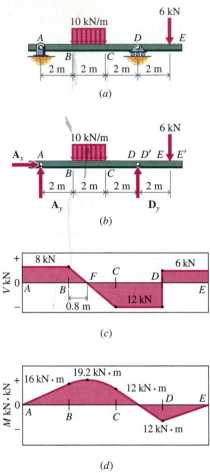

(a)

(b)

(c)

(d)

**Fig. 8-18**

---

## SOLUTION

A free-body diagram, or load diagram, for the beam is shown in Fig. 8-18b. The reaction $D$ is determined by summing moments about point $A$.

$$+\circlearrowleft \Sigma M_A = D_y(6) - 10(2)(3) - 6(8) = 0 \qquad D_y = 18 \text{ kN}\uparrow$$

The reaction components at $A$ are determined by summing forces in the $x$- and $y$-directions. Thus,

$$+\uparrow \Sigma F_y = A_y - 10(2) - 6 + D_y$$
$$= A_y - 20 - 6 + 18 = 0 \qquad A_y = 8 \text{ kN}\uparrow$$
$$+\rightarrow \Sigma F_x = A_x = 0 \qquad\qquad A_x = 0$$

The shear and moment diagrams for the beam can be drawn directly from the load diagram without writing the shear and moment equations. The shear diagram, shown in Fig. 8-18c, is drawn below the load diagram. The shear at the left end of the beam is zero. At support $A$, the shear changes abruptly by an amount equal to the magnitude of the reaction $A_y$ and in the direction of $A_y$. Thus,

$$V_A = 0 + 8 = 8 \text{ kN}$$

The slope of the shear diagram is equal to the load $w$; therefore, the slope of the diagram between $A$ and $B$ is zero since there is no distributed load $w$ in this interval of the beam. The change in shear from $A$ to $B$ is equal to the area under the load diagram; thus,

$$V_B = V_A + \Delta V = 8 + 0 = 8 \text{ kN}$$

A distributed load $w = 10$ kN/m is applied in the interval of the beam between $B$ and $C$; therefore, the slope of the shear diagram between $B$ and $C$ is $-10$ kN/m. The change in shear from $B$ to $C$ is equal to the area under the load diagram; thus,

$$V_C = V_B + \Delta V = 8 + (-10)(2) = -12 \text{ kN}$$

The shear diagram for intervals $CD$ and $DE$ of the beam are drawn in the same way as interval $AB$ since there is no distributed load $w$ in either interval. Thus,

$$V_D = V_C + \Delta V = -12 + 0 = -12 \text{ kN}$$
$$V_{D'} = V_D + \Delta V = -12 + 18 = 6 \text{ kN}$$
$$V_E = V_{D'} + \Delta V = 6 + 0 = 6 \text{ kN}$$
$$V_{E'} = V_E + \Delta V = 6 - 6 = 0$$

The moment diagram is drawn below the shear diagram, as shown in Fig. 8-18$d$. The moment at the left support of the beam is zero. The slope of the moment diagram is equal to the shear $V$; therefore, the slope of the moment diagram is constant ($V = 8$ kN) from $A$ to $B$. The change in moment from $A$ to $B$ is equal to the area under the shear diagram; thus,

$$M_B = M_A + \Delta M = 0 + 8(2) = 16 \text{ kN} \cdot \text{m}$$

The slope of the moment diagram decreases uniformly from 8 kN at $B$ to $-12$ kN at $C$. The slope is zero at point $F$ which is located $d = \dfrac{V}{W} = \dfrac{8}{10} = 0.8$ m to the right of point $B$. The change in moment from $B$ to $F$ is equal to the area under the shear diagram; thus,

$$M_F = M_B + \Delta M = 16 + \frac{1}{2}(8)(0.8) = 19.2 \text{ kN} \cdot \text{m}$$

Similarly, the moment at $C$ is

$$M_C = M_F + \Delta M = 19.2 + \frac{1}{2}(-12)(1.2) = 12 \text{ kN} \cdot \text{m}$$

The shear is constant (but different) in the two remaining intervals of the beam; therefore,

$$M_D = M_C + \Delta M = 12 + (-12)(2) = -12 \text{ kN} \cdot \text{m}$$
$$M_E = M_D + \Delta M = -12 + (6)(2) = 0$$

The moment at the right end of the beam must be zero since it is a free end. If $M_D$ were not zero, it would indicate that an error has occurred.

When a uniformly distributed load acts on a section of the beam, the shear diagram is a straight line with a negative slope.

Closing of the shear diagram provides a check of the equilibrium equation $\Sigma F_y = 0$ for the complete beam.

When a uniformly distributed load acts on a section of the beam, the moment diagram is a parabola.

Closing of the moment diagram provides a check of the equilibrium equation $\Sigma M_E = 0$ at the right end of the beam.

A beam is loaded and supported as shown in Fig. 8-19a. Draw complete shear and moment diagrams for the beam.

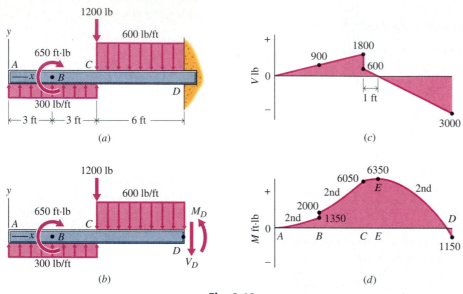

Fig. 8-19

## SOLUTION

A free-body diagram, or load diagram, for the beam is shown in Fig. 8-19b. The reactions at $D$ are computed from the equations of equilibrium. It is not necessary to compute the reactions for a cantilever beam in order to draw shear and moment diagrams from the load diagram, but the reactions provide a convenient check. From the equilibrium equation $\Sigma F_y = 0$,

$$V_D = 300(6) - 1200 - 600(6) = -3000 \text{ lb}$$

From the equilibrium equation $\Sigma M_D = 0$,

$$M_D = 300(6)(9) + 650 - 1200(6) - 600(6)(3) = -1150 \text{ ft} \cdot \text{lb}$$

The shear and moment diagrams can be drawn directly from the load diagram without writing the shear and moment equations. The shear diagram, shown in Fig. 8-19c, is drawn below the diagram. The shear at the end of the beam is zero. The slope of the shear diagram is equal to the load $w$; therefore, the slope of the diagram between $A$ and $C$ is $+300$ lb/ft. The change in shear from $A$ to $C$ is equal to the area under the load diagram; thus,

$$V_C = V_A + \Delta V = 0 + 300(6) = 1800 \text{ lb}$$

A concentrated load $\mathbf{P}$ is applied at $C$; therefore, the shear changes abruptly by the magnitude of the load and in the direction of the load. Thus,

$$V_{C'} = V_C + P = 1800 - 1200 = 600 \text{ lb}$$

A uniform load $w$ is applied to the beam between $C'$ and $D$. Since the slope of the shear diagram is equal to the load $w$, the slope of the diagram between $C'$ and $D$ is $-600$ lb/ft. The change in shear from $C'$ to $D$ is equal to the area under the load diagram; thus,

$$V_D = V_{C'} + \Delta V = 600 + (-600)(6) = -3000 \text{ lb}$$

This shear is the same as the reaction at $D$, which provides a check.

The moment diagram is drawn below the shear diagram, as shown in Fig. 8-19$d$. The moment at the left end of the beam is zero. The slope of the moment diagram is equal to the shear $V$; therefore, the slope of the diagram increases uniformly from 0 at $A$ to 900 lb at $B$, where a concentrated moment **C** is applied. The change in moment from $A$ to $B$ is equal to the area under the shear diagram; thus,

$$M_B = M_A + \Delta M = 0 + \frac{1}{2}(900)(3) = 1350 \text{ ft} \cdot \text{lb}$$

In the interval $AB$ of the beam,

$$\frac{dM}{dx} = V = 300x \qquad\qquad (a)$$

because of the positive (upward) 300 lb/ft distributed load on the beam. Thus,

$$M = 150x^2 + C_1 = 150x^2 \qquad\qquad (b)$$

since $C_1 = 0$ because of the boundary condition $M = 0$ at $x = 0$. Equations $a$ and $b$ show that a constant distributed load distribution produces a linear shear distribution and a second-order moment distribution in the interval $AB$ of the beam. The degree (2nd) of the moment equation between $A$ and $B$ is indicated on the moment diagram. At $B$, the moment changes abruptly by an amount equal to the moment of the couple ($+650$ ft $\cdot$ lb) applied to the beam at $B$. This moment does not appear on the shear diagram, so some procedure should be established to recall its presence on the beam when the moment diagram is being drawn by using shear diagram information. Thus,

$$M_{B'} = M_B + C = 1350 + 650 = 2000 \text{ ft} \cdot \text{lb}$$

From $B'$ to $C$, the slope of the moment diagram increases uniformly from 900 lb at $B'$ to 1800 lb at $C$. Again, the change in moment from $B'$ to $C$ is equal to the area under the shear diagram; thus,

$$M_C = M_{B'} + \Delta M = 2000 + \frac{1}{2}(900 + 1800)(3) = 6050 \text{ ft} \cdot \text{lb}$$

The slope of the moment diagram changes abruptly at $C$ from $+1800$ lb to $+600$ lb. It then decreases uniformly from $+600$ lb at $C$ to $-3000$ lb at $D$. The slope of the moment diagram is zero at point $E$, which is 1 ft to the right of point $C$. From the areas under the shear diagram, moments at $E$ and $D$ are

$$M_E = M_C + \Delta M = 6050 + \frac{1}{2}(600)(1) = 6350 \text{ ft} \cdot \text{lb}$$

$$M_D = M_E + \Delta M = 6350 + \frac{1}{2}(-3000)(5) = -1150 \text{ ft} \cdot \text{lb}$$

This moment is the same as the reaction at $D$, which provides a check.

A beam is loaded and supported as shown in Fig. 8-20a. Draw complete shear and moment diagrams for the beam.

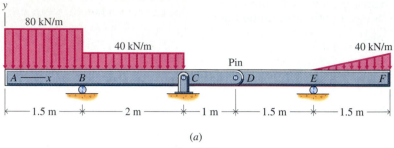

(a)

**Fig. 8-20**

## SOLUTION

A free-body diagram, or load diagram, for the beam is shown in Fig. 8-20b. The reactions at B, C, and E are computed by using the equations of equilibrium for the entire beam. Since the pin at D will not transmit a moment, applying the equilibrium equation $\Sigma M_D = 0$ to the right portion of the beam yields $R_E = 50$ kN. Once $R_E$ is known, the equilibrium equation $\Sigma M_B = 0$ yields $R_C = -35$ kN. Finally, the equilibrium equation $\Sigma F_y = 0$ yields $R_B = 215$ kN.

Once reactions $R_B$, $R_C$, and $R_E$ are known, the shear and moment diagrams can be drawn directly from the load diagram as shown in Fig. 8-20c. The shear at the left end of the beam is zero. The slope of the shear diagram is equal to the load $w$; therefore, the slope of the diagram between A and B is $-80$ kN/m. The change in shear from A to B is equal to the area under the load diagram; thus,

$$V_B = V_A + \Delta V = 0 + (-80)(1.5) = -120 \text{ kN}$$

The shear changes abruptly at support B. Thus,

$$V_{B'} = V_B + R_B = -120 + 215 = +95 \text{ kN}$$

The slope of the shear diagram between B and C is $-40$ kN/m. The change in shear from B' to C is equal to the area under the load diagram; thus,

$$V_C = V_{B'} + \Delta V = 95 + (-40)(2) = +15 \text{ kN}$$

The shear changes abruptly at support C. Thus,

$$V_{C'} = V_C + R_C = 15 + (-35) = -20 \text{ kN}$$

The beam is not loaded between C and E; therefore, the slope of the shear diagram is zero and

$$V_E = V_{C'} + \Delta V = -20 + (0)(2.5) = -20 \text{ kN}$$

The shear changes abruptly at support E. Thus,

$$V_{E'} = V_E + R_E = -20 + 50 = 30 \text{ kN}$$

The slope of the shear diagram decreases uniformly from 0 at E to $-40$ kN/m at F. Thus, from the area under the load diagram,

$$V_F = V_{E'} + \Delta V = 30 + \frac{1}{2}(-40)(1.5) = 0$$

(b)

(c)

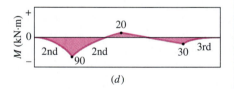

(d)

The shear at the right end of the beam must be zero since it is a free end. If the shear at $F$ is not zero, it indicates that an error has occurred.

The moment diagram, shown in Fig. 8-20d, is drawn below the shear diagram. The moment at the left end of the beam is zero. The slope of the moment diagram is equal to the shear $V$; therefore, the slope of the moment diagram decreases uniformly from 0 at $A$ to $-120$ kN at support $B$. The change in moment from $A$ to $B$ is equal to the area under the shear diagram; thus,

$$M_B = M_A + \Delta M = 0 + \frac{1}{2}(-120)(1.5) = -90 \text{ kN} \cdot \text{m}$$

At support $B$ the slope of the moment diagram changes abruptly from $-120$ kN to $+95$ kN; it then decreases uniformly from $+95$ kN at support $B$ to $+15$ kN at support $C$. From the area under the shear diagram, the moment at $C$ is

$$M_C = M_B + \Delta M = -90 + \frac{1}{2}(95 + 15)(2) = +20 \text{ kN} \cdot \text{m}$$

At support $C$, the slope of the moment diagram changes abruptly from $+15$ kN to $-20$ kN; it then remains constant at $-20$ kN between supports $C$ and $E$. From the area under the shear diagram, the moment at $E$ is

$$M_E = M_C + \Delta M = +20 + (-20)(2.5) = -30 \text{ kN} \cdot \text{m}$$

At support $E$, the slope of the moment diagram changes abruptly from $-20$ kN to $+30$ kN. It then decreases parabolically from $+30$ kN at support $E$ to 0 at end $F$ of the beam. From the area under the shear diagram, the moment at $F$ is

$$M_F = M_E + \Delta M = -30 + \frac{2}{3}(30)(1.5) = 0$$

The moment at the right end of the beam must be zero since it is a free end. If the moment at $F$ were not zero, it would indicate that an error has occurred. Also,

$$M_D = M_C + \Delta M = 20 - 20(1) = 0$$

The moment at hinge $D$ must be zero. If $M_D$ were not zero, it would indicate that an error has occurred.

Calculus would suggest that the locations of maximum shear and maximum moment are where $dV/dx = w = 0$ and $dM/dx = V = 0$, respectively. However, the functions $w(x)$ and $V(x)$ usually are not smooth, continuous functions, and the maximum values of $V$ and $M$ often occur at points where concentrated loads or couples are applied to the beam rather than where $w(x) = 0$ or where $V(x) = 0$. Both possibilities need to be checked.

Sections where the bending moment is zero, called points of inflection or contraflexure, can be located by equating the expression for $M$ to zero. The flexural (bending) stress is zero at such sections, and if a beam must be spliced, the splice should be located at or near a point of inflection if there is one.

Closing of the shear diagram provides a check of the equilibrium equation $\Sigma F_y = 0$ for the complete beam.

Observe that the pin connection at $D$ can transmit a shear but no moment.

Observe that maximum and minimum moments occur at points of zero shear ($dM/dx = V = 0$).

# PROBLEMS

In Problems 8-51 to 8-68, beams are loaded and supported as shown in the accompanying figures. Draw complete shear and moment diagrams for

## Introductory Problems

**8-51*** The beam shown in Fig. P8-51.

**Fig. P8-51**

**8-52*** The beam shown in Fig. P8-52.

**Fig. P8-52**

**8-53** The beam shown in Fig. P8-53.

**Fig. P8-53**

**8-54** The beam shown in Fig. P8-54.

**Fig. P8-54**

**8-55*** The beam shown in Fig. P8-55.

**Fig. P8-55**

**8-56*** The beam shown in Fig. P8-56.

**Fig. P8-56**

**8-57** The beam shown in Fig. P8-57.

**Fig. P8-57**

**8-58** The beam shown in Fig. P8-58.

**Fig. P8-58**

**8-59*** The beam shown in Fig. P8-59.

**Fig. P8-59**

**8-60** The beam shown in Fig. P8-60.

**Fig. P8-60**

## Intermediate Problems

**8-61\*** The beam shown in Fig. P8-61.

**Fig. P8-61**

**8-62\*** The beam shown in Fig. P8-62.

**Fig. P8-62**

**8-63** The beam shown in Fig. P8-63.

**Fig. P8-63**

**8-64** The beam shown in Fig. P8-64.

**Fig. P8-64**

## Challenging Problems

**8-65\*** The beam shown in Fig. P8-65.

**Fig. P8-65**

**8-66\*** The beam shown in Fig. P8-66.

**Fig. P8-66**

**8-67** The beam shown in Fig. P8-67.

**Fig. P8-67**

**8-68** The beam shown in Fig. P8-68.

**Fig. P8-68**

## 8-6   FLEXIBLE CABLES

Flexible cables are used for suspension bridges and aerial tramways, for power transmission and telephone lines, for guy wires on radio and television towers, and for many other engineering applications. A cable is said to be perfectly flexible when it offers no resistance to bending. Actual cables are not perfectly flexible; however, the resistance they offer to bending is generally so small that any bending effects can be neglected in the analysis of the cable without introducing serious error. Once it is assumed that the cable offers no resistance to bending, the resultant internal force on any cross section must act along a tangent to the cable at that cross section.

In previous applications, cables were assumed to be straight two-force members capable of transmitting only axial tensile forces. When transverse loads are applied to a cable, it cannot remain straight but sags. Sag is defined as the difference in elevation between the lowest point on the cable and a support. When the supports are not at the same elevation, the sag measured from one support will be different from the sag measured from the other support. The span of a cable is defined as the horizontal distance between supports.

Flexible cables may be subjected to a series of distinct concentrated loads, or they may be subjected to loads that are uniformly distributed over the horizontal span of the cable or uniformly distributed over the length of the cable. The weights of cars and their contents on an aerial tramway is an example of a cable subjected to a series of concentrated loads. The weight of a suspension-bridge roadway is an example of a load that is uniformly distributed along the horizontal span of the cable. The weight of a power transmission cable of constant cross section is an example of a load that is uniformly distributed along the length of the cable.

In the following discussion of cables it will be assumed that the cables are perfectly flexible and inextensible. Relationships between the length, span, and sag of the cable, the tension in the cable, and the loads applied to the cable will be determined from equilibrium considerations.

### 8-6.1   Cables Subjected to Concentrated Loads

A cable subjected to concentrated loads $P_1$, $P_2$, and $P_3$ at discrete points along its length is depicted in Fig. 8-21$a$. The cable is anchored to rigid walls at ends $A$ and $B$ with pins. If the loads are much larger than the weight of the cable, the weight of the cable can be neglected in the analysis and the segments of the cable can be considered as straight two-force bars.

Assume for the following discussion that loads $P_1$, $P_2$, and $P_3$ together with distances $x_1$, $x_2$, $x_3$, and span $a$ are known. The distances $y_1$, $y_2$, and $y_3$ are unknowns to be determined. A free-body diagram of the cable is shown in Fig. 8-21$b$. Since the distances $y_1$, $y_2$, and $y_3$ are unknown, the slopes of the cable segments at ends $A$ and $B$ are not known; therefore, the reactions at $A$ and $B$ are represented by two components each. Since four unknowns are involved, the three equations of equilibrium for this free-body diagram (Fig. 8-21$b$) are not sufficient to determine the reactions at $A$ and $B$. The information that can be determined is as follows:

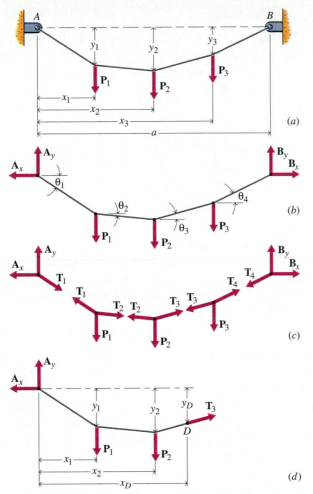

**Figure 8-21** Flexible cable supporting a system of concentrated loads.

From the equilibrium equation $\Sigma M_A = 0$:

$$+\curvearrowleft B_y a - P_1 x_1 - P_2 x_2 - P_3 x_3 = 0$$

$$B_y = \frac{1}{a}(P_1 x_1 + P_2 x_2 + P_3 x_3)$$

From the equilibrium equation $\Sigma F_y = 0$:

$$+\uparrow A_y + B_y - P_1 - P_2 - P_3 = 0$$
$$A_y = P_1 + P_2 + P_3 - B_y$$
$$= P_1 + P_2 + P_3 - \frac{1}{a}(P_1 x_1 + P_2 x_2 + P_3 x_3)$$

From the equilibrium equation $\Sigma F_x = 0$:

$$+\rightarrow B_x - A_x = 0$$
$$B_x = A_x$$

Previously, in the discussion of frames and trusses, additional equations were obtained by considering the equilibrium of a portion of the structure. For a cable, if the weight of the cable is neglected, the internal forces in the different segments of the cable can be represented

**Fig. 8-21**

as shown in Fig. 8-21c. From this series of free-body diagrams and the equilibrium equation $\Sigma F_x = 0$, it is observed that

$$A_x = T_1 \cos \theta_1 = T_2 \cos \theta_2 = T_3 \cos \theta_3 = T_4 \cos \theta_4 = B_x$$

This equation indicates that the horizontal component of the tensile force at any point in the cable is constant and equal to the horizontal component of the pin reaction at the supports. The maximum tension $T$ will occur in the segment with the largest angle of inclination $\theta$ since $\cos \theta$ will be minimum in this segment. Such a segment must be adjacent to one of the two supports. The equilibrium equation $\Sigma F_x = 0$ and $\Sigma F_y = 0$ applied to the pins at supports $A$ and $B$ yield

$$
\begin{array}{ll}
A_x = T_1 \cos \theta_1 & B_x = T_4 \cos \theta_4 \\
A_y = T_1 \sin \theta_1 & B_y = T_4 \sin \theta_4
\end{array}
$$

from which

$$
\begin{array}{ll}
T_1 = \sqrt{A_x^2 + A_y^2} & T_4 = \sqrt{B_x^2 + B_y^2} \\
\theta_1 = \tan^{-1} \dfrac{A_y}{A_x} & \theta_4 = \tan^{-1} \dfrac{B_y}{B_x}
\end{array}
\qquad (a)
$$

If either the maximum tension or the maximum slope is specified for a given problem, Eqs. *a* can be used to determine the unknown horizontal components of the support reactions. Once either $A_x$ or $B_x$ is known, all the remaining unknowns, $T_1$, $T_2$, $T_3$, $T_4$, $y_1$, $y_2$, and $y_3$ can be determined by using the free-body diagrams shown in Fig. 8-21c. The procedure is illustrated in Example Problem 8-11.

The unknown horizontal components of the support reactions can also be determined if the vertical distance from a support to any point along the cable is known. For example, consider the free-body diagram shown in Fig. 8-21d, where it is assumed that the distance $y_D$ is known. The equilibrium equation $\Sigma M_D = 0$ yields

$$A_y x_D - P_1(x_D - x_1) - P_2(x_D - x_2) - A_x y_D = 0$$

from which

$$A_x = \frac{1}{y_D}[A_y x_D - P_1(x_D - x_1) - P_2(x_D - x_2)]$$

Finally, problems of this type can be solved by specifying the length of the cable. In this case, the length of each segment of the cable is written in terms of the vertical distances $y_1$, $y_2$, and $y_3$ and the horizontal distances $x_1$, $x_2$, $x_3$, and span $a$. The required additional equation is then obtained by equating the sum of the individual segment lengths to the total length $L$:

$$L = \sqrt{x_1^2 + y_1^2} + \sqrt{(x_2 - x_1)^2 + (y_2 - y_1)^2}$$
$$+ \sqrt{(x_3 - x_2)^2 + (y_2 - y_3)^2} + \sqrt{(a - x_3)^2 + y_3^2}$$

Since the terms for the segment lengths in this equation involve square roots of the unknown vertical distances, any solution is extremely tedious and time-consuming if performed by hand calculation. Computer solution is recommended for this formulation of cable problems.

A cable supports concentrated loads of 500 lb and 200 lb as shown in Fig. 8-22a. If the maximum tension in the cable is 1000 lb, determine

**a.** The support reactions $A_x$, $A_y$, $D_x$, and $D_y$.
**b.** The tensions $T_1$, $T_2$, and $T_3$ in the three segments of the cable.
**c.** The vertical distances $y_B$ and $y_C$ from the level of support $A$.
**d.** The length $L$ of the cable.

(a)

(b)

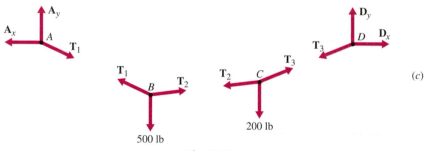

(c)

**Fig. 8-22**

## SOLUTION

**a.** A free-body diagram for the cable is shown in Fig. 8-22b. From the equilibrium equation $\Sigma M_D = 0$,

$$A_y = \frac{1}{34}[500(22) + 200(10)] = 382.4 \cong 382 \text{ lb} \qquad \text{Ans.}$$

The maximum tension of 1000 lb will occur in interval $AB$ of the cable. Thus, from the free-body diagram at point $A$ of the cable (Fig. 8-22c),

$$A_x = \sqrt{T_1^2 - A_y^2} = \sqrt{1000^2 - 382.4^2} = 924.0 = 924 \text{ lb} \qquad \text{Ans.}$$

Since the horizontal component of the tensile force at any point in the cable is constant, the maximum tension in the cable occurs in the segment with the largest angle of inclination.

Also,

$$\theta_1 = \tan^{-1} \frac{A_y}{A_x} = \tan^{-1} \frac{382.4}{924.0} = 22.48°$$

Return now to the free-body diagram for the entire cable shown in Fig. 8-22b. From the equilibrium equation $\Sigma F_x = 0$,

$$D_x = A_x = 924.0 = 924 \text{ lb} \qquad\qquad \text{Ans.}$$

From the equilibrium equation $\Sigma F_y = 0$,

$$D_y = 500 + 200 - A_y = 700 - 382.4 = 317.6 \cong 318 \text{ lb} \qquad\qquad \text{Ans.}$$

b. Free-body diagrams at points $B$ and $C$ (Fig. 8-22c) can be used to determine the tensions $T_2$ and $T_3$. From the equilibrium equation $\Sigma F_x = 0$ and $\Sigma F_y = 0$,

At point B:

$$T_{2x} = T_2 \cos \theta_2 = T_1 \cos \theta_1 = 1000 \cos 22.48° = A_x = 924.0 \text{ lb}$$
$$T_{2y} = T_2 \sin \theta_2 = 500 - T_1 \sin \theta_1 = 500 - 1000 \sin 22.48° = 117.64 \text{ lb}$$
$$T_2 = \sqrt{T_{2x}^2 + T_{2y}^2} = \sqrt{924.0^2 + 117.64^2} = 931.4 \cong 931 \text{ lb} \qquad\qquad \text{Ans.}$$
$$\theta_2 = \tan^{-1} \frac{T_{2y}}{T_{2x}} = \tan^{-1} \frac{117.64}{924.0} = 7.256°$$

At point $C$:

$$T_{3x} = T_3 \cos \theta_3 = T_2 \cos \theta_2 = 931.4 \cos 7.256° = A_x = 924.0 \text{ lb}$$
$$T_{3y} = T_3 \sin \theta_3 = 200 + T_2 \sin \theta_2 = 200 + 931.4 \sin 7.256° = 317.6 \text{ lb}$$
$$T_3 = \sqrt{T_{3x}^2 + T_{3y}^2} = \sqrt{924.0^2 + 317.6^2} = 977.1 \cong 977 \text{ lb} \qquad\qquad \text{Ans.}$$
$$\theta_3 = \tan^{-1} \frac{T_{3y}}{T_{3x}} = \tan^{-1} \frac{317.6}{924.0} = 18.969°$$

As a check at point $D$:

$$T_3 = \sqrt{D_x^2 + D_y^2} = \sqrt{924.0^2 + 317.6^2} = 977.1 \cong 977 \text{ lb} \qquad\qquad \text{Ans.}$$

c. Once the angles are known, the vertical distances $y_B$ and $y_C$ are

$$y_B = 12 \tan \theta_1 = 12 \tan 22.48° = 4.966 \cong 4.97 \text{ ft} \qquad\qquad \text{Ans.}$$
$$y_C = 10 \tan \theta_3 = 10 \tan 18.969° = 3.439 \cong 3.44 \text{ ft} \qquad\qquad \text{Ans.}$$

d. The cable length $L$ is obtained from the segment lengths as

$$L = \sqrt{12^2 + 4.966^2} + \sqrt{12^2 + (4.966 - 3.439)^2} + \sqrt{10^2 + 3.439^2}$$
$$= 35.66 \cong 35.7 \text{ ft} \qquad\qquad \text{Ans.}$$

# PROBLEMS

## Introductory Problems

**8-69\*** A cable supports two vertical loads as shown in Fig. P8-69. If the maximum tension in the cable is 2000 lb, determine

**a.** The horizontal and vertical components of the reactions at supports $A$ and $D$.
**b.** The vertical distances $y_B$ and $y_C$.
**c.** The length $L$ of the cable.

**Fig. P8-69**

**8-70\*** A cable supports two vertical loads as shown in Fig. P8-70. The sag at point $B$ of the cable is 2 m. Determine

**a.** The horizontal and vertical components of the reactions at supports $A$ and $D$.
**b.** The tensions in the three segments of the cable.
**c.** The length $L$ of the cable.

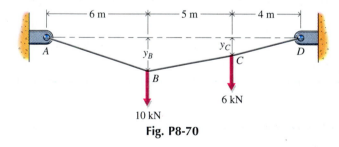

**Fig. P8-70**

**8-71** Solve Problem 8-69 if the maximum tension in the cable is 1500 lb.

**8-72** Solve Problem 8-70 if the sag at point $B$ of the cable is 3 m.

**8-73\*** A cable supports three vertical loads as shown in Fig. P8-73. The sag at point $C$ of the cable is 4 ft. Determine

**a.** The horizontal and vertical components of the reactions at supports $A$ and $E$.
**b.** The tensions in the four segments of the cable.
**c.** The length $L$ of the cable.

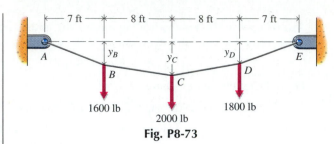

**Fig. P8-73**

**8-74\*** A cable supports three vertical loads as shown in Fig. P8-74. If the maximum tension in the cable is 50 kN, determine

**a.** The horizontal and vertical components of the reactions at supports $A$ and $D$.
**b.** The vertical distances $y_B$, $y_C$, and $y_D$.
**c.** The length $L$ of the cable.

**Fig. P8-74**

**8-75** Solve Problem 8-73 if the sag at point $C$ of the cable is 6 ft.

**8-76** Solve Problem 8-74 if the maximum tension in the cable is 35 kN.

## Intermediate Problems

**8-77\*** A cable supports two vertical loads as shown in Fig. P8-77. If the maximum tension in the cable is 1000 lb, determine

**a.** The horizontal and vertical components of the reactions at supports $A$ and $D$.

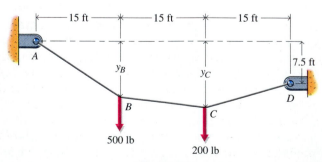

**Fig. P8-77**

**448**

**b.** The vertical distances $y_B$ and $y_C$.
**c.** The length $L$ of the cable.

**8-78** A cable is loaded and supported as shown in Fig. P8-78. Determine

**a.** The horizontal and vertical components of the reactions at supports $A$ and $D$.
**b.** The tensions in the three segments of the cable.
**c.** The vertical distance $y_C$.
**d.** The length $L$ of the cable.

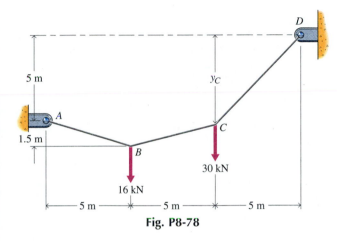

**Fig. P8-78**

### Challenging Problems

**8-79*** A cable is loaded and supported as shown in Fig. P8-79. The vertical distance $y_C$ is 4 ft. Determine
**a.** The horizontal and vertical components of the reactions at supports $A$ and $D$.
**b.** The tensions in the three segments of the cable.
**c.** The vertical distance $y_B$.
**d.** The length $L$ of the cable.

**Fig. P8-79**

**8-80*** A cable is loaded and supported as shown in Fig. P8-80. If the maximum tension in the cable is 40 kN, determine

**a.** The horizontal and vertical components of the reactions at supports $A$ and $D$.
**b.** The vertical distances $y_B$ and $y_C$.
**c.** The length $L$ of the cable.

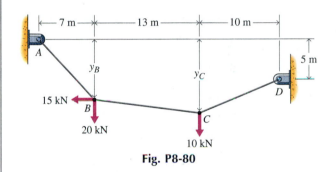

**Fig. P8-80**

**8-81** Solve Problem 8-79 if the direction of the 600-lb load at point $C$ is reversed.

**8-82** Solve Problem 8-80 if the direction of the 15-kN load at point $B$ is reversed.

## 8-6.2 Cables with Loads Uniformly Distributed Along the Horizontal

A cable carrying a load **W** distributed uniformly over a distance $a$ is illustrated in Fig. 8-23a. Such a cable could be analyzed by using the procedures discussed in the previous section; however, the process would be time-consuming and tedious due to the large number of loads involved. Such a cable can also be analyzed by representing the large number of discrete loads as a uniformly distributed load $w(x) = W/a$ along the horizontal. Little error is introduced by this simplification if the weight of the cable is small in comparison to the weight being supported by the cable. The loads on the cables of a suspension bridge closely approximate this type of loading since the weights of the cables are usually small in comparison to the weight of the roadway. The weight of the roadway is uniformly distributed along the length of the roadway.

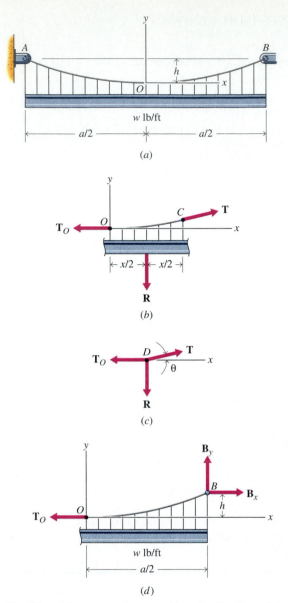

**Figure 8-23**   Flexible cable supporting a uniformly distributed load.

Equations that express relationships between the length $L$, span $a$, and sag $h$ of a cable with supports at the same height, the tension $T$ in the cable, and the distributed load $w(x)$ applied to the cable can be developed by using equilibrium considerations and the free-body diagram of a portion of the cable shown in Fig. 8-23$b$. In this diagram, the lowest point on the cable (point $O$) has been taken as the origin of the $xy$-coordinate system.

The segment of cable shown in Fig. 8-23$b$ is subjected to three forces; the tension $\mathbf{T}_O$ in the cable at point $O$, the tension $\mathbf{T}$ in the cable at an arbitrary point $C$, located a distance $x$ from the origin, and the resultant $\mathbf{R}$ ($R = wx$) of the distributed load whose line of action is located a distance $x/2$ from point $O$. Since segment $OC$ of the cable is in equilibrium under the action of these three forces, the three forces must be concurrent at point $D$, as shown in Fig. 8-23$c$. The two equilibrium equations $\Sigma F_x = 0$ and $\Sigma F_y = 0$ for this concurrent force system yield

$$+\rightarrow\Sigma F_x = T \cos \theta - T_O = 0 \qquad T_x = T \cos \theta = T_O \qquad (a)$$
$$+\uparrow\Sigma F_y = T \sin \theta - R = 0 \qquad T_y = T \sin \theta = R = wx \qquad (b)$$

Equation $a$ shows that the horizontal component $T_x$ of the tension $T$ at any point in the cable is constant and equal to the tension $T_O$ at the lowest point on the cable. Solving Eqs. $a$ and $b$ for $T$ and $\theta$ yields

$$T = \sqrt{T_O^2 + w^2 x^2} \qquad \text{(8-5)}$$

$$\theta = \tan^{-1} \frac{wx}{T_O} \qquad \text{(8-6)}$$

Equation 8-5 shows that the tension is minimum at the lowest point in the cable (where $x = 0$) and maximum at the supports (where $x = a/2$). Thus,

$$T_{\max} = \sqrt{T_O^2 + \frac{w^2 a^2}{4}} \qquad \text{(8-7)}$$

The shape of the curve can be determined by using Eq. 8-6, which gives the slope. Thus,

$$\frac{dy}{dx} = \tan \theta = \frac{wx}{T_O} \qquad (c)$$

Integrating Eq. $c$ yields

$$y = \frac{wx^2}{2T_O} + C$$

The constant $C$ is determined by using the boundary conditions resulting from the choice of axes. For the $xy$-coordinate system being used, $y$ is equal to zero when $x$ is equal to zero; therefore, $C$ is equal to zero, and the equation of the loaded cable is

$$y = \frac{wx^2}{2T_O} = kx^2 \qquad \text{(8-8)}$$

Equation 8-8 indicates that the shape of the loaded cable is a parabola with its vertex at the lowest point of the cable. By applying the boundary condition $y = h$ (the sag) where $x = a/2$, an expression for the tension $T_O$ is obtained in terms of the applied load $w$, the span $a$, and the sag $h$. Thus

$$T_O = \frac{wa^2}{8h} \qquad \text{(8-9)}$$

If Eq. 8-9 is substituted into Eq. 8-7, the maximum tension in the cable $T_{\max}$ is obtained in terms of the applied load $w$, the span $a$, and the sag $h$. Thus,

$$T_{max} = \sqrt{\frac{w^2 a^4}{64 h^2} + \frac{w^2 a^2}{4}}$$
$$= \frac{wa}{8h}\sqrt{a^2 + (4h)^2} \qquad \text{(8-10)}$$

Equations 8-9 and 8-10 can also be determined by using the free-body diagram of the right half of the cable shown in Fig. 8-23$d$. From the equilibrium equation $\Sigma M_B = 0$,

$$+\lcurvearrowleft M_B = \frac{wa}{2}\left(\frac{a}{4}\right) - T_O(h) = 0 \qquad T_O = \frac{wa^2}{8h} \qquad \text{(8-9)}$$

From the equilibrium equation $\Sigma F_x = 0$,

$$+\rightarrow \Sigma F_x = B_x - T_O = 0 \qquad B_x = T_O = \frac{wa^2}{8h}$$

From the equilibrium equation $\Sigma F_y = 0$,

$$+\uparrow \Sigma F_y = B_y - \frac{wa}{2} = 0 \qquad B_y = \frac{wa}{2}$$

Thus,

$$T_{\max} = B = \sqrt{B_x^2 + B_y^2} = \sqrt{\left(\frac{wa^2}{8h}\right)^2 + \left(\frac{wa}{2}\right)^2}$$

$$= \frac{wa}{8h}\sqrt{a^2 + (4h)^2} \qquad \text{(8-10)}$$

For any curve, the length $dL$ of an arc of the curve can be obtained from the equation

$$dL = \sqrt{(dx)^2 + (dy)^2} = \sqrt{1 + \left(\frac{dy}{dx}\right)^2}\, dx \qquad \text{(d)}$$

Thus, for the cable, from Eqs. 8-8 and 8-9,

$$y = kx^2 = \left(\frac{w}{2T_O}\right)x^2 = \left(\frac{4h}{a^2}\right)x^2$$

$$\frac{dy}{dx} = 2kx = \left(\frac{8h}{a^2}\right)x$$

$$L = \int_{\text{span}} dL = 2\int_0^{a/2} \sqrt{1 + 4k^2x^2}\, dx$$

$$= \left[x\sqrt{1 + 4k^2\,x^2} + \frac{1}{2k}\ln\left(2kx + \sqrt{1 + 4k^2x^2}\right)\right]_0^{a/2}$$

$$= \frac{a}{2}\sqrt{1 + k^2a^2} + \frac{1}{2k}\ln\left(ka + \sqrt{1 + k^2a^2}\right)$$

Finally, substituting $k = 4h/a^2$ yields

$$L = \frac{1}{2}\sqrt{a^2 + 16h^2} + \frac{a^2}{8h}\ln\frac{1}{a}(4h + \sqrt{a^2 + 16h^2} \qquad \text{(8-11)}$$

When the supports have different elevations, as shown in Fig. 8-24, the location of the lowest point on the cable (the origin of the $xy$-coordinate system) is not known and must be determined. The equation of the loaded cable (Eq. 8-8) remains valid since the free-body diagram (Fig. 8-23$b$) used in its development would be identical for a cable with supports at different elevations as shown in Fig. 8-24. If the sags $h_A$ and $h_B$ of the cable with respect to supports $A$ and $B$ are known, Eq. 8-8 can be used to locate the origin of the coordinate system. Thus, from Eq. 8-8,

$$y = kx^2$$

$$k = \frac{y}{x^2} = \frac{h_B}{x_B^2} = \frac{h_A}{x_A^2} \qquad (e)$$

Since $x_A = x_B - a$, Eq. $e$ can be written

$$k = \frac{y}{x^2} = \frac{h_B}{x_B^2} = \frac{h_A}{(x_B - a)^2}$$

which leads to the quadratic equation

$$(h_B - h_A)x_B^2 - 2ah_B x_B + h_B a^2 = 0$$

Solving for $x_B$ yields

$$x_B = \frac{h_B - \sqrt{h_B h_A}}{h_B - h_A} a \qquad (8\text{-}12)$$

Also, since $x_A = x_B - a$,

$$x_A = \frac{h_A - \sqrt{h_B h_A}}{h_B - h_A} a \qquad (8\text{-}13)$$

The horizontal distance from support $A$ to the origin $O$ of the coordinate system is shown as distance $d$ in Fig. 8-24. At support $A$, $x_A = -d$. Thus,

$$d = \frac{\sqrt{h_B h_A} - h_A}{h_B - h_A} a \qquad (8\text{-}14)$$

The tension $T_O$ in the cable can be found by using Eq. 8-8. Thus,

$$T_O = \frac{wx^2}{2y} = \frac{wx_A^2}{2h_A} = \frac{wx_B^2}{2h_B} \qquad (8\text{-}15)$$

Equation 8-5 shows that the tension in the cable varies from a minimum value $T_O$ at the lowest point in the cable to a maximum value at the highest (most distant) support. By substituting the magnitudes of $x_A$ and $x_B$ in Eq. 8-5 and using Eq. 8-15, we find that the tensions in the cable at the supports are

$$T_A = wd_A\left[1 + \frac{d_A^2}{4h_A^2}\right]^{1/2} \qquad (8\text{-}16a)$$

$$T_B = wd_B\left[1 + \frac{d_B^2}{4h_B^2}\right]^{1/2} \qquad (8\text{-}16b)$$

where $d_A = |x_A|$ and $d_B = |x_B|$.

**Figure 8-24**  Flexible cable with supports at different elevations.

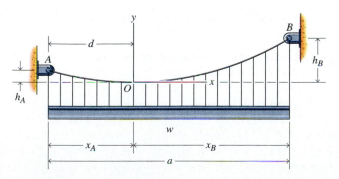

The length of the curve between the lowest point $O$ and a support can be determined by using Eq. $d$. Thus, for support $B$,

$$
\begin{aligned}
L_{OB} &= \int_0^{d_B} \sqrt{1 + 4k^2x^2}\, dx \\
&= \left[ \frac{x}{2}\sqrt{1 + 4k^2x^2} + \frac{1}{4k}\ln\left(2kx + \sqrt{1 + 4k^2x^2}\right) \right]_0^{d_B} \\
&= \frac{d_B}{2}\sqrt{1 + 4k^2d_B^2} + \frac{1}{4k}\ln\left(2kd_B + \sqrt{1 + 4k^2d_B^2}\right)
\end{aligned}
$$

Finally, substituting $k = h_B/d_B^2$ yields

$$
L_{OB} = \frac{1}{2}\sqrt{d_B^2 + 4h_B^2} + \frac{d_B^2}{4h_B}\ln\frac{1}{d_B}(2h_B + \sqrt{d_B^2 + 4h_B^2}) \tag{8-17}
$$

Similarly,

$$
L_{OA} = \frac{1}{2}\sqrt{d_A^2 + 4h_A^2} + \frac{d_A^2}{4h_A}\ln\frac{1}{d_A}(2h_A + \sqrt{d_A^2 + 4h_A^2}) \tag{8-18}
$$

The total length $L$ of the cable is the sum of the arc lengths $L_{OA}$ and $L_{OB}$ measured from the origin to each support.

A cable supports a uniformly distributed load of 300 lb/ft along the horizontal as shown in Fig. 8-25a. Determine

a. The minimum tension in the cable.
b. The tension in the cable at the supports and the angle it makes with the horizontal.
c. The length $L$ of the cable.

$w = 300$ lb/ft
800 ft

(a)

$B_y$
$B$
$B_x$
120 ft
$y$
$T_O$
$O$
$x$
200 ft — 200 ft
$W$

(b)

**Fig. 8-25**

---

## SOLUTION

Parts $a$ and $b$ of the problem will be solved by using the free-body diagram shown in Fig. 8-25b. Thus,

$$W = w(400) = 300(400) = 120{,}000 \text{ lb} = 120 \text{ kip}$$

a. From the equilibrium equation $\Sigma M_B = 0$,

$$+\curvearrowleft M_B = W(200) - T_O(120) = 120(200) - T_O(120) = 0$$
$$T_O = 200 \text{ kip} \qquad \text{Ans.}$$

> Since the supports are at the same elevation, the low point in the cable occurs at the middle of the span.

b. From the equilibrium equation $\Sigma F_x = 0$,

$$+\rightarrow \Sigma F_x = B_x - T_O = B_x - 200 = 0 \quad B_x = 200 \text{ kip}$$

From the equilibrium equation $\Sigma F_y = 0$,

$$+\uparrow \Sigma F_y = B_y - W = B_y - 120 = 0 \quad B_y = 120 \text{ kip}$$
$$T_A = T_B = T_{max} = \sqrt{B_x^2 + B_y^2}$$
$$= \sqrt{(200)^2 + (120)^2} = 233 \text{ kip} \qquad \text{Ans.}$$
$$\theta_x = \tan^{-1}\frac{B_y}{B_x} = \tan^{-1}\frac{120}{200} = 30.96 \cong 31.0° \qquad \text{Ans.}$$

c. The length of the cable is determined by using Eq. 8-11. Thus,

$$L = \frac{1}{2}\sqrt{a^2 + 16h^2} + \frac{a^2}{8h} \ln \frac{1}{a}(4h + \sqrt{a^2 + 16h^2})$$
$$= \frac{1}{2}\sqrt{(800)^2 + 16(120)^2}$$
$$+ \frac{(800)^2}{8(120)} \ln \frac{1}{800}\left[4(120) + \sqrt{(800)^2 + 16(120)^2}\right]$$
$$= 845.7 \cong 846 \text{ ft} \qquad \text{Ans.}$$

The cable shown in Fig. 8-26a supports a pipeline as it passes over a river. The support at the middle of the river is 25 m above the supports on the two sides. The lowest points on the cable are 45 m below the middle support. Determine

a. The minimum tension in the cable.
b. The maximum tension in the cable.
c. The length L of the cable.

(a)

(b)

**Fig. 8-26**

## SOLUTION

The lowest point in the cable (origin of the $xy$-coordinate system) can be located by using Eq. 8-14. Thus,

$$d = \frac{\sqrt{h_B h_A} - h_A}{h_B - h_A} a = \frac{\sqrt{45(20)} - 20}{45 - 20} (300) = 120.0 \text{ m}$$

Once the distance $d$ is known, the free-body diagram shown in Fig. 8-26b can be used to solve parts $a$ and $b$ of the problem. The distance from the origin of coordinates to support $B$ is

$$x_B = a - d = 300 - 120 = 180 \text{ m}$$
$$W = w(x_B) = 5(180) = 900 \text{ kN}$$

a. From the equilibrium equation $\Sigma M_B = 0$,

$$+ \curvearrowleft M_B = W\left(\frac{x_B}{2}\right) - T_O(h_B) = 900(90) - T_O(45) = 0$$
$$T_O = T_{min} = 1800 \text{ kN} \qquad \text{Ans.}$$

> If the ends of the cable are not at the same level, it is not possible to use symmetry to locate the lowest point on the cable.

**b.** From the equilibrium equation $\Sigma F_x = 0$,

$$+\rightarrow\Sigma F_x = B_x - T_O = B_x - 1800 = 0 \qquad B_x = 1800 \text{ kN}$$

From the equilibrium equation $\Sigma F_y = 0$,

$$+\uparrow\Sigma F_y = B_y - W = B_y - 900 = 0 \qquad B_y = 900 \text{ kN}$$
$$T_A = T_B = T_{max} = \sqrt{B_x^2 + B_y^2}$$
$$= \sqrt{(1800)^2 + (900)^2} = 2010 \text{ kN} \qquad \text{Ans.}$$

**c.** The length of the cable is determined by using Eqs. 8-17 and 8-18. Thus,

$$L_{OB} = \frac{1}{2}\sqrt{d_B^2 + 4h_B^2} + \frac{d_B^2}{4h_B}\ln\frac{1}{d_B}\left(2h_B + \sqrt{d_B^2 + 4h_B^2}\right)$$
$$= \frac{1}{2}\sqrt{(180)^2 + 4(45)^2}$$
$$+ \frac{(180)^2}{4(45)}\ln\frac{1}{180}\left[2(45) + \sqrt{(180)^2 + 4(45)^2}\right]$$
$$= 187.24 \text{ m}$$

Similarly,

$$L_{OA} = \frac{1}{2}\sqrt{d_A^2 + 4h_A^2} + \frac{d_A^2}{4h_A}\ln\frac{1}{d_A}\left(2h_A + \sqrt{d_A^2 + 4h_A^2}\right) \qquad \textbf{(8-18)}$$
$$= \frac{1}{2}\sqrt{(120)^2 + 4(20)^2}$$
$$+ \frac{(120)^2}{4(20)}\ln\frac{1}{120}\left[2(20) + \sqrt{(120)^2 + 4(20)^2}\right]$$
$$= 122.19 \text{ m} \qquad \text{Ans.}$$

Therefore,

$$L = 2(L_{OA} + L_{OB}) = 2(122.19 + 187.24) = 618.9 \cong 619 \text{ m} \qquad \text{Ans.}$$

# PROBLEMS

### Introductory Problems

**8-83*** A cable with supports at the same elevation has a span of 600 ft and a sag of 60 ft. The cable supports a uniformly distributed load of 750 lb/ft along the horizontal. Determine

**a.** The maximum tension in the cable.
**b.** The length of the cable.

**8-84*** A cable with supports at the same elevation is being designed to carry a uniformly distributed load of 1.50 kN/m along the horizontal. The sag of the cable at midspan is to be 4 m. If the maximum tension in the cable must be limited to 12 kN, determine

**a.** The maximum allowable span for the cable.
**b.** The required length for the cable.

**8-85** The load carried by each cable of the suspension bridge shown in Fig. P8-85 is 2000 lb per foot of horizontal length. The span of the bridge is 1000 ft, and the sag at midspan is 100 ft. Determine

**a.** The tension in the cable at the middle of the span.
**b.** The tension in the cable at the supports and the angle it makes with the horizontal.
**c.** The length of the cable.

**Fig. P8-85**

**8-86** The center span of the suspension bridge shown in Fig. P8-86 is 500 m. The sag at the center of the span is 50 m. The cables have been designed to resist a maximum tension of 5000 kN. Determine

a.  The design load per horizontal meter of roadway.
b.  The length of the cable in the center span.

**Fig. P8-86**

**8-87\*** A cable with supports at the same elevation carries a uniformly distributed load of 500 lb/ft along the horizontal. If the maximum tension in the cable is 100 kip and the sag at midspan is 40 ft, determine

a.  The horizontal distance between supports.
b.  The length of the cable.

**8-88\*** A cable with supports at the same elevation has a span of 400 m. The cable supports a uniformly distributed load of 6 kN/m along the horizontal. The maximum tension in the cable is 5000 kN. Determine

a.  The angle between the cable and the horizontal at a support.
b.  The sag in the cable at the middle of the span.
c.  The length of the cable.

**8-89** A cable with supports at the same elevation has a span of 600 ft. The length of the cable is 630 ft. If the maximum tension in the cable resulting from a uniformly distributed load $w$ along the horizontal is 360 kip, determine

a.  The sag in the cable at the middle of the span.
b.  The angle between the cable and the horizontal at a support.
c.  The magnitude of the distributed load $w$.

**Intermediate Problems**

**8-90\*** The cable shown in Fig. P8-90 carries a uniformly distributed load of 25 kN per meter of horizontal length. The left support of the cable is located 50 m below the right support. The angle between the cable and the horizontal at the left support is 10°. Determine

a.  The tension in the cable at the left support.
b.  The tension in the cable at the right support.
c.  The angle between the cable and the horizontal at the right support.
d.  The length of the cable.

**Fig. P8-90**

**8-91** The left support of the cable shown in Fig. P8-91 is located 40 ft below the right support. The cable is horizontal at the left support. If the maximum tension in the cable resulting from a uniformly distributed load $w$ along the horizontal is 400 kip, determine

a.  The angle between the cable and the horizontal at the right support.
b.  The magnitude of the distributed load $w$.
c.  The length of the cable.

**Fig. P8-91**

**Challenging Problems**

**8-92\*** The left support of the cable shown in Fig. P8-92 is located 10 m below the right support. The lowest point on the cable is 13 m below the right support. If the maximum tension in the cable resulting from a uniformly distributed load $w$ along the horizontal is 400 kN, determine

a.  The angle between the cable and the horizontal at the right support.
b.  The magnitude of the distributed load $w$.
c.  The length of the cable.

**Fig. P8-92**

**8-93\*** A cable carries a uniformly distributed load of 500 lb per foot of horizontal length. The supports are 500 ft apart and the left support is 30 ft lower than the right support. The lowest point on the cable is 25 ft below the left support. Determine

a. The minimum tension in the cable.
b. The maximum tension in the cable.
c. The angle between the cable and the horizontal at the right support.
d. The length of the cable.

**8-94** The right support of a cable is located 45 m below the left support. The horizontal distance between supports is 250 m. The lowest point on the cable is 15 m below the right support. If the maximum tension in the cable resulting from a uniformly distributed load $w$ along the horizontal is 850 kN, determine

a. The angle between the cable and the horizontal at the right support.
b. The magnitude of the distributed load $w$.
c. The length of the cable.

## 8-6.3 Cables with Loads Uniformly Distributed Along Their Length

In the previous two sections, the weight of the cable was small in comparison to the loads being supported by the cable. For power transmission lines, telephone lines, and guy wires on radio and television towers, the weight of the cable is the only significant load being applied. In these applications, the weight is uniformly distributed along the length of the cable. When the supports are at the same elevation and the sag ratio is small (a taut cable), the curve assumed by the cable may be regarded with small error as being a parabola, since a uniformly distributed load along the cable does not differ significantly from the same load uniformly distributed along the horizontal. When the sag ratio is large ($h/a > 0.10$), the parabolic formulas of the previous section should not be used.

A uniform cable having a weight per unit length $w$ over its length is illustrated in Fig. 8-27a. Equations that express relationships between the length $L$, span $a$, and sag $h$ of the cable, the tension $T$ in the cable, and the weight $w$ of the cable can be developed by using equilibrium considerations and the free-body diagram of a portion of the cable shown in Fig. 8-27b. In this diagram, the lowest point on the cable (point $O$) has been taken as the origin of the $xy$-coordinate system.

The segment of cable shown in Fig. 8-27b is subjected to three forces; the tension $\mathbf{T}_O$ in the cable at point $O$, the tension $\mathbf{T}$ in the cable at an arbitrary point $C$, which is located a distance $s$ from the origin, and the resultant $\mathbf{R}$ ($R = ws$) of the distributed load, whose line of action is located at an unknown distance $x$ from point $O$. The weight acts at the centroid of the curve formed by the cable. The angle that $\mathbf{T}$ makes with the horizontal is denoted by $\theta$. Since segment $OC$ of the cable is in equilibrium under the action of these three forces, the three forces must be concurrent at point $D$ as shown in Fig. 8-27c. The two equilibrium equations $\Sigma F_x = 0$ and $\Sigma F_y = 0$ for this concurrent force system yield

$$+\rightarrow\Sigma F_x = T \cos \theta - T_O = 0 \qquad T_x = T \cos \theta = T_O \qquad (a)$$
$$+\uparrow\Sigma F_y = T \sin \theta - R = 0 \qquad T_y = T \sin \theta = R = ws \qquad (b)$$

Equation $a$ shows that the horizontal component $T_x$ of the tension $T$ at any point in the cable is constant and equal to the tension $T_O$ at the

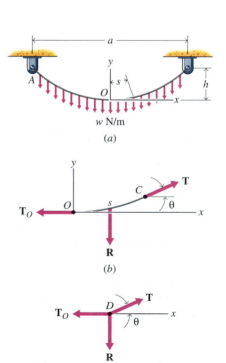

**Figure 8-27** Flexible cable supporting a load uniformly distributed along its length.

459

lowest point on the cable. Solving Eqs. $a$ and $b$ for $T$ and $\theta$ yields

$$T = [T_O^2 + w^2 s^2]^{1/2} \tag{8-19}$$

$$\theta = \tan^{-1} \frac{ws}{T_O} \tag{8-20}$$

Equation 8-19 shows that the tension is minimum at the lowest point in the cable (where $s = 0$) and maximum at the supports (where $s$ is greatest). The distance $s$ remains to be determined.

The shape of the curve can be determined by using Eq. 8-20, which gives the slope. Thus,

$$\frac{dy}{dx} = \tan \theta = \frac{ws}{T_O} \tag{c}$$

Recall the relationship

$$(ds)^2 = (dx)^2 + (dy)^2$$

from which

$$\frac{dy}{dx} = \left[\left(\frac{ds}{dx}\right)^2 - 1\right]^{1/2} \tag{d}$$

Substituting Eq. $d$ into Eq. $c$ yields

$$\left[\left(\frac{ds}{dx}\right)^2 - 1\right]^{1/2} = \frac{ws}{T_O}$$

from which

$$\frac{ds}{dx} = \left[1 + \left(\frac{ws}{T_O}\right)^2\right]^{1/2}$$

Solving for $dx$ yields

$$dx = \frac{(T_O/w)dx}{[(T_O/w)^2 + s^2]^{1/2}} \tag{e}$$

Integrating Eq. $e$ yields

$$x = \left(\frac{T_O}{w}\right) \ln \left\{ s + \left[\left(\frac{T_O}{w}\right)^2 + s^2\right]^{1/2} \right\} + C$$

The integration constant $C$ can be determined by substituting the boundary condition $s = 0$ when $x = 0$. Thus,

$$C = -\left(\frac{T_O}{w}\right) \ln \left(\frac{T_O}{w}\right)$$

therefore,

$$x = \left(\frac{T_O}{w}\right) \ln \frac{s + [(T_O/w)^2 + s^2]^{1/2}}{(T_O/w)} \tag{8-21}$$

Equation 8-21 can also be written in exponential form as

$$\left(\frac{T_O}{w}\right) e^{wx/T_O} = s + \left[\left(\frac{T_O}{w}\right)^2 + s^2\right]^{1/2}$$

Solving for $s$ yields

$$s = \left(\frac{T_O}{w}\right)\left[\frac{e^{wx/T_O} - e^{-wx/T_O}}{2}\right] = \left(\frac{T_O}{w}\right)\sinh\left(\frac{wx}{T_O}\right) \qquad (8\text{-}22)$$

Once the distance $s$ is known in terms of $x$, Eq. 8-19 yields

$$T = T_O[1 + \sinh^2(wx/T_O)]^{1/2} = T_O\cosh(wx/T_O) \qquad (8\text{-}23)$$

Similarly, Eq. 8-22 can be substituted into Eq. $c$ to obtain

$$\frac{dy}{dx} = \sinh\left(\frac{wx}{T_O}\right)$$

which can be integrated to give

$$y = \left(\frac{T_O}{w}\right)\cosh\left(\frac{wx}{T_O}\right) + C$$

Substituting the boundary condition $y = 0$ at $x = 0$ yields

$$C = -\left(\frac{T_O}{w}\right)$$

Thus,

$$y = \left(\frac{T_O}{w}\right)\left[\cosh\left(\frac{wx}{T_O}\right) - 1\right] \qquad (8\text{-}24)$$

Equation 8-24 is the Cartesian equation of a catenary.

The tension $T$ at an arbitrary point along the cable can be expressed in terms of $T_O$ and the $y$-coordinate of the point by solving Eq. 8-24 for $\cosh(wx/T_O)$ and substituting the result into Eq. 8-23. The final result is

$$T = T_O + wy \qquad (8\text{-}25)$$

Equation 8-24 can also be used to determine $T_O$ if both coordinates of a point on the cable are known. For example, the coordinates at a support for a cable with span $a$ and sag $h$ and supports at the same elevation are $x = a/2$ and $y = h$. Thus,

$$h = \left(\frac{T_O}{w}\right)\left[\cosh\left(\frac{wa}{2T_O}\right) - 1\right] \qquad (8\text{-}26)$$

Unfortunately, Eq. 8-26 cannot be solved directly for $T_O$; therefore, values for $T_O$ must be obtained by using trial-and-error solutions, iterative procedures, numerical solutions, or graphs and tables.

A cable with supports at the same elevation has a span of 800 ft. The cable weighs 5 lb/ft and the sag at midspan is 200 ft. Determine

**a.** The tension in the cable at midspan.
**b.** The tension in the cable at a support.
**c.** The length of the cable.

## SOLUTION

**a.** The tension in the cable at midspan can be determined by using Eq. 8-26. Thus,

$$h = \left(\frac{T_O}{w}\right)\left[\cosh\left(\frac{wa}{2T_O}\right) - 1\right]$$

$$200 = \frac{T_O}{5}\left[\cosh\frac{5(800)}{2T_O} - 1\right]$$

A trial-and-error solution for the above equation is shown in the following table. Equation 8-9 for a parabolic cable was used to obtain a first estimate. Thus $T_O = wa^2/8h = 5(800)^2/8(200) = 2000$ lb.

The equation for $T_O$ cannot be solved directly; therefore, values for $T_O$ must be obtained by using trial-and-error methods, iterative procedures, or numerical solutions.

| $T_O$ | $\dfrac{T_O}{5}$ | $\cosh\dfrac{2000}{T_O}$ | $\dfrac{T_O}{5}\left[\cosh\dfrac{2000}{T_O} - 1\right]$ | $h$ | % Error |
|---|---|---|---|---|---|
| 2000 | 400 | 1.54308 | 217.23 | 200 | +8.62 |
| 2100 | 420 | 1.48885 | 205.32 | 200 | +2.66 |
| 2200 | 440 | 1.44248 | 194.69 | 200 | −2.66 |
| 2150 | 430 | 1.46478 | 199.86 | 200 | −0.07 |

Thus,

$$T_O = 2150 \text{ lb} \qquad\qquad \text{Ans.}$$

The tension $T_O$ in the cable at midspan can also be determined by using the Newton-Raphson Method. Equation 8-26 is first written in the form

$$f(T_O) = T_O\left[\cosh\left(\frac{2000}{T_O}\right) - 1\right] - 1000 = 0$$

Using the Newton-Raphson Method

$$(T_O)_{n+1} = (T_O)_n + \delta_n$$

$$\delta_n = -\frac{f(T_O)_n}{f'(T_O)_n}$$

$$f'(T_O) = \left[\cosh\left(\frac{2000}{T_O}\right) - 1\right] + T_O\left(-\frac{2000}{T_O^2}\right)\sinh\left(\frac{2000}{T_O}\right)$$

Using the initial value $(T_O)_0 = 2000$

| $n$ | $(T_O)_n$ | $f(T_O)_n$ | $f'(T_O)_n$ | $\delta_n$ |
|---|---|---|---|---|
| 0 | 2000 | 86.1614 | −0.6321 | 136.3053 |
| 1 | 2136.31 | 6.6034 | −0.5390 | 12.2505 |
| 2 | 2148.56 | 0.0440 | −0.5317 | 0.0845 |
| 3 | 2148.64 | 0.0000 | | |

Thus,

$$T_O = 2149 \text{ lb} \qquad \text{Ans.}$$

**b.** Once the tension at midspan is known, Eq. 8-25 can be used to determine the tension at a support. Thus,

$$T_B = T_{max} = T_O + wh = 2150 + 5(200) = 3150 \text{ lb} \qquad \text{Ans.}$$

**c.** The distance $s$ along the cable from the low point to a support can be determined by using Eq. 8-22. Thus,

$$s = \left(\frac{T_O}{w}\right) \sinh\left(\frac{wx}{T_O}\right)$$

$$s_B = \frac{2150}{5} \sinh \frac{5(400)}{2150} = 460.2 \text{ ft}$$

Therefore

$$L = 2s_B = 2(460.2) = 920.4 \cong 920 \text{ ft} \qquad \text{Ans.}$$

The equivalent results obtained by using the parabolic equations are

$$T_B = \frac{1}{2}wa\left[1 + \left(\frac{a}{4h}\right)^2\right]^{1/2}$$

$$= \frac{1}{2}(5)(800)\left[1 + \left(\frac{800}{4(200)}\right)^2\right]^{1/2} = 2830 \text{ lb}$$

The length of the cable is determined by using Eq. 8-11. Thus,

$$L = \frac{1}{2}\sqrt{a^2 + 16h^2} + \frac{a^2}{8h} \ln \frac{1}{a}\left(4h + \sqrt{a^2 + 16h^2}\right)$$

$$= \frac{1}{2}\sqrt{(800)^2 + 16(200)^2}$$

$$+ \frac{(800)^2}{8(200)} \ln \frac{1}{800}\left[4(200) + \sqrt{(800)^2 + 16(200)^2}\right]$$

$$= 918.2 \cong 918 \text{ ft} \qquad \text{Ans.}$$

The results obtained by using the equations for parabolic and catenary cables are summarized in the following table.

| Quantity | Catenary | Parabolic |
|----------|----------|-----------|
| $T_O$ | 2150 | 2000 |
| $T_{max}$ | 3150 | 2830 |
| $L$ | 920 | 918 |

# PROBLEMS

## Introductory Problems

**8-95*** A flexible cable with supports at the same elevation has a span of 400 ft. If the length of the cable is 500 ft, determine the sag midway between the supports.

**8-96*** A flexible cable with supports at the same elevation has a span of 300 m. If the sag midway between the supports is 80 m, determine the length of the cable.

**8-97** A flexible cable with supports at the same elevation has a span of 900 ft. The cable weighs 0.75 lb/ft and is 1000 ft long. Determine

a. The sag midway between the supports.
b. The maximum tension in the cable.

**8-98** A flexible cable with supports at the same elevation has a span of 250 m. The cable weighs 12 N/m. If the sag midway between the supports is 50 m, determine

a. The length of the cable.
b. The maximum tension in the cable.

**8-99*** A cable with supports at the same elevation has a span of 550 ft. The cable weighs 4 lb/ft and the sag at midspan is 100 ft. Determine

a. The tension in the cable at midspan.
b. The tension in the cable at a support.
c. The length of the cable.

**8-100*** A cable with supports at the same elevation has a span of 300 m. The cable weighs 75 N/m and the sag at midspan is 75 m. Determine

a. The tension in the cable at midspan.
b. The tension in the cable at a support.
c. The length of the cable.

**8-101** A segment of an electric power transmission line with supports at the same elevation is shown in Fig. P8-101. A horizontal force of 3000 lb was applied to the line before it was attached to the towers. The weight of the line is 1.5 lb/ft. Determine

a. The maximum tension in the cable.

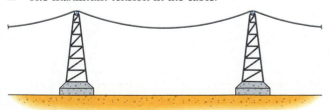

**Fig. P8-101**

b. The sag at the middle of the span.
c. The length of the cable.

## Intermediate Problems

**8-102*** A tugboat is used to tow a barge as shown in Fig. P8-102. The weight of the cable, after accounting for the buoyant force of the water on the submerged cable, is 120 N/m. The horizontal component of the force exerted on the barge by the cable is 40 kN. Determine

a. The maximum tension in the cable.
b. The sag in the cable at its lowest point.
c. The length of the cable.

**Fig. P8-102**

**8-103*** The flexible cable shown in Fig. P8-103 weighs 5 lb/ft. The cable is horizontal at the left support. Determine

a. The tension in the cable at support $A$.
b. The tension in the cable at support $B$.
c. The length of the cable.

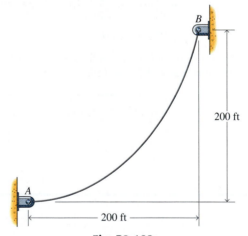

**Fig. P8-103**

**8-104** The flexible cable used to tether the balloon shown in Fig. P8-104 weighs 7.0 N/m and is 100 m long. The balloon exerts a buoyant vertical force of 2000 N on the cable at end $B$. Determine

a. The force exerted by the wind on the balloon.
b. The height $h$ of the balloon.
c. The horizontal distance $d$ from the anchor $A$ to the balloon.

**Fig. P8-106**

**Fig. P8-104**

## Challenging Problems

**8-105\*** An electric power transmission cable weighing 2.5 lb/ft is supported by towers on two sides of a river valley as shown in Fig. P8-105. Determine

a.  The tension in the cable at its lowest point.
b.  The maximum tension in the cable.
c.  The length of the cable.

**Fig. P8-105**

**8-106\*** The cable shown in Fig. P8-106 weighs 30 N/m. If the sag $h_A$ in the cable is 200 m, determine

a.  The tension in the cable at its lowest point.
b.  The tension in the cable at support $A$.
c.  The tension in the cable at support $B$.
d.  The length of the cable.

**8-107** Three pairs of flexible cables, spaced at 120° intervals, are used to stabilize a 1000-ft television tower. The tower and one pair of cables is shown in Fig. P8-107. The weight of the cables is 3.0 lb/ft. The horizontal component of the force exerted by each cable is 6000 lb. At anchor $A$, the angle between cable $AC$ and the horizontal must be 40°. Determine

a.  The maximum tension in cable $AC$.
b.  The horizontal distance $d$ from the tower to anchor $A$.
c.  The length of cable $AC$.

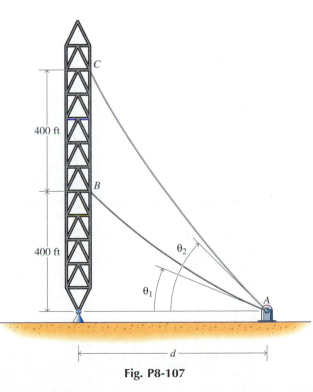

**Fig. P8-107**

## Computer Problems

**C8-108**   An electric power transmission cable weighing 30 N/m is supported by towers on two sides of a river valley as shown in Fig. P8-108.

**a.**   Plot $h_B$, the sag in the cable, as a function of the length of the cable $L$ ($605 \leq L \leq 630$ m).

**b.**   Plot $T_{max}$, the maximum tension in the cable, as a function of the length of the cable $L$ ($605 \leq L \leq 630$ m).

**C8-109**   An electric power transmission cable weighing 2.5 lb/ft is supported by towers on two sides of a river valley as shown in Fig. P8-109. Design specifications require that the maximum tension in the cable be less than 12,000 lb and that the minimum clearance above the river be at least 75 ft. Determine the minimum allowable height of tower $B$.

Fig. P8-108

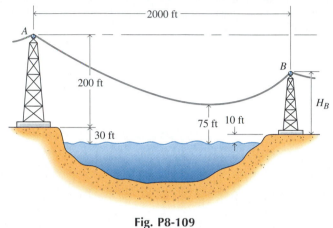

Fig. P8-109

## SUMMARY

When a structural member or machine component is subjected to a system of external loads, a system of internal resisting forces develops within the member to balance the external forces. The resultant of the internal forces on a specific plane within a body can be determined by assuming that the plane separates the body into two parts. Since the body is in equilibrium, each part of the body must also be in equilibrium under the action of the internal forces that develop on the plane. The internal force distribution on the plane can be represented on a free-body diagram of either part of the body by a resultant force **R** at a point on the plane (usually the centroid) and a resultant couple **C.** In most cases, it will be convenient to represent the resultant force **R** and the resultant couple **C** with rectangular components. Thus, with the $x$-axis normal to the plane at the point, the resultant force **R** is represented by a normal force $\mathbf{N}_x$ and two mutually perpendicular in-plane shearing forces $\mathbf{V}_y$ and $\mathbf{V}_z$. The resultant couple **C** is represented by a twisting moment (torque), $\mathbf{M}_x$ and two mutually perpendicular bending moments $\mathbf{M}_y$ and $\mathbf{M}_z$. Generally, it will be possible to solve for the six unknown force and moment components by using the equations

$$\Sigma F_x = 0 \qquad \Sigma F_y = 0 \qquad \Sigma F_z = 0$$
$$\Sigma M_x = 0 \qquad \Sigma M_y = 0 \qquad \Sigma M_z = 0$$

Axial force diagrams provide a graphical representation of the variation of internal axial force along the length of an axially loaded member. In a similar manner, torque diagrams provide a graphical representation of the variation in resisting torque transmitted by transverse cross sections of a shaft subjected to a series of twisting moments. Shear and bending moment diagrams provide similar graphical representations of the variation of shear force and bending moment along the length of a member. From such diagrams, the magnitudes and locations of maximum values of axial force, torque, shear force, or bending moment are easily established. Shear-force and bending-moment diagrams can be established directly from a free-body diagram (load diagram) for a beam by using several simple mathematical relationships that exist between loads and shears and between shears and moments. The equations developed from a free-body diagram of an elemental length of a beam are

$$dV = w\,dx \qquad \text{or} \qquad \frac{dV}{dx} = w \tag{8-1}$$

$$dM = V\,dx \qquad \text{or} \qquad \frac{dM}{dx} = V \tag{8-3}$$

Flexible cables are used for suspension bridges, aerial tramways, power lines, and many other engineering applications. A cable is said to be perfectly flexible when it offers no resistance to bending. Actual cables are not perfectly flexible; however, the resistance they offer to bending is generally so small that any bending effects can be neglected in the analysis of the cable without introducing serious error. Once it is assumed that the cable offers no resistance to bending, the resultant internal force on any cross section must act along a tangent to the cable at that cross section. Typical types of loads applied to flexible cables include a series of distinct vertical concentrated loads, uniformly distributed vertical loads over the horizontal span of the cable, and uniformly distributed vertical loads (such as the weight of the cable) over the length of the cable. For these three types of loading, the horizontal component of the internal force on a cross section is constant for the length of the cable.

# REVIEW PROBLEMS

8-110*   Determine the internal resisting forces and moment transmitted by section $aa$ in the bracket shown in Fig. P8-110.

8-111*   A simple two-bar mechanism supports a 500-lb load as shown in Fig. P8-111. Determine the internal resisting forces and moment transmitted

a.   By cross section $aa$ in bar $AC$.
b.   By the cross section at support $A$.

Fig. P8-110

Fig. P8-111

**8-112**  The hook shown in Fig. P8-112 supports a 10-kN load. Determine the internal resisting forces and moment transmitted by

**a.** Section *aa*.
**b.** Section *bb*.
**c.** Section *cc*.

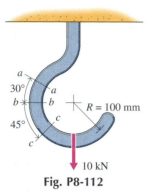

**Fig. P8-112**

**8-113**  An 800-lb automobile engine is supported by an engine hoist as shown in Fig. P8-113. Determine the internal resisting forces and moment transmitted by cross section *aa* in beam *ABC*.

**Fig. P8-113**

**8-114\***  Two bars, a pulley, and a cable are used to support a block as shown in Fig. P8-114. The two bars have negligible weight. The mass of the pulley is 50 kg and the mass of the block is 100 kg. Determine the internal resisting forces and moment transmitted by cross section *aa* in bar *AB*.

**Fig. P8-114**

**8-115\***  A beam is loaded and supported as shown in Fig. P8-115. Using the coordinate axes shown, write equations for the shear force *V* and bending moment *M* for any section of the beam in the interval 4 ft $< x <$ 8 ft.

**Fig. P8-115**

**8-116\***  Draw complete shear and moment diagrams for the beam shown in Fig. P8-116.

**Fig. P8-116**

**8-117**  Draw complete shear and moment diagrams for the beam shown in Fig. P8-117.

**Fig. P8-117**

**8-118** Draw complete shear and moment diagrams for the beam shown in Fig. P8-118.

**Fig. P8-118**

**8-119*** The load carried by each cable of the suspension bridge shown in Fig. P8-119 is 2500 lb per foot of horizontal length. The span of the bridge is 1500 ft, and the sag at midspan is 150 ft. Determine

a. The tension in the cable at the middle of the span.
b. The tension in the cable at the supports and the angle it makes with the horizontal.
c. The length of the cable.

**Fig. P8-119**

**8-120*** The cable shown in Fig. P8-120 weighs 35 N/m. Determine the maximum tension in the cable and the sag $h_A$ in the cable if

a. The length of the cable is 33 m.
b. The length of the cable is 45 m.
c. The length of the cable is 60 m.

**Fig. P8-120**

**8-121** A 500-lb block $W$ is supported by a boom $CB$ and a flexible cable $AB$ as shown in Fig. P8-121. The boom weighs 250 lb, and the cable weighs 2.5 lb/ft. Determine the maximum tension in the cable, the length $L$ of the cable, and the sag $h$ at midspan. Assume that the weight of the cable is uniformly distributed along its horizontal length.

**Fig. P8-121**

469

*R*ock climbing is a sport that depends heavily on friction. Friction forces developed when the pins used to support the climbing ropes are driven into cracks in the rock often support the climber's weight during difficult parts of the climb.

# FRICTION

## 9-1 INTRODUCTION

 Thus far in the book, the forces between two surfaces in contact have been either perfectly smooth or perfectly rough. Whenever tangential or friction forces were needed, they were assumed to supply whatever force was necessary to ensure equilibrium. This is the concept of a rough surface. However, no surface is perfectly rough any more than it is perfectly smooth. In the present chapter, it is desired to identify the origin and the nature of these tangential forces and to quantify the concept of roughness.

A perfectly smooth or frictionless surface that exerts only normal forces on bodies is a useful model for a large number of situations. However, frictional forces that act tangential to the surface are present in the contact between all real surfaces. Whether the friction forces are large or small depends on a number of things including the types of materials in contact.

Friction forces act to oppose the tendency of contacting surfaces to slip relative to one another and can be either good or bad. Without friction it would be impossible to walk or ride a bicycle, drive a car, or pick up objects. Even in some machine applications such as brakes and belt drives, a design consideration is to maximize the friction. In many other machine applications, however, friction is undesirable. Friction causes energy loss and wears down sliding surfaces in contact. In these cases, a primary design consideration is to minimize friction.

Two main types of friction are commonly encountered in engineering practice: dry friction and fluid friction. As its name suggests, **dry friction** or **Coulomb[1] friction** describes the tangential component

[1]Charles A. Coulomb (1736–1806), was a French physicist and military engineer. His study of the laws of friction was published in 1781, and his work on fluid resistance in 1799.

**Figure 9-1**   Free-body diagram for a block at rest on a rough horizontal surface.

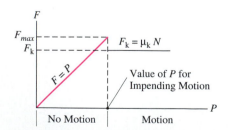

**Figure 9-2**   Variation of the friction force between two bodies.

of the contact force that exists when two dry surfaces slide or tend to slide relative to one another. Coulomb friction is the primary concern of this chapter and will be studied in considerable detail in Section 9-2.

**Fluid friction** describes the tangential component of the contact force that exists between adjacent layers in a fluid that are moving at different velocities relative to each other, as in the thin layer of oil between bearing surfaces. The tangential forces developed between the adjacent fluid layers oppose the relative motion and depend primarily on the relative velocity between the two layers. Fluid friction is one of the primary concerns in the study of fluid mechanics and is more properly treated in a course in fluid mechanics.

## 9-2   CHARACTERISTICS OF COULOMB FRICTION

In order to investigate the behavior of frictional forces, consider a simple experiment consisting of a solid block of mass $m$ resting on a rough horizontal surface and acted on by a horizontal force $\mathbf{P}$ as shown in Fig. 9-1a. A free-body diagram for the block is shown in Fig. 9-1b. The free-body diagram indicates that the block is subjected to the action of three forces, the weight $\mathbf{W} = m\mathbf{g}$, the applied force $\mathbf{P}$, and the reaction $\mathbf{R}$ of the support surface. In order for the block to be in equilibrium, the three forces must be concurrent. The reaction $\mathbf{R}$ can be represented on the free-body diagram by a normal component ($\mathbf{N} = \mathbf{W} = m\mathbf{g}$) and a horizontal or friction component ($\mathbf{F} = \mathbf{P}$) on the contact surface as shown in Fig. 9-1c. When the horizontal force $\mathbf{P}$ is zero, no horizontal component of force $\mathbf{F}$ will be required for equilibrium and friction will not exert a force on the surface. As the force $\mathbf{P}$ increases, the friction force $\mathbf{F}$ also increases, as shown in the graph of Fig. 9-2. The friction force cannot increase indefinitely, however, and it eventually reaches its maximum value $F_{max}$. The maximum value is also called the **limiting value of static friction.**

The condition when the friction force is at its maximum value is called the condition of **impending motion.** That is, if $P$ increases beyond the point $P = F_{max}$, then friction can no longer supply the amount of force necessary for equilibrium. Therefore, the block will no longer be in equilibrium, but will start moving in the direction of the force $P$. When the block starts moving, the friction force $F$ normally decreases in magnitude by about 20 to 25 percent. From this point on, the block will slide with increasing speed while the friction force (the kinetic friction force) remains approximately constant (Fig. 9-2).

Repeating the experiment with a second block of mass $m_2 = 2m$ would produce similar results, but the limiting force at which the block starts to move would be observed to be twice as great. Repeating the experiment with two blocks of different sizes but the same mass and material would yield the same limiting force for both blocks. That is, the value of limiting friction is proportional to the normal force at the contact surface

$$F_{max} = \mu_s N \tag{9-1}$$

The constant of proportionality $\mu_s$ is called the **coefficient of static friction,** and it depends on the types of material in contact. However, $\mu_s$ is observed to be relatively independent of both the normal force and the area of contact.

In some instances, as the horizontal force **P** is increased, the friction force **F** will be sufficient to permit the line of action of the reaction **R** of Fig. 9-1b to reach corner A, as shown in Fig. 9-3a, before impending motion by slipping begins. In this case, the body will tip before it slips, since the couple (see Fig. 9-3b) consisting of forces **W** and **N** will be smaller than the couple consisting of forces **P** and **F**.

In order to understand how $\mu_s$ can be independent of the area of contact, one must consider where the friction forces come from. It is generally believed that dry friction results primarily from the roughness between two surfaces and to a lesser extent from attraction between the molecules of the two surfaces. Even two surfaces that are considered to be smooth have small irregularities, as the (idealized) enlargement of the contact surfaces of Fig. 9-4 shows. Therefore, contact between the block and the surface takes place only over a few very small areas of the common surface. The friction force **F** is then the resultant of the tangential component of the forces acting at each of these tiny contact points just as normal force **N** is the resultant of the normal components of the forces acting at each of the contact points. (Normal and tangential here are relative to the overall contact plane and not the individual tiny contact points.) Increasing the number of contact points just means that the normal and frictional components at each point are proportionately smaller, but their sums **F** and **N** do not change. Therefore, $\mu_s$ will not change either.

Before going on it must be noted that since the normal force **N** is the resultant of a distributed force, it acts through the centroid of that force distribution. If the force distribution is uniform, **N** will act at the center of the surface. In general, however, **N** will not act at the center of the surface or through the center of the body. Since the actual distribution of forces is generally not known, the location of **N** is determined using moment equilibrium.

In many friction problems, it is easily recognized that a body is in no danger of tipping over. Since only force equilibrium is considered, it does not matter where the normal force is drawn on the free-body diagram. Even so, the student should not get into the habit of showing the force as always acting through the center of the body.

It must also be noted that friction is a resistive force. That is, friction always acts to opposite motion; it never acts to create motion. Equation 9-1 only tells how much friction is available, $F_{avail} = F_{max} = \mu_s N$, to prevent motion. No matter how much frictional force is available on a surface, however, the *frictional force actually exerted is never*

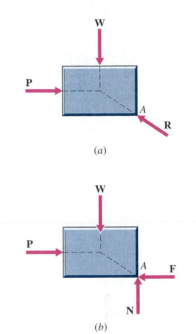

*(a)*

*(b)*

**Figure 9-3** Free-body diagram for a block on a rough horizontal surface at the condition of impending tipping.

**Figure 9-4** Dry friction results from the roughness between two surfaces.

*greater than that required to satisfy the equations of equilibrium*

$$F \le \mu_s N \qquad (9\text{-}2)$$

where *the equality holds only at the point of impending motion.*

Once the block starts to slip relative to the surface, the friction force will decrease to

$$F = \mu_k N \qquad (9\text{-}3)$$

where $\mu_k$ is called the **coefficient of kinetic friction.** This coefficient is again independent of the normal force and is also independent of the speed of the relative motion—at least for low speeds. Of course, the presence of any oil or moisture on the surface can change the problem from one of dry friction, where the friction force is independent of the speed of the body, to one of fluid friction, where the friction force is a function of the speed. At higher speeds, the effect of lubrication by an intervening fluid film (such as oil, surface moisture, or even air) can become appreciable.

These results are summarized by

---

### Coulomb's Laws of Dry Friction

The direction of the friction force on a surface is such as to oppose the tendency of the one surface to slide relative to the other. It is the relative motion or the impending relative motion of one body relative to another that is important.

The friction force is never greater than just sufficient to prevent motion.

For the static equilibrium case in which the two surfaces are stationary with respect to one another, the normal and tangential components of the contact force satisfy

$$F \le \mu_s N$$

where the equality holds for the case of **impending motion** in which the contacting surfaces are on the verge of sliding relative to each other.

For the case where two contacting surfaces are sliding over each other, the normal and tangential components of the contact force satisfy

$$F = \mu_k N$$

where $\mu_k < \mu_s$.

---

Of course, Coulomb's laws only apply when $N$ is positive, that is, when the surfaces are being pressed together.

The values of $\mu_s$ and $\mu_k$ must be determined experimentally for each pair of contacting surfaces. Average values of $\mu_s$ for various types of materials are given in Table 9-1. Reported values for $\mu_s$ vary widely, however, depending on the exact nature of the contacting surfaces. Values for $\mu_k$ are generally 20 to 25 percent less than those reported for $\mu_s$. Values for $\mu_k$ are not listed in Table 9-1 since the uncertainty in $\mu_s$ is much larger (by as much as 100 percent in some cases) than the difference between it and $\mu_k$.

**TABLE 9-1 COEFFICIENTS OF FRICTION FOR COMMON SURFACES**

| Materials | $\mu_s$ |
| --- | --- |
| Metal on metal | 0.5 |
| Wood on metal | 0.5 |
| Wood on wood | 0.4 |
| Leather on wood | 0.4 |
| Rubber on metal | 0.5 |
| Rubber on wood | 0.5 |
| Rubber on pavement | 0.7 |

Because of the uncertainty in the values of $\mu_s$ and $\mu_k$, Table 9-1 should be used to get only a rough estimate of the magnitude of the friction forces. If more accurate values are needed, experiments should be performed using the actual surfaces being studied.

Since the coefficients of friction are the ratio of two forces, they are dimensionless quantities and can be used with either SI or USCS units.

In many simple friction problems it is convenient to use the resultant of friction and normal forces rather than their separate components. In the case of the block in Fig. 9-1 (which is redrawn in Fig. 9-5), this leaves only three forces acting on the block. Moment equilibrium is established simply by making the three forces concurrent and only force equilibrium need be considered. Since $\mathbf{N}$ and $\mathbf{F}$ are rectangular components of the resultant $\mathbf{R}$, the magnitude and direction of the resultant are given by

$$R = \sqrt{N^2 + F^2} \quad \text{and} \quad \tan\theta = \frac{F}{N} \tag{9-4}$$

At the point of impending motion, Eq. 9-4 becomes

$$R = \sqrt{N^2 + F_{\max}^2} = \sqrt{N^2 + (\mu_s N)^2} = N\sqrt{1 + \mu_s^2} \tag{9-5a}$$

$$\tan\phi_s = \frac{F}{N} = \frac{\mu_s N}{N} = \mu_s \tag{9-5b}$$

where the angle $\phi_s$, the angle between the resultant and the normal to the surface, is called the **angle of static friction.** For a given normal force $\mathbf{N}$, if the friction force is less than the maximum, $F < \mu_s N$, then the angle of the resultant will be less than the angle of static friction, $\theta < \phi_s$. In no case can the angle of the resultant $\theta$ be greater than $\phi_s$ for a body in equilibrium. A similar relation is obtained in the case of kinetic friction,

$$\tan\phi_k = \mu_k$$

where $\phi_k$ is called the **angle of kinetic friction.**

When a block sits on an inclined surface and is acted on only by gravity, the resultant of the normal and friction force must be collinear, as shown in Fig. 9-6. But the angle between the resultant and the normal to the surface can never be greater than the angle of static friction $\phi_s$. Therefore, the steepest inclination $\theta$ for which the block will be in equilibrium is equal to the angle of static friction. This angle is called the **angle of respose.**

There are three typical types of friction problems encountered in engineering analysis:

**Figure 9-5** Free-body diagram for a block at rest on a rough horizontal surface.

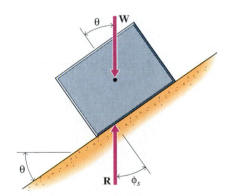

**Figure 9-6** Free-body diagram for a block at rest on a rough inclined surface.

1. **Impending Motion Is Not Assumed.** This first case is just the type of equilibrium problem solved in the previous chapters of the book. The friction force $F_{req}$ ($\neq \mu_s N$) and the normal force $N$ are drawn on the free-body diagram and are determined using force equilibrium. The normal force should not be drawn as acting through the center of the body. Instead, its location is determined using moment equilibrium. The amount of friction required for equilibrium $F_{req}$ and the location of the normal force are then checked against their maximum values. Three possibilities exist:

   a. If the amount of friction required for equilibrium $F_{req}$ is smaller than or equal to the maximum amount of friction available, $F_{avail} = F_{max} = \mu_s N$, and the location of the normal force is on the body, then the body is in equilibrium. In this case, the actual friction force supplied by the surface is

   $$F_{actual} = F_{req}$$
   $$< F_{avail} = \mu_s N$$

   That is, the surface supplies just enough friction force (resistance) to keep the body from moving.

   b. If the amount of friction required for equilibrium $F_{req}$ is smaller than or equal to the maximum amount of friction available, $F_{avail} = F_{max} = \mu_s N$, but the location of the normal force is not somewhere on the body, then the body is not in equilibrium and will tip over.

   c. If the amount of friction required for equilibrium $F_{req}$ is greater than the maximum amount of friction available, $F_{avail} = F_{max} = \mu_s N$, then the body is not in equilibrium and will slide. In this case, the actual friction force supplied by the surface is the kinetic friction force $F_{actual} = \mu_k N$.

2. **Impending Slipping Is Known to Occur at All Surfaces of Contact.** Since impending slipping is known to occur at all surfaces of contact, the magnitude of the friction forces can be shown as $\mu_s N$ on the free-body diagrams. The equations of equilibrium are then written, and

   a. If all the applied forces are given but $\mu_s$ is unknown, the equations of equilibrium can be solved for $N$ and $\mu_s$. This $\mu_s$ is the smallest coefficient for which the body will be in equilibrium.

   b. If the coefficient of friction is given but one of the applied forces is unknown, the equations of equilibrium can be solved for $N$ and the unknown applied force.

3. **Impending Motion Is Known to Exist but the Type of Motion or Surface of Slip Is Not Known.** Since it is not known whether the body tips or slips, the free-body diagrams must be drawn as in case 1 above. That is, the friction forces must not be shown as $\mu_s N$ on the free-body diagrams. At this point the three equations of equilibrium will contain more than three unknowns. Assumptions must be made about the type of motion that is about to occur until the number of equations equals the number of unknowns. The equations are then solved for the remaining unknowns and checked against the assumptions made about slipping or tipping. If $F_{req}$

comes out greater than $F_{avail} = \mu_s N$ at some surface or if the location of the normal force is not on the body, then the assumptions must be changed and the problem solved again.

In both of the last two cases (2b and 3), the friction force is treated as if it is a known force. Care must be taken to be sure that it opposes the tendency of the other forces to cause motion. The result will not include a negative sign to indicate that the friction is in the wrong direction. If the direction of the friction force is drawn incorrectly, incorrect answers will result. The direction is easily determined by pretending for a moment that friction does not exist. Then apply the friction in such a direction as to opposite the motion that would occur in the absence of friction.

# CONCEPTUAL EXAMPLE 9-1: ANGLE OF REPOSE

*A skier is at the point of experiencing impending motion on a ski slope as illustrated in Fig. CE9-1a. Determine the coefficient of static friction between a ski and the slope if angle θ (the angle of repose) is 5°.*

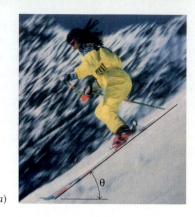

(a)

## SOLUTION

A free-body diagram of the system shows the weight **W** of the skier and his equipment being resisted by a resultant contact force **R** that is equal, opposite, and collinear with the weight **W**, as shown in Fig. CE9-1b. Normal and tangential components of the weight **W** and the resultant force **R** are shown in Fig. CE9-1c. From the free-body diagram of Fig. CE9-1c it is observed that equilibrium is satisfied if

$$N = W \cos \theta$$
$$F = W \sin \theta$$

For the case of impending motion, the normal and tangential components of the resultant force **R** satisfy the equation

$$F = \mu_s N$$

Thus,

$$\mu_s = \frac{F}{N} = \frac{W \sin \theta}{W \cos \theta} = \tan \theta$$

(b)

An angle of repose of 5° would indicate that the coefficient of static friction between a ski and the slope would be

$$\mu_s = \tan \theta = \tan 5° = 0.087$$

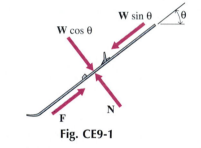

(c)

**Fig. CE9-1**

## CONCEPTUAL EXAMPLE 9-2: MAXIMUM FRICTIONAL FORCE

*A small boy has the option of pushing or pulling his small sister on a sled as shown in Figs. CE9-2a and CE9-2b, respectively. Which option will require the smaller force?*

(a)          (b)

(c)          (d)

**Fig. CE9-2**

---

### SOLUTION

Free-body diagrams of the sled for the two options are shown in Figs. CE9-2c (pushing) and CE9-2d (pulling).

In Fig. CE9-2c, for the case where the sled is being pushed, the normal force $N_1$ exerted on the sled runners by the snowy surface is

$$N_1 = W + P_1 \sin \theta$$

In Fig. CE9-2d, for the case where the sled is being pulled, the normal force $N_2$ exerted on the sled runners by the snowy surface is

$$N_2 = W - P_2 \sin \theta$$

Clearly, force $\mathbf{N}_2$ is less than force $\mathbf{N}_1$; therefore, since the frictional force $\mathbf{F}$ required to move the sled depends on the normal force $\mathbf{N}$,

$$F_2 = \mu N_2 = P_2 \cos \theta < F_1 = \mu N_1 = P_1 \cos \theta$$

Thus, $P_2 < P_1$, so pulling the sled requires less force than pushing the sled.

A 20-lb piece of electronic equipment is placed on a wooden skid that weighs 10 lb and that rests on a concrete floor (Fig. 9-7). Assume a coefficient of static friction of 0.45 and determine the minimum pushing force along the handle necessary to cause the skid to start sliding across the floor.

**P**

30°

**Fig. 9-7**

## SOLUTION

The free-body diagram of the skid is drawn in Fig. 9-8. Since motion is known to be impending (to the right), the friction force has magnitude $A_f = \mu_s A_n$ and is drawn pointing to the left so as to oppose the impending motion. (The symbol $A_n$ was chosen for the normal force rather than $N$ so as to avoid any confusion with the abbreviation for Newton.) The equilibrium equations are

$$+\rightarrow \Sigma F_x = P \cos 30° - 0.45\, A_n = 0$$
$$+\uparrow \Sigma F_y = A_n - P \sin 30° - 30 = 0$$

These equations can be rewritten in the form

$$P \sin 30° = A_n - 30 \qquad (a)$$
$$P \cos 30° = 0.45\, A_n \qquad (b)$$

Now dividing Eq. $a$ by Eq. $b$ gives

$$\tan 30° = \frac{A_n - 30}{0.45\, A_n}$$

or

$$A_n = 40.53 \text{ lb}$$

This value can be substituted into either Eq. $a$ or Eq. $b$ to get

$$P = 21.06 \text{ lb} \cong 21.1 \text{ lb} \qquad \text{Ans.}$$

the minimum force necessary to start the skid sliding.

**P**     **W**

$A_f = \mu_s A_n$

$A_n$

**Fig. 9-8**

> The condition that motion by slipping is impending means that $A_f = \mu_s A_n$.

> The frictional force between two bodies always opposes the tendency to move.

In Example Problem 9-1, assume that a pulling force instead of a pushing force is applied to the handle. Determine the minimum force necessary to start the skid in motion now.

**Fig. 9-9**

---

### SOLUTION

The free-body diagram for this case is shown in Fig. 9-9. Note that now the direction of impending motion is to the left, and so the friction must act to the right. The equilibrium equations are now

$$+\rightarrow \Sigma F_x = -P \cos 30° + 0.45\, A_n = 0$$
$$+\uparrow \Sigma F_y = A_n - P \sin 30° - 30 = 0$$

Solving as before gives

$$A_n = 23.81 \text{ lb}$$
$$P = 12.38 \text{ lb} \qquad \text{Ans.}$$

(Note that this time the force $P$ is only 41 percent of the total weight whereas in Example 9-1 the force required was 70 percent of the total weight.)

> The condition that motion by slipping is impending means that $A_f = \mu_s A_n$.

> The frictional force between two bodies always opposes the tendency to move.

A 20-kg triangular block sits on top of a 10-kg rectangular block as shown in Fig. 9-10a. The coefficient of friction is 0.40 at all surfaces. Determine the maximum horizontal force **P** for which motion will not occur.

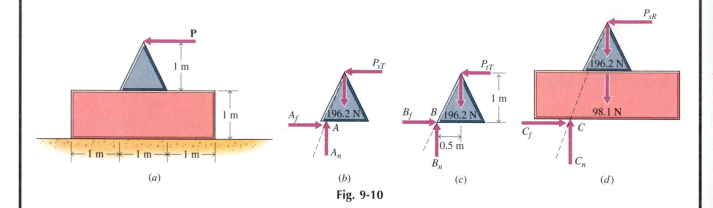

**Fig. 9-10**

## SOLUTION

Three types of motion are possible: The triangular block can slip, the triangular block can tip, or the rectangular block can slip. First, consider equilibrium of the triangular block. Three forces act on the block; the weight $W_T = m_T g = 20(9.81) = 196.2$ N, the applied force $P$, and the reaction force $A$ (with components $A_f$ and $A_n$) exerted on the triangular block by the rectangular block. For slipping, the line of action of force $A$ must intersect the contact surface between the two blocks somewhere between corner $B$ and the vertical centerline of the block, as shown in Fig. 9-10b. Summing forces in the x- and y-directions yields

$$+\uparrow\Sigma F_y = A_n - 196.2 = 0 \qquad A_n = 196.2 \text{ N}$$
$$+\rightarrow\Sigma F_x = A_f - P_{sT} = 0$$
$$P_{sT} = A_f = A_f(\max) = \mu_s A_n = 0.40(196.2) = 78.48 \text{ N}$$

> The three forces acting on the triangular block must be concurrent if the block is in equilibrium; therefore, the line of action of force $A$ passes through the apex of the block.

For tipping, the line of action of force $B$ must intersect the left corner of the triangular block as shown in Fig. 9-10c. Summing moments about point $B$ at the left corner yields

$$+ \downarrow\Sigma M_B = P_{tT}(1) - 196.2(0.5) = 0$$
$$P_{tT} = 98.1 \text{ N} > P_{sT} = 78.48 \text{ N}$$

Therefore, the triangular block will not tip.

Finally, consider a system consisting of both blocks as shown in Fig. 9-10d. The weight of the rectangular block is $W_R = m_R g = 10(9.81) = 98.1$ N. Summing forces in the x- and y-directions yields

$$+\uparrow\Sigma F_y = C_n - 196.2 - 98.1 = 0 \qquad C_n = 294.3 \text{ N}$$
$$+\rightarrow\Sigma F_x = C_f - P_{sR} = 0$$
$$P_{sR} = C_f = C_f(\max)$$
$$= \mu_s C_n = 0.40(294.3) = 117.7 \text{ N} > P_{sT} = 78.48 \text{ N}$$

> The triangular block will slip before it tips if the line of action of force $A$ intersects the contact surface.

> Alternatively, for the triangular block, $(b/2)/h = 0.5/1 = 0.50 > 0.40$; therefore, the block slips before it tips.

Therefore:

$$P_{\max} = P_{sT} = 78.48 \text{ N} \cong 78.5 \text{ N} \qquad\qquad \text{Ans.}$$

The block in Fig. 9-11a has a mass of 100 kg. The coefficient of friction between the block and the inclined surface is 0.20. Determine

a.  If the system is in equilibrium when **P** = 600 N.
b.  The minimum force **P** to prevent motion.
c.  The minimum force **P** for which the system is in equilibrium.

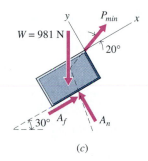

(a)  (b)  (c)

**Fig. 9-11**

## SOLUTION

a.  Three forces act on the block; the weight $W = mg = 100(9.81) = 981$ N, the applied force $P = 600$ N, and the reaction force $A$ (with components $A_f$ and $A_n$) exerted on the block by the inclined surface as shown in Fig. 9-11b. The three forces must be concurrent if the block is in equilibrium. The reaction force is obtained by summing forces parallel and perpendicular to the inclined surface. Thus,

> The equation $A_f = \mu_s A_n$ cannot be assumed since motion is not known to be impending or occurring.

$$+\nwarrow\Sigma F_y = P \sin 20° + A_n - W \cos 30°$$
$$= 600 \sin 20° + A_n - 981 \cos 30° = 0 \quad A_n = 644.36$$
N
$$+\nearrow\Sigma F_x = P \cos 20° - A_f - W \sin 30°$$
$$= 600 \cos 20° - A_f - 981 \sin 30° = 0 \quad A_f = 73.32 \text{ N}$$
$$A_f(\text{max}) = \mu_s A_n = 0.20(644.36) = 128.87 \text{ N} > A_f(\text{req}) = 73.32 \text{ N}$$

Therefore, the block is in equilibrium

b.  The minimum force **P** will be required when motion of the block down the incline is impending. For this condition, the frictional force $A_f$ will tend to resist this motion as shown in Fig. 9-11c. Equilibrium exists when

> The frictional force between two bodies always opposes the tendency to move.

$$+\nearrow\Sigma F_x = P_{min} \cos 20° + A_f - W \sin 30°$$
$$= P_{min} \cos 20° + 0.20 A_n - 981 \sin 30° = 0$$
$$+\nwarrow\Sigma F_y = P_{min} \sin 20° + A_n - W \cos 30°$$
$$= P_{min} \sin 20° + A_n - 981 \cos 30° = 0$$

> The equation $F = \mu_s N$ gives the maximum frictional force that can develop between the two surfaces.

Solving simultaneously yields

$$A_n = 723.73 \text{ N} \cong 724 \text{ N}$$
$$P_{min} = 367.94 \text{ N} \cong 368 \text{ N} \quad \text{Ans.}$$

c.  The maximum force **P** will be required when motion of the block up the incline is impending. For this condition, the frictional force $A_f$ will tend to

(d)

(e)    (f)    (g)

**Fig. 9-11**

resist this motion as shown in Fig. 9-11$d$. Equilibrium exists when

$$+\nearrow\Sigma F_x = P_{max}\cos 20° - A_f - W\sin 30°$$
$$= P_{max}\cos 20° - 0.2A_n - 981\sin 30° = 0$$
$$+\nwarrow\Sigma F_y = P_{max}\sin 20° + A_n - W\cos 30°$$
$$= P_{max}\sin 20° + A_n - 981\cos 30° = 0$$

Solving simultaneously yields

$$A_n = 625.52 \text{ N} \cong 626 \text{ N}$$
$$P_{max} = 655.11 \text{ N} \cong 655 \text{ N} \qquad\qquad \text{Ans.}$$

Force triangles for parts $a$, $b$, and $c$ of the problem are shown in Figs. 9-11$e$, $f$, and $g$, respectively.

A 20-lb homogeneous box has tipped and is resting against a 40-lb homogeneous box (Fig. 9-12). The coefficient of friction between box A and the floor is 0.7; between box B and the floor, 0.4. Treat the contact surface between the two boxes as smooth and determine if the boxes are in equilibrium.

**Fig. 9-12**

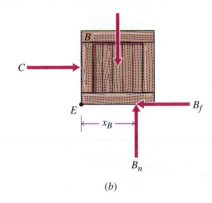

**Fig. 9-13**

## SOLUTION

The free-body diagram of box A is drawn in Fig. 9-13a. The equilibrium equations

$$+\rightarrow \Sigma F_x = A_f - C = 0$$
$$+\uparrow \Sigma F_y = A_n - 20.0 = 0$$
$$\downarrow + \Sigma M_D = 12.00C - (7.392)(20) = 0$$

are solved to get

$$C = 12.320 \text{ lb} \qquad A_n = 20.000 \text{ lb} \qquad A_f = 12.320 \text{ lb}$$

The friction force available at this surface is

$$F_{\max} = \mu_s A_n = (0.7)(20) = 14.00 \text{ lb}$$

Since the friction force required (12.32 lb) is less than the friction force available (14.00 lb), box A is in equilibrium.

The free-body diagram of box B is drawn in Fig. 9-13b. The equilibrium equations for this box

$$+\rightarrow \Sigma F_x = 12.320 - B_f = 0$$
$$+\uparrow \Sigma F_y = B_n - 40 = 0$$
$$\downarrow + \Sigma M_E = x_B B_n - (12.00)(12.230) - (12.00)(40.0) = 0$$

give

$$B_f = 12.320 \text{ lb} \qquad B_n = 40.00 \text{ lb} \qquad x_B = 15.669 \text{ in.}$$

The friction force available at this surface is

$$F_{\max} = \mu_s B_n = (0.4)(40.00) = 16.00 \text{ lb}$$

Again the friction force available (16.00 lb) is greater than the friction force required (12.32 lb) and box B is also in equilibrium.

Thus, both boxes are in equilibrium. Ans.

> The contact surface between the boxes is smooth; therefore, friction forces do not exist.

> The equation $A_f = \mu_s A_n$ can not be assumed since motion is not known to be impending or occurring.

> The equation $F = \mu_s N$ gives the maximum frictional force that can develop between two surfaces.

> Since $x_B < 24$ in., the line of action of force B intersects the contact surface; therefore, box B will slip before it tips.

The wheels of the refrigerator of Fig. 9-14 are stuck and will not turn. The refrigerator weighs 600 N. Assume a coefficient of friction between the wheels and the floor of 0.6 and determine the force necessary to cause the refrigerator to just start to move (impending motion). Also determine the maximum height $h$ at which the force can be applied without causing the refrigerator to tip over.

Fig. 9-14                    Fig. 9-15

## SOLUTION

The free-body-diagram of the refrigerator is drawn in Fig. 9-15. The equilibrium equations

$$+\rightarrow \Sigma F_x = P - A_f - B_f = 0$$
$$+\uparrow \Sigma F_y = A_n + B_n - 600 = 0$$
$$\downarrow + \Sigma M_B = (0.4)(600) - hP - 0.8\, A_n = 0$$

give

$$A_n + B_n = 600$$
$$P = A_f + B_f = 0.6(A_n + B_n) = 360 \text{ N} \qquad \text{Ans.}$$

and

$$h = \frac{240 - 0.8 A_n}{360}$$

When $h = 0$, $A_n = B_n = 300$ N, and the wheels share the load of the weight equally. As $h$ increases, $A_s$ gets smaller. However, the force at $A$ cannot be negative, so the condition for impending motion by tipping is $A_n = 0$. Thus,

$$h < \frac{240}{360} = 0.667 \text{ m} \qquad \text{Ans.}$$

At the point of impending tipping, none of the weight is carried by the wheels at $A$; it has all been shifted to the wheels at the right side of the refrigerator.

> The condition that motion by slipping is impending means that $A_f = \mu_s A_n$ and $B_f = \mu_s B_n$.

> For impending tipping, when all weight is being carried by the wheels at $B$, $0.4/h = \tan \phi = \mu_s = 0.60$ which verifies that $h = 0.667$ m.

# EXAMPLE PROBLEM 9-7

The pickup of Fig. 9-16 is traveling at a constant speed of 50 mi/h and is carrying a 60-lb box in the back. The box projects 1 ft above the cab of the pickup. The wind resistance on the box can be approximated as a uniformly distributed force of 25 lb/ft on the exposed edge of the box. Calculate the minimum coefficient of friction required to keep the box from sliding on the bed of the pickup. Also determine whether or not the box will tip over.

truck
drawer
Fridge

whatever
$\mu$ is smaller

How fast

max

**Fig. 9-16**

**Fig. 9-17**

## SOLUTION

The free-body diagram of the box is drawn in Fig. 9-17. The equilibrium equations for the box

$$+\rightarrow \Sigma F_x = (25)(1) - C_f = 0$$
$$+\uparrow \Sigma F_y = C_n - 60 = 0$$
$$\downarrow + \Sigma M_A = (1.5)(60) - (3.5)(25)(1) - x_C C_n = 0$$

are solved to get

$$C_f = 25.0 \text{ lb} \quad C_n = 60.0 \text{ lb}$$

and

$$x_C = 0.042 \text{ ft} = 0.500 \text{ in.}$$

The friction and normal force must act on the box, so $x_C$ must be a number between 0 and 3 ft. If the solution had given $x_C$ to be negative, then no normal force on the bottom of the box could satisfy moment equilibrium and the box would tip over. Thus, the required coefficient of friction is

$$\mu_s = \frac{C_f}{C_n} = \frac{25.0}{60.0} = 0.417 \qquad \text{Ans.}$$

and since $x_C$ is positive, the box will not tip. Ans.

# PROBLEMS

## Introductory Problems

**9-1\*** Determine the horizontal force **P** required to start moving the 250-lb block shown in Fig. P9-1 up the inclined surface. The coefficient of friction between the inclined surface and the block is $\mu_s = 0.30$.

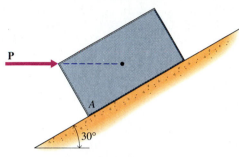

**Fig. P9-1**

**9-2\*** The 20-N force shown in Fig. P9-2 is required to produce impending motion of the block on the inclined surface. If the mass of the block is 10 kg, determine the coefficient of friction between the block and the surface.

**Fig. P9-2**

**9-3** Workers are pulling a 400-lb crate up an incline as shown in Fig. P9-3. The coefficient of friction between the crate and the surface is 0.20, and the rope on which the workers are pulling is horizontal.

**a.** Determine the force **P** that the workers must exert to start sliding the crate up the incline.
**b.** If one of the workers lets go of the rope for a moment, determine the minimum force the other workers must exert to keep the crate from sliding back down the incline.

**Fig. P9-3**

**9-4** Two blocks with masses $m_A = 20$ kg and $m_B = 80$ kg are connected with a flexible cable that passes over a frictionless pulley as shown in Fig. P9-4. The coefficient of friction between the blocks is 0.25. If motion of the blocks is impending, determine the coefficient of friction between block $B$ and the inclined surface and the tension in the cable between the two blocks.

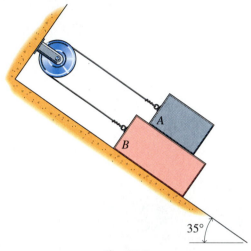

**Fig. P9-4**

**9-5\*** Determine the minimum weight $W$ required to hold the 100-lb block of Fig. P9-5 in equilibrium on the inclined surface if the coefficient of friction between the block and the surface is 0.50.

**Fig. P9-5**

**9-6\*** Two blocks with masses $m_A = 50$ kg and $m_B = 100$ kg are connected with a flexible cable that passes over a frictionless pulley as shown in Fig. P9-6. Determine the

**Fig. P9-6**

force **P** required to start moving the blocks if the coefficient of friction between the blocks and the surface is 0.10.

**9-7** A homogeneous, triangular block of weight $W$ has height $h$ and base width $b$ as shown in Fig. P9-7. Develop an expression for the coefficient of friction between the block and the surface for which impending motion by slipping and tipping occur simultaneously if

**a.** Force **P** acts in the direction shown on the figure.
**b.** The direction of force **P** is reversed.

**Fig. P9-7**

**9-8** Three blocks with masses $m_A = 25$ kg, $m_B = 50$ kg, and $m_C = 100$ kg are connected with a flexible cable that passes over a frictionless pulley as shown in Fig. P9-8. Determine the force **P** required to start moving the blocks if the coefficient of friction is 0.50 between blocks $A$ and $B$, 0.35 between block $B$ and the surface, and 0.26 between block $C$ and the surface.

**Fig. P9-8**

**9-9\*** A light, inextensible cord passes over a frictionless pulley as shown in Fig. P9-9. One end of the rope is attached to block $B$; a force **P** is applied to the other end.

**Fig. P9-9**

Block $A$ weighs 600 lb, and block $B$ weighs 1000 lb. The coefficient of friction between the blocks is 0.33, and the coefficient of friction between block $B$ and the floor is 0.25. Determine

**a.** If the system would be in equilibrium for $P = 400$ lb.
**b.** The maximum force **P** for which the system is in equilibrium.

**9-10\*** A force of 500 N is applied to the end of the cord that is attached to block $A$ of Fig. P9-10. Block $A$ weighs 2000 N, and block $B$ weighs 4000 N. The coefficient of friction between the blocks is 0.30, and the coefficient of friction between $B$ and the floor is 0.20. Determine

**a.** If the system would be in equilibrium for $P = 2500$ N.
**b.** The maximum force **P** for which the system is in equilibrium.

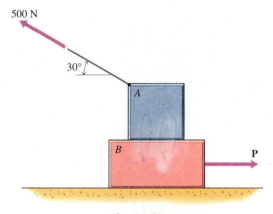

**Fig. P9-10**

**9-11** The weights of cylinder $A$ and block $B$ of Fig. P9-11 are $W_A = 100$ lb and $W_B = 200$ lb. Determine the minimum coefficient of friction for which block $B$ is in equilibrium on the inclined surface.

**Fig. P9-11**

**9-12**  The masses of blocks A and B of Fig. P9-12 are $m_A$ = 40 kg and $m_B$ = 85 kg. The coefficient of friction is 0.25 for both surfaces. Determine the force **P** and the tension in the cord when motion of block B down the inclined surface is impending.

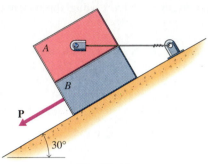

**Fig. P9-12**

**9-13\***  The 200-lb block of Fig. P9-13 is sitting on a 30° inclined surface. The coefficient of friction between the block and the surface is 0.50. A light, inextensible cord is attached to the block, passes around a frictionless pulley, and is attached to a second block of mass M. Determine the minimum and maximum masses, $M_{min}$ and $M_{max}$, such that the system is in equilibrium. Is impending motion by slipping or by tipping?

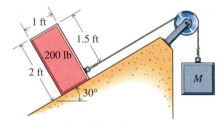

**Fig. P9-13**

**9-14\***  A block of mass M rests on a 50-kg block which in turn rests on an inclined plane as shown in Fig. P9-14. The coefficient of friction is 0.40 between the blocks and 0.30 between the 50-kg block and the floor. The pulley is frictionless, and the weight of the cord can be neglected. Determine the minimum mass $M_{min}$ necessary to prevent slipping.

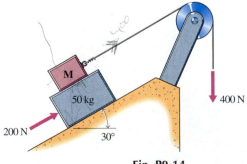

**Fig. P9-14**

**9-15**  Two blocks are connected with a flexible cable that passes over a frictionless pulley as shown in Fig. P9-15. The weight of block A is 25 lb, and the coefficient of friction is 0.20 for both inclined surfaces. Determine the maximum and minimum weights for block B if the system remains in equilibrium.

**Fig. P9-15**

**9-16**  The masses of blocks A and B of Fig. P9-16 are $m_A$ = 50 kg and $m_B$ = 75 kg. If the coefficient of friction is 0.25 for both surfaces, determine the force **P** required to cause impending motion of block B.

**Fig. P9-16**

**9-17\***  The weights of block A and cylinder B of Fig. P9-17 are $W_A$ = 250 lb and $W_B$ = 100 lb. If the coefficient of friction is 0.35 for all surfaces, determine the minimum force **P** required for equilibrium.

**Fig. P9-17**

**9-18\*** A boy is pulling a sled with a box (at constant velocity) up an inclined surface as shown in Fig. P9-18. The mass of the box and sled is 50 kg; the mass of the boy is 40 kg. If the coefficient of kinetic friction between the sled runners and the icy surface is 0.05, determine the minimum coefficient of static friction needed between the boy's shoes and the icy surface.

Fig. P9-18

**9-19** The 250-lb cylinder shown in Fig. P9-19 is in equilibrium on the inclined surface. Determine the weight $W$ and the minimum coefficient of friction between the inclined surface and the cylinder.

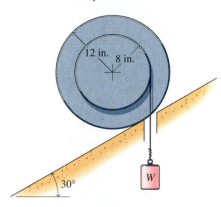

Fig. P9-19

**9-20** A 30-kg box is sitting on an inclined surface as shown in Fig. P9-20. If the coefficient of friction between the box and the surface is 0.50 and the angle $\alpha = 60°$, determine the maximum force **T** for which the box will be in equilibrium. Is impending motion by tipping or by slipping?

Fig. P9-20

**9-21\*** A 120-lb girl is walking up a 48-lb uniform beam as shown in Fig. P9-21. Determine how far up the beam the girl can walk before the beam starts to slip if

**a.** The coefficient of friction is 0.20 at all surfaces.
**b.** The coefficient of friction at the bottom end of the beam is increased to 0.40 by placing a piece of rubber between the beam and the floor.

Fig. P9-21

**9-22\*** A 75-kg man starts climbing a ladder that leans against a wall as shown in Fig. P9-22. If the weight of the ladder is negligible, determine how far up the ladder the man can climb before the ladder starts to slip if

**a.** The coefficient of friction is 0.25 at both surfaces.
**b.** The coefficient of friction at the bottom end of the ladder is increased to 0.40 by placing a piece of rubber between the ladder and the floor.

Fig. P9-22

**9-23** Determine the force **P** required to cause impending motion of the two 100-lb cylinders up the inclined surface shown in Fig. P9-23. The coefficient of friction is 0.15 for all contacting surfaces. The pin in cylinder $A$ is frictionless. Do the cylinders slip or roll?

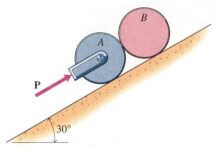

**Fig. P9-23**

**9-24** The masses of blocks $A$ and $B$ of Fig. P9-24 are $m_A = 100$ kg and $m_B = 50$ kg. If the coefficient of friction is 0.15 for block $A$ and 0.25 for block $B$, determine the maximum angle $\theta$ for equilibrium of the blocks.

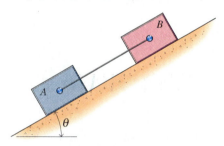

**Fig. P9-24**

**9-25\*** Block $A$ is pressed against the floor when a 65-lb force is applied to the handle shown in Fig. P9-25. The pin is frictionless, and the weight of the handle can be neglected. The coefficient of friction is 0.20 at all surfaces. Determine the minimum weight $W_A$ necessary to prevent slipping.

**Fig. P9-25**

**9-26\*** A worker is lifting a 35-kg uniform beam (see Fig. P9-26) with a force that is perpendicular to the beam. The length of the beam is 5 m and the coefficient of friction between the beam and the ground is 0.20. Determine the height $h$ at which the beam will begin to slip.

**Fig. P9-26**

**9-27** A lightweight rope is wrapped around a 100-lb drum, passes over a frictionless pulley, and is attached to a weight $W$ (see Fig. P9-27). The coefficient of friction between the drum and the surfaces is 0.50. Determine the maximum amount of weight that can be supported by this arrangement.

**Fig. P9-27**

**9-28** A lightweight rope is wrapped around a drum as shown in Fig. P9-28. The coefficient of friction between the drum and the ground is 0.30. Determine the maximum angle $\theta$ such that the drum does not slip. Also determine the tension in the cable for this angle if the drum weighs 100 N.

**Fig. P9-28**

**9-29\*** A lightweight rope is wrapped around a 25-lb drum that is 3 ft in diameter (see Fig. P9-29). The coefficient of friction between the drum and the ground is 0.30. Determine whether or not the drum is in equilibrium.

**Fig. P9-29**

**9-30** The broom shown in Fig. P9-30 weighs 8 N and is held up by the two cylinders, which are wedged between the broom handle and the side rails. The coefficient of friction between the broom and cylinders and between the cylinders and side rails is 0.30. The side rails are at an angle of $\theta = 60°$ to the horizontal. The weight of the cylinders may be neglected. Determine whether or not this system is in equilibrium. If the system is in equilibrium, determine the force exerted on the broom handle by the rollers.

**Fig. P9-30**

## Intermediate Problems

**9-31** The cylinder shown in Fig. P9-31 weighs 200 lb. If the coefficient of friction is 0.25 for all surfaces, determine the maximum weight $W$ that can be supported without causing the cylinder to turn.

**Fig. P9-31**

**9-32** The rods of Fig. P9-32 are lightweight and all pins are frictionless. The coefficient of friction between the 40-kg slider block and the floor is 0.40. Determine

**a.** The maximum force **P** for which motion does not occur if the force **P** is horizontal ($\theta = 0$).

**b.** The angle $\theta$ that gives the absolute greatest force **P** for which motion does not occur.

**Fig. P9-32**

**9-33\*** The weight of block $A$ in Fig. P9-33 is 125 lb. The weights of the lever and cylinder can be neglected, and the pin at $D$ is frictionless. If the coefficient of friction is 0.50 at $B$ and 0.30 for all other surfaces, determine the force **P** required to cause impending motion of block $A$ to the left.

**Fig. P9-33**

**9-34\*** A 100-kg cylinder rests against a wall and a plate (see Fig. P9-34). The coefficient of friction is 0.30 at both surfaces. The plate rests on rollers, and friction between the plate and the floor is negligible. Neglect the weight of the plate and determine the minimum force **P** necessary to move the plate.

**Fig. P9-34**

493

**9-35** A 100-lb uniform beam 16 ft long lies against a corner as shown in Fig. P9-35. Determine the maximum force **P** for which the beam will be in equilibrium if the coefficient of friction $\mu_s$ is

**a.** 0.60 at both surfaces.
**b.** 0.75 at the bottom surface and 0.40 at the corner surface.

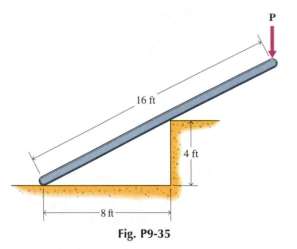

Fig. P9-35

**9-36** The masses of blocks $A$ and $B$ of Fig. P9-36 are $m_A = 40$ kg and $m_B = 85$ kg. If the coefficient of friction is 0.25 for both surfaces, determine the force **P** required to cause impending motion of block $B$.

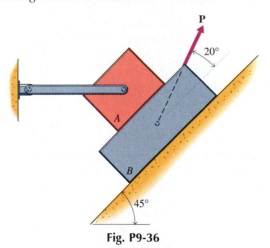

Fig. P9-36

## Challenging Problems

**9-37\*** The weights of block $A$ and cylinder $B$ of Fig. P9-37 are $W_A = 100$ lb and $W_B = 140$ lb. If the coefficient of

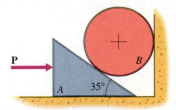

Fig. P9-37

static friction is 0.25 for all surfaces, determine the force **P** required to cause impending motion of the block to the right.

**9-38\*** The automobile shown in Fig. P9-38 has a mass of 1500 kg. The coefficient of friction between the rubber tires and the pavement is 0.70. Determine the maximum incline $\theta$ that the automobile can drive up if the automobile has

**a.** A rear-wheel drive.
**b.** A front-wheel drive.

Fig. P9-38

**9-39** The weights of block $A$ and cylinder $B$ of Fig. P9-39 are $W_A = 250$ lb and $W_B = 100$ lb. Determine the minimum coefficient of friction for which block $B$ is in equilibrium on the inclined surface.

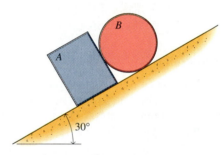

Fig. P9-39

**9-40** Two blocks are connected with a rigid link as shown in Fig. P9-40. Determine the minimum force **P** required to prevent motion of the blocks if $m_A = 75$ kg and $m_B = 25$ kg. The coefficient of friction between the blocks and the surface is 0.15.

Fig. P9-40

**9-41\*** Three identical cylinders are stacked as shown in Fig. P9-41. The cylinders each weigh 100 lb and are 10 in. in diameter. The coefficient of friction is 0.40 at all surfaces. Determine the maximum force **P** that the cylinders can support without moving.

**Fig. P9-41**

**9-42\*** A 50-kg uniform plank rests on rough supports at $A$ and $B$ as shown in Fig. P9-42. The coefficient of friction is 0.60 at both surfaces. If a man weighing 800 N pulls on the rope with a force $P = 400$ N, determine

a. The minimum and maximum angles $\theta_{\min}$ and $\theta_{\max}$ for which the system will be in equilibrium;

b. The minimum coefficient of friction that must exist between the man's shoes and the ground for the two cases in part *a*.

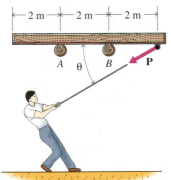

**Fig. P9-42**

**9-43** When a drawer is pulled by only one of the handles, it tends to twist and rub as shown (highly exaggerated) in Fig. P9-43. The weight of the drawer and its contents is 2 lb and is uniformly distributed. The coefficient of friction is 0.60 between the sides of the drawer and the sides of the dresser and 0.10 between the bottom of the drawer and the

**Fig. P9-43**

side rails the drawer rides on. Determine the minimum amount of force necessary to pull the drawer out.

**9-44** An ill-fitting window is about 10 mm narrower than its frame (see Fig. P9-44). The window weighs 40 N, and the coefficient of friction between the window and the frame is 0.20. Determine the amount of force **P** that must be applied at the lower corner to keep the window from lowering.

**Fig. P9-44**

## Computer Problems

**C9-45** The simple mechanism of Fig. P9-45 is often used to hold the handles of brooms, mops, shovels, and other such tools. The weight of the tool causes the two otherwise free cylinders to become wedged into the corner between the handle and the rails. Although no amount of downward force will cause the handle to slip, the tool can be removed easily by lifting it upward and pulling it forward. The coefficient of friction is the same between the cylinder and the broom handle and between the cylinder and the side rail. If the broom shown weighs 1.5 lb and the weight of the two small cylinders may be neglected,

**Fig. P9-45**

a. Plot $\mu_s$, the minimum coefficient of friction for which the system is in equilibrium, as a function of the rail angle $\theta$ ($5° \le \theta \le 75°$).
b. Plot $A_n$, the normal component of the force exerted on the broom handle by one of the cylinders, as a function of the angle $\theta$ ($5° \le \theta \le 75°$).

**Fig. P9-45**

**C9-46** The simple mechanism of Fig. P9-46 is often used to hold notes or signs on bulletin boards. The weight of the paper causes the otherwise free cylinder to become wedged into the corner between the paper and the rail. Although no amount of downward force will cause the paper to slip out, the paper can be removed easily by lifting it upward and pulling it sideways. The coefficient of friction is the same between the cylinder and the paper and between the cylinder and the front rail. The small amount of friction between the back side of the paper and the back of the device may be neglected.

a. Plot $\mu_s$, the minimum coefficient of friction for which the system is in equilibrium, as a function of the rail angle $\theta$ ($5° \le \theta \le 75°$).

15 N

**Fig. P9-46**

b. If a downward force of 15 N is applied to the paper, plot $A_n$, the normal component of the force exerted on the paper by the cylinder, as a function of the angle $\theta$ ($5° \le \theta \le 75°$).

**C9-47** A 50-lb cylinder rests against a wall and a plate as shown in Fig. P9-47. The coefficient of friction is the same at both surfaces of contact with the cylinder.

a. Plot **P**, the maximum force that may be applied to the plate without moving the plate, as a function of the coefficient of friction $\mu_s$ ($0.05 \le \mu_s \le 0.80$).
b. On the same graph, plot $(A_f)_{actual}$ and $(B_f)_{actual}$, the actual amounts of friction that act at points $A$ and $B$, and $(A_f)_{available}$ and $(B_f)_{available}$, the maximum amounts of friction available for equilibrium at points $A$ and $B$.

**Fig. P9-47**

**C9-48** A 10-kg cylinder rests on a thin, lightweight piece of cardboard as shown in Fig. P9-48. The coefficient of friction is the same at all surfaces.

a. Plot **P**, the maximum force that may be applied to the cardboard without moving it, as a function of the coefficient of friction $\mu_s$ ($0.05 \le \mu_s \le 0.80$).
b. On the same graph, plot $(A_f)_{actual}$ and $(B_f)_{actual}$, the actual amounts of friction that act at points $A$ and $B$, and $(A_f)_{available}$ and $(B_f)_{available}$, the maximum amounts of friction available for equilibrium at points $A$ and $B$.
c. What happens to the system at $\mu_s \cong 0.364$?
d. What does the solution for $\mu_s > 0.364$ mean?

**Fig. P9-48**

## 9-3  ANALYSIS OF SYSTEMS INVOLVING DRY FRICTION

As observed in the previous section, friction is often encountered in engineering applications in simple situations. Just as often, however, friction is encountered in situations involving more complex applications. Some of these machine applications, which will be studied here, are (Fig. 9-18) wedges, screws, journal bearings, thrust bearings, and belts. As was the case with machine and nonrigid frames in Chapter 7, it will be necessary to consider equilibrium of the component parts of the machines—even when only external forces or reactions are desired.

### 9-3.1  Wedges

A wedge is just a block that has two flat faces that make a small angle with each other. Wedges are often used in pairs as shown in Fig. 9-18 to raise heavy loads. Depending on the angle between the two surfaces of the wedge, the weight being lifted (the output force) can be many times that of the force $P$ (the input force) applied to the wedge. Also, a properly designed wedge will stay in place and support the load even after the force $P$ is removed.

Wedge problems can often be solved using a semigraphical approach. Wedges are almost always constrained against rotation so that only force equilibrium need be considered. Also, the number of forces acting on a wedge is usually small (the friction and normal forces are usually combined into a single resultant force as in Fig. 9-19a), so force equilibrium can be expressed as a force polygon (Fig. 9-19b). The law of sines and the law of cosines can then be used to relate the forces and angles.

For the case of impending motion, the resultant of the normal and friction force is drawn at the angle of static friction (Fig. 9-20a) and the magnitude of the resultant or some other force determined. If motion is not impending, the resultant is drawn with whatever magnitude and at whatever angle $\phi_1$ is required for equilibrium. This angle is then compared with the angle of static friction $\phi_1 \le \phi_{1s}$ (Fig. 9-20b) to determine whether or not equilibrium exists.

Like other machines, wedges are typically characterized by their **mechanical advantage** (M.A.), or the ratio of their output and input forces. In the case of wedges, the mechanical advantage is defined as

Wedges

Screw

Thrust Bearing

Journal Bearing

Belt

**Figure 9-18**  Machine applications involving friction.

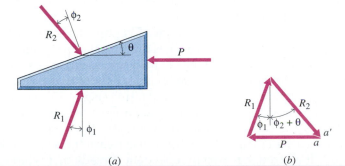

(a)                    (b)

**Figure 9-19**  Free-body diagram and force polygon for a wedge.

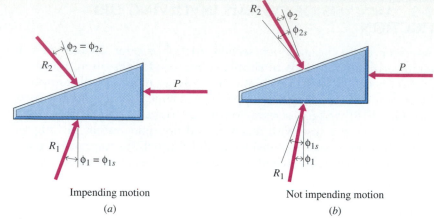

Impending motion     Not impending motion

(a)                          (b)

**Figure 9-20**   Free-body diagrams for a wedge when motion is impending and not impending.

the ratio

$$\text{M.A.} = \frac{\text{direct force}}{\text{wedge force}} \qquad \textbf{(9-6)}$$

The numerator of Eq. 9-6 is the force that must be applied directly to some object to accomplish a desired task. In the case of the wedge in Fig. 9-18, this is just the weight of the object being raised. The denominator of Eq. 9-6 is the force that must be applied to the wedge to accomplish the same task. For the wedge of Fig. 9-18, this is $P$. Clearly, a well-designed wedge should have a mechanical advantage greater than one.

A wedge with a large mechanical advantage may not be the best overall design, however. A common design criterion for wedges is that the wedge remain in place after being forced under the load. A wedge that must be forcibly removed is called **self-locking.**

### 9-3.2   Square-Threaded Screws

Square-threaded screws are essentially wedges that have been wound around a cylindrical shaft. These simple devices can be found in nearly every facet of our lives. Screws are used as fasteners to hold machinery together. Screws are used in jacks to raise heavy loads and on the feet of heavy appliances such as refrigerators to level them. Screws are also used in vises and clamps to squeeze objects together. In each of these cases and many more like them, friction on the threads keeps the screws from turning and loosening.

For example, consider the simple C-clamp of Fig. 9-21. When a twisting moment $M$ is applied to the screw, the clamp tightens and exerts an axial force $W$ on whatever is held in the clamp. As the screw turns and tightens, however, a small segment of the screw's thread will travel around and up the groove in the frame (Fig. 9-22). The distance that the screw moves in the axial direction during one revolution (from point $A$ to $A'$) is called the lead of the screw. For a single-threaded screw, the lead is the same as the distance between the adjacent threads (Fig. 9-23). If a screw has two independent threads that wind around it, the lead would be twice the distance between the adjacent threads.

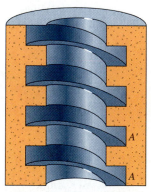

**Figure 9-21**   Twisting moment applied to the screw of a simple C-clamp.

**Figure 9-22**   Lead for a single-threaded screw.

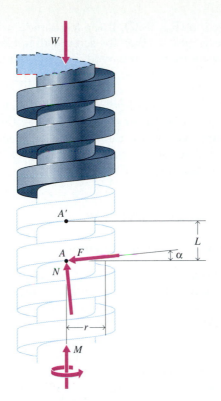

**Figure 9-23** Normal and friction forces on a screw.

**Figure 9-24** Free-body diagram for a partial section of a screw.

**Figure 9-25** Free-body diagram for a screw being advanced.

During each complete turn of the screw a small segment of the screw's thread will travel a distance $2\pi r$ around the shaft while advancing the distance $L$. It is as if the small segment of the screw thread is being pushed up a wedge or inclined plane of angle $\alpha = \tan^{-1}(L/2\pi r)$ (Fig. 9-24). In addition to the normal and friction forces on the thread, the free-body diagram for a typical segment of the screw includes a portion of the axial force $dW$ and a force $dP$ due to the twisting moment, $dM = r\,dP$. If the equilibrium equations for each little segment of the screw are added together, the resulting set of equations would be the same as the equilibrium equations for the free-body diagram shown in Fig. 9-25 in which $W = \int dW$ is the total axial force, $P = \int dP = (\int dM/r) = M/r$ is the total pushing force due to the moment $M$, and $F = \int dF$ and $N = \int dN$ are the total friction and normal forces, respectively.

If the twisting moment $M$ is just sufficient to turn the screw, then $R$, the resultant of the friction and normal forces, will act at the angle of static friction $\phi_s$ to the normal and force equilibrium is neatly expressed by the force triangle (Fig. 9-26) from which

$$\tan(\alpha + \phi_s) = \frac{M/r}{W}$$

Therefore,

$$M = r\,W\tan(\alpha + \phi_s) \qquad \textbf{(9-7)}$$

is the minimum torque $M$ necessary to advance the screw against a load $W$.

When the twisting moment $M$ is removed or reduced to near zero, the screw will tend to unwind and the friction force will change di-

**Figure 9-26** Force triangle when the minimum moment needed to advance the screw is being applied.

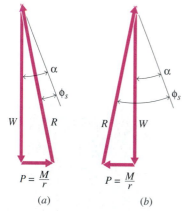

**Figure 9-27** Free-body diagram for a screw as it unwinds.

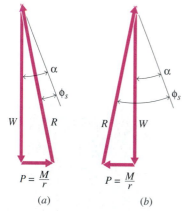

**Figure 9-28** Free-body diagrams for a screw when the lead angle is greater than or less than the angle of static friction.

**Figure 9-29** Schematic illustration of a journal bearing.

rection (Fig. 9-27). If the lead angle $\alpha$ is greater than the angle of static friction $\phi_s$, then the screw will not be in equilibrium when the twisting moment is removed, but will require a twisting moment of

$$M = rW \tan (\alpha - \phi_s) \qquad (9\text{-}8a)$$

to maintain equilibrium (Fig. 9-28a). However, if the lead angle $\alpha$ is less than the angle of static friction $\phi_s$, then the screw will be in equilibrium even when the twisting moment is removed. This condition is called self-locking and is a design criterion in most screw designs. In this case a reverse moment of

$$M = rW \tan (\phi_s - \alpha) \qquad (9\text{-}8b)$$

is required to remove the screw (Fig. 9-28b).

### 9-3.3  Journal Bearings

Journal bearings are used to support the sides of rotating shafts, as shown in Fig. 9-29. Journal bearings are usually lubricated to reduce friction, and the analysis of lubricated journal bearings is a topic for fluid mechanics. Some journal bearings, however, are not lubricated or are poorly lubricated. The analysis that follows can be used to estimate the amount of friction in these bearings.

Suppose a torque $M$ is applied to the shaft of Fig. 9-29 to cause it to rotate. Because of the lateral load $L$ on the shaft, there will be contact between the shaft and bearing. As the shaft rotates, friction between the rotating shaft and the bearing causes the shaft to climb a small distance up the inner surface of the bearing. Since slip is occurring, the friction force is that of kinetic friction, $F = \mu_k N$, and the resultant of the friction and normal forces act at the angle of kinetic friction $\phi_k$ to the normal.

Applying force equilibrium to the free-body diagram of Fig. 9-30a gives that $R$ must be a vertical force equal in magnitude to the load $L$. Then, for moment equilibrium,

$$M = Lr \sin \phi_k \qquad (9\text{-}9a)$$

where $r \sin \phi_k$ is the distance between the lines of action of $L$ and $R$ (Fig. 9-30b). Even for dry journal bearings, however, the coefficient of friction $\mu_k = \tan \phi_k$ is generally small, so $\sin \phi_k \cong \tan \phi_k = \mu_k$. Therefore, the torque required to rotate a shaft at constant speed is approximately

$$M \cong \mu_k Lr \qquad (9\text{-}9b)$$

**Figure 9-30** Free-body diagram for a journal bearing.

## 9-3.4  Thrust Bearings

Thrust bearings are used to support shafts acted on by axial loads. Thrust bearings include both the collar bearing and end bearing shown in Fig. 9-31. With collar bearings, friction forces act on the annular region of contact between the collar and the bearing. With end bearings, friction forces act on a circular area if the shaft is solid or on an annular region if the shaft is hollow. The analysis for clutch plates and disk brakes, which also involve friction on circular or annular regions, is identical to that for end bearings except that the kinetic coefficient of friction $\mu_k$ must be replaced with the static coefficient of friction $\mu_s$ since friction is only impending and not actually occurring.

The end bearing of Fig. 9-32 supports a hollow circular shaft that is subjected to an axial force $P$. If the contact pressure (normal force per unit area) between the shaft and the bearing is $p$, then the normal force on a small element of area $dA$ is $dN = p\,dA$; the friction force on $dA$ is $dF = \mu_k(p\,dA)$; and the moment of the friction force about the axis of the shaft is $dM = r\,dF = \mu_k pr\,dA$. Integrating these quantities over the contact area gives the total axial load carried by the bearing

$$P = \int p\,dA = \int_{R_1}^{R_2} p(2\pi r\,dr) \qquad (9\text{-}10a)$$

and the total moment

$$M = \int_A dM = \int_A \mu_k pr\,dA = \mu_k \int_{R_1}^{R_2} pr(2\pi r\,dr) \qquad (9\text{-}10b)$$

For new surfaces that are flat and well-supported, the pressure $p$ is essentially constant over the contact area, so the total axial load carried by the bearing is

$$P = p \int_{R_1}^{R_2} 2\pi r\,dr = p(R_2^2 - R_1^2) \qquad (9\text{-}11)$$

and the total moment is given by

$$M = \mu_k p \int_{R_1}^{R_2} 2\pi r^2\,dr = \frac{2}{3}\mu_k \pi p(R_2^3 - R_1^3)$$

$$= \frac{2\mu_k P(R_2^3 - R_1^3)}{3(R_2^2 - R_1^2)} \qquad (9\text{-}12a)$$

For a solid circular shaft of radius $R$, Eq. 9-12a simplifies to

$$M = \frac{2}{3}\mu_k PR \qquad (9\text{-}12b)$$

As the shaft rotates in the bearing, however, the surfaces will gradually wear and the pressure will likely decrease with radial distance over the bearing surface. That is, during each rotation of the shaft, a small element of area travels a distance of $2\pi r$. Therefore, outer portions of the shaft, which travel farther, probably wear down faster than inner portions of the shaft, and contact will probably not be as strong there. Assuming that the pressure in a worn bearing decreases with radius according to $rp = C$, where $C$ is a constant, gives the total axial force carried by the bearing as

$$P = \int p\,dA = \int_{R_1}^{R_2} \left(\frac{C}{r}\right)(2\pi r\,dr) = 2\pi C(R_2 - R_1) \qquad (9\text{-}13)$$

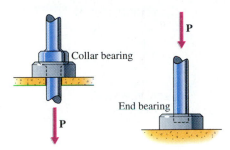

**Figure 9-31**  Schematic illustrations of thrust bearings.

**Figure 9-32**  Friction force in a thrust bearing.

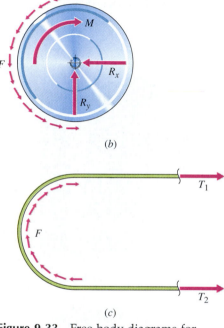

(a)

(b)

(c)

**Figure 9-33**  Free-body diagrams for a flat belt in contact with a circular drum.

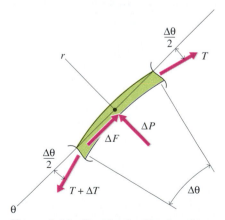

**Figure 9-34**  Free-body diagram for a small segment of a flat belt.

and the total moment as

$$M = \mu_k C \int_{R_1}^{R_2} 2\pi r \, dr = \mu_k \pi C (R_2^2 - R_1^2)$$
$$= \frac{\mu_k P (R_2 + R_1)}{2} \tag{9-14a}$$

For a solid circular shaft of radius $R$, Eq. 9-14a simplifies to

$$M = \frac{1}{2}\mu_k PR \tag{9-14b}$$

which is just 3/4 as much as for new surfaces.

### 9-3.5   Flat Belts and V-Belts

Many types of power machinery rely on belt drives to transfer power from one piece of equipment to another. Without friction, the belts would slip on their pulleys and no power transfer would be possible. Maximum torque is applied to the pulley when the belt is at the point of impending slip, and that is the case discussed here.

Although the analysis presented is for flat belts, it also applies to any shape belt as well as circular ropes, as long as the only contact between the belt and the pulley is on the bottom surface of the belt. This section ends with a brief discussion of V-belts, which indicates the kind of modifications required when friction acts on the sides of the belt instead of the bottom.

Figure 9-33a shows a flat belt passing over a circular drum. The tensions in the belt on either side of the drum are $T_1$ and $T_2$ and the bearing reaction is $R$. Friction in the bearing is neglected for this analysis, but a torque $M$ is applied to the drum to keep it from rotating. If there is no friction between the belt and the drum, the two tensions must be equal, $T_1 = T_2$, and no torque is required for moment equilibrium, $M = 0$. If there is friction between the belt and the drum, however, then the two tensions need not be equal and a torque $M = r (T_2 - T_1)$ is needed to satisfy moment equilibrium. Assuming that $T_2 > T_1$, this means that friction must exert a counterclockwise moment on the drum (Fig. 9-33b) and the drum will exert an opposite frictional resistance on the belt (Fig. 9-33c). Because the friction force depends on the normal force and the normal force varies around the drum, care must be taken in adding up the total frictional resistance.

The free-body diagram (Fig. 9-34) of a small segment of the belt includes the friction force $\Delta F$ and the normal force $\Delta P$. The tension in the belt increases from $T$ on one side of the segment to $T + \Delta T$ on the other side. Equilibrium in the radial direction gives

$$\nwarrow \Sigma F_r = \Delta P - T \sin\left(\frac{\Delta\theta}{2}\right) - (T + \Delta T) \sin\left(\frac{\Delta\theta}{2}\right) = 0$$

or

$$\Delta P = 2T \sin\left(\frac{\Delta\theta}{2}\right) + \Delta T \sin\left(\frac{\Delta\theta}{2}\right) \tag{a}$$

# CONCEPTUAL EXAMPLE 9-3: FRONT-WHEEL VERSUS REAR-WHEEL DRIVE

*The car shown in Fig. CE9-3a weighs 13 kN. When the car is moving at a constant speed of 50 km/h it must overcome a drag force $F_D$ of 450 N. If the car encounters a stretch of ice-covered pavement where the coefficient of friction between the pavement and the rubber tires is 0.065, would a front-wheel or a rear-wheel drive car be better able to maintain the speed?*

(a)

## SOLUTION

Free-body diagrams for front-wheel drive and rear-wheel drive cars are shown in Figs. CE9-3b and CE9-3c, respectively. In both cases, the 450-N drag force is opposed by frictional forces between the car tires and the pavement. Summing moments about point $A$ for the front-wheel drive car shown in Fig. CE9-3b yields

$$+\curvearrowleft \Sigma M_A = N_F(3.000) + 0.450(0.750) - 13(1.750) = 0$$

from which

$$N_F = 7.4708 \text{ kN} \cong 7.471 \text{ kN}$$

(b)

Summing moments about point $B$ for the rear-wheel drive car shown in Fig. CE9-3c yields

$$+\curvearrowleft \Sigma M_B = -N_R(3.000) + 0.450(0.750) + 13(1.250) = 0$$

from which

$$N_R = 5.5292 \text{ kN} \cong 5.529 \text{ kN}$$

Since frictional forces are developed only at the drive wheels,

For the front-wheel drive car: $\mu_{\text{req}} = \dfrac{F_F}{N_F} = \dfrac{0.450}{7.471} = 0.058 < 0.065$

For the rear-wheel drive car: $\mu_{\text{req}} = \dfrac{F_R}{N_R} = \dfrac{0.450}{5.529} = 0.081 > 0.065$

(c)

**Fig. CE9-3**

Thus, the front-wheel drive car would be able to maintain the speed, since the required coefficient of friction is less than that available, while the rear-wheel drive car could not maintain the speed, since the required coefficient of friction is more than that available.

while equilibrium in the circumferential ($\theta$-) direction gives

$$\nwarrow\Sigma F_\theta = (T + \Delta T) \cos\left(\frac{\Delta\theta}{2}\right) - T \cos\left(\frac{\Delta\theta}{2}\right) - \Delta F = 0$$

or

$$\Delta T \cos\left(\frac{\Delta\theta}{2}\right) = \Delta F \qquad\qquad (b)$$

In the limit as $\Delta\theta \to 0$ the normal force $\Delta P$ on the small segment of the belt must vanish according to Eq. $a$. But when the normal force vanishes ($\Delta P \to 0$), there can be no friction on the belt either ($\Delta F \to 0$). Therefore, the change in tension across the small segment of the belt must also vanish ($\Delta T \to 0$) in the limit as $\Delta\theta \to 0$ according to Eq. $b$.

Assuming that slip is impending gives $\Delta F = \mu_s \Delta P$, and Eqs. $a$ and $b$ can be combined to give

$$\Delta T \cos (\Delta\theta/2) = \mu_s\, 2T \sin (\Delta\theta/2) + \mu_s\, \Delta T \sin (\Delta\theta/2) \qquad (c)$$

which, after dividing through by $\Delta\theta$, is

$$\frac{\Delta T}{\Delta\theta} \cos (\Delta\theta/2) = \mu_s T \frac{\sin (\Delta\theta/2)}{\Delta\theta/2} + \frac{\mu_s \Delta T}{2} \frac{\sin (\Delta\theta/2)}{\Delta\theta/2} \qquad (d)$$

Finally, taking the limits as $\Delta\theta \to 0$ and recalling that

$$\lim_{\Delta\theta\to 0} \frac{\Delta T}{\Delta\theta} \to \frac{dT}{d\theta} \qquad \lim_{x\to 0} \cos x \to x \qquad \lim_{x\to 0} \frac{\sin x}{x} \to 1$$

gives

$$\frac{dT}{d\theta} = \mu_s T \qquad\qquad (e)$$

Equation $e$ can be rearranged in the form

$$dT/T = \mu_s\, d\theta \qquad\qquad \textbf{(9-15)}$$

which, since the coefficient of friction is a constant, can be immediately integrated from $\theta_1$, where the tension is $T_1$, to $\theta_2$, where the tension is $T_2$, to get

$$\ln (T_2/T_1) = \mu_s(\theta_2 - \theta_1) = \mu_s\beta \qquad\qquad \textbf{(9-16}a\textbf{)}$$

or

$$T_2 = T_1 e^{\mu_s\beta} \qquad\qquad \textbf{(9-16}b\textbf{)}$$

where $\beta = \theta_2 - \theta_1$ is the central angle of the drum for which the belt is in contact with the drum. The angle of wrap $\beta$ must be measured in radians and must obviously be positive. Angles greater than $2\pi$ radians are possible and simply mean that the belt is wrapped more than one complete revolution around the drum.

It must be emphasized that Eq. 9-16 assumes impending slip at all points along the belt surface and therefore gives the maximum change in tension that the belt can have. Since the exponential function of a positive value is always greater than 1, Eq. 9-16 gives that $T_2$ (the tension in the belt on the side toward which slip tends to occur) will always be greater than $T_1$ (the tension in the belt on the side away from which slip tends to occur). Of course, if slip is not known to be impending, then Eq. 9-16 does not apply and $T_2$ may be larger or smaller than $T_1$.

V-belts as shown in Fig. 9-35a are handled similarly to the above. A view of the belt cross section (Fig. 9-35b), however, shows that there are now two normal forces and there will also be two frictional forces (acting along the edges of the belt and pointing into the plane of the figure). Equilibrium in the circumferential ($\theta$-) direction now gives

$$\Delta T \cos (\Delta\theta/2) = 2\Delta F$$

while equilibrium in the radial direction gives

$$2\Delta P \sin (\alpha/2) = 2T \sin (\Delta\theta/2) + \Delta T \sin (\Delta\theta/2)$$

Continuing as above results finally in

$$T_2 = T_1 e^{(\mu_s)_{enh}\beta} \qquad\qquad (9\text{-}17)$$

in which $(\mu_s)_{enh} = [\mu_s/\sin (\alpha/2)] > \mu_s$ is an *enhanced coefficient of friction*. That is, V-belts always give a larger $T_2$ than flat belts for a given coefficient of friction $\mu_s$ and a given angle of wrap $\beta$.

Equations 9-16 and 9-17 can also be used when slipping is actually occurring by replacing the static coefficient of friction $\mu_s$ with the kinetic coefficient of friction $\mu_k$.

(a)

(b)

**Figure 9-35**  Cross section of a V-belt showing normal forces on the inclined surfaces.

A wedge is to be used to slide the 3000 N safe of Fig. 9-36a across the floor. Determine the minimum force **P** necessary if the coefficient of friction is 0.35 at all surfaces and the weight of the wedges may be neglected.

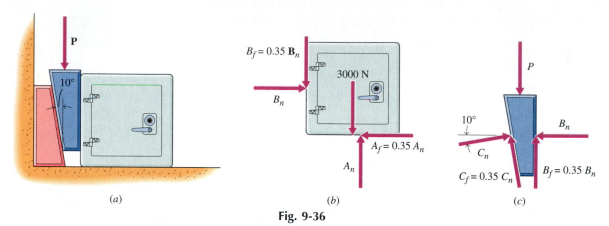

(a)        (b)        (c)

**Fig. 9-36**

## SOLUTION

First draw the free-body diagram of the safe (Fig. 36b). Clearly, the safe will tend to move to the right and the friction force $A_f$ must act to the left to oppose the motion. However, the direction of the friction force $B_f$ is not as easy to ascertain. Even though the safe is not moving up or down, it appears to be moving up relative to the wedge. Hence, the friction force $B_f$ must act downward on the safe to oppose this relative motion.

It may be easier to see the correct direction for this friction force on the equilibrium diagram of the wedge (Fig. 9-36c). The motion of the wedge is downward and the friction force $B_f$ must act upward to oppose the motion. And if the safe exerts an upward frictional force on the wedge, then the wedge must exert an equal frictional force downward on the safe.

The equilibrium equations for the safe are

$$+\rightarrow \Sigma F_x = B_n - 0.35A_n = 0$$
$$+\uparrow \Sigma F_y = A_n - 0.35B_n - 3000 = 0$$

which are solved to get

$$A_n = 3419 \ N \quad \text{and} \quad B_n = 1197 \ N$$

Equilibrium equations for the wedge are

$$+\rightarrow \Sigma F_x = C_n \cos 10° - 0.35C_n \sin 10° - 1197 = 0$$
$$+\uparrow \Sigma F_y = C_n \sin 10° + 0.35C_n \cos 10° + (0.35)(1197) - P = 0$$

which give

$$C_n = 1295 \ N \quad \text{and} \quad P = 1090 \ N \qquad \text{Ans.}$$

Since motion by slipping is impending, the friction force on each sliding surface of the wedge is the maximum available and it acts in a direction to oppose the motion.

## SOLUTION 2. Using the force equilibrium triangle

The free-body diagrams are drawn as above (Figs. 9-36b and 9-36c). Then the force equilibrium triangles are drawn for the safe (Fig. 9-36d) and for the wedge

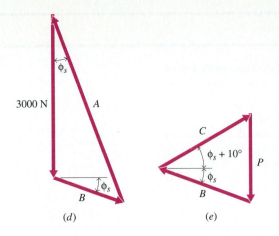

(d)                    (e)

(Fig. 9-36e). In these diagrams, the normal and frictional forces have been combined into a single resultant force, which is drawn at the angle of static friction

$$\phi_s = \tan^{-1} 0.35 = 19.29°$$

since motion is impending.

Using the law of sines on the first force triangle (Fig. 9-36d)

$$\frac{B}{\sin (19.29°)} = \frac{3000}{\sin [90° - 2(19.29°)]}$$

gives immediately

$$B = 1268 \text{ N}$$

Then using the law of sines on the second force triangle (Fig. 9-36e)

$$\frac{P}{\sin [2(19.29°) + 10°]} = \frac{1268}{\sin (90° - 19.29° - 10°)}$$

gives

$$P = 1090 \text{ N} \qquad\qquad \text{Ans.}$$

> Wedge problems can be solved using force triangles. Normal and frictional forces on each contact surface are replaced by resultants which act at the angle of static friction.

507

A wedge is used to raise a 350-lb refrigerator (Fig. 9-37a). The coefficient of friction is 0.2 at all surfaces.

a.  Determine the minimum force **P** needed to insert the wedge.
b.  Determine if the system would still be in equilibrium if **P = 0.**
c.  If the system is not in equilibrium when **P = 0,** determine the force necessary to keep the wedge in place, or if the system is in equilibrium when **P = 0,** determine the force necessary to remove the wedge.

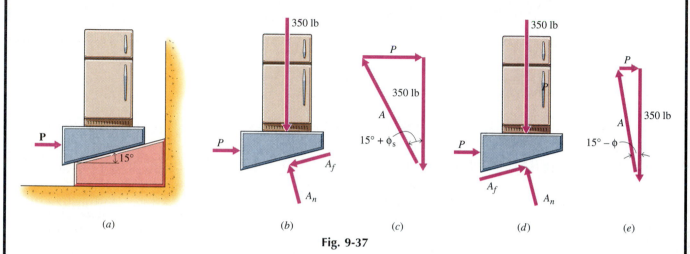

**Fig. 9-37**

## SOLUTION

a.  First draw the free-body diagram of the upper wedge and the refrigerator (Fig. 9-37b) and its force equilibrium triangle (Fig. 9-37c). The normal and frictional forces are combined into a single resultant force acting at the angle of static friction

$$\phi_s = \tan^{-1} 0.2 = 11.31°$$

relative to the normal force or $15° + \phi_s$ relative to the vertical direction. The force triangle is just a right triangle, so

$$\tan (15° + 11.31°) = \frac{P}{350}$$

and

$$P = 173.1 \text{ lb} \qquad \qquad \text{Ans.}$$

b.  For $P$ very small the wedge will tend to move to the left, and the friction will have to act to the right to oppose this motion (Fig. 9-37d). In the force equilibrium triangle (Fig. 9-37e), the resultant force is drawn at an angle $\phi$ relative to the normal force or $15° - \phi$ relative to the vertical direction. Since motion is not known to be impending, the angle $\phi$ is not necessarily equal to $\phi_s$. In this case the angle $\phi$ is merely an unknown to be determined as part of the solution.

From the equilibrium force triangle, when $P = 0$

$$15 - \phi = \tan^{-1} \frac{P}{350} = 0$$

> Since motion by slipping is impending, the friction force on each sliding surface of the wedge is the maximum available and it acts in a direction to oppose the motion.

> Wedge problems can be solved using force triangles. Normal and frictional forces on each contact surface are replaced by resultants which act at the angle of static friction.

or $\phi = 15°$. But the angle of the resultant $\phi$ can never be greater than the angle of static friction $\phi_s = 11.31°$. Therefore, the wedge will not be in equilibrium if the force $P$ is removed.                                                   Ans.

c.  Since the wedge will not stay in place by itself, a force $P$ to the right will be required to hold the wedge in place. The minimum force necessary to hold the wedge in place is attained when $\phi = \phi_s = 11.31°$. The force equilibrium triangle (Fig. 9-37e) is again a right triangle, so

$$\tan (15° - 11.31°) = \frac{P}{350}$$

and

$$P = 22.6 \text{ lb} \qquad\qquad\qquad \text{Ans.}$$

## EXAMPLE PROBLEM 9-10

In the C-clamp of Fig. 9-21, the screw has a mean radius of 3 mm and a single thread with a pitch of 2 mm. If the coefficient of friction is 0.2, determine

a.  The minimum twisting moment necessary to produce a clamping force of 600 N.
b.  The minimum twisting moment necessary to release the clamp when the clamping force is 600 N.
c.  The minimum coefficient of friction for which the clamp is self-locking.

## SOLUTION

The screw has a single thread, so the lead is equal to the pitch

$$\alpha = \tan^{-1}\left(\frac{L}{2\pi r}\right) = \tan^{-1}\left(\frac{2}{6\pi}\right) = 6.06°$$

The angle of static friction is

$$\phi_s = \tan^{-1} 0.2 = 11.31°$$

a.  Since the twisting moment is just sufficient to tighten the screw, the free-body diagram and force triangle of Figs. 9-25 and 9-26 apply and

$$M = rW \tan (\alpha + \phi_s) = 0.003(600) \tan 17.37°$$
$$= 0.563 \text{ N} \cdot \text{m} \qquad\qquad \text{Ans.}$$

b.  Now the twisting moment is just sufficient to loosen the screw, so the free-body diagram and force triangle of Figs. 9-27 and 9-28b apply, and

$$M = rW \tan (\phi_s - \alpha) = 0.003(600) \tan 5.25°$$
$$= 0.1654 \text{ N} \cdot \text{m} \qquad\qquad \text{Ans.}$$

c.  When $M = 0$, $\phi_s \geq \phi = \alpha$. The minimum coefficient of friction corresponds to $\phi = \phi_s = 6.06°$, and hence

$$\mu_s = \tan 6.06° = 0.106 \qquad\qquad \text{Ans.}$$

The distance a screw advances during one turn is called the lead $L$ of the screw.

The pitch angle $\alpha$ for a screw can be determined by using the triangle formed when a single thread is visualized as being unwound for one turn of the screw.

## EXAMPLE PROBLEM 9-11

A 1-in.-diameter shaft rotates inside a journal bearing. The coefficient of friction between the shaft and the bearing is 0.12, and the shaft has a lateral load of 120 lb applied to it. Determine the torque required to rotate the shaft and the angle through which the shaft will climb up the bearing.

### SOLUTION

The moment required is

$$M = \mu_k \, Lr = 0.12(120)(1) = 14.40 \text{ lb} \cdot \text{in.} \qquad \text{Ans.}$$

The angle that the shaft will climb is given by

$$\phi_k = \tan^{-1} \mu_k = \tan^{-1} 0.12 = 6.84° \qquad \text{Ans.}$$

(Note that for this angle, $\tan 6.84° = 0.1200$ while $\sin 6.84° = 0.1191$, and therefore the approximation $M = Lr \sin \phi_k \cong Lr \tan \phi_k \cong Lr \, \mu_k$ is within the accuracy of the knowledge of the coefficient of friction and other approximations of the problem.)

Journal bearings have low coefficients of friction; therefore, the approximation $\sin \phi_k \cong \tan \phi_k \cong \mu_k$ is valid.

The pulley of Fig. 9-38a consists of a 100-mm-diameter wheel that fits loosely over a 15-mm-diameter axle. The axle is rigidly supported at its ends and does not turn. The static and dynamic coefficients of friction between the pulley and axle are 0.4 and 0.35, respectively, and the weight of the pulley may be neglected. Determine the force in the rope that is necessary

a. To just start raising the 250-N load.
b. To raise the 250-N load at a constant rate.
c. To lower the 250-N load at a constant rate.

250 N

$T$

$(a)$

$M$

$A_x$

$A_y$

250 N          $T$

$(b)$

**Fig. 9-38**

## SOLUTION

a. The free-body diagram of the pulley is shown in Fig. 9-38b. The moment $M$ is due to axle friction and for the case of impending motion is given by $M = \mu_s Lr$, where $L$ is the resultant contact force between the axle and pulley and $r$ is the radius of the axle. The moment acts in a counterclockwise direction to oppose the impending motion. The equations of equilibrium give

$$+\rightarrow \Sigma F_x = A_x = 0 \qquad +\uparrow \Sigma F_y = A_y - T - 250 = 0$$
$$\curvearrowright +\Sigma M_A = (50)(250) - 50T + (0.4)\, A_y\, (7.5) = 0$$

Solution of these equations gives

$$A_x = 0 \text{ N} \qquad A_y = 532 \text{ N} \qquad T = 281.9 \text{ N} \cong 282 \text{ N} \qquad \text{Ans.}$$

> The frictional force between two bodies always opposes the tendency to move.

b. Now motion is occurring and the moment due to axle friction is given by $M = \mu_k Lr$. Replacing the coefficient of static friction with the coefficient of kinetic friction and re-solving the above equations gives

$$A_x = 0 \text{ N} \qquad A_y = 528 \text{ N} \qquad T = 277.7 \text{ N} \cong 278 \text{ N} \qquad \text{Ans.}$$

c. Now a counterclockwise motion is occurring and the moment due to axle friction must act in a clockwise direction to oppose the motion. The magnitude of the moment is again given by $M = \mu_k Lr$. Changing the sign on the third term in the moment equation above and re-solving the equations gives

$$A_x = 0 \text{ N} \qquad A_y = 475 \text{ N} \qquad T = 225.1 \text{ N} \cong 225 \text{ N} \qquad \text{Ans.}$$

A collar bearing supports an axial load of 15 lb. The inside and outside diameters of the bearing surface are 1 in. and 2 in., respectively. If the coefficient of kinetic friction is 0.15, compute the moment necessary to overcome friction for a new and for a worn-in bearing.

**SOLUTION**

For a new bearing the moment is

$$M = \frac{2\mu_k P(R_2^3 - R_1^3)}{3(R_2^2 - R_1^2)}$$
$$= \frac{2(0.15)(15)(2^3 - 1^3)}{3(2^2 - 1^2)}$$
$$= 350 \text{ lb} \cdot \text{in.} \qquad \text{Ans.}$$

For a worn-in bearing the moment would be

$$M = \frac{\mu_k P(R_2^2 - R_1^2)}{2(R_2 - R_1)}$$
$$= \frac{(0.15)(15)(2^2 - 1^2)}{2(2 - 1)}$$
$$= 3.38 \text{ lb} \cdot \text{in.} \qquad \text{Ans.}$$

> In a collar bearing, friction forces act on the annular region of contact between the collar and the bearing.

The machine of Fig. 9-39 consists of a polishing disk that spins clockwise to shine a waxed floor. The coefficient of friction between the disk and the floor is 0.3, and the polishing unit weighs 175 N. Determine the forces that must be applied to the handle of the floor polisher to counteract the frictional moment of the polishing disk.

**SOLUTION**

Assuming that the pressure is distributed uniformly over the polishing disk, the frictional moment is given by

$$M = \frac{2}{3}\mu_k\, WR = \frac{2}{3}(0.3)(175)(250) = 8750 \text{ N} \cdot \text{mm}$$

and this must equal the moment of the couple applied to the handle

$$350\, P = 8750 \text{ N} \cdot \text{mm}$$

Therefore

$$P = 25.0 \text{ N} \qquad \text{Ans.}$$

Fig. 9-39

An 80-lb child is sitting on a swing suspended by a rope that passes over a tree branch (Fig. 9-40). The coefficient of friction between the rope and the branch (which can be modeled as a flat belt over a drum) is 0.5, and the weight of the rope can be ignored. Determine the minimum force that must be applied to the other end of the rope to keep the child suspended.

**Fig. 9-40**

## SOLUTION

The rope is wrapped one-half turn or $\pi$ radians about the branch. The child is sitting on the side that motion is impending toward, so the tension in that side of the rope is designated $T_2$ and $T_1 = P$. Therefore,

$$80 \text{ lb} = P\, e^{0.5\pi}$$

or

$$P = 16.63 \text{ lb} \qquad \text{Ans.}$$

The tendency of the rope is to slide toward the side where the tension is the largest. The frictional force opposes this tendency to slide.

A car is prevented from moving by pulling on a rope that is wrapped $n + \frac{1}{4}$ times around a tree (Fig. 9-41). The coefficient of friction between the rope and the tree is 0.35, and the force exerted by the car is 3600 N. If it is desired that the force exerted on the rope be no more than 125 N, determine $n$, the number of times the rope must be wrapped around the tree.

**Fig. 9-41**

**SOLUTION**

The angle of twist necessary to hold the car is found from

$$3600 = 125 \, e^{0.35\beta}$$

or

$$\beta = \frac{\ln(3600/125)}{0.35} = 9.60 \text{ radians}$$

which is 1.528 times around the tree. Any angle less than this will require a resisting force greater than 125 N, whereas any angle greater than this will require less force. Thus

$$n = 2 \qquad \qquad \text{Ans.}$$

will be sufficient.

> If the rope is wrapped $n$ times around the tree, the angle $\beta$ equals $2\pi n$ radians.

---

Suppose the child of Example Problem 9-15 is holding the other end of the rope. Determine the minimum force he must exert on the rope to keep himself suspended.

**SOLUTION**

The free-body diagrams of the child and the tree limb are drawn in Fig. 9-42a and 9-42b, respectively. The rope is still wrapped one-half turn or $\pi$ radians about the branch so that

$$T_2 = T_1 \, e^{0.5\pi}$$

Vertical equilibrium of the child gives

$$+\uparrow \Sigma F_y = T_2 + T_1 - 80 = 0$$

Combining these two equations gives

$$80 - T_1 = T_1 \, (4.81)$$

or

$$T_1 = 13.77 \text{ lb} \qquad \qquad \text{Ans.}$$

$T_2 \quad T_1$      $C$

$T_2 \quad T_1$

80 lb

(a)      (b)

**Fig. 9-42**

> The ratio of tensions given by Eq. 9-16 can also be used for circular belts and ropes.

# PROBLEMS

## Wedge Friction

### Introductory Problems

**9-49*** The weights of blocks $A$ and $B$ of Fig. P9-49 are $W_A = 1000$ lb and $W_B = 100$ lb. If the coefficient of friction is 0.30 at all surfaces, determine the force $P$ required to produce impending motion of block $A$.

**Fig. P9-49**

**9-50*** The masses of blocks $A$ and $B$ of Fig. P9-50 are $m_A = 100$ kg and $m_B = 10$ kg. If the coefficient of friction is 0.25 at all surfaces, determine the force $P$ required to produce impending motion of block $A$.

**Fig. P9-50**

**9-51** The weights of blocks $A$ and $B$ of Fig. P9-51 are $W_A = 250$ lb and $W_B = 500$ lb. If the coefficient of friction is 0.30

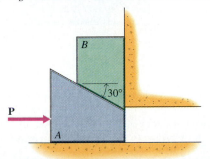

**Fig. P9-51**

at all surfaces, determine the force $P$ required to produce impending motion of block $A$.

**9-52** A 2500-kg block is being lifted with two 8° wedges as shown in Fig. P9-52. If the coefficient of friction is 0.10 at all surfaces, determine the force $P$ required to start lifting the block.

**Fig. P9-52**

**9-53*** The weights of blocks $A$ and $B$ of Fig. P9-53 are $W_A = 500$ lb and $W_B = 1000$ lb. If the coefficient of friction is 0.25 at the contact surface between blocks $A$ and $B$ and 0.10 at all other surfaces, determine the force $P$ required to produce impending motion of block $B$.

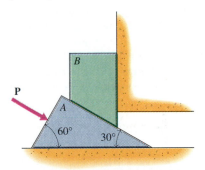

**Fig. P9-53**

**9-54*** The masses of blocks $A$ and $C$ of Fig. P9-54 are $m_A = 800$ kg and $m_C = 300$ kg. The mass of wedge $B$ is negligible. If the coefficient of friction is 0.20 at all surfaces, determine the force $P$ required to produce impending motion of wedge $B$.

**Fig. P9-54**

**9-55** The plunger of a door latch is held in place by a spring as shown in Fig. P9-55. Friction on the sides of the plunger may be ignored. If a force of 2 lb is required to just start closing the door and the coefficient of friction between the plunger and the striker plate is 0.25, determine the force exerted on the plunger by the spring.

**Fig. P9-55**

## Intermediate Problems

**9-56\*** A wedge is being forced under an 80-kg drum as shown in Fig. P9-56. The coefficient of friction between the wedge and the drum is 0.10, while the coefficient of friction is 0.30 at all other surfaces. Assuming a wedge angle $\theta$ of 25° and that the weight of the wedge may be neglected, determine the minimum force **P** necessary to insert the wedge.

**Fig. P9-56**

**9-57\*** A wedge rests between a 75-lb drum and a wall as shown in Fig. P9-57. The coefficient of friction between the drum and the floor is 0.15, while the coefficient of friction is 0.50 at all other surfaces. Assuming a wedge angle $\theta$ of 40°, determine the minimum weight of the wedge that will cause motion.

**Fig. P9-57**

**9-58** The wedge of Problem 9-56 is to be designed so that slip occurs at all surfaces simultaneously. Determine the required wedge angle $\theta$ and the corresponding force **P**.

**9-59** The wedge of Problem 9-57 is to be designed so that slip occurs at all surfaces simultaneously. Determine the required wedge angle $\theta$ and the corresponding wedge weight $W$.

## Challenging Problems

**9-60\*** A pair of wedges is used to lift a crate as shown in Fig. P9-60. The crate weighs 3000 N, the wedge angle $\theta$ is 15°, and the coefficient of friction is 0.25 at all surfaces. The weight of the wedges is negligible. Determine

**a.** The force **P** necessary to insert the wedge.
**b.** If the system would be in equilibrium if the force **P** were removed.
**c.** The force **P** necessary to remove the wedge (or to prevent the wedge from slipping out, depending on the answer to part b).
**d.** The maximum angle $\theta$ for which the system would be in equilibrium if the force **P** were removed.

**Fig. P9-60**

**9-61\*** A pair of wedges is used to lift a crate as shown in Fig. P9-61. The crate weighs 4000 lb, the wedge angle $\theta$ is 18°, and the coefficient of friction is 0.15 at all surfaces. The weight of the wedges is negligible. Determine

**a.** The force **P** necessary to insert the wedge.

**b.** If the system would be in equilibrium if the force **P** were removed.

**c.** The force **P** necessary to remove the wedge or to prevent the wedge from slipping out depending on the answer to part *b*.

**d.** The maximum angle $\theta$ for which the system would be in equilibrium if the force **P** were removed.

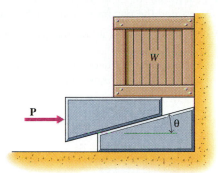

Fig. P9-61

**9-62** A pair of wedges is used to lift a crate as shown in Fig. P9-62. The crate weighs 1200 N, the wedge angle $\theta$ is 10°, and the coefficient of friction is 0.35 for the contacting surface between the wedges and 0.10 at all other surfaces. The weight of the wedges is negligible. Determine

**a.** The force **P** necessary to insert the wedge.

**b.** The force **P** necessary to prevent the wedge from slipping out.

**c.** The maximum angle $\theta$ for which the system would be in equilibrium if the force **P** were removed.

Fig. P9-62

**9-63** A pair of wedges is used to move a crate as shown in Fig. P9-63. The crate weighs 2500 lb, the wedge angle $\theta$ is 12°, and the coefficient of friction is 0.20 for the contacting surface between the wedges and 0.10 at all other surfaces. The weight of the wedges is negligible. Determine

**a.** The force **P** necessary to insert the wedge for the given angle $\theta$.

**b.** The maximum angle $\theta$ for which **P** is smaller than the force needed to push the crate directly.

**c.** The minimum angle $\theta$ such that no value of **P** will cause the crate to move.

Fig. P9-63

**9-64** A pair of wedges is used to move a crate as shown in Fig. P9-64. The crate weighs 3000 N, the wedge angle $\theta$ is 15°, and the coefficient of friction is 0.25 for the contacting surface between the wedges and 0.10 at all other surfaces. The weight of the wedges is negligible. Determine

**a.** The force **P** necessary to insert the wedge for the given angle $\theta$.

**b.** The maximum angle $\theta$ for which **P** is smaller than the force needed to push the crate directly.

**c.** The minimum angle $\theta$ such that no value of **P** will cause the crate to move.

Fig. P9-64

**517**

## Square-Threaded Screws Friction

**9-65\*** The clamp of Fig. P9-65 is used to hold a cover in place. The screw of the clamp has a single thread with a mean radius of $\frac{1}{4}$ in. and a pitch of 0.10 in. The coefficient of friction is 0.40, and the required clamping force is 80 lb. Determine the twisting moment necessary to tighten the clamp and whether or not the clamp will stay in place if the twisting moment is removed.

**Fig. P9-65**

**9-66\*** In the old-fashioned printing press of Fig. P9-66, a square-threaded screw is used to press the paper against the type bed. The single-threaded screw has a mean radius of 20 mm and a lead of 100 mm. The coefficient of friction is 0.15, and the clamping force necessary to guarantee clear printing on all parts of the paper is 400 N. Determine the twisting moment necessary to operate the press and whether or not a moment must be applied to release the press.

**Fig. P9-66**

**9-67** The twisting moment of Problem 9-65 is applied by means of a force-couple applied to a rod through the head of the screw. Determine the minimum length rod necessary if the applied forces are not to exceed 8 lb and the clamping force does not exceed 150 lb.

**9-68** The twisting moment of Problem 9-66 is applied by means of a single force on a long handle attached to the head of the screw. Determine the minimum length handle necessary such that the force required on the handle is 5 percent of the force applied by the press on the paper.

**9-69\*** The screw jack of Fig. P9-69 uses a square-threaded screw having a coefficient of friction of 0.35 and a mean radius of 1 in. Determine the maximum lead such that

a. The jack will stay in place when the moment is removed.
b. The moment necessary to lower the jack is 20 percent of the moment necessary to raise the jack.

**Fig. P9-69**

**9-70\*** The upside-down screw jack shown in Fig. P9-70 is often used as a foot leveler on refrigerators and other heavy appliances. Suppose the weight carried by a single foot screw is 800 N and the screw has a mean radius of 3 mm and a lead of 1 mm. The coefficient of friction is 0.3 between the screw and the refrigerator frame, and friction between the screw head and the floor may be ignored. Determine the minimum moment necessary to raise and the minimum moment to lower the corner of the refrigerator.

**Fig. P9-70**

**9-71** The gear puller shown in Fig. P9-71 has a double-threaded square-threaded screw with a mean radius of $\frac{1}{4}$ in., a pitch of 0.1 in., and a coefficient of friction of 0.30. The gear is wedged onto its shaft and requires a force of 100 lb to remove it. Determine the moment that must be applied to the gear puller screw to remove the gear.

**Fig. P9-71**

**9-72** The turnbuckle shown in Fig. P9-72 is used to secure a support cable on a sailboat. The double-threaded screws each have a mean radius of 4 mm and a pitch of 2 mm and are threaded in opposite directions. The coefficient of friction is 0.2 and the tension in the cables is 600 N. Determine the minimum moment necessary to tighten and the minimum moment necessary to loosen the turnbuckle.

**Fig. P9-72**

## Journal Bearing Friction

**9-73\*** A grinding wheel weighing 2 lb is supported by a journal bearing at each end of the axle as shown in Fig. P9-73. The coefficient of friction between the axle and bearing is $\mu_k = 0.10$, and the diameter of the axle is 0.75 in. Determine the torque required to rotate the wheel at a constant speed.

**Fig. P9-73**

**9-74\*** A pulley consists of a 120-mm-diameter wheel that fits loosely over an 18-mm-diameter shaft as shown in Fig. P9-74. If the coefficient of friction between the pulley and the shaft is $\mu_k = 0.17$ and the weight of the pulley is 5 N, determine the torque required to raise a 2.5-kg load at a constant speed.

**Fig. P9-74**

**9-75** The pulley of Fig. P9-75 consists of a 4-in.-diameter wheel that fits loosely over a $\frac{1}{2}$ in.-diameter shaft. A rope passes over the pulley and is attached to a 50-lb weight. The static and kinetic coefficients of friction between the pulley and the shaft are 0.35 and 0.25, respectively. The weight of the pulley is 2 lb. The rope being pulled makes an angle $\theta = 90°$ with the horizontal. Determine the force necessary to

a. Just start raising the weight.
b. Raise the weight at a constant speed.
c. Just start lowering the weight.
d. Lower the weight at a constant speed.

**Fig. P9-75**

**9-76** Repeat Problem 9-75 for $\theta = 0°$ and a 110-mm-diameter, 1-kg pulley on a 12-mm-diameter shaft raising/lowering a 225-N weight.

**9-77\*** Repeat Problem 9-75 for $\theta = 30°$.

**9-78** The two-pan balance scale of Fig. P9-78 consists of a 400-mm-long bar that fits loosely over a 20-mm-diameter shaft. The center of mass of the bar coincides with the center of the shaft. The scale functions by placing an item to be weighed (for example, a mini-TV) in the right pan and then placing known weights in the left pan until the scale is in equilibrium. However, because of bearing friction, the system can be in equilibrium for more than a single value of weight in the left pan. If the coefficient of friction between the bar and the shaft is 0.30, determine the range of weights that can be placed in the left pan to balance an 80-N object placed in the right pan.

**Fig. P9-78**

## Thrust Bearing Friction

**9-79*** An axial load of 160 lb is applied to the end bearing of Fig. P9-79. Contact between the end bearing and the seat takes place over an annular ring of radius 4 in. to 7 in. For a coefficient of friction of 0.20, determine the torque needed to rotate the shaft for

**a.** A new bearing surface.
**b.** A worn-in bearing surface.

**Fig. P9-79**

**9-80*** An axial load of 750 N is applied to the 60-mm-diameter end bearing of Fig. P9-80. For a coefficient of friction of 0.20, determine the torque needed to rotate the shaft for

**a.** A new bearing surface.
**b.** A worn-in bearing surface.

**Fig. P9-80**

**9-81** The electric auto polisher shown in Fig. P9-81 has an 8-in.-diameter polishing pad. The coefficient of friction between the pad and the auto surface is 0.30. Determine the forces that must be applied to the handles to hold the polisher steady when it is pressed against the surface with a force of 20 lb.

**Fig. P9-81**

**9-82** An electric disk sander (see Fig. P9-82) is pressed against a surface with a force of 50 N. The coefficient of friction between the sandpaper and the surface is 0.6 and the diameter of the sanding disk is 200 mm. Determine the torque that the motor must develop to overcome the developed friction.

**Fig. P9-82**

**9-83\*** Recalculate the forces for Problem 9-81 for the case in which the polishing pad is attached to a flexible backing surface and the pressure distribution on the disk is as shown in Fig. P9-83.

**Fig. P9-83**

**9-84** Recalculate the torque for Problem 9-82 for the case in which the polishing pad is attached to a flexible backing surface and the pressure distribution on the disk is as shown in Fig. P9-84.

**Fig. P9-84**

## Belt Friction

**9-85\*** A rope attached to a 500-lb block passes over a frictionless pulley and is wrapped for one full turn around a fixed post as shown in Fig. P9-85. If the coefficient of friction between the rope and the post is 0.25, determine

**Fig. P9-85**

a. The minimum force **P** that must be used to keep the block from falling.
b. The minimum force **P** that must be used to begin to raise the block.

**9-86\*** A rope attached to a 220-kg block passes over a fixed drum as shown in Fig. P9-86. If the coefficient of friction between the rope and the drum is 0.30, determine

a. The minimum force **P** that must be used to keep the block from falling.
b. The minimum force **P** that must be used to begin to raise the block.

**Fig. P9-86**

**9-87** A rope attached to a 35-lb crate passes around two fixed pegs as shown in Fig. P9-87. The 45-lb crate is attached to a wall by a second cord. The coefficient of friction between the two crates is 0.25; between the crate and the floor, 0.25; and between the rope and the pegs, 0.20. Determine the minimum force $P_{min}$ that must be used to cause motion.

**Fig. P9-87**

**9-88\*** A conveyor belt is driven with the 200-mm-diameter multiple-pulley drive shown in Fig. P9-88. Couples $C_A$ and $C_B$ are applied to the system at pulleys $A$ and $B$, respectively. The angle of contact between the belt and a pulley is 225° for each pulley, and the coefficient of friction is 0.30. Determine the maximum force $T$ that can be developed by the drive and the magnitudes of input couples $C_A$ and $C_B$.

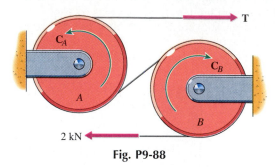

Fig. P9-88

**9-89** The electric motor shown in Fig. P9-89 weighs 30 lb and delivers 50 in. lb of torque to pulley $A$ of a furnace blower by means of a V-belt. The effective diameters of the 36° pulleys are 5 in. The coefficient of friction is 0.30. Determine the minimum distance $a$ to prevent slipping of the belt if the rotation of the motor is clockwise.

Fig. P9-89

**9-90** The band brake of Fig. P9-90 is used to control the rotation of a drum. The coefficient of friction between the belt and the drum is 0.35 and the weight of the handle is 15 N. If a force of 200 N is applied to the end of the handle, determine the maximum torque for which no motion occurs if the torque is applied

**a.** Clockwise.
**b.** Counterclockwise.

Fig. P9-90

**9-91\*** The band brake of Fig. P9-91 is used to control the rotation of a drum. The coefficient of friction between the belt and the drum is 0.25, and the weight of the handle is 5 lb. Determine the force $P$ that must be applied to the end of the handle to resist a maximum torque of 200 in. · lb if the torque is applied

**a.** Clockwise.
**b.** Counterclockwise.

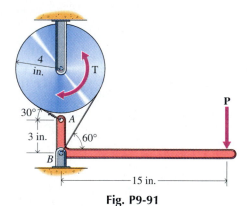

Fig. P9-91

**9-92\*** The band brake of Fig. P9-92 is used to control the rotation of a drum. The coefficient of friction between the belt and the drum is 0.30 and the weight of the handle is 15 N. Determine the force $P$ that must be applied to the end of the handle to resist a maximum torque of 75 N · m if the torque is applied

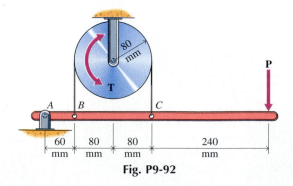

Fig. P9-92

a. Clockwise.
b. Counterclockwise.

**9-93** A uniform belt 36 in. long and weighing 1 lb hangs over a small fixed peg as shown in Fig. P9-93. If the coefficient of friction between the belt and the peg is 0.40, determine the maximum distance $d$ between the two ends for which the belt will not slip off the peg.

**Fig. P9-93**

**9-94** The band wrench of Fig. P9-94 is used to unscrew an oil filter from a car. (The filter acts as if it were a wheel with a resisting torque of 40 lb · ft.) Determine the minimum coefficient of friction between the band and the filter that will prevent slippage. The weight of the handle can be neglected.

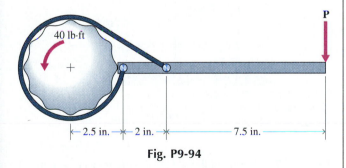

**Fig. P9-94**

**9-95*** The band brake of Fig. P9-95 is used to control the rotation of a drum. The weight of the handle is 2 lb. Determine the minimum coefficient of friction between the belt and drum for which the band brake is self-locking, that is, for which no force on the handle is necessary to restrain arbitrarily large torques on the drum.

**Fig. P9-95**

**9-96*** A 100-N crate is sitting on a 200-N crate as shown in Fig. P9-96. The rope that joins the crates passes around a frictionless pulley and a fixed drum. The coefficient of friction is 0.30 at all surfaces. Determine the minimum force **P** that must be applied to the 200-N crate to start motion.

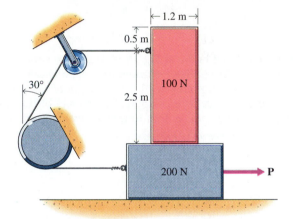

**Fig. P9-96**

**9-97** A rope connecting two blocks is wrapped $\frac{3}{4}$-turn around a fixed peg as shown in Fig. P9-97. The coefficient of friction between the peg and the rope is 0.15; between the 65-lb block and the floor, 0.40. Determine the minimum and maximum weight of block $B$ for which no motion occurs.

**Fig. P9-97**

523

**9-98** A rope is attached to the top corner of a 450-N block, passes over a fixed drum, and is attached to another block as shown in Fig. P9-98. The coefficient of friction between the 450-N block and the floor is 0.30; between the rope and the drum, 0.60. Determine the maximum mass of block $B$ for which no motion occurs.

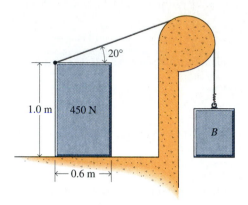

**Fig. P9-98**

**9-99\*** The band brake of Fig. P9-95 is used to control the rotation of a drum. The coefficient of friction between the belt and the 40° V-pulley is 0.10. The weight of the handle is 2 lb. If a force of 50 lb is applied to the end of the handle, determine the maximum torque for which no motion occurs if the torque is applied

**a.** Clockwise.
**b.** Counterclockwise.

**Fig. P9-95**

**9-100** The band brake of Fig. P9-90 is used to control the rotation of a drum. The coefficient of friction between the belt and the 36° V-pulley is 0.15, and the weight of the handle is 15 N. If a force of 200 N is applied to the end of the handle, determine the maximum torque for which no motion occurs if the torque is applied

**a.** Clockwise.
**b.** Counterclockwise.

**Fig. P9-90**

## Computer Problems

**C9-101** The hand brake of Fig. P9-101 is used to control the rotation of a drum. The coefficient of friction between the belt and the drum is $\mu_s = 0.15$, and the weight of the handle may be neglected. If a counterclockwise torque of 5 ft · lb is applied to the drum

**a.** Plot **P**, the minimum force that must be applied to the handle to prevent motion as a function of $a$, the location of the pivot point (0.5 in. $\leq a \leq$ 5.8 in.).
**b.** Repeat the problem for a clockwise torque of 5 ft · lb. What happens to the system at $a \cong 4.6$ in.? What does the solution mean for $a > 4.6$ in.?

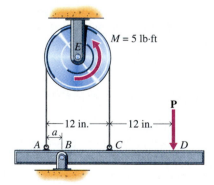

**Fig. P9-101**

**C9-102** The hand brake of Fig. P9-102 is used to control the rotation of a drum. The coefficient of friction between the belt and the drum is $\mu_s = 0.35$, and the weight of the handle may be neglected. If a clockwise torque of 475 N · m is applied to the drum

**a.** Plot **P**, the minimum force that must be applied to the handle to prevent motion, as a function of $a$, the location of the pivot point (0.5 mm $\leq a \leq$ 500 mm).
**b.** Repeat the problem for a counterclockwise torque of 475 N · m.

150 mm

$M = 475$ N·m

800 mm

$a$

$A$  $B$  $C$

$P$

**Fig. P9-102**

a. Plot **P**, the minimum force that must be applied to the wrench to loosen the filter, as a function of $a$ (1 in. ≤ $a$ ≤ 5 in.).
b. Plot $T_B$, the tension of the handle at $B$, as a function of $a$ (1 in. ≤ $a$ ≤ 5 in.).
c. Plot $A$, the normal force exerted on the filter by the handle of the wrench, as a function of $a$ (1 in. ≤ $a$ ≤ 5 in.).

**C9-103** The band wrench of Fig. P9-103 is used to unscrew an oil filter from a car. (The filter acts as if it were a wheel with a resisting torque of 40 lb · ft.) The coefficient of friction between the band and the filter is $\mu_s = 0.15$, and the friction between the end of the wrench and the filter may be neglected.

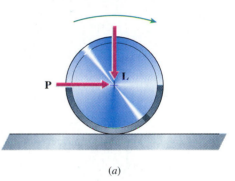

5.5 in.

10 in.

$P$

40 lb·ft

$D$  $A$  $B$  $C$

$a$

**Fig. P9-103**

## 9-4 ROLLING RESISTANCE

Rolling resistance refers to the forces that cause a rolling wheel to gradually slow down and stop. Rolling resistance, however, is not due to Coulomb friction and cannot be described with a coefficient of friction in the same sense as wedges and bearings. Therefore, it is worth examining the origin and nature of rolling resistance in a little more detail.

Several effects including air resistance and bearing friction combine to cause a rolling wheel to gradually slow down and stop. In what follows, it will be assumed that these sources of resistance have been minimized and that the primary source of resistance is the interaction between the wheel and the surface on which it is rolling.

Figure 9-43a shows a wheel that is being rolled at a constant speed across a flat horizontal surface by a force $P$ at its center. The free-body diagram of the wheel is shown in Fig. 9-43b. The force $L$ represents the weight of the wheel plus any vertical load that might be applied to its axle. If the wheel and the surface are both perfectly rigid, the normal force will act through the center of the wheel. Summing moments about the center of the wheel then gives $F = 0$. That is, no Coulomb friction acts on the bottom of the wheel. Furthermore, summing forces in the horizontal direction gives $P = 0$. That is, no pushing force is required to keep the wheel rolling at constant speed.

The apparent contradiction with our common experience is caused by assuming the wheel and surface are perfectly rigid. All real materials deform to some extent. The softer the materials and/or the heavier the load, the more the deformation will be.

The real situation, then, is probably more like Fig. 9-44a, which shows a rubber wheel rolling across a rubber surface. The load $L$ will cause both the surface and the wheel to be dented. In addition, since the wheel is being pushed to the right, the surface will probably stretch and be pushed up a little in front of the wheel. The free-body diagram

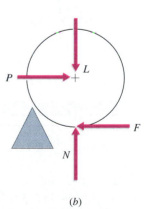

(a)

(b)

**Figure 9-43** Free-body diagram for a rigid wheel rolling at a constant speed on a flat, rigid, horizontal surface.

525

**Figure 9-44** Free-body diagram for a deformable wheel rolling at a constant speed on a flat, deformable, horizontal surface.

for this situation is shown in Fig. 9-44$b$, in which $R$ represents the resultant of all contact forces between the wheel and the surface.

Equilibrium of moments about the center of the wheel requires that $R$ act through the center of the wheel as shown. Then, summing moments about $A$, the point through which $R$ acts, gives

$$\circlearrowleft +\Sigma M_A = La - Pb = 0$$

In most cases of interest the deformation is small so that $b$ is very nearly equal to the radius of the wheel $r$. Therefore, the force $P$ required to keep the wheel rolling at a constant speed is given by

$$P = aL/r \qquad \text{(9-18)}$$

The distance $a$ is usually termed the coefficient of rolling resistance.[2] It is not a dimensionless coefficient as are $\mu_s$ and $\mu_k$, but has a dimension of length. Like $\mu_s$ and $\mu_k$, the coefficient of rolling resistance depends on the properties of the contacting surfaces. Since it arises due to deformation of the surface rather than from the roughness of the surface, however, it is not related to $\mu_s$ and $\mu_k$. In fact, there is even considerable disagreement over whether or not $a$ is even a constant.

Bearing these things in mind, Table 9-2 presents a list of rolling coefficients. The reported values vary by 50 percent or more, however, and no great accuracy should be expected from this procedure.

**TABLE 9-2   COEFFICIENTS OF ROLLING RESISTANCE**

| Materials | $a$ (in.) | $a$ (mm) |
| --- | --- | --- |
| Hardened steel on hardened steel | 0.0004 | 0.01 |
| Mild steel on mild steel | 0.015 | 0.38 |
| Steel on wood | 0.08 | 2 |
| Steel on soft ground | 5 | 130 |
| Pneumatic tires on pavement | 0.025 | 0.6 |
| Pneumatic tires on dirt road | 0.05 | 1.3 |

[2]Some people prefer to keep friction coefficients dimensionless and refer to the ratio $a/r$ as the coefficient of rolling resistance.

# CONCEPTUAL EXAMPLE 9-4: ROLLING RESISTANCE

*A typical value for the coefficient of rolling resistance for a rubber tire on a paved surface is 0.025 in. Determine the angle θ for which motion would impend for the 28-in.-diameter wheel and tire shown in Fig. CE9-4.*

**Fig. CE9-4**

---

## SOLUTION

Rolling resistance is a measure of the interaction between a wheel and a surface (on which the wheel is rolling) that tends to cause the wheel to gradually slow down and stop. The concept can be illustrated by considering a wheel in equilibrium on an inclined surface. A free-body diagram of the wheel shows the weight **W** of the wheel being resisted by a resultant contact force **R** that is equal, opposite, and collinear with the weight **W**, as shown in Fig. CE9-4*a*. As shown in Fig. CE9-4*b*, the line of action of the resultant **R** is displaced from a normal to the surface as a result of deformations of the wheel and the contact surface. When motion is impending, the displacement reaches a distance *a* (defined as the coefficient of rolling resistance).

Under this condition, summing moments about point *A* yields

$$+\curvearrowleft \Sigma M_A = W \sin \theta \, (r) - W \cos \theta \, (a) = 0$$

from which

$$a = \frac{W \sin \theta}{W \cos \theta} \, (r) = r \tan \theta$$

Thus, the angle θ for which motion would impend for the 28-in.-diameter wheel and tire is

$$\theta = \tan^{-1} \frac{a}{r} = \tan^{-1} \frac{0.025}{28/2} = 0.1023°$$

The two-wheel cart of Fig. 9-45a has 500-mm-diameter bicycle tires and a mass of 15 kg. The cart is carrying 95 kg of rocks on a dirt road. The wheels fit loosely over a 15-mm-diameter fixed axle. The coefficient of kinetic friction between the wheel and the axle is 0.25, and the center of mass of the cart and the rocks is 100 mm in front of the axle. Determine the force (both the horizontal and vertical components) that must be applied to the handle of the cart to pull it at a constant speed.

(a)

## SOLUTION

Both the axle friction and the rolling friction depend on the normal load on the axle. Therefore, the first step is to draw a free-body diagram of the cart minus the wheels (Fig. 9-45b) to bring the axle load into the equations. The equilibrium equations give

$$+\rightarrow \Sigma F_x = T_x - P = 0$$
$$+\uparrow \Sigma F_y = L - 1078 + T_y = 0$$
$$\downarrow + \Sigma M_O = 1T_y - (0.1)(1078) - M = 0$$

where $L$ is the total vertical load on the axle, $P$ is the total horizontal force on the axle, and $M = 0.25L(0.015/2)$ is the total moment on the axle due to both wheels. Solution of these equations gives

$$T_x = P \qquad T_y = 109.6 \text{ N} \qquad L = 968 \text{ N}$$

(b)

Next, draw a free-body diagram of one of the wheels (Fig. 9-45c). The forces $L_f$ and $P_f$ are the components of the vertical and horizontal forces that are carried by the wheel on the front side of the cart and $M_f = 0.25L_f(0.015/2)$ is the torque exerted on the front wheel by bearing friction. The ground reaction acts at point $A$, which is ahead of the axle by the distance $a = 1.3$ mm according to Table 9-2. Summing moments about point $A$ gives

$$\downarrow + \Sigma M_A = M_f + aL_f - rP_f = 0$$

or

(c)

**Fig. 9-45**

$$P_f = \frac{aL_f}{r} + \frac{M_f}{r} = \frac{1.3L_f}{250} + \frac{0.25L_f(7.5)}{250}$$
$$= (0.00520 + 0.00750)L_f = 0.01270L_f$$

The first term in the foregoing equation represents the force necessary to overcome rolling resistance, and the second term represents the force necessary to overcome axle friction. Analysis of the back wheel similarly gives

$$P_b = 0.01270L_b$$

The total horizontal force is the sum of the horizontal forces on both wheels

$$P = P_f + P_b = 0.01270(L_f + L_b)$$

and the total vertical force is the sum of the vertical forces on both wheels

$$L = L_f + L_b = 968 \text{ N}$$

so finally,

$$T_x = P = 0.01270(968) = 12.30 \text{ N} \qquad \text{Ans.}$$

The reaction $R$ on the wheel is the resultant of all distributed forces exerted on the wheel by the dirt road.

# PROBLEMS

**9-104\*** A horizontal force of 170 N is required to push a 180-kg refrigerator across a level linoleum floor. The refrigerator rolls on four 50-mm-diameter plastic wheels. Neglect all friction except rolling resistance and determine the coefficient of rolling friction between the plastic wheels and the linoleum floor.

**9-105\*** Determine the horizontal force required to push a 2500-lb automobile on level pavement. The auto has four 23-in.-diameter tires. Neglect all friction except rolling resistance.

**9-106** A 1200-kg automobile is observed to roll at a constant speed down a 1° incline. The auto has four 550-mm-diameter tires. Neglect all friction except rolling resistance and determine the coefficient of rolling friction between the tires and the surface.

**9-107** Determine the angle of incline down which a 50-ton railroad car (see Fig. P9-107) would roll at constant speed. The car has eight 20-in.-diameter steel wheels, which roll on steel rails. Neglect all friction except rolling resistance.

**Fig. P9-107**

**9-108\*** An overhead crane has two 75-mm-diameter steel wheels that ride on steel rails as shown in Fig. P9-108.

**Fig. P9-108**

Determine the horizontal force necessary to push the crane when it is carrying a load of mass 2000 kg. Neglect all friction except rolling resistance.

**9-109\*** A horizontal force of 200 lb is required to push an 800-lb piano across a level carpeted floor. The piano rolls on four 2-in.-diameter steel wheels. Neglect all friction except rolling resistance and determine the coefficient of rolling friction between the steel wheels and the carpeted floor.

**9-110** A 35-kg child rides on a 10-kg bicycle with 500-mm-diameter wheels. Determine the horizontal force necessary to push the child and bicycle

**a.** On a level sidewalk.
**b.** Through level sand.

Neglect all friction except rolling resistance.

**9-111\*** The wheels of Problem 9-105 rotate on 1.5-in.-diameter shafts. If the coefficient of friction between the wheel and the shaft is 0.15, determine the force necessary to push the automobile and the ratio of the rolling friction to the axle friction.

**9-112\*** The wheels of Problem 9-109 rotate on 5-mm-diameter axles. If the coefficient of friction between the wheel and the axle is 0.40, determine the force necessary to push the crane and the ratio of the rolling friction to the axle friction.

**9-113** A 600-lb safe rolls on four 2-in.-diameter steel wheels. The wheels rotate on $\frac{1}{4}$-in.-diameter axles. If the coefficient of friction between the wheel and the axle is 0.45, determine the force necessary to push the safe and the ratio of the rolling friction to the axle friction.

**9-114** The wheels of Problem 9-110 rotate on 5-mm-diameter axles. If the coefficient of friction between the wheel and the axle is 0.20, determine the force necessary to push the bicycle and the ratio of the rolling friction to the axle friction.

# SUMMARY

Tangential forces due to friction are always present at the interface between two contacting bodies, and these forces always act in a direction to oppose the tendency of the contacting surfaces to slip relative to one another. In some types of machine elements, such as bearings and skids, it is desired to minimize friction, while in other types, such as brakes and belt drives, it is desired to maximize friction.

Two main types of friction are commonly encountered in engineering practice: dry friction and fluid friction. Dry friction describes the force that opposes the slip of one solid surface on another. Dry friction depends on the normal force pressing the surfaces together and on the nature of the materials in contact. Fluid friction describes the force that opposes the slip of two layers of fluid moving at different speeds. Fluid friction depends on the relative speed of the moving fluid layers.

The strength of friction forces is limited, as shown in Fig. 9-2. The condition when a friction force is at its maximum value is called impending motion. Beyond that point, friction can no longer supply the amount of force required for equilibrium. The value of limiting friction is proportional to the normal force $N$ at the contact surface

$$F_{\max} = \mu_s N \qquad (9\text{-}1)$$

The constant of proportionality $\mu_s$ is called the coefficient of static friction; it depends on the types of material in contact but is independent of both the normal force and the area of contact.

The frictional force actually exerted is never greater than that required to satisfy the equations of equilibrium

$$F \le \mu_s N \qquad (9\text{-}2)$$

The equality in Eq. 9-2 holds only at the point of impending motion. Once a body starts to slip relative to its supporting surface, the friction force will decrease slightly and

$$F = \mu_k N \qquad (9\text{-}3)$$

where $\mu_k$ is the coefficient of kinetic friction. This coefficient is usually about 20 to 25 percent smaller than $\mu_s$ and is again independent of the normal force. It is also independent of the speed of the relative motion—at least for low speeds. The presence of any moisture or oil on the surface, however, can change the problem from one of dry friction where the friction force is independent of the speed of the body to one of fluid friction where the friction force is a function of the speed.

Some components in engineering systems where frictional forces play an important role include wedges, screws, journal bearings, thrust bearings, and belts. The procedure followed in analyzing all these components is the same as that used previously in Chapter 7 for pin-connected machines and frames.

Rolling resistance is not due to Coulomb friction and cannot be described with the usual coefficients of friction. Rolling resistance arises because of the deformation of the wheel and/or the surface on which it is rolling.

The magnitude of the force $P$ required to keep a wheel rolling at

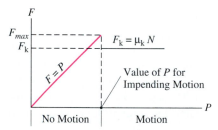

**Figure 9-2**   Variation of the friction force between two bodies.

a constant speed is given by

$$P = a\,L/r \qquad\qquad \textbf{(9-18)}$$

where $r$ is the radius of the wheel and $L$ is the load supported by the wheel. The coefficient of rolling resistance $a$ is not a dimensionless coefficient as are $\mu_s$ and $\mu_k$, but has a dimension of length. Although $a$ depends on the properties of the contacting surfaces, it arises because of the deformation of the surface rather than the roughness of the surface. Therefore, the coefficient of rolling resistance is not directly related to $\mu_s$ and $\mu_k$.

# REVIEW PROBLEMS

**9-115\*** A device for lifting rectangular objects such as bricks and concrete blocks is shown in Fig. P9-115. Determine the minimum coefficient of static friction between the contacting surfaces required to make the device work.

**Fig. P9-115**

**9-116\*** A post is held up by a rope that passes around a frictionless pulley and is attached to a mass $M$ as shown in Fig. P9-116. Find the mass $M$ in terms of $\mu_s$, $\theta$, and $m_p$ (the mass of the post) for the case of impending slip at the floor.

**Fig. P9-116**

**9-117** A uniform plank 8 ft long is balanced on a corner using a horizontal force of 75 lb as shown in Fig. P9-117. If the plank weighs 45 lb, determine

**a.** The angle $\theta$ for equilibrium.
**b.** The minimum coefficient of friction for which the plank is in equilibrium.

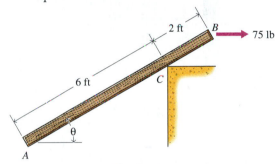

**Fig. P9-117**

**9-118** A bumper jack, used to raise the wheel of an automobile in order to change a tire, is shown in Fig. P9-118. The mean diameter of the screw is 30 mm, and the pitch of the screw is 6 mm. The coefficient of friction between the

**Fig. P9-118**

screw and the nut is 0.30. Determine the force which must be applied to the crank in order to

a. Raise the automobile.
b. Lower the automobile.

**9-119\*** The device shown in Fig. P9-119 is used to raise boxes and crates between floors of a factory. The frame, which slides on the 4-in.-diameter vertical post, weighs 50 lb. If the coefficient of friction between the post and the frame is 0.10, determine the force **P** required to raise a 150-lb box.

**Fig. P9-119**

**9-120\*** The masses of blocks $A$ and $B$ of Fig. P9-120 are $m_A = 50$ kg and $m_B = 25$ kg. If the coefficient of static friction is 0.15 for both surfaces, determine the force **P** required to cause impending motion of block $B$.

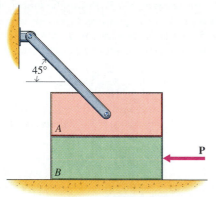

**Fig. P9-120**

**9-121** The 200-lb crate of Fig. P9-121 is being moved by a rope that passes over a smooth pulley. The coefficient of friction between the crate and the floor is 0.30.

a. Assume that $h = 4$ ft and determine the force **P** necessary to produce impending motion.

b. Determine the value of $h$ for which impending motion by slipping and by tipping would occur simultaneously.

**Fig. P9-121**

**9-122** The static coefficient of friction between the brake pad and the brake drum of Fig. P9-122 is 0.40. When a force of 350 N is being applied to the brake arm, determine the couple required to initiate rotation of the drum if the direction of rotation is (a) clockwise and (b) counterclockwise.

**Fig. P9-122**

**9-123\*** In Fig. P9-123, box $A$ weighs 10 lb while box $B$ weighs 20 lb. The coefficient of friction between box $A$ and the inclined surface is 0.45; between box $B$ and the horizontal surface, 0.5. The pulleys are all frictionless. Deter-

**Fig. P9-123**

mine the maximum weight of box *C* such that no motion will occur. Which motion would occur first?

**9-124\*** The brake shown in Fig. P9-124 is used to control the motion of block *B*. If the mass of block *B* is 25 kg and the kinetic coefficient of friction between the brake drum and brake pad is 0.30, determine the force **P** required for a constant-velocity descent.

**Fig. P9-124**

**9-125** Blocks *A* and *B* of Fig. P9-125 weigh 50 lb and 500 lb, respectively. The static coefficients of friction are 0.25 for all contacting surfaces. Determine the force **P** required to initiate movement of the blocks.

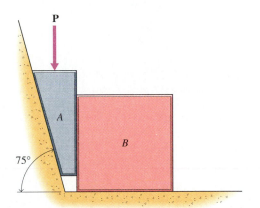

**Fig. P9-125**

**9-126** A rope connecting two blocks passes around a frictionless pulley and a fixed drum (Fig. P9-126). The coefficient of friction between the rope and the drum and between the 10-kg block and the wall is 0.30; the pin supporting the 30-kg block is frictionless. Determine the maximum force **P** for which no motion occurs.

**Fig. P9-126**

**9-127\*** A rotating shaft with a belt-type brake is shown in Fig. P9-127. The static coefficient of friction between the brake drum and the brake belt is 0.20. When a 75-lb force **P** is applied to the brake arm, rotation of the shaft is prevented. Determine the maximum torque that can be resisted by the brake

**a.** If the shaft is tending to rotate counterclockwise.
**b.** If the shaft is tending to rotate clockwise.

**Fig. P9-127**

**9-128\*** Blocks *A* and *B* of Fig. P9-128 have 20- and 50-kg masses, respectively. The static coefficients of friction be-

**Fig. P9-128**

**Fig. P9-129**

**Fig. P9-128**

9-130 The rope connecting the two blocks of Fig. P9-130 passes over a fixed drum. The coefficient of friction between the left block and the floor is 0.50; between the right block and the floor, 0.40; and between the rope and the drum, 0.30. Determine the minimum and maximum mass of the left block for which motion does not occur.

tween blocks $A$ and $B$ and between block $B$ and the horizontal surface are 0.25. The static coefficient of friction between the rope and the fixed pegs is 0.20. Determine the force **P** required to initiate motion of block $B$.

**9-129** Determine an expression for the torque $M$ required to rotate a shaft carrying an axial load $P$ and supported by a conical end bearing as shown in Fig. P9-129. The coefficient of friction is $\mu$.

**Fig. P9-130**

*T*he strengths of the beams and columns that make up the skeleton of a modern building depend on the second moments of the area of their cross sections.

Chapter **10**

# SECOND MOMENTS OF AREA AND MOMENTS OF INERTIA

## 10-1 INTRODUCTION

In Chapter 5 the centroid for an area was located by using an integral of the form $\int_A x\,dA$. In the analysis of stresses and deflections in beams and shafts, an expression of the form $\int_A x^2\,dA$ is frequently encountered, in which $dA$ represents an element of area and $x$ represents the distance from the element to some axis in, or perpendicular to, the plane of the area. An expression of the form $\int_A x^2\,dA$ is known as the second moment of the area. In the analysis of the angular motion of rigid bodies, an expression of the form $\int_m r^2\,dm$ is encountered, in which $dm$ represents an element of mass and $r$ represents the distance from the element to some axis. Euler[1] gave the name "moment of inertia" to expressions of the form $\int_m r^2\,dm$. Because of the similarity between the two types of integrals, both have become widely known as moments of inertia. In this text, the integrals involving areas will generally be referred to as "second moments of area" and less frequently as "area moments of inertia." Integrals involving masses will be referred to as "mass moments of inertia" or simply "moments of inertia."

Methods used to determine both second moments of area and mass moments of inertia will be discussed in later sections of this chapter.

## 10-2 SECOND MOMENT OF PLANE AREAS

The second moment of an area with respect to an axis will be denoted by the symbol $I$ for an axis in the plane of the area and by the symbol $J$ for an axis perpendicular to the plane of the area. The particular axis about which the second moment is taken will be denoted by subscripts.

[1]Leonhard Euler (1707–1783), a noted Swiss mathematician and physicist.

**Figure 10-1**  Location of an element of area $dA$ with respect to a Cartesian coordinate system.

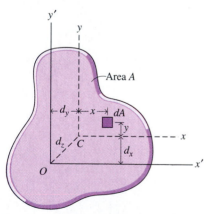

**Figure 10-2**  Area $A$ in the $xy$-plane.

Thus, the second moments of the area $A$ shown in Fig. 10-1 with respect to $x$- and $y$-axes in the plane of the area are

$$I_x = \int_A y^2 \, dA \quad \text{and} \quad I_y = \int_A x^2 \, dA \qquad (10\text{-}1)$$

The quantities $I_x$ and $I_y$ are sometimes referred to as rectangular second moments of area $A$.

Similarly, the second moment of the area $A$ shown in Fig. 10-1 with respect to a $z$-axis, which is perpendicular to the plane of the area at the origin $O$ of the $xy$-coordinate system, is

$$J_z = \int_A r^2 \, dA = \int_A (x^2 + y^2) \, dA \qquad (10\text{-}2)$$

Thus,

$$J_z = \int_A x^2 \, dA + \int_A y^2 \, dA = I_y + I_x \qquad (10\text{-}3)$$

The quantity $J_z$ is known as the polar second moment of the area $A$.

The second moment of an area can be visualized as the sum of a number of terms each consisting of an area multiplied by a distance squared. Thus, the units of a second moment are a length raised to the fourth power ($L^4$). Common units are $mm^4$ and $in.^4$. Also, the sign of each term summed to obtain the second moment is positive, since either a positive or negative distance squared is positive. Therefore, the second moment of an area is always positive.

## 10-2.1  Parallel-Axis Theorem for Second Moments of Area

When the second moment of an area has been determined with respect to a given axis, the second moment with respect to a parallel axis can be obtained by means of the parallel axis theorem. If one of the axes (say the $x$-axis) passes through the centroid of the area, as shown in Fig. 10-2, the second moment of the area about a parallel $x'$-axis located a distance $d_x$ from the $x$-axis is

$$I_{x'} = \int_A (y + d_x)^3 \, dA = \int_A y^2 \, dA + 2d_x \int_A y \, dA + d_x^2 \int_A dA$$

Since $d_x$ is the same for every element of area $dA$, the first integral is the second moment of the area with respect to the $x$-axis $I_x$ and the last integral is the total area $A$. Therefore,

$$I_{x'} = I_x + 2d_x \int_A y \, dA + d_x^2 A \qquad (a)$$

The integral $\int_A y \, dA$ is the first moment of the area with respect to the $x$-axis. Since the $x$-axis passes through the centroid $C$ of the area, the first moment is zero and Eq. $a$ becomes

$$I_{x'} = I_{xC} + d_x^2 A \qquad (10\text{-}4)$$

where $I_{xC}$ is the second moment of the area with respect to the $x$-axis through the centroid and $d_x$ is the distance between the $x$ and $x'$-axes.

In a similar manner it can be shown that

$$J_{z'} = J_{zC} + (d_x^2 + d_y^2)A = J_{zC} + d_z^2 A \qquad (10\text{-}5)$$

where $J_{zC}$ is the polar second moment of the area with respect to the z-axis through the centroid and $d_z$ is the distance between the z- and z'-axes.

The parallel-axis theorem states that the second moment of an area with respect to any axis in the plane of the area is equal to the second moment of the area with respect to a parallel axis through the centroid of the area added to the product of the area and the square of the distance between the two axes. The theorem also indicates that the second moment of an area with respect to an axis through the centroid of the area is less than that for any parallel axis since

$$I_{xC} = I_{x'} - d_y^2 A \qquad (10\text{-}6)$$

As a point of caution, note that the parallel-axis theorem (Eq. 10-4) is valid only for transfers to or from a centroidal axis. That is, if $x'$ and $x''$ are two parallel axes (neither of which passes through the centroid of an area) then

$$\begin{aligned}
I_{x''} = I_{xC} + d_{x2}^2 A &= (I_{x'} - d_{x1}^2 A) + d_{x2}^2 A \\
&= I_{x'} + (d_{x2}^2 - d_{x1}^2)A \\
&\neq I_{x'} + (d_{x2} - d_{x1})^2 A
\end{aligned}$$

## 10-2.2    Second Moments of Areas by Integration

Rectangular and polar second moments of area were defined in Section 10-2. When the second moment of a plane area with respect to a line is determined by using Eqs. 10-1 and 10-2, it is possible to select the element of area $dA$ in various ways and to express the area of the element in terms of either Cartesian or polar coordinates. In some cases, an element of area with dimensions $dA = dy\,dx$, as shown in Fig. 10-1, may be required. This type of element has the slight advantage that it can be used for calculating both $I_x$ and $I_y$ but it has the greater disadvantage of requiring double integration. Most problems can be solved with less work by choosing elements of the type shown in Figs. 10-3a and 10-3b. The following should be considered when selecting an element of area $dA$ for a specific problem.

1. If all parts of the element of area are the same distance from the axis, the second moment can be determined directly by using Eqs. 10-1 or 10-2. Thus, the element shown in Fig. 10-1 can be used to determine either $I_x$ or $I_y$ directly, but a double integration is required. The element shown in Fig. 10-3a can be used to determine $I_y$ directly since the dimension $x$ is constant for the element. The element shown in Fig. 10-3a is not suitable for determining $I_x$ directly since the y-dimension is not constant for the element. Similarly, the element shown in Fig. 10-3b is suitable for determining $I_x$ directly but not $I_y$. A single integration would be required with elements of the type shown in Fig. 10-3.

2. If the second moment of the element of area with respect to the axis about which the second moment of the area is to be found is known, the second moment of the area can be found by summing the second moments of the individual elements that make up the

(a)

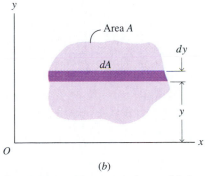

(b)

**Figure 10-3**    Elements of area $dA$ for integration about the x- and y-axes.

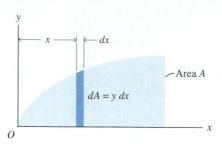

**Figure 10-4**   Integration using an element of area $dA$ with a known second moment.

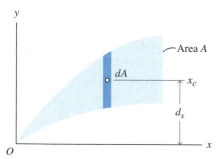

**Figure 10-5**   Element of area $dA$ in the $xy$-plane.

area. For example, if the second moment $dI_x$ for the rectangular area $dA$ in Fig. 10-4 is known, the second moment $I_x$ for the complete area $A$ is simply $I_x = \int_A dI_x$.

3. If both the location of the centroid of the element and the second moment of the element about its centroidal axis parallel to the axis of interest for the complete area are known, the parallel-axis theorem can often be used to simplify the solution of a problem. For example, consider the area shown in Fig. 10-5. If both the distance $d_x$ and the second moment $dI_{xC}$ for the rectangular element $dA$ are known, then by the parallel-axis theorem, the second moment of the area of the rectangular element about the $x$-axis is $dI_x = dI_{xC} + d_x^2(dA)$. The second moment for the complete area $A$ is then simply $I_x = \int_A dI_x$.

From the previous discussion it is evident that either single or double integration may be required for the determination of second moments of area, depending on the element of area $dA$ selected. When double integration is used, all parts of the element will be the same distance from the moment axis, and the second moment of the element can be written directly. Special care must be taken in establishing the limits for the two integrations to see that the correct area is included. If a strip element is selected, the second moment can usually be obtained by a single integration, but the element must be properly selected in order for its second moment about the reference axis to be either known or readily calculated by using the parallel-axis theorem. The following example problems illustrate the procedure for determining the second moments of area by integration.

Determine the second moment for the rectangle shown in Fig. 10-6a with respect to

a. The base of the rectangle.
b. An axis through the centroid parallel to the base.
c. An axis through the centroid normal to the area.

(a)

## SOLUTION

a. An element of area $dA = b \, dy$, as shown in Fig. 10-6a, will be used. Since all parts of the element are located a distance $y$ from the $x$-axis, Eq. 10-1 can be used directly for the determination of the second moment $I_x$ about the base of the rectangle. Thus,

$$I_x = \int_A y^2 \, dA = \int_0^h y^2 \, b \, dy = \left[ \frac{by^3}{3} \right]_0^h = \frac{bh^3}{3} \qquad \text{Ans.}$$

This result will be used frequently in later examples, when elements of the type shown in Fig. 10-4 are used to determine second moments about the $x$-axis.

> The second moment $I_x$ can be computed with a single integration if a thin strip element parallel to the $x$-axis is used.

b. The parallel-axis theorem (Eq. 10-6) will be used to determine the second moment $I_{xC}$ about an axis that passes through the centroid of the rectangle (see Fig. 10-6b) and is parallel to the base. Thus,

$$I_{xC} = I_x - d_x^2 \, A = \frac{bh^3}{3} - \left(\frac{h}{2}\right)^2 (bh) = \frac{bh^3}{12} \qquad \text{Ans.}$$

This result will be used frequently in later examples when elements of the type shown in Fig. 10-5 are used to determine second moments about the $x$-axis.

c. The second moment $I_{yC}$ for the rectangle can be determined in an identical manner. It can also be obtained from the preceding solution by interchanging $b$ and $h$; that is,

$$I_{yC} = \frac{hb^3}{12}$$

The polar second moment $J_{zC}$ about the $z$-axis through the centroid of the rectangle is given by Eq. 10-3 as

$$J_{zC} = I_{xC} + I_{yC} = \frac{bh^3}{12} + \frac{hb^3}{12} = \frac{bh}{12}(h^2 + b^2) \qquad \text{Ans.}$$

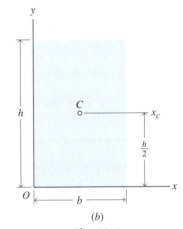

(b)

**Fig. 10-6**

> If an area has an axis of symmetry, the centroid is located on that axis; if the area has two axes of symmetry, the centroid is located at the point of intersection of the two axes.

Determine the second moment of area for the circle shown in Fig. 10-7 with respect to a diameter of the circle.

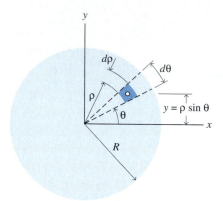

**Fig. 10-7**

---

### SOLUTION

Polar coordinates are convenient for this problem. An element of area $dA = \rho \, d\theta \, d\rho$, as shown in Fig. 10-7, will be used. If the $x$-axis is selected as the diameter about which the second moment of area is to be determined, then $y = \rho \sin \theta$. Application of Eq. 10-1 yields

Polar coordinates are usually more efficient when circular boundaries are involved.

$$I_x = \int_A y^2 \, dA = \int_0^{2\pi} \int_0^R (\rho \sin \theta)^2 (\rho \, d\theta \, d\rho)$$

$$= \int_0^{2\pi} \int_0^R \rho^3 \sin^2 \theta \, d\rho \, d\theta = \frac{R^4}{4} \left[ \frac{\theta}{2} - \frac{\sin 2\theta}{4} \right]_0^{2\pi} = \frac{\pi R^4}{4} \qquad \text{Ans.}$$

Determine the polar second moment of area for the circle shown in Fig. 10-8 with respect to

a. An axis through the center of the circle and normal to the plane of the area.
b. An axis through the edge of the circle and normal to the plane of the area.

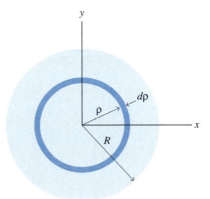

**Fig. 10-8**

## SOLUTION

a. Polar coordinates are convenient for this problem. An element of area $dA = 2\pi\rho \, d\rho$, as shown in Fig. 10-8, will be used. Since all parts of the element are located a constant distance $\rho$ from the center of the circle, Eq. 10-2 can be used directly for the determination of the polar second moment $J_z$ about an axis through the center of the circle and normal to the plane of the area. Thus,

$$J_z = \int_A r^2 \, dA = \int_0^R \rho^2 \, (2\pi\rho \, d\rho) = \int_0^R 2\pi\rho^3 \, d\rho = \frac{\pi R^4}{2} \qquad \text{Ans.}$$

> The second moment $J_z$ can be computed with a single integration if a thin annular element at a constant distance $\rho$ from the z-axis is used.

This result could have been obtained from the solution of Example Problem 10-2 and use of Eq. 10-3. Thus,

$$J_z = I_x + I_y = \frac{\pi R^4}{4} + \frac{\pi R^4}{4} = \frac{\pi R^4}{2}$$

b. The parallel-axis theorem (Eq. 10-5) will be used to determine the polar second moment $J_{z'}$ about an axis that passes through the edge of the circle and is normal to the plane of the area. Thus,

$$J_{z'} = J_{zC} + d_z^2 \, A = \frac{\pi R^4}{2} + R^2(\pi R^2) = \frac{3\pi R^4}{2} \qquad \text{Ans.}$$

Determine the second moment of area for the shaded region of Fig. 10-9 with respect to

a.   The $x$-axis.
b.   An axis through the origin of the $xy$-coordinate system and normal to the plane of the area.

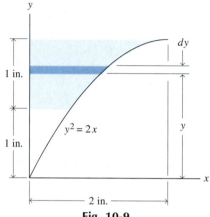

**Fig. 10-9**

---

## SOLUTION

a.   An element of area $dA = x\, dy = (y^2/2)\, dy$, as shown in Fig. 10-3b, will be used. Since all parts of the element are located a distance $y$ from the $x$-axis, Eq. 10-1 can be used directly for the determination of the second moment $I_x$ about the $x$-axis. Thus,

$$I_x = \int_A y^2\, dA = \int_A y^2 \left(\frac{y^2}{2}\, dy\right) = \int_1^2 \frac{y^4}{2}\, dy = \left[\frac{y^5}{10}\right]_1^2 = 3.10 \text{ in.}^4 \qquad \text{Ans.}$$

> The second moment $I_x$ can be computed with a single integration if a thin strip element parallel to the $x$-axis is used.

b.   The same element of area can be used to obtain the second moment $I_y$ if the result of Example Problem 10-1 is used as the known value for $dI_y$. Thus,

$$dI_y = \frac{bh^3}{3} = \frac{dy(x)^3}{3} = \frac{dy(y^2/2)^3}{3} = \frac{y^6}{24}\, dy$$

Summing all such elements yields

$$I_y = \int_A dI_y = \int_1^2 \frac{y^6}{24}\, dy = \left[\frac{y^7}{168}\right]_1^2 = \frac{127}{168} = 0.756 \text{ in.}^4$$

Once $I_x$ and $I_y$ are known for the area, the polar second moment for an axis through the origin of the $xy$-coordinate system and normal to the plane of the area is obtained by using Eq. 10-3. Thus,

$$J_z = I_x + I_y = 3.10 + 0.756 = 3.856 \cong 3.86 \text{ in.}^4 \qquad \text{Ans.}$$

# PROBLEMS

## Introductory Problems

**10-1*** Determine the second moment of area for the isosceles triangle shown in Fig. P10-1 with respect to

**a.** The base of the triangle (the $x$-axis).
**b.** An axis through the centroid parallel to the base.

**Fig. P10-1**

**10-2*** Determine the second moment of area for the triangle shown in Fig. P10-2 with respect to

**a.** The base of the triangle (the $x$-axis).
**b.** An axis through the centroid parallel to the base.

**Fig. P10-2**

**10-3** Determine the second moment of area for the quarter circle shown in Fig. P10-3 with respect to

**a.** The $x$-axis.
**b.** An axis through the centroid and normal to the plane

**Fig. P10-3**

**10-4** Determine the second moment of area for the half circle shown in Fig. P10-4 with respect to

**a.** The $x$-axis.
**b.** An axis through the centroid parallel to the $x$-axis.

**Fig. P10-4**

**10-5*** Determine the second moment of area for the shaded region shown in Fig. P10-5 with respect to

**a.** The $x$-axis.
**b.** The $y$-axis.

$$y^2 = x$$

16 in.

**Fig. P10-5**

**10-6*** Determine the second moment of area for the shaded region shown in Fig. P10-6 with respect to

**a.** The $x$-axis.
**b.** The $y$-axis.

$$100y = x^2$$

50 mm

**Fig. P10-6**

**10-7** Determine the second moment of area for the shaded region shown in Fig. P10-7 with respect to

**a.** The $x$-axis.
**b.** The $y$-axis.

Fig. P10-7

**10-8** Determine the second moment of area for the shaded region shown in Fig. P10-8 with respect to

**a.** The $x$-axis.
**b.** The $y$-axis.

Fig. P10-8

**10-9*** Determine the second moment of area for the shaded region shown in Fig. P10-9 with respect to

**a.** The $x$-axis.
**b.** The $y$-axis.

Fig. P10-9

**10-10*** Determine the second moment of area for the shaded region shown in Fig. P10-10 with respect to

**a.** The $x$-axis.
**b.** The $y$-axis.

Fig. P10-10

**10-11** Determine the second moment of area for the shaded region shown in Fig. P10-11 with respect to

**a.** The $x$-axis.
**b.** The $y$-axis.

Fig. P10-11

**10-12** Determine the second moment of area for the shaded region shown in Fig. P10-12 with respect to

**a.** The $x$-axis.
**b.** The $y$-axis.

Fig. P10-12

## Intermediate Problems

**10-13\*** Determine the second moment of area for the sector of a circle shown in Fig. P10-13 with respect to

a. The $x$-axis.
b. An axis through the origin of the $xy$-coordinate system and normal to the plane of the area.

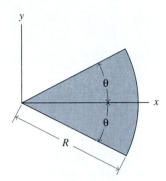

**Fig. P10-13**

**10-14** Determine the second moment of area for the shaded region shown in Fig. P10-14 with respect to

a. The $x$-axis.
b. An axis through the centroid parallel to the $x$-axis.

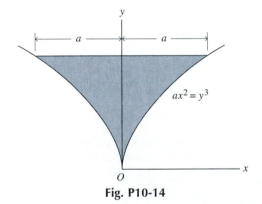

**Fig. P10-14**

## Challenging Problems

**10-15\*** Determine the second moment of area for the ellipse shown in Fig. P10-15 with respect to

a. The $x$-axis.
b. The $y$-axis.

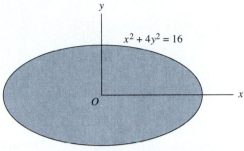

**Fig. P10-15**

**10-16** Determine the second moment of area for the shaded region shown in Fig. P10-16 with respect to

a. The $x$-axis.
b. The $y$-axis.

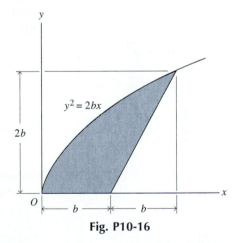

**Fig. P10-16**

## 10-2.3   Radius of Gyration of Areas

Since the second moment of an area has the dimensions of length to the fourth power, it can be expressed as the area $A$ multiplied by a length $k$ squared. Thus, from Eqs. 10-1 and 10-2

$$I_x = \int_A y^2 \, dA = A\,k_x^2 \qquad k_x = \sqrt{\frac{I_x}{A}}$$

$$I_y = \int_A x^2 \, dA = A\,k_y^2 \qquad k_y = \sqrt{\frac{I_y}{A}} \qquad \text{(10-7)}$$

$$I_z = \int_A r^2 \, dA = A\,k_z^2 \qquad k_z = \sqrt{\frac{I_z}{A}}$$

and from Eq. 10-3

$$k_z^2 = k_x^2 + k_y^2 \qquad \text{(10-8)}$$

The distance $k$ is called the radius of gyration. The subscript denotes the axis about which the second moment of area is taken. The radius of gyration of an area with respect to an axis can be visualized as the distance from the axis to the point where the area can be assumed to be concentrated in order to produce the same second moment of area with respect to the axis as does the actual area.

The parallel-axis theorem for second moments of area was discussed in Section 10-2.1. A corresponding relation exists between the radii of gyration of the area with respect to two parallel axes, one of which passes through the centroid of the area. Thus, from Eqs. 10-4 and 10-7,

$$k_{x'}^2 = k_{xC}^2 + d_x^2$$

and

$$\qquad \text{(10-9)}$$

$$k_{y'}^2 = k_{yC}^2 + d_y^2$$

Similarly for polar second moments of area and radii of gyration,

$$k_{z'}^2 = k_{zC}^2 + (d_x^2 + d_y^2) = k_{zC}^2 + d_z^2 \qquad \text{(10-10)}$$

Solving Eqs. 10-9 and 10-10 for rectangular and polar radii of gyration for arbitrary and centroidal $x$-, $y$-, and $z$-axes yields

$$\begin{aligned}
k_{x'} &= \sqrt{k_{xC}^2 + d_x^2} & k_{xC} &= \sqrt{k_{x'}^2 - d_x^2} \\
k_{y'} &= \sqrt{k_{yC}^2 + d_y^2} & k_{yC} &= \sqrt{k_{y'}^2 - d_y^2} \\
k_{z'} &= \sqrt{k_{zC}^2 + d_z^2} & k_{zC} &= \sqrt{k_{z'}^2 - d_z^2}
\end{aligned} \qquad \text{(10-11)}$$

The following example illustrates the concepts discussed in this section.

*The radius of gyration of an area is used in engineering for determining the buckling strength of columns. Two column cross sections with equal areas are shown in Fig. CE10-1. Which cross section offers the greatest resistance to buckling (largest radius of gyration) with respect to the x- and y-axes shown on the figures?*

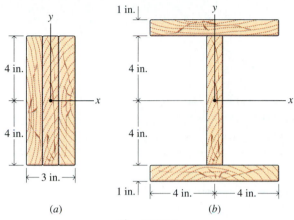

(a)                                          (b)

**Fig. CE10-1**

## SOLUTION

The radius of gyration of an area with respect to a given axis is the distance from the axis at which the area can be conceived to be concentrated and have the same second moment of area with respect to the axis as does the actual (or distributed) area. For both cross sections, $A = 3(8)(1) = 24$ in.$^2$.

For section $a$:

$$I_x = \frac{1}{12}(3)(8)^3 = 128 \text{ in.}^4 \qquad k_x = \sqrt{\frac{I}{A}} = \sqrt{\frac{128}{24}} = 2.31 \text{ in.}$$

$$I_y = \frac{1}{12}(8)(3)^3 = 18 \text{ in.}^4 \qquad k_y = \sqrt{\frac{I}{A}} = \sqrt{\frac{18}{24}} = 0.866 \text{ in.}$$

For section $b$:

$$I_x = \frac{1}{12}(8)(10)^3 - \frac{1}{12}(7)(8)^3 = 368 \text{ in.}^4 \qquad k_x = \sqrt{\frac{I}{A}} = \sqrt{\frac{368}{24}} = 3.92 \text{ in.}$$

$$I_y = \frac{1}{12}(2)(8)^3 + \frac{1}{12}(8)(1)^3 = 86 \text{ in.}^4 \qquad k_y = \sqrt{\frac{I}{A}} = \sqrt{\frac{86}{24}} = 1.89 \text{ in.}$$

These results indicate that section $b$ would offer a greater resistance to buckling about both the $x$- and $y$-axes. This result explains why most columns in large buildings have a cross section similar to the one shown in Fig. CE10-1$b$.

For the shaded area shown in Fig. 10-10a, determine

**a.** The radii of gyration $k_x$, $k_y$, and $k_z$.
**b.** The radius of gyration for an axis passing through the centroid and parallel to the $y$-axis.

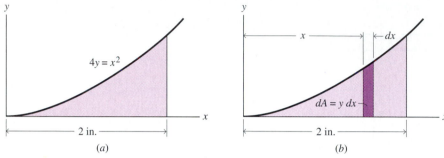

**Fig. 10-10**

---

## SOLUTION

**a.** The quantities required for the determination of $k_x$, $k_y$, and $k_z$ are the area $A$ and the second moments $I_x$ and $I_y$. Since none of these quantities is readily available from known solutions, they will be determined by integration using the element of area shown in Fig. 10-10b. For area $A$,

$$A = \int_A y \, dx = \int_0^2 \frac{x^2}{4} \, dx = \left[ \frac{x^3}{12} \right]_0^2 = \frac{2}{3} \text{ in.}^2$$

For the second moment $I_x$, the results of Example Problem 10-1 can be used. In Example Problem 10-1 it was shown that the second moment $I$ about the base of a rectangle having a base $b$ and a height $h$ was $I = \frac{1}{3} bh^3$. Thus, for the shaded element of area $dA$ of Fig. 10-10b, which has a base $dx$ and a height $y$,

$$dI_x = \frac{1}{3}(dx)(y)^3 = \frac{1}{3}(dx)(x^2/4)^3 = \frac{x^6}{192} \, dx$$

and

$$I_x = \int_A dI_x = \int_0^2 \frac{x^6}{192} \, dx = \left[ \frac{x^7}{1344} \right]_0^2 = \frac{2}{21} \text{ in.}^4$$

For the second moment $I_y$,

$$I_y = \int_A x^2 \, dA = \int_A x^2 y \, dx = \int_0^2 x^2 \left( \frac{x^2}{4} \right) dx = \int_0^2 \frac{x^4}{4} \, dx = \left[ \frac{x^5}{20} \right]_0^2 = \frac{8}{5} \text{ in.}^4$$

Once $A$, $I_x$, and $I_y$ are known, the radii of gyration $k_x$ and $k_y$ are obtained from Eqs. 10-7. Thus,

$$k_x = \left[ \frac{I_x}{A} \right]^{1/2} = \left[ \frac{2/21}{2/3} \right]^{1/2} = 0.3780 = 0.378 \text{ in.} \qquad \text{Ans.}$$

$$k_y = \left[ \frac{I_y}{A} \right]^{1/2} = \left[ \frac{8/5}{2/3} \right]^{1/2} = 1.5492 \cong 1.549 \text{ in.} \qquad \text{Ans.}$$

> The second moment $I_x$ can be computed with a single integration if a thin vertical strip element and the known results for a rectangle are used.

The polar radius of gyration $k_z$ obtained from $k_x$ and $k_y$ by using Eq. 10-8 is

$$k_z = \sqrt{k_x^2 + k_y^2} = \sqrt{(0.3780)^2 + (1.5492)^2} = 1.595 \text{ in.} \qquad \text{Ans.}$$

b.  In order to determine the radius of gyration $k_{yC}$, the distance between the $y$-axis and the centroid of the area must be determined. Thus,

$$A x_C = \int_A x \, dA = \int_0^2 x \left( \frac{x^2}{4} \right) dx = \int_0^2 \frac{x^3}{4} \, dx = \left[ \frac{x^4}{16} \right]_0^2 = 1.000 \text{ in.}^3$$

from which

$$d_y = x_C = \frac{\int_A x \, dA}{A} = \frac{1.000}{2/3} = 1.500 \text{ in.}$$

The radius of gyration $k_{yC}$ is then obtained by using Eqs. 10-11. Thus,

$$k_{yC} = \sqrt{k_y^2 - d_y^2} = \sqrt{(1.5492)^2 - (1.500)^2} = 0.387 \text{ in.} \qquad \text{Ans.}$$

Determine the radii of gyration of the area of the isosceles triangle shown in Fig. 10-11a with respect to

**a.** Horizontal and vertical centroidal axes.
**b.** The x- and y-axes shown on the figure.

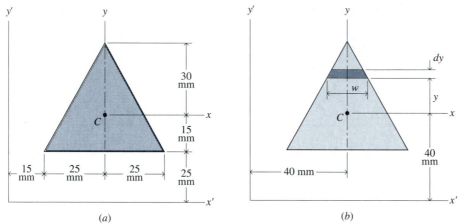

(a)                                              (b)

**Fig. 10-11**

## SOLUTION

**a.** The quantities required for the determination $k_{xC}$ and $k_{yC}$ are the area $A$ and the second moments $I_{xC}$ and $I_{yC}$. The area $A$ for the triangle is

$$A = \frac{1}{2}bh = \frac{1}{2}(50)(45) = 1125 \text{ mm}^2$$

The second moments $I_{xC}$ and $I_{yC}$ will be determined, by integration, using the element of area $dA = w\,dy$ shown in Fig. 10-11b. From similar triangles,

$$\frac{w}{30-y} = \frac{50}{45} \quad \text{from which} \quad w = \frac{50}{45}(30-y)$$

$$dA = w\,dy = \frac{50}{45}(30-y)\,dy$$

Since all parts of the element $dA = w\,dy$ are located a distance $y$ from the $x_C$-axis, the definition of a second moment can be used to determine $I_{xC}$. Thus,

$$I_{xC} = \int_A y^2\,dA = \frac{50}{45}\int_{-15}^{+30}(30-y)y^2\,dy$$

$$= \frac{50}{45}\left[\left(\frac{30y^3}{3} - \frac{y^4}{4}\right)\right]_{-15}^{+30} = 126{,}563 \text{ mm}^4$$

For the second moment $I_{yC}$, the results of Example Problem 10-1 can be used. In Example Problem 10-1 it was shown that the second moment $I$ about a centroidal axis parallel to the base of a rectangle having a base $b$ and a height $h$ was $I = bh^3/12$. Thus, for the shaded element of Fig. 10-11b, which has a base $dy$ and a height $w$,

$$dI_{yC} = \frac{1}{12}(dy)(w)^3 = \frac{1}{12}(dy)\left[\frac{50}{45}(30-y)\right]^3 = \frac{1}{12}\left[\frac{50}{45}(30-y)\right]^3 dy$$

> The second moment $I_x$ can be computed with a single integration if a thin strip element parallel to the x-axis is used.

> The width of the element can be expressed as a function of $y$ by observing that the entire triangle is similar to the triangle that lies above the shaded element.

From which

$$I_{yC} = \int_A dI_{yC} = \int_{-15}^{+30} \frac{1}{12}\left[\frac{50}{45}(30-y)\right]^3 dy$$

$$= -\frac{1}{48}\left(\frac{45}{50}\right)\left\{\left[\frac{50}{45}(30-y)\right]^4\right\}_{-15}^{+30} = 117{,}188 \text{ mm}^4$$

From the definition of a radius of gyration,

$$k_{xC} = \sqrt{\frac{I_{xC}}{A}} = \frac{126{,}563}{1125} = 10.607 \text{ mm} \cong 10.61 \text{ mm} \qquad \text{Ans.}$$

$$k_{yC} = \sqrt{\frac{I_{yC}}{A}} = \frac{117{,}188}{1125} = 10.206 \text{ mm} \cong 10.21 \text{ mm} \qquad \text{Ans.}$$

**b.** The radii of gyration $k_x$ and $k_y$ can be determined by using the parallel axis theorem for radii of gyration. Thus,

$$k_x = \sqrt{k_{xC}^2 + d_x^2} = \sqrt{(10.607)^2 + (40)^2} = 41.4 \text{ mm} \qquad \text{Ans.}$$

$$k_y = \sqrt{k_{yC}^2 + d_y^2} = \sqrt{(10.206)^2 + (40)^2} = 41.3 \text{ mm} \qquad \text{Ans.}$$

# PROBLEMS

### Introductory Problems

**10-17*** Determine the radii of gyration for the rectangular area shown in Fig. P10-17 with respect to

**a.** The $x$- and $y$-axes shown on the figure.
**b.** The horizontal and vertical centroidal axes.

**Fig. P10-17**

**10-18*** Determine the radii of gyration for the triangular area shown in Fig. P10-18 with respect to

**a.** The $x$- and $y$-axes shown on the figure.
**b.** The horizontal and vertical centroidal axes.

**Fig. P10-18**

**10-19** Determine the radii of gyration for the half circle shown in Fig. P10-19 with respect to

**a.** The $x$-axis shown on the figure.
**b.** An axis through the centroid parallel to the $x$-axis.

**Fig. P10-19**

553

**10-20** Determine the radii of gyration for the half ring shown in Fig. P10-20 with respect to

a. The $y$-axis shown on the figure.
b. An axis through the origin of the $xy$-coordinate system and normal to the plane of the area.

**Fig. P10-20**

**10-21\*** Determine the radii of gyration for the shaded area shown in Fig. P10-21 with respect to the $x$- and $y$-axes shown on the figure.

$$y = a \cos \frac{\pi x}{2L}$$

**Fig. P10-21**

**10-22** Determine the radii of gyration for the shaded area shown in Fig. P10-22 with respect to the $x$- and $y$-axes shown on the figure.

$$y = a \sin \frac{\pi x}{L}$$

**Fig. P10-22**

## Intermediate Problems

**10-23\*** Determine the radii of gyration for the circular area shown in Fig. P10-23 with respect to

a. The $x$- and $y$-axes shown on the figure.
b. The horizontal and vertical centroidal axes.

4 in.

5 in.

6 in.

**Fig. P10-23**

**10-24\*** Determine the radii of gyration for the area of the isosceles triangle shown in Fig. P10-24 with respect to

a. The $x$- and $y$-axes shown on the figure.
b. The horizontal and vertical centroidal axes.

45 mm

15 mm    50 mm    25 mm

**Fig. P10-24**

**10-25** Determine the polar radius of gyration for the circular area shown in Fig. P10-25 with respect to an axis through the origin of the $xy$-coordinate system and normal to the plane of the area.

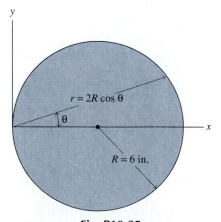

$r = 2R \cos \theta$

$R = 6$ in.

**Fig. P10-25**

**10-26** Determine the radius of gyration for the shaded area shown in Fig. P10-26 with respect to an axis through the centroid of the area and normal to the plane of the area.

$$y = \frac{x^2}{2b}$$

$$\frac{b}{2}$$

$$b$$

**Fig. P10-26**

**10-27\*** Determine the polar radius of gyration for the shaded area shown in Fig. P10-27 with respect to an axis through the origin of the $xy$-coordinate system and normal to the plane of the area.

$$y = b$$

$$y^2 = \frac{b^2 x}{a}$$

$$b$$

$$a$$

**Fig. P10-27**

### Challenging Problems

**10-28\*** Determine the radii of gyration for the shaded area shown in Fig. P10-28 with respect to

a.  The $x$- and $y$-axes shown on the figure.
b.  The horizontal and vertical centroidal axes.

$$y^2 = 25x$$

50 mm

100 mm

**Fig. P10-28**

**10-29\*** Determine the radii of gyration for the shaded area shown in Fig. P10-29 with respect to

a.  The $x$- and $y$-axes shown on the figure.
b.  The horizontal and vertical centroidal axes.

$$y = 2$$

$$y = \frac{x^2}{8}$$

2 in.

4 in.

**Fig. P10-29**

**10-30** Determine the radii of gyration for the shaded area shown in Fig. P10-30 with respect to the $x$- and $y$-axes shown on the figure.

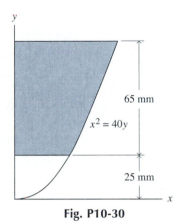

65 mm

$$x^2 = 40y$$

25 mm

**Fig. P10-30**

**10-31** Determine the radii of gyration for the shaded area shown in Fig. P10-31 with respect to the $x$- and $y$-axes shown on the figure.

$$y^2 = 3x$$

2 in.          3 in.

**Fig. P10-31**

**10-32\*** Determine the polar radius of gyration for the shaded area shown in Fig. P10-32 with respect to an axis through the origin of the $xy$-coordinate system and normal to the plane of the area.

**Fig. P10-32**

## 10-2.4 Second Moments of Composite Areas

The second moments $I_x$, $I_y$, and $I_z$ of an area $A$ with respect to any set of $x$-, $y$-, and $z$-coordinate axes were defined as

$$I_x = \int_A y^2 \, dA \qquad I_y = \int_A x^2 \, dA \qquad I_z = \int_A r^2 \, dA$$

Frequently in engineering practice, an irregular area $A$ will be encountered that can be broken up into a series of simple areas $A_1$, $A_2$, $A_3, \ldots, A_n$ for which the integrals have been evaluated and tabulated. The second moment of the irregular area, the *composite area*, with respect to any axis is equal to the sum of the second moments of the separate parts of the area with respect to the specified axis. For example,

$$I_x = \int_A y^2 \, dA$$

$$= \int_{A_1} y^2 \, dA_1 + \int_{A_2} y^2 \, dA_2 + \int_{A_3} y^2 \, dA_3 + \cdots + \int_{A_n} y^2 \, dA_n$$

$$= I_{x1} + I_{x2} + I_{x3} + \cdots + I_{xn}$$

When an area such as a hole is removed from a larger area, its second moment must be subtracted from the second moment of the larger area to obtain the resulting second moment. Thus, for the case of a square plate with a hole

$$I_\blacksquare = I_\blacksquare + I_\boxdot$$

Therefore,

$$I_\blacksquare = I_\blacksquare - I_\boxdot$$

Table 10-1 contains a listing of the values of the integrals for frequently encountered shapes such as rectangles, triangles, circles, and semicircles. Tables listing second moments of area and other properties for the cross sections of common structural shapes are found in engineering handbooks and in data books prepared by industrial organizations such as the American Institute of Steel Construction. Properties of a few selected shapes are listed in Tables 10-2$a$ and 10-2$b$ for use in solving problems.

## TABLE 10-1   SECOND MOMENTS OF PLANE AREAS

$$I_x = \frac{bh^3}{12}$$

$$I_{x'} = \frac{bh^3}{3}$$

$$A = bh$$

$$I_x = \frac{bh^3}{36}$$

$$I_{x'} = \frac{bh^3}{12}$$

$$A = \frac{1}{2}bh$$

$$I_x = \frac{\pi R^4}{4}$$

$$I_{x'} = \frac{5\pi R^4}{4}$$

$$A = \pi R^2$$

$$I_x = \frac{\pi R^4}{8} - \frac{8R^4}{9\pi}$$

$$I_y = \frac{\pi R^4}{8}$$

$$I_{x'} = \frac{\pi R^4}{8}$$

$$A = \frac{1}{2}\pi R^2$$

$$I_x = \frac{\pi R^4}{16} - \frac{4R^4}{9\pi}$$

$$I_{x'} = \frac{\pi R^4}{16}$$

$$A = \frac{1}{4}\pi R^2$$

$$I_x = \frac{R^4}{4}\left(\theta - \frac{1}{2}\sin 2\theta\right)$$

$$I_y = \frac{R^4}{4}\left(\theta + \frac{1}{2}\sin 2\theta\right)$$

$$x_C = \frac{2}{3}\frac{R\sin\theta}{\theta}$$

$$A = \theta R^2$$

TABLE 10-2*A*    PROPERTIES OF ROLLED-STEEL SHAPES (U.S. CUSTOMARY UNITS)

| Shape | Area in.$^2$ | Depth in. | Width in. | Flange Thickness in. | Web Thickness in. | Axis XX | | Axis YY | |
|---|---|---|---|---|---|---|---|---|---|
| | | | | | | $I_X$ in.$^4$ | $k_X$ in. | $I_Y$ in.$^4$ | $k_Y$ in. |
| W24 × 84 | 24.7 | 24.10 | 9.020 | 0.770 | 0.470 | 2370 | 9.79 | 94.4 | 1.95 |
| W18 × 76 | 22.3 | 18.21 | 11.035 | 0.680 | 0.425 | 1330 | 7.73 | 152 | 2.61 |
| W14 × 43 | 12.6 | 13.66 | 7.995 | 0.530 | 0.305 | 428 | 5.82 | 45.2 | 1.89 |
| W12 × 30 | 8.79 | 12.34 | 6.520 | 0.440 | 0.260 | 238 | 5.21 | 20.3 | 1.52 |
| W10 × 22 | 6.49 | 10.17 | 5.750 | 0.360 | 0.240 | 118 | 4.27 | 11.4 | 1.33 |
| S18 × 70 | 20.6 | 18.00 | 6.251 | 0.691 | 0.711 | 926 | 6.71 | 24.1 | 1.08 |
| S12 × 50 | 14.7 | 12.00 | 5.477 | 0.659 | 0.687 | 305 | 4.55 | 15.7 | 1.03 |
| S10 × 35 | 10.3 | 10.00 | 4.944 | 0.491 | 0.594 | 147 | 3.78 | 8.36 | 0.901 |

| Shape | Area in.$^2$ | Depth in. | Width in. | Flange Thickness in. | Web Thickness in. | Axis XX | | Axis YY | | |
|---|---|---|---|---|---|---|---|---|---|---|
| | | | | | | $I_X$ in.$^4$ | $k_X$ in. | $I_Y$ in.$^4$ | $k_Y$ in. | $x_C$ in. |
| C15 × 50 | 14.7 | 15.00 | 3.716 | 0.650 | 0.716 | 404 | 5.24 | 11.0 | 0.867 | 0.798 |
| C12 × 30 | 8.82 | 12.00 | 3.170 | 0.501 | 0.510 | 162 | 4.29 | 5.14 | 0.763 | 0.674 |
| C10 × 30 | 8.82 | 10.00 | 3.033 | 0.436 | 0.673 | 103 | 3.42 | 3.94 | 0.669 | 0.649 |

| Shape | Area in.$^2$ | Axis XX | | | Axis YY | | |
|---|---|---|---|---|---|---|---|
| | | $I_X$ in.$^4$ | $k_X$ in. | $y_C$ in. | $I_Y$ in.$^4$ | $k_Y$ in. | $x_C$ in. |
| L8 × 8 × 1 | 15.00 | 89.0 | 2.44 | 2.37 | 89.0 | 2.44 | 2.37 |
| L8 × 6 × 1 | 13.00 | 80.8 | 2.49 | 2.65 | 38.8 | 1.73 | 1.65 |
| L8 × 4 × 1 | 11.00 | 69.6 | 2.52 | 3.05 | 11.6 | 1.03 | 1.05 |
| L6 × 6 × 1 | 11.00 | 35.5 | 1.80 | 1.86 | 35.5 | 1.80 | 1.86 |
| L6 × 4 × 3/4 | 6.94 | 24.5 | 1.88 | 2.08 | 8.68 | 1.12 | 1.08 |
| L5 × 3 1/2 × 3/4 | 5.81 | 13.9 | 1.55 | 1.75 | 5.55 | 0.977 | 0.996 |
| L4 × 4 × 3/4 | 5.44 | 7.67 | 1.19 | 1.27 | 7.67 | 1.19 | 1.27 |
| L4 × 3 × 1/2 | 3.25 | 5.05 | 1.25 | 1.33 | 2.42 | 0.864 | 0.827 |
| L3 × 3 × 1/2 | 2.75 | 2.22 | 0.898 | 0.932 | 2.22 | 0.898 | 0.932 |

In some instances the second moments $I_{xC}$, $I_{yC}$, and $I_{zC}$ of a composite shape with respect to centroidal $x$-, $y$-, and $z$-axes of the composite may be required. These quantities can be determined by first evaluating the second moments $I_{x'}$, $I_{y'}$, and $I_{z'}$ of the composite with respect to any convenient set of parallel $x'$-, $y'$-, and $z'$-axes and then transferring these second moments to the centroidal axes by using the parallel-axis theorem.

## TABLE 10-2*B*    PROPERTIES OF ROLLED-STEEL SHAPES (SI UNITS)

W-SHAPE    S-SHAPE    C-SHAPE    L-SHAPE

| Shape | Area mm² | Depth mm | Width mm | Flange Thickness mm | Web Thickness mm | Axis XX | | Axis YY | |
|---|---|---|---|---|---|---|---|---|---|
| | | | | | | $I_X$ 10⁶ mm⁴ | $k_X$ mm | $I_Y$ 10⁶ mm⁴ | $k_Y$ mm |
| W610 × 125 | 15,935 | 612 | 229 | 19.6 | 11.9 | 985 | 249 | 39.3 | 49.5 |
| W457 × 113 | 14,385 | 463 | 280 | 17.3 | 10.8 | 554 | 196 | 63.3 | 66.3 |
| W356 × 64 | 8,130 | 347 | 203 | 13.5 | 7.7 | 178 | 148 | 18.8 | 48.0 |
| W305 × 45 | 5,670 | 313 | 166 | 11.2 | 6.6 | 99.1 | 132 | 8.45 | 38.6 |
| W254 × 33 | 4,185 | 258 | 146 | 9.1 | 6.1 | 49.1 | 108 | 4.75 | 33.8 |
| S457 × 104 | 13,290 | 457 | 159 | 17.6 | 18.1 | 358 | 170 | 10.0 | 27.4 |
| S305 × 74 | 9,485 | 305 | 139 | 16.7 | 17.4 | 127 | 116 | 6.53 | 26.2 |
| S254 × 52 | 6,645 | 254 | 126 | 12.5 | 15.1 | 61.2 | 96.0 | 3.48 | 22.9 |

| Shape | Area mm² | Depth mm | Width mm | Flange Thickness mm | Web Thickness mm | Axis XX | | Axis YY | | |
|---|---|---|---|---|---|---|---|---|---|---|
| | | | | | | $I_X$ 10⁶ mm⁴ | $k_X$ mm | $I_Y$ 10⁶ mm⁴ | $k_Y$ mm | $x_C$ mm |
| C381 × 74 | 9485 | 381 | 94.4 | 16.5 | 18.2 | 168 | 133 | 4.58 | 22.0 | 20.3 |
| C305 × 45 | 5690 | 305 | 80.5 | 12.7 | 13.0 | 67.4 | 109 | 2.14 | 19.4 | 17.1 |
| C254 × 45 | 5690 | 254 | 77.0 | 11.1 | 17.1 | 42.9 | 86.9 | 1.64 | 17.0 | 16.5 |

| Shape | Area mm² | Axis XX | | | Axis YY | | |
|---|---|---|---|---|---|---|---|
| | | $I_X$ 10⁶ mm⁴ | $k_X$ mm | $y_C$ mm | $I_Y$ 10⁶ mm⁴ | $k_Y$ mm | $x_C$ mm |
| L203 × 203 × 25.4 | 9675 | 37.0 | 62.0 | 60.2 | 37.0 | 62.0 | 60.2 |
| L203 × 152 × 25.4 | 8385 | 33.6 | 63.2 | 67.3 | 16.1 | 43.9 | 41.9 |
| L203 × 102 × 25.4 | 7095 | 29.0 | 64.0 | 77.5 | 4.83 | 26.2 | 26.7 |
| L152 × 152 × 25.4 | 7095 | 14.8 | 45.7 | 47.2 | 14.8 | 45.7 | 47.2 |
| L152 × 102 × 19.1 | 4475 | 10.2 | 47.8 | 52.8 | 3.61 | 28.4 | 27.4 |
| L127 × 89 × 19.1 | 3750 | 5.79 | 39.4 | 44.5 | 2.31 | 24.8 | 25.3 |
| L102 × 102 × 19.1 | 3510 | 3.19 | 30.2 | 32.3 | 3.19 | 30.2 | 32.3 |
| L102 × 76 × 12.7 | 2095 | 2.10 | 31.8 | 33.8 | 1.01 | 21.9 | 21.0 |
| L76 × 76 × 12.7 | 1775 | 0.924 | 22.8 | 23.7 | 0.924 | 22.8 | 23.7 |

A beam having the cross section shown in Fig. 10-12a is constructed by gluing a 2 × 6-in. wooden plank 10 ft long to a second 2 × 6-in. wooden plank also 10 ft long. Determine the second moment of the cross-sectional area with respect to

a.   The $x$-axis.
b.   The $y$-axis.
c.   The $x_C$-axis, which passes through the centroid of the area and is parallel to the $x$-axis.

(a)

**SOLUTION**

a.   As shown in Fig. 10-12b, the cross-sectional area can be divided into two simple rectangles. Since both rectangles have an edge along the $x$-axis, their moments of inertia are just $bh^3/3$. Therefore, the moment of inertia of the entire area with respect to the $x$-axis is

$$I_x = I_{x1} + I_{x2} = \frac{1}{3}(4)(2)^3 + \frac{1}{3}(2)(8)^3 = 352.0 \text{ in.}^4 \qquad \text{Ans.}$$

> The second moment of a composite area with respect to a given axis equals the sum of the second moments of the separate parts of the area with respect to the same axis.

b.   Using the same division of areas as in part $a$, the first area has an edge along the $y$-axis and its moment of inertia is given by $bh^3/3$. However, the parallel-axis theorem is needed for the second rectangle, since neither the centroid of the rectangle nor either edge is along the $y$-axis. Therefore, the moment of inertia of the entire area with respect to the $y$-axis is

$$I_y = I_{y1} + I_{y2}$$
$$= \frac{1}{3}(2)(4)^3 + \left[ \frac{1}{12}(8)(2)^3 + 16(5)^2 \right] = 448 \text{ in.}^4 \qquad \text{Ans.}$$

(b)

**Fig. 10-12**

c.   The centroid will be located using the same division of areas as above

$$d(8 + 16) = 1(8) + 4(16) \qquad d = 3 \text{ in.}$$

Then, using the parallel-axis theorem for both rectangles, the moment of inertia of the entire area with respect to the $x_C$-axis is

$$I_{xC} = I_{xC1} + I_{xC2}$$
$$= \left[ \frac{1}{12}(4)(2)^3 + 8(3 - 1)^2 \right] + \left[ \frac{1}{12}(2)(8)^3 + 16(4 - 3)^2 \right] = 136.0 \text{ in.}^4 \quad \text{Ans.}$$

Note that, since the $x_C$-axis passes through the centroid of the entire area, the moments of inertia $I_x$ and $I_{xC}$ are related by the parallel-axis theorem

$$I_x = I_{xC} + Ad^2$$
$$= 136.0 + 24(3)^2 = 352 \text{ in.}^4$$

Determine the second moment of the shaded area shown in Fig. 10-13a with respect to

a. The x-axis.
b. The y-axis.
c. An axis through the origin $O$ of the $xy$-coordinate system and normal to the plane of the area.

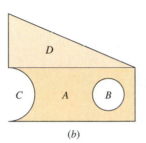

(a)　　　　　　　　　　(b)

**Fig. 10-13**

## SOLUTION

As shown in Fig. 10-13b, the shaded area can be divided into a $100 \times 200$-mm rectangle ($A$), with a 60-mm-diameter circle ($B$) and a 100-mm-diameter half circle ($C$) removed, and a $100 \times 200$-mm triangle ($D$). The second moments for these four areas, with respect to the $x$- and $y$-axes, can be obtained by using information from Table 10-1, as follows:

a. For the rectangle (shape $A$),

$$I_{x1} = \frac{bh^3}{3} = \frac{200(100^3)}{3} = 66.667(10^6) \text{ mm}^4$$

For the circle (shape $B$),

$$I_{x2} = I_{xC} + d_x^2 A = \frac{\pi R^4}{4} + d_x^2(\pi R^2)$$

$$= \frac{\pi(30^4)}{4} + (50^2)(\pi)(30^2) = 7.705(10^6) \text{ mm}^4$$

For the half circle (shape $C$),

$$I_{x3} = I_{xC} + d_x^2 A = \frac{\pi R^4}{8} + d_x^2\left(\frac{\pi R^2}{2}\right)$$

$$= \frac{\pi(50^4)}{8} + (50)^2\left[\frac{\pi(50)^2}{2}\right] = 12.272(10^6) \text{ mm}^4$$

For the triangle (shape $D$),

$$I_{x4} = I_{xC} + d_x^2 A$$
$$= \frac{bh^3}{36} + d_x^2\left(\frac{bh}{2}\right)$$
$$= \frac{200(100^3)}{36} + \left(100 + \frac{100}{3}\right)^2\left[\frac{200(100)}{2}\right] = 183.333(10^6) \text{ mm}^4$$

For the composite area,

$$I_x = I_{x1} - I_{x2} - I_{x3} + I_{x4}$$
$$= 66.667(10^6) - 7.705(10^6) - 12.272(10^6) + 183.333(10^6)$$
$$= 230.023(10^6) \cong 230(10^6) \text{ mm}^4 \qquad \text{Ans.}$$

When a hole is present in a larger area, its second moment must be subtracted from the second moment of the larger area in order to obtain the required second moment.

**b.** For the rectangle (shape $A$),

$$I_{y1} = \frac{b^3 h}{3} = \frac{200^3(100)}{3} = 266.667(10^6) \text{ mm}^4$$

For the circle (shape $B$),

$$I_{y2} = I_{yC} + d_y^2 A$$
$$= \frac{\pi R^4}{4} + d_y^2(\pi R^2)$$
$$= \frac{\pi(30^4)}{4} + (150^2)(\pi)(30^2) = 64.253(10^6) \text{ mm}^4$$

For the half circle (shape $C$),

$$I_{y3} = \frac{\pi R^4}{8} = \frac{\pi(50^4)}{8} = 2.454(10^6) \text{ mm}^4$$

For the triangle (shape $D$),

$$I_{y4} = \frac{bh^3}{12} = \frac{100(200^3)}{12} = 66.667(10^6) \text{ mm}^4$$

For the composite area,

$$I_y = I_{y1} - I_{y2} - I_{y3} + I_{y4}$$
$$= 266.667(10^6) - 64.253(10^6) - 2.454(10^6) + 66.667(10^6)$$
$$= 266.627(10^6) \cong 267(10^6) \text{ mm}^4 \qquad \text{Ans.}$$

**c.** For the composite area,

$$J_z = I_x + I_y$$
$$= 230.023(10^6) + 266.627(10^6)$$
$$= 496.650(10^6) \cong 497(10^6) \text{ mm}^4 \qquad \text{Ans.}$$

A column with the cross section shown in Fig. 10-14a is constructed from a W24 × 84 wide-flange section and a C12 × 30 channel section. Determine the second moments and radii of gyration of the cross-sectional area with respect to horizontal and vertical axes through the centroid of the cross section.

C12 x 30

← W24 x 84

(a)

y

4.51 in. ∗ 4.51 in.

12.05 in.

C

12.05 in.

$A = 24.7$ in.$^2$
$I_x = 2370$ in.$^4$
$I_y = 94.4$ in.$^4$

(b)

0.674 in.

y

6.00 in. ∗ 6.00 in.

11.886 in.

C x'

0.510 in.

$A = 8.82$ in.$^2$
$I_y = 162$ in.$^4$
$I_{x'} = 5.19$ in.$^4$

(c)

y

C $x_C$
x

3.128 in.

(d)

**Fig. 10-14**

## SOLUTION

Properties and dimensions for the structural shapes can be obtained from Table 10-2a. The properties and dimensions for the wide-flange section are shown in Fig. 10-14b. The properties and dimensions for the channel section are shown in Fig. 10-14c. In Figs. 10-14b and c, the x-axis passes through the centroid of the wide-flange section and a parallel x′-axis passes through the centroid of the channel. A centroidal $x_C$-axis for the composite section (see Fig. 10-14d) can be located by using the principle of moments as applied to areas. The total area $A_T$ for the composite section is

$$A_T = A_{WF} + A_{CH} = 24.7 + 8.82 = 33.52 \text{ in.}^2$$

The moment of the composite area about the x-axis is

$$A_T(y_C)_T = A_{WF}(y_C)_{WF} + A_{CH}(y_C)_{CH} = 24.7(0) + 8.82(11.886) = 104.835 \text{ in.}^3$$

The distance $(y_C)_T$ from the x-axis to the centroid of the composite section is

$$(y_C)_T = \frac{104.835}{33.52} = 3.128 \text{ in.}$$

The second moment $(I_{xC})_{WF}$ for the wide-flange section about the centroidal $x_C$-axis of the composite section is determined by using the parallel-axis theorem. Thus,

$$(I_{xC})_{WF} = (I_x)_{WF} + (y_C)^2_{WF} A_{WF} = 2370 + (3.128)^2(24.7) = 2611.7 \text{ in.}^4$$

C12 x 30

← W24 x 84

(a)

A = 24.7 in.$^2$
$I_x$ = 2370 in.$^4$
$I_y$ = 94.4 in.$^4$

(b)

A = 8.82 in.$^2$
$I_y$ = 162 in.$^4$
$I_{x'}$ = 5.19 in.$^4$

(c)

(d)

**Fig. 10-14**

Similarly, the second moment $(I_{xC})_{CH}$ for the channel about the centroidal $x_C$-axis of the composite section is

$$(I_{xC})_{CH} = (I_{x'})_{CH} + [(y_C)_{CH} - (y_C)_T]^2 \, A_{CH}$$
$$= 5.14 + (11.886 - 3.128)^2 \, (8.82) = 681.7 \text{ in.}^4$$

For the composite area,

$$(I_{xC})_T = (I_{xC})_{WF} + (I_{xC})_{CH}$$
$$= 2611.7 + 681.7 = 3293.4 \cong 3290 \text{ in.}^4 \qquad \text{Ans.}$$

The $y$-axis passes through the centroid of both areas; therefore, the second moment $(I_{yC})_T$ for the composite section is

$$(I_{yC})_T = (I_{yC})_{WF} + (I_{yC})_{CH}$$
$$= 94.4 + 162 = 256.4 \cong 256 \text{ in.}^4 \qquad \text{Ans.}$$

The radius of gyration about the $x_C$-axis for the composite section is

$$(k_{xC})_T = \left[\frac{(I_{xC})_T}{A_T}\right]^{1/2} = \left(\frac{3293.4}{33.52}\right)^{1/2} = 9.912 \cong 9.91 \text{ in.} \qquad \text{Ans.}$$

The radius of gyration about the $y_C$-axis for the composite section is

$$(k_{yC})_T = \left[\frac{(I_{yC})_T}{A_T}\right]^{1/2} = \left(\frac{256.4}{33.52}\right)^{1/2} = 2.766 \cong 2.77 \text{ in.} \qquad \text{Ans.}$$

# PROBLEMS

Determine the second moments with respect to $x$- (horizontal) and $y$- (vertical) axes through the centroid of

## Introductory Problems

**10-33\*** The shaded area shown in Fig. P10-33.

1 in.

4 in.

1 in. →

1 in.

4 in.

**Fig. P10-33**

**10-34\*** The shaded area shown in Fig. P10-34.

20 mm

80 mm

20 mm

← 40 mm →

20 mm

**Fig. P10-34**

**10-35** The shaded area shown in Fig. P10-35.

2 in.

6 in.

2 in. | 2 in. | 2 in.

**Fig. P10-35**

**10-36** The shaded area shown in Fig. P10-36.

60 mm | 160 mm | 60 mm

140 mm

60 mm

**Fig. P10-36**

**10-37\*** The two $10 \times 1$-in. steel plates that are welded to the flanges of an S18 × 70 I-beam as shown in Fig. P10-37.

1 in.

18 in.

10 in.

1 in.

**Fig. P10-37**

**10-38\*** The two $250 \times 25$-mm steel plates and two C254 × 45 channels that are welded together to form the cross section shown in Fig. P10-38.

25 mm

254 mm

250 mm

25 mm

**Fig. P10-38**

**10-39** The four L8 × 8 × 1 angles that are riveted together to form the cross section shown in Fig. P10-39.

Fig. P10-39

**10-40** The four C305 × 45 channels that are welded together to form the cross section shown in Fig. P10-40.

Fig. P10-40

**10-41\*** The 10 × 1/2-in. steel plate and four L5 × 3 1/2 × 3/4 in. angles that are riveted together to form the cross section shown in Fig. P10-41.

Fig. P10-41

**10-42\*** The shaded area shown in Fig. P10-42.

Fig. P10-42

**10-43** The three S10 × 35 I-beams that are welded together to form the cross section shown in Fig. P10-43.

Fig. P10-43

**10-44** The shaded area shown in Fig. P10-44.

Fig. P10-44

## Intermediate Problems

Determine the second moments with respect to the $x$- and $y$-axes shown on the figure for

**10-45\*** The shaded area shown in Fig. P10-45 if $d = 2$ in.

**Fig. P10-45**

**10-46\*** The shaded area shown in Fig. P10-46 if $d = 30$ mm.

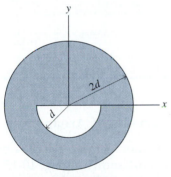

**Fig. P10-46**

**10-47** The shaded area shown in Fig. P10-47.

**Fig. P10-47**

**10-48** The shaded area shown in Fig. P10-48.

**Fig. P10-48**

## Challenging Problems

Determine the second moments with respect to

**a.** The $x$- and $y$-axes shown on the figure.
**b.** The $x$- and $y$-axes through the centroid of the area.

**10-49\*** For the shaded area shown in Fig. P10-49.

**Fig. P10-49**

**10-50\*** For the shaded area shown in Fig. P10-50.

**Fig. P10-50**

**10-51** For the shaded area shown in Fig. P10-51.

Fig. P10-51

**10-53*** For the shaded area shown in Fig. P10-53.

Fig. P10-53

**10-52** For the shaded area shown in Fig. P10-52.

Fig. P10-52

**10-54** For the shaded area shown in Fig. P10-54.

Fig. P10-54

## 10-2.5 Mixed Second Moments of Areas

The mixed second moment (commonly called the area product of inertia) $dI_{xy}$ of the element of area $dA$ shown in Fig. 10-15 with respect to the $x$- and $y$-axes is defined as the product of the two coordinates of the element multiplied by the area of the element; thus,

$$dI_{xy} = xy \, dA$$

The mixed second moment (area product of inertia) of the total area $A$ about the $x$- and $y$-axes is the sum of the mixed second moments of the elements of the area; thus,

$$I_{xy} = \int_A xy \, dA \qquad (10\text{-}12)$$

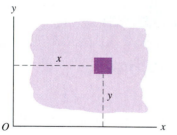

**Figure 10-15** Element of area $dA$ in the $xy$-plane.

The dimensions of the mixed second moment are the same as for the rectangular or polar second moments, but since the product $xy$ can be either positive or negative, the mixed second moment can be positive, negative, or zero. Recall that rectangular or polar second moments are always positive.

The mixed second moment of an area with respect to any two orthogonal axes is zero when either of the axes is an axis of symmetry. This statement can be demonstrated by means of Fig. 10-16, which is symmetrical with respect to the $y$-axis. For every element of area $dA$ on one side of the axis of symmetry, there is a corresponding element of area $dA'$ on the opposite side of the axis such that the mixed second moments of $dA$ and $dA'$ will be equal in magnitude but opposite in sign. Thus, they will cancel each other in the summation, and the resulting mixed second moment for the total area will be zero.

The parallel-axis theorem for mixed second moments can be derived from Fig. 10-17, in which the $x$- and $y$-axes pass through the centroid C of the area and are parallel to the $x'$- and $y'$-axes. The mixed second moment with respect to the $x'$- and $y'$-axes is

$$I_{x'y'} = \int_A x'y' \, dA = \int_A (d_y + x)(d_x + y) \, dA$$

$$= d_x d_y \int_A dA + d_y \int_A y \, dA + d_x \int_A x \, dA + \int_A xy \, dA$$

**Figure 10-16**   Elements of area $dA$ and $dA'$ on opposite sides of an axis of symmetry.

since $d_x$ and $d_y$ are the same for every element of area $dA$. The second and third integrals in the preceding equation are zero since $x$ and $y$ are centroidal axes. The last integral is the mixed second moment with respect to the centroidal axes. Consequently, the mixed second moment about a pair of axes parallel to a pair of centroidal axes is

$$I_{x'y'} = I_{xyC} + d_x d_y A \qquad (10\text{-}13)$$

where the subscript C indicates that the $x$- and $y$-axes are centroidal axes. The parallel-axis theorem for mixed second moments (area products of inertia) can be stated as follows: The mixed second moment of an area with respect to any two perpendicular axes $x$ and $y$ in the plane of the area is equal to the mixed second moment of the area with respect to a pair of centroidal axes parallel to the $x$- and $y$-axes added to the product of the area and the two centroidal locations from the $x$- and $y$-axes. The parallel-axis theorem for mixed second moments is used most frequently in determining mixed second moments for composite areas. Values from Eq. 10-12 for some of the shapes commonly used in these calculations are listed in Table 10-3.

The determination of the mixed second moment (area product of inertia) is illustrated in the next two example problems.

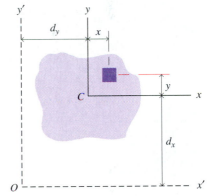

**Figure 10-17**   Element of area $dA$ in the $xy$-plane.

**TABLE 10-3 MIXED SECOND MOMENTS OF PLANE AREAS**

$$I_{xy} = 0 \qquad\qquad I_{x'y'} = \frac{b^2h^2}{4}$$

$$I_{xy} = -\frac{b^2h^2}{72} \qquad\qquad I_{x'y'} = \frac{b^2h^2}{24}$$

$$I_{xy} = \frac{b^2h^2}{72} \qquad\qquad I_{x'y'} = -\frac{b^2h^2}{24}$$

$$I_{xy} = \frac{(9\pi - 32)R^4}{72\pi} \qquad\qquad I_{x'y'} = \frac{R^4}{8}$$

$$I_{xy} = -\frac{(9\pi - 32)R^4}{72\pi} \qquad\qquad I_{x'y'} = -\frac{R^4}{8}$$

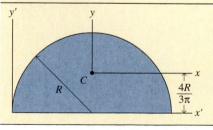

$$I_{xy} = 0 \qquad\qquad I_{x'y'} = \frac{2R^4}{3}$$

Determine the mixed second moment (area product of inertia) of the shaded area shown in Fig. 10-18a with respect to the $x$- and $y$-axes by using

**a.** Double integration.
**b.** The parallel-axis theorem and single integration.

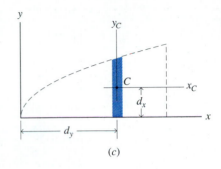

(a)  (b)  (c)

**Fig. 10-18**

## SOLUTION

**a.** The second mixed moment (area product of inertia) $dI_{xy}$ of the element of area $dA = dy\, dx$ shown in Fig. 10-18a with respect to the $x$- and $y$-axes is defined as $dI_{xy} = xy\, dA$. Therefore,

$$I_{xy} = \int_A xy\, dA = \int_0^4 \int_0^{\sqrt{x}} xy\, dy\, dx$$

$$= \int_0^4 \left[ \frac{y^2}{2} \right]_0^{\sqrt{x}} x\, dx$$

$$= \int_0^4 \frac{x^2}{2}\, dx = \left[ \frac{x^3}{6} \right]_0^4 = 10.67 \text{ in.}^4 \qquad \text{Ans.}$$

**b.** The mixed second moment (area product of inertia) $dI_{xy}$ of the element of area $dA = y\, dx$ shown in Fig. 10-18b with respect to axes through the centroid of the element parallel to the $x$- and $y$-axes is zero. Thus, the mixed second moment of the element with respect to the $x$- and $y$-axes is

$$dI_{xy} = dI_{xyC} + d_x d_y\, dA = 0 + \frac{y}{2} x\, (y\, dx) = \frac{x^2}{2}\, dx$$

> The mixed second moment of an area with respect to any two orthogonal axes is zero when either of the axes is an axis of symmetry.

Therefore,

$$I_{xy} = \int_0^4 dI_{xy} = \int_0^4 \frac{x^2}{2}\, dx = \left[ \frac{x^3}{6} \right]_0^4 = 10.67 \text{ in.}^4 \qquad \text{Ans.}$$

## EXAMPLE PROBLEM 10-11

Determine the mixed second moment (area product of inertia) of the quadrant of a circle shown in Fig. 10-19 with respect to

a. The $x$- and $y$-axes.
b. A pair of axes through the centroid of the area and parallel to the $x$- and $y$-axes.

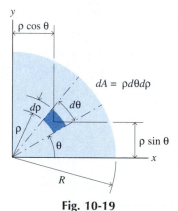

**Fig. 10-19**

---

### SOLUTION

a. The mixed second moment (area product of inertia) $dI_{xy}$ of the element of area $dA = \rho \, d\theta \, d\rho$ shown in Fig. 10-19 with respect to the $x$- and $y$-axes is defined as $dI_{xy} = xy \, dA$. Therefore,

$$dI_{xy} = (\rho \cos \theta)(\rho \sin \theta)(\rho \, d\theta \, d\rho)$$

and

$$I_{xy} = \int_A dI_{xy} = \int_0^R \int_0^{\pi/2} (\rho \cos \theta)(\rho \sin \theta) \, \rho \, d\theta \, d\rho$$

$$= \int_0^R \left[ \frac{\sin^2 \theta}{2} \right]_0^{\pi/2} \rho^3 \, d\rho$$

$$= \int_0^R \frac{\rho^3}{2} \, d\rho = \left[ \frac{\rho^4}{8} \right]_0^R = \frac{R^4}{8} \qquad \text{Ans.}$$

> Polar coordinates are usually more efficient when circular boundaries are involved.

b. Once the mixed second moment is known with respect to a pair of axes, the mixed second moment with respect to a parallel set of axes through the centroid of the area can be found by using the parallel-axis theorem. For the quarter circle, $d_x = d_y = 4R/3\pi$; therefore,

$$I_{xyC} = I_{xy} - d_x d_y \, dA$$

$$= \frac{R^4}{8} - \left( \frac{4R}{3\pi} \right)\left( \frac{4R}{3\pi} \right)\left( \frac{\pi R^2}{4} \right) = \frac{(9\pi - 32)R^4}{72\pi} \qquad \text{Ans.}$$

> The parallel-axis theorem for second moments is valid only for transfers to or from a parallel axis through the centroid of the area.

# PROBLEMS

Determine the mixed second moment with respect to the x- and y-axes

## Introductory Problems

**10-55\*** For the rectangle shown in Fig. P10-55.

**Fig. P10-55**

**10-56\*** For the angle section shown in Fig. P10-56.

**Fig. P10-56**

**10-57** For the shaded area shown in Fig. P10-57.

**Fig. P10-57**

**10-58** For the U-section shown in Fig. P10-58.

**Fig. P10-58**

**10-59\*** For the T-section shown in Fig. P10-59.

**Fig. P10-59**

**10-60\*** For the Z-section shown in Fig. P10-60.

**Fig. P10-60**

**10-61** For the shaded area shown in Fig. P10-61.

**Fig. P10-61**

**10-62** For the shaded area shown in Fig. P10-62.

**Fig. P10-62**

## Intermediate Problems

**10-63*** Determine the mixed second moment of the shaded area shown in Fig. P10-63 with respect to

a. The $x$- and $y$-axes.
b. A pair of axes through the centroid of the area and parallel to the $x$- and $y$-axes.

**Fig. P10-63**

**10-64*** Determine the mixed second moment of the shaded area in Fig. P10-64 with respect to

a. The $x$- and $y$-axes.
b. A pair of axes through the centroid of the area and parallel to the $x$- and $y$-axes.

**Fig. P10-64**

**10-65** Determine the mixed second moment of the shaded area shown in Fig. P10-65 with respect to the $x$- and $y$-axes.

**Fig. P10-65**

**10-66** Determine the mixed second moment of the shaded area shown in Fig. P10-66 with respect to the $x$- and $y$-axes.

**Fig. P10-66**

**10-67*** Determine the mixed second moment of the shaded area shown in Fig. P10-67 with respect to the $x$- and $y$-axes.

**10-68** Determine the mixed second moment of the shaded area shown in Fig. P10-68 with respect to the $x$- and $y$-axes.

**Fig. P10-67**

**Fig. P10-68**

## 10-3 PRINCIPAL SECOND MOMENTS

The second moment of the area $A$ in Fig. 10-20 with respect to the $x'$-axis through $O$ will, in general, vary with the angle $\theta$. The $x$- and $y$-axes used to obtain the polar second moment $J_z$ about a $z$-axis through $O$ (Eq. 10-2) were any pair of orthogonal axes in the plane of the area passing through $O$; therefore,

$$J_z = I_x + I_y = I_{x'} + I_{y'}$$

where $x'$ and $y'$ are any pair of orthogonal axes through $O$. Since the sum of $I_{x'}$ and $I_{y'}$ is a constant, $I_{x'}$ will be maximum and the corresponding $I_{y'}$ will be minimum for one particular value of $\theta$.

The set of axes for which the second moments are maximum and minimum are called the principal axes of the area through point $O$ and are designated as the $u$- and $v$-axes. The second moments of the area with respect to these axes are called the principal second moments of the area (principal area moments of inertia) and are designated $I_u$ and $I_v$. There is only one set of principal axes for any point in an area unless all axes have the same second moment, such as the diameters of a circle. Principal axes are important in mechanics of materials courses, which deal with stresses and deformations in beams and columns.

The principal second moments of an area can be determined by expressing $I_{x'}$ as a function of $I_x$, $I_y$, $I_{xy}$, and $\theta$ and setting the derivative of $I_{x'}$ with respect to $\theta$ equal to zero to obtain the values of $\theta$ that give the maximum and minimum second moments. From Fig. 10-20,

$$dI_{x'} = y'^2 \, dA = (y \cos \theta - x \sin \theta)^2 \, dA$$

Therefore,

$$I_{x'} = \int_A dI_{x'}$$

$$= \cos^2 \theta \int_A y^2 \, dA + \sin^2 \theta \int_A x^2 \, dA - 2 \sin \theta \cos \theta \int_A xy \, dA$$

$$= I_x \cos^2 \theta + I_y \sin^2 \theta - 2I_{xy} \sin \theta \cos \theta \qquad (10\text{-}14)$$

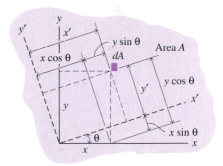

**Figure 10-20** Element of area $dA$ in the $xy$-plane.

Equation 10-14 can be expressed in terms of the double angle $2\theta$ by using the trigonometric identities

$$\sin 2\theta = 2 \sin \theta \cos \theta$$
$$\cos 2\theta = \cos^2 \theta - \sin^2 \theta$$

Thus,

$$I_{x'} = \frac{1}{2}(I_x + I_y) + \frac{1}{2}(I_x - I_y) \cos 2\theta - I_{xy} \sin 2\theta \quad \text{(10-15)}$$

The angle $2\theta$ for which $I_{x'}$ is a maximum (or a minimum) can be obtained by setting the derivative of $I_{x'}$ with respect to $\theta$ equal to zero; thus,

$$\frac{dI_{x'}}{d\theta} = -(I_x - I_y) \sin 2\theta - 2I_{xy} \cos 2\theta = 0$$

from which

$$\tan 2\theta_p = \frac{-2I_{xy}}{(I_x - I_y)} \quad \text{(10-16)}$$

where $\theta_p$ represents the two values of $\theta$ that locate the principal axes $u$ and $v$. Equation 10-16 gives two values of $2\theta_p$ that are 180° apart, and thus two values of $\theta_p$ that are 90° apart. The principal second moments can be obtained by substituting these values of $\theta_p$ in Eq. 10-14. Thus,

$$I_{u,v} = \frac{I_x + I_y}{2} \pm \left[ \left( \frac{I_x - I_y}{2} \right)^2 + I_{xy}^2 \right]^{1/2} \quad \text{(10-17)}$$

The mixed second moment (area product of inertia) of the element of area in Fig. 10-20 with respect to the $x'$- and $y'$-axes is

$$dI_{x'y'} = x'y' \, dA = (x \cos \theta + y \sin \theta)(y \cos \theta - x \sin \theta) \, dA$$

Therefore,

$$I_{x'y'} = \int_A dI_{x'y'}$$

$$= \sin \theta \cos \theta \int_A (y^2 - x^2) \, dA + (\cos^2 \theta - \sin^2 \theta) \int_A xy \, dA$$

$$= (I_x - I_y) \sin \theta \cos \theta + I_{xy}(\cos^2 \theta - \sin^2 \theta) \quad \text{(10-18)}$$

or, in terms of the double angle $2\theta$,

$$I_{x'y'} = I_{xy} \cos 2\theta + \frac{1}{2}(I_x - I_y) \sin 2\theta \quad \text{(10-19)}$$

The mixed second moment $I_{x'y'}$ will be zero when

$$\tan 2\theta = -\frac{2I_{xy}}{(I_x - I_y)} \quad \text{(a)}$$

Equation $a$ is the same as Eq. 10-16. This fact indicates that mixed second moments with respect to principal axes are zero. Since mixed second moments are zero with respect to any axis of symmetry, it follows that any axis of symmetry must be a principal axis for any point on the axis.

The following example problem illustrates the procedure for determining second moments of area with respect to the principal axes.

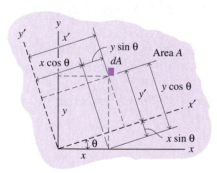

**Figure 10-20**   Element of area $dA$ in the $xy$-plane.

Determine the maximum and minimum second moments for the triangular area shown in Fig. 10-21 with respect to axes through the centroid of the area.

48 mm

24 mm

73.15°

16.85°

(a)          (b)

**Fig. 10-21**

## SOLUTION

The second moments $I_x$, $I_y$, and $I_{xy}$ can be determined by using the properties listed in Tables 10-1 and 10-4. Thus,

$$I_x = \frac{bh^3}{36} = \frac{24(48)^3}{36} = 73{,}728 \text{ mm}^4$$

$$I_y = \frac{hb^3}{36} = \frac{48(24)^3}{36} = 18{,}432 \text{ mm}^4$$

$$I_{xy} = \frac{b^2h^2}{72} = \frac{(24)^2(48)^2}{72} = 18{,}432 \text{ mm}^4$$

The principal angles $\theta_p$ are

$$\theta_p = \frac{1}{2} \tan^{-1}\left[-\frac{2I_{xy}}{I_x - I_y}\right]$$

$$= \frac{1}{2} \tan^{-1}\left[-\frac{2(18432)}{73728 - 18432}\right] = -16.85° \text{ or } 73.15°$$

With $\theta_p = -16.85°$, Eq. 10-14 yields

$$I_{x'} = I_x \cos^2 \theta + I_y \sin^2 \theta - 2I_{xy} \sin \theta \cos \theta$$
$$= 73{,}728 \cos^2 (-16.85°) + 18{,}432 \sin^2 (-16.85°)$$
$$- 2(18{,}432) \sin (-16.85°) \cos (-16.85°) = 79{,}309 \text{ mm}^4$$

With $\theta_p = 73.15°$, Eq. 10-14 yields

$$I_{x'} = I_x \cos^2 \theta + I_y \sin^2 \theta - 2I_{xy} \sin \theta \cos \theta$$
$$= 73{,}728 \cos^2 (73.15°) + 18{,}432 \sin^2 (73.15°)$$
$$- 2(18{,}432) \sin (73.15°) \cos (73.15°) = 12{,}851 \text{ mm}^4$$

Therefore, with respect to the x-axis,

$$I_u = I_{max} = 79{,}300 \text{ mm}^4 \text{ at } \theta_p = -16.85° \qquad \text{Ans.}$$
$$I_v = I_{min} = 12{,}850 \text{ mm}^4 \text{ at } \theta_p = 73.15° \qquad \text{Ans.}$$

> Second moments associated with principal axes are the maximum and minimum values for all axes in the plane of the area that pass through the point.

The principal second moments can also be determined by using Eq. 10-17:

$$I_{u,v} = \frac{I_x + I_y}{2} \pm \left[\left(\frac{I_x - I_y}{2}\right)^2 + I_{xy}^2\right]^{1/2}$$

$$= \frac{73{,}728 + 18{,}432}{2} \pm \left[\left(\frac{73{,}728 - 18{,}432}{2}\right)^2 + (18{,}432)^2\right]^{1/2}$$

$$= 46{,}080 \pm 33{,}229 = 79{,}309 \text{ mm}^4 \quad \text{and} \quad 12{,}851 \text{ mm}^4$$

The orientations of the principal axes are shown in Fig. 10-21b.

# PROBLEMS

Determine the maximum and minimum second moments of area with respect to axes through the origin of the *xy*-coordinate system and show the orientations of the principal axes on a sketch for

## Introductory Problems

**10-69*** The rectangle shown in Fig. P10-69.

**Fig. P10-69**

**10-70*** The angle section shown in Fig. P10-70.

**Fig. P10-70**

**10-71** The T-section shown in Fig. P10-71.

**Fig. P10-71**

**10-72** The Z-section shown in Fig. P10-72.

**Fig. P10-72**

## Intermediate Problems

**10-73*** The triangle shown in Fig. P10-73.

**Fig. P10-73**

**10-74** The triangle shown in Fig. P10-74.

**Fig. P10-74**

## 10-3.1 Mohr's Circle for Second Moments of Areas

The equations for locating principal axes and determining maximum and minimum second moments for an area were developed in Section 10-3. The German engineer Otto Mohr (1835–1918) developed a graphic interpretation of these equations that provides a visual picture of the relationships between the second moments, mixed moments, and principal moments of an area. This method, commonly called Mohr's circle, involves the construction of a circle in such a manner that the coordinates of each point on the circle represent the rectangular second moment $I_{x'}$ and the mixed second moment $I_{x'y'}$ for one orientation of the $x'$- and $y'$-axes. The proof that the second moments $I_{x'}$ and $I_{x'y'}$ for a given value of $\theta$ can be represented as a point on a circle follows from the general expressions for $I_{x'}$ and $I_{x'y'}$ developed in Section 10-3. Recall Eqs. 10-15 and 10-19 and rewrite in the following form

$$I_{x'} - \frac{I_x + I_y}{2} = \frac{I_x - I_y}{2} \cos 2\theta - I_{xy} \sin 2\theta$$

$$I_{x'y'} = \frac{I_x - I_y}{2} \sin 2\theta + I_{xy} \cos 2\theta$$

Squaring both equations, adding, and simplifying yields

$$\left(I_{x'} - \frac{I_x + I_y}{2}\right)^2 + I_{x'y'}^2 = \left(\frac{I_x - I_y}{2}\right)^2 + I_{xy}^2$$

This is the equation of a circle in terms of the variables $I_{x'}$ and $I_{x'y'}$. The circle is centered on the $I_{x'}$ axis at a distance $(I_x + I_y)/2$ from the $I_{x'y'}$ axis, and the radius of the circle is given by the expression

$$R = \left[\left(\frac{I_x - I_y}{2}\right)^2 + I_{xy}^2\right]^{1/2}$$

Construction of Mohr's circle is initiated by plotting two points $A$ and $B$ as shown in Fig. 10-22a. The coordinates of point $A$ are $(I_x, I_{xy})$. The coordinates of point $B$ are $(I_y, -I_{xy})$. In this figure, it has been assumed that $I_x$ is greater than $I_y$ and that $I_{xy}$ is positive. Rectangular second moments (area moments of inertia) are plotted along the horizontal axis. Mixed second moments (area products of inertia) are plotted along the vertical axis. Rectangular second moments are always positive; therefore, they are always plotted to the right of the origin. Mixed second moments can either be positive or negative: positive values are plotted above the horizontal axis; negative values are plotted below the horizontal axis. A line connecting $A$ and $B$ is a diameter of Mohr's circle, as shown on Fig. 10-22a. On a physical body, the angle between the $x$- and $y$-axes is 90°. On Mohr's circle, points $A$ and $B$ on the circle, which represent quantities associated with the $x$- and $y$-axes, are 180° apart. Therefore, all angles on Mohr's circle are double those on the physical body. The diameter intersects the horizontal axis at point $C$. The distance from the origin $O$ of the coordinate system to the center $C$ of Mohr's circle is

$$OC = \frac{I_x + I_y}{2}$$

(a)

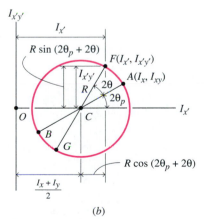

(b)

**Figure 10-22** Mohr's circle for second moments.

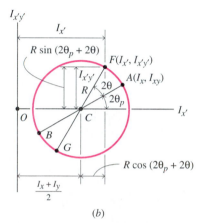

**Figure 10-22**  Mohr's circle for second moments.

The radius $R$ of the circle is

$$R = OA = \left[\left(\frac{I_x - I_y}{2}\right)^2 + I_{xy}^2\right]^{1/2}$$

Maximum and minimum values of $I_{x'}$ occur at points $D$ and $E$, respectively. At point $D$,

$$I_{max} = OC + R = \frac{I_x + I_y}{2} + \left[\left(\frac{I_x - I_y}{2}\right)^2 + I_{xy}^2\right]^{1/2} \qquad (a)$$

At point $E$,

$$I_{min} = OC - R = \frac{I_x - I_y}{2} - \left[\left(\frac{I_x - I_y}{2}\right)^2 + I_{xy}^2\right]^{1/2} \qquad (b)$$

On Mohr's circle, angle $ACD$ locates $I_{max}$ with respect to $I_x$. Thus,

$$ACD = \tan^{-1}\frac{I_{xy}}{(I_x - I_y)/2} = \tan^{-1}\frac{2I_{xy}}{(I_x - I_y)} \quad \text{clockwise} \qquad (c)$$

The rectangular second moment $I_{x'}$ and the mixed second moment $I_{x'y'}$ for an axis passing through a point $O$ in a body and oriented at an angle $\theta$ with respect to the $x$-axis can be obtained by using the Mohr's circle shown in Fig. 10-22b. Thus,

$$\begin{aligned}
I_{x'} &= OC + R\cos(2\theta_p + 2\theta)\\
&= OC + R\cos 2\theta_p \cos 2\theta - R\sin 2\theta_p \sin 2\theta\\
&= \tfrac{1}{2}(I_x + I_y) + \tfrac{1}{2}(I_x - I_y)\cos 2\theta - I_{xy}\sin 2\theta \qquad (d)\\
I_{x'y'} &= R\sin(2\theta_p + 2\theta)\\
&= R\cos 2\theta_p \sin 2\theta + R\sin 2\theta_p \cos 2\theta\\
&= \tfrac{1}{2}(I_x - I_y)\sin 2\theta + I_{xy}\cos 2\theta \qquad (e)
\end{aligned}$$

Equations $a$, $b$, $c$, $d$, and $e$, obtained using Mohr's circle, are identical to Eqs. 10-15, 10-16, 10-17, and 10-19, obtained by theoretical analysis. Figure 10-23 represents an area and a set of axes for which the data used to construct Fig. 10-22 are valid.

The following example problem illustrates the procedure for locating the principal axes and determining the maximum and minimum second moments of area by using Mohr's circle.

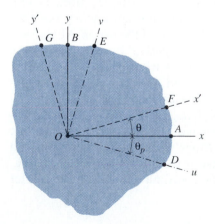

**Figure 10-23**  Principal axes and their orientation.

Determine the maximum and minimum second moments for the cross-sectional area of the unequal-leg angle, shown in Fig. 10-24a, with respect to axes through the centroid $C$ of the area.

Fig. 10-24

---

### SOLUTION

The location of the centroid of the angle with respect to the $x'$- and $y'$-axes of Fig. 10-24a are

$$x_C = \frac{60(10)(30) + 70(10)(5)}{60(10) + 70(10)} = 16.54 \text{ mm}$$

$$y_C = \frac{80(10)(40) + 50(10)(5)}{80(10) + 50(10)} = 26.54 \text{ mm}$$

The area is divided into two rectangles $A$ and $B$. The location of the centroid of each of these areas with respect to the centroid of the composite area is indicated on Fig. 10-24b. The values $I_x$, $I_y$, and $I_{xy}$ for the two areas are obtained by applying the parallel-axis theorem to the expressions for $I_{xC}$, $I_{yC}$, and $I_{xyC}$ from Tables 10-1 and 10-3. Thus,

$$I_x = \frac{bh^3}{12} + A(d_x^2)^2$$

$$I_x = I_{xA} + I_{xB}$$
$$= \frac{80(10)^3}{12} + 80(10)(13.46)^2 + \frac{50(10)^3}{12} + 50(10)(21.54)^2 = 808(10^3) \text{ mm}^4$$

$$I_y = I_{yA} + I_{yB}$$
$$= \frac{10(80)^3}{12} + 10(80)(11.54)^2 + \frac{10(50)^3}{12} + 10(50)(18.46)^2 = 388(10^3) \text{ mm}^4$$

$$I_{xy} = I_{xyC} + d_x d_y A = 0 + d_x d_y A$$

$$I_{xy} = I_{xyA} + I_{xyB}$$
$$= 13.46(-11.54)(10)(80) + (-21.54)(18.46)(50)(10) = -323(10^3) \text{ mm}^4$$

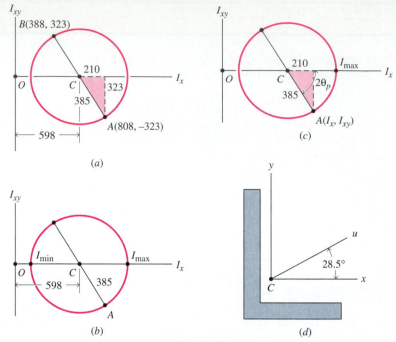

Fig. 10-25

Mohr's circle, shown in Fig. 10-25a, is constructed using these values. Point A (808, −323) is located at $I_x$ and $I_{xy}$ (which is negative), and point B (388, 323) is located at $I_y$ and $-I_{xy}$. A line drawn between points A and B locates point C (the center of the circle). The distance OC is

$$OC = \frac{I_x + I_y}{2} = \frac{(808 + 388)(10^3)}{2} = 598(10^3) \text{ mm}^4$$

The radius of the circle is

$$R = \sqrt{(210)^2 + (323)^2} \; (10^3) = 385(10^3) \text{ mm}^4$$

Thus, as shown in Fig. 10-25b, the principal second moments are

$$I_{max} = OC + R = (598 + 385)(10^3) = 983(10^3) \text{ mm}^4 \qquad \text{Ans.}$$
$$I_{min} = OC - R = (598 - 385)(10^3) = 213(10^3) \text{ mm}^4 \qquad \text{Ans.}$$

Twice the angle from the x-axis to the principal axis with the maximum second moment is shown on Fig. 10-25c as $2\theta_p$. From the information provided on Mohr's circle,

$$\theta_p = \frac{1}{2} \cos^{-1} \frac{210}{385} = 28.5° \qquad \text{counterclockwise from the x-axis}$$

Results of this Mohr circle analysis are shown in Fig. 10-25d. These values were obtained analytically from the geometry of Mohr's circle. They could also have been measured directly from the figure; however, the accuracy of the results obtained by scaling distances from the figure would depend on the scale used and the care employed in constructing the figure.

Each point on Mohr's circle represents $I_x$ and $I_{xy}$ for one particular set of x and y axes through the point.

All angles on Mohr's circle are double those on the physical body.

The two points where Mohr's circle intersects the horizontal axis represent maximum and minimum second moments with respect to axes through the centroid of the angle.

582

# PROBLEMS

Use Mohr's circle to determine the maximum and minimum second moments of area with respect to axes through the origin of the $xy$-coordinate system and the orientations of the principal axes for

## Introductory Problems

**10-75\*** The composite section shown in Fig. P10-75.

**Fig. P10-75**

**10-76\*** The T-section shown in Fig. P10-76.

**Fig. P10-76**

**10-77** The composite section shown in Fig. P10-77.

**Fig. P10-77**

**10-78** The Z-section shown in Fig. P10-78.

**Fig. P10-78**

## Intermediate Problems

**10-79\*** The triangle shown in Fig. P10-79.

**Fig. P10-79**

**10-80** The triangle shown in Fig. P10-80.

**Fig. P10-80**

**Figure 10-26**   Element of mass $dm$ at a distance $r$ from axis $OO$.

## 10-4   MOMENTS OF INERTIA

In analyses of the motion of rigid bodies, expressions are often encountered that involve the product of the mass of a small element of the body and the square of its distance from a line of interest. This product is called the second moment of the mass of the element or more frequently the *moment of inertia* of the element. Thus, the moment of inertia $dI$ of an element of mass $dm$ about the axis $OO$ shown in Fig. 10-26 is defined as

$$dI = r^2 \, dm$$

The moment of inertia of the entire body about axis $OO$ is defined as

$$I = \int_m r^2 \, dm \qquad (10\text{-}20)$$

Since both the mass of the element and the distance squared from the axis to the element are always positive, the moment of inertia of a mass is always a positive quantity.

Moments of inertia have the dimensions of mass multiplied by length squared, $ML^2$. Common units for the measurement of moment of inertia in the SI system are $kg \cdot m^2$. In the U.S. Customary system, force, length, and time are selected as the fundamental quantities, and mass has the dimensions $FT^2L^{-1}$. Therefore, moment of inertia has the units $lb \cdot s^2 \cdot ft$. If the mass of the body $W/g$ is expressed in slugs ($lb \cdot s^2/ft$), the units for measurement of moment of inertia in the U.S. Customary system are $slug \cdot ft^2$.

The moments of inertia of a body with respect to an $xyz$-coordinate system can be determined by considering an element of mass as shown in Fig. 10-27. From the definition of moment of inertia,

$$dI_x = r_x^2 \, dm = (y^2 + z^2) \, dm$$

Similar expressions can be written for the $y$- and $z$-axes. Thus,

$$I_x = \int_m r_x^2 \, dm = \int_m (y^2 + z^2) \, dm$$

$$I_y = \int_m r_y^2 \, dm = \int_m (z^2 + x^2) \, dm \qquad (10\text{-}21)$$

$$I_z = \int_m r_z^2 \, dm = \int_m (x^2 + y^2) \, dm$$

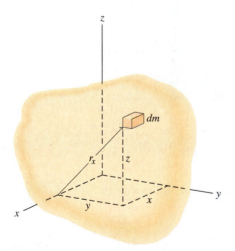

**Figure 10-27**   Location of an element of mass $dm$ with respect to a Cartesian coordinate system.

### 10-4.1   Radius of Gyration

The definition of moment of inertia (Eq. 10-20) indicates that the dimensions of moment of inertia are mass multiplied by a length squared. As a result, the moment of inertia of a body can be expressed as the product of the mass $m$ of the body and a length $k$ squared. This length $k$ is defined as the *radius of gyration* of the body. Thus, the moment of inertia $I$ of a body with respect to a given line can be expressed as

$$I = mk^2 \qquad or \qquad k = \sqrt{I/m} \qquad (10\text{-}22)$$

The radius of gyration of the mass of a body with respect to any axis can be viewed as the distance from the axis to the point where the total mass must be concentrated to produce the same moment of inertia with respect to the axis as does the actual (or distributed) mass.

The radius of gyration for masses is very similar to the radius of gyration for areas discussed in Section 10-2.3. The radius of gyration for masses is not the distance from the given axis to any fixed point in the body such as the mass center. The radius of gyration of the mass of a body with respect to any axis is always greater than the distance from the axis to the mass center of the body. There is no useful physical interpretation for a radius of gyration; it is merely a convenient means of expressing the moment of inertia of the mass of a body in terms of its mass and a length.

## 10-4.2   Parallel-Axis Theorem for Moments of Inertia

The parallel-axis theorem for moments of inertia is very similar to the parallel-axis theorem for second moments of area discussed in Section 10-2.1. Consider the body shown in Fig. 10-28a, which has an $xyz$-coordinate system with its origin at the mass center $G$ of the body and a parallel $x'y'z'$-coordinate system with its origin at point $O'$. Observe in Fig. 10-28b that

$$x' = x_G + x$$
$$y' = y_G + y$$
$$z' = z_G + z$$

The distance $d_x$ (see Fig. 10-28a) between the $x'$- and $x$-axes is

$$d_x = \sqrt{y_G^2 + z_G^2}$$

the moment of inertia of the body about an $x'$-axis that is parallel to the $x$-axis through the mass center (see Fig. 10-28b) is by definition

$$I_{x'} = \int_m r_{x'}^2 \, dm$$

$$= \int_m [(y')^2 + (z')^2] \, dm$$

$$= \int_m [(y_G + y)^2 + (z_G + z)^2] \, dm$$

$$= \int_m (y^2 + z^2) \, dm + y_G^2 \int_m dm + 2y_G \int_m y \, dm + z_G^2 \int_m dm + 2z_G \int z \, dm$$

However,

$$\int_m (y^2 + z^2) \, dm = I_{xG}$$

and, since the $x$- and $y$-axes pass through the mass center $G$ of the body,

$$\int_m y \, dm = 0 \quad \int_m z \, dm = 0$$

Therefore,

$$I_{x'} = I_{xG} + (y_G^2 + z_G^2)m = I_{xG} + d_x^2 m$$
$$I_{y'} = I_{yG} + (z_G^2 + x_G^2)m = I_{yG} + d_y^2 m \quad \text{(10-23)}$$
$$I_{z'} = I_{zG} + (x_G^2 + y_G^2)m = I_{zG} + d_z^2 m$$

Equation 10-23 is the parallel-axis theorem for moments of inertia. The subscript $G$ indicates that the $x$-axis passes through the mass center $G$ of the body. Thus, if the moment of inertia of a body with respect to an axis passing through its mass center is known, the moment

(a)

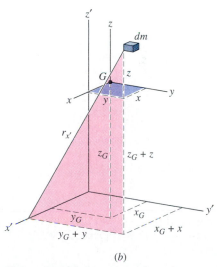

(b)

**Figure 10-28**   Location of an element of mass $dm$ with respect to a Cartesian coordinate system.

of inertia of the body with respect to any parallel axis can be found, without integrating, by use of Eqs. 10-23.

A similar relationship exists between the radii of gyration for the two axes. Thus, if the radii of gyration for the two parallel axes are denoted by $k_x$ and $k_{x'}$, the foregoing equation may be written

$$k_{x'}^2 m = k_{xG}^2 m + d_x^2 m$$

Hence,

$$k_{x'}^2 = k_{xG}^2 + d_x^2$$
$$k_{y'}^2 = k_{yG}^2 + d_y^2 \qquad (10\text{-}24)$$
$$k_{z'}^2 = k_{zG}^2 + d_z^2$$

(*Note:* Equations 10-23 and 10-24 are valid only for transfers to or from *xyz*-axes passing through the mass center of the body. They are not valid for two arbitrary axes.)

### 10-4.3   Moments of Inertia by Integration

When integration methods are used to determine the moment of inertia of a body with respect to an axis, the mass of the body can be divided into elements in various ways. Depending on the way the element is chosen, single, double, or triple integration may be required. The geometry of the body usually determines whether Cartesian or polar coordinates are used. In any case, the elements of mass should always be selected so that

1. All parts of the element are the same distance from the axis with respect to which the moment of inertia is to be determined.
2. If condition 1 is not satisfied, the element should be selected so that the moment of inertia of the element with respect to the axis about which the moment of inertia of the body is to be found is known. The moment of inertia of the body can then be found by summing the moments of inertia of the elements.
3. If the location of the mass center of the element is known and the moment of inertia of the element with respect to an axis through its mass center and parallel to the given axis is known, the moment of inertia of the element can be determined by using the parallel-axis theorem. The moment of inertia of the body can then be found by summing the moments of inertia of the elements.

When triple integration is used, the element always satisfies the first requirement, but this condition is not necessarily satisfied by elements used for single or double integration.

In some instances, a body can be regarded as a system of particles. The moment of inertia of a system of particles with respect to a line of interest is the sum of the moments of inertia of the particles with respect to the given line. Thus, if the masses of the particles of a system are denoted by $m_1, m_2, m_3, \ldots, m_n$, and the distances of the particles from a given line are denoted by $r_1, r_2, r_3, \ldots, r_n$, the moment of inertia of the system can be expressed as

$$I = \Sigma mr^2 = m_1 r_1^2 + m_2 r_2^2 + m_3 r_3^2 + \cdots + m_n r_n^2$$

Moments of inertia for thin plates are relatively easy to determine. For example, consider the thin plate shown in Fig. 10-29. The plate has a uniform density $\rho$, a uniform thickness $t$, and a cross-sectional area $A$. The moments of inertia about $x$-, $y$-, and $z$-axes are by definition

$$I_{xm} = \int_m y^2 \, dm = \int_V y^2\rho \, dV = \int_A y^2\rho t \, dA = \rho t \int_A y^2 \, dA = \rho t I_{xA}$$

$$I_{ym} = \int_m x^2 \, dm = \int_V x^2\rho \, dV = \int_A x^2\rho t \, dA = \rho t \int_A x^2 \, dA = \rho t I_{yA} \quad \textbf{(10-25)}$$

$$I_{zm} = \int_m (x^2 + y^2) \, dm = \rho t I_{yA} + \rho t I_{xA} = \rho t (I_{yA} + I_{xA})$$

where the subscripts $m$ and $A$ denote moments of inertia and second moments of area, respectively. Since the equations for the moments of inertia of thin plates contain the expressions for the second moments of area, the results listed in Table 10-1 for second moments of areas can be used for moments of inertia by simply multiplying the results listed in the table by $\rho t$.

For the general three-dimensional body, moments of inertia with respect to $x$-, $y$-, and $z$-axes are

$$I_x = \int_m r_x^2 \, dm = \int_m (y^2 + z^2) \, dm$$

$$I_y = \int_m r_y^2 \, dm = \int_m (z^2 + x^2) \, dm \quad \textbf{(10-21)}$$

$$I_z = \int_m r_z^2 \, dm = \int_m (x^2 + y^2) \, dm$$

**Figure 10-29**   Element of mass $dm$ in a thin plate.

If the density of the body is uniform, the element of mass $dm$ can be expressed in terms of the element of volume $dV$ of the body as $dm = \rho \, dV$. Equations 10-21 then become

$$I_x = \rho \int_V (y^2 + z^2) \, dV$$

$$I_y = \rho \int_V (z^2 + x^2) \, dV \quad \textbf{(10-26)}$$

$$I_z = \rho \int_V (x^2 + y^2) \, dV$$

If the density of the body is not uniform, it must be expressed as a function of position and retained within the integral sign.

The specific element of volume to be used depends on the geometry of the body. For the general three-dimensional body, the differential element $dV = dx \, dy \, dz$, which requires a triple integration, is usually used. For bodies of revolution, circular plate elements, which require only a single integration, can be used. For some problems, cylinder elements and polar coordinates are useful. Procedures for determining moments of inertia are illustrated in the following example problems.

Determine the moment of inertia of a homogeneous right circular cylinder with respect to the axis of the cylinder.

**Fig. 10-30**

## SOLUTION

The moment of inertia of the cylinder can be determined from the definition of moment of inertia (Eq. 10-20) by selecting a cylindrical tube type of element as shown in Fig. 10-30a. Thus,

$$dI_{zm} = r^2 \, dm = r^2(\rho \, dV) = r^2\rho(2\pi rh \, dr) = 2\pi\rho hr^3 \, dr$$

> This problem can be solved with a single integration if a thin annular element at a constant distance $r$ from the $z$-axis is used.

Therefore,

$$I_z = \int_m dI_{zm} = \int_0^R 2\pi\rho hr^3 \, dr = \left[\frac{\pi\rho hr^4}{2}\right]_0^R = \frac{1}{2}\pi\rho hR^4$$

Alternatively, a thin circular plate type of element, such as the one shown in Fig. 10-30b, can be used. The moment of inertia for this type of element is given by Eq. 10-25 as

$$dI_{zm} = \rho t(I_{yA} + I_{xA})$$

> This problem can also be solved with a single integration if a thin circular disk element is used.

Substituting the second moment for a circular area from Table 10-1 yields

$$dI_{zm} = \rho\left(\frac{\pi R^4}{4} + \frac{\pi R^4}{4}\right) dz = \frac{1}{2}\pi\rho R^4 \, dz$$

Therefore,

$$I_z = \int_m dI_{zm} = \int_0^h \frac{1}{2}\pi\rho R^4 \, dz = \left[\frac{1}{2}\pi\rho R^4 z\right]_0^h = \frac{1}{2}\pi\rho hR^4$$

The mass of the cylinder is

$$m = \rho V = \rho(\pi R^2 h) = \rho\pi R^2 h$$

Therefore,

$$I_z = \frac{1}{2}(\rho\pi R^2 h)R^2 = \frac{1}{2}mR^2 \qquad\qquad \text{Ans.}$$

Determine the moment of inertia for the homogeneous rectangular prism shown in Fig. 10-31a with respect to

a.  Axis $y$ through the mass center of the prism.
b.  Axis $y'$ along an edge of the prism.
c.  Axis $x$ through the centroid of an end of the prism.

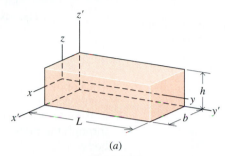

(a)

**Fig. 10-31**

## SOLUTION

a.  A thin rectangular plate type of element, such as the one shown in Fig. 10-31b, will be used. The moment of inertia for this type of element is given by Eq. 10-25 as

$$dI_{ym} = \rho t(I_{zA} + I_{xA})$$

Substituting the second moments for a rectangular area from Table 10-1 yields

> Moments of inertia for thin plates can be determined by using known results for second moments of area.

$$dI_{ym} = \rho \left( \frac{hb^3}{12} + \frac{bh^3}{12} \right) dy = \rho \frac{bh}{12} (b^2 + h^2)\, dy$$

Therefore,

$$I_y = \int_m dI_{ym} = \int_0^L \rho \frac{bh}{12} (b^2 + h^2)\, dy$$

$$= \rho \frac{bh}{12} \left[ (b^2 + h^2)\, y \right]_0^L = \frac{\rho bhL}{12} (b^2 + h^2)$$

The mass of the prism is

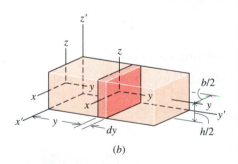

(b)

$$m = \rho V = \rho(bhL) = \rho bhL$$

Therefore,

$$I_y = \frac{\rho bhL}{12} (b^2 + h^2) = \frac{1}{12} m(b^2 + h^2) \qquad \text{Ans.}$$

b.  The parallel-axis theorem (Eq. 10-23) can be used to determine the moment of inertia about the $y'$-axis along an edge of the prism. Thus,

$$I_{y'} = I_{yG} + (z_G^2 + x_G^2)m$$

$$= \frac{1}{12} m(b^2 + h^2) + \left( \frac{h^2}{4} + \frac{b^2}{4} \right) m$$

$$= \frac{1}{3} m(b^2 + h^2)$$

**c.** The moment of inertia about an $x$-axis through the mass center of the thin rectangular plate type of element shown in Fig. 10-31$b$ is given by Eq. 10-25 as

$$dI_{xm} = \rho t I_{xA}$$

Substituting the second moment for a rectangular area from Table 10-1 yields

$$dI_{xG} = \rho \frac{bh^3}{12} \, dy = \frac{\rho bh^3}{12} \, dy$$

The parallel-axis theorem (Eq. 10-23) with $d_x = y$ then gives the moment of inertia for the thin rectangular plate element about the $x$-axis shown in Fig. 10-31$b$ as

$$dI_x = dI_{xG} + d_x^2 m = \frac{\rho bh^3}{12} \, dy + y^2(\rho bh \, dy) = \frac{\rho bh}{12} \, (h^2 + 12y^2) \, dy$$

$$I_x = \int_m dI_x = \int_0^L \frac{\rho bh}{12} \, (h^2 + 12y^2) \, dy$$

$$= \frac{\rho bh}{12} \left[ h^2 y + 4y^3 \right]_0^L = \frac{\rho bhL}{12} \, (h^2 + 4L^2)$$

(b)

**Fig. 10-31**

> The parallel-axis theorem for moments of inertia is valid only for transfers to or from parallel axes through the mass center of the body.

But

$$m = \rho bhL$$

Therefore,

$$I_x = \frac{\rho bhL}{12} \, (h^2 + 4L^2) = \frac{1}{12} \, m \, (h^2 + 4L^2) \qquad \text{Ans.}$$

---

# PROBLEMS

### Introductory Problems

**10-81\*** Determine the amount of inertia of the homogeneous right circular cone, shown in Fig. P10-81, with respect to the axis of the cone.

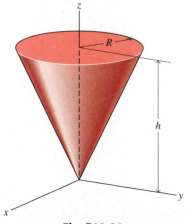

**Fig. P10-81**

**10-82\*** Determine the moment of inertia of the homogenous right circular cone, shown in Fig. P10-81, with respect to an axis perpendicular to the axis of the cone at the apex of the cone.

**10-83** Determine the moment of inertia of the homogeneous right circular cone, shown in Fig. P10-81, with respect to an axis perpendicular to the axis of the cone at the base of the cone.

**10-84** Determine the amount of inertia of a solid homogeneous sphere of radius $R$ with respect to a diameter of the sphere.

**10-85\*** Determine the moment of inertia of a solid homogeneous cylinder of radius $R$ and length $L$ with respect to a diameter in the base of the cylinder.

**10-86\*** Determine the moment of inertia of the solid homogeneous hemisphere shown in Fig. P10-86 with respect to the $x$-axis shown on the figure.

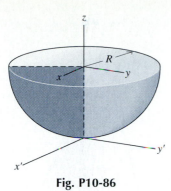

**Fig. P10-86**

**10-87** Determine the moment of inertia of the solid homogeneous hemisphere shown in Fig. P10-86 with respect to the $x'$-axis shown on the figure.

**10-88** A homogeneous solid revolution is formed by revolving the area shown in Fig. P10-88 around the $y$-axis. Determine the moment of inertia of the body with respect to the $y$-axis.

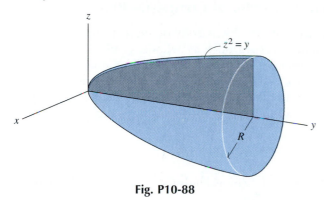

**Fig. P10-88**

**10-89*** A homogeneous solid of revolution is formed by revolving the area shown in Fig. P10-88 around the $y$-axis. Determine the moment of inertia of the body with respect to the $x$-axis.

**10-90** A homogeneous octant of a sphere is formed by rotating the quarter circle shown in Fig. P10-90 for 90° around the $z$-axis. Determine the moment of inertia of the body with respect to a $y$-axis through the mass center of the body.

**Fig. P10-90**

## Intermediate Problems

**10-91*** Determine the moment of inertia of the solid homogeneous triangular prism shown in Fig. P10-91 with respect to the $x$-axis shown on the figure.

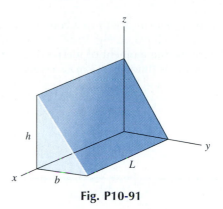

**Fig. P10-91**

**10-92*** Determine the moment of inertia of the solid homogeneous triangular prism shown in Fig. P10-91 with respect to a $y$-axis through the mass center of the prism.

**10-93** Determine the moment of inertia of the solid homogeneous tetrahedron shown in Fig. P10-93 with respect to the $x$-axis shown on the figure.

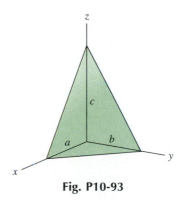

**Fig. P10-93**

**10-94** Determine the moment of inertia of the solid homogeneous tetrahedron shown in Fig. P10-93 with respect to a $y$-axis through the mass center of the body.

**10-95*** A homogeneous octant of a cone is formed by rotating the triangle shown in Fig. P10-95 for 90° around the z-axis. Determine the moment of inertia of the body with respect to the y-axis shown on the figure.

**10-96** A homogeneous octant of a cone is formed by rotating the triangle shown in Fig. P10-95 for 90° around the z-axis. Determine the moment of inertia of the body with respect to an x-axis through the mass center of the body.

**Fig. P10-95**

### 10-4.4   Moment of Inertia of Composite Bodies

Frequently in engineering practice, a body of interest can be broken up into a number of simple shapes, such as cylinders, spheres, plates, and rods, for which the moments of inertia have been evaluated and tabulated. The moment of inertia of the *composite body* with respect to any axis is equal to the sum of the moments of inertia of the separate parts of the body with respect to the specified axis. For example,

$$I_x = \int_m (y^2 + z^2) \, dm$$

$$= \int_{m_1} (y^2 + z^2) \, dm_1 + \int_{m_2} (y^2 + z^2) \, dm_2 + \cdots + \int_{m_n} (y^2 + z^2) \, dm_n$$

$$= I_{x1} + I_{x2} + I_{x3} + \cdots + I_{xn}$$

When one of the component parts is a hole, its moment of inertia must be subtracted from the moment of inertia of the larger part to obtain the moment of inertia for the composite body. Table 10-4 contains a listing of the moments of inertia for some frequently encountered shapes such as rods, plates, cylinders, spheres, and cones.

## TABLE 10-4 MOMENTS OF INERTIA OF COMMON SHAPES

**Slender rod**

$$I_x = 0$$
$$I_y = I_z = \frac{1}{12} mL^2$$

**Thin rectangular plate**

$$I_x = \frac{1}{12}m(b^2 + h^2)$$
$$I_y = \frac{1}{12}mb^2$$
$$I_z = \frac{1}{12}mh^2$$

**Thin circular plate**

$$I_x = \frac{1}{2}mR^2$$
$$I_y = I_z = \frac{1}{4}mR^2$$

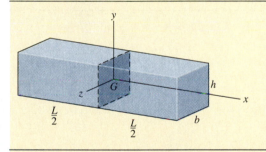

**Rectangular prism**

$$V = bhL$$

$$I_x = \frac{1}{12}m(b^2 + h^2)$$
$$I_y = \frac{1}{12}m(b^2 + L^2)$$
$$I_z = \frac{1}{12}m(h^2 + L^2)$$

**Right circular cone**

$$V = \frac{1}{3}\pi R^2 h$$
$$x_G = \frac{3}{4}h$$

$$I_x = \frac{3}{10}mR^2$$
$$I_y = I_z = \frac{3}{20}m(R^2 + 4h^2)$$
$$I_{yG} = I_{zG} = \frac{3}{80}m(4R^2 + h^2)$$

**TABLE 10-4** (CONTINUED)

Circular cylinder

$$V = \pi R^2 L$$

$$I_x = \frac{1}{2}mR^2$$

$$I_y = I_z = \frac{1}{12}m(3R^2 + L^2)$$

Hemisphere

$$V = \frac{2}{3}\pi R^3$$

$$x_G = \frac{3}{8}R$$

$$I_x = I_y = I_z = \frac{2}{5}mR^2$$

$$I_{yG} = I_{zG} = \frac{83}{320}mR^2$$

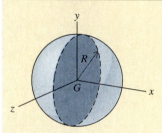

Sphere

$$V = \frac{4}{3}\pi R^3$$

$$I_x = I_y = I_z = \frac{2}{5}mR^2$$

Determine the moment of inertia of the cast-iron flywheel shown in Fig. 10-32 with respect to the axis of rotation of the flywheel. The specific weight of the cast iron is 460 lb/ft³.

**Fig. 10-32**

## SOLUTION

The rim and hub of the flywheel are hollow cylinders, and the spokes are rectangular prisms. The density of the cast iron is

$$\rho = \frac{w}{8} = \frac{460}{32.2} = 14.29 \text{ slugs/ft}^3$$

With all dimensions converted to feet, the moment of inertia of the rim is

$$I_R = 1/2\, m_0 R_0^2 - 1/2\, m_i R_i^2$$

$$= \frac{1}{2}\left[\pi\left(\frac{34}{12}\right)^2\left(\frac{16}{12}\right)(14.29)\right]\left(\frac{34}{12}\right)^2 - \frac{1}{2}\left[\pi\left(\frac{30}{12}\right)^2\left(\frac{16}{12}\right)(14.29)\right]\left(\frac{30}{12}\right)^2$$

$$= 1929 - 1169 = 760 \text{ slug} \cdot \text{ft}^2$$

The moment of inertia of the hub is

$$I_H = 1/2\, m_0 R_0^2 - 1/2\, m_i R_i^2$$

$$= \frac{1}{2}\left[\pi\left(\frac{10}{12}\right)^2\left(\frac{12}{12}\right)(14.29)\right]\left(\frac{10}{12}\right)^2 - \frac{1}{2}\left[\pi\left(\frac{6}{12}\right)^2\left(\frac{12}{12}\right)(14.29)\right]\left(\frac{6}{12}\right)^2$$

$$= 10.82 - 1.40 = 9.42 \text{ slug} \cdot \text{ft}^2$$

The moment of inertia of each spoke is

$$I_S = I_G + d^2 m$$

$$= \frac{1}{12}\left[\frac{3}{12}\left(\frac{4}{12}\right)\left(\frac{20}{12}\right)(14.29)\right]\left[\left(\frac{3}{12}\right)^2 + \left(\frac{20}{12}\right)^2\right]$$
$$+ \left(\frac{20}{12}\right)^2\left[\frac{3}{12}\left(\frac{4}{12}\right)\left(\frac{20}{12}\right)(14.29)\right]$$

$$= 0.4698 + 5.5131 = 5.9829 \cong 5.98 \text{ slug} \cdot \text{ft}^2$$

The total moment of inertia for the flywheel is

$$I = I_R + I_H + 6I_S$$
$$= 760 + 9.42 + 6(5.98) = 805 \text{ slug} \cdot \text{ft}^2 \qquad \textbf{Ans.}$$

The moment of inertia of a composite body with respect to a given axis equals the sum of the moments of inertia of the separate parts of the body with respect to the same axis.

# PROBLEMS

## Introductory Problems

**10-97\*** A composite body is constructed by attaching a steel ($w = 490$ lb/ft$^3$) hemisphere to an aluminum ($w = 175$ lb/ft$^3$) right circular cone as shown in Fig. P10-97. Determine the moment of inertia of the composite body with respect to the $y$-axis shown on the figure.

**Fig. P10-97**

**10-98\*** A composite body consists of a rectangular brass ($\rho = 8.75$ Mg/m$^3$) block attached to a steel ($\rho = 7.87$ Mg/m$^3$) cylinder as shown in Fig. P10-98. Determine the moment of inertia of the composite body with respect to the $y$-axis shown on the figure.

**Fig. P10-98**

**10-99** A composite body is constructed by attaching a steel ($w = 490$ lb/ft$^3$) hemisphere to an aluminum ($w = 175$ lb/ft$^3$) right circular cone as shown in Fig. P10-97. Determine the moment of inertia of the composite body with respect to the $x$-axis shown on the figure.

**10-100** A composite body consists of a cylinder attached to a rectangular block as shown in Fig. P10-98. Determine the moment of inertia of the composite body with respect to the $x$-axis shown on the figure if the body is made of cast iron ($\rho = 7.37$ Mg/m$^3$).

## Intermediate Problems

**10-101\*** Two steel ($w = 490$ lb/ft$^3$) cylinders and a brass ($w = 546$ lb/ft$^3$) sphere form the composite body shown in Fig. P10-101. Determine the moment of inertia of the composite body with respect to the $x$-axis shown on the figure.

**Fig. P10-101**

**10-102\*** Determine the moment of inertia of the composite body shown in Fig. P10-102 with respect to the $x$-axis shown on the figure. The density of the material is 7.87 Mg/m$^3$.

**Fig. P10-102**

**10-103** Two steel ($w = 490$ lb/ft$^3$) cylinders and an aluminum ($w = 173$ lb/ft$^3$) sphere form the composite body shown in Fig. P10-101. Determine the moment of inertia of the composite body with respect to the $z$-axis shown on the figure.

**10-104** Determine the moment of inertia of the composite body shown in Fig. P10-102 with respect to the $z$-axis shown on the figure. The density of the material is 7.87 Mg/m$^3$.

## Challenging Problems

**10-105\*** Determine the moment of inertia of the composite body shown in Fig. P10-105 with respect to the $y$-axis shown on the figure. The specific weight of the material is 175 lb/ft$^3$.

Fig. P10-105

**10-106*** Determine the moment of inertia of the composite body shown in Fig. P10-106 with respect to the *y*-axis shown on the figure. The density of the material is 7.87 Mg/m³.

Fig. P10-106

**10-107** Determine the moment of inertia of the composite body shown in Fig. P10-107 with respect to the *y*-axis shown on the figure. The specific weight of the material is 553 lb/ft³.

Fig. P10-107

**10-108** Determine the moment of inertia of the composite body shown in Fig. P10-106 with respect to the *x*-axis shown on the figure. The density of the material is 8.75 Mg/m³.

**Figure 10-33**   Location of an element of mass $dm$ with respect to the $xz$- and $yz$-planes.

## 10-4.5   Product of Inertia

In analyses of the motion of rigid bodies, expressions are sometimes encountered that involve the product of the mass of a small element and the coordinate distances from a pair of orthogonal coordinate planes. This product, which is similar to the mixed second moment of an area, is called the *product of inertia* of the element. For example, the product of inertia of the element shown in Fig. 10-33 with respect to the $xz$- and $yz$-planes is by definition

$$dI_{xy} = xy\,dm \tag{10-27}$$

The sum of the products of inertia of all elements of mass of the body with respect to the same orthogonal planes is defined as the product of inertia of the body. The three products of inertia for the body shown in Fig. 10-33 are

$$
\begin{aligned}
I_{xy} &= \int_m xy\,dm \\[2mm]
I_{yz} &= \int_m yz\,dm \\[2mm]
I_{zx} &= \int_m zx\,dm
\end{aligned}
\tag{10-28}
$$

Products of inertia, like moments of inertia, have the dimensions of mass multiplied by a length squared, $ML^2$. Common units for the measurement of product of inertia in the SI system are kg · m². In the U.S. Customary system, common units are slug · ft².

The product of inertia of a body can be positive, negative, or zero, since the two coordinate distances have independent signs. The product of inertia will be positive for coordinates with the same sign and negative for coordinates with opposite signs. The product of inertia will be zero if either of the planes is a plane of symmetry, since pairs of elements on opposite sides of the plane of symmetry will have positive and negative products of inertia that will add to zero in the summation process.

The integration methods used to determine moments of inertia apply equally well to products of inertia. Depending on the way the element is chosen, single, double, or triple integration may be required. Moments of inertia for thin plates were related to second moments of area for the same plate. Likewise, products of inertia can be related to the mixed second moments for the plates. If the plate has a uniform density $\rho$, a uniform thickness $t$, and a cross-sectional area $A$, the products of inertia are by definition

$$
\begin{aligned}
I_{xym} &= \int_m xy\,dm = \int_V xy\,\rho\,dV = \int_A xy\,\rho t\,dA = \rho t \int_A xy\,dA = \rho t\,I_{xyA} \\[2mm]
I_{yzm} &= \int_m yz\,dm = 0 \\[2mm]
I_{zxm} &= \int_m zx\,dm = 0
\end{aligned}
\tag{10-29}
$$

where the subscripts $m$ and $A$ denote products of inertia of mass and mixed second moments of area, respectively. The products of inertia $I_{yzm}$ and $I_{zxm}$ for a thin plate are zero since the $x$- and $y$-axes are assumed to lie in the midplane of the plate (plane of symmetry).

A parallel-axis theorem for products of inertia can be developed that is very similar to the parallel-axis theorem for mixed second moments of area discussed in Section 10-2.5. Consider the body shown in Fig. 10-34, which has an $xyz$-coordinate system with its origin at the mass center $G$ of the body and a parallel $x'y'z'$-coordinate system with its origin at point $O'$. Observe in the figure that

$$x' = x_G + x$$

$$y' = y_G + y$$

$$z' = z_G + z$$

The product of inertia $I_{x'y'}$ of the body with respect to the $x'z'$- and $y'z'$-planes is by definition

$$I_{x'y'} = \int_m x'y'\, dm = \int_m (x_G + x)(y_G + y)\, dm$$

$$= \int_m x_G y_G\, dm + \int_m x_G y\, dm + \int_m y_G x\, dm + \int_m xy\, dm$$

Since $x_G$ and $y_G$ are the same for every element of mass $dm$,

$$I_{x'y'} = x_G y_G \int_m dm + x_G \int_m y\, dm + y_G \int_m x\, dm + \int_m xy\, dm$$

However,

$$\int_m xy\, dm = I_{xy}$$

and, since the $x$- and $y$-axes pass through the mass center $G$ of the body,

$$\int_m y\, dm = 0 \qquad \int_m z\, dm = 0$$

Therefore,

$$I_{x'y'} = I_{xyG} + x_G y_G m$$

$$I_{y'z'} = I_{yzG} + y_G z_G m \qquad\qquad \text{(10-30)}$$

$$I_{z'x'} = I_{zxG} + z_G x_G m$$

Equations 10-30 are the parallel-axis theorem for products of inertia. The subscript $G$ indicates that the $x$- and $y$-axes pass through the mass center $G$ of the body. Thus, if the product of inertia of a body with respect to a pair of orthogonal planes that pass through its mass-center is known, the product of inertia of the body with respect to any other pair of parallel planes can be found, without integrating, by use of Eqs. 10-30.

Procedures for determining products of inertia are illustrated in the following example problems.

**Figure 10-34**  Location of an element of mass $dm$ with respect to the $xy$-, $yz$- and $zx$-planes.

Determine the product of inertia $I_{xy}$ for the homogeneous quarter cylinder shown in Fig. 10-35a.

Fig. 10-35

## SOLUTION

All parts of the element of mass $dm$, shown in Fig. 10-35b, are located at the same distances $x$ and $y$ from the $xz$- and $yz$-planes; therefore, the product of inertia $dI_{xy}$ for the element is by definition

$$dI_{xy} = xy \, dm$$

Summing the elements for the entire body yields

$$
\begin{aligned}
I_{xy} &= \int_m dI_{xy} = \int_m xy \, dm = \int_V xy\rho \, dV \\
&= \int_0^R \int_0^{\sqrt{R^2 - x^2}} \rho xy \, (h \, dy \, dx) \\
&= \int_0^R \rho hx \left[ \frac{y^2}{2} \right]_0^{\sqrt{R^2 - x^2}} dx \\
&= \int_0^R \frac{1}{2} \rho h (R^2 x - x^3) \, dx \\
&= \frac{1}{2} \rho h \left[ \frac{R^2 x^2}{2} - \frac{x^4}{4} \right]_0^R = \frac{1}{2} \rho h \left( \frac{R^4}{4} \right) = \frac{1}{8} \rho R^4 h
\end{aligned}
$$

Alternatively, the thin-plate element shown in Fig. 10-35c could be used to determine $I_{xy}$. From Eq. 10-29 and the data from Table 10-3,

$$dI_{xym} = \rho t \, dI_{xyA} = \frac{1}{8} \rho R^4 \, dz$$

Therefore,

$$I_{xym} = \rho t \int_A dI_{xyA} = \int_0^h \frac{1}{8} \rho R^4 \, dz = \frac{1}{8} \rho R^4 h$$

Since the mass of the body is

$$m = \rho V = \rho \left( \frac{1}{4} \rho R^2 h \right) = \frac{1}{4} \rho \pi R^2 h$$

the product of inertia $I_{xy}$ can be written as

$$I_{xy} = \frac{1}{2\pi} \left( \frac{1}{4} \rho \pi R^2 h \right) R^2 = \frac{1}{2\pi} m R^2 \qquad \text{Ans.}$$

This problem can be solved with a single integration if a thin-plate element is used.

Moments of inertia for thin plates can be determined by using known results for second moments of area.

Determine the products of inertia $I_{xy}$, $I_{yz}$, and $I_{zx}$ for the homogeneous flat-plate steel ($\rho = 7870$ kg/m$^3$) washer shown in Fig. 10-36. The hole is located at the center of the plate.

**Fig. 10-36**

___

## SOLUTION

The products of inertia are zero for the planes of symmetry through the mass centers of the plate and hole. Since the $xy$-, $yz$-, and $zx$-planes shown in Fig. 10-36 are parallel to these planes of symmetry, the parallel-axis theorem for products of inertia (Eqs. 10-30) can be used to determine the required products of inertia. The masses of the plate, hole, and washer are

$$m_P = \rho V = \rho bht = 7870(0.280)(0.250)(0.060) = 33.05 \text{ kg}$$

$$m_H = \rho V = \rho \pi R^2 t = 7870\pi (0.050)^2(0.060) = 3.71 \text{ kg}$$

$$m_W = m_P - m_H = 33.05 - 3.71 = 29.34 \text{ kg}$$

From Eqs. 10-30,

$$I_{xy} = I_{xyG} + x_G y_G \, m$$
$$= 0 + (-0.125)(0.140)(29.34) = -0.513 \text{ kg} \cdot \text{m}^2 \qquad \text{Ans.}$$

$$I_{yz} = I_{yzG} + y_G z_G \, m$$
$$= 0 + (0.140)(0.030)(29.34) = 0.1232 \text{ kg} \cdot \text{m}^2 \qquad \text{Ans.}$$

$$I_{zx} = I_{zxG} + z_G x_G \, m$$
$$= 0 + (0.030)(-0.125)(29.34) = -0.1100 \text{ kg} \cdot \text{m}^2 \qquad \text{Ans.}$$

The product of inertia of a body with respect to two perpendicular planes is zero if either of the planes is a plane of symmetry.

The parallel-axis theorem for products of inertia is valid only for transfers to or from parallel axes through the mass center of the body.

# PROBLEMS

## Introductory Problems

**10-109*** Determine the products of inertia $I_{xy}$, $I_{yz}$, and $I_{zx}$ for the homogeneous rectangular block shown in Fig. P10-109.

**Fig. P10-109**

**10-110*** Determine the products of inertia $I_{xy}$, $I_{yz}$, and $I_{zx}$ for the homogeneous half cylinder shown in Fig. P10-110.

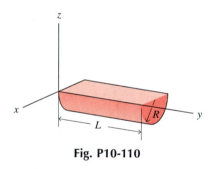

**Fig. P10-110**

**10-111** Determine the products of inertia $I_{xy}$, $I_{yz}$, and $I_{zx}$ for the homogeneous steel ($w = 490$ lb/ft$^3$) bracket shown in Fig. P10-111.

**Fig. P10-111**

**10-112** Determine the product of inertia $I_{xy}$ for the homogeneous octant of a sphere shown in Fig. P10-112.

**Fig. P10-112**

## Intermediate Problems

**10-113*** Determine the products of inertia $I_{xy}$ and $I_{zx}$ for the homogeneous triangular block shown in Fig. P10-113.

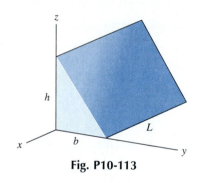

**Fig. P10-113**

**10-114** Determine the products of inertia $I_{xy}$ and $I_{zx}$ for the homogeneous right circular quarter cone shown in Fig. P10-114.

**Fig. P10-114**

# Challenging Problems

**10-115\*** Determine the products of inertia $I_{yz}$ and $I_{zx}$ for the homogeneous body shown in Fig. P10-115.

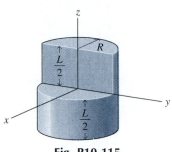

**Fig. P10-115**

**10-116** Determine the products of inertia $I_{xy}$, $I_{yz}$, and $I_{zx}$ for the homogeneous steel ($\rho = 7870 \text{ kg/m}^3$) block shown in Fig. P10-116.

**Fig. P10-116**

## 10-5 PRINCIPAL MOMENTS OF INERTIA

In some instances, in the dynamic analysis of bodies, principal axes and maximum and minimum moments of inertia, which are similar to maximum and minimum second moments of an area, must be determined. Again, the problem is one of transforming known or easily calculated moments and products of inertia with respect to one coordinate system (such as an *xyz*-coordinate system along the edges of a rectangular prism) to a second $x'y'z'$-coordinate system through the same origin $O$ but inclined with respect to the *xyz*-system.

For example, consider the body shown in Fig. 10-37, where the $x'$-axis is oriented at angles $\theta_{x'x}$, $\theta_{x'y}$, and $\theta_{x'z}$ with respect to the $x$-, $y$-, and $z$-axes, respectively. The moment of inertia $I_{x'}$ is by definition

$$I_{x'} = \int_m r^2 \, dm$$

The distance $d$ from the origin of coordinates to the element $dm$ is given by the expression

$$d^2 = x^2 + y^2 + z^2 = x'^2 + y'^2 + z'^2 = x'^2 + r^2$$

Therefore,

$$r^2 = x^2 + y^2 + z^2 - x'^2$$

and since

$$x' = x \cos \theta_{x'x} + y \cos \theta_{x'y} + z \cos \theta_{x'z}$$
$$r^2 = x^2 + y^2 + z^2 - (x \cos \theta_{x'x} + y \cos \theta_{x'y} + z \cos \theta_{x'z})^2$$

Recall that

$$\cos^2 \theta_{x'x} + \cos^2 \theta_{x'y} + \cos^2 \theta_{x'z} = 1$$

Therefore,

$$r^2 = (x^2 + y^2 + z^2)(\cos^2 \theta_{x'x} + \cos^2 \theta_{x'y} + \cos^2 \theta_{x'z})$$
$$- (x \cos \theta_{x'x} + y \cos \theta_{x'y} + z \cos \theta_{x'z})^2$$

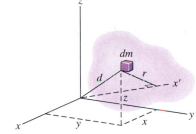

**Figure 10-37** Location of an element of mass $dm$ with respect to a Cartesian coordinate system.

**603**

which reduces to

$$r^2 = (y^2 + z^2) \cos^2 \theta_{x'x} + (z^2 + x^2) \cos^2 \theta_{x'y} + (x^2 + y^2) \cos^2 \theta_{x'z}$$
$$- 2xy \cos \theta_{x'x} \cos \theta_{x'y} - 2yz \cos \theta_{x'y} \cos \theta_{x'z} - 2zx \cos \theta_{x'z} \cos \theta_{x'x}$$

Therefore,

$$I_{x'} = \int_m r^2 \, dm = \cos^2 \theta_{x'x} \int_m (y^2 + z^2) \, dm + \cos^2 \theta_{x'y} \int_m (z^2 + x^2) \, dm$$

$$+ \cos^2 \theta_{x'z} \int_m (x^2 + y^2) \, dm - \cos \theta_{x'x} \cos \theta_{x'y} \int_m 2xy \, dm$$

$$- \cos \theta_{x'y} \cos \theta_{x'z} \int_m 2yz \, dm - \cos \theta_{x'z} \cos \theta_{x'x} \int_m 2zx \, dm$$

From Eqs. 10-21 and 10-28,

$$I_x = \int_m (y^2 + z^2) \, dm \qquad I_{xy} = \int_m xy \, dm$$

$$I_y = \int_m (z^2 + x^2) \, dm \qquad I_{yz} = \int_m yz \, dm$$

$$I_z = \int_m (x^2 + y^2) \, dm \qquad I_{zx} = \int_m zx \, dm$$

Therefore,

$$I_{x'} = I_x \cos^2 \theta_{x'x} + I_y \cos^2 \theta_{x'y} + I_z \cos^2 \theta_{x'z} - 2I_{xy} \cos \theta_{x'x} \cos \theta_{x'y}$$
$$- 2I_{yz} \cos \theta_{x'y} \cos \theta_{x'z} - 2I_{zx} \cos \theta_{x'z} \cos \theta_{x'x} \qquad \textbf{(10-31a)}$$

In a similar fashion the product of inertia

$$I_{x'y'} = \int_m x'y' \, dm$$

can be expressed in terms of $I_x$, $I_y$, $I_z$, $I_{xy}$, $I_{yz}$, and $I_{zx}$ as

$$I_{x'y'} = -I_x \cos \theta_{x'x} \cos \theta_{y'x} - I_y \cos \theta_{x'y} \cos \theta_{y'y} - I_z \cos \theta_{x'z} \cos \theta_{y'z}$$
$$+ I_{xy}(\cos \theta_{x'x} \cos \theta_{y'y} + \cos \theta_{x'y} \cos \theta_{y'x})$$
$$+ I_{yz}(\cos \theta_{x'y} \cos \theta_{y'z} + \cos \theta_{x'z} \cos \theta_{y'y})$$
$$+ I_{zx}(\cos \theta_{x'z} \cos \theta_{y'x} + \cos \theta_{x'x} \cos \theta_{y'z}) \qquad \textbf{(10-31b)}$$

If the original $xyz$-axes are principal axes (such as those shown for the figures in Table 10-4),

$$I_{xy} = I_{yz} = I_{zx} = 0$$

and Eqs. 10-31 reduce to

$$I_{x'} = I_x \cos^2 \theta_{x'x} + I_y \cos^2 \theta_{x'y} + I_z \cos^2 \theta_{x'z} \qquad \textbf{(10-32a)}$$

and

$$I_{x'y'} = -I_x \cos \theta_{x'x} \cos \theta_{y'x} - I_y \cos \theta_{x'y} \cos \theta_{y'y}$$
$$- I_z \cos \theta_{x'z} \cos \theta_{y'z} \qquad \textbf{(10-32b)}$$

Equation 10-31a for moments of inertia is the three-dimensional equivalent of Eq. 10-14 for second moments of area. By using a similar but much more complicated procedure than the one used with Eq. 10-14 to locate principal axes and determine maximum and minimum second moments of area, principal axes can be located and maximum and minimum moments of inertia can be determined. The procedure yields the following equations:

$$(I_x - I_P) \cos \theta_{Px} - I_{xy} \cos \theta_{Py} - I_{zx} \cos \theta_{Pz} = 0$$

$$(I_y - I_P) \cos \theta_{Py} - I_{yz} \cos \theta_{Pz} - I_{xy} \cos \theta_{Px} = 0 \qquad (10\text{-}33)$$

$$(I_z - I_P) \cos \theta_{Pz} - I_{zx} \cos \theta_{Px} - I_{yz} \cos \theta_{Py} = 0$$

This set of equations has a nontrivial solution only if the determinant of the coefficients of the direction cosines is equal to zero. Expansion of the determinant yields the following cubic equation for determining the principal moments of inertia of the body for the particular origin of coordinates being used:

$$I_P^3 - (I_x + I_y + I_z)I_P^2 + (I_xI_y + I_yI_z + I_zI_x - I_{xy}^2 - I_{yz}^2 - I_{zx}^2)I_P$$
$$- (I_xI_yI_z - I_xI_{yz}^2 - I_yI_{zx}^2 - I_zI_{xy}^2 - 2I_{xy}I_{yz}I_{zx}) = 0 \qquad (10\text{-}34)$$

Equation 10-34 yields three values $I_1$, $I_2$, and $I_3$ for the principal moments of inertia. One value is the maximum moment of inertia of the body for the origin of coordinates being used, a second value is the minimum moment of inertia of the body for the origin of coordinates being used, and the third value is an intermediate value of the moment of inertia of the body that has no particular significance.

The direction cosines for the principal inertia axes can be obtained by substituting the three values $I_1$, $I_2$, and $I_3$ obtained from Eq. 10-34, in turn, into Eqs. 10-33 and using the additional relation

$$\cos^2 \theta_{Px} + \cos^2 \theta_{Py} + \cos^2 \theta_{Pz} = 1$$

Equations 10-33 and 10-34 are valid for bodies of any shape. The procedure for locating principal axes and determining maximum and minimum moments of inertia is illustrated in the following example problem.

Locate the principal axes and determine the maximum and minimum moments of inertia for the rectangular steel ($w = 490$ lb/ft$^3$) block shown in Fig. 10-38.

Fig. 10-38

## SOLUTION

The moments and products of inertia for the block are given by Eqs. 10-23 and 10-30 as

$$I_x = I_{xG} + (y_G^2 + z_G^2)m \qquad I_{xy} = I_{xyG} + x_G y_G m$$
$$I_y = I_{yG} + (z_G^2 + x_G^2)m \qquad I_{yz} = I_{yzG} + y_G z_G m$$
$$I_z = I_{zG} + (x_G^2 + y_G^2)m \qquad I_{zx} = I_{zxG} + z_G x_G m$$

The mass of the block is

$$m = \rho V = \frac{w}{g}bhL = \frac{490}{32.2}\left(\frac{8}{12}\right)\left(\frac{4}{12}\right)\left(\frac{16}{12}\right) = 4.509 \text{ slugs}$$

Thus, from the results listed in Table 10-4,

$$I_x = \frac{1}{12}m(b^2 + h^2) + \left[\left(\frac{h}{2}\right)^2 + \left(\frac{b}{2}\right)^2\right]m$$
$$= \frac{1}{3}m(b^2 + h^2) = \frac{1}{3}(4.509)\left[\left(\frac{8}{12}\right)^2 + \left(\frac{4}{12}\right)^2\right] = 0.835 \text{ slug} \cdot \text{ft}^2$$

$$I_y = \frac{1}{12}m(b^2 + L^2) + \left[\left(\frac{b}{2}\right)^2 + \left(\frac{L}{2}\right)^2\right]m$$
$$= \frac{1}{3}m(b^2 + L^2) = \frac{1}{3}(4.509)\left[\left(\frac{8}{12}\right)^2 + \left(\frac{16}{12}\right)^2\right] = 3.340 \text{ slug} \cdot \text{ft}^2$$

$$I_z = \frac{1}{12}m(h^2 + L^2) + \left[\left(\frac{h}{2}\right)^2 + \left(\frac{L}{2}\right)^2\right]m$$
$$= \frac{1}{3}m(h^2 + L^2) = \frac{1}{3}(4.509)\left[\left(\frac{4}{12}\right)^2 + \left(\frac{16}{12}\right)^2\right] = 2.839 \text{ slug} \cdot \text{ft}^2$$

$$I_{xy} = 0 + \left(\frac{L}{2}\right)\left(\frac{h}{2}\right)m = \frac{1}{4}mLh = \frac{1}{4}(4.509)\left(\frac{16}{12}\right)\left(\frac{4}{12}\right) = 0.501 \text{ slug} \cdot \text{ft}^2$$

$$I_{yz} = 0 + \left(\frac{h}{2}\right)\left(\frac{b}{2}\right)m = \frac{1}{4}mhb = \frac{1}{4}(4.509)\left(\frac{4}{12}\right)\left(\frac{8}{12}\right) = 0.251 \text{ slug} \cdot \text{ft}^2$$

$$I_{zx} = 0 + \left(\frac{b}{2}\right)\left(\frac{L}{2}\right)m = \frac{1}{4}mbL = \frac{1}{4}(4.509)\left(\frac{8}{12}\right)\left(\frac{16}{12}\right) = 1.002 \text{ slug} \cdot \text{ft}^2$$

The principal moments of inertia are

$$I_P^3 - (I_x + I_y + I_z)I_P^2 + (I_xI_y + I_yI_z + I_zI_x - I_{xy}^2 - I_{yz}^2 - I_{zx}^2)I_P$$
$$- (I_xI_yI_z - I_xI_{yz}^2 - I_yI_{zx}^2 - I_zI_{xy}^2 - 2I_{xy}I_{yz}I_{zx}) = 0$$

The three principal moments of inertia are the roots of Eq. 10-34.

Substituting values for the moments and products of inertia gives

$$I_P^3 - 7.014I_P^2 + 13.324I_P - 3.548 = 0$$

which has the solution

$$I_1 = I_{max} = 3.451 \text{ slug} \cdot \text{ft}^2 \qquad \text{Ans.}$$
$$I_2 = I_{int} = 3.246 \text{ slug} \cdot \text{ft}^2$$
$$I_3 = I_{min} = 0.317 \text{ slug} \cdot \text{ft}^2 \qquad \text{Ans.}$$

The principal directions are obtained by substituting the principal moments of inertia, in turn, into Eqs. 10-33. With $I_P = I_1 = I_{max} = 3.451$ slug $\cdot$ ft$^2$:

Once the three-principal moments of inertia are known, Eqs. 10-33 can be used to determine the directions of the principal axes.

$$(I_x - I_P) \cos \theta_{Px} - I_{xy} \cos \theta_{Py} - I_{zx} \cos \theta_{Pz} = 0$$
$$(I_y - I_P) \cos \theta_{Py} - I_{yz} \cos \theta_{Pz} - I_{xy} \cos \theta_{Px} = 0$$
$$(I_z - I_P) \cos \theta_{Pz} - I_{zx} \cos \theta_{Px} - I_{yz} \cos \theta_{Py} = 0$$

$$-2.616 \cos \theta_{1x} - 0.501 \cos \theta_{1y} - 1.002 \cos \theta_{1z} = 0$$
$$-0.501 \cos \theta_{1x} - 0.111 \cos \theta_{1y} - 0.251 \cos \theta_{1z} = 0 \qquad (a)$$
$$-1.002 \cos \theta_{1x} - 0.251 \cos \theta_{1y} - 0.612 \cos \theta_{1z} = 0$$

Equations $a$ together with the required relationship for direction cosines

$$\cos^2 \theta_{1x} + \cos^2 \theta_{1y} + \cos^2 \theta_{1z} = 1$$

have the solution

$$\cos \theta_{1x} = 0.0891 \qquad or \qquad \theta_{1x} = 84.9°$$
$$\cos \theta_{1y} = -0.9643 \qquad \theta_{1y} = 164.6° \qquad \text{Ans.}$$
$$\cos \theta_{1z} = 0.2495 \qquad \theta_{1z} = 75.6°$$

With $I_P = I_2 = I_{int} = 3.246$ slug $\cdot$ ft$^2$:

$$-2.411 \cos \theta_{2x} - 0.501 \cos \theta_{2y} - 1.002 \cos \theta_{2z} = 0$$
$$-0.501 \cos \theta_{2x} + 0.094 \cos \theta_{2y} - 0.251 \cos \theta_{2z} = 0$$
$$-1.002 \cos \theta_{2x} - 0.251 \cos \theta_{2y} - 0.407 \cos \theta_{2z} = 0$$
$$\cos^2 \theta_{2x} + \cos^2 \theta_{2y} + \cos^2 \theta_{2z} = 1$$

the solution becomes

$$\cos \theta_{2x} = -0.4105 \qquad or \qquad \theta_{2x} = 114.2°$$
$$\cos \theta_{2y} = 0.1925 \qquad \theta_{2y} = 78.9° \qquad \text{Ans.}$$
$$\cos \theta_{2z} = 0.8913 \qquad \theta_{2z} = 27.0°$$

With $I_P = I_3 = I_{min} = 0.317$ slug $\cdot$ ft$^2$:

$$0.518 \cos \theta_{3x} - 0.501 \cos \theta_{3y} - 1.002 \cos \theta_{3z} = 0$$
$$-0.501 \cos \theta_{3x} + 3.023 \cos \theta_{3y} - 0.251 \cos \theta_{3z} = 0$$
$$-1.002 \cos \theta_{3x} - 0.251 \cos \theta_{3y} + 2.522 \cos \theta_{3z} = 0$$
$$\cos^2 \theta_{3x} + \cos^2 \theta_{3y} + \cos^2 \theta_{3z} = 1$$

the solution becomes

$$\cos \theta_{3x} = 0.9075 \qquad or \qquad \theta_{3x} = 24.8°$$
$$\cos \theta_{3y} = 0.1818 \qquad \theta_{3y} = 79.5° \qquad \text{Ans.}$$
$$\cos \theta_{3z} = 0.3786 \qquad \theta_{3z} = 67.8°$$

Thus, the unit vectors associated with the three principal directions are

$$\mathbf{n}_1 = 0.0891\mathbf{i} - 0.9643\mathbf{j} + 0.2495\mathbf{k} \qquad \mathbf{n}_1 \cdot \mathbf{n}_2 = 0$$
$$\mathbf{n}_2 = -0.4105\mathbf{i} + 0.1925\mathbf{j} + 0.8913\mathbf{k} \qquad \mathbf{n}_2 \cdot \mathbf{n}_3 = 0$$
$$\mathbf{n}_3 = 0.9075\mathbf{i} + 0.1818\mathbf{j} + 0.3786\mathbf{k} \qquad \mathbf{n}_3 \cdot \mathbf{n}_1 = 0$$

which verifies that the three principal axes are orthogonal.

# PROBLEMS

## Introductory Problems

**10-117\*** Locate the principal axes and determine the maximum and minimum moments of inertia for the triangular steel ($w = 490$ lb/ft³) block shown in Fig. P10-117 if $b = 8$ in., $h = 6$ in., and $L = 10$ in.

**Fig. P10-117**

**10-118\*** Locate the principal axes and determine the maximum and minimum moments of inertia for the steel ($\rho = 7870$ kg/m³) half cylinder shown in Fig. P10-118 if $R = 100$ mm and $L = 150$ mm.

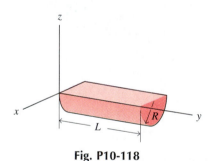

**Fig. P10-118**

**10-119** Locate the principal axes and determine the maximum and minimum moments of inertia for the cast iron ($w = 460$ lb/ft³) tetrahedron shown in Fig. P10-119 if $a = 6$ in., $b = 8$ in., and $c = 10$ in.

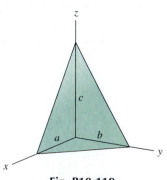

**Fig. P10-119**

**10-120** Locate the principal axes and determine the maximum and minimum moments of inertia for the brass ($\rho = 8750$ kg/m³) sphere shown in Fig. P10-120. The radius of the sphere is 200 mm. The surface of the sphere is tangent to the three coordinate planes.

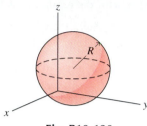

**Fig. P10-120**

## Intermediate Problems

**10-121\*** Locate the principal axes and determine the maximum and minimum moments of inertia for the steel ($w = 460$ lb/ft³) angle bracket shown in Fig. P10-121.

**Fig. P10-121**

**10-122** Locate the principal axes and determine the maximum and minimum moments of inertia for the steel ($\rho = 7870$ kg/m³) washer shown in Fig. P10-122.

**Fig. P10-122**

## Challenging Problems

**10-123\*** Locate the principal axes and determine the maximum and minimum moments of inertia for the aluminum ($w = 175$ lb/ft³) cylinder shown in Fig. P10-123 if $R = 4$ in. and $L = 6$ in.

**Fig. P10-123**

**10-124** Locate the principal axes and determine the maximum and minimum moments of inertia for the brass ($\rho =$ 8750 kg/m$^3$) block shown in Fig. P10-124.

**Fig. P10-124**

## Computer Problems

**C10-125** The principle moments of inertia are the roots of the cubic equation

$$I_P^3 - aI_P^2 + bI_P - c = 0$$

in which

$$a = I_x + I_y + I_z$$
$$b = I_xI_y + I_yI_z + I_zI_x - I_{xy}^2 - I_{yz}^2 - I_{zx}^2$$
$$c = I_xI_yI_z - I_xI_{yz}^2 - I_yI_{zx}^2 - I_zI_{xy}^2 - 2I_{xy}I_{yz}I_{zx}$$

Write a program to compute and print the three principal moments of inertia ($I_1 \geq I_2 \geq I_3$) and the maximum products of inertia ($I_{12}, I_{23}, I_{31}$).

**C10-126** The direction cosines for the principal inertia axes are obtained by solving the system of equations

$$(I_x - I_P)\ell_P - I_{xy}m_P \qquad - I_{zx}n_P = 0$$
$$-I_{xy}\ell_P + (I_y - I_P)m_P - I_{yz}n_P = 0$$
$$-I_{zx}\ell_P - I_{yz}m_P \qquad + (I_z - I_P)n_P = 0$$

and using the additional relation

$$\ell_P^2 + m_P^2 + n_P^2 = 1$$

where $I_P$ ($p = 1, 2, 3$) is one of the principal moments obtained by solving the cubic equation

$$I_P^3 - aI_P^2 + bI_P - c = 0$$

and $\ell_P = \cos\theta_{Px}$, $m_P = \cos\theta_{Py}$, and $n_P = \cos\theta_{Pz}$. Write a program to compute and print the three direction cosines $\ell_P$, $m_P$, and $n_P$ for each of the principal moments $I_P(p = 1, 2, 3)$.

**C10-127** A common approximation for the moment of inertia of a cylinder about an axis parallel to the axis of the cylinder is $I \cong ma^2$ where $a$ is the distance between the two axes as shown in Fig. P10-127. Plot the percent error in using this approximation as a function of $a/r$ ($1 \leq a/r \leq 6$).

**Fig. P10-127**

# SUMMARY

Previously, the centroid of an area $A$ was located by using the integrals $\int_A y \, dA$ and $\int_A x \, dA$, which are the first moments of the area with respect to the $x$- and $y$-axes, respectively. Integrals of the form $\int_A x^2 \, dA$ and $\int_A xy \, dA$ yield quantities known as second moments of the area and mixed second moments of the area, respectively.

When the second moment of an area has been determined with respect to a given axis, the second moment with respect to a parallel axis can be obtained by means of the parallel-axis theorem. If one of the axes (say the $x$-axis) passes through the centroid of the area, the second moment of the area about a parallel $x'$-axis is given by Eq. 10-4 as

$$I_{x'} = I_{xC} + d_x^2 A$$

Similarly from Eqs. 10-4, 10-5, and 10-13,

$$I_{y'} = I_{yC} + d_y^2 A$$

$$J_{z'} = J_{zC} + (d_x^2 + d_y^2)A = J_{zC} + d_z^2 A$$

$$I_{x'y'} = I_{xyC} + d_x d_y A$$

The parallel-axis theorem is valid only for transfers to or from centroidal axes.

The second moment of an area $A$ with respect to an $x'$-axis through the origin $O$ of an $xy$-coordinate system will, in general, vary with the angle $\theta$ between the $x$- and $x'$-axes. The set of $x'$- and $y'$-axes for which the second moments are maximum and minimum are called principal axes. Second moments of the area with respect to these axes, called the principal second moments, are designated $I_u$ and $I_v$. In terms of the second moments $I_x$, $I_y$, and the mixed second moment $I_{xy}$, the principal second moments $I_u$ and $I_v$ are

$$I_{u,v} = \frac{I_x + I_y}{2} \pm \left[ \left( \frac{I_x - I_y}{2} \right)^2 + I_{xy}^2 \right]^{1/2} \qquad \textbf{(10-17)}$$

The angle $\theta_P$ for which second moments are maximum or minimum is determined from the expression

$$\tan 2\theta_P = -\frac{2 \, I_{xy}}{(I_x - I_y)} \qquad \textbf{(10-16)}$$

In analyses of the motion of rigid bodies, second moments of the mass of an element are often encountered. These second moments are commonly known as moments of inertia and products of inertia. Moments of inertia and products of inertia of a body with respect to an $xyz$-coordinate system are given by Eqs. 10-21 and 10-28 as

$$I_x = \int_m (y^2 + z^2) \, dm \qquad I_{xy} = \int_m xy \, dm$$

$$I_y = \int_m (z^2 + x^2) \, dm \qquad I_{yz} = \int_m yz \, dm$$

$$I_z = \int_m (x^2 + y^2) \, dm \qquad I_{zx} = \int_m zx \, dm$$

For a body which has an $xyz$-coordinate system with its origin at the mass center $G$ of the body and a parallel $x'y'z'$-coordinate system with its origin at point $O'$, the parallel-axis theorem for moments and products of inertia, Eqs. 10-23 and 10-30, yield

$$I_x = I_{xG} + (y_G^2 + z_G^2)m \qquad I_{xy} = I_{xyG} + x_G y_G m$$

$$I_y = I_{yG} + (z_G^2 + x_G^2)m \qquad I_{yz} = I_{yzG} + y_G z_G m$$

$$I_z = I_{zG} + (x_G^2 + y_G^2)m \qquad I_{zx} = I_{zxG} + z_G x_G m$$

Equations 10-23 and 10-30 are valid only for transfers to or from axes passing through the mass center $G$ of the body.

For some dynamic analyses of bodies, maximum and minimum moments of inertia must be determined. Maximum and minimum moments of inertia can be determined and principal axes can be located by using the following equations:

$$(I_x - I_P) \cos \theta_{Px} - I_{xy} \cos \theta_{Py} - I_{zx} \cos \theta_{Pz} = 0$$

$$(I_y - I_P) \cos \theta_{Py} - I_{yz} \cos \theta_{Pz} - I_{xy} \cos \theta_{Px} = 0 \qquad \textbf{(10-33)}$$

$$(I_z - I_P) \cos \theta_{Pz} - I_{zx} \cos \theta_{Px} - I_{yz} \cos \theta_{Py} = 0$$

Equations 10-33 have a solution if the determinant of the coefficients of the cosine terms is equal to zero. Expansion of the determinant yields the following cubic equation for determining the principal moments of inertia of the body for the particular origin of coordinates being used:

$$I_P^3 - (I_x + I_y + I_z)I_P^2 + (I_x I_y + I_y I_z + I_z I_x - I_{xy}^2 - I_{yz}^2 - I_{zx}^2)I_P$$
$$- (I_x I_y I_z - I_x I_{yz}^2 - I_y I_{zx}^2 - I_z I_{xy}^2 - 2I_{xy}I_{yz}I_{zx}) = 0 \qquad \textbf{(10-34)}$$

Equation 10-34 yields three values $I_1$, $I_2$, and $I_3$ for the principal moments of inertia. One value is the maximum moment of inertia of the body for the origin of coordinates being used, the second value is the minimum moment of inertia, and the third value is an intermediate value of the moment of inertia that has no particular significance. The direction cosines for the principal axes are obtained by substituting the three values $I_1$, $I_2$, and $I_3$ obtained from Eq. 10-34, in turn, into Eqs. 10-33 and using the additional relation

$$\cos^2 \theta_{Px} + \cos^2 \theta_{Py} + \cos^2 \theta_{Pz} = 1$$

# REVIEW PROBLEMS

**10-128*** Determine the second moments $I_x$ and $I_y$ for the shaded area shown in Fig. P10-128.

**Fig. P10-128**

**10-129*** Determine the second moments $I_x$ and $I_y$ for the shaded area shown in Fig. P10-129.

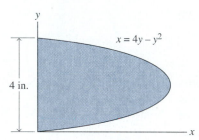

**Fig. P10-129**

**10-130** Determine the second moments $I_x$ and $I_y$ for the shaded area shown in Fig. P10-130.

**Fig. P10-130**

**10-131** Determine the radii of gyration $k_x$ and $k_y$ for the shaded area shown in Fig. P10-131.

**Fig. P10-131**

**10-132*** Determine the mixed second moment $I_{xy}$ for the shaded area shown in Fig. P10-130.

**10-133*** Determine the second moments $I_x$, $I_y$, and $I_{xy}$ for the composite area shown in Fig. P10-133.

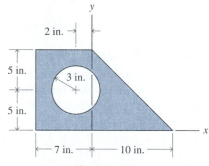

**Fig. P10-133**

**10-134** Determine the second moments $I_x$ and $I_y$ for the cross section shown in Fig. P10-134. Four C381 × 74 channels are welded together to form the cross section.

**Fig. P10-134**

**10-135** For the shaded composite area shown in Fig. P10-135, determine

**a.** The second moment of the area with respect to the $x$-axis.

**b.** The second moment of the area with respect to a centroidal $x$-axis.

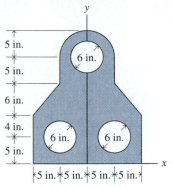

**Fig. P10-135**

**10-136*** For the shaded composite area shown in Fig. P10-136, determine

**a.** The mixed second moment $I_{xy}$ for the area.

**b.** The maximum and minimum second moments of area with respect to axes through the origin of the $xy$-coordinate system and their orientations with respect to the $x$-axis.

**Fig. P10-136**

**10-137*** Determine the moment of inertia $I_y$ for the frustrum of a homogeneous cone shown in Fig. P10-137.

**Fig. P10-137**

**10-138** Determine the product of inertia $I_{xy}$ for the homogeneous tetrahedron shown in Fig. P10-138.

**Fig. P10-138**

**10-139*** Determine the moment of inertia about an axis which passes through points $A$ and $B$ of the homogeneous circular cylinder shown in Fig. P10-139. Line $AB$ lies in the $xy$-plane and passes through the mass center $G$ of the cylinder.

**Fig. P10-139**

**10-140** A composite body consists of a homogeneous sphere and a homogeneous bent bar as shown in Fig. P10-140. The mass of the sphere is 35 kg. The masses of sections $OA$ and $AB$ of the bar are 16 kg and 8 kg, respectively. Determine

**a.** The moment of inertia $I_z$ for the composite body.

**b.** The product of inertia $I_{yz}$ for the composite body.

**Fig. P10-140**

*Virtual work principles can be used to determine the stability of equilibrium for various positions of a complicated machine or structure.*

# METHOD OF VIRTUAL WORK

## 11-1 INTRODUCTION

Static equilibrium problems in Chapter 6 ("Equilibrium of Rigid Bodies"), Chapter 7 ("Trusses, Frames, and Machines"), and Chapter 8 ("Internal Forces in Structural Members") were solved by using the problem-solving method outlined in Chapter 1. Once the problem was clearly identified, free-body diagrams (one or more as required for the specific problem) were drawn to identify all forces (applied forces and support reactions) involved in the problem. The free-body diagram was then used with the appropriate equations of equilibrium to solve for the required unknowns. All of these types of problems can also be solved by using a method based on work and potential energy concepts. The method provides a convenient means for determining an unknown force whenever displacements of applied forces can be determined in terms of the displacement of an unknown force. Since most equilibrium problems involve bodies at rest, the displacements involved in the process are virtual[1] (imaginary or fictitious) displacements. Thus, the method has become known as the method of virtual displacements or the method of virtual work. The method is often used when systems of connected bodies are studied. Frequently, when the equations of equilibrium are used to solve such problems, the system must be dismembered in order to determine a particular reaction component. It is not necessary to investigate the internal forces of a system when the method of virtual work is used.

The method of virtual work, as applied to the solution of equilibrium problems, is developed in the following sections.

[1]*Webster's New World Dictionary of the American Language* describes the word *virtual* as "being so in effect or essence, although not in actual fact or name."

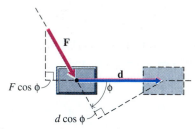

**Figure 11-1**   Motion of a particle along a straight line.

## 11-2   DEFINITION OF WORK AND VIRTUAL WORK

In everyday life, the word *work* is applied to any form of activity that requires the exertion of muscular or mental effort. In mechanics, the term is used in a much more restricted sense.

### 11-2.1   Work of a Force

In mechanics, a force does work only when the particle to which the force is applied moves. For example, when a constant force **F** is applied to a particle that moves a distance **d** in a straight line, as shown in Fig. 11-1, the work done on the particle by the force **F** is defined by the scalar product

$$U = \mathbf{F} \cdot \mathbf{d} = Fd \cos \phi$$
$$= F_x d_x + F_y d_y + F_z d_z \tag{11-1}$$

where $\phi$ is the angle between the vectors **F** and **d**. Equation 11-1 is usually interpreted as follows:

> The work done by a force **F** is the product of the magnitude of the force ($F$) and the magnitude of the rectangular component of the displacement in the direction of the force ($d \cos \phi$) (see Fig. 11-1).

However, $\cos \phi$ can also be associated with the force **F** instead of with the displacement **d.** Then Eq. 11-1 would be interpreted as follows:

> The work done by a force **F** is the product of the magnitude of the displacement ($d$) and the magnitude of the rectangular component of the force in the direction of the displacement ($F \cos \phi$) (see Fig. 11-1).

Since work is defined as the scalar product of two vectors, work is a scalar quantity with only magnitude and algebraic sign. When the sense of the displacement and the sense of the force component in the direction of the displacement are the same ($0 \leq \phi < 90°$), the work done by the force is positive. When the sense of the displacement and the sense of the force component in the direction of the displacement are opposite ($90° < \phi \leq 180°$), the work done by the force is negative. When the direction of the force is perpendicular to the direction of the displacement ($\phi = 90°$), the component of the force in the direction of the displacement is zero ($d = 0$) and the work done by the force is zero. Of course, the work done by the force is also zero if the displacement is zero ($d = 0$).

Work has the dimensions of force times length. In the SI system of units, this combination of dimensions is called a joule (1 J = 1 N · m)[2]. In the U.S. Customary system of units, there is no special unit for work. It is expressed simply as ft · lb.

If the force is not constant or if the displacement is not in a straight line, Eq. 11-1 gives the work done by the force only during an infini-

---

[2]It may be noted that work and moment of a force have the same dimensions: They are both force times length. However, work and moment are two totally different concepts, and the special unit joule should be used only to describe work. The moment of a force must always be expressed as N · m.

tesimal part of the displacement, $d\mathbf{r}$ (see Fig. 11-2):

$$dU = \mathbf{F} \cdot d\mathbf{r} = F\, ds \cos \phi = F_t\, ds$$
$$= F_x\, dx + F_y\, dy + F_z\, dz \qquad \textbf{(11-2)}$$

where $d\mathbf{r} = ds\, \mathbf{e}_t = dx\, \mathbf{i} + dy\, \mathbf{j} + dz\, \mathbf{k}$. The total work done by the force as the particle moves from position 1 to position 2 is obtained by integrating Eq. 11-2 along the path of the particle

$$U_{1 \to 2} = \int_1^2 dU = \int_{s_1}^{s_2} F_t\, ds$$
$$= \int_{x_1}^{x_2} F_x\, dx = \int_{y_1}^{y_2} F_y\, dy = \int_{z_1}^{z_2} F_z\, dz \qquad \textbf{(11-3)}$$

For the special case in which the force $\mathbf{F}$ is constant (both in magnitude and direction), the force components can be taken outside the integral signs in Eq. 11-3. Then Eq. 11-13 gives

$$U_{1 \to 2} = F_x \int_{x_1}^{x_2} dx + F_y \int_{y_1}^{y_2} dy + F_z \int_{z_1}^{z_2} dz$$
$$= F_x(x_2 - x_1) + F_y(y_2 - y_1) + F_z(z_2 - z_1) \qquad \textbf{(11-4)}$$

Note that the evaluation of the work done by a constant force depends on the coordinates at the end points of the particle's path but not on the actual path traveled by the particle. For the constant force $\mathbf{F}$ shown in Fig. 11-3, it doesn't matter if the particle moves along path $a$ from position 1 to position 2, along path $b$, or along some other path. The work done by the force is always the same. Forces for which the work done is independent of the path are called conservative forces.

The weight $W$ of a particle is a particular example of a constant force. When bodies move near the surface of the earth, the force of the earth's gravity is essentially constant ($F_x = 0$, $F_y = 0$, and $F_z = -W$). Therefore, the work done on a particle by its weight is $-W(z_2 - z_1)$. When $z_2 > z_1$ the particle moves upward (opposite the gravitational force), and the work done by gravity is negative. When $z_2 < z_1$ the particle moves downward (in the direction of the gravitational force), and the work done by gravity is positive.

Examples of forces that do work when a body moves from one position to another include the weight of the body, friction between the body and other surfaces, and externally applied loads. Examples of forces which do no work include forces at fixed points ($ds = 0$) and forces acting in a direction perpendicular to the displacement ($\cos \phi = 0$).

## 11-2.2 Work of a Couple

The work done by a couple is obtained by calculating the work done by each force of the couple separately and adding them together. For example, consider a couple $\mathbf{C}$ acting on a rigid body as shown in Fig. 11-4$a$. During some small time $dt$ the body translates and rotates. If the displacement of point $A$ is $d\mathbf{r} = ds\, \mathbf{e}_t$, choose a second point $B$ such that the line $AB$ is perpendicular to $d\mathbf{r}$. Then the motion that takes $A$ to $A'$ will take $B$ to $B'$. This motion may be considered in two parts: first a translation that takes the line $AB$ to $A'\hat{B}$, followed by a rotation $d\theta$ about $A'$ that takes $\hat{B}$ to $B'$ (see Fig. 11-4$b$).

**Figure 11-2** Motion of a particle along a curved path.

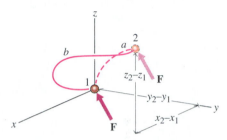

**Figure 11-3** Motion of a particle along a curved path.

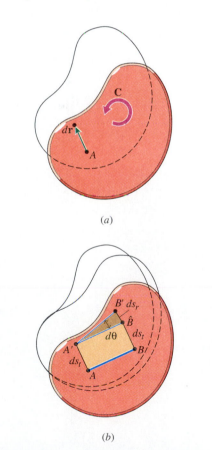

**Figure 11-4** Action of a couple on a rigid body.

(c)

Now represent the couple by a pair of forces of magnitude $F = C/b$ in the direction perpendicular to the line $AB$ (see Fig. 11-4c). During the translational part of the motion, one force will do positive work $F\,ds_t$ and the other will do negative work $-F\,ds_t$; therefore, the sum of the work done on the body by the pair of forces during the translational part of the motion is zero. During the rotational part of the motion, $A'$ is a fixed point and the force applied at $A'$ does no work. The work done by the force at $B$ is $dU = F\,ds_r \cong Fb\,d\theta$, where $d\theta$ is in radians and $C = Fb$ is the magnitude of the moment of the couple. Therefore, when a body is simultaneously translated and rotated, the couple does work only as a result of the rotation.

The total work done by the couple during the differential motion is

$$dU = F\,ds_t - F\,ds_t + Fb\,d\theta = C\,d\theta \qquad (11\text{-}5)$$

The work is positive if the angular displacement $d\theta$ is in the same direction as the sense of rotation of the couple and negative if the displacement is in the opposite direction. No work is done if the couple is translated or rotated about an axis parallel to the plane of the couple.

The work done on the body by the couple as the body rotates through a finite angle $\Delta\theta = \theta_2 - \theta_1$ is obtained by integrating Eq. 11-5:

$$U_{1\to2} = \int_1^2 dU = \int_{\theta_1}^{\theta_2} C\,d\theta \qquad (11\text{-}6)$$

If the couple is constant, then $C$ can be taken outside the integral sign and Eq. 11-6 becomes

$$U_{1\to2} = C\int_{\theta_1}^{\theta_2} d\theta = C(\theta_2 - \theta_1) = C\,\Delta\theta \qquad (11\text{-}7)$$

If the body rotates in space, the component of the infinitesimal angular displacement $d\boldsymbol{\theta}$ in the direction of the couple $\mathbf{C}$ is required. For this case, the work done is determined by using the dot product relationship,

$$dU = \mathbf{C} \cdot d\boldsymbol{\theta} = M\,d\theta\cos\phi \qquad (11\text{-}8)$$

where $M$ is the magnitude of the moment of the couple, $d\theta$ is the magnitude of the infinitesimal angular displacement, and $\phi$ is the angle between $\mathbf{C}$ and $d\boldsymbol{\theta}$. For planar motion (in the $xy$-plane), $\mathbf{C} = C\,\mathbf{k}$, $d\boldsymbol{\theta} = d\theta\,\mathbf{k}$, $\mathbf{C} \cdot d\boldsymbol{\theta} = C\,d\theta$, and Eq. 11-8 gives the same result as Eq. 11-7.

Since work is a scalar quantity, the work done on a rigid body by a system of external forces and couples is the algebraic sum of work done by the individual forces and couples.

### 11-2.3   Virtual Work

When a body being acted on by a force $\mathbf{F}$ moves through an infinitesimal linear displacement $d\mathbf{r}$ as described in Section 11-2.1, the body is not in equilibrium. In studying the equilibrium of bodies by the method of virtual work, it is necessary to introduce fictitious displacements, called virtual displacements. An infinitesimal virtual linear displacement will be represented by the first-order differential $\delta\mathbf{s}$ rather than $d\mathbf{s}$. The work done by a force $\mathbf{F}$ acting on a body during a virtual displacement $\delta s$ is called virtual work $\delta U$ and is represented mathematically as

$$\delta U = \mathbf{F} \cdot \delta \mathbf{s} \qquad \text{or} \qquad \delta U = F \, \delta s \cos \alpha \qquad \textbf{(11-9)}$$

where $F$ and $\delta s$ are the magnitudes of the force $\mathbf{F}$ and virtual displacement $\delta \mathbf{s}$, respectively, and $\alpha$ is the angle between $\mathbf{F}$ and $\delta \mathbf{s}$.

A virtual displacement may also be a rotation of the body. The virtual work done by a couple $\mathbf{C}$ during an infinitesimal virtual angular displacement $\delta \boldsymbol{\theta}$ of the body is

$$\delta U = \mathbf{C} \cdot \delta \boldsymbol{\theta} \qquad \text{or} \qquad \delta U = M \, \delta \theta \cos \alpha \qquad \textbf{(11-10)}$$

where $M$ and $\delta \theta$ are the magnitudes of the couple $\mathbf{C}$ and virtual displacement $\delta \boldsymbol{\theta}$, respectively, and $\alpha$ is the angle between $\mathbf{C}$ and $\delta \boldsymbol{\theta}$. Since the infinitesimal virtual displacements $\delta s$ and $\delta \theta$ in Eqs. 11-9 and 11-10 refer to fictitious movements, the equations cannot be integrated.

A 500-lb block $A$ is held in equilibrium on an inclined surface with a cable and weight system and a force $\mathbf{P}$ as shown in Fig. 11-5a. When the force $\mathbf{P}$ is removed, block $A$ slides down the incline at a constant velocity for a distance of 10 ft. The coefficient of friction between block $A$ and the inclined surface is 0.20. Determine

a. The work done by the cable on block $A$.
b. The work done by gravity on block $A$.
c. The work done by the surface of the incline on block $A$.
d. The total work done by all forces on block $A$.
e. The work done by the cable on block $B$.
f. The work done by gravity on block $B$.
g. The total work done by all forces on block $B$.

(a)

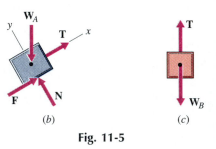

(b)                    (c)

**Fig. 11-5**

## SOLUTION

Free-body diagrams for blocks $A$ and $B$ are shown in Fig. 11-5b. Four forces act on block $A$: the cable tension $\mathbf{T}$, the weight $\mathbf{W}_A$, and the normal and frictional forces $\mathbf{N}$ and $\mathbf{F}$ at the surface of contact between the block and the inclined surface. Two forces act on block $B$: the cable tension $\mathbf{T}$ and the weight $\mathbf{W}_B$. Since the blocks move at a constant velocity, they are in equilibrium and the equilibrium equations applied to block $A$ yield

$$N = W_A \cos 30° = 500 \cos 30° = 433 \text{ lb}$$
$$F = \mu N = 0.20(433) = 86.6 \text{ lb}$$
$$T = W_A \sin 30° - F = 500 \sin 30° - 86.6 = 163.4 \text{ lb}$$

The equilibrium equations applied to block $B$ yield

$$W_B = T = 163.4 \text{ lb}$$

Since all of the forces are constant in both magnitude and direction during the movements of the blocks, the work can be computed by using Eq. 11-4 (in two dimensions, so that $z_2 - z_1 = 0$)

$$U_{1\to2} = F_x(x_2 - x_1) + F_y(y_2 - y_1)$$

Therefore, for block $A$, $x_2 - x_1 = -10$ ft, $y_2 - y_1 = 0$ ft, and

a.  $U_T = 163.4(-10) = -1634 \text{ ft} \cdot \text{lb}$  Ans.
b.  $U_W = (-500 \cos 60°)(-10) = 2500 \text{ ft} \cdot \text{lb}$  Ans.

Alternatively, the weight force acts vertically downward and block $A$ drops a vertical distance $h = 10 \sin 30°$ so that

$U_W = W_A h = 500(10 \sin 30°) = 2500 \text{ ft} \cdot \text{lb}$

c.  $U_F = 86.6(-10) = -866 \text{ ft} \cdot \text{lb}$  Ans.

$$U_N = 0(-10) + 433(0) = 0 \text{ ft} \cdot \text{lb}$$  Ans.

d.  $U_{\text{total}} = U_T + U_W + U_F + U_N = -1634 + 2500 - 866 + 0 = 0$  Ans.

and for block B, $x_2 - x_2 = 0$ ft, $y_2 - y_1 = 10$ ft, and

e.  $U_T = 163.4(10) = 1634 \text{ ft} \cdot \text{lb}$  Ans.
f.  $U_W = 163.4(-10) = -1634 \text{ ft} \cdot \text{lb}$  Ans.
g.  $U_{\text{total}} = U_T + U_W = -1634 + 1634 = 0$  Ans.

> The work done by a force $F$ during a linear displacement $ds$ in the direction of the force is $dU = F\,ds$.

> The total work done by a system of forces is the algebraic sum of the work done by the individual forces.

A constant couple $\mathbf{C} = 25\mathbf{i} + 35\mathbf{j} - 50\mathbf{k}$ N · m acts on a rigid body. The unit vector associated with the fixed axis of rotation of the body for an infinitesimal angular displacement $d\theta$ is $\mathbf{e}_\theta = 0.667\mathbf{i} + 0.333\mathbf{j} + 0.667\mathbf{k}$. Determine the work done on the body by the couple during an angular displacement of 2.5 rad.

**SOLUTION**

The magnitude (moment $M$) of the couple is

$$M = \sqrt{(25)^2 + (35)^2 + (-50)^2} = 65.95 \text{ N · m}$$

The unit vector associated with the couple is

$$\mathbf{e}_C = \frac{+25}{65.95}\mathbf{i} + \frac{+35}{65.95}\mathbf{j} + \frac{-50}{65.95}\mathbf{k} = 0.379\mathbf{i} + 0.531\mathbf{j} - 0.758\mathbf{k}$$

The cosine of the angle between the axis of the couple and the axis of rotation of the body is

$$
\begin{aligned}
\cos\alpha &= \mathbf{e}_C \cdot \mathbf{e}_\theta \\
&= (0.379\mathbf{i} + 0.531\mathbf{j} - 0.758\mathbf{k}) \cdot (0.667\mathbf{i} + 0.333\mathbf{j} + 0.667\mathbf{k}) \\
&= -0.0760
\end{aligned}
$$

Therefore,

$$\alpha = 94.36°$$

The work done by the couple during the finite rotation of 2.5 rad can be obtained by using Eq. 11-8. Thus,

$$dU = M \cos\alpha \, d\theta = 65.95(-0.0760) \, d\theta = -5.0122 \, d\theta$$

$$U = \int_0^{2.5} dU = \int_0^{2.5} (M \cos\alpha) \, d\theta$$

$$= -5.0122 \int_0^{2.5} d\theta = -12.53 \text{ N · m} \qquad \text{Ans.}$$

> The work done by the moment $M$ of a couple during an angular displacement $d\theta$ in the plane of the couple is $dU = M \, d\theta$.

Alternatively,

$$U = \int_0^{2.5} \mathbf{C} \cdot d\boldsymbol{\theta}$$

$$= \int_0^{2.5} (25\mathbf{i} + 35\mathbf{j} - 50\mathbf{k}) \cdot (0.667\mathbf{i} + 0.333\mathbf{j} + 0.667\mathbf{k}) \, d\theta$$

$$= \int_0^{2.5} -5.02 \, d\theta = -12.55 \text{ N · m} \qquad \text{Ans.}$$

> The work done by a couple $\mathbf{C}$ during an infinitesimal angular displacement $d\theta$ is given by the dot product $dU = \mathbf{C} \cdot d\boldsymbol{\theta}$.

The slight difference between this answer and the previous answer is due to the way the numbers were rounded off before being multiplied together.

# PROBLEMS

## Introductory Problems

**11-1\*** Determine the work done by a locomotive in drawing a freight train for 1 mi at constant speed on a level track if the locomotive exerts a constant force of 6000 lb on the train.

**11-2\*** A horse tows a canal boat with a rope that makes an angle of 10° with the towpath. If the tension in the rope is 800 N, how much work does the horse do while pulling the boat for 300 m along the canal?

**11-3** A box is dragged across a floor by using a rope that makes an angle of 30° with the horizontal. Determine the work done if the tension in the rope is 50 lb and the horizontal distance moved is 12 ft.

**11-4** A child on a sled is pulled by using a rope that makes an angle of 40° with the horizontal. Determine the work done if the tension in the rope is 75 N and the horizontal distance moved is 30 m.

**11-5\*** A constant force acting on a particle can be expressed in Cartesian vector form as $\mathbf{F} = 8\mathbf{i} - 6\mathbf{j} + 2\mathbf{k}$ lb. Determine the work done by the force on the particle if the displacement of the particle can be expressed in Cartesian vector form as $\mathbf{s} = 5\mathbf{i} + 4\mathbf{j} + 6\mathbf{k}$ ft.

**11-6** A constant force acting on a particle can be expressed in Cartesian vector form as $\mathbf{F} = 3\mathbf{i} + 5\mathbf{j} - 4\mathbf{k}$ N. Determine the work done by the force on the particle if the displacement of the particle can be expressed in Cartesian vector form as $\mathbf{s} = 4\mathbf{i} - 2\mathbf{j} + 3\mathbf{k}$ m.

## Intermediate Problems

**11-7\*** A 175-lb man climbs a flight of stairs 12 ft high. Determine

**a.** The work done by the man.
**b.** The work done on the man by gravity.

**11-8\*** A box with a mass of 600 kg is dragged up an incline 12 m long and 4 m high by using a cable that is parallel to the incline. The force in the cable is 2500 N. Determine

**a.** The work done on the box by the cable.
**b.** The work done on the box by gravity.

**11-9** A constant couple $\mathbf{C} = 120\mathbf{i} + 75\mathbf{j} - 150\mathbf{k}$ ft · lb acts on a rigid body. The unit vector associated with the fixed axis of rotation of the body for an infinitesimal angular displacement $d\boldsymbol{\theta}$ is $\mathbf{e}_\theta = 0.600\mathbf{i} + 0.300\mathbf{j} - 0.742\mathbf{k}$. Determine the work done on the body by the couple during an angular displacement of 0.75 rad.

**11-10** A constant couple $\mathbf{C} = 200\mathbf{i} + 300\mathbf{j} + 350\mathbf{k}$ N · m acts on a rigid body. The unit vector associated with the fixed axis of rotation of the body for an infinitesimal angular displacement $d\boldsymbol{\theta}$ is $\mathbf{e}_\theta = 0.250\mathbf{i} + 0.350\mathbf{j} - 0.903\mathbf{k}.$ Determine the work done on the body by the couple during an angular displacement of 1.5 rad.

## Challenging Problems

**11-11\*** A 100-lb block is pushed at constant speed for a distance of 20 ft along a level floor by a force that makes an angle of 35° with the horizontal, as shown in Fig. P11-11. The coefficient of friction between the block and the floor is 0.35. Determine

**a.** The work done on the block by the force.
**b.** The work done on the block by gravity.
**c.** The work done on the block by the floor.

**Fig. P11-11**

**11-12\*** A block with a mass of 100 kg slides down an inclined surface that is 5 m long and makes an angle of 25° with the horizontal. A man pushes horizontally on the block so that it slides down the incline at a constant speed. The coefficient of friction between the block and the inclined surface is 0.15. Determine

**a.** The work done on the block by the man.
**b.** The work done on the block by gravity.
**c.** The work done on the block by the inclined surface.

**11-13** A crate weighing 300 lb is supported by a rope that is 40 ft long. A man pushes the crate 8 ft horizontally and holds it there. Determine

**a.** The work done on the crate by the man.
**b.** The work done on the crate by the rope.
**c.** The work done on the crate by gravity.

**11-14** A steel bar of uniform cross section is 2 m long and has a mass of 25 kg. The bar is supported in a vertical position by a horizontal pin at the top end of the bar. Determine the work done on the bar by gravity as the bar rotates 60° about the pin in a vertical plane.

## 11-3  PRINCIPLE OF VIRTUAL WORK AND EQUILIBRIUM

The principle of virtual work can be stated as follows:

> If the virtual work done by all external forces (or couples) acting on a particle, a rigid body, or a system of connected rigid bodies with ideal (frictionless) connections and supports is zero for all virtual displacements of the system, the system is in equilibrium.

The principle of virtual work can be expressed mathematically as

$$\delta U = \sum_{i=1}^{m} \mathbf{F}_i \cdot \delta \mathbf{s}_i \; + \; \sum_{j=1}^{n} \mathbf{C}_j \cdot \delta \boldsymbol{\theta}_j = 0 \qquad \textbf{(11-11)}$$

### 11-3.1  Equilibrium of a Particle

Consider the particle shown in Fig. 11-6, which is acted on by several forces $\mathbf{F}_1, \mathbf{F}_2, ..., \mathbf{F}_n$. The work done on the particle by these forces during an arbitrary virtual displacement $\delta \mathbf{s}$ is

$$\begin{aligned}\delta U &= \mathbf{F}_1 \cdot \delta \mathbf{s} + \mathbf{F}_2 \cdot \delta \mathbf{s} + \cdots + \mathbf{F}_n \cdot \delta \mathbf{s} \\ &= (\mathbf{F}_1 + \mathbf{F}_2 + \cdots + \mathbf{F}_n) \cdot \delta \mathbf{s} = \Sigma \mathbf{F} \cdot \delta \mathbf{s} = \mathbf{R} \cdot \delta \mathbf{s} \qquad \textbf{(11-12)}\end{aligned}$$

where $\mathbf{R}$ is the resultant of the forces acting on the particle.

Expressing the resultant $\mathbf{R}$ and the virtual displacement $\delta \mathbf{s}$ in Cartesian vector form and computing the vector scalar product yields

$$\begin{aligned}\delta U = \mathbf{R} \cdot \delta \mathbf{s} &= (\Sigma F_x \mathbf{i} + \Sigma F_y \mathbf{j} + \Sigma F_z \mathbf{k}) \cdot (\delta x \, \mathbf{i} + \delta y \, \mathbf{j} + \delta z \, \mathbf{k}) \\ &= \Sigma F_x \, \delta x + \Sigma F_y \, \delta y + \Sigma F_z \, \delta z \qquad \textbf{(11-13)}\end{aligned}$$

Applying the principle of virtual work by combining Eqs. 11-11 and 11-13 yields

$$\delta U = \mathbf{R} \cdot \delta \mathbf{s} = \Sigma F_x \, \delta x + \Sigma F_y \, \delta y + \Sigma F_z \, \delta z = 0 \qquad \textbf{(11-14)}$$

By considering virtual displacements ($\delta x$, $\delta y$, and $\delta z$) taken one at a time in each of the three mutually perpendicular coordinate directions,

$$\Sigma F_x = 0 \qquad \Sigma F_y = 0 \qquad \Sigma F_z = 0$$

Thus, the virtual work equation $\delta U = 0$ is simply an alternative statement of the equilibrium equations for a particle. The principle of virtual work does not simplify the solution of problems involving equilibrium of a particle since the equations $\delta U = 0$ and $\Sigma F = 0$ are equivalent.

### 11-3.2  Equilibrium of a Rigid Body

If a rigid body is in equilibrium, all particles forming the body must be in equilibrium. Therefore, according to the principle of virtual work, the total virtual work of all forces, both internal and external, acting on all of the particles must be zero. Since the internal forces between particles occur as equal, opposite, collinear pairs, the work done by each pair of forces sum to zero during any virtual displacement of the rigid body. Thus, only the external forces do work during any virtual displacement of the body. Any system of forces acting on a rigid body can be replaced by a resultant force $\mathbf{R}$ and a resultant couple $\mathbf{C}$. Therefore, the work done on a rigid body by the external forces during an

**Figure 11-6**  Virtual displacement of a particle acted on by several forces.

*A postal patron tries to close the door on a mail box after inserting a package. If the door is blocked as shown in Fig. CE11-1a, what force must the door hinge be able to resist if P = 45 lb, a = 10 in., and b = 15 in.*

(a)

(b)

(c)

**Fig. CE11-1**

## SOLUTION

If the door is in equilibrium, the virtual work done by all external forces acting on the door must be zero for all virtual displacements of the door. In Fig. CE11-1b, a small virtual rotation $\delta\theta$ of the door about the hinge at $A$ is assumed. As a result of this rotation, $\delta x = \delta y = 0$ at hinge $A$, $\delta y = b\ \delta\theta$ at support $B$, and $\delta x = -a\ \delta\theta$ at door handle $C$. As a result, the virtual work principle $\delta U = 0$ gives

$$\delta U = A_x(0) + A_y(0) + P(-a\ \delta\theta) + B_y(b\ \delta\theta) = 0$$

or

$$B_y(b) - P(a) = 0 \qquad B_y = \frac{a}{b}P = \frac{10}{15}(45) = 30\ \text{lb}\uparrow$$

which is simply a statement of the equilibrium equation $\Sigma M_A = 0$.

Similarly, for the virtual displacement $\delta x$ of Fig. CE11-1c,

$$\delta U = A_x(\delta x) + A_y(0) + P(\delta x) + B_y(0) = 0$$

or

$$A_x + P = 0 \qquad A_x = -P = -45\ \text{lb} = 45\ \text{lb}\leftarrow$$

which is simply a statement of the equilibrium equation $\Sigma F_x = 0$.

Finally, for the virtual displacement $\delta y$ of Fig. CE11-1d,

$$\delta U = A_x(0) + A_y(\delta y) + P(0) + B_y(\delta y) = 0$$

or

$$A_y + B_y = 0 \qquad A_y = -B_y = -30\ \text{lb} = 30\ \text{lb}\downarrow$$

which is simply a statement of the equilibrium equation $\Sigma F_y = 0$. Finally,

$$A = \sqrt{(A_x)^2 + (A_y)^2} = \sqrt{(45)^2 + (30)^2} = 54.1\ \text{lb}$$
$$\mathbf{A} = 54.1\ \text{lb} \nearrow 33.7°$$

Ans.

(d)

arbitrary linear virtual displacement $\delta \mathbf{s}$ and an arbitrary angular virtual displacement $\delta \boldsymbol{\theta}$ is

$$\delta U = \mathbf{R} \cdot \delta \mathbf{s} + \mathbf{C} \cdot \delta \boldsymbol{\theta} \qquad \text{(11-15)}$$

Expressing the resultant force $\mathbf{R}$, the resultant couple $\mathbf{C}$, and virtual displacements $\delta \mathbf{r}$ and $\delta \boldsymbol{\theta}$ in Cartesian vector form and computing the vector scalar products yields

$$\begin{aligned}
\delta U &= \mathbf{R} \cdot \delta \mathbf{s} + \mathbf{C} \cdot \delta \boldsymbol{\theta} \\
&= (\Sigma F_x \mathbf{i} + \Sigma F_y \mathbf{j} + \Sigma F_z \mathbf{k}) \cdot (\delta x \mathbf{i} + \delta y \mathbf{j} + \delta z \mathbf{k}) \\
&\quad + (\Sigma M_x \mathbf{i} + \Sigma M_y \mathbf{j} + \Sigma M_z \mathbf{k}) \cdot (\delta \theta_x \mathbf{i} + \delta \theta_y \mathbf{j} + \delta \theta_z \mathbf{k}) \\
&= \Sigma F_x \, \delta x + \Sigma F_y \, \delta y + \Sigma F_z \, \delta z + \Sigma M_x \, \delta \theta_x + \Sigma M_y \, \delta \theta_y + \Sigma M_z \, \delta \theta_z
\end{aligned}$$
$$\text{(11-16)}$$

Applying the principle of virtual work by combining Eqs. 11-11 and 11-16 yields

$$\delta U = \Sigma F_x \, \delta x + \Sigma F_y \, \delta y + \Sigma F_z \, \delta z + \Sigma M_x \, \delta \theta_x + \Sigma M_y \, \delta \theta_y + \Sigma M_z \, \delta \theta_z = 0$$
$$\text{(11-17)}$$

By considering virtual linear displacements ($\delta x$, $\delta y$, and $\delta z$) and virtual angular displacements ($\delta \theta_x$, $\delta \theta_y$, and $\delta \theta_z$) taken one at a time,

$$\Sigma F_x = 0 \qquad \Sigma F_y = 0 \qquad \Sigma F_z = 0$$
$$\Sigma M_x = 0 \qquad \Sigma M_y = 0 \qquad \Sigma M_z = 0$$

Therefore, the virtual work equation $\delta U = 0$ is simply an alternative statement of the equilibrium equations for a rigid body. The principle of virtual work does not simplify the solution of problems involving equilibrium of a single rigid body since the equation $\delta U = 0$ is equivalent to the equilibrium equations $\Sigma F = 0$ and $\Sigma M = 0$.

## 11-3.3   Equilibrium of an Ideal System of Connected Rigid Bodies

The principle of virtual work can also be used to study systems of connected rigid bodies. Frequently it is possible to solve such problems by using the complete system rather than individual free-body diagrams of each member of the system.

When the system remains connected during the virtual displacement, only the work of forces external to the system need be considered, since the net work done by the internal forces at connections between members during any virtual displacement is zero because the forces exist as equal, opposite, collinear pairs. Such a condition exists when the connection is a smooth pin, a smooth roller, or an inextensible link or cable. When the reaction exerted by a support is to be determined, the restraint is replaced by a force and the body is given a virtual displacement with a component in the direction of the force. The virtual work done by the reaction and all other forces acting on the body is computed. If several forces are to be determined, the system of bodies can be given a series of separate virtual displacements in which only one of the unknown forces does virtual work during each displacement.

The problems in this chapter will be limited to a single degree of freedom, systems for which the virtual displacements of all points can be expressed in terms of a single variable (displacement).

A beam is loaded and supported as shown in Fig. 11-7a. Use the method of virtual work to determine the reaction at support B. Neglect the weight of the beam.

Fig. 11-7

## SOLUTION

A free-body diagram for the beam is shown in Fig. 11-7b. The 100-lb/ft distributed load has been replaced by its resultant $R = w\ell = 100(8) = 800$ lb. The beam is given a counterclockwise virtual angular displacement $\delta\theta$ about support A, which produces the virtual linear displacements $\delta y_B = 22\ \delta\theta$, $\delta y_C = 15\ \delta\theta$, and $\delta y_D = 4\ \delta\theta$ at support B and load points C and D, respectively, as shown in the displacement diagram (Fig. 11-7c). The virtual work done as a result of these linear and angular virtual displacements is given by Eqs. 11-9 and 11-10 as

$$\delta U = F\ \delta s \cos\alpha \quad \text{and} \quad \delta U = M\ \delta\theta \cos\alpha$$

Thus,

$$\delta U_B = B_y\ \delta y_B \cos 0° = B_y\ (22\ \delta\theta) \cos 0° = 22 B_y\ \delta\theta$$
$$\delta U_C = F_C\ \delta y_C \cos 180° = 500\ (15\ \delta\theta) \cos 180° = -7500\ \delta\theta$$
$$\delta U_D = F_D\ \delta y_D \cos 180° = 800\ (4\ \delta\theta) \cos 180° = -3200\ \delta\theta$$
$$\delta U_M = M\ \delta\theta \cos 180° = 300\ \delta\theta \cos 180° = -300\ \delta\theta$$

The total work on the beam is zero when the beam is in equilibrium. Thus,

$$\delta U_{\text{total}} = \delta U_B + \delta U_C + \delta U_D + \delta U_M = (22 B_y - 7500 - 3200 - 300)\ \delta\theta = 0$$

Since $\delta\theta \neq 0$

$$(22 B_y - 7500 - 3200 - 300) = 0 \qquad \mathbf{B}_y = 500\ \text{lb}\uparrow \qquad \text{Ans.}$$

Solutions to problems of this type are much simpler if the equations of equilibrium are used. As an example, for this problem,

$$+\downarrow\Sigma M_A = 0 \qquad B_y(22) - 500(15) - 300 - 800(4) = 0 \qquad \mathbf{B}_y = 500\ \text{lb}\uparrow \quad \text{Ans.}$$

> A force does positive virtual work if its linear virtual displacement has the same direction as the force.

> A couple does positive virtual work if its angular virtual displacement has the same direction as the sense of rotation of the couple.

The slender bar shown in Fig. 11-8a is 7.2 m long and has a mass of 100 kg. The bar rests against smooth surfaces at supports A and B. Use the method of virtual work to determine the magnitude of the force **F** required to maintain the bar in the equilibrium position shown in the figure.

(a)

(b)

**Fig. 11-8**

(c)

## SOLUTION

A free-body diagram of the bar is shown in Fig. 11-8b, and a virtual displacement diagram is shown in Fig. 11-8c. The virtual displacements of interest can be expressed in terms of the displacement $\delta x_B$ in the direction of the force **F**. Thus,

$$(x_B - \delta x_B)^2 + (y_A + \delta y_A)^2 = x_B^2 + y_A^2 = L^2 \tag{a}$$

from which

$$x_B^2 - 2x_B\delta x_B + \delta x_B^2 + y_A^2 + 2y_A\delta y_A + \delta y_A^2 = x_B^2 + y_A^2 \tag{b}$$

Since the virtual displacements may be considered to be very small, the terms $\delta x_B^2$ and $\delta y_A^2$ can be neglected and Eq. b simplifies to

$$\delta y_A = \frac{x_B}{y_A}\,\delta x_B$$

The y-component of the virtual displacement of the center of gravity is

$$\delta y_G = \frac{1}{2}\,\delta y_A = \frac{1}{2}\frac{x_B}{y_A}\,\delta x_B$$

For the given geometry

$$y_A = \sqrt{L^2 - x_B^2} = \sqrt{(7.2)^2 - (4)^2} = 5.987 \text{ m}$$

Therefore,

$$\delta y_G = \frac{1}{2}\frac{x_B}{y_A}\,\delta x_B = \frac{1}{2}\frac{4}{5.987}\,\delta x_B = 0.3341\,\delta x_B$$

Forces **A** and **B** undergo no displacements in the directions of the forces as a result of displacement $\delta x_B$; therefore, they do no work. The work of the force

> The length L of the bar remains constant; therefore, the Pythagorean theorem can be used to relate the virtual displacements at ends A and B of the bar.

**627**

**F** and the work of the weight **W** on the bar as a result of displacement $\delta x_B$ are as follows:

Since $F$ and $\delta x_B$ are in the same direction,

$$\delta U_F = F \, \delta x_B$$

Since $W$ and $\delta y$ are in opposite directions,

$$\delta U_W = W \, (-\delta y_G) = -0.3341W \, \delta x_B$$

For the bar to be in equilibrium, the total work on the bar must be zero. Thus,

$$\delta U_{\text{total}} = \delta U_F + \delta U_W = F \, \delta x_B - 0.3341W \, \delta x_B = (F - 0.3341W) \, \delta x_B = 0$$

Since $\delta x_B \neq 0$,

$$F - 0.3341W = 0$$
$$F = 0.3341 \, W = 0.3341(mg) = 0.3341(100)(9.81) = 328 \text{ N} \qquad \text{Ans.}$$

The relationships between the virtual displacements can also be obtained by using an angular displacement $\delta\theta$ as shown in Fig. 11-8d. Thus,

$$x_B = -L \cos \theta \qquad \text{and} \qquad \delta x_B = L \sin \theta \, \delta\theta = 0.8315L \, \delta\theta$$
$$y_A = L \sin \theta \qquad \text{and} \qquad \delta y_A = L \cos \theta \, \delta\theta = 0.5556L \, \delta\theta$$
$$\delta y_G = \frac{1}{2} \, \delta y_A = 0.2778L \, \delta\theta$$

$$\delta U_F = F \, \delta x_B = F(0.8315L \, \delta\theta) = 0.8315FL \, \delta\theta$$
$$\delta U_W = W(-\delta y_G) = W(-0.2778L \, \delta\theta) = -0.2778WL \, \delta\theta$$
$$\delta U_{\text{total}} = \delta U_F + \delta U_W = (0.8315F - 0.2778W) \, L \, \delta\theta = 0$$

Since $L$ and $\delta\theta \neq 0$

$$0.8315F - 0.2778W = 0$$
$$F = 0.3341W = 0.3341(mg) = 0.3341(100)(9.81) = 328 \text{ N} \qquad \text{Ans.}$$

(d)

A force does positive virtual work if its linear virtual displacement has the same direction as the force.

Trigonometric relations and standard mathematical rules of differentiation can be used to relate virtual displacements at different points on the bar.

A system of pin-connected bars supports a system of loads shown in Fig. 11-9a. Use the method of virtual work to determine the horizontal component of the reaction at support G. Neglect the weights of the bars.

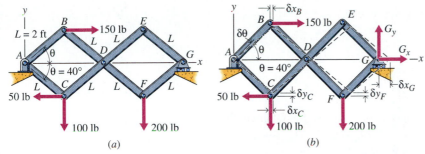

**Fig. 11-9**

## SOLUTION

A virtual displacement diagram of the system with the loads superimposed is shown in Fig. 11-9b. The virtual displacements $\delta x$ and $\delta y$ at each of the load points and at the support G can be expressed in terms of a virtual angular displacement $\delta\theta$ about support A. Thus,

Trigonometric relations and standard mathematical rules of differentiation can be used to relate virtual displacements at different points on the system of bars.

$$x_B = L \cos \theta \qquad \delta x_B = -L \sin \theta \, \delta\theta = -2(\sin 40°) \, \delta\theta = -1.2856 \, \delta\theta$$
$$x_C = L \cos \theta \qquad \delta x_C = -L \sin \theta \, \delta\theta = -2(\sin 40°) \, \delta\theta = -1.2856 \, \delta\theta$$
$$x_G = 4L \cos \theta \qquad \delta x_G = -4L \sin \theta \, \delta\theta = -4(2)(\sin 40°) \, \delta\theta = -5.1423 \, \delta\theta$$

$$y_C = -L \sin \theta \qquad \delta y_C = -L \cos \theta \, \delta\theta = -2(\cos 40°) \, \delta\theta = -1.5321 \, \delta\theta$$
$$y_F = -L \sin \theta \qquad \delta y_F = -L \cos \theta \, \delta\theta = -2(\cos 40°) \, \delta\theta = -1.5321 \, \delta\theta$$

The virtual work done by each of the forces can then be computed by carefully noting the directions of the forces and the directions of the associated virtual displacements. Thus,

A force does positive virtual work if its linear virtual displacement has the same direction as the force.

$$\delta U_{Bx} = B_x(\delta x_B) = 150(-1.2856 \, \delta\theta) = -192.84 \, \delta\theta$$
$$\delta U_{Cx} = C_x(\delta x_C) = 50(+1.2856 \, \delta\theta) = +64.28 \, \delta\theta$$
$$\delta U_{Cy} = C_y(\delta y_C) = 100(+1.5321 \, \delta\theta) = +153.21 \, \delta\theta$$
$$\delta U_{Fy} = F_y(\delta y_F) = 200(+1.5321 \, \delta\theta) = +306.42 \, \delta\theta$$
$$\delta U_{Gx} = G_x(\delta x_G) = G_x(-5.1423 \, \delta\theta) = -5.1423G_x \, \delta\theta$$

Force $G_y$ does not undergo a virtual displacement in the direction of the force as a result of the virtual angular displacement $\delta\theta$; therefore, it does no work. For the system to be in equilibrium, the total work on the system must be zero. Thus

$$\delta U_{total} = \delta U_{Bx} + \delta U_{Cx} + \delta U_{Cy} + \delta U_{Fy} + \delta U_{Gx}$$
$$= (-192.84 + 64.28 + 153.21 + 306.42 - 5.1423G_x) \, \delta\theta$$
$$= (331.07 - 5.1423G_x) \, \delta\theta = 0$$

Since $\delta\theta \neq 0$,

$$331.07 - 5.1423G_x = 0$$
$$G_x = \frac{331.07}{5.1423} = 64.4 \text{ lb} \qquad\qquad \text{Ans.}$$

The two-bar mechanism shown in Fig. 11-10a is used to support a body W, which has a mass of 50 kg. A couple $C = 1500$ N · m is also applied to member CDE as shown in the figure. Use the method of virtual work to determine the magnitude of the force exerted by the link at support A to maintain the mechanism in equilibrium in the position shown in the figure. Neglect the weights of the bars.

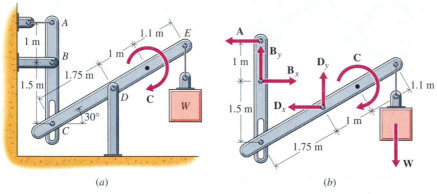

(a)                                                    (b)

**Fig. 11-10**

### SOLUTION

A free-body diagram mechanism is shown in Fig. 11-10b and a virtual displacement diagram of the mechanism is shown in Fig. 11-10c. The magnitudes of the virtual displacements $\delta x_A$ and $\delta y_E$ can be expressed in terms of a virtual angular displacement $\delta\theta$ about support B. Thus,

$$\delta x_A = 1 \, \delta\theta = 1.000 \, \delta\theta$$

Note from the small diagram shown in Fig. 11-10d that

$$1.5 \, \delta\theta = 1.75 \, \delta\phi \sin 30°$$

therefore,

$$\delta\phi = \frac{1.5 \, \delta\theta}{1.75(0.5000)} = 1.7143 \, \delta\theta$$

From the small diagram shown in Fig. 11-10e,

$$\delta y_E = 2.1 \, \delta\phi \cos 30° = 2.1(1.7143 \, \delta\theta) \cos 30° = 3.118 \, \delta\theta$$

Forces $B_x$, $B_y$, $D_x$, and $D_y$ undergo no virtual displacements in the directions of the forces as a result of the virtual angular displacement $\delta\theta$; therefore, they do no work. The work of the force $A$, the couple $C$, and the weight $W$ on the mechanism, as a result of the virtual angular displacement $\delta\theta$, is

$$\delta U_A = A \, \delta x_A \cos 180° = A(1.000 \, \delta\theta)(-1) = -1.000A \, \delta\theta$$
$$\delta U_C = C \, \delta\phi \cos 0° = 1500(1.7143 \, \delta\theta)(1) = 2571 \, \delta\theta$$
$$\delta U_W = W \, \delta y_E \cos 0° = mg \, \delta y_E \cos 0°$$
$$= 50(9.81)(3.118 \, \delta\theta)(1) = 1529 \, \delta\theta$$

The total work must be zero for the mechanism to be in equilibrium. Thus,

$$\delta U_{total} = \delta U_A + \delta U_C + \delta U_W = -1.000A \, \delta\theta + 2571 \, \delta\theta + 1529 \, \delta\theta$$
$$= (-1.000A + 2571 + 1529) \, \delta\theta = 0$$

A force does positive virtual work if its linear virtual displacement has the same direction as the force.

A couple does positive virtual work if its angular virtual displacement has the same direction as the sense of rotation of the couple.

(c)    (d)    (e)

Since $\delta\theta \neq 0$,

$$-1.000A + 2571 + 1529 = 0$$

Therefore,

$$A_x = 4100 \text{ N} = 4.10 \text{ kN} \qquad \text{Ans.}$$

# PROBLEMS

Use the method of virtual work to solve the following problems.

## Introductory Problems

**11-15*** Two carts carry boxes $A$ and $B$ as shown in Fig. P11-15. The two carts are connected by an inextensible cable that passes over an ideal pulley. Determine the angle $\theta$ for equilibrium if the weights of boxes $A$ and $B$ are 500 and 1000 lb, respectively.

**Fig. P11-15**

**11-16*** The cable and pulley system shown in Fig. P11-16 is in equilibrium. If the mass of block $B$ is 200 kg, determine the mass of block $A$.

**Fig. P11-16**

**11-17** A beam is loaded and supported as shown in Fig. P11-17. Determine the reaction at support $B$. Neglect the weight of the beam.

**Fig. P11-17**

**631**

**11-18** A beam is loaded and supported as shown in Fig. P11-18. Determine the reaction at support A. Neglect the weight of the beam.

**Fig. P11-18**

**11-19\*** A system of beams is loaded and supported as shown in Fig. P11-19. Determine the reaction at support C. Neglect the weight of the beams.

**Fig. P11-19**

**11-20\*** A system of beams is loaded and supported as shown in Fig. P11-20. Determine the reaction at support A. Neglect the weight of the beams.

**Fig. P11-20**

**11-21** Two spheres are supported with slender rods as shown in Fig. P11-21. Determine the angle $\theta$ for equilibrium if the weights of spheres A and B are 25 and 10 lb, respectively.

**Fig. P11-21**

**11-22** A three-bar frame is loaded and supported as shown in Fig. P11-22. Determine the angle $\theta$ for equilibrium if $\mathbf{P} = 500$ N, $\mathbf{M} = 350$ N $\cdot$ m, $L = 1$ m, and the masses of blocks A and B are 50 and 75 kg, respectively. The weights of the bars are negligible.

**Fig. P11-22**

**Intermediate Problems**

**11-23\*** Determine the force transmitted by member CD of the truss shown in Fig. P11-23.

**Fig. P11-23**

**11-24\*** Determine the force transmitted by member BC of the truss shown in Fig. P11-24.

3 m

2.5 m — 2.5 m — 2.5 m

4 kN     3 kN     2 kN

**Fig. P11-24**

**11-25** A frame is loaded and supported as shown in Fig. P11-25. If the weight of block *A* is 500 lb and angle $\theta$ equals 45°, determine the magnitude of force **F** required to maintain equilibrium.

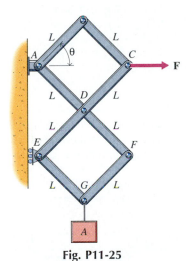

**Fig. P11-25**

**11-26** A frame is loaded and supported as shown in Fig. P11-26. Determine the magnitude of the force **F** required to maintain equilibrium if **P** = 250 N and $\theta$ = 45°.

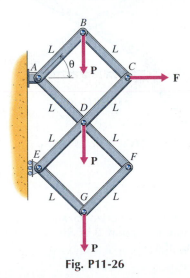

**Fig. P11-26**

**11-27*** The slender bar shown in Fig. P11-27 weighs 150 lb. If **P** = 75 lb and **M** = 50 ft · lb, determine the magnitude of the force **F** required to maintain the bar in the equilibrium position shown in the figure.

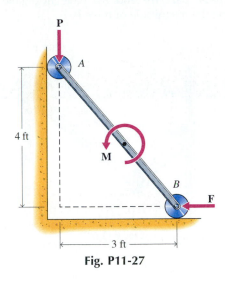

**Fig. P11-27**

**11-28** Determine the force transmitted by cable *CD* of the A-frame shown in Fig. P11-28.

**Fig. P11-28**

## Challenging Problems

**11-29\*** The two-bar linkage shown in Fig. P11-29 is in equilibrium. The collars at supports $A$ and $C$ slide on smooth rods. Bars $AB$ and $BC$ have negligible weight. Determine the magnitude of force $\mathbf{P}$.

**Fig. P11-29**

**11-30\*** Two blocks are suspended on a continuous inextensible cord as shown in Fig. P11-30. Determine the angle $\theta$ for equilibrium if the masses of blocks $A$ and $B$ are 50 and 40 kg, respectively.

**Fig. P11-30**

**Fig. P11-31**

**11-32** The mass of block $A$ shown in Fig. P11-32 is 250 kg. Determine the magnitude of the force $\mathbf{F}$ required to support the block in the equilibrium position shown in the figure.

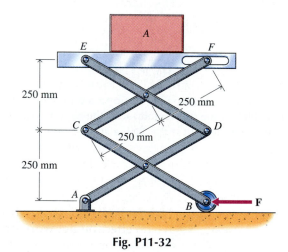

**Fig. P11-32**

**11-31** Determine the horizontal and vertical components of the reaction at support $B$ of the two-bar mechanism shown in Fig. P11-31.

**11-33\*** Determine the horizontal and vertical components of the reaction at support $B$ of the three-bar mechanism shown in Fig. P11-33.

**Fig. P11-33**

**11-34** Determine the horizontal and vertical components of the reaction at support $A$ of the two-bar frame shown in Fig. P11-34.

**Fig. P11-34**

## 11-4 POTENTIAL ENERGY AND EQUILIBRIUM

The potential energy of a body is a measure of its capacity for doing work. The potential energy can be defined quantitatively as the amount of work that the body is capable of doing against outside forces as it passes from a given position or configuration to some standard or reference position or configuration. A compressed spring is capable of doing work by virtue of the relative positions (configuration) of its particles. Potential energy is expressed in the same units as work: in joules (J) if SI units are used and in ft · lb or in. · lb if U.S. Customary units are used.

**Conservative Force Systems** When the work done by a body depends only on the initial and final configurations of the system and not at all on the paths taken by the parts of the system while moving from the initial state to the final state, the system is a conservative force system. Conservative force systems occur frequently in engineering. A common example is that of a system consisting of the earth and an elevated body (rigid or deformable). The work done on the body during any displacement is equal to the earth pull (weight) of the body times the vertical displacement of the center of gravity of the body. Any intermediate positions occupied by the body as it moves from its initial position to its final position have no effect on net work done on the body.

Another example is that of an elastic body such as a spring. If the body is elastic, the energy possessed by the body in a given strained condition (that is, for a given configuration of its particles) is independent of the movements of the particles that occurred while the body was being put in the given strained condition. The standard configuration may be arbitrarily chosen, but, for convenience, it is usually chosen so that the potential energy of the body is initially positive or zero. Thus, in the case of the earth and an elevated body, the earth is considered fixed and the standard configuration occurs when the body is in contact with the earth.

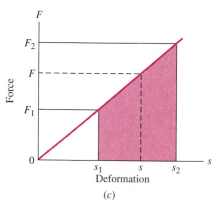

**Figure 11-11**   Deformation of a
spring under the action of a force **F.**

**Nonconservative Force Systems**   The most common case of a nonconservative system is one in which the system does work against frictional forces. Any rigid body under the action of a force system in which friction does not occur (or may be considered negligible) is a conservative system provided no change in state or condition of the body except that of position or configuration takes place.

### 11-4.1   Elastic Potential Energy

A deformable body that changes shape under load but resumes its original shape when the loads are removed is known as an elastic body. The spring shown in Fig. 11-11$a$ is an example of an elastic body that is widely used in engineering applications. When a tensile force is applied to the spring, the length of the spring increases. Similarly, when a compressive force is applied to the spring, the length decreases. The magnitude $F$ of the force that must be applied to the spring to stretch it by an amount $s$ is given by the expression

$$F = ks \tag{11-18}$$

where $s$ is the deformation of the spring from its unloaded position and $k$ is a constant known as the modulus of the spring. Any spring whose behavior is governed by Eq. 11-18 is an ideal linear elastic spring.

The work done in stretching an ideal spring from an initial unstretched position to a stretched position $s$ can be determined from Eq. 11-1. Since the force $F$ and displacement $s$ are in the same direction, the angle $\alpha$ is zero and Eq. 11-1 becomes

$$U = \int_0^s ks\, ds = \frac{1}{2} ks^2 \tag{11-19}$$

For this case, the work done by the force $F$ as it stretches the spring can be represented as the shaded triangular area under the load versus deformation curve shown in Fig. 11-11$b$. This area also represents the elastic potential energy $V_e$ stored in the spring as a result of the change in shape of the spring.

In a similar manner, the work done in stretching an ideal spring from an initial position $s_1$ to a further extended position $s_2$ can be determined from Eq. 11-1. Thus,

$$U = V_e = \int_{s_1}^{s_2} ks\, ds = \frac{1}{2} k(s_2^2 - s_1^2) \tag{11-20}$$

In this case, the work done by the force $F$ as it stretches the spring can be represented as the shaded trapezoidal area under the load versus deformation curve shown in Fig. 11-11$c$. It is important to note here that Eq. 11-19 is valid only if the deflection of the spring is measured from its undeformed position.

As a spring is being deformed (either stretched or compressed), the force on the spring and the displacement are in the same direction; therefore, the work done on the spring is positive, which increases its potential energy. If a spring is initially deformed and gradually released, the force and displacement are in opposite directions, the work is negative, and the potential energy decreases. The force of a spring on a body is opposite to the force of the body on the spring. Since both the body and the end of the spring have the same displacement, the work done by the spring on the body results in an equal decrease in the potential energy of the spring.

A torsional spring, which is used to resist rotation rather than linear displacement, can also store and release potential energy. The magnitude of the torque (twisting moment) $T$ that must be applied to a torsional spring to produce a rotation $\theta$ is given by the expression

$$T = k\theta \tag{11-21}$$

where $T$ is the torque, $\theta$ is the angular deformation of the spring from its unloaded position, and $k$ is a constant known as the modulus of the spring. The work $U$ done by the torque $T$, which is equal to the potential energy $V_e$ stored in the spring, is given by the expression

$$U = V_e = \int_0^\theta T \, d\theta = \int_0^\theta k\theta \, d\theta = \frac{1}{2} k\theta^2 \tag{11-22}$$

which is analogous to the expression for the linear spring.

During a virtual displacement $\delta s$ of a spring, the virtual work $\delta U$ done on the spring and the change in virtual elastic potential energy $\delta V_e$ of the spring are given by the expression

$$\delta U = \delta V_e = F \, \delta s = ks \, \delta s \tag{11-23}$$

## 11-4.2 Gravitational Potential Energy

The work done by the weight of a body is obtained by multiplying the weight $W$ of the body by the vertical displacement $dh$ of the center of gravity of the body. Thus,

$$U = W \, dh \tag{11-24}$$

If the displacement is toward the center of the earth, the work is positive; if it is away from the center of the earth, the work is negative. The work is independent of the path followed by the body in getting from its initial position to its final position; it depends only on the vertical displacement $h$ of the center of gravity of the body. Thus,

$$U = V_g = \int_0^h W \, dh = Wh = mgh \tag{11-25}$$

The gravitational potential energy possessed by the body is available to do work as the body returns to a lower position. The work done by the weight $W$ as it displaces downward will be positive. This downward movement will produce a decrease in the gravitational potential energy of the body equal to the work done. Thus,

$$\Delta U = W \, \Delta h = -\Delta V_g \tag{11-26}$$

During a virtual displacement $\delta h$ of the body, the virtual work $\delta U$ done by the weight of the body, and the change in virtual gravitational potential energy $\delta V_g$ of the body are given by the expression

$$\delta U = -\delta V_g = W \, \delta h \tag{11-27}$$

## 11-4.3 The Principle of Potential Energy

Since the work done by a linear spring on a body equals the negative of the change in elastic potential energy of the spring and the work done by the weight $W$ of a body equals the negative of the change in gravitational potential energy of the body, the virtual-work equation can be written as

$$\delta U + \delta V_e + \delta V_g = \delta U + \delta V = 0 \qquad \text{or} \qquad \delta U = -\delta V \quad \text{(11-28)}$$

where $V = V_e + V_g$ stands for the total potential energy of the system and $\delta U$ represents the virtual work done on the system during a virtual displacement by all external forces other than spring forces and gravitational forces. Thus, the principle of virtual work can be restated as follows:

> A system is in equilibrium if the virtual work done on the system by the external forces equals the change in elastic and gravitational potential energy of the system for all possible virtual displacements.

**System Having One Degree of Freedom**   For a system whose position can be defined in terms of a single variable (say $s$), we can write

$$V = V(s) \qquad \delta V = \frac{dV}{ds} \delta s$$

Since $\delta s \neq 0$, the principle of virtual work for equilibrium of the system, as represented by Eq. 11-28, yields

$$\frac{dV}{ds} = 0 \qquad \qquad \text{(11-29)}$$

Thus, the derivative of the potential energy of a conservative system in equilibrium with respect to its position variable is zero. This property is referred to as the principle of potential energy. A system whose position can be described by a single variable is known as a single-degree-of-freedom system.

**System Having *n* Degrees of Freedom**   If the position of a system depends on several independent variables, the total potential energy stored in the system will be a function of the $n$ independent variables. Thus,

$$V = V(s_1, s_2,\ldots, s_n)$$

The change in potential energy $\delta V$ of such a system is obtained by using the "chain rule" of differential calculus. Thus,

$$\delta V = \frac{\partial V}{\partial s_1} \delta s_1 + \frac{\partial V}{\partial s_2} \delta s_2 + \cdots + \frac{\partial V}{\partial s_n} \delta s_n = 0$$

Since $\delta s_1, \delta s_2,\ldots, \delta s_n$ are independent and not equal to zero,

$$\frac{\partial V}{\partial s_1} = 0 \qquad \frac{\partial V}{\partial s_2} = 0 \qquad \cdots \qquad \frac{\partial V}{\partial s_n} = 0 \qquad \text{(11-30)}$$

Equations 11-30 are mathematical expressions of the principle of potential energy for a conservative system with $n$ degrees of freedom.

## 11-5   STABILITY OF EQUILIBRIUM

The principle of potential energy used in Section 11-4 to determine the equilibrium position of a body or a system of bodies can also be used to determine the state of equilibrium of the body or the system of bodies. The three states of equilibrium (stable, neutral, and unstable) are illustrated in Fig. 11-12.

### 11-5.1 Stable Equilibrium

The ball in Fig. 11-12*a* is in a stable equilibrium position because gravity will cause the ball to return to the bottom of the pit once any perturbing force is removed. In this state of stable equilibrium, the work done on the ball during any small displacement is negative since the weight of the ball acts opposite to a component of the displacement; therefore, the potential energy of the ball increases. When the disturbing force is removed, the weight of the ball does positive work as the ball returns to its position of stable equilibrium at the bottom of the pit. Thus, it is evident that a configuration of stable equilibrium occurs at a point of minimum potential energy. Mathematically, the potential energy is minimum for a single-degree-of-freedom system when the second derivative of the potential energy with respect to the displacement variable is positive. Thus, for stable equilibrium

$$\frac{dV}{ds} = 0 \quad \text{and} \quad \frac{d^2V}{ds^2} > 0 \qquad \textbf{(11-31)}$$

### 11-5.2 Neutral Equilibrium

The ball shown in Fig. 11-12*b* is in a neutral equilibrium position on the horizontal plane because it will remain at any new position to which it is displaced once the disturbing force is removed. In this state of neutral equilibrium, no work is done on the ball during the displacement since the weight of the ball acts perpendicular to the direction of the displacement; therefore, the potential energy of the ball remains constant. Thus, the conditions for neutral equilibrium are

$$\frac{dV}{ds} = 0 \quad \text{and} \quad \frac{d^2V}{ds^2} = \frac{d^3V}{ds^3} = \cdots = 0 \qquad \textbf{(11-32)}$$

### 11-5.3 Unstable Equilibrium

The ball shown in Fig. 11-12*c* is in an unstable equilibrium position at the top of a hill because if the ball is perturbed, gravity will cause it to move away from its original equilibrium position. In this state of unstable equilibrium, the work done on the ball during any small displacement is positive, since the weight of the ball acts in the same direction as a component of the displacement; therefore, the potential energy of the ball decreases. When the disturbing force is removed, the potential energy continues to decrease as the weight of the body causes the body to continue to move away from the position of unstable equilibrium. Thus, a configuration of unstable equilibrium occurs at a point of maximum potential energy. Mathematically, the potential energy is maximum when the second derivative of the potential energy with respect to the displacement variable is negative. Thus, for unstable equilibrium

$$\frac{dV}{ds} = 0 \quad \text{and} \quad \frac{d^2V}{ds^2} < 0 \qquad \textbf{(11-33)}$$

If both the first and the second derivatives of $V$ are zero, the sign of a higher-order derivative must be used to determine the type of equi-

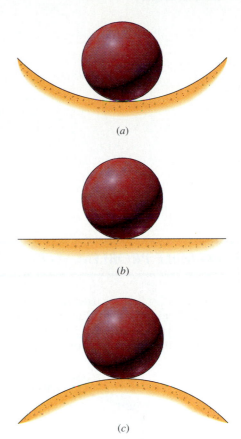

(a)

(b)

(c)

**Figure 11-12** Stable, neutral, and unstable equilibrium.

librium. The equilibrium will be stable if the order of the first nonzero derivative is even and the value of the derivative is positive. The equilibrium will be neutral if all derivatives are zero. In all other cases the equilibrium will be unstable.

If the system possesses several degrees of freedom, the potential energy $V$ depends on several variables, and a more advanced treatment involving the theory of functions of several variables is required to determine whether $V$ is minimum. Therefore, stability criteria for multiple-degree-of-freedom systems are beyond the scope of this text.

The use of the potential energy of a system for the determination of static equilibrium and stability is illustrated in the following example problems.

## EXAMPLE PROBLEM 11-7

Two bars ($L = 2$ ft) are used to support a weight ($W = 100$ lb) as shown in Fig. 11-13a. Determine the value of $k$ for equilibrium at $\theta = 30°$. Assume that the weights of the bars are negligible and that the spring is unstretched when $\theta = 0°$. Show that the equilibrium is stable.

(a)                                    (b)

**Fig. 11-13**

### SOLUTION

A free-body diagram for the system is shown in Fig. 11-13b. From the diagram it is evident that

$$y_B = -L \sin \theta \quad \text{and} \quad x_C = 2L \cos \theta$$

Therefore, the potential energies of the weight, spring, and system with respect to a reference position at $\theta = 0°$ are

$$V_W = W y_B = W(-L \sin \theta) = -WL \sin \theta$$
$$V_S = 1/2 k s^2 = 1/2 k (2L - x_C)^2 = 1/2 k (2L - 2L \cos \theta)^2 = 2kL^2(1 - \cos \theta)^2$$
$$V = V_W + V_S = -WL \sin \theta + 2kL^2(1 - \cos \theta)^2$$

For equilibrium (Eq. 11-29),

$$\frac{dV}{d\theta} = -WL \cos \theta + 4kL^2(1 - \cos \theta) \sin \theta = 0$$

Solving for $k$ yields

$$k = \frac{W}{4L(1 - \cos \theta) \tan \theta} = \frac{100}{4(2)(1 - \cos 30°) \tan 30°} = 161.6 \text{ lb/ft} \quad \text{Ans.}$$

From Eq. 11-31,

$$\frac{d^2V}{d\theta^2} = WL \sin\theta + 4kL^2 \cos\theta + 4kL^2 \sin^2\theta - 4kL^2 \cos^2\theta$$

With $W = 100$ lb, $L = 2$ ft, $k = 161.6$ lb/ft, and $\theta = 30°$,

$$\frac{d^2V}{d\theta^2} = +1046 > 0$$

Therefore, the equilibrium is stable.

A one-degree-of-freedom mechanical system is in equilibrium when the derivative of its total potential energy is zero.

A small displacement away from a stable equilibrium position produces an increase in potential energy.

## EXAMPLE PROBLEM 11-8

A mass ($m = 100$ kg) and a spring ($k = 50$ kN/m) are attached to a crank, which rotates about a frictionless pin at support $O$ as shown in Fig. 11-14a. The spring is unstretched when $\theta = 0°$. Determine the angle $\theta$ for equilibrium and show that the equilibrium is stable.

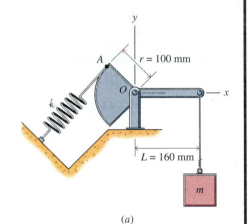

(a)

### SOLUTION

The potential energies of the mass, spring, and system with respect to a reference position at $\theta = 0°$ (see Fig. 11-14b) are

$$V_m = mg(-h) = mg(-L \sin\theta) = -mgL \sin\theta$$
$$V_S = 1/2ks^2 = 1/2k(r\theta)^2 = 1/2kr^2\theta^2$$
$$V = V_m + V_S = -mgL \sin\theta + 1/2kr^2\theta^2$$

For equilibrium (Eq. 11-29)

$$\frac{dV}{d\theta} = -mgL \cos\theta + kr^2\theta = 0$$

Therefore,

$$\frac{\cos\theta}{\theta} = \frac{kr^2}{mgL} = \frac{50(10^3)(0.100)^2}{100(9.81)(0.160)} = 3.1855 \qquad (a)$$

Equation $a$ can be solved by using trial and error or by using a numerical method. Solving by using the Newton–Raphson method yields

$$f(\theta) = \cos\theta - 3.1855\,\theta = 0$$
$$f'(\theta) = -\sin\theta - 3.1855 = 0$$
$$\delta = -\frac{f(\theta)}{f'(\theta)} = \frac{\cos\theta - 3.1855\,\theta}{\sin\theta + 3.1855}$$
$$\theta_{n+1} = \theta_n + \delta(\theta_n) \qquad \theta \text{ in radians}$$

| $n$ | $\theta_n$ | $\delta(\theta_n)$ |
|-----|-----------|--------------------|
| 0 | 0 | 0.3139 |
| 1 | 0.3139 | −0.0140 |
| 2 | 0.2999 | $9.94(10^{-6})$ |

(b)

**Fig. 11-14**

Thus,

$$\theta = 0.2999 \text{ rad} = 17.18°$$       Ans.

From Eq. 11.31

$$\frac{d^2V}{d\theta^2} = mgL \sin \theta + kr^2$$

$$= 100(9.81)(0.160)(\sin 17.18°) + 50(10^3)(0.100)^2$$
$$= +546 > 0$$

Therefore, the equilibrium is stable.       Ans.

## EXAMPLE PROBLEM 11-9

A system of bars and springs is used to support a weight $W$ as shown in Fig. 11-15a. Determine the value of $W$ for equilibrium and investigate the stability of equilibrium. The weights of the bars are negligible and the springs are unstretched when $\theta = 0°$. Assume that the spring forces remain horizontal.

(a)

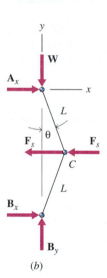
(b)

**Fig. 11-15**

### SOLUTION

A free-body diagram for the system is shown in Fig. 11-15b. From the diagram it is evident that

$$x_C = L \sin \theta \quad \text{and} \quad y_A = 2L - 2L \cos \theta$$

Therefore, the potential energies of the weight, springs, and system with respect to a reference position at $\theta = 0°$ are

$$V_W = Wy_A = W(2L - 2L \cos \theta) = -2WL(1 - \cos \theta)$$
$$V_S = 1/2ks^2 = 1/2k(x_C)^2 = 1/2k(L \sin \theta)^2 = 1/2kL^2 \sin^2 \theta$$
$$V = V_W + 2\,V_S = -2\,WL(1 - \cos \theta) + kL^2 \sin^2 \theta$$

For equilibrium (Eq. 11-29),

$$\frac{dV}{d\theta} = -2\,WL \sin \theta + 2kL^2 \sin \theta \cos \theta = 0$$

**642**

or
$$\sin \theta \, (- W + kL \cos \theta) = 0$$

Thus, the system is in equilibrium when

$$\sin \theta = 0 \quad \text{or} \quad \theta = 0$$
$$-W + kL \cos \theta = 0 \quad \text{or} \quad W = kL \cos \theta$$

From Eq. 11-31,

$$\frac{d^2V}{d\theta^2} = -2 \, WL \cos \theta - 2kL^2 \sin^2 \theta + 2kL^2 \cos^2 \theta$$

For $\theta = 0°$,

$$\frac{d^2V}{d\theta^2} = -2WL + 2kL^2 = 2L(-W + kL)$$

With $W < kL$,

$$\frac{d^2V}{d\theta^2} > 0 \qquad \text{the equilibrium is stable}$$

With $W = kL$,

$$\frac{d^2V}{d\theta^2} = 0 \qquad \text{the equilibrium is neutral}$$

With $W > kL$,

$$\frac{d^2V}{d\theta^2} < 0 \qquad \text{the equilibrium is unstable}$$

For $W = kL \cos \theta$,

$$\frac{d^2V}{d\theta^2} = -2 \, WL \cos \theta - 2kL^2 \sin^2 \theta + 2kL^2 \cos^2 \theta$$

$$= -2 \, WL \left(\frac{W}{kL}\right) - 2kL^2 \left[1 - \left(\frac{W}{kL}\right)^2\right] + 2kL^2\left(\frac{W}{kL}\right)^2 = \frac{2}{k} \, (W^2 - k^2L^2)$$

With $W > kL$,

$$\frac{d^2V}{d\theta^2} > 0 \qquad \text{the equilibrium is stable}$$

With $W = kL$,

$$\frac{d^2V}{d\theta^2} = 0 \qquad \text{the equilibrium is neutral}$$

With $W < kL$,

$$\frac{d^2V}{d\theta^2} < 0 \qquad \text{the equilibrium is unstable}$$

A small displacement away from a stable equilibrium position produces an increase in potential energy.

A small displacement away from a neutral equilibrium position produces no change in potential energy.

A small displacement away from an unstable equilibrium position produces a decrease in potential energy.

# PROBLEMS

## Introductory Problems

**11-35\***  A body ($W$ = 120 lb) is supported by a flexible cable and a spring ($k$ = 80 lb/in.) as shown in Fig. P11-35. The pulley over which the cable passes is weightless and frictionless. If a force **P** of 100 lb is slowly applied to the body, what is the change in potential energy of

**a.**  The spring?
**b.**  The system consisting of the body, the cable, and the spring?

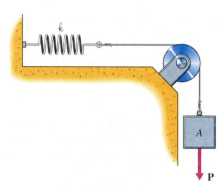

**Fig. P11-35**

**11-36\***  A sphere ($m$ = 10 kg) is supported at the end of a slender rod ($L$ = 300 mm) as shown in Fig. P11-36. The torsional spring ($k$ = 300 N · m/radian) at support $A$ is neutral when $\theta = 0°$. Determine the potential energy of the system at angles of 30°, 60°, and 90°.

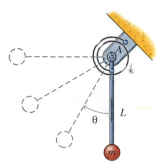

**Fig. P11-36**

**11-37**  A moment **M** is slowly increased as it is applied to the disk shown in Fig. P11-37. Determine the change in potential energy of the system with respect to a reference at $\theta = 0°$ as the disk is rotated to the position $\theta = 45°$. The spring ($k$ = 10 lb/in.) is unstretched at $\theta = 0°$.

**Fig. P11-37**

**11-38**  A force **F** is slowly increased as it is applied to bar $AB$ of Fig. P11-38. Determine the change in potential energy of the system with respect to a reference at $\theta = 0°$ as the bar is rotated to the position $\theta = 30°$. The bar is in equilibrium at $\theta = 0°$ with **F** = 0. The spring constant $k$ equals 500 N/m.

**Fig. P11-38**

**11-39*** A system of bars and a spring are used to support a weight $W$ as shown in Fig. P11-39. The weights of the bars are negligible and the spring is unstretched when $\theta = 25°$. Determine the angle $\theta$ for equilibrium if $W = 50$ lb, $L = 12$ in., and $k = 12$ lb/in.

**Fig. P11-39**

**11-40*** Two bars and a torsional spring are used to support a mass $m$ as shown in Fig. P11-40. The masses of the bars are negligible and the spring is neutral when $\theta = 0°$. Determine the angle $\theta$ for equilibrium if $m = 25$ kg, $L = 1.25$ m, and $k = 150$ N · m rad.

**Fig. P11-40**

**11-41** A two-bar mechanism supports a weight ($W = 100$ lb) as shown in Fig. P11-41. The length of bar $AB$ is 3 ft, the unstretched length of the spring is 1.5 ft, and the spring stiffness $k = 30$ lb/ft. Determine the angle $\theta$ for equilibrium of the mechanism. Assume that the weights of the bars are negligible.

**Fig. P11-41**

**11-42** A two-bar mechanism supports a body ($m = 75$ kg) as shown in Fig. P11-42. The unstretched length of the spring is 600 mm. Determine the spring stiffness $k$ required to limit the angle $\theta$ for equilibrium to 35°. Assume that the masses of the bars are negligible.

**Fig. P11-42**

**11-43*** The uniform bar $AB$ shown in Fig. P11-43 weighs 30 lb. If the spring is unstretched when $\theta = 0°$, determine the angle $\theta$ for equilibrium if $k = 15$ lb/ft and $L = 3$ ft.

**Fig. P11-43**

**11-44** The uniform bar $AB$ shown in Fig. P11-44 has a mass of 10 kg. The spring is unstretched when $\theta = 0°$. If $k = 50$ N/m and $L = 2$m, determine the angle $\theta$ for equilibrium.

**645**

**Fig. P11-44**

## Intermediate Problems

**11-45\*** The uniform bar $AB$ shown in Fig. P11-45 weighs 100 lb. If the spring has an unstretched length of 5 ft, determine the spring constant $k$ required for equilibrium at the position indicated in the figure.

**Fig. P11-45**

**11-46\*** Bar $AB$ of Fig. P11-46 with weight $W$ and length $L$ has a uniform cross section. Rotation of the bar is resisted by a torsional spring with spring constant $k$. The spring is neutral when $\theta = 0°$. Determine

a. The minimum stiffness $k$ for the spring, in terms of $W$ and $L$, if the bar is to be in stable equilibrium when $\theta = 0°$.

b. The equilibrium angle $\theta$ and the stability of the equilibrium if $W = 100$ N, $L = 1$ m, and $k = 500$ N/m.

**Fig. P11-46**

**11-47** A block is supported by the three-bar frame shown in Fig. P11-47. The weights of the bars are negligible. The spring can act in either tension or compression and is unstretched when $\theta = 0°$. Determine the maximum weight $W$ for the block, in terms of $k$ and $L$, if the system is to be in stable equilibrium in the position shown in the figure.

**Fig. P11-47**

**11-48** A block of mass $m$ is supported by the three-bar frame shown in Fig. P11-48. The masses of the bars are negligible and the torsional springs are neutral when $\theta = 0°$. Determine the minimum stiffness $k$ for the springs, in terms of mass $m$ and length $L$, if the system is to be in stable equilibrium in the position shown in the figure.

**Fig. P11-48**

**11-49\*** The spring in Fig. P11-49 has an unstretched length of 2 ft. Determine the angle $\theta$ for equilibrium when $W_D = W_E = W_F = 80$ lb, $L = 2$ ft, and $k = 350$ lb/ft. Assume that the weights of the bars are negligible.

**Fig. P11-49**

**11-50** The spring in Fig. P11-50 is unstretched when $\theta = 0°$. Find the spring constant $k$ required for equilibrium at an angle $\theta = 25°$ if $m_B = 50$ kg, $m_D = 25$ kg, and $L = 1.5$ m. Assume that the masses of the bars are negligible.

**Fig. P11-50**

## Challenging Problems

**11-51\*** The spring in Fig. P11-51 is unstretched when $\theta = \theta_0 = 45°$. Find the angle $\theta$ for equilibrium if $W = 20$ lb, $L = 18$ in., $a = 8$ in., and $k = 12$ lb/in. Assume that the weights of the bars are negligible.

**Fig. P11-51**

**11-52\*** The spring in Fig. P11-52 is unstretched when $\theta = \theta_0 = 15°$. Find the angle $\theta$ for equilibrium if $m_B = 60$ kg, $m_D = 25$ kg, $L = 1$ m, and $k = 750$ N/m. Assume that the masses of the bars are negligible.

**Fig. P11-52**

**11-53** The spring in Fig. P11-53 has an unstretched length of 2 ft. Find the spring constant $k$ required for equilibrium at an angle $\theta = 30°$ if $W_B = 300$ lb, $W_D = 500$ lb, and $L = 4$ ft. Assume that the weights of the bars are negligible.

**Fig. P11-53**

**11-54** The uniform bar $AB$ shown in Fig. P11-54 has a mass of 20 kg and a length of 2 m. If the spring is unstretched at an angle $\theta = 30°$, determine the spring constant $k$ for equilibrium at an angle $\theta = 60°$.

**Fig. P11-54**

**11-55\*** Bar $AB$ of Fig. P11-55 with weight $W$ and length $L$ has a uniform cross section. The spring has an unstretched length $L$. Determine the possible positions of equilibrium and the stability of equilibrium in each of these positions if $W = 20$ lb, $L = 2$ ft, and $k = 28$ lb/ft.

**Fig. P11-55**

**11-56** The weight of bar $BC$ of Fig. P11-56 is negligible. The spring has an unstretched length $L$. Determine the possible positions of equilibrium and stability of equilibrium in each of these positions if $m = 50$ kg, $L = 1.5$ m, and $k = 600$ N/m.

**Fig. P11-56**

## SUMMARY

Static equilibrium problems are normally solved by isolating the body of interest, drawing a free-body diagram, and solving the equations $\mathbf{R} = \mathbf{C} = 0$ for the desired unknowns. A second method for solving static equilibrium problems makes use of the principle of virtual work. In mechanics, a force does work only when the body to which the force is applied moves. When a constant force $\mathbf{F}$ is applied to a body that moves a distance $\mathbf{d}$ in a straight line, the work $U$ done on the particle by the force $\mathbf{F}$ is defined by the scalar product

$$U = \mathbf{F} \cdot \mathbf{d} = Fd \cos \phi \qquad \textbf{(11-1)}$$

where $\phi$ is the angle between the vectors $\mathbf{F}$ and $\mathbf{d}$. If the force is not constant or if the displacement is not in a straight line, Eq. 11-1 gives the work done by the force only during an infinitesimal part of the displacement $d\mathbf{r}$. Thus,

$$dU = \mathbf{F} \cdot d\mathbf{r} = F\, ds \cos \phi = F_t\, ds \qquad \textbf{(11-2)}$$

The work done by a couple is obtained by calculating the work done by each force of the couple separately and adding their works together. Thus,

$$dU = \mathbf{C} \cdot d\boldsymbol{\theta} = M\, d\theta \cos \phi \qquad \textbf{(11-8)}$$

where $M$ is the magnitude of the moment of the couple, $d\theta$ is the magnitude of the infinitesimal angular displacement, and $\phi$ is the angle between $\mathbf{C}$ and $d\boldsymbol{\theta}$. The work is positive if the angular displacement $d\theta$ is in the same direction as the sense of rotation of the couple and negative if the displacement is in the opposite direction. No work is done if the couple is translated or rotated about an axis parallel to the plane of the couple. When a body being acted on by a force $\mathbf{F}$ moves through an infinitesimal linear displacement $d\mathbf{s}$, the body is not in equilibrium. In studying the equilibrium of bodies by the method of virtual work, fictitious displacements called virtual displacements ($\delta\mathbf{s}$ rather than $d\mathbf{s}$) are introduced. The work done by a force $\mathbf{F}$ acting on a body during a virtual displacement $\delta\mathbf{s}$ is called virtual work $\delta U$ and is represented mathematically as

$$\delta U = \mathbf{F} \cdot \delta\mathbf{s} \quad \text{or} \quad \delta U = F\, \delta s \cos \alpha \qquad \textbf{(11-9)}$$
$$\delta U = \mathbf{C} \cdot \delta\boldsymbol{\theta} \quad \text{or} \quad \delta U = M\, \delta s \cos \alpha \qquad \textbf{(11-10)}$$

The principle of virtual work states that a particle, a rigid body, or a system of connected rigid bodies is in equilibrium if the virtual work done by all external forces and couples acting on the system is zero for all possible virtual displacements of the system.

The principle of virtual work can be expressed mathematically as

$$\delta U = \sum_{i=1}^{m} \mathbf{F}_i \cdot \delta \mathbf{s}_i + \sum_{j=1}^{n} \mathbf{C}_j \cdot \delta \boldsymbol{\theta}_j = 0 \qquad \text{(11-11)}$$

If the virtual work done by all external forces acting on the body for all possible virtual displacements of the body is zero, then

$$\delta U = \Sigma F_x \ \delta_x + \Sigma F_y \ \delta y + \Sigma F_z \ \delta z + \Sigma M_x \ \delta \theta_x + \Sigma M_y \ \delta \theta_y + \Sigma M_z \ \delta \theta_z = 0 \qquad \text{(11-17)}$$

Applying the virtual work principle by taking virtual linear and angular displacements one at a time yields expressions that are identical to the six scalar equations of equilibrium; namely,

$$\Sigma F_x = 0 \qquad \Sigma F_y = 0 \qquad \Sigma F_z = 0$$
$$\Sigma M_x = 0 \qquad \Sigma M_y = 0 \qquad \Sigma M_z = 0$$

Therefore, the virtual work equation $\delta U = 0$ is simply an alternative statement of the equilibrium equations for a rigid body.

The potential energy of a body is a measure of its capacity for doing work. A deformed elastic member (such as a spring) stores energy $V_e$, which is potentially available to do work on some other body. Similarly, an elevated body possesses gravitational potential energy $V_g$, which is available to do work on some other body. In terms of potential energy, the virtual-work equation can be written as

$$\delta U + \delta V_e + \delta V_g = 0 \qquad \text{or} \qquad \delta U = -(\delta V_e + \delta V_g) = -\delta V \quad \text{(11-28)}$$

Equation 11-28 indicates that the virtual work done by the external forces on a system in equilibrium is equal to the change in elastic and gravitational potential energy of the system for all possible virtual displacements consistent with the constraints.

For a system whose position can be defined in terms of a single variable (single-degree-of-freedom system),

$$V = V(s) \qquad \delta V = \frac{dV}{ds} \delta s$$

Since $\delta s \neq 0$, Eq. 11-28 indicates that the system is in equilibrium when

$$\frac{dV}{ds} = 0 \qquad \text{(11-29)}$$

The potential energy method can be extended to determine whether the equilibrium is stable or unstable. Mathematically, the potential energy is minimum for a single-degree-of-freedom system when the second derivative of the potential energy with respect to the displacement variable is positive. For stable equilibrium (at a position of minimum potential energy)

$$\frac{dV}{ds} = 0 \qquad \text{and} \qquad \frac{d^2V}{ds^2} > 0 \qquad \text{(11-31)}$$

For unstable equilibrium (at a position of maximum potential energy)

$$\frac{dV}{ds} = 0 \quad \text{and} \quad \frac{d^2V}{ds^2} < 0 \qquad (11\text{-}33)$$

If both the first and the second derivatives of the potential energy $V$ are zero, the sign of a higher-order derivative must be used to determine the type of equilibrium. The equilibrium will be stable if the first nonzero derivative is even and positive. The equilibrium will be neutral if all derivatives are zero. In all other cases the equilibrium will be unstable. If the system possesses several degrees of freedom, the potential energy $V$ depends on several variables, and a more advanced treatment involving the theory of functions of several variables is required to determine when the potential energy $V$ is minimum. Stability criteria for such systems is beyond the scope of this text.

## REVIEW PROBLEMS

Use the method of virtual work to solve the following problems.

**11-57*** A system of beams supports three concentrated loads as shown in Fig. P11-57. Determine the reaction at support $B$.

**Fig. P11-57**

**11-58*** A pin-connected truss is loaded and supported as shown in Fig. P11-58. Determine

a. The reaction at support $E$.
b. The force in member $CD$.

**Fig. P11-58**

**11-59** Determine the equilibrium position (angles $\theta_1$ and $\theta_2$) for the system of pin-connected bars shown in Fig. P11-59. Bars $AB$ and $BC$ weigh 50 and 30 lb, respectively, and $P = 20$ lb.

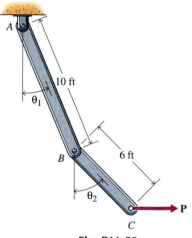

**Fig. P11-59**

**11-60** The masses of bar $AB$, bar $BC$, and body $W$ of Fig. P11-60 are 20 kg, 20 kg, and 10 kg, respectively. Determine the magnitude of force $P$ if the system is in equilibrium in the position shown.

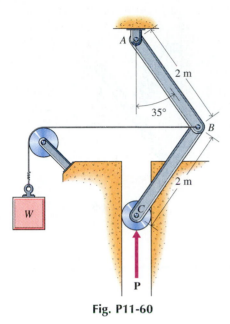

**Fig. P11-60**

**11-61\*** A slider crank mechanism with a spring attached to the slider is shown in Fig. P11-61. Bars $AB$ and $BC$ weigh 10 and 20 lb, respectively. The spring has a modulus $k$ of 50 lb/in. and is unstretched when $\theta = 0°$. If the system is in equilibrium when $\theta = 30°$, determine the magnitude of the moment $M$ that must be applied to crank $AB$. Friction between the slider and the horizontal surfaces is negligible.

**Fig. P11-61**

**11-62\*** The two springs in Fig. P11-62 are identical and have unstretched lengths of 0.5 m. Determine the spring constant $k$ required for equilibrium at an angle $\theta = 30°$ if $m_A = m_B = m_C = 20$ kg and $L = 1.5$ m. Assume that the masses of the bars are negligible.

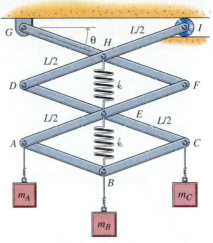

**Fig. P11-62**

**11-63** The two springs shown in Fig. P11-63 can act in either tension or compression and are unstretched when $\theta = 0°$. Determine the possible positions of equilibrium and the stability of equilibrium in each of these positions if $W = 25$ lb, $L = 6$ ft, and $k = 6.0$ lb/ft.

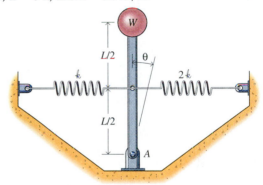

**Fig. P11-63**

**11-64** Determine the horizontal and vertical components of the reaction at support $A$ of the frame shown in Fig. P11-64.

**Fig. P11-64**

**11-65\*** The three-bar frame shown in Fig. P11-65 supports a block $W$, which has a weight of 200 lb. The spring $AC$ is unstretched when $\theta = 0°$. The three bars have negligible masses. If the system is in equilibrium when $\theta = 30°$, determine

a. The modulus $k$ of the spring.
b. The stability of this equilibrium position.

**Fig. P11-65**

**11-66** The uniform bar $AB$ shown in Fig. P11-66 has a mass of 20 kg, and the block $W$ has a mass of 50 kg. Determine the angle $\theta$ for equilibrium if the spring ($k = 5$ kN/m) is unstretched when $\theta = 30°$.

**Fig. P11-66**

# VECTOR OPERATIONS

## A-1  INTRODUCTION

Physical quantities of interest in mechanics can be represented by two kinds of quantities: scalars and vectors. The mathematical operations required for their use in statics and dynamics problems will be presented in this appendix.

### A-1.1  Scalar Quantities

Scalar quantities can be completely described with a magnitude. Examples of scalar quantities in mechanics are mass, density, length, area, volume, speed, energy, time, and temperature. In this book and in the companion text on dynamics, symbols representing scalar quantities will be printed in lightface italic type $A$. In mathematical operations, scalars follow the rules in elementary algebra.

### A-1.2  Vector Quantities

A vector quantity has both a magnitude and a direction (line of action and sense) and obeys the parallelogram law of addition, which will be described in Section A-2 of this appendix. Examples of vector quantities in mechanics are force, displacement, velocity, and acceleration. A vector quantity can be represented graphically by using a directed line segment (arrow), as shown in Fig. A-1. The magnitude of the quantity is represented by the length of the arrow. The direction is specified by using the angle $\theta$ between the arrow and some known reference direction. The sense is indicated by the arrowhead at the tip of the line segment. Symbols representing vector quantities will be printed in boldface type **A** to distinguish them from scalar quantities. Symbols representing magnitudes $|\mathbf{A}|$ of vector quantities will be printed in lightface italic type $A$. In all handwritten work, it is important to distinguish between scalar and vector quantities, since in mathematical operations vector quantities do not follow the rules of elementary algebra. A small arrow over the symbol for a vector quantity $\vec{A}$ is often used in handwritten work to take the place of boldface type in printed material.

**Figure A-1**  Representing a vector quantity with a directed line segment.

Vectors can be classified into three types: free, sliding, or fixed.

1. A free vector has a specific magnitude, slope, and sense, but its line of action does not pass through a unique point in space.
2. A sliding vector has a specific magnitude, slope, and sense, and its line of action passes through a unique point in space. The point of application of a sliding vector can be anywhere along its line of action.
3. A fixed vector has a specific magnitude, slope, and sense, and its line of action passes through a unique point in space. The point of application of a fixed vector is confined to a fixed point on its line of action.

## A-2   ELEMENTARY VECTOR OPERATIONS

### A-2.1   Addition of Vectors

Two free vectors **A** and **B** can be translated so that their tails meet at a common point, as shown in Fig. A-2. The parallelogram law of vector addition states that the two vectors **A** and **B** are equivalent to a vector **R**, which is the diagonal of a parallelogram constructed by using vectors **A** and **B** as the two adjacent sides (see Fig. A-2). The vector **R** is the diagonal that passes through the point of intersection of vectors **A** and **B**. Vector **R** is known as the resultant of the two vectors **A** and **B**. Vector quantities must obey this parallelogram law of addition. The resultant **R** can be represented mathematically by the vector equation

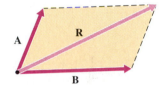

**Figure A-2**   Parallelogram law for vector addition.

$$\mathbf{R} = \mathbf{A} + \mathbf{B} \qquad \text{(A-1)}$$

The plus sign used in conjunction with vector quantities **A** and **B** (boldface type) indicates vector (parallelogram law) addition, not scalar (algebraic) addition.

It is obvious from Fig. A-2 that the two vectors **A** and **B** can be added in a head-to-tail fashion, as shown in the top part of Fig. A-3, to obtain the vector sum **R**. Similarly, the two vectors can be added in the manner shown in the bottom part of Fig. A-3 to obtain **R**, since the two triangles shown in Fig. A-3 are the two different halves of the parallelogram shown in Fig. A-2. The procedure of adding vectors in a head-to-tail fashion is known as the triangle law. Note from Fig. A-3 that the order of addition of the vectors does not affect their sum; therefore,

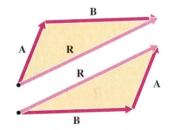

**Figure A-3**   Triangle law for vector addition.

$$\mathbf{A} + \mathbf{B} = \mathbf{B} + \mathbf{A} \qquad \text{(A-2)}$$

Equation A-2 establishes the fact that vector addition is commutative.

The sum of three or more vectors **A** + **B** + **C** + ··· can be obtained by first adding the vectors **A** and **B**, and then adding the vector **C** to the vector sum (**A** + **B**). The procedure, which is illustrated in Fig. A-4, can be expressed mathematically as

$$\mathbf{A} + \mathbf{B} + \mathbf{C} = (\mathbf{A} + \mathbf{B}) + \mathbf{C} \qquad \text{(A-3)}$$

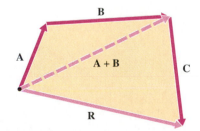

**Figure A-4**   Triangle law for vector addition.

Similarly, the sum of four vectors can be obtained by adding the fourth vector to the sum of the first three. Thus, the sum of any number of vectors can be obtained by applying the parallelogram law, as expressed mathematically by Eq. A-3, to successive pairs of vectors until all the given vectors are replaced by a single resultant vector **R**.

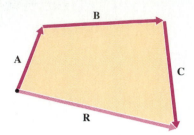

**Figure A-5**   Tip-to-tail procedure for vector addition.

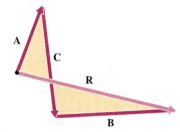

**Figure A-6**   Tip-to-tail procedure for vector addition.

**Figure A-7**   Lengths and angles for an oblique triangle *ABC*.

The sum of three or more vectors $\mathbf{A} + \mathbf{B} + \mathbf{C} + \cdots$ can also be obtained by repeated application of the triangle law. In Fig. A-5 the sum of three vectors is obtained directly by connecting the three vectors, in sequence, in a tip-to-tail fashion. The resultant $\mathbf{R}$ is the vector that connects the tail of the first vector with the tip of the last one. The procedure, illustrated in Fig. A-5, can be applied to any number of vectors and is known as the polygon rule for vector addition. The sketch of the vectors connected in a tip-to-tail fashion is known as a vector polygon. As illustrated in Fig. A-6, the resultant $\mathbf{R}$ is independent of the order in which the vectors $\mathbf{A}$, $\mathbf{B}$, and $\mathbf{C}$ are introduced into the vector polygon. This fact, which can be expressed mathematically as

$$\mathbf{A} + \mathbf{B} + \mathbf{C} = (\mathbf{A} + \mathbf{B}) + \mathbf{C} = \mathbf{A} + (\mathbf{B} + \mathbf{C}) \tag{A-4}$$

shows that vector addition is associative.

The resultant $\mathbf{R}$ of two or more vectors can be determined graphically by drawing either the parallelogram or the vector polygon to scale. In practice, magnitude $R$ of the resultant is determined algebraically by applying the cosine law for a general triangle, as illustrated in Fig. A-7. Thus,

$$R = \sqrt{A^2 + B^2 - 2AB \cos \theta} \tag{A-5}$$

The angles between resultant $\mathbf{R}$ and vectors $\mathbf{A}$ and $\mathbf{B}$ can be determined by using the law of sines for a general triangle. Thus, for the triangle shown in Fig. A-7,

$$\frac{\sin \alpha}{A} = \frac{\sin \beta}{B} = \frac{\sin \theta}{R} \tag{A-6}$$

## A-2.2   Subtraction of Vectors

The difference between vectors $\mathbf{A}$ and $\mathbf{B}$ is defined by the relation

$$\mathbf{A} - \mathbf{B} = \mathbf{A} + (-\mathbf{B}) \tag{A-7}$$

The vector $-\mathbf{B}$ has the same magnitude and direction angle $\theta$ as vector $\mathbf{B}$ but is of opposite sense. Two vectors $\mathbf{A}$ and $\mathbf{B}$ are shown in Fig. A-8a. Their sum and difference are illustrated in Figs. A-8b and A-8c, respectively. A zero or null vector is obtained when a vector is subtracted from itself; that is,

$$\mathbf{A} - \mathbf{A} = \mathbf{0} \tag{A-8}$$

It is important to realize that the zero that occurs on the right-hand side of this vector equation must be a vector.

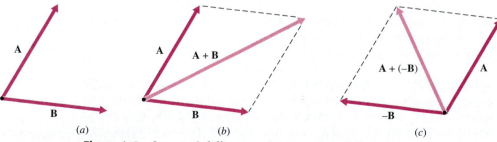

**Figure A-8**   Sum and difference of vectors $\mathbf{A}$ and $\mathbf{B}$.

The minus sign used in conjunction with vectors **A** and **B** (boldface type) indicates vector (parallelogram law) subtraction and not scalar subtraction. Since vector subtraction is defined as a special case of vector addition, all of the rules for vector addition discussed in Section A-2.1 apply to vector subtraction.

### A-2.3   Multiplication of Vectors by Scalars

The product of a vector **A** and a scalar $m$ is a vector $m\mathbf{A}$, whose magnitude is $mA$ (a positive number) and whose direction has the same sense as **A** if $m$ is positive and a sense opposite to **A** if $m$ is negative. Operations involving the products of scalars $m$ and $n$ and vectors **A** and **B** include the following:

$$(m + n)\mathbf{A} = m\mathbf{A} + n\mathbf{A}$$
$$m(\mathbf{A} + \mathbf{B}) = m\mathbf{A} + m\mathbf{B} \qquad \text{(A-9)}$$
$$m(n\mathbf{A}) = (mn)\mathbf{A} = n(m\mathbf{A})$$

## A-3   VECTORS REFERRED TO RECTANGULAR CARTESIAN COORDINATES

### A-3.1   Unit Vectors

A vector of unit magnitude is called a unit vector. Unit vectors directed along the $x$-, $y$-, and $z$-axes of a Cartesian coordinate system are normally given the symbols **i**, **j**, and **k**, respectively. The sense of these unit vectors is indicated analytically by using a plus sign if the unit vector points in a positive $x$-, $y$-, or $z$-direction and a minus sign if the unit vector points in a negative $x$-, $y$-, or $z$-direction. The unit vectors shown in Fig. A-9 are positive. In this text, the symbol **e** with an appropriate subscript is used to denote a unit vector in a direction other than a coordinate direction.

Any vector quantity can be written as the product of its magnitude (a positive number) and a unit vector in the direction of the given vector. Thus, for a vector **A** in the positive $n$ direction

$$\mathbf{A} = |\mathbf{A}|\mathbf{e}_n = A\mathbf{e}_n \qquad \text{(A-10)}$$

For a vector **B** in the negative $n$ direction

$$\mathbf{B} = |\mathbf{B}|(-\mathbf{e}_n) = -B\mathbf{e}_n \qquad \text{(A-11)}$$

It is obvious from Eq. A-10 that the unit vector $\mathbf{e}_n$ can be expressed as

$$\mathbf{e}_n = \frac{\mathbf{A}}{|\mathbf{A}|} = \frac{\mathbf{A}}{A} \qquad \text{(A-12)}$$

The Cartesian coordinate axes shown in Fig. A-9 are arranged as a right-hand system. In a right-hand system, if the fingers of the right hand are curled about the $z$-axis in a direction from the positive $x$-axis toward the positive $y$-axis, then the thumb points in the positive $z$-direction, as shown in Fig. A-10. All the vector relationships developed in the remainder of this appendix will utilize a right-hand coordinate system.

**Figure A-9**   Unit vectors in Cartesian coordinates.

**Figure A-10**   Coordinate axes in a right-hand system.

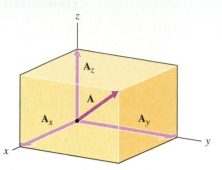

**Figure A-11**  Rectangular components $A_x$, $A_y$, and $A_z$ of a vector **A**.

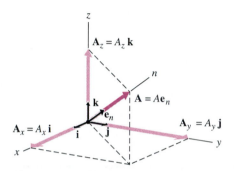

**Figure A-12**  Representing the rectangular components of a vector **A** with unit vectors **i**, **j**, and **k**.

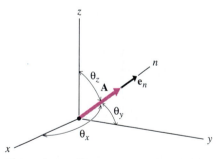

**Figure A-13**  Representing the direction of a vector in space with direction angles.

## A-3.2  Cartesian Components of a Vector

The process of adding two or more vectors to obtain a resultant vector **R** was discussed in Section A-2.1. The reverse process of resolving a given vector **A** into two or more components will now be discussed. The vector **A** can be resolved into any number of components provided the components sum to the given vector **A** by the parallelogram law. Mutually perpendicular components, called rectangular components, along the $x$-, $y$-, and $z$-coordinate axes are most useful. Any vector **A** can be resolved into rectangular components $A_x$, $A_y$, and $A_z$ by constructing a rectangular parallelepiped (see Fig. A-11) with vector **A** as the diagonal and rectangular components $A_x$, $A_y$, and $A_z$ along the $x$-, $y$-, and $z$-axes, respectively, as the edges. The three components $A_x$, $A_y$, and $A_z$ can be written in Cartesian vector form, as illustrated in Fig. A-12, by using Eq. A-10. Thus

$$\mathbf{A}_x = A_x\mathbf{i} \qquad \mathbf{A}_y = A_y\mathbf{j} \qquad \mathbf{A}_z = A_z\mathbf{k} \qquad \text{(A-13)}$$

Furthermore, from Eqs. A-1, A-10, and A-13,

$$\mathbf{A} = \mathbf{A}_x + \mathbf{A}_y + \mathbf{A}_z = A_x\mathbf{i} + A_y\mathbf{j} + A_z\mathbf{k} = A\mathbf{e}_n \qquad \text{(A-14)}$$

The magnitudes of the rectangular components are related to the magnitude of vector **A** by the expressions

$$A_x = A \cos \theta_x \qquad A_y = A \cos \theta_y \qquad A_z = A \cos \theta_z \qquad \text{(A-15)}$$

The terms $\cos \theta_x$, $\cos \theta_y$, and $\cos \theta_z$ are called the direction cosines of the vector **A**. These cosines are also the direction cosines of the unit vector $\mathbf{e}_n$ since its line of action, as illustrated in Fig. A-13, coincides with the line of action of vector **A**.

The magnitude of vector **A** can be expressed in terms of the magnitudes of its rectangular components by the Pythagorean theorem as

$$A = \sqrt{A_x^2 + A_y^2 + A_z^2} \qquad \text{(A-16)}$$

Also, from Eq. A-15, the direction cosines of a vector **A** can be expressed in terms of the magnitude of vector **A** and the magnitudes of its rectangular components as

$$\cos \theta_x = \frac{A_x}{A} \qquad \cos \theta_y = \frac{A_y}{A} \qquad \cos \theta_z = \frac{A_z}{A} \qquad \text{(A-17)}$$

Finally, the unit vector $\mathbf{e}_n$ can be written in terms of vector **A** and its rectangular components by using Eqs. A-14 and A-15. Thus,

$$\mathbf{e}_n = \frac{\mathbf{A}}{A} = \frac{A_x}{A}\mathbf{i} + \frac{A_y}{A\mathbf{j}} + \frac{A_z}{A}\mathbf{k}$$
$$= \cos \theta_x \, \mathbf{i} + \cos \theta_y \, \mathbf{j} + \cos \theta_z \, \mathbf{k} \qquad \text{(A-18)}$$

Since the magnitude of $\mathbf{e}_n = 1$

$$\cos^2 \theta_x + \cos^2 \theta_y + \cos^2 \theta_z = 1 \qquad \text{(A-19)}$$

## A-3.3  Position Vectors

A directed line segment **r,** known as a position vector, can be used to locate a point in space relative to the origin of a coordinate system or relative to another point in space. For example, consider the two points

$A$ and $B$ shown in Fig. A-14. The location of point $A$ can be specified by using the position vector $\mathbf{r}_A$ drawn from the origin of coordinates to point $A$. Position vector $\mathbf{r}_A$ can be written in Cartesian vector form as

$$\mathbf{r}_A = x_A\mathbf{i} + y_A\mathbf{j} + z_A\mathbf{k}$$

Similarly, the location of point $B$ with respect to the origin of coordinates can be specified by using the position vector $\mathbf{r}_B$, which can be written in Cartesian vector form as

$$\mathbf{r}_B = x_B\mathbf{i} + y_B\mathbf{j} + z_B\mathbf{k}$$

Finally, the position of point $B$ with respect to point $A$ can be specified by using the position vector $\mathbf{r}_{B/A}$ where the subscript $B/A$ indicates $B$ with respect to $A$. Observe in Fig. A-14 that

$$\mathbf{r}_B = \mathbf{r}_A + \mathbf{r}_{B/A}$$

Therefore,

$$\begin{aligned}\mathbf{r}_{B/A} &= \mathbf{r}_B - \mathbf{r}_A \\ &= (x_B\mathbf{i} + y_B\mathbf{j} + z_B\mathbf{k}) - (x_A\mathbf{i} + y_A\mathbf{j} + z_A\mathbf{k}) \\ &= (x_B - x_A)\mathbf{i} + (y_B - y_A)\mathbf{j} + (z_B - z_A)\mathbf{k} \end{aligned} \qquad \text{(A-20)}$$

**Figure A-14**   Position vector $\mathbf{r}_{B/A}$ in terms of vectors $\mathbf{r}_A$ and $\mathbf{r}_B$.

## EXAMPLE PROBLEM A-1

Determine the magnitude and direction of the position vector from point $A$ to point $B$ if the coordinates of points $A$ and $B$ are $(3, 4, 5)$ and $(6, -3, -2)$, respectively.

### SOLUTION

The position vector $\mathbf{r}_{B/A}$ is given by Eq. A-20 as

$$\begin{aligned}\mathbf{r}_{B/A} &= (x_B - x_A)\mathbf{i} + (y_B - y_A)\mathbf{j} + (z_B - z_A)\mathbf{k} \\ &= (6 - 3)\mathbf{i} + (-3 - 4)\mathbf{j} + (-2 - 5)\mathbf{k} \\ &= 3\mathbf{i} - 7\mathbf{j} - 7\mathbf{k}\end{aligned}$$

The magnitude of $\mathbf{r}_{B/A}$ is determined by using Eq. A-22. Thus,

$$\begin{aligned}r_{B/A} &= \sqrt{(x_B - x_A)^2 + (y_B - y_A)^2 + (z_B - z_A)^2} \\ &= \sqrt{(3)^2 + (-7)^2 + (-7)^2} = 10.344\end{aligned} \qquad \text{Ans.}$$

The direction is determined by using Eqs. A-23. Thus,

$$\theta_x = \cos^{-1}\frac{x_B - x_A}{r_{B/A}} = \cos^{-1}\frac{3}{10.344} = 73.1° \qquad \text{Ans.}$$

$$\theta_y = \cos^{-1}\frac{y_B - y_A}{r_{B/A}} = \cos^{-1}\frac{-7}{10.344} = 132.6° \qquad \text{Ans.}$$

$$\theta_z = \cos^{-1}\frac{z_B - z_A}{r_{B/A}} = \cos^{-1}\frac{-7}{10.344} = 132.6° \qquad \text{Ans.}$$

## A-4   ADDITION OF CARTESIAN VECTORS

The vector operations of addition and subtraction are greatly simplified when more than two vectors are involved if the vectors are expressed in Cartesian vector form. For example, consider the two vectors **A** and **B**, which can be written as

$$\mathbf{A} = A_x\mathbf{i} + A_y\mathbf{j} + A_z\mathbf{k}$$
$$\mathbf{B} = B_x\mathbf{i} + B_y\mathbf{j} + B_z\mathbf{k}$$

The sum of the two vectors is

$$\mathbf{R} = \mathbf{A} + \mathbf{B} = (A_x\mathbf{i} + A_y\mathbf{j} + A_z\mathbf{k}) + (B_x\mathbf{i} + B_y\mathbf{j} + B_z\mathbf{k})$$
$$= (A_x + B_x)\mathbf{i} + (A_y + B_y)\mathbf{j} + (A_z + B_z)\mathbf{k} \qquad \text{(A-21)}$$

Thus, the resultant vector **R** has components that represent the scalar sums of the rectangular components of **A** and **B**. The process represented by Eq. A-21 can be extended to any number of vectors. Thus,

$$\mathbf{R}_x = (A_x + B_x + C_x + \cdots)\mathbf{i} = R_x\mathbf{i}$$
$$\mathbf{R}_y = (A_y + B_y + C_y + \cdots)\mathbf{j} = R_y\mathbf{j}$$
$$\mathbf{R}_z = (A_z + B_z + C_z + \cdots)\mathbf{k} = R_z\mathbf{k}$$
$$\mathbf{R} = \mathbf{R}_x + \mathbf{R}_y + \mathbf{R}_z = R_x\mathbf{i} + R_y\mathbf{j} + R_z\mathbf{k}$$

The magnitude and direction of the resultant **R** are then obtained by using Eqs. A-16 and A-17. Thus,

$$R = \sqrt{(R_x)^2 + (R_y)^2 + (R_z)^2} \qquad \text{(A-22)}$$

$$\theta_x = \cos^{-1}\frac{R_x}{R} \qquad \theta_y = \cos^{-1}\frac{R_y}{R} \qquad \theta_z = \cos^{-1}\frac{R_z}{R} \qquad \text{(A-23)}$$

### EXAMPLE PROBLEM A-2

Determine the magnitude and direction of the resultant of the following three vectors $\mathbf{A} = 3\mathbf{i} + 7\mathbf{j} + 8\mathbf{k}$, $\mathbf{B} = 4\mathbf{i} - 5\mathbf{j} + 3\mathbf{k}$, and $\mathbf{C} = 2\mathbf{i} + 3\mathbf{j} - 4\mathbf{k}$. Express the resultant **R** and the unit vector $\mathbf{e}_R$ associated with the resultant in Cartesian vector form.

### SOLUTION

The $x$-, $y$-, and $z$-components of the resultant **R** are

$$R_x = A_x + B_x + C_x = 3 + 4 + 2 = 9$$
$$R_y = A_y + B_y + C_y = 7 - 5 + 3 = 5$$
$$R_z = A_z + B_z + C_z = 8 + 3 - 4 = 7$$

The magnitude of the resultant is determined by using Eq. A-22. Thus,

$$R = \sqrt{(R_x)^2 + (R_y)^2 + (R_z)^2} = \sqrt{(9)^2 + (5)^2 + (7)^2} = 12.450 \qquad \text{Ans.}$$

The direction is determined by using Eqs. A-23. Thus,

$$\theta_x = \cos^{-1}\frac{R_x}{R} = \cos^{-1}\frac{9}{12.450} = 43.7° \qquad \text{Ans.}$$

$$\theta_y = \cos^{-1}\frac{R_y}{R} = \cos^{-1}\frac{5}{12.450} = 66.3° \qquad \text{Ans.}$$

$$\theta_z = \cos^{-1}\frac{R_z}{R} = \cos^{-1}\frac{7}{12.450} = 55.8° \qquad \text{Ans.}$$

The resultant can be expressed in Cartesian vector form as

$$\mathbf{R} = R_x\mathbf{i} + R_y\mathbf{j} + R_z\mathbf{k} = 9\mathbf{i} + 5\mathbf{j} + 7\mathbf{k} \qquad \text{Ans.}$$

The unit vector $\mathbf{e}_R$ associated with the resultant $\mathbf{R}$ is given by Eq. A-18 as

$$\mathbf{e}_R = \frac{\mathbf{R}}{R} = \frac{9\mathbf{i} + 5\mathbf{j} + 7\mathbf{k}}{12.450} = 0.723\mathbf{i} + 0.402\mathbf{j} + 0.562\mathbf{k} \qquad \text{Ans.}$$

## A-5  MULTIPLICATION OF CARTESIAN VECTORS

Two types of multiplication involving vector quantities are performed; namely, the scalar or dot product (written as $\mathbf{A} \cdot \mathbf{B}$) and the vector or cross product (written as $\mathbf{A} \times \mathbf{B}$). These two types of multiplication have entirely different properties and are used for different purposes.

### A-5.1  Dot or Scalar Product

The dot or scalar product of two intersecting vectors is defined as the product of the magnitudes of the vectors and the cosine of the angle between them. Thus, by definition, for the vectors $\mathbf{A}$ and $\mathbf{B}$ shown in Fig. A-15,

$$\mathbf{A} \cdot \mathbf{B} = \mathbf{B} \cdot \mathbf{A} = AB \cos\theta \qquad \text{(A-24)}$$

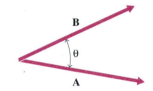

**Figure A-15**  Two intersecting vectors.

where $0° \le \theta \le 180°$. This type of vector multiplication yields a scalar, not a vector. For $0° \le \theta < 90°$, the scalar is positive. For $90° < \theta \le 180°$, the scalar is negative. When $\theta = 90°$, the two vectors are perpendicular and the scalar is zero.

The dot product can be used to obtain the rectangular scalar component of a vector in a specified direction. For example, the rectangular scalar component of vector $\mathbf{A}$ along the $x$-axis is

$$A_x = \mathbf{A} \cdot \mathbf{i} = A(1) \cos\theta_x = A \cos\theta_x$$

Similarly, the rectangular scalar component of vector $\mathbf{A}$ in a direction $n$ (see Fig. A-16a) is

$$A_n = \mathbf{A} \cdot \mathbf{e}_n = A \cos\theta_n \qquad \text{(A-25)}$$

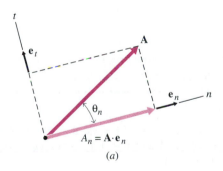

(a)

where $\mathbf{e}_n$ is the unit vector associated with the direction $n$. The vector component of $\mathbf{A}$ in the direction $n$ (see Fig. A-16b) is then given by Eq. A-10 as

$$\mathbf{A}_n = (\mathbf{A} \cdot \mathbf{e}_n)\mathbf{e}_n \qquad \text{(A-26)}$$

The component of vector $\mathbf{A}$ perpendicular to the direction $n$ lies in the plane containing $\mathbf{A}$ and $n$, as shown in Fig. A-16b, and can be obtained from the expression

$$\mathbf{A}_t = \mathbf{A} - \mathbf{A}_n \qquad \text{(a)}$$

(b)

**Figure A-16**  Dot or scalar product of two vectors.

Once an expression for $\mathbf{A}_t$ has been obtained by using Eq. $a$, the magnitude and direction of $\mathbf{A}_t$ can be determined by using Eqs. A-22 and A-18.

If two vectors $\mathbf{A}$ and $\mathbf{B}$ are written in Cartesian vector form, the scalar product becomes

$$
\begin{aligned}
\mathbf{A} \cdot \mathbf{B} &= (A_x\mathbf{i} + A_y\mathbf{j} + A_z\mathbf{k}) \cdot (B_x\mathbf{i} + B_y\mathbf{j} + B_z\mathbf{k}) \\
&= A_xB_x(\mathbf{i} \cdot \mathbf{i}) + A_xB_y(\mathbf{i} \cdot \mathbf{j}) + A_xB_z(\mathbf{i} \cdot \mathbf{k}) \\
&\quad + A_yB_x(\mathbf{j} \cdot \mathbf{i}) + A_yB_y(\mathbf{j} \cdot \mathbf{j}) + A_yB_z(\mathbf{j} \cdot \mathbf{k}) \\
&\quad + A_zB_x(\mathbf{k} \cdot \mathbf{i}) + A_zB_y(\mathbf{k} \cdot \mathbf{j}) + A_zB_z(\mathbf{k} \cdot \mathbf{k})
\end{aligned}
$$

Since $\mathbf{i}, \mathbf{j},$ and $\mathbf{k}$ are orthogonal,

$$
\begin{aligned}
\mathbf{i} \cdot \mathbf{j} = \mathbf{j} \cdot \mathbf{k} = \mathbf{k} \cdot \mathbf{i} = (1)(1) \cos 90° = 0 \\
\mathbf{i} \cdot \mathbf{i} = \mathbf{j} \cdot \mathbf{j} = \mathbf{k} \cdot \mathbf{k} = (1)(1) \cos 0° = 1
\end{aligned}
$$

Therefore,

$$
\mathbf{A} \cdot \mathbf{B} = A_xB_x + A_yB_y + A_zB_z \tag{A-27}
$$

Note also the special case

$$
\mathbf{A} \cdot \mathbf{A} = A^2 \cos 0° = A^2 = A_x^2 + A_y^2 + A_z^2
$$

which verifies Eq. A-16 obtained by using the Pythagorean theorem.

By combining Eqs. A-24 and A-27, an expression for the angle between two vectors is obtained. Thus,

$$
\mathbf{A} \cdot \mathbf{B} = A_xB_x + A_yB_y + A_zB_z = AB \cos \theta
$$

or

$$
\cos \theta = \frac{\mathbf{A} \cdot \mathbf{B}}{AB} = \frac{A_xB_x + A_yB_y + A_zB_z}{AB} \tag{A-28}
$$

In mechanics, the scalar product is used to determine the rectangular component of a vector (force, moment, velocity, acceleration, etc.) along a line and to find the angle between two vectors (two forces, a force and a line, a force and an acceleration, etc.).

Determine the rectangular components of vector $\mathbf{A} = -312\mathbf{i} + 72\mathbf{j} - 228\mathbf{k}$ parallel and perpendicular to a line whose direction is given by the unit vector $\mathbf{e}_n = -0.60\mathbf{i} + 0.80\mathbf{j}$. Express the unit vector $\mathbf{e}_t$ associated with the perpendicular component of vector $\mathbf{A}$ in Cartesian vector form.

## SOLUTION

The magnitude of the component of vector $\mathbf{A}$ parallel to the line is given by Eq. A-27 as

$$A_n = \mathbf{A} \cdot \mathbf{e}_n = (-312\mathbf{i} + 72\mathbf{j} - 228\mathbf{k}) \cdot (-0.60\mathbf{i} + 0.80\mathbf{j})$$
$$= -312(-0.60) + 72(0.80) - 228(0) = 224.8$$

Thus, from Eq. A-26

$$\mathbf{A}_n = (\mathbf{A} \cdot \mathbf{e}_n)\mathbf{e}_n = 224.8(-0.60\mathbf{i} + 0.80\mathbf{j})$$
$$= -146.9\mathbf{i} + 195.8\mathbf{j} \qquad \text{Ans.}$$

From Eq. A-21

$$\mathbf{A} = \mathbf{A}_n + \mathbf{A}_t$$

Therefore,

$$\mathbf{A}_t = \mathbf{A} - \mathbf{A}_n = (-312\mathbf{i} + 72\mathbf{j} - 228\mathbf{k}) - (-146.9\mathbf{i} + 195.8\mathbf{j})$$
$$= -165.1\mathbf{i} - 123.8\mathbf{j} - 228\mathbf{k} \qquad \text{Ans.}$$

The magnitude of vector $\mathbf{A}_t$ is determined by using Eq. A-22. Thus,

$$A_t = |\mathbf{A}_t| = \sqrt{(-165.1)^2 + (-123.8)^2 + (-228)^2} = 307.5$$

The unit vector $\mathbf{e}_t$ is given by Eq. A-18 as

$$\mathbf{e}_t = \frac{\mathbf{A}_t}{A_t} = \frac{-165.1\mathbf{i} - 123.8\mathbf{j} - 228\mathbf{k}}{307.5}$$

$$= -0.537\mathbf{i} - 0.403\mathbf{j} - 0.741\mathbf{k} \qquad \text{Ans.}$$

Determine the angle $\theta$ between the vectors

$$\mathbf{A} = 8\mathbf{i} + 9\mathbf{j} + 7\mathbf{k} \quad \text{and} \quad \mathbf{B} = 6\mathbf{i} - 5\mathbf{j} + 3\mathbf{k}$$

### SOLUTION

The magnitudes of vectors **A** and **B** are determined by using Eq. A-22. Thus,

$$A = \sqrt{(A_x)^2 + (A_y)^2 + (A_z)^2} = \sqrt{(8)^2 + (9)^2 + (7)^2} = 13.928$$
$$B = \sqrt{(B_x)^2 + (B_y)^2 + (B_z)^2} = \sqrt{(6)^2 + (-5)^2 + (3)^2} = 8.367$$

The dot product of vectors **A** and **B** is given by Eq. A-27 as

$$\mathbf{A} \cdot \mathbf{B} = A_x B_x + A_y B_y + A_z B_z = 8(6) + 9(-5) + 7(3) = 24$$

The angle $\theta$ between the two vectors is given by Eq. A-28 as

$$\theta = \cos^{-1}\frac{\mathbf{A} \cdot \mathbf{B}}{AB} = \cos^{-1}\frac{24}{13.928(8.367)} = 78.1° \qquad \text{Ans.}$$

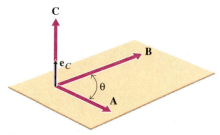

**Figure A-17** Cross or vector product of two vectors.

## A-5.2   Cross or Vector Product

The cross or vector product of two intersecting vectors **A** and **B,** by definition, yields a vector **C** that has a magnitude that is the product of the magnitudes of vectors **A** and **B** and the sine of the angle $\theta$ between them and a direction that is perpendicular to the plane containing the vectors **A** and **B.** Thus, by definition, for the vectors **A** and **B** shown in Fig. A-17,

$$\mathbf{C} = \mathbf{A} \times \mathbf{B} = (AB \sin \theta)\mathbf{e}_C \qquad \text{(A-29)}$$

where $0 \le \theta \le 180°$ and $\mathbf{e}_C$ is a unit vector in a direction perpendicular to the plane containing the vectors **A** and **B.** The sense of $\mathbf{e}_C$ is obtained by using the right-hand rule; that is, if the fingers of the right hand are curled from **A** toward **B** about an axis perpendicular to the plane containing vectors **A** and **B,** then the thumb points in the direction of vector **C.** Because of this definition of the unit vector $\mathbf{e}_C$, the cross product is not commutative. In fact,

$$\mathbf{A} \times \mathbf{B} = -\mathbf{B} \times \mathbf{A}$$

It is apparent from Eq. A-29, however, that

$$\mathbf{A} \times s\mathbf{B} = s(\mathbf{A} \times \mathbf{B})$$

and by combining Eqs. A-29 and the law of addition that

$$\mathbf{A} \times (\mathbf{B} + \mathbf{C}) = (\mathbf{A} \times \mathbf{B}) + (\mathbf{A} \times \mathbf{C})$$

If two vectors **A** and **B** are written in Cartesian vector form, the cross product becomes

$$\begin{aligned}
\mathbf{A} \times \mathbf{B} &= (A_x\mathbf{i} + A_y\mathbf{j} + A_z\mathbf{k}) \times (B_x\mathbf{i} + B_y\mathbf{j} + B_z\mathbf{k}) \\
&= A_x B_x(\mathbf{i} \times \mathbf{i}) + A_x B_y(\mathbf{i} \times \mathbf{j}) + A_x B_z(\mathbf{i} \times \mathbf{k}) \\
&\quad + A_y B_x(\mathbf{j} \times \mathbf{i}) + A_y B_y(\mathbf{j} \times \mathbf{j}) + A_y B_z(\mathbf{j} \times \mathbf{k}) \\
&\quad + A_z B_x(\mathbf{k} \times \mathbf{i}) + A_z B_y(\mathbf{k} \times \mathbf{j}) + A_z B_z(\mathbf{k} \times \mathbf{k})
\end{aligned} \qquad (a)$$

Since **i**, **j**, and **k** are orthogonal,

$$\mathbf{i} \times \mathbf{i} = [(1)(1) \sin 0°]\mathbf{k} = \mathbf{0}$$
$$\mathbf{i} \times \mathbf{j} = [(1)(1) \sin 90°]\mathbf{k} = \mathbf{k}$$

Similarly,

$$
\begin{array}{llll}
\mathbf{i} \times \mathbf{i} = \mathbf{0} & \quad \mathbf{j} \times \mathbf{i} = -\mathbf{k} & \quad \mathbf{k} \times \mathbf{i} = \mathbf{j} & \\
\mathbf{i} \times \mathbf{j} = \mathbf{k} & \quad \mathbf{j} \times \mathbf{j} = \mathbf{0} & \quad \mathbf{k} \times \mathbf{j} = -\mathbf{i} & \quad \textbf{(A-30)} \\
\mathbf{i} \times \mathbf{k} = -\mathbf{j} & \quad \mathbf{j} \times \mathbf{k} = \mathbf{i} & \quad \mathbf{k} \times \mathbf{k} = \mathbf{0} &
\end{array}
$$

Equations A-30 can be represented graphically by arranging the unit vectors **i**, **j**, and **k** in a circle in a counterclockwise order, as shown in Fig. A-18. The product of two unit vectors will be positive if they follow each other in counterclockwise order, and negative if they follow each other in clockwise order. Substituting Eqs. A-30 into Eq. *a* yields

$$
\begin{aligned}
\mathbf{A} \times \mathbf{B} &= A_x B_y \mathbf{k} + A_x B_z(-\mathbf{j}) + A_y B_x(-\mathbf{k}) \\
&\quad + A_y B_z \mathbf{i} + A_z B_x \mathbf{j} + A_z B_y(-\mathbf{i}) \\
&= (A_y B_z - A_z B_y)\mathbf{i} + (A_z B_x - A_x B_z)\mathbf{j} \qquad \textbf{(A-31)} \\
&\quad + (A_x B_y - A_y B_x)\mathbf{k}
\end{aligned}
$$

which is the expanded form of the determinant

$$
\mathbf{A} \times \mathbf{B} = \begin{vmatrix} \mathbf{i} & \mathbf{j} & \mathbf{k} \\ A_x & A_y & A_z \\ B_x & B_y & B_z \end{vmatrix} = \mathbf{C} \qquad \textbf{(A-32)}
$$

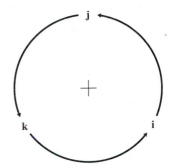

**Figure A-18**   Unit vectors **i**, **j**, and **k** arranged in a circle in counterclockwise order.

Note carefully the arrangement of terms in the determinant that places the unit vectors **i**, **j**, **k** in the first row, the components $A_x$, $A_y$, $A_z$ of **A** in the second row, and the components $B_x$, $B_y$, $B_z$ of **B** in the third row. If the two bottom rows of the determinant are interchanged, the sign of the determinant will change. Thus,

$$\mathbf{A} \times \mathbf{B} = -\mathbf{B} \times \mathbf{A}$$

In mechanics, the cross product is used to find the moment of a force about a point. The direction of the moment vector is the axis about which the force tends to rotate a body, and the magnitude of the moment is the strength of the tendency to rotate the body.

## EXAMPLE PROBLEM A-5

If **A** = $-3.75\mathbf{i} - 2.50\mathbf{j} + 1.50\mathbf{k}$ and **B** = $32\mathbf{i} + 44\mathbf{j} + 64\mathbf{k}$, determine the magnitude and direction of the vector **C** if **C** = **A** × **B**. Express the unit vector $\mathbf{e}_C$ associated with the vector **C** in Cartesian vector form.

## SOLUTION

The vector cross product **A** × **B** is obtained by using Eq. A-32. Thus,

$$
\mathbf{C} = \mathbf{A} \times \mathbf{B} = \begin{vmatrix} \mathbf{i} & \mathbf{j} & \mathbf{k} \\ A_x & A_y & A_z \\ B_x & B_y & B_z \end{vmatrix} = \begin{vmatrix} \mathbf{i} & \mathbf{j} & \mathbf{k} \\ -3.75 & -2.50 & 1.50 \\ 32 & 44 & 64 \end{vmatrix}
$$

$$= -226\mathbf{i} + 288\mathbf{j} - 85\mathbf{k} \qquad \text{Ans.}$$

The magnitude of vector **C** is determined by using Eq. A-22. Thus,

$$C = |\mathbf{C}| = \sqrt{C_x^2 + C_y^2 + C_z^2} = \sqrt{(-226)^2 + (288)^2 + (-85)^2} = 375.8 \quad \text{Ans.}$$

The direction is determined by using Eqs. A-23. Thus,

$$\theta_x = \cos^{-1} \frac{C_x}{|\mathbf{C}|} = \cos^{-1} \frac{-226}{375.8} = 126.9° \qquad \text{Ans.}$$

$$\theta_y = \cos^{-1} \frac{C_y}{|\mathbf{C}|} = \cos^{-1} \frac{288}{375.8} = 40.0° \qquad \text{Ans.}$$

$$\theta_z = \cos^{-1} \frac{C_z}{|\mathbf{C}|} = \cos^{-1} \frac{-85}{375.8} = 103.1° \qquad \text{Ans.}$$

The unit vector $\mathbf{e}_C$ is given by Eq. A-18 as

$$\mathbf{e}_C = \frac{\mathbf{C}}{|\mathbf{C}|} = \frac{-266\mathbf{i} + 288\mathbf{j} - 85\mathbf{k}}{375.8}$$
$$= -0.601\mathbf{i} + 0.766\mathbf{j} - 0.226\mathbf{k} \qquad \text{Ans.}$$

## A-5.3   Triple Scalar Product

The triple scalar product involves the dot product of vector **A** and the cross product of vectors **B** and **C**. The triple scalar product is written as

$$\mathbf{A} \cdot (\mathbf{B} \times \mathbf{C}) \qquad \text{or} \qquad (\mathbf{B} \times \mathbf{C}) \cdot \mathbf{A} \qquad \text{(A-33)}$$

Expressing vectors **A, B,** and **C** in Cartesian vector form and expanding yields

$$\begin{aligned}
\mathbf{A} \cdot (\mathbf{B} \times \mathbf{C}) &= (A_x\mathbf{i} + A_y\mathbf{j} + A_z\mathbf{k}) \cdot \\
&\quad [(B_x\mathbf{i} + B_y\mathbf{j} + B_z\mathbf{k}) \times (C_x\mathbf{i} + C_y\mathbf{j} + C_z\mathbf{k})] \\
&= A_x(B_yC_z - B_zC_y) + A_y(B_zC_x - B_xC_z) \qquad \text{(A-34)} \\
&\quad + A_z(B_xC_y - B_yC_x)
\end{aligned}$$

which is the expanded form of the determinant

$$\mathbf{A} \cdot (\mathbf{B} \times \mathbf{C}) = \begin{vmatrix} A_x & A_y & A_z \\ B_x & B_y & B_z \\ C_x & C_y & C_z \end{vmatrix} \qquad \text{(A-35)}$$

This product is a scalar; hence the name triple scalar product. However, the vectors cannot be indiscriminately interchanged since vector cross products are not commutative. It is true, however, that

$$\mathbf{A} \cdot (\mathbf{B} \times \mathbf{C}) = (\mathbf{A} \times \mathbf{B}) \cdot \mathbf{C} = (\mathbf{B} \times \mathbf{C}) \cdot \mathbf{A} = \mathbf{B} \cdot (\mathbf{C} \times \mathbf{A})$$

Since the cross product is used to find the moment of a force about a point and the scalar product is used to find the component of a vector along a line, the triple scalar product represents the component of the moment of a force along a line. That is, the triple scalar product represents the tendency of a force to rotate a body about a line.

## A-5.4   Triple Vector Product

The triple vector product involves the cross product of a vector **A** with the result of a cross product of vectors **B** and **C**. Thus, the triple vector product is written as

$$\mathbf{A} \times (\mathbf{B} \times \mathbf{C})$$

If the vectors are written in Cartesian vector form, Eq. A-31 gives the cross product $(\mathbf{B} \times \mathbf{C})$ as

$$\mathbf{B} \times \mathbf{C} = (B_x\mathbf{i} + B_y\mathbf{j} + B_z\mathbf{k}) \times (C_x\mathbf{i} + C_y\mathbf{j} + C_z\mathbf{k})$$
$$= (B_yC_z - B_zC_y)\mathbf{i} + (B_zC_x - B_xC_z)\mathbf{j} + (B_xC_y - B_yC_x)\mathbf{k}$$

Similarly, Eq. A-31 gives the cross product $\mathbf{A} \times (\mathbf{B} \times \mathbf{C})$ as

$$\mathbf{A} \times (\mathbf{B} \times \mathbf{C}) = (A_x\mathbf{i} + A_y\mathbf{j} + A_z\mathbf{k}) \times [(B_yC_z - B_zC_y)\mathbf{i}$$
$$+ (B_zC_x - B_xC_z)\mathbf{j} + (B_xC_y - B_yC_x)\mathbf{k}]$$
$$= [A_y(B_xC_y - B_yC_x) - A_z(B_zC_x - B_xC_z)]\mathbf{i}$$
$$+ [A_z(B_yC_z - B_zC_y) - A_x(B_xC_y - B_yC_x)]\mathbf{j}$$
$$+ [A_x(B_zC_x - B_xC_z) - A_y(B_yC_z - B_zC_y)]\mathbf{k}$$
$$= (A_yC_y + A_zC_z)B_x\mathbf{i} + (A_zC_z + A_xC_x)B_y\mathbf{j}$$
$$+ (A_xC_x + A_yC_y)B_z\mathbf{k} - (A_yB_y + A_xB_z)C_x\mathbf{i}$$
$$- (A_zB_z + A_xB_x)C_y\mathbf{j} - (A_xB_x + A_yB_y)C_z\mathbf{k}$$

Adding and subtracting the vector

$$A_xB_xC_x\mathbf{i} + A_yB_yC_y\mathbf{j} + A_zB_zC_z\mathbf{k}$$

and regrouping yields

$$\mathbf{A} \times (\mathbf{B} \times \mathbf{C}) = (A_xC_x + A_yC_y + A_zC_z)(B_x\mathbf{i} + B_y\mathbf{j} + B_z\mathbf{k})$$
$$- (A_xB_x + A_yB_y + A_zB_z)(C_x\mathbf{i} + C_y\mathbf{j} + C_z\mathbf{k})$$

However, from Eq. A-27,

$$A_xC_x + A_yC_y + A_zC_z = \mathbf{A} \cdot \mathbf{C}$$
$$A_xB_x + A_yB_y + A_zB_z = \mathbf{A} \cdot \mathbf{B}$$

Therefore

$$\mathbf{A} \times (\mathbf{B} \times \mathbf{C}) = (\mathbf{A} \cdot \mathbf{C})\mathbf{B} - (\mathbf{A} \cdot \mathbf{B})\mathbf{C} \qquad \text{(A-36)}$$

In a similar fashion it can be established that

$$(\mathbf{A} \times \mathbf{B}) \times \mathbf{C} = (\mathbf{A} \cdot \mathbf{C})\mathbf{B} - (\mathbf{B} \cdot \mathbf{C})\mathbf{A} \qquad \text{(A-37)}$$

The triple vector product is used in dynamics to determine the acceleration of points in a rotating body.

## EXAMPLE PROBLEM A-6

Determine the scalar triple product $\mathbf{A} \cdot (\mathbf{B} \times \mathbf{C})$ and the vector triple product $\mathbf{A} \times (\mathbf{B} \times \mathbf{C})$ for the vectors

$$\mathbf{A} = 3\mathbf{i} + 5\mathbf{j} + 8\mathbf{k}$$
$$\mathbf{B} = 4\mathbf{i} - 5\mathbf{j} + 3\mathbf{k}$$
$$\mathbf{C} = 2\mathbf{i} + 3\mathbf{j} - 4\mathbf{k}$$

### SOLUTION

The scalar triple product $\mathbf{A} \cdot (\mathbf{B} \times \mathbf{C})$ can be expressed in determinate form as given by Eq. A-35. Thus,

$$\mathbf{A} \cdot (\mathbf{B} \times \mathbf{C}) = \begin{vmatrix} A_x & A_y & A_z \\ B_x & B_y & B_z \\ C_x & C_y & C_z \end{vmatrix} = \begin{vmatrix} 3 & 5 & 8 \\ 4 & -5 & 3 \\ 2 & 3 & -4 \end{vmatrix}$$

$$= 3(20 - 9) - 5(-16 - 6) + 8(12 + 10)$$
$$= 319 \qquad \text{Ans.}$$

The triple vector product $\mathbf{A} \times (\mathbf{B} \times \mathbf{C})$ is given by Eq. A-36 as

$$\mathbf{A} \times (\mathbf{B} \times \mathbf{C}) = (\mathbf{A} \cdot \mathbf{C})\mathbf{B} - (\mathbf{A} \cdot \mathbf{B})\mathbf{C}$$
$$= [(3\mathbf{i} + 5\mathbf{j} + 8\mathbf{k}) \cdot (2\mathbf{i} + 3\mathbf{j} - 4\mathbf{k})](4\mathbf{i} - 5\mathbf{j} + 3\mathbf{k})$$
$$\quad - [(3\mathbf{i} + 5\mathbf{j} + 8\mathbf{k}) \cdot (4\mathbf{i} - 5\mathbf{j} + 3\mathbf{k})](2\mathbf{i} + 3\mathbf{j} - 4\mathbf{k})$$
$$= [3(2) + 5(3) + 8(-4)](4\mathbf{i} - 5\mathbf{j} + 3\mathbf{k})$$
$$\quad - [3(4) + 5(-5) + 8(3)](2\mathbf{i} + 3\mathbf{j} - 4\mathbf{k})$$
$$= -11(4\mathbf{i} - 5\mathbf{j} + 3\mathbf{k}) - 11(2\mathbf{i} + 3\mathbf{j} - 4\mathbf{k})$$
$$= -66\mathbf{i} + 22\mathbf{j} + 11\mathbf{k} \qquad \text{Ans.}$$

# ANSWERS TO SELECTED PROBLEMS

## CHAPTER 1

| | |
|---|---|
| **1-1** | $m = 18.65$ slug |
| **1-2** | $W = 6.62$ kN |
| **1-5** | $g = 27.8$ ft/s$^2$ |
| **1-6** | $r = 3.38(10^6)$ m |
| **1-9** | $g_e = 32.16$ ft/s$^2$ |
| | $g_p = 32.38$ ft/s$^2$ |
| **1-10** | $F = 8.90(10^{-7})$ N |
| **1-13** | $W = 767$ lb |
| **1-14** | $W = 486$ N |
| **1-17** | **a.** $W = 209$ lb |
| | **b.** $W = 190.3$ lb |
| **1-19** | $g = 81.6$ ft/s$^2$ |
| **1-20** | **a.** $W = 7.03$ N |
| | **b.** $W = 482$ N |
| **1-23** | $h = 1640$ mi |
| **1-25** | $W = 187.4$ lb |
| **1-26** | $m = 272$ kg |
| **1-29** | $v = 33.5$ m/s |
| **1-30** | $FC = 32.9$ mi/gal |
| **1-33** | $v = 0.617$ km/min |
| **1-34** | $A = 1235$ acres |
| **1-41** | $p = 241$ kPa |
| **1-43** | $\rho = 7860$ kg/m$^3$ |
| **1-44** | $\gamma = 0.1626$ lb/in.$^3$ |
| **1-47** | $G = L^3/MT^2$ |
| **1-48** | $E = M/LT^2$ |
| **1-51** | $k = 1$ (dimensionless) |
| **1-52** | Yes |
| **1-55** | $P = ML/T^2; I = L^4$ |
| **1-56** | $P = ML/T^2; I = ML^2$ |
| **1-59** | $x = L; a = 1/T; b = T; \alpha = 1$ (dimensionless) |
| **1-63** | **a.** 0.015 (%D $= -1.961$) |
| | **b.** 0.035 (%D $= +0.865$) |
| | **c.** 0.057 (%D $= +0.707$) |
| **1-64** | **a.** 0.84 (%D $= +0.310$) |
| | **b.** 0.47 (%D $= -0.613$) |
| | **c.** 0.66 (%D $= -0.662$) |
| **1-67** | **a.** 26.4 (%D $= +0.0227$) |
| | **b.** 74.8 (%D $= -0.0388$) |
| | **c.** 55.3 (%D $= -0.0651$) |
| **1-68** | **a.** 375 (%D $= +0.01867$) |
| | **b.** 826 (%D $= -0.0581$) |
| | **c.** 349 (%D $= -0.0945$) |
| **1-71** | **a.** 63,750 (%D $= +0.00596$) |
| | **b.** 27,380 (%D $= -0.00950$) |
| | **c.** 55,130 (%D $= +0.0001814$) |
| **1-72** | **a.** 937,300 (%D $= +0.001707$) |
| | **b.** 274,900 (%D $= -0.00655$) |
| | **c.** 339,900 (%D $= +0.00824$) |
| **1-78** | $g = 11.64$ m/s$^2$ |

| | |
|---|---|
| **1-79** | $F_H = 139.3$ lb |
| | $F_L = 171.2$ lb |
| **1-82** | $d = 0.1546(10^{12})$ m |
| **1-83** | $F = 48.9$ lb |
| **1-86** | $c = M/T$ |
| **1-87** | $a = M; c = M/T$ |

## CHAPTER 2

| | |
|---|---|
| **2-1** | $\mathbf{R} = 150$ lb∠36.9° |
| **2-2** | $\mathbf{R} = 98.8$ N∠28.3° |
| **2-5** | $\mathbf{R} = 826$ lb∠51.3° |
| **2-6** | $\mathbf{R} = 31.2$ kN∠19.91° |
| **2-9** | $\mathbf{R} = 884$ lb∠0.300° |
| **2-11** | $\mathbf{R} = 100.1$ lb∠50.5° |
| **2-12** | $\mathbf{R} = 151.7$ N∠30.5° |
| **2-15** | $\mathbf{R} = 821$ lb∠83.0° |
| **2-17** | $\mathbf{R} = 1361$ lb∠43.7° |
| **2-18** | $\mathbf{R} = 115.3$ kN∠67.2° |
| **2-21** | $\mathbf{R} = 23.2$ kip∠71.9° |
| **2-22** | $\mathbf{R} = 468$ N∠35.1° |
| **2-25** | $\mathbf{R} = 782$ lb∠59.0° |
| **2-26** | $\mathbf{R} = 1140$ N∠77.3° |
| **2-29** | $\mathbf{R} = 273$ lb∠58.4° |
| **2-31** | $F_u = 437$ lb |
| | $F_v = 539$ lb |
| **2-32** | $F_u = 1286$ N |
| | $F_v = 1000$ N |
| **2-35** | $F_u = 395$ lb |
| | $F_v = 318$ lb |
| **2-36** | $F_u = 900$ N |
| | $F_v = 466$ N |
| **2-39** | $F_u = 707$ lb |
| | $F_v = 948$ lb |
| **2-40** | $F_u = 383$ N |
| | $F_v = 273$ N |
| **2-43** | $F_2 = 4260$ lb |
| | $F_3 = 1213$ lb |
| **2-45** | $F_x = 693$ lb |
| | $F_y = 400$ lb |
| **2-46** | $F_x = 697$ N |
| | $F_y = 1495$ N |
| **2-49** | $F_x = -440$ lb |
| | $F_y = -177.9$ lb |
| **2-50** | $F_x = 167.7$ N |
| | $F_y = -108.9$ N |
| **2-53** | $F_x = 268$ lb |
| | $F_y = -134.2$ lb |
| **2-54** | $F_x = -627$ N |
| | $F_y = -251$ N |
| **2-55** | **a.** $F_x = -583$ lb |
| | $F_y = 694$ lb |

| | |
|---|---|
| | $F_z = 423$ lb |
| | **b.** $\mathbf{F} = -583\mathbf{i} + 694\mathbf{j} + 423\mathbf{k}$ lb |
| **2-56** | **a.** $F_x = 369$ N |
| | $F_y = 638$ N |
| | $F_z = 516$ N |
| | **b.** $\mathbf{F} = 369\mathbf{i} + 638\mathbf{j} + 516\mathbf{k}$ N |
| **2-59** | **a.** $F_x = 255$ lb |
| | $F_y = 424$ lb |
| | $F_z = 339$ lb |
| | **b.** $\mathbf{F} = 255\mathbf{i} + 424\mathbf{j} + 339\mathbf{k}$ lb |
| **2-60** | **a.** $F_x = -5.61$ kN |
| | $F_y = -11.22$ kN |
| | $F_z = 16.84$ kN |
| | **b.** $\mathbf{F} = -5.61\mathbf{i} - 11.22\mathbf{j} +$ $16.84\mathbf{k}$ kN |
| **2-63** | **a.** $F_{1x} = -557$ lb |
| | $F_{1y} = 278$ lb |
| | $F_{1z} = 650$ lb |
| | **b.** $\mathbf{F}_1 = -557\mathbf{i} + 278\mathbf{j} + 650\mathbf{k}$ lb |
| | **c.** $\alpha = 30.7°$ |
| **2-64** | **a.** $F_{1x} = 25.4$ kN |
| | $F_{1y} = -15.21$ kN |
| | $F_{1z} = 5.07$ kN |
| | **b.** $\mathbf{F}_1 = 25.4\mathbf{i} - 15.21\mathbf{j} +$ $5.07\mathbf{k}$ kN |
| | **c.** $\alpha = 30.6°$ |
| **2-67** | $\mathbf{R} = 846$ lb∠47.2° |
| **2-68** | $\mathbf{R} = 433$ N∠18.05° |
| **2-71** | $\mathbf{R} = 1696$ lb∠10.22° |
| **2-72** | $\mathbf{R} = 25.2$ kN∠21.3° |
| **2-74** | $\mathbf{R} = 3.69$ kN∠52.3° |
| **2-76** | $\mathbf{R} = 15.77$ kN∠78.8° |
| **2-78** | $R = 28.6$ kN |
| | $\theta_x = 82.2°$ |
| | $\theta_y = 69.6°$ |
| | $\theta_z = 22.0°$ |
| **2-79** | $R = 279$ lb |
| | $\theta_x = 66.1°$ |
| | $\theta_y = 66.1°$ |
| | $\theta_z = 34.9°$ |
| **2-82** | $R = 135.7$ N |
| | $\theta_x = 43.2°$ |
| | $\theta_y = 61.8°$ |
| | $\theta_z = 60.4°$ |
| **2-83** | $R = 1718$ lb |
| | $\theta_x = 51.9°$ |
| | $\theta_y = 54.8°$ |
| | $\theta_z = 57.6°$ |
| **2-85** | $R = 42.1$ kip |
| | $\theta_x = 76.6°$ |
| | $\theta_y = 42.2°$ |
| | $\theta_z = 50.9°$ |
| **2-87** | $\mathbf{R} = 669$ lb∠73.0° |

# CHAPTER 4

**2-88**   $R = 642$ N$\angle 10.79°$

**2-91**   a. $R = 223$ lb$\angle 22.5°$
     b. $F_x = 206$ lb
       $F_y = -85.2$ lb

**2-94**   $R = 408$ lb$\angle 75.2°$

**2-96**   $R = 10.37$ kN
     $\theta_x = 47.1°$
     $\theta_y = 84.8°$
     $\theta_z = 43.4°$

## CHAPTER 3

**3-1**   $P = 8.78$ lb
    $N = 23.4$ lb

**3-2**   $P = 398$ N
    $N = 719$ N

**3-5**   $F_{AC} = 869$ lb$\nwarrow$
    $F_{BC} = 424$ lb$\nearrow$
    $F_{CD} = 1000$ lb$\downarrow$

**3-6**   $F_{AC} = 7.36$ kN
    $F_{BC} = 12.74$ kN
    $F_{CD} = 7.36$ kN

**3-8**   $F_x = 2.86$ kN
    $F_y = 0.250$ kN

**3-10**   $F_4 = 620$ N
     $\theta = -12.28°$

**3-13**   $T_{AB} = 12.73$ lb
     $T_{BC} = 9.06$ lb
     $T_{CD} = 12.73$ lb
     $\alpha = 6.34°$

**3-14**   $N_L = 1304$ N
     $N_R = 829$ N

**3-16**   $F_A = 390$ N
     $F_B = 753$ N
     $F_{AB} = 195.0$ N$\angle 2.42°$

**3-19**   $F_4 = 165.8$ lb
     $\theta_x = 150.3°$
     $\theta_y = 113.2°$
     $\theta_z = 74.4°$

**3-20**   $F_4 = 766$ N
     $\theta_x = 157.2°$
     $\theta_y = 112.5°$
     $\theta_z = 86.8°$

**3-23**   $F_B = 33.4$ lb
     $F_C = 11.62$ lb
     $F_D = 41.5$ lb

**3-24**   $F_C = 3.65$ kN
     $F_D = 1.277$ kN
     $F_E = 3.65$ kN

**3-31**   $F_4 = 7.29$ kip
     $\theta = 75.9°$

**3-32**   $F_4 = 1283$ N
     $\theta = -5.71°$

**3-35**   $F_1 = 160.0$ lb
     $F_2 = 90.0$ lb
     $T = 120.0$ lb
     $\theta = 36.9°$

**3-36**   $T_A = 3.75$ kN
     $T_B = 69.8$ kN
     $T_C = 70.7$ kN

## CHAPTER 4

**4-1**   $M_A = 750$ ft $\cdot$ lb$\downarrow$

**4-2**   $M_A = 135.0$ N $\cdot$ m$\curvearrowleft$
    $M_B = 90.0$ N $\cdot$ m$\curvearrowleft$
    $M_C = 180.0$ N $\cdot$ m$\curvearrowleft$

**4-5**   a. $M_A = 640$ in. $\cdot$ lb$\curvearrowleft$
    b. $M_B = 1125$ in. $\cdot$ lb$\downarrow$

**4-6**   a. $M_A = 89.6$ N $\cdot$ m$\curvearrowleft$
    b. $M_B = 98.4$ N $\cdot$ m$\curvearrowleft$

**4-9**   a. $M_O = 600$ in. $\cdot$ lb$\downarrow$
    b. $M_O = 600$ in. $\cdot$ lb$\curvearrowleft$
    c. $M_A = 1358$ in. $\cdot$ lb$\curvearrowleft$

**4-10**   $M_A = 60$ N $\cdot$ m$\downarrow$
     $M_B = 100$ N $\cdot$ m$\downarrow$

**4-13**   $M_{A1} = 3380$ in. $\cdot$ lb$\downarrow$
     $M_{A1} = 2550$ in. $\cdot$ lb$\downarrow$
     $M_{B1} = 5330$ in. $\cdot$ lb$\downarrow$
     $M_{B2} = 1867$ in. $\cdot$ lb$\curvearrowleft$

**4-14**   $M_{A1} = 63.0$ N $\cdot$ m$\curvearrowleft$
     $M_{A2} = 92.5$ N $\cdot$ m$\downarrow$
     $M_{B1} = 81.3$ N $\cdot$ m$\curvearrowleft$
     $M_{B2} = 117.8$ N $\cdot$ m$\downarrow$

**4-17**   $M_A = 3.40$ in. $\cdot$ kip$\downarrow$

**4-18**   $M_A = 80.8$ N $\cdot$ m$\curvearrowleft$

**4-21**   $M_B = 9470$ in. $\cdot$ lb$\curvearrowleft$

**4-23**   $M_O = 1047$ in. $\cdot$ lb$\downarrow$
     $M_A = 1286$ in. $\cdot$ lb$\downarrow$

**4-24**   a. $M_A = 1.129$ kN $\cdot$ m$\curvearrowleft$
     b. $M_B = 1.344$ kN $\cdot$ m$\downarrow$

**4-27**   $M_A = 375$ in. $\cdot$ lb$\curvearrowleft$
     $M_B = 990$ in. $\cdot$ lb$\curvearrowleft$

**4-28**   $M_A = 256$ N $\cdot$ m$\downarrow$
     $M_B = 103.7$ N $\cdot$ m$\downarrow$

**4-31**   $M_O = 751\mathbf{k}$ in. $\cdot$ lb

**4-32**   $M_O = -215\mathbf{k}$ N $\cdot$ m

**4-37**   a. $M_A = 12.34\mathbf{k}$ in. $\cdot$ kip
     b. $M_B = -14.14\mathbf{k}$ in. $\cdot$ kip

**4-38**   a. $M_O = -7.07\mathbf{k}$ kN $\cdot$ m
     b. $M_A = -1.358\mathbf{k}$ kN $\cdot$ m

**4-41**   a. $M_A = 5.06\mathbf{k}$ in. $\cdot$ kip
     b. $M_B = -5.22\mathbf{k}$ in. $\cdot$ kip

**4-42**   a. $M_A = 63.0\mathbf{k}$ N $\cdot$ m
     b. $M_A = -134.7\mathbf{k}$ N $\cdot$ m

**4-45**   $M_C = -14.90\mathbf{i} - 12.42\mathbf{j} - 18.63\mathbf{k}$ in. $\cdot$ kip

**4-47**   $M_O = 7.80\mathbf{i} - 6.24\mathbf{j}$ in. $\cdot$ kip

**4-49**   $M_A = -13.40\mathbf{i} + 8.37\mathbf{j} - 0.893\mathbf{k}$ in. $\cdot$ kip
     $\theta_x = 147.9°$
     $\theta_y = 58.0°$
     $\theta_z = 93.2°$

**4-50**   $M_O = -215\mathbf{i} - 9.57\mathbf{j} + 129.2\mathbf{k}$ N $\cdot$ m
     $\theta_x = 149.0°$
     $\theta_y = 92.2°$
     $\theta_z = 59.1°$

**4-52**   $M_B = -190.4\mathbf{i} + 220\mathbf{j} + 214\mathbf{k}$ N $\cdot$ m

**4-53**   $M_B = -4.35\mathbf{i} + 2.37\mathbf{j} - 0.510\mathbf{k}$ in. $\cdot$ kip

**4-56**   $M_O = -119.9\mathbf{i} + 139.9\mathbf{j} - 210\mathbf{k}$ N $\cdot$ m

     $\theta_x = 115.4°$
     $\theta_y = 59.9°$
     $\theta_z = 138.7°$

**4-57**   $M_O = -8.14\mathbf{i} + 7.83\mathbf{j} + 8.14\mathbf{k}$ in. $\cdot$ kip
     $\theta_x = 125.8°$
     $\theta_y = 55.8°$
     $\theta_z = 54.2°$

**4-58**   $M_O = -42.6\mathbf{i} - 106.5\mathbf{j} + 159.8\mathbf{k}$ N $\cdot$ m

**4-61**   $M_{BC} = 2.78$ in. $\cdot$ kip

**4-62**   $M_{OC} = 142.2$ N $\cdot$ m

**4-65**   $M_{DE} = -3.63$ in. $\cdot$ kip

**4-66**   $M_{DE} = 67.1$ N $\cdot$ m

**4-69**   $M_{AC} = -2.40$ ft $\cdot$ kip

**4-70**   $M_{AB} = 86.6$ N $\cdot$ m

**4-72**   $M_{CD} = 91.1$ N $\cdot$ m

**4-73**   $M_{OC} = 1584\mathbf{j} + 915\mathbf{k}$ in. $\cdot$ lb

**4-76**   $M_{OB} = -60.1\mathbf{i} - 32.8\mathbf{k}$ N $\cdot$ m

**4-77**   $M_{BC} = -1292\mathbf{i} + 1939\mathbf{j}$ ft $\cdot$ lb

**4-78**   $M_{OC} = -32.9\mathbf{i} - 54.8\mathbf{j} - 5.48\mathbf{k}$ N $\cdot$ m

**4-81**   $M = 3030\mathbf{k}$ in. $\cdot$ lb
     $d = 8.66$ in.

**4-82**   $M = 168.1\mathbf{k}$ N $\cdot$ m
     $d = 221$ mm

**4-85**   $C = 2.50\mathbf{k}$ in. $\cdot$ kip

**4-86**   $C = 28.5\mathbf{k}$ N $\cdot$ m

**4-89**   $C = 5.60\mathbf{i} + 4.80\mathbf{k}$ in. $\cdot$ kip
     $\theta_x = 40.6°$
     $\theta_y = 90.0°$
     $\theta_z = 49.4°$

**4-90**   $C = 11.61$ N $\cdot$ m
     $\theta_x = 74.4°$
     $\theta_y = 64.5°$
     $\theta_z = 30.5°$

**4-93**   a. $C = 3.56$ in. $\cdot$ kip
      $\theta_x = 90.0°$
      $\theta_y = 145.9°$
      $\theta_z = 124.1°$
     b. $C_{OA} = -2.56$ in. $\cdot$ kip

**4-94**   a. $C = 140.2$ N $\cdot$ m
      $\theta_x = 55.2°$
      $\theta_y = 62.4°$
      $\theta_z = 47.3°$
     b. $C_{OA} = 56.7$ N $\cdot$ m

**4-97**   $F_A = 50\mathbf{j}$ lb
     $C_A = -300\mathbf{k}$ in. $\cdot$ lb

**4-98**   $F_B = -300\mathbf{j}$ N
     $C_B = -318\mathbf{k}$ N $\cdot$ m

**4-100**   $F_B = 540\mathbf{i} - 405\mathbf{j}$ N
     $C_B = -211\mathbf{k}$ N $\cdot$ m

**4-103**   $F_B = 380\mathbf{i} - 816\mathbf{j}$ lb
     $C_B = -10.06\mathbf{k}$ ft $\cdot$ kip

**4-105**   $F_B = 39.4\mathbf{i} - 30.8\mathbf{j}$ lb
     $C_B = 990\mathbf{k}$ in. $\cdot$ lb

**4-106**   $F_B = -212\mathbf{i} + 212\mathbf{j}$ N
     $C_B = 251\mathbf{k}$ N $\cdot$ m

**4-109**   a. $F_B = -210\mathbf{i} - 450\mathbf{j} + 300\mathbf{k}$ lb
     $M_B = 2.40\mathbf{i} + 4.20\mathbf{j} + 7.98\mathbf{k}$ in. $\cdot$ kip
     b. $\theta_x = 75.1°$
      $\theta_y = 63.2°$

$\theta_z = 31.2°$

**4-110**  a. $\mathbf{F}_C = 275\mathbf{i} - 250\mathbf{j} + 325\mathbf{k}$ N
$\mathbf{M}_C = -162.6\mathbf{i} - 89.4\mathbf{j} + 68.8\mathbf{k}$ N · m
b. $\theta_x = 145.2°$
$\theta_y = 116.9°$
$\theta_z = 69.7°$

**4-113**  $\mathbf{R} = 634$ lb∠14.8°
$d_R = 13.19$ in.↘

**4-114**  $\mathbf{R} = 1034$ N∠18.36°
$d_R = 135.6$ mm↖

**4-117**  $\mathbf{R} = 95.0$ lb↑
$x_R = 26.0$ in.→

**4-118**  a. $\mathbf{R} = 12.37$ kN⭦76.0°
b. $d_R = 3.23$ m↗

**4-119**  $\mathbf{R} = 901$ N⭦30.9°
$\mathbf{C}_B = 1.681$ kN · m↓

**4-120**  $\mathbf{R} = 203$ lb⭦11.35°
$\mathbf{C}_A = 4.46$ in. · kip↓

**4-123**  $\mathbf{R} = 589$ lb∠26.2°
$\mathbf{M}_O = 15.93$ in. · kip↰

**4-124**  $\mathbf{R} = 92.4$ N↘82.9°
$d_R = 76.0$ mm↙

**4-127**  $\mathbf{R} = 950$ lb↓
$x_R = 1.158$ ft
$y_R = 3.37$ ft

**4-128**  $\mathbf{R} = 1600$ N↓
$x_R = 68.8$ mm
$y_R = 241$ mm

**4-131**  $\mathbf{R} = 40.0\mathbf{i}$ lb
$y_R = -8.13$ ft
$z_R = -1.50$ ft

**4-133**  $F_1 = 100.0$ lb
$F_2 = 100.0$ lb
$F_3 = 60.0$ lb

**4-134**  $F_1 = 50.0$ N
$F_2 = 160.0$ N
$F_3 = 90.0$ N

**4-137**  $\mathbf{R} = 238\mathbf{i} - 47.0\mathbf{j} + 238\mathbf{k}$ lb
$\mathbf{C} = -953\mathbf{j} - 1071\mathbf{k}$ ft · lb

**4-139**  $\mathbf{R} = 75.0\mathbf{i} + 40.0\mathbf{j} + 50.0\mathbf{k}$ lb
$\mathbf{C} = 900\mathbf{i} + 720\mathbf{j} - 1350\mathbf{k}$ ft · lb

**4-140**  $\mathbf{R} = -20.6\mathbf{i} + 195.7\mathbf{j} - 241\mathbf{k}$ N
$\mathbf{C} = -168.8\mathbf{i} + 80.0\mathbf{j} + 33.7\mathbf{k}$ N · m

**4-143**  $\mathbf{R} = 300\mathbf{i} + 240\mathbf{j} + 150\mathbf{k}$ lb
$\mathbf{C} = 6.66\mathbf{i} - 2.40\mathbf{j} - 5.16\mathbf{k}$ in. · kip

**4-144**  $\mathbf{R} = 225\mathbf{i} + 475\mathbf{j} + 250\mathbf{k}$ N
$\mathbf{C} = 63.5\mathbf{i} - 82.5\mathbf{j} + 24.0\mathbf{k}$ N · m

**4-147**  $\mathbf{R} = 175\mathbf{i} + 350\mathbf{j} - 100\mathbf{k}$ lb
$\mathbf{C} = -6.45\mathbf{i} + 7.60\mathbf{j} + 3.15\mathbf{k}$ in. · kip

**4-149**  $\mathbf{R} = 50\mathbf{j} + 155\mathbf{k}$ lb
$\mathbf{C} = 216\mathbf{i} + 671\mathbf{k}$ in. · lb
$x_R = 13.67$ in.
$z_R = 4.41$ in.

**4-150**  $\mathbf{R} = 500\mathbf{j}$ N
$\mathbf{C} = -37.5\mathbf{j}$ N · m
$x_R = 25$ mm
$z_R = -60$ mm

**4-153**  a. $\mathbf{M} = 11.47$ ft · kip↰
b. $\mathbf{M} = 86.8$ ft · kip↓

**4-155**  $\mathbf{R} = 750\mathbf{i}$ lb

**4-156**  $\mathbf{R} = 458$ N↘19.11°
$d = 639$ mm↗

**4-159**  $\mathbf{C} = -1600\mathbf{i} - 1200\mathbf{j} - 1540\mathbf{k}$ in. · lb

**4-160**  a. $\mathbf{M}_O = -320\mathbf{i} + 448\mathbf{j} - 200\mathbf{k}$ N · m
b. $M_{OA} = -377$ N · m

# CHAPTER 5

**5-1**  $x_G = -0.0481$ in.
$y_G = -3.33$ in.

**5-2**  $x_G = 11.60$ mm
$y_G = 29.0$ mm

**5-5**  $x_G = 7.39$ in.
$y_G = 9.02$ in.
$z_G = 4.07$ in.

**5-6**  $x_G = 225$ mm
$y_G = 263$ mm
$z_G = 173.6$ mm

**5-9**  $x_G = -0.1667$ ft
$y_G = 1.167$ ft
$z_G = 5.33$ ft

**5-11**  $x_C = 4.00$ in.
$y_C = 2.67$ in.

**5-12**  $x_C = 133.3$ mm
$y_C = 100.0$ mm

**5-15**  $x_C = 3b/5$
$y_C = 3b^{1/2}/8$

**5-17**  $x_C = 3a/10$
$y_C = 3b/4$

**5-18**  $x_C = 2a/5$
$y_C = a/2$

**5-21**  $x_C = r - (4r/3\pi)$
$y_C = r - (4r/3\pi)$

**5-22**  $x_C = (10 - 3\pi)r/(12 - 3\pi)$
$y_C = (10 - 3\pi)r/(12 - 3\pi)$

**5-25**  $x_C = 10.94$ in.
$y_C = 7.43$ in.

**5-27**  $x_C = r/\pi$
$y_C = r/\pi$
$z_C = h/4$

**5-28**  $x_C = y_C = 0$
$z_C = 18r/25$

**5-31**  $x_C = 3.18$ in.
$y_C = 6.68$ in.

**5-33**  $x_C = 1.500$ in.
$y_C = 2.00$ in.

**5-34**  $x_C = 22.0$ mm
$y_C = 60.0$ mm

**5-37**  $x_C = 0$
$y_C = 4.00$ in.

**5-39**  $x_C = 0.431$ in.
$y_C = 0.431$ in.

**5-41**  $x_C = 4.23$ in.
$y_C = 8.33$ in.

**5-43**  $x_C = 5.61$ in.
$y_C = 5.61$ in.

**5-44**  $x_C = 104.1$ mm
$y_C = 110.6$ mm

**5-47**  $x_C = 2.09$ in.

$y_C = 7.44$ in.
$z_C = 5.83$ in.

**5-49**  a. $x_C = 2.35$ in.
$y_C = 3.54$ in.
$z_C = 1.606$ in.
b. $x_G = 3.19$ in.
$y_G = 3.75$ in.
$z_G = 0.850$ in.

**5-50**  a. $x_C = 0$
$y_C = 0$
$z_C = 229$ mm
b. $x_G = 0$
$y_G = 0$
$z_G = 197.0$ mm

**5-53**  $x_G = 10.86$ in.
$y_G = 9.95$ in.
$z_G = 3.50$ in.

**5-55**  $A = 4\pi R^2$
$V = 4\pi R^3/3$

**5-56**  $A = \pi R[R^2 + (R^2 + h^2)^{1/2}]$
$V = \pi R^2 h/3$

**5-59**  $A = 1276$ in.$^2$
$V = 1679$ in.$^3$

**5-61**  $A = 3740$ in.$^2$
$V = 8130$ in.$^3$

**5-62**  $A = 2.19(10^6)$ mm$^2$
$V = 138.0(10^6)$ mm$^3$

**5-65**  $W = 6740$ lb

**5-67**  $V = 4.87$ in.$^3$

**5-69**  $\mathbf{R} = 1350$ lb↓
$d = 3.83$ ft→

**5-70**  $\mathbf{R} = 2.25$ kN↓
$d = 2.67$ m→

**5-73**  $\mathbf{R} = 3775$ lb↓
$d = 8.06$ ft→

**5-74**  $\mathbf{R} = 4.50$ kN↓
$d = 4.35$ m→

**5-77**  $\mathbf{R} = 8530$ lb↓
$d = 9.60$ ft→

**5-78**  $\mathbf{R} = 1867$ N↓
$d = 3.57$ m→

**5-81**  $\mathbf{R} = 2.50(10^6)$ lb

**5-82**  $\mathbf{R} = 63.7$ MN

**5-85**  $\mathbf{R} = 12.23$ kip
$d_P = 2.33$ ft

**5-87**  $\mathbf{R} = 9.59$ kip
$d_P = 2.59$ ft

**5-88**  $\mathbf{R} = 277$ kN
$d_P = 1.347$ m

**5-92**  $\mathbf{R} = 116.6$ kN
$d_P = 0.640$ m

**5-93**  $\mathbf{R} = 12.62$ kip
$d_P = 3.27$ ft

**5-101**  $x_C = 0.900$ in.
$y_C = 0.900$ in.

**5-104**  $y_C = -64.3$ mm

**5-106**  $x_C = 0$
$y_C = 122.0$ mm
$z_C = 0$
$x_G = 0$
$y_G = 206$ mm
$z_G = 0$

**5-107**  $\mathbf{R} = 108.0$ kip↘33.7°
$d_O = 9.15$ ft↗

**6-31** $\mathbf{A} = 250$ lb↑
$M_A = 750$ ft · lb↺

**6-32** $\mathbf{A} = 300$ N→
$\mathbf{B} = 424$ N⬊45°

**6-35** $\mathbf{A} = 607$ lb↑
$\mathbf{E} = 468$ lb↑

**6-36** $\mathbf{A} = 2.54$ kN⬃9.01°
$\mathbf{B} = 3.42$ kN⬊60.0°

**6-39** $\mathbf{A} = 202$ lb⬊33.0°
$T = 107.7$ lb

**6-40** $\mathbf{A} = 374$ N∡68.9°
$\mathbf{B} = 635$ N←

**6-43** $\mathbf{A} = 84.7$ lb∡45.0°
$\mathbf{B} = 60.7$ lb⬃9.37°

**6-44** $\mathbf{A} = 1349$ N∡13.74°
$T_B = 1413$ N

**6-47** $\mathbf{A} = 2940$ lb↑
$\mathbf{B} = 3040$ lb↑

**6-48** $\mathbf{A} = 9.00$ kN↑
$M_A = 39.6$ kN · m↓

**6-51** $\mathbf{A} = 72.0$ lb⬃5.32°
$F_{CD} = 59.6$ lb (C)

**5-52** $\mathbf{A} = 7.00$ kN↑
$\mathbf{B} = 2.00$ kN←

**6-55** $P = 55.6$ lb

**6-56** $T = 828$ N

**6-59** $P = 10.46$ lb
$F_F = 14.94$ lb
$F_R = 24.1$ lb

**6-60** $P = 9.81$ kN
$F_F = 8.06$ kN
$F_R = 8.93$ kN

**6-62** $\mathbf{A} = 312$ N∡56.8°
$\mathbf{E} = 197.6$ N⬊30.0°

**6-65** $\mathbf{R}_A = 64.6\mathbf{i} + 129.2\mathbf{j} + 92.3\mathbf{k}$ lb
$\mathbf{C}_A = -1185\mathbf{i} + 2649\mathbf{k}$ in. · lb
$T_C = 180.2$ lb

**6-73** $\mathbf{R}_A = 50\mathbf{k}$ lb
$\mathbf{C}_A = 1150\mathbf{i} + 350\mathbf{j}$ in. · lb

**6-74** $\mathbf{T}_A = 1561\mathbf{k}$ N
$\mathbf{T}_B = 1394\mathbf{k}$ N
$\mathbf{T}_C = 1951\mathbf{k}$ N

**6-77** $\mathbf{A} = 450\mathbf{j} + 450\mathbf{k}$ lb
$\mathbf{B} = -450\mathbf{i} - 525\mathbf{j}$ lb
$\mathbf{C} = 400\mathbf{i} - 350\mathbf{k}$ lb

**6-78** $\mathbf{A} = -1900\mathbf{i} + 500\mathbf{j}$ N
$\mathbf{B} = 1580\mathbf{i} - 800\mathbf{j}$ N
$\mathbf{C} = 720\mathbf{i} + 200\mathbf{k}$ N

**6-81** $\mathbf{P} = -150.0\mathbf{k}$ lb
$\mathbf{R}_A = 928\mathbf{i} - 116.7\mathbf{j}$ lb
$\mathbf{R}_B = -278\mathbf{i} - 200\mathbf{j} + 267\mathbf{k}$ lb

**6-82** $\mathbf{P} = -750\mathbf{k}$ N
$\mathbf{R}_A = 231\mathbf{i} + 519\mathbf{k}$ N
$\mathbf{R}_B = 519\mathbf{i} + 231\mathbf{k}$ N

**6-85** $T_C = 180.2$ lb
$\mathbf{A} = 64.6\mathbf{i} + 129.2\mathbf{j} + 92.3\mathbf{k}$ lb
$\mathbf{C}_A = -1185\mathbf{i} + 2649\mathbf{k}$ in. · lb

**6-87** $T_C = 144.8$ lb
$\mathbf{A} = 107.0\mathbf{j} + 50.0\mathbf{k}$ lb
$\mathbf{B} = -83.3\mathbf{i} - 15.31\mathbf{j} + 25.0\mathbf{k}$ lb

**6-88** $T_C = 3.51$ kN

$\mathbf{R}_A = -0.537\mathbf{i} + 3.00\mathbf{j} + 1.403\mathbf{k}$ kN
$\mathbf{B} = 2.00\mathbf{i}$ kN

**6-91** $\mathbf{R}_A = 903\mathbf{i} - 220\mathbf{j} + 13.9\mathbf{k}$ lb
$T_B = 548$ lb
$T_C = 636$ lb

**6-95** $\mathbf{A} = 353$ lb∡45.0°
$\mathbf{B} = 333$ lb↑
$\mathbf{T} = 499$ lb⬀60.0°

**6-96** $\mathbf{A} = 5.48$ kN⬊47.3°
$\mathbf{B} = 5.26$ kN↑

**6-98** $\mathbf{A} = 639$ N∡17.48°
$T = 384$ N

**6-101** $\mathbf{T} = 750\mathbf{j}$ in. · lb
$\mathbf{A} = 97.2\mathbf{i} - 155.6\mathbf{k}$ lb
$\mathbf{B} = 152.8\mathbf{i} - 44.4\mathbf{k}$ lb

**6-102** $\mathbf{A} = 803$ N⬊61.6°
$\mathbf{B} = 855$N∡63.4°

**6-105** $T = 83.6$ lb
$\mathbf{A} = 26.3\mathbf{j} + 51.9\mathbf{k}$ lb
$\mathbf{B} = 20.2\mathbf{i} + 107.0\mathbf{j} + 57.6\mathbf{k}$ lb

# CHAPTER 7

**7-1** $T_{AB} = 1000$ lb (T)
$T_{AC} = 1500$ lb (T)
$T_{BC} = 1732$ lb (C)

**7-2** $T_{AB} = 3.93$ kN (T)
$T_{AC} = 5.52$ kN (T)
$T_{BC} = 9.20$ kN (C)

**7-5** $T_{AB} = 560$ lb (C)
$T_{AD} = 1536$ lb (T)
$T_{BC} = 1920$ lb (C)
$T_{BD} = 1600$ lb (T)
$T_{CD} = 1536$ lb (T)

**7-6** $T_{AB} = 2.38$ kN (C)
$T_{AD} = 0.567$ kN (T)
$T_{BC} = 4.00$ kN (C)
$T_{BD} = 3.23$ kN (C)
$T_{CD} = 4.62$ kN (T)

**7-9** $T_{AB} = 866$ lb (C)
$T_{AC} = 1000$ lb (T)
$T_{BC} = 0$
$T_{BD} = 1732$ lb (C)
$T_{CD} = 1000$ lb (T)

**7-10** $T_{AB} = 4.04$ kN (C)
$T_{AE} = 2.02$ kN (T)
$T_{BC} = 2.31$ kN (C)
$T_{BE} = 0.577$ kN (T)
$T_{CD} = 5.20$ kN (C)
$T_{CE} = 0.577$ kN (C)
$T_{DE} = 2.60$ kN (T)

**7-13** $T_{AB} = 1200$ lb (T)
$T_{AE} = 1039$ lb (C)
$T_{EF} = 2250$ lb (C)

**7-14** $T_{CD} = 10.61$ kN (T)
$T_{DE} = 7.50$ kN (T)
$T_{DF} = 19.50$ kN (C)

**7-17** $T_{AB} = 5000$ lb (T)
$T_{AE} = 1000$ lb (T)
$T_{BC} = 3330$ lb (T)
$T_{BD} = 2000$ lb (T)

**7-19** $T_{BE} = 1667$ lb (C)
$T_{CD} = 2670$ lb (C)
$T_{DE} = 2670$ lb (C)
$T_{BC} = 1451$ lb (C)
$T_{BG} = 161.3$ lb (T)
$T_{CG} = 1298$ lb (T)

**7-20** $T_{CD} = 6.19$ kN (C)
$T_{CG} = 6.87$ kN (T)
$T_{DG} = 2.43$ kN (C)

**7-23** $T_{AB} = T_{FG} = 2290$ lb (C)
$T_{AC} = T_{EG} = 2320$ lb (C)
$T_{BC} = T_{EF} = 0$
$T_{BD} = T_{DF} = 907$ lb (C)
$T_{CD} = T_{DE} = 2320$ lb (C)

**7-24** $T_{DE} = 16.00$ kN (C)
$T_{EI} = 10.56$ kN (T)
$T_{HI} = 8.07$ kN (T)

**7-27** $T_{BC} = 400$ lb (C)
$T_{BG} = 200$ lb (T)
$T_{CG} = 178.9$ lb (T)

**7-28** $T_{CG} = 2.49$ kN (C)
$T_{FG} = 13.30$ kN (T)

**7-43** $T_{CD} = 2730$ lb (C)
$T_{DE} = 138.6$ lb (T)
$T_{EF} = 2660$ lb (T)

**7-44** $T_{CD} = 8.63$ kN (C)
$T_{DE} = 2.71$ kN (T)
$T_{EF} = 7.00$ kN (T)

**7-47** $T_{BC} = 556$ lb (C)
$T_{CF} = 89.0$ lb (T)
$T_{EF} = 445$ lb (T)

**7-48** $T_{CD} = 9.19$ kN (C)
$T_{CG} = 0.750$ kN (C)
$T_{GH} = 4.95$ kN (T)

**7-51** $T_{CD} = 267$ lb (C)
$T_{CF} = 83.3$ lb (C)
$T_{FG} = 333$ lb (T)

**7-52** $T_{CD} = 12.00$ kN (C)
$T_{CE} = 0$
$T_{EF} = 12.00$ kN (T)

**7-55** $P_{max} = 1000$ lb

**7-56** $P_{max} = 4.99$ kN

**7-59** $T_{CD} = 1510$ lb (C)
$T_{CK} = 448$ lb (C)
$T_{LK} = 1620$ lb (T)

**7-60** $T_{DE} = 9.37$ kN (C)
$T_{EJ} = 3.89$ kN (C)
$T_{HJ} = 9.20$ kN (C)

**7-63** $T_{DE} = 10,000$ lb (C)
$T_{DJ} = 8000$ lb (C)
$T_{JK} = 7000$ lb (T)

**7-65** $T_{CD} = 1617$ lb (C)
$T_{CE} = 400$ lb (C)
$T_{EF} = 924$ lb (T)

**7-67** $T_{DE} = 8750$ lb (C)
$T_{EG} = 1000$ lb (T)
$T_{FG} = 3610$ lb (T)
$T_{FH} = 4000$ lb (T)

**7-73** $\mathbf{A} = 50.0\mathbf{i} + 191.7\mathbf{j} + 93.8\mathbf{k}$ lb
$\mathbf{B} = -100.0\mathbf{i}$ lb
$\mathbf{C} = 50.0\mathbf{i} - 141.7\mathbf{j} + 56.3\mathbf{k}$ lb
$T_{AB} = 114.6$ lb (C)
$T_{AD} = 114.6$ lb (C)

$T_{BC} = 52.1$ lb (C)

$T_{BD} = 141.4$ lb (T)

$T_{CD} = 114.6$ lb (C)

**7-74**   $A = -750i + 469j + 375k$ N

$B = -938j$ N

$C = 750i + 469j + 375k$ N

$T_{AB} = T_{BC} = 530$ N (C)

$T_{AD} = T_{CD} = 600$ N (C)

$T_{BD} = 1200$ N (T)

**7-77**   $A = 263i + 150.0j$ lb

$C = 525i$ lb

$D = -788i - 750j + 1650k$ lb

$T_{AB} = 263$ lb (C)

$T_{AC} = 150.0$ lb (C)

$T_{AD} = 0$

$T_{BC} = 797$ lb (C)

$T_{BD} = 1196$ lb (T)

$T_{CD} = 1061$ lb (T)

**7-78**   $A = 230i + 800k$ N

$B = 250i - 500j$ N

$D = -480i$ N

$T_{AB} = 0$

$T_{AC} = 230$ N (C)

$T_{AD} = 800$ N (C)

$T_{BC} = 559$ N (C)

$T_{BD} = 0$

$T_{CD} = 933$ N (T)

**7-81**   $A = -3.75i$ lb

$B = 116.3i + 15.00j + 120.0k$ lb

$C = 7.50i$ lb

$F = -120.0i$ lb

$T_{AB} = 7.50$ lb (C)

$T_{AD} = 8.39$ lb (T)

$T_{AF} = 0$

$T_{BC} = 7.50$ lb (T)

$T_{BD} = 116.3$ lb (C)

$T_{BF} = 120.0$ lb (C)

$T_{CE} = 10.61$ lb (C)

$T_{CF} = 0$

$T_{DE} = 72.5$ lb (C)

$T_{DF} = 72.1$ lb (T)

$T_{EF} = 100.0$ lb (T)

**7-83**   $B = 48.2$ lb ∠57.4°

$C = 85.7$ lb ∡45.0°

**7-84**   $A = 1269$ N ⬎23.2°

$B = 1537$ N ⬑40.6°

**7-87**   $F = 589$ lb↓

**7-88**   $B = 834$ N←

$C = 2410$ N ⬏55.0°

$D = 2970$ N ∡41.7°

**7-91**   $A = 167.7$ lb ⬎63.4°

$B = 424$ lb ⬑45.0°

$C = 335$ lb ⬎26.6°

**7-92**   $F_{BC} = 675$ N (T)

$F_{AD} = 375$ N (C)

**7-95**   $A = 200$ lb ⬑56.4°

$D = 180.0$ lb ⬎52.0°

**7-96**   $B = 13.75$ lb↓

$D = 8.75$ N↓

**7-99**   $A = 1179$ lb ∡232.0°

$C = 686$ lb ⬑19.48°

$D = 924$ lb ⬏67.5°

**7-100**   $B = 3.10$ kN ⬑29.9°

$C = 3.40$ kN ⬐59.9°

$D = 3.59$ kN ∡41.5°

$E = 1.966$ kN ⬏30.1°

**7-103**   $A = 25.3$ lb ∡81.9°

$B = 10.85$ lb ⬏39.8°

$C = 18.67$ lb ⬐75.2°

**7-104**   $A = 1535$ N ⬐26.6°

$B = 1541$ N ∡82.1°

$D = 1717$ N ⬑22.6°

**7-107**   $A = 70.5$ lb ⬐7.13°

$B = 61.3$ lb↓

$C = 99.0$ lb ⬑45.0°

$M_C = 245$ ft · lb↓

**7-109**   $A = 5990$ lb ∡39.0°

$B = 4650$ lb ⬏8.75°

$C = 5920$ lb ∡17.37°

**7-110**   $A = 736$ N↑

$T = 1568$ N→

$C = 2300$ N←

$E = 1041$ N ⬐45.0°

**7-113**   $A = 353$ lb ⬑10.89°

$C = 353$ lb ∡70.9°

$E = 462$ lb ⬐60.0°

**7-119**   $T_{AB} = 2500$ lb (C)

$T_{AF} = 5660$ lb (T)

$T_{BC} = 3540$ lb (C)

$T_{BD} = 2500$ lb (T)

$T_{BF} = 0$

$T_{CD} = 3540$ lb (C)

$T_{DE} = 6500$ lb (C)

$T_{DF} = 5660$ lb (T)

$T_{EF} = 0$

**7-120**   $T_{AB} = 9.81$ kN (T)

$T_{AC} = 8.50$ kN (C)

$T_{BC} = 6.22$ kN (C)

$T_{CD} = 7.62$ kN (C)

**7-125**   $B = 1166$ lb ⬐59.0°

$F = 1131$ lb ⬑45.0°

$G = 320$ lb ∡38.7°

**7-126**   $A = 1414$ N ⬐45.0°

$B = 3270$ N ⬑66.6°

$C = 3310$ N ⬏65.0°

$D = 1972$ N ∡30.5°

**7-129**   $P = 138.6$ lb ⬏40.0°

$A = 72.6$ lb ⬐77.3°

$B = 138.6$ lb ⬑60.0°

$C = 72.6$ lb ⬐42.7°

# CHAPTER 8

**8-1**   **a.** $F_{AB} = 35.0$ kip (T)

$F_{BC} = 2.00$ kip (C)

$F_{CD} = 18.00$ kip (T)

**8-2**   **a.** $F_{AB} = 25.0$ kN (C)

$F_{BC} = 50.0$ kN (T)

$F_{CD} = 50.0$ kN (C)

**8-5**   $T_A = 100$ ft · lb↓

$T_B = 50$ ft · lb↗

$T_C = 250$ ft · lb↓

**8-7**   **a.** $T_{AB} = 80.0$ ft · kip↓

$T_{BC} = 20.0$ ft · kip↓

$T_{CD} = 20.0$ ft · kip↓

$T_{DE} = 45.0$ ft · kip↓

**8-8**   $T_{max} = 30.0$ kN · m↓

**8-11**   $T_A = 10$ ft · kip↗

$T_B = 7$ ft · kip↓

$T_C = 47$ ft · kip↗

$T_D = 13$ ft · kip↓

$T_E = 12$ ft · kip↗

**8-12**   $F_{max} = 70.0$ kN (T)

**8-15**   $P = 300$ lb

$V = 500$ lb

$M = 3600$ in. · lb

**8-16**   $P = 225$ N

$V = 390$ N

$M = 137.9$ N · m

**8-19**   $P = 85.1$ lb

$V = 206$ lb

$M = 789$ ft · lb

**8-20**   $P = 1.543$ kN

$V = 2.57$ kN

$M = 0.300$ kN · m

**8-23**   $P = 96.3$ lb

$V = 265$ lb

$M = 5290$ in. · lb

**8-24**   $P = 354$ N

$V = 493$ N

$M = 25.7$ N · m

**8-27**   **a.** $P = 877$ lb

$V = 280$ lb

$M = 700$ ft · lb

**b.** $P = 750$ lb

$V = 13.36$ lb

$M = 1000$ in. · lb

**8-28**   $P = 0$

$V = 450$ N

$M = 203$ N · m

**8-31**   **a.** $P = 2080$ lb

$V = 1562$ lb

$M = 5000$ ft · lb

**b.** $P = 520$ lb

$V = 80$ lb

$M = 2080$ ft · lb

**8-32**   **a.** $P = 1248$ N

$V = 69.3$ N

$M = 25.0$ N · m

**b.** $P = 750$ N

$V = 500$ N

$M = 100.0$ N · m

**8-35**   $V = -900x + 6000$ lb

$M = -450x^2 + 6000x$ ft · lb

**8-36**   $V = -5x + 21$ kN

$M = -2.5x^2 + 21x - 18$ kN · m

**8-39**   $V = -300x + 2500$ lb

$M = -150x^2 + 2500x - 11{,}250$ ft · lb

**8-41**   $V = -750x + 4875$ lb

$M = -375x^2 + 4875x - 6000$ ft · lb

**8-42**   $V = -6x + 36.75$ kN

$M = -3x^2 + 36.75x - 102$ kN · m

**8-45**   $V = -500x + 6000$ lb

$M = -250x^2 + 6000x - 27{,}000$ ft · lb

**8-47**   **a.** $V = -1000x + 5500$ lb

**673**

$M = -500x^2 + 5500x$ ft $\cdot$ lb

**b.** $V = -6500$ lb

$M = -6500x + 72{,}000$ ft $\cdot$ lb

**c.** $V_{max} = -6500$ lb (12 ft$\leq x\leq$ 16 ft)

$M_{max} = -32{,}000$ ft $\cdot$ lb (at $x = 16$ ft)

**8-51** $V_{4ft} = 6185$ lb and 1185 lb

$V_{9ft} = 1185$ lb and $-6815$ lb

$M_{4ft} = 24{,}740$ ft $\cdot$ lb

$M_{9ft} = 30{,}665$ ft $\cdot$ lb

**8-52** $V_{4m} = 9.09$ kN and $-0.91$ kN

$V_{8m} = -0.91$ kN and $-30.9$ kN

$V_{11m} = -30.9$ kN and 20.0 kN

$M_{4m} = 36.4$ kN $\cdot$ m

$M_{8m} = 32.7$ kN $\cdot$ m

$M_{11m} = -60.0$ kN $\cdot$ m

**8-55** $V_{8ft} = 3370$ lb and 870 lb

$V_{13ft} = 870$ lb

$M_{8ft} = 26{,}960$ ft $\cdot$ lb

$M_{13ft} = 31{,}310$ ft $\cdot$ lb

**8-56** $V_{4m} = -20$ kN

$V_{6m} = -20$ kN and $-60$ kN

$M_{4m} = 160$ kN $\cdot$ m

$M_{6m} = 120$ kN $\cdot$ m

**8-59** $V_{6ft} = 2600$ lb and $-400$ lb

$V_{12ft} = -400$ lb

$M_{6ft} = 15{,}600$ ft $\cdot$ lb

$M_{9ft} = 14{,}400$ ft $\cdot$ lb and 12,600 ft $\cdot$ lb

$M_{12ft} = 11{,}400$ ft $\cdot$ lb

**8-61** $V_0 = -1000$ lb and 1600 lb

$V_{10ft} = -400$ lb and $-1000$ lb

$M_0 = -4000$ ft $\cdot$ lb

$M_{10ft} = 2000$ ft $\cdot$ lb and 5000 ft $\cdot$ lb

**8-62** $V_0 = 0$ and 21 kN

$V_{6m} = -15$ kN and 3 kN

$M_0 = -24$ kN $\cdot$ m

$M_{6m} = -6$ kN $\cdot$ m

**8-65** $V_0 = -3000$ lb and 5000 lb

$V_{3ft} = 2000$ lb

$V_{6ft} = 2000$ lb and $-3000$ lb

$V_{9ft} = -3000$ lb and 6000 lb

$M_0 = -4500$ ft $\cdot$ lb

$M_{3ft} = 6000$ ft $\cdot$ lb

$M_{6ft} = 12{,}000$ ft $\cdot$ lb and 3000 ft $\cdot$ lb

$M_{3ft} = -6000$ ft $\cdot$ lb

**8-66** $V_0 = -7.5$ kN and 16.75 kN

$V_{2m} = 1.75$ kN

$V_{3m} = 1.75$ kN and $-13.25$ kN

$V_{4m} = -13.25$ kN and 6.75 kN

$M_0 = -3.75$ kN $\cdot$ m

$M_{2m} = 14.75$ kN $\cdot$ m

$M_{3m} = 16.50$ kN $\cdot$ m and 6.50 kN $\cdot$ m

$M_{4m} = -6.75$ kN $\cdot$ m

**8-69** **a.** $\mathbf{A}_x = 1810$ lb$\leftarrow$

$\mathbf{A}_y = 350$ lb$\uparrow$

$\mathbf{D}_x = 1810$ lb$\rightarrow$

$\mathbf{D}_y = 850$ lb$\uparrow$

**b.** $y_B = 1.933$ ft

$y_C = 2.35$ ft

**c.** $L = 30.7$ ft

**8-70** **a.** $\mathbf{A}_x = 22.8$ kN$\leftarrow$

$\mathbf{A}_y = 7.60$ kN$\uparrow$

$\mathbf{D}_x = 22.8$ kN$\rightarrow$

$\mathbf{D}_y = 8.40$ kN$\uparrow$

**b.** $T_{AB} = 24.0$ kN

$T_{BC} = 22.9$ kN

$T_{CD} = 24.3$ kN

**c.** $L = 15.62$ m

**8-73** **a.** $\mathbf{A}_x = 6730$ lb$\leftarrow$

$\mathbf{A}_y = 2650$ lb$\uparrow$

$\mathbf{E}_x = 6730$ lb$\rightarrow$

$\mathbf{E}_y = 2750$ lb$\uparrow$

**b.** $T_{AB} = 7230$ lb

$T_{BC} = 6810$ lb

$T_{CD} = 6790$ lb

$T_{DE} = 7270$ lb

**c.** $L = 31.3$ ft

**8-74** **a.** $\mathbf{A}_x = 46.4$ kN$\leftarrow$

$\mathbf{A}_y = 18.6$ kN$\uparrow$

$\mathbf{E}_x = 46.4$ kN$\rightarrow$

$\mathbf{E}_y = 13.4$ kN$\uparrow$

**b.** $y_B = 1.202$ m

$y_C = 1.771$ m

$y_D = 0.866$ m

**c.** $L = 15.48$ m

**8-77** **a.** $\mathbf{A}_x = 842$ lb$\leftarrow$

$\mathbf{A}_y = 540$ lb$\uparrow$

$\mathbf{D}_x = 842$ lb$\rightarrow$

$\mathbf{D}_y = 159.7$ lb$\uparrow$

**b.** $y_B = 9.63$ ft

$y_C = 10.35$ ft

**c.** $L = 48.1$ ft

**8-79** **a.** $\mathbf{A}_x = 4460$ lb$\leftarrow$

$\mathbf{A}_y = 1070$ lb$\uparrow$

$\mathbf{D}_x = 3860$ lb$\rightarrow$

$\mathbf{D}_y = 1930$ lb$\uparrow$

**b.** $T_{AB} = 4590$ lb

$T_{BC} = 4560$ lb

$T_{CD} = 4320$ lb

**c.** $y_B = 6.92$ ft

**d.** $L = 31.5$ ft

**8-80** **a.** $\mathbf{A}_x = 24.2$ kN$\leftarrow$

$\mathbf{A}_y = 22.0$ kN$\uparrow$

$\mathbf{D}_x = 39.2$ kN$\rightarrow$

$\mathbf{D}_y = 7.98$ kN$\uparrow$

**b.** $y_B = 6.37$ m

$y_C = 7.04$ m

**c.** $L = 32.7$ m

**8-83** **a.** $T_{max} = 606$ kip

**b.** $L = 616$ ft

**8-84** **a.** $a = 12.58$ m

**b.** $L = 15.42$ m

**8-87** **a.** $a = 229$ ft

**b.** $L = 247$ ft

**8-88** **a.** $\theta_x = 13.88°$

**b.** $h = 24.7$ m

**c.** $L = 404$ m

**8-90** **a.** $T_A = 12{,}130$ kN

**b.** $T_B = 13{,}300$ kN

**c.** $\theta_B = 26.1°$

**d.** $L = 158.6$ m

**8-92** **a.** $\theta_B = 21.1°$

**b.** $w = 2.13$ kN/m

**c.** $L = 101.8$ m

**8-93** **a.** $T_o = 405$ kip

**b.** $T_{max} = 432$ kip

**c.** $\theta_x = 20.2°$

**d.** $L = 509$ ft

**8-95** $h = 132.7$ ft

**8-96** $L = 351$ m

**8-99** **a.** $T_o = 1575$ lb

**b.** $T_B = 1975$ lb

**c.** $L = 596$ ft

**8-100** **a.** $T_o = 12.09$ kN

**b.** $T_B = 17.71$ kN

**c.** $L = 345$ m

**8-102** **a.** $T_{max} = 41.8$ kN

**b.** $h = 15.11$ m

**c.** $L = 203$ m

**8-103** **a.** $T_A = 619$ lb

**b.** $T_B = 1619$ lb

**c.** $L = 299$ ft

**8-105** **a.** $T_o = 9630$ lb

**b.** $T_{max} = 10{,}250$ lb

**c.** $L = 2032$ ft

**8-106** **a.** $T_o = 2260$ N

**b.** $T_A = 8260$ N

**c.** $T_B = 4660$ N

**d.** $L = 400$ m

**8-110** $P = 260$ N

$V = 150$ N

$M = 99.6$ N $\cdot$ m

**8-111** **a.** $P = 0$

$V = 1544$ lb

$M = 31.8$ in. $\cdot$ kip

**b.** $P = 1350$ lb

$V = 750$ lb

$M = 65.5$ in. $\cdot$ kip

**8-114** $P = 809$ N

$V = 638$ N

$M = 191.3$ N $\cdot$ m

**8-115** $V = -500x + 4000$ lb

$M = -250x^2 + 4000x - 16{,}000$ ft $\cdot$ lb

**8-116** $V_o = 29$ kN

$V_{2m} = 13$ kN and 8 kN

$V_{4m} = -8$ kN

$V_{6m} = -8$ kN and $-13$ kN

$M_{2m} = 42$ kN $\cdot$ m and 47 kN $\cdot$ m

$M_{4m} = 47$ kN $\cdot$ m

$M_{6m} = 31$ kN $\cdot$ m and 26 kN $\cdot$ m

$M_{4m} = -6.75$ kN $\cdot$ m

**8-119** **a.** $T_O = 4690$ kip

**b.** $T_B = 5050$ kip

$\theta_B = 21.8°$

**c.** $L = 1539$ ft

**8-120** **a.** $T_{max} = 898$ N

$h = 6.01$ m

**b.** $T_{max} = 851$ N

$h = 15.08$ m

**c.** $T_{max} = 1077$ N

$h = 23.9$ m

# CHAPTER 9

**9-1**    $P = 265$ lb
**9-2**    $\mu = 0.830$
**9-5**    $W = 35.4$ lb
**9-6**    $P = 407$ N
**9-9**    **a.** System is in equilibrium
     **b.** $P_{max} = 800$ lb
**9-10**    **a.** System is not in equilibrium
     **b.** $P_{max} = 1583$ N
**9-13**    $M_{min} = 0.832$ slug
     $M_{max} = 5.80$ slug
**9-14**    $M_{min} = 48.2$ kg
**9-17**    $P = 268$ lb
**9-18**    $\mu = 0.556$
**9-21**    **a.** $x = 3.21$ ft
     **b.** $x = 6.26$ ft
**9-22**    **a.** $d = 2.33$ m
     **b.** $d = 3.60$ m
**9-25**    $W_A = 181.8$ N
**9-26**    $h = 1.071$ m
**9-29**    The drum is in equilibrium
**9-33**    $P = 87.0$ lb
**9-34**    $P_{min} = 210$ N
**9-37**    $P = 260$ lb
**9-38**    **a.** $\theta = 20.0°$
     **b.** $\theta = 21.9°$
**9-41**    $P = 100$ lb
**9-42**    **a.** $\theta_{min} = 19.92°$
     $\theta_{max} = 37.8°$
     **b.** $\mu_{min} = 0.567$
     $\mu_{max} = 0.570$
**9-49**    $P = 227$ lb
**9-50**    $P = 96.0$ N
**9-53**    $P = 1447$ lb
**9-54**    $P = 1354$ N
**9-56**    $P = 794$ N
**9-57**    $W_{min} = 43.9$ lb
**9-60**    **a.** $P = 1665$ N
     **b.** System is not in equilibrium
     **c.** $P = 50.5$ N
     **d.** $\phi_{max} = 14.04°$
**9-61**    **a.** $P = 2660$ lb
     **b.** System is not in equilibrium
     **c.** $P = 667$ lb
     **d.** $\phi_{max} = 8.53°$
**9-65**    $M = 9.52$ in. $\cdot$ lb
     Clamp will stay in place
**9-66**    $M = 8.59$ N $\cdot$ m
     No moment needed to release
**9-69**    **a.** $L_{max} = 2.20$ in.
     **b.** $L = 1.368$ in.
**9-70**    $M_R = 0.861$ N $\cdot$ m
     $M_L = 0.583$ N $\cdot$ m
**9-73**    $M = 0.0750$ in. $\cdot$ lb
**9-74**    $M = 1.516$ N $\cdot$ m
**9-77**    **a.** $P = 54.0$ lb
     **b.** $P = 52.8$ lb
     **c.** $P = 46.3$ lb
     **d.** $P = 47.3$ lb
**9-79**    **a.** $M = 180.4$ in. $\cdot$ lb
     **b.** $M = 176.0$ in. $\cdot$ lb
**9-80**    **a.** $M = 3.00$ N $\cdot$ m

     **b.** $M = 2.25$ N $\cdot$ m
**9-83**    $P = 2.59$ lb
**9-85**    **a.** $P_{min} = 103.9$ lb
     **b.** $P_{min} = 2410$ lb
**9-86**    **a.** $P_{min} = 1347$ N
     **b.** $P_{min} = 3460$ N
**9-88**    $T = 21.1$ kN
     $C_A = 1.460$ kN $\cdot$ m
     $C_B = 0.450$ kN $\cdot$ m
**9-91**    **a.** $P = 10.51$ lb
     **b.** $P = 1.852$ lb
**9-92**    **a.** $P = 479$ N
     **b.** $P = 805$ N
**9-95**    **b.** $\mu = 0.350$
     **b.** Brake is not self-locking
**9-96**    $P_{min} = 159.0$ N
**9-99**    **a.** $T_{max} = 8340$ in. $\cdot$ lb
     **b.** $T_{max} = 632$ in. $\cdot$ lb
**9-104**    $a = 2.41$ mm
**9-105**    $F = 5.43$ lb
**9-108**    $F = 198.8$ N
**9-109**    $a = 0.25$ in.
**9-111**    $F = 29.9$ lb
     Ratio $= 0.222$
**9-112**    $F = 722$ N
     Ratio $= 0.380$
**9-115**    $\mu_{min} = 1/3$
**9-116**    $M = \mu_s m_p /(\sin \theta + \mu_s \cos \theta)$
**9-119**    $P = 211$ lb
**9-120**    $P = 210$ N
**9-123**    $W_C = 12.00$ lb (tipping)
**9-124**    $P = 144.4$ N
**9-127**    **a.** $T_{max} = 766$ in. $\cdot$ lb
     **b.** $T_{max} = 3150$ in. $\cdot$ lb
**9-128**    $P_{min} = 414$ N

# CHAPTER 10

**10-1**    **a.** $I_x = bh^3/12$
     **b.** $I_{xC} = bh^3/36$
**10-2**    **a.** $I_x = bh^3/12$
     **b.** $I_{xC} = bh^3/36$
**10-5**    **a.** $I_x = 205$ in.$^4$
     **b.** $I_y = 780$ in.$^4$
**10-6**    **a.** $I_x = 37,200$ mm$^4$
     **b.** $I_y = 625,000$ mm$^4$
**10-9**    **a.** $I_x = 4a^3/9$
     **b.** $I_y = (\pi^2 - 4)a$
**10-10**    **a.** $I_x = 2/9$
     **b.** $I_y = \pi^2/4 - 2$
**10-13**    **a.** $I_x = R^4[\theta - (1/2)\sin 2\theta]/4$
     **b.** $J_z = R^4\theta/2$
**10-15**    **a.** $I_x = 8\pi$
     **b.** $I_y = 32\pi$
**10-17**    **a.** $k_x = 3.46$ in.
     $k_y = 1.732$ in.
     **b.** $k_{xC} = 1.732$ in.
     $k_{yC} = 0.866$ in.
**10-18**    **a.** $k_x = 51.0$ mm
     $k_y = 81.6$ mm
     **b.** $k_{xC} = 29.4$ mm
     $k_{yC} = 47.1$ mm

**10-21**    $k_x = 0.471a$
     $k_y = 0.435L$
**10-23**    **a.** $k_{xC} = 2.00$ in.
     **b.** $k_x = 5.39$ in.
     $k_y = 6.32$ in.
**10-24**    **a.** $k_{xC} = 10.61$ mm
     $k_{yC} = 10.21$ mm
     **b.** $k_x = 41.4$ mm
     $k_y = 41.3$ mm
**10-27**    $k_z = [(21b^2 + 5a^2)/35]^{1/2}$
**10-28**    **a.** $k_x = 22.4$ mm
     $k_y = 65.5$ mm
     **b.** $k_{xC} = 12.18$ mm
     $k_{yC} = 26.2$ mm
**10-29**    **a.** $k_x = 1.309$ in.
     $k_y = 1.789$ in.
     **b.** $k_{xC} = 0.524$ in.
     $k_{yC} = 0.975$ in.
**10-32**    $k_z = 33.9$ mm
**10-33**    $I_{xC} = 56.0$ in.$^4$
     $I_{yC} = 11.0$ in.$^4$
**10-34**    $I_{xC} = 6.93(10^6)$ mm$^4$
     $I_{yC} = 1.157(10^6)$ mm$^4$
**10-37**    $I_{xC} = 2730$ in.$^4$
     $I_{yC} = 190.8$ in.$^4$
**10-38**    $I_{xC} = 330(10^6)$ mm$^4$
     $I_{yC} = 115.7(10^6)$ mm$^4$
**10-41**    $I_{xC} = 436$ in.$^4$
     $I_{yC} = 148.7$ in.$^4$
**10-42**    $I_{xC} = 172.2(10^6)$ mm$^4$
     $I_{yC} = 37.7(10^6)$ mm$^4$
**10-45**    $I_x = 163.4$ in.$^4$
     $I_y = 188.5$ in.$^4$
**10-46**    $I_x = 9.86(10^6)$ mm$^4$
     $I_y = 9.86(10^6)$ mm$^4$
**10-49**    **a.** $I_x = 1739$ in.$^4$
     $I_y = 134.9$ in.$^4$
     **b.** $I_{xC} = 647$ in.$^4$
     $I_{yC} = 134.9$ in.$^4$
**10-50**    **a.** $I_x = 543.9(10^6)$ mm$^4$
     $I_y = 543.9(10^6)$ mm$^4$
     **b.** $I_{xC} = 543.9(10^6)$ mm$^4$
     $I_{yC} = 104.8(10^6)$ mm$^4$
**10-53**    **a.** $I_x = 2210$ in.$^4$
     $I_y = 2210$ in.$^4$
     **b.** $I_{xC} = 2180$ in.$^4$
     $I_{yC} = 2180$ in.$^4$
**10-55**    $I_{xy} = b^2h^2/4$
**10-56**    $I_{xy} = 6.25(10^6)$ mm$^4$
**10-59**    $I_{xy} = 360$ in.$^4$
**10-60**    $I_{xyC} = -108.3(10^6)$ mm$^4$
**10-63**    **a.** $I_{xy} = b^2h^2/8$
     **b.** $I_{xyC} = b^2h^2/72$
**10-64**    **a.** $I_{xy} = 5b^2h^2/24$
     **b.** $I_{xyC} = -b^2h^2/72$
**10-67**    $I_{xy} = 7790$ in.$^4$
**10-69**    $I_{max} = 1070$ in.$^4$ @ $-28.5°$
     $I_{min} = 116.2$ in.$^4$ @ $61.5°$
**10-70**    $I_{max} = 35.7(10^6)$ mm$^4$ @ $45.0°$
     $I_{min} = 21.8(10^6)$ mm$^4$ @ $-45.0°$
**10-73**    $I_{max} = 372$ in.$^4$ @ $-16.33$
     $I_{min} = 8.61$ in.$^4$ @ $73.7°$
**10-75**    $I_{max} = 12.67$ in.$^4$ @ $26.6°$
     $I_{min} = 2.67$ in.$^4$ @ $-63.4°$

**10-76**  $I_{max} = 148.6(10^6)$ mm$^4$ @ $-28.2°$
$I_{min} = 10.13(10^6)$ mm$^4$ @ $61.8°$

**10-79**  $I_{max} = 1568$ in.$^4$ @ $-16.84°$
$I_{min} = 254$ in.$^4$ @ $73.2°$

**10-81**  $I_z = 3mR^2/10$

**10-82**  $I_y = 3m(R^2 + 4h^2)/20$

**10-85**  $I_y = m(3R^2 + 4L^2)/12$

**10-86**  $I_x = 2mR^2/5$

**10-89**  $I_x = mR^2(1 + 3R^2)/6$

**10-91**  $I_x = m(h^2 + b^2)/6$

**10-92**  $I_{yG} = m(2h^2 + 3L^2)/36$

**10-95**  $I_y = m(3b^2 + 2h^2)/20$

**10-97**  $I_y = 0.267$ slug · ft$^2$

**10-98**  $I_y = 1.004$ kg · m$^2$

**10-101**  $I_x = 19.22$ slug · ft$^2$

**10-102**  $I_x = 22.9$ kg · m$^2$

**10-105**  $I_y = 0.555$ slug · ft$^2$

**10-106**  $I_y = 1.616$ kg · m$^2$

**10-109**  $I_{xy} = mbL/4$
$I_{yz} = mhL/4$
$I_{zx} = mbh/4$

**10-110**  $I_{xy} = -mRL/2$
$I_{yz} = -2mRL/3\pi$
$I_{zx} = 4mR^2/3\pi$

**10-113**  $I_{xy} = -mbL/6$
$I_{yz} = mbh/12$

**10-115**  $I_{yz} = 0$
$I_{zx} = -mRL/9\pi$

**10-117**  $I_{max} = 0.695$ slug · ft$^2$
$\theta_x = 68.6°$
$\theta_y = 81.5°$
$\theta_z = 23.1°$
$I_{int} = 0.665$ slug · ft$^2$
$\theta_x = 71.0°$
$\theta_y = 25.2°$
$\theta_z = 105.9°$
$I_{min} = 0.1078$ slug · ft$^2$
$\theta_x = 29.2°$
$\theta_y = 113.5°$
$\theta_z = 106.4°$

**10-118**  $I_{max} = 0.402$ kg · m$^2$
$\theta_x = 101.3°$
$\theta_y = 76.5°$
$\theta_z = 17.8°$
$I_{int} = 0.378$ kg · m$^2$
$\theta_x = 52.9°$
$\theta_y = 37.4°$
$\theta_z = 94.1°$
$I_{min} = 0.0546$ kg · m$^2$
$\theta_x = 39.3°$

$\theta_y = 124.0°$
$\theta_z = 72.7°$

**10-121**  $I_{max} = 1.773$ slug · ft$^2$
$\theta_x = 63.1°$
$\theta_y = 70.7°$
$\theta_z = 34.1°$
$I_{int} = 1.627$ slug · ft$^2$
$\theta_x = 73.9°$
$\theta_y = 33.9°$
$\theta_z = 118.9°$
$I_{min} = 0.388$ slug · ft$^2$
$\theta_x = 32.1°$
$\theta_y = 116.7°$
$\theta_z = 106.5°$

**10-123**  $I_{max} = 0.0419$ slug · ft$^2$
$\theta_x = 60.2°$
$\theta_y = 90.0°$
$\theta_z = 29.8°$
$I_{int} = 0.0346$ slug · ft$^2$
$\theta_x = 90.0°$
$\theta_y = 0°$
$\theta_z = 90.0°$
$I_{min} = 0.0322$ slug · ft$^2$
$\theta_x = 29.8°$
$\theta_y = 90.0°$
$\theta_z = 119.8°$

**10-128**  **a.**  $k_x = 9.43$ mm
$k_y = 25.3$ mm
**b.**  $k_{xC} = 4.99$ mm
$k_{yC} = 10.84$ mm

**10-129**  **a.**  $I_x = 51.2$ in.$^4$
**b.**  $I_y = 39.0$ in.$^4$

**10-132**  $I_{xy} = 0.210(10^6)$ mm$^4$

**10-133**  $I_x = 2400$ in.$^4$
$I_y = 1800$ in.$^4$
$I_{xy} = -526$ in.$^4$

**10-136**  $I_{max} = 2390(10^6)$ mm$^4$ @ $-21.0°$
$I_{min} = 301(10^6)$ mm$^4$ @ $69.0°$

**10-137**  $I_y = 372ma^2/1120$

**10-139**  $I_{x'} = (6R^2 + 5L^2)mR^2/(4R^2 + L^2)6$

# CHAPTER 11

**11-1**  $U = 31.7(10^6)$ ft · lb

**11-2**  $U = 236$ kJ

**11-5**  $U = 28.0$ ft · lb

**11-7**  **a.**  $U = 2100$ ft · lb
**b.**  $U = -2100$ ft · lb

**11-8**  **a.**  $U = 30.0$ kJ
**b.**  $U = -23.5$ kJ

**11-11**  **a.**  $U = 927$ ft · lb
**b.**  $U = 0$
**c.**  $U = -927$ ft · lb

**11-12**  **a.**  $U = -1314$ J
**b.**  $U = 2070$ J
**c.**  $U = -729$ J

**11-15**  $\theta = 31.2°$

**11-16**  $m_A = 100.0$ kg

**11-19**  $C = 550$ lb↑

**11-20**  $A = 500$ N↑
$M_A = 750$ N · m↓

**11-23**  $T_{CD} = 708$ lb C

**11-24**  $T_{BC} = 9.42$ kN T

**11-27**  $F = 125.0$ lb

**11-29**  $P = 198.8$ lb

**11-30**  $\theta = 38.7°$

**11-33**  $B_x = 30.0$ lb→
$B_y = 8.33$ lb↓

**11-35**  **a.**  $V_s = 213$ in. · lb
**b.**  $V_T = 62.5$ in. · lb

**11-36**  $V_T = 45.1$ J (at 30°)
$V_T = 179.2$ J (at 60°)
$V_T = 400$ J (at 90°)

**11-39**  $\theta = 32.1°$

**11-40**  $\theta = 20.6°$

**11-43**  $\theta = 0°$ and 48.2°

**11-45**  $k = 37.5$ lb/ft

**11-46**  **a.**  $k_{min} = WL/2$
**b.**  $\theta = 0°$ (unstable)
$\theta = 34.7°$ (stable)

**11-49**  $\theta = 44.4°$

**11-51**  $\theta = 56.7°$

**11-52**  $\theta = 44.5°$

**11-55**  $\theta = 0°$ (unstable)
$\theta = 18.45°$ (stable)

**11-57**  $B = 1038$ lb↑

**11-58**  **a.**  $E = 6.50$ kN↑
**b.**  $T_{CD} = 8.00$ kN C

**11-61**  $M = 3080$ in. · lb

**11-62**  $k = 2550$ N/m

**11-65**  **a.**  $k = 4370$ N/m
**b.**  $\theta = 30°$ (unstable)

## PHOTO CREDITS

**Chapter 1**    Opener: Courtesy NASA. Page 5: Will & Deni McIntyre/Tony Stone Images/New York, Inc. Page 7: Pitkin Studio. Page 14: Courtesy Bureau International des Poids et Mesures, France. Page 15 (top): Courtesy National Institute of Standards and Measures. Page 15 (bottom): Courtesy Bureau International des Poids et Mesures, France. Page 25: Courtesy Texas Instruments.

**Chapter 2**    Opener: Guy Marche/FPG International.

**Chapter 3**    Opener: Darrell Wong/Tony Stone Images/New York, Inc.

**Chapter 4**    Opener: Hugh Sitton/Tony Stone Images/New York, Inc.

**Chapter 5**    Opener: Hank Delespinasse/The Image Bank. Page 192: Terranova International/Photo Researchers. Page 227: Mark Wagner/Tony Stone Images/New York, Inc.

**Chapter 6**    Opener: Tony Stone Images/New York, Inc.

**Chapter 7**    Opener: Dale Boyer/Tony Stone Images/New York, Inc.

**Chapter 8**    Opener: Donnovan Reese/Tony Stone Images/New York, Inc.

**Chapter 9**    Opener: Zefa/Mathis/The Stock Market. Page 476: Scott Markewitz/FPG International.

**Chapter 10**    Opener: Chris Baker/Tony Stone Images/New York, Inc.

**Chapter 11**    Opener: Chris Baker/Tony Stone Images/New York, Inc.